Praise for the first edition

"Quantum field theory is an extraordinarily beautiful subject, but it can be an intimidating one. The profound and deeply physical concepts it embodies can get lost, to the beginner, amidst its technicalities. In this book, Zee imparts the wisdom of an experienced and remarkably creative practitioner in a user-friendly style. I wish something like it had been available when I was a student."
—Frank Wilczek, Massachusetts Institute of Technology

"Finally! Zee has written a ground-breaking quantum field theory text based on the course I made him teach when I chaired the Princeton physics department. With utmost clarity he gives the eager student a light-hearted and easy-going introduction to the multifaceted wonders of quantum field theory. I wish I had this book when I taught the subject."
—Marvin L. Goldberger, President, Emeritus, California Institute of Technology

"This book is filled with charming explanations that students will find beneficial."
—Ed Witten, Institute for Advanced Study

"This book is perhaps the most user-friendly introductory text to the essentials of quantum field theory and its many modern applications. With his physically intuitive approach, Professor Zee makes a serious topic more reachable for beginners, reducing the conceptual barrier while preserving enough mathematical details necessary for a firm grasp of the subject."
—Bei Lok Hu, University of Maryland

"Like the famous Feynman Lectures on Physics, this book has the flavor of a good blackboard lecture. Zee presents technical details, but only insofar as they serve the larger purpose of giving insight into quantum field theory and bringing out its beauty."
—Stephen M. Barr, University of Delaware

"This is a fantastic book—exciting, amusing, unique, and very valuable."
—Clifford V. Johnson, University of Durham

"Tony Zee explains quantum field theory with a clear and engaging style. For budding or seasoned condensed matter physicists alike, he shows us that field theory is a nourishing nut to be cracked and savored."
—Matthew P. A. Fisher, Kavli Institute for Theoretical Physics

"I was so engrossed that I spent all of Saturday and Sunday one weekend absorbing half the book, to my wife's dismay. Zee has a talent for explaining the most abstruse and arcane concepts in an elegant way, using the minimum number of equations (the jokes and anecdotes help). . . . I wish this were available when I was a graduate student. Buy the book, keep it by your bed, and relish the insights delivered with such flair and grace."
—N. P. Ong, Princeton University

What readers are saying

"Funny, chatty, physical: QFT education transformed!! This text stands apart from others in so many ways that it's difficult to list them all. . . . The exposition is breezy and chatty. The text is never boring to read, and is at times very, very funny. Puns and jokes abound, as do anecdotes. . . . A book which is much easier, and more fun, to read than any of the others. Zee's skills as a popular physics writer have been used to excellent effect in writing this textbook. . . . Wholeheartedly recommended."
  —M. Haque

"A readable, and rereadable instant classic on QFT. . . . At an introductory level, this type of book—with its pedagogical (and often very funny) narrative—is priceless. [It] is full of fantastic insights akin to reading the Feynman lectures. I have since used *QFT in a Nutshell* as a review for [my] year-long course covering all of Peskin and Schroder, and have been pleasantly surprised at how Zee is able to preemptively answer many of the open questions that eluded me during my course. . . . I value *QFT in a Nutshell* the same way I do the Feynman lectures. . . . It's a text to teach an understanding of physics."
  —Flip Tanedo

"One of those books a person interested in theoretical physics simply must own! A real scientific masterpiece. I bought it at the time I was a physics sophomore and that was the best choice I could have made. It was this book that triggered my interest in quantum field theory and crystallized my dreams of becoming a theoretical physicist. . . . The main goal of the book is to make the reader gain real intuition in the field. Amazing . . . amusing . . . real fun. What also distinguishes this book from others dealing with a similar subject is that it is written like a tale. . . . I feel enormously fortunate to have come across this book at the beginning of my adventure with theoretical physics. . . . Definitely the best quantum field theory book I have ever read."
  —Anonymous

"I have used *Quantum Field Theory in a Nutshell* as the primary text. . . . I am immensely pleased with the book, and recommend it highly. . . . Don't let the 'damn the torpedoes, full steam ahead' approach scare you off. Once you get used to seeing the physics quickly, I think you will find the experience very satisfying intellectually."
  —Jim Napolitano

"This is undoubtedly the best book I have ever read about the subject. Zee does a fantastic job of explaining quantum field theory, in a way I have never seen before, and I have read most of the other books on this topic. If you are looking for quantum field theory explanations that are clear, precise, concise, intuitive, and fun to read—this is the book for you."
  —Anonymous

"One of the most artistic and deepest books ever written on quantum field theory. Amazing . . . extremely pleasant . . . a lot of very deep and illuminating remarks. . . . I recommend the book by Zee to everybody who wants to get a clear idea what good physics is about."
—Slava Mukhanov

"Perfect for learning field theory on your own—by far the clearest and easiest to follow book I've found on the subject."
—Ian Z. Lovejoy

"A beautifully written introduction to the modern view of fields . . . breezy and enchanting, leading to exceptional clarity without sacrificing depth, breadth, or rigor of content. . . . [It] passes my test of true greatness: I wish it had been the first book on this topic that I had found."
—Jeffrey D. Scargle

"A breeze of fresh air . . . a real literary gem which will be useful for students who make their first steps in this difficult subject and an enjoyable treat for experts, who will find new and deep insights. Indeed, the *Nutshell* is like a bright light source shining among tall and heavy trees—the many more formal books that exist—and helps seeing the forest as a whole! . . . I have been practicing QFT during the past two decades and with all my experience I was thrilled with enjoyment when I read some of the sections."
—Joshua Feinberg

"This text not only teaches up-to-date quantum field theory, but also tells readers how research is actually done and shows them how to think about physics. [It teaches things that] people usually say 'cannot be learned from books.' [It is] in the same style as *Fearful Symmetry* and *Einstein's Universe*. All three books . . . are classics."
—Yu Shi

"I belong to the [group of] enthusiastic laymen having enough curiosity and insistence . . . but lacking the mastery of advanced math and physics. . . . I really could not see the forest for the trees. But at long last I got this book!"
—Makay Attila

"More fun than any other QFT book I have read. The comparisons to Feynman's writings made by several of the reviewers seem quite apt. . . . His enthusiasm is quite infectious. . . . I doubt that any other book will spark your interest like this one does."
—Stephen Wandzura

"I'm having a blast reading this book. It's both deep and entertaining; this is a rare breed, indeed. I usually prefer the more formal style (big Landau fan), but I have to say that when Zee has the talent to present things his way, it's a definite plus."
—Pierre Jouvelot

"Required reading for QFT: [it] heralds the introduction of a book on quantum field theory that you can sit down and read. My professor's lectures made much more sense as I followed along in this book, because concepts were actually EXPLAINED, not just worked out."
    —Alexander Scott

"Not your father's quantum field theory text: I particularly appreciate that things are motivated physically before their mathematical articulation. . . . Most especially though, the author's 'heuristic' descriptions are the best I have read anywhere. From them alone the essential ideas become crystal clear."
    —Dan Dill

Quantum Field Theory in a Nutshell

Quantum Field Theory in a Nutshell

SECOND EDITION

A. Zee

PRINCETON UNIVERSITY PRESS · PRINCETON AND OXFORD

Published by Princeton University Press, 41 William Street,
Princeton, New Jersey 08540
In the United Kingdom: Princeton University Press,
6 Oxford Street, Woodstock, Oxfordshire OX20 1TW

Library of Congress Cataloging-in-Publication Data

Zee, A.
   Quantum field theory in a nutshell / A. Zee.—2nd ed.
      p.   cm.
   Includes bibliographical references and index.
   ISBN 978-0-691-14034-6 (hardcover : alk. paper)   1. Quantum field theory.   I. Title.
   QC174.45.Z44 2010
   530.14′3—dc22   2009015469

British Library Cataloging-in-Publication Data is available

This book has been composed in Scala LF with ZzTEX
by Princeton Editorial Associates, Inc., Scottsdale, Arizona

Printed on acid-free paper.

press.princeton.edu

Printed in the United States of America

10 9

To my parents,
who valued education above all else

Contents

Ⅰ Part I: Motivation and Foundation

Ⅱ Part II: Dirac and the Spinor

III   Part III: Renormalization and Gauge Invariance

IV   Part IV: Symmetry and Symmetry Breaking

V   Part V: Field Theory and Collective Phenomena

VI   Part VI: Field Theory and Condensed Matter

VII   Part VII: Grand Unification

VIII   Part VIII: Gravity and Beyond

N   Part N

Preface to the First Edition

As a student, I was rearing at the bit, after a course on quantum mechanics, to learn quantum field theory, but the books on the subject all seemed so formidable. Fortunately, I came across a little book by Mandl on field theory, which gave me a taste of the subject enabling me to go on and tackle the more substantive texts. I have since learned that other physicists of my generation had similar good experiences with Mandl.

In the last three decades or so, quantum field theory has veritably exploded and Mandl would be hopelessly out of date to recommend to a student now. Thus I thought of writing a book on the essentials of modern quantum field theory addressed to the bright and eager student who has just completed a course on quantum mechanics and who is impatient to start tackling quantum field theory.

I envisaged a relatively thin book, thin at least in comparison with the many weighty tomes on the subject. I envisaged the style to be breezy and colloquial, and the choice of topics to be idiosyncratic, certainly not encyclopedic. I envisaged having many short chapters, keeping each chapter "bite-sized."

The challenge in writing this book is to keep it thin and accessible while at the same time introducing as many modern topics as possible. A tough balancing act! In the end, I had to be unrepentantly idiosyncratic in what I chose to cover. Note to the prospective book reviewer: You can always criticize the book for leaving out your favorite topics. I do not apologize in any way, shape, or form. My motto in this regard (and in life as well), taken from the Ricky Nelson song "Garden Party," is "You can't please everyone so you gotta please yourself."

This book differs from other quantum field theory books that have come out in recent years in several respects.

I want to get across the important point that the usefulness of quantum field theory is far from limited to high energy physics, a misleading impression my generation of theoretical physicists were inculcated with and which amazingly enough some recent textbooks on quantum field theory (all written by high energy physicists) continue to foster. For instance, the study of driven surface growth provides a particularly clear, transparent, and physical example of the importance of the renormalization group in quantum field theory. Instead of being entangled in all sorts of conceptual irrelevancies such as divergences, we have the obviously physical notion of changing the ruler used to measure the fluctuating surface. Other examples include random matrix theory and Chern-Simons gauge theory in quantum Hall fluids. I hope that condensed matter theory students will find this book helpful in getting a first taste of quantum field theory. The book is divided into eight parts,[1] with two devoted more or less exclusively to condensed matter physics.

I try to give the reader at least a brief glimpse into contemporary developments, for example, just enough of a taste of string theory to whet the appetite. This book is perhaps also exceptional in incorporating gravity from the beginning. Some topics are treated quite differently than in traditional texts. I introduce the Faddeev-Popov method to quantize electromagnetism and the language of differential forms to develop Yang-Mills theory, for example.

The emphasis is resoundingly on the conceptual rather than the computational. The only calculation I carry out in all its gory details is that of the magnetic moment of the electron. Throughout, specific examples rather than heavy abstract formalism will be favored. Instead of dealing with the most general case, I always opt for the simplest.

I had to struggle constantly between clarity and wordiness. In trying to anticipate and to minimize what would confuse the reader, I often find that I have to belabor certain points more than what I would like.

I tried to avoid the dreaded phrase "It can be shown that . . . " as much as possible. Otherwise, I could have written a much thinner book than this! There are indeed thinner books on quantum field theory: I looked at a couple and discovered that they hardly explain anything. I must confess that I have an almost insatiable desire to explain.

As the manuscript grew, the list of topics that I reluctantly had to drop also kept growing. So many beautiful results, but so little space! It almost makes me ill to think about all the stuff (bosonization, instanton, conformal field theory, etc., etc.) I had to leave out. As one colleague remarked, the nutshell is turning into a coconut shell!

Shelley Glashow once described the genesis of physical theories: "Tapestries are made by many artisans working together. The contributions of separate workers cannot be discerned in the completed work, and the loose and false threads have been covered over." I regret that other than giving a few tidbits here and there I could not go into the fascinating history of quantum field theory, with all its defeats and triumphs. On those occasions when I refer to original papers I suffer from that disconcerting quirk of human psychology of tending to favor my own more than decorum might have allowed. I certainly did not attempt a true bibliography.

---

[1] Murray Gell-Mann used to talk about the eightfold way to wisdom and salvation in Buddhism (M. Gell-Mann and Y. Ne'eman, *The Eightfold Way*). Readers familiar with contemporary Chinese literature would know that the celestial dragon has eight parts.

The genesis of this book goes back to the quantum field theory course I taught as a beginning assistant professor at Princeton University. I had the enormous good fortune of having Ed Witten as my teaching assistant and grader. Ed produced lucidly written solutions to the homework problems I assigned, to the extent that the next year I went to the chairman to ask "What is wrong with the TA I have this year? He is not half as good as the guy last year!" Some colleagues asked me to write up my notes for a much needed text (those were the exciting times when gauge theories, asymptotic freedom, and scores of topics not to be found in any texts all had to be learned somehow) but a wiser senior colleague convinced me that it might spell disaster for my research career. Decades later, the time has come. I particularly thank Murph Goldberger for urging me to turn what expository talents I have from writing popular books to writing textbooks. It is also a pleasure to say a word in memory of the late Sam Treiman, teacher, colleague, and collaborator, who as a member of the editorial board of Princeton University Press persuaded me to commit to this project. I regret that my slow pace in finishing the book deprived him of seeing the finished product.

Over the years I have refined my knowledge of quantum field theory in discussions with numerous colleagues and collaborators. As a student, I attended courses on quantum field theory offered by Arthur Wightman, Julian Schwinger, and Sidney Coleman. I was fortunate that these three eminent physicists each has his own distinctive style and approach.

The book has been tested "in the field" in courses I taught. I used it in my field theory course at the University of California at Santa Barbara, and I am grateful to some of the students, in particular Ted Erler, Andrew Frey, Sean Roy, and Dean Townsley, for comments. I benefitted from the comments of various distinguished physicists who read all or parts of the manuscript, including Steve Barr, Doug Eardley, Matt Fisher, Murph Goldberger, Victor Gurarie, Steve Hsu, Bei-lok Hu, Clifford Johnson, Mehran Kardar, Ian Low, Joe Polchinski, Arkady Vainshtein, Frank Wilczek, Ed Witten, and especially Joshua Feinberg. Joshua also did many of the exercises.

Talking about exercises: You didn't get this far in physics without realizing the absolute importance of doing exercises in learning a subject. It is especially important that you do most of the exercises in this book, because to compensate for its relative slimness I have to develop in the exercises a number of important points some of which I need for later chapters. Solutions to some selected problems are given.

I will maintain a web page http://theory.kitp.ucsb.edu/~zee/nuts.html listing all the errors, typographical and otherwise, and points of confusion that will undoubtedly come to my attention.

I thank my editors, Trevor Lipscombe, Sarah Green, and the staff of Princeton Editorial Associates (particularly Cyd Westmoreland and Evelyn Grossberg) for their advice and for seeing this project through. Finally, I thank Peter Zee for suggesting the cover painting.

Preface to the Second Edition

What one fool could understand, another can.
—R. P. Feynman[1]

Appreciating the appreciators

It has been nearly six years since this book was published on March 10, 2003. Since authors often think of books as their children, I may liken the flood of appreciation from readers, students, and physicists to the glorious report cards a bright child brings home from school. Knowing that there are people who appreciate the care and clarity crafted into the pedagogy is a most gratifying feeling. In working on this new edition, merely looking at the titles of the customer reviews on Amazon.com would lighten my task and quicken my pace: "Funny, chatty, physical. QFT education transformed!," "A readable, and re-readable instant classic on QFT," "A must read book if you want to understand essentials in QFT," "One of the most artistic and deepest books ever written on quantum field theory," "Perfect for learning field theory on your own," "Both deep and entertaining," "One of those books a person interested in theoretical physics simply must own," and so on.

In a *Physics Today* review, Zvi Bern, a preeminent younger field theorist, wrote:

> Perhaps foremost in his mind was how to make *Quantum Field Theory in a Nutshell* as much fun as possible. . . . I have not had this much fun with a physics book since reading *The Feynman Lectures on Physics*. . . . [This is a book] that no student of quantum field theory should be without. *Quantum Field Theory in a Nutshell* is the ideal book for a graduate student to curl up with after having completed a course on quantum mechanics. But, mainly, it is for anyone who wishes to experience the sheer beauty and elegance of quantum field theory.

A classical Chinese scholar famously lamented "He who knows me are so few!" but here Zvi read my mind.

Einstein proclaimed, "Physics should be made as simple as possible, but not any simpler." My response would be "Physics should be made as fun as possible, but not

---

[1] R. P. Feynman, *QED: The Strange Theory of Light and Matter*, p. xx.

any funnier." I overcame the editor's reluctance and included jokes and stories. And yes, I have also written a popular book *Fearful Symmetry* about the "sheer beauty and elegance" of modern physics, which at least in that book largely meant quantum field theory. I want to share that sense of fun and beauty as much as possible. I've heard some people say that "Beauty is truth" but "Beauty is fun" is more like it.

I had written books before, but this was my first textbook. The challenges and rewards in writing different types of book are certainly different, but to me, a university professor devoted to the ideals of teaching, the feeling of passing on what I have learned and understood is simply incomparable. (And the nice part is that I don't have to hand out final grades.) It may sound corny, but I owe it, to those who taught me and to those authors whose field theory texts I studied, to give something back to the theoretical physics community. It is a wonderful feeling for me to meet young hotshot researchers who had studied this text and now know more about field theory than I do.

How I made the book better: The first text that covers the twenty-first century

When my editor Ingrid Gnerlich asked me for a second edition I thought long and hard about how to make this edition better than the first. I have clarified and elaborated here and there, added explanations and exercises, and done more "practical" Feynman diagram calculations to appease those readers of the first edition who felt that I didn't calculate enough. There are now three more chapters in the main text. I have also made the "most accessible" text on quantum field theory even more accessible by explaining stuff that I thought readers who already studied quantum mechanics should know. For example, I added a concise review of the Dirac delta function to chapter I.2. But to the guy on Amazon.com who wanted complex analysis explained, sorry, I won't do it. There is a limit. Already, I gave a basically self-contained coverage of group theory.

More excitingly, and to make my life more difficult, I added, to the existing eight parts (of the celestial dragon), a new part consisting of four chapters, covering field theoretic happenings of the last decade or so. Thus I can say that this is the first text since the birth of quantum field theory in the late 1920s that covers the twenty-first century.

Quantum field theory is a mature but certainly not a finished subject, as some students mistakenly believe. As one of the deepest constructs in theoretical physics and all encompassing in its reach, it is bound to have yet unplumbed depths, secret subterranean connections, and delightful surprises. While many theoretical physicists have moved past quantum field theory to string theory and even string field theory, they often take the limit in which the string description reduces to a field description, thus on occasion revealing previously unsuspected properties of quantum field theories. We will see an example in chapter N.4.

My friends admonished me to maintain, above all else, the "delightful tone" of the first edition. I hope that I have succeeded, even though the material contained in part N is "hot off the stove" stuff, unlike the long-understood material covered in the main text. I also added a few jokes and stories, such as the one about Fermi declining to trace.

As with the first edition, I will maintain a web site http://theory.kitp.ucsb.edu/~zee/ nuts2.html listing the errors, typographical or otherwise, that will undoubtedly come to my attention.

Encouraging words

In the quote that started this preface, Feynman was referring to himself, and to you! Of course, Feynman didn't simply understand the quantum field theory of electromagnetism, he also invented a large chunk of it. To paraphrase Feynman, I wrote this book for fools like you and me. If a fool like me could write a book on quantum field theory, then surely you can understand it.

As I said in the preface to the first edition, I wrote this book for those who, having learned quantum mechanics, are eager to tackle quantum field theory. During a sabbatical year (2006–07) I spent at Harvard, I was able to experimentally verify my hypothesis that a person who has mastered quantum mechanics could understand my book on his or her own without much difficulty. I was sent a freshman who had taught himself quantum mechanics in high school. I gave him my book to read and every couple of weeks or so he came by to ask a question or two. Even without these brief sessions, he would have understood much of the book. In fact, at least half of his questions stem from the holes in his knowledge of quantum mechanics. I have incorporated my answers to his field theoretic questions into this edition.

As I also said in the original preface, I had tested some of the material in the book "in the field" in courses I taught at Princeton University and later at the University of California at Santa Barbara. Since 2003, I have been gratified to know that it has been used successfully in courses at many institutions.

I understand that, of the different groups of readers, those who are trying to learn quantum field theory on their own could easily get discouraged. Let me offer you some cheering words. First of all, that is very admirable of you! Of all the established subjects in theoretical physics, quantum field theory is by far the most subtle and profound. By consensus it is much much harder to learn than Einstein's theory of gravity, which in fact should properly be regarded as part of field theory, as will be made clear in this book. So don't expect easy cruising, particularly if you don't have someone to clarify things for you once in a while. Try an online physics forum. Do at least some of the exercises. Remember: "No one expects a guitarist to learn to play by going to concerts in Central Park or by spending hours reading transcriptions of Jimi Hendrix solos. Guitarists practice. Guitarists play the guitar until their fingertips are calloused. Similarly, physicists solve problems."[2] Of course, if you don't have the prerequisites, you won't be able to understand this or any other field theory text. But if you have mastered quantum mechanics, keep on trucking and you will get there.

---

[2] N. Newbury et al., *Princeton Problems in Physics with Solutions*, Princeton University Press, Princeton, 1991.

The view will be worth it, I promise. My thesis advisor Sidney Coleman used to start his field theory course saying, "Not only God knows, I know, and by the end of the semester, you will know." By the end of this book, you too will know how God weaves the universe out of a web of interlocking fields. I would like to change Dirac's statement "God is a mathematician" to "God is a quantum field theorist."

Some of you steady truckers might want to ask what to do when you get to the end. During my junior year in college, after my encounter with Mandl, I asked Arthur Wightman what to read next. He told me to read the textbook by S. S. Schweber, which at close to a thousand pages was referred to by students as "the monster" and which could be extremely opaque at places. After I slugged my way to the end, Wightman told me, "Read it again." Fortunately for me, volume I of Bjorken and Drell had already come out. But there is wisdom in reading a book again; things that you miss the first time may later leap out at you. So my advice is "Read it again." Of course, every physics student also knows that different explanations offered by different books may click with different folks. So read other field theory books. Quantum field theory is so profound that most people won't get it in one pass.

On the subject of other field theory texts: James Bjorken kindly wrote in my much-used copy of Bjorken and Drell that the book was obsolete. Hey BJ, it isn't. Certainly, volume I will never be passé. On another occasion, Steve Weinberg told me, referring to his field theory book, that "I wrote the book that I would have liked to learn from." I could equally well say that "I wrote the book that *I* would have liked to learn from." Without the least bit of hubris, I can say that I prefer my book to Schweber's. The moral here is that if you don't like this book you should write your own.

I try not to do clunky

I explained my philosophy in the preface to the first edition, but allow me a few more words here. I will teach you how to calculate, but I also have what I regard as a higher aim, to convey to you an enjoyment of quantum field theory in all its splendors (and by "all" I mean not merely quantum field theory as defined by some myopic physicists as applicable only to particle physics). I try to erect an elegant and logically tight framework and put a light touch on a heavy subject.

In spite of the image conjured up by Zvi Bern of some future field theorist curled up in bed reading this book, I expect you to grab pen and paper and work. You could do it in bed if you want, but work you must. I intentionally did not fill in all the steps; it would hardly be a light touch if I do every bit of algebra for you. Nevertheless, I have done algebra when I think that it would help you. Actually, I love doing algebra, particularly when things work out so elegantly as in quantum field theory. But I don't do clunky. I do not like clunky-looking equations. I avoid spelling everything out and so expect you to have a certain amount of "sense." As a small example, near the end of chapter I.10 I suppressed the spacetime dependence of the fields φ_a and $\delta\varphi_a$. If you didn't realize, after some 70 pages, that fields are functions of where you are in spacetime, you are quite lost, my friend. My plan is to "keep you on your toes" and I purposely want you to feel puzzled occasionally. I have faith that the sort of person who would be reading this book can always figure it out after a bit of thought. I realize that there are at least three distinct groups of readers, but let me say to the students, "How do you expect to do research if you have to be spoon-fed from line to line in a textbook?"

Nuts who do not appreciate the *Nutshell*

In the original preface, I quoted Ricky Nelson on the impossibility of pleasing everyone and so I was not at all surprised to find on Amazon.com a few people whom one of my friends calls "nuts who do not appreciate the *Nutshell*." My friends advise me to leave these people alone but I am sufficiently peeved to want to say a few words in my defense, no matter how nutty the charge. First, I suppose that those who say the book is too mathematical cancel out those who say the book is not mathematical enough. The people in the first group are not informed, while those in the second group are misinformed.

Quantum field theory does not have to be mathematical. I know of at least three Field Medalists who enjoyed the book. A review for the American Mathematical Society offered this deep statement in praise of the book: "It is often deeper to know why something is true rather than to have a proof that it is true." (Indeed, a Fields Medalist once told me that top mathematicians secretly think like physicists and after they work out the broad outline of a proof they then dress it up with epsilons and deltas. I have no idea if this is true only for one, for many, or for all Fields Medalists. I suspect that it is true for many.)

Then there is the person who denounces the book for its lack of rigor. Well, I happen to know, or at least used to know, a thing or two about mathematical rigor, since I wrote my senior thesis with Wightman on what I would call "fairly rigorous" quantum field theory. As we like to say in the theoretical physics community, too much rigor soon leads to rigor mortis. Be warned. Indeed, as Feynman would tell students, if this ain't rigorous enough for you the math department is just one building over. So read a more rigorous book. It is a free country.

More serious is the impression that several posters on Amazon.com have that the book is too elementary. I humbly beg to differ. The book gives the impression of being elementary but in fact covers more material than many other texts. If you master everything in the *Nutshell*, you would know more than most professors of field theory and could start doing research. I am not merely making an idle claim but could give an actual proof. All the ingredients that went into the spinor helicity formalism that led to a deep field theoretic discovery described in part N could be found in the first edition of this book. Of course, reading a textbook is not enough; you have to come up with the good ideas.

As for he who says that the book does not look complicated enough and hence can't be a serious treatment, I would ask him to compare a modern text on electromagnetism with Maxwell's treatises.

Thanks

In the original preface and closing words, I mentioned that I learned a great deal of quantum field theory from Sidney Coleman. His clarity of thought and lucid exposition have always inspired me. Unhappily, he passed away in 2007. After this book was published, I visited Sidney on different occasions, but sadly, he was already in a mental fog.

In preparing this second edition, I am grateful to Nima Arkani-Hamed, Yoni Ben-Tov, Nathan Berkovits, Marty Einhorn, Joshua Feinberg, Howard Georgi, Tim Hsieh, Brendan Keller, Joe Polchinski, Yong-shi Wu, and Jean-Bernard Zuber for their helpful comments. Some of them read parts or all of the added chapters. I thank especially Zvi Bern and Rafael Porto for going over the chapters in part N with great care and for many useful suggestions. I also thank Craig Kunimoto, Richard Neher, Matt Pillsbury, and Rafael Porto for teaching me the black art of composing equations on the computer. My editor at Princeton University Press, Ingrid Gnerlich, has always been a pleasure to talk to and work with. I also thank Kathleen Cioffi and Cyd Westmoreland for their meticulous work in producing this book. Last but not least, I am grateful to my wife Janice for her encouragement and loving support.

Convention, Notation, and Units

For the same reason that we no longer use a certain king's feet to measure distance, we use natural units in which the speed of light c and the Dirac symbol $\hbar$ are both set equal to 1. Planck made the profound observation that in natural units all physical quantities can be expressed in terms of the Planck mass $M_{\text{Planck}} \equiv 1/\sqrt{G_{\text{Newton}}} \simeq 10^{19}\text{Gev}$. The quantities c and $\hbar$ are not so much fundamental constants as conversion factors. In this light, I am genuinely puzzled by condensed matter physicists carrying around Boltzmann's constant k, which is no different from the conversion factor between feet and meters.

Spacetime coordinates x^μ are labeled by Greek indices ($\mu = 0, 1, 2, 3$) with the time coordinate x^0 sometimes denoted by t. Space coordinates x^i are labeled by Latin indices ($i = 1, 2, 3$) and $\partial_\mu \equiv \partial/\partial x^\mu$. We use a Minkowski metric $\eta^{\mu\nu}$ with signature $(+, -, -, -)$ so that $\eta^{00} = +1$. We write $\eta^{\mu\nu}\partial_\mu\varphi\partial_\nu\varphi = \partial_\mu\varphi\partial^\mu\varphi = (\partial\varphi)^2 = (\partial\varphi/\partial t)^2 - \sum_i(\partial\varphi/\partial x^i)^2$. The metric in curved spacetime is always denoted by $g^{\mu\nu}$, but often I will also use $g^{\mu\nu}$ for the Minkowski metric when the context indicates clearly that we are in flat spacetime.

Since I will be talking mostly about relativistic quantum field theory in this book I will without further clarification use a relativistic language. Thus, when I speak of momentum, unless otherwise specified, I mean energy and momentum. Also since $\hbar = 1$, I will not distinguish between wave vector k and momentum, and between frequency ω and energy.

In local field theory I deal primarily with the Lagrangian density $\mathcal{L}$ and not the Lagrangian $L = \int d^3x\,\mathcal{L}$. As is common practice in the literature and in oral discussion, I will often abuse terminology and simply refer to $\mathcal{L}$ as the Lagrangian. I will commit other minor abuses such as writing 1 instead of I for the unit matrix. I use the same symbol φ for the Fourier transform $\varphi(k)$ of a function $\varphi(x)$ whenever there is no risk of confusion, as is almost always the case. I prefer an abused terminology to cluttered notation and unbearable pedantry.

The symbol $*$ denotes complex conjugation, and $\dagger$ hermitean conjugation: The former applies to a number and the latter to an operator. I also use the notation c.c. and h.c. Often when there is no risk of confusion I abuse the notation, using $\dagger$ when I should use $*$. For instance, in a path integral, bosonic fields are just number-valued fields, but nevertheless I write $\varphi^\dagger$ rather than φ^*. For a matrix M, then of course $M^\dagger$ and M^* should be carefully distinguished from each other.

I made an effort to get factors of 2 and π right, but some errors will be inevitable.

Q uantum Field Theory in a Nutshell

Part I | Motivation and Foundation

I.1 | Who Needs It?

Who needs quantum field theory?

Quantum field theory arose out of our need to describe the ephemeral nature of life.

No, seriously, quantum field theory is needed when we confront simultaneously the two great physics innovations of the last century of the previous millennium: special relativity and quantum mechanics. Consider a fast moving rocket ship close to light speed. You need special relativity but not quantum mechanics to study its motion. On the other hand, to study a slow moving electron scattering on a proton, you must invoke quantum mechanics, but you don't have to know a thing about special relativity.

It is in the peculiar confluence of special relativity and quantum mechanics that a new set of phenomena arises: Particles can be born and particles can die. It is this matter of birth, life, and death that requires the development of a new subject in physics, that of quantum field theory.

Let me give a heuristic discussion. In quantum mechanics the uncertainty principle tells us that the energy can fluctuate wildly over a small interval of time. According to special relativity, energy can be converted into mass and vice versa. With quantum mechanics and special relativity, the wildly fluctuating energy can metamorphose into mass, that is, into new particles not previously present.

Write down the Schrödinger equation for an electron scattering off a proton. The equation describes the wave function of one electron, and no matter how you shake and bake the mathematics of the partial differential equation, the electron you follow will remain one electron. But special relativity tells us that energy can be converted to matter: If the electron is energetic enough, an electron and a positron ("the antielectron") can be produced. The Schrödinger equation is simply incapable of describing such a phenomenon. Nonrelativistic quantum mechanics must break down.

You saw the need for quantum field theory at another point in your education. Toward the end of a good course on nonrelativistic quantum mechanics the interaction between radiation and atoms is often discussed. You would recall that the electromagnetic field is

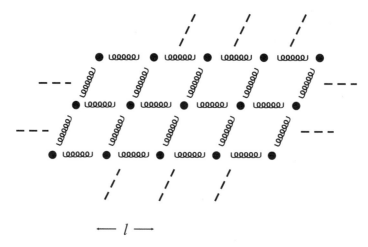

Figure I.1.1

treated as a field; well, it is a field. Its Fourier components are quantized as a collection of harmonic oscillators, leading to creation and annihilation operators for photons. So there, the electromagnetic field is a quantum field. Meanwhile, the electron is treated as a poor cousin, with a wave function $\Psi(x)$ governed by the good old Schrödinger equation. Photons can be created or annihilated, but not electrons. Quite aside from the experimental fact that electrons and positrons could be created in pairs, it would be intellectually more satisfying to treat electrons and photons, as they are both elementary particles, on the same footing.

So, I was more or less right: Quantum field theory is a response to the ephemeral nature of life.

All of this is rather vague, and one of the purposes of this book is to make these remarks more precise. For the moment, to make these thoughts somewhat more concrete, let us ask where in classical physics we might have encountered something vaguely resembling the birth and death of particles. Think of a mattress, which we idealize as a 2-dimensional lattice of point masses connected to each other by springs (fig. I.1.1). For simplicity, let us focus on the vertical displacement [which we denote by $q_a(t)$] of the point masses and neglect the small horizontal movement. The index a simply tells us which mass we are talking about. The Lagrangian is then

$$L = \tfrac{1}{2}\left(\sum_a m\dot{q}_a^2 - \sum_{a,b} k_{ab}q_a q_b - \sum_{a,b,c} g_{abc}q_a q_b q_c - \cdots\right) \tag{1}$$

Keeping only the terms quadratic in q (the "harmonic approximation") we have the equations of motion $m\ddot{q}_a = -\sum_b k_{ab}q_b$. Taking the q's as oscillating with frequency ω, we have $\sum_b k_{ab}q_b = m\omega^2 q_a$. The eigenfrequencies and eigenmodes are determined, respectively, by the eigenvalues and eigenvectors of the matrix k. As usual, we can form wave packets by superposing eigenmodes. When we quantize the theory, these wave packets behave like particles, in the same way that electromagnetic wave packets when quantized behave like particles called photons.

Since the theory is linear, two wave packets pass right through each other. But once we include the nonlinear terms, namely the terms cubic, quartic, and so forth in the q's in (1), the theory becomes anharmonic. Eigenmodes now couple to each other. A wave packet might decay into two wave packets. When two wave packets come near each other, they scatter and perhaps produce more wave packets. This naturally suggests that the physics of particles can be described in these terms.

Quantum field theory grew out of essentially these sorts of physical ideas.

It struck me as limiting that even after some 75 years, the whole subject of quantum field theory remains rooted in this harmonic paradigm, to use a dreadfully pretentious word. We have not been able to get away from the basic notions of oscillations and wave packets. Indeed, string theory, the heir to quantum field theory, is still firmly founded on this harmonic paradigm. Surely, a brilliant young physicist, perhaps a reader of this book, will take us beyond.

Condensed matter physics

In this book I will focus mainly on relativistic field theory, but let me mention here that one of the great advances in theoretical physics in the last 30 years or so is the increasingly sophisticated use of quantum field theory in condensed matter physics. At first sight this seems rather surprising. After all, a piece of "condensed matter" consists of an enormous swarm of electrons moving nonrelativistically, knocking about among various atomic ions and interacting via the electromagnetic force. Why can't we simply write down a gigantic wave function $\Psi(x_1, x_2, \cdots, x_N)$, where x_j denotes the position of the jth electron and N is a large but finite number? Okay, Ψ is a function of many variables but it is still governed by a nonrelativistic Schrödinger equation.

The answer is yes, we can, and indeed that was how solid state physics was first studied in its heroic early days (and still is in many of its subbranches).

Why then does a condensed matter theorist need quantum field theory? Again, let us first go for a heuristic discussion, giving an overall impression rather than all the details. In a typical solid, the ions vibrate around their equilibrium lattice positions. This vibrational dynamics is best described by so-called phonons, which correspond more or less to the wave packets in the mattress model described above.

This much you can read about in any standard text on solid state physics. Furthermore, if you have had a course on solid state physics, you would recall that the energy levels available to electrons form bands. When an electron is kicked (by a phonon field say) from a filled band to an empty band, a hole is left behind in the previously filled band. This hole can move about with its own identity as a particle, enjoying a perfectly comfortable existence until another electron comes into the band and annihilates it. Indeed, it was with a picture of this kind that Dirac first conceived of a hole in the "electron sea" as the antiparticle of the electron, the positron.

We will flesh out this heuristic discussion in subsequent chapters in parts V and VI.

Marriages

To summarize, quantum field theory was born of the necessity of dealing with the marriage of special relativity and quantum mechanics, just as the new science of string theory is being born of the necessity of dealing with the marriage of general relativity and quantum mechanics.

I.2 | Path Integral Formulation of Quantum Physics

The professor's nightmare: a wise guy in the class

As I noted in the preface, I know perfectly well that you are eager to dive into quantum field theory, but first we have to review the path integral formalism of quantum mechanics. This formalism is not universally taught in introductory courses on quantum mechanics, but even if you have been exposed to it, this chapter will serve as a useful review. The reason I start with the path integral formalism is that it offers a particularly convenient way of going from quantum mechanics to quantum field theory. I will first give a heuristic discussion, to be followed by a more formal mathematical treatment.

Perhaps the best way to introduce the path integral formalism is by telling a story, certainly apocryphal as many physics stories are. Long ago, in a quantum mechanics class, the professor droned on and on about the double-slit experiment, giving the standard treatment. A particle emitted from a source S (fig. I.2.1) at time $t = 0$ passes through one or the other of two holes, A_1 and A_2, drilled in a screen and is detected at time $t = T$ by a detector located at O. The amplitude for detection is given by a fundamental postulate of quantum mechanics, the superposition principle, as the sum of the amplitude for the particle to propagate from the source S through the hole A_1 and then onward to the point O and the amplitude for the particle to propagate from the source S through the hole A_2 and then onward to the point O.

Suddenly, a very bright student, let us call him Feynman, asked, "Professor, what if we drill a third hole in the screen?" The professor replied, "Clearly, the amplitude for the particle to be detected at the point O is now given by the sum of three amplitudes, the amplitude for the particle to propagate from the source S through the hole A_1 and then onward to the point O, the amplitude for the particle to propagate from the source S through the hole A_2 and then onward to the point O, and the amplitude for the particle to propagate from the source S through the hole A_3 and then onward to the point O."

The professor was just about ready to continue when Feynman interjected again, "What if I drill a fourth and a fifth hole in the screen?" Now the professor is visibly losing his

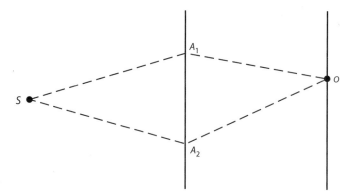

Figure I.2.1

patience: "All right, wise guy, I think it is obvious to the whole class that we just sum over all the holes."

To make what the professor said precise, denote the amplitude for the particle to propagate from the source S through the hole A_i and then onward to the point O as $A(S \rightarrow A_i \rightarrow O)$. Then the amplitude for the particle to be detected at the point O is

$$A(\text{detected at } O) = \sum_i A(S \rightarrow A_i \rightarrow O) \tag{1}$$

But Feynman persisted, "What if we now add another screen (fig. I.2.2) with some holes drilled in it?" The professor was really losing his patience: "Look, can't you see that you just take the amplitude to go from the source S to the hole A_i in the first screen, then to the hole B_j in the second screen, then to the detector at O, and then sum over all i and j?"

Feynman continued to pester, "What if I put in a third screen, a fourth screen, eh? What if I put in a screen and drill an infinite number of holes in it so that the screen is no longer there?" The professor sighed, "Let's move on; there is a lot of material to cover in this course."

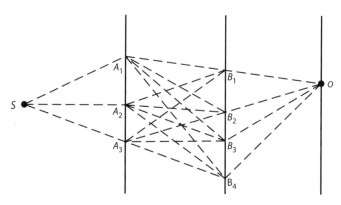

Figure I.2.2

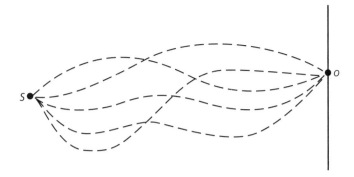

Figure I.2.3

But dear reader, surely you see what that wise guy Feynman was driving at. I especially enjoy his observation that if you put in a screen and drill an infinite number of holes in it, then that screen is not really there. Very Zen! What Feynman showed is that even if there were just empty space between the source and the detector, the amplitude for the particle to propagate from the source to the detector is the sum of the amplitudes for the particle to go through each one of the holes in each one of the (nonexistent) screens. In other words, we have to sum over the amplitude for the particle to propagate from the source to the detector following all possible paths between the source and the detector (fig. I.2.3).

$\mathcal{A}$(particle to go from S to O in time T) =

$$\sum_{\text{(paths)}} \mathcal{A} \text{ (particle to go from } S \text{ to } O \text{ in time } T \text{ following a particular path)} \tag{2}$$

Now the mathematically rigorous will surely get anxious over how $\sum_{\text{(paths)}}$ is to be defined. Feynman followed Newton and Leibniz: Take a path (fig. I.2.4), approximate it by straight line segments, and let the segments go to zero. You can see that this is just like filling up a space with screens spaced infinitesimally close to each other, with an infinite number of holes drilled in each screen.

Fine, but how to construct the amplitude $\mathcal{A}$(particle to go from S to O in time T following a particular path)? Well, we can use the unitarity of quantum mechanics: If we know the amplitude for each infinitesimal segment, then we just multiply them together to get the amplitude of the whole path.

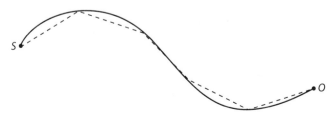

Figure I.2.4

In quantum mechanics, the amplitude to propagate from a point q_I to a point q_F in time T is governed by the unitary operator e^{-iHT}, where H is the Hamiltonian. More precisely, denoting by $|q\rangle$ the state in which the particle is at q, the amplitude in question is just $\langle q_F | e^{-iHT} | q_I \rangle$. Here we are using the Dirac bra and ket notation. Of course, philosophically, you can argue that to say the amplitude is $\langle q_F | e^{-iHT} | q_I \rangle$ amounts to a postulate and a definition of H. It is then up to experimentalists to discover that H is hermitean, has the form of the classical Hamiltonian, et cetera.

Indeed, the whole path integral formalism could be written down mathematically starting with the quantity $\langle q_F | e^{-iHT} | q_I \rangle$, without any of Feynman's jive about screens with an infinite number of holes. Many physicists would prefer a mathematical treatment without the talk. As a matter of fact, the path integral formalism was invented by Dirac precisely in this way, long before Feynman.[1]

A necessary word about notation even though it interrupts the narrative flow: We denote the coordinates transverse to the axis connecting the source to the detector by q, rather than x, for a reason which will emerge in a later chapter. For notational simplicity, we will think of q as 1-dimensional and suppress the coordinate along the axis connecting the source to the detector.

Dirac's formulation

Let us divide the time T into N segments each lasting $\delta t = T/N$. Then we write

$$\langle q_F | e^{-iHT} | q_I \rangle = \langle q_F | e^{-iH\delta t} e^{-iH\delta t} \cdots e^{-iH\delta t} | q_I \rangle$$

Our states are normalized by $\langle q' | q \rangle = \delta(q' - q)$ with δ the Dirac delta function. (Recall that δ is defined by $\delta(q) = \int_{-\infty}^{\infty} (dp/2\pi) e^{ipq}$ and $\int dq \, \delta(q) = 1$. See appendix 1.) Now use the fact that $|q\rangle$ forms a complete set of states so that $\int dq \, |q\rangle\langle q| = 1$. To see that the normalization is correct, multiply on the left by $\langle q''|$ and on the right by $|q'\rangle$, thus obtaining $\int dq \, \delta(q'' - q) \delta(q - q') = \delta(q'' - q')$. Insert 1 between all these factors of $e^{-iH\delta t}$ and write

$$\langle q_F | e^{-iHT} | q_I \rangle$$
$$= (\prod_{j=1}^{N-1} \int dq_j) \langle q_F | e^{-iH\delta t} | q_{N-1} \rangle \langle q_{N-1} | e^{-iH\delta t} | q_{N-2} \rangle \cdots \langle q_2 | e^{-iH\delta t} | q_1 \rangle \langle q_1 | e^{-iH\delta t} | q_I \rangle \tag{3}$$

Focus on an individual factor $\langle q_{j+1} | e^{-iH\delta t} | q_j \rangle$. Let us take the baby step of first evaluating it just for the free-particle case in which $H = \hat{p}^2/2m$. The hat on $\hat{p}$ reminds us that it is an operator. Denote by $|p\rangle$ the eigenstate of $\hat{p}$, namely $\hat{p} |p\rangle = p |p\rangle$. Do you remember from your course in quantum mechanics that $\langle q|p \rangle = e^{ipq}$? Sure you do. This

---

[1] For the true history of the path integral, see p. xv of my introduction to R. P. Feynman, *QED: The Strange Theory of Light and Matter.*

just says that the momentum eigenstate is a plane wave in the coordinate representation. (The normalization is such that $\int (dp/2\pi)|p\rangle\langle p| = 1$. Again, to see that the normalization is correct, multiply on the left by $\langle q'|$ and on the right by $|q\rangle$, thus obtaining $\int (dp/2\pi)e^{ip(q'-q)} = \delta(q'-q)$.) So again inserting a complete set of states, we write

$$\langle q_{j+1}|e^{-i\delta t(\hat{p}^2/2m)}|q_j\rangle = \int \frac{dp}{2\pi} \langle q_{j+1}|e^{-i\delta t(\hat{p}^2/2m)}|p\rangle\langle p|q_j\rangle$$

$$= \int \frac{dp}{2\pi} e^{-i\delta t(p^2/2m)}\langle q_{j+1}|p\rangle\langle p|q_j\rangle$$

$$= \int \frac{dp}{2\pi} e^{-i\delta t(p^2/2m)}e^{ip(q_{j+1}-q_j)}$$

Note that we removed the hat from the momentum operator in the exponential: Since the momentum operator is acting on an eigenstate, it can be replaced by its eigenvalue. Also, we are evidently working in the Heisenberg picture.

The integral over p is known as a Gaussian integral, with which you may already be familiar. If not, turn to appendix 2 to this chapter.

Doing the integral over p, we get (using (21))

$$\langle q_{j+1}|e^{-i\delta t(\hat{p}^2/2m)}|q_j\rangle = \left(\frac{-im}{2\pi\delta t}\right)^{\frac{1}{2}} e^{[im(q_{j+1}-q_j)^2]/2\delta t}$$

$$= \left(\frac{-im}{2\pi\delta t}\right)^{\frac{1}{2}} e^{i\delta t(m/2)[(q_{j+1}-q_j)/\delta t]^2}$$

Putting this into (3) yields

$$\langle q_F|e^{-iHT}|q_I\rangle = \left(\frac{-im}{2\pi\delta t}\right)^{\frac{N}{2}} \left(\prod_{k=1}^{N-1}\int dq_k\right) e^{i\delta t(m/2)\sum_{j=0}^{N-1}[(q_{j+1}-q_j)/\delta t]^2}$$

with $q_0 \equiv q_I$ and $q_N \equiv q_F$.

We can now go to the continuum limit $\delta t \to 0$. Newton and Leibniz taught us to replace $[(q_{j+1}-q_j)/\delta t]^2$ by $\dot{q}^2$, and $\delta t \sum_{j=0}^{N-1}$ by $\int_0^T dt$. Finally, we define the integral over paths as

$$\int Dq(t) = \lim_{N\to\infty} \left(\frac{-im}{2\pi\delta t}\right)^{\frac{N}{2}} \left(\prod_{k=1}^{N-1}\int dq_k\right)$$

We thus obtain the path integral representation

$$\langle q_F|e^{-iHT}|q_I\rangle = \int Dq(t)\, e^{i\int_0^T dt\,\frac{1}{2}m\dot{q}^2} \tag{4}$$

This fundamental result tells us that to obtain $\langle q_F|e^{-iHT}|q_I\rangle$ we simply integrate over all possible paths $q(t)$ such that $q(0) = q_I$ and $q(T) = q_F$.

As an exercise you should convince yourself that had we started with the Hamiltonian for a particle in a potential $H = \hat{p}^2/2m + V(\hat{q})$ (again the hat on $\hat{q}$ indicates an operator) the final result would have been

$$\langle q_F|e^{-iHT}|q_I\rangle = \int Dq(t)\, e^{i\int_0^T dt[\frac{1}{2}m\dot{q}^2 - V(q)]} \tag{5}$$

We recognize the quantity $\frac{1}{2}m\dot{q}^2 - V(q)$ as just the Lagrangian $L(\dot{q}, q)$. The Lagrangian has emerged naturally from the Hamiltonian! In general, we have

$$\langle q_F | e^{-iHT} | q_I \rangle = \int Dq(t)\, e^{i\int_0^T dt\, L(\dot{q},q)} \tag{6}$$

To avoid potential confusion, let me be clear that t appears as an integration variable in the exponential on the right-hand side. The appearance of t in the path integral measure $Dq(t)$ is simply to remind us that q is a function of t (as if we need reminding). Indeed, this measure will often be abbreviated to Dq. You might recall that $\int_0^T dt\, L(\dot{q}, q)$ is called the action $S(q)$ in classical mechanics. The action S is a functional of the function $q(t)$.

Often, instead of specifying that the particle starts at an initial position q_I and ends at a final position q_F, we prefer to specify that the particle starts in some initial state I and ends in some final state F. Then we are interested in calculating $\langle F | e^{-iHT} | I \rangle$, which upon inserting complete sets of states can be written as

$$\int dq_F \int dq_I \langle F | q_F \rangle \langle q_F | e^{-iHT} | q_I \rangle \langle q_I | I \rangle,$$

which mixing Schrödinger and Dirac notation we can write as

$$\int dq_F \int dq_I \Psi_F(q_F)^* \langle q_F | e^{-iHT} | q_I \rangle \Psi_I(q_I).$$

In most cases we are interested in taking $|I\rangle$ and $|F\rangle$ as the ground state, which we will denote by $|0\rangle$. It is conventional to give the amplitude $\langle 0 | e^{-iHT} | 0 \rangle$ the name Z.

At the level of mathematical rigor we are working with, we count on the path integral $\int Dq(t)\, e^{i\int_0^T dt[\frac{1}{2}m\dot{q}^2 - V(q)]}$ to converge because the oscillatory phase factors from different paths tend to cancel out. It is somewhat more rigorous to perform a so-called Wick rotation to Euclidean time. This amounts to substituting $t \to -it$ and rotating the integration contour in the complex t plane so that the integral becomes

$$Z = \int Dq(t)\, e^{-\int_0^T dt[\frac{1}{2}m\dot{q}^2 + V(q)]}, \tag{7}$$

known as the Euclidean path integral. As is done in appendix 2 to this chapter with ordinary integrals we will always assume that we can make this type of substitution with impunity.

The classical world emerges

One particularly nice feature of the path integral formalism is that the classical limit of quantum mechanics can be recovered easily. We simply restore Planck's constant $\hbar$ in (6):

$$\langle q_F | e^{-(i/\hbar)HT} | q_I \rangle = \int Dq(t)\, e^{(i/\hbar)\int_0^T dt\, L(\dot{q},q)}$$

and take the $\hbar \to 0$ limit. Applying the stationary phase or steepest descent method (if you don't know it see appendix 3 to this chapter) we obtain $e^{(i/\hbar)\int_0^T dt\, L(\dot{q}_c, q_c)}$, where $q_c(t)$ is the "classical path" determined by solving the Euler-Lagrange equation $(d/dt)(\delta L/\delta \dot{q}) - (\delta L/\delta q) = 0$ with appropriate boundary conditions.

Appendix 1

For your convenience, I include a concise review of the Dirac delta function here. Let us define a function $d_K(x)$ by

$$d_K(x) \equiv \int_{-\frac{K}{2}}^{\frac{K}{2}} \frac{dk}{2\pi} e^{ikx} = \frac{1}{\pi x} \sin \frac{Kx}{2} \tag{8}$$

for arbitrary real values of x. We see that for large K the even function $d_K(x)$ is sharply peaked at the origin $x = 0$, reaching a value of $K/2\pi$ at the origin, crossing zero at $x = 2\pi/K$, and then oscillating with ever decreasing amplitude. Furthermore,

$$\int_{-\infty}^{\infty} dx \, d_K(x) = \frac{2}{\pi} \int_0^{\infty} \frac{dx}{x} \sin \frac{Kx}{2} = \frac{2}{\pi} \int_0^{\infty} \frac{dy}{y} \sin y = 1 \tag{9}$$

The Dirac delta function is defined by $\delta(x) = \lim_{K \to \infty} d_K(x)$. Heuristically, it could be thought of as an infinitely sharp spike located at $x = 0$ such that the area under the spike is equal to 1. Thus for a function $s(x)$ well-behaved around $x = a$ we have

$$\int_{-\infty}^{\infty} dx \, \delta(x - a)s(x) = s(a) \tag{10}$$

(By the way, for what it is worth, mathematicians call the delta function a "distribution," not a function.)

Our derivation also yields an integral representation for the delta function that we will use repeatedly in this text:

$$\delta(x) = \int_{-\infty}^{\infty} \frac{dk}{2\pi} e^{ikx} \tag{11}$$

We will often use the identity

$$\int_{-\infty}^{\infty} dx \, \delta(f(x))s(x) = \sum_i \frac{s(x_i)}{|f'(x_i)|} \tag{12}$$

where x_i denotes the zeroes of $f(x)$ (in other words, $f(x_i) = 0$ and $f'(x_i) = df(x_i)/dx$.) To prove this, first show that $\int_{-\infty}^{\infty} dx \, \delta(bx)s(x) = \int_{-\infty}^{\infty} dx \, \frac{\delta(x)}{|b|} s(x) = s(0)/|b|$. The factor of $1/b$ follows from dimensional analysis. (To see the need for the absolute value, simply note that $\delta(bx)$ is a positive function. Alternatively, change integration variable to $y = bx$: for b negative we have to flip the integration limits.) To obtain (12), expand around each of the zeroes of $f(x)$.

Another useful identity (understood in the limit in which the positive infinitesimal ε tends to zero) is

$$\frac{1}{x + i\varepsilon} = \mathcal{P}\frac{1}{x} - i\pi\delta(x) \tag{13}$$

To see this, simply write $1/(x + i\varepsilon) = x/(x^2 + \varepsilon^2) - i\varepsilon/(x^2 + \varepsilon^2)$, and then note that $\varepsilon/(x^2 + \varepsilon^2)$ as a function of x is sharply spiked around $x = 0$ and that its integral from $-\infty$ to ∞ is equal to π. Thus we have another representation of the Dirac delta function:

$$\delta(x) = \frac{1}{\pi} \frac{\varepsilon}{x^2 + \varepsilon^2} \tag{14}$$

Meanwhile, the principal value integral is defined by

$$\int dx \, \mathcal{P}\frac{1}{x} f(x) = \lim_{\varepsilon \to 0} \int dx \, \frac{x}{x^2 + \varepsilon^2} f(x) \tag{15}$$

Appendix 2

I will now show you how to do the integral $G \equiv \int_{-\infty}^{+\infty} dx\, e^{-\frac{1}{2}x^2}$. The trick is to square the integral, call the dummy integration variable in one of the integrals y, and then pass to polar coordinates:

$$G^2 = \int_{-\infty}^{+\infty} dx\, e^{-\frac{1}{2}x^2} \int_{-\infty}^{+\infty} dy\, e^{-\frac{1}{2}y^2} = 2\pi \int_{0}^{+\infty} dr\, r e^{-\frac{1}{2}r^2}$$

$$= 2\pi \int_{0}^{+\infty} dw\, e^{-w} = 2\pi$$

Thus, we obtain

$$\int_{-\infty}^{+\infty} dx\, e^{-\frac{1}{2}x^2} = \sqrt{2\pi} \tag{16}$$

Believe it or not, a significant fraction of the theoretical physics literature consists of varying and elaborating this basic Gaussian integral. The simplest extension is almost immediate:

$$\int_{-\infty}^{+\infty} dx\, e^{-\frac{1}{2}ax^2} = \left(\frac{2\pi}{a}\right)^{\frac{1}{2}} \tag{17}$$

as can be seen by scaling $x \to x/\sqrt{a}$.

Acting on this repeatedly with $-2(d/da)$ we obtain

$$\langle x^{2n}\rangle \equiv \frac{\int_{-\infty}^{+\infty} dx\, e^{-\frac{1}{2}ax^2} x^{2n}}{\int_{-\infty}^{+\infty} dx\, e^{-\frac{1}{2}ax^2}} = \frac{1}{a^n}(2n-1)(2n-3)\cdots 5\cdot 3\cdot 1 \tag{18}$$

The factor $1/a^n$ follows from dimensional analysis. To remember the factor $(2n-1)!! \equiv (2n-1)(2n-3)\cdots 5\cdot 3\cdot 1$ imagine $2n$ points and connect them in pairs. The first point can be connected to one of $(2n-1)$ points, the second point can now be connected to one of the remaining $(2n-3)$ points, and so on. This clever observation, due to Gian Carlo Wick, is known as Wick's theorem in the field theory literature. Incidentally, field theorists use the following graphical mnemonic in calculating, for example, $\langle x^6 \rangle$: Write $\langle x^6 \rangle$ as $\langle xxxxxx \rangle$ and connect the x's, for example

$$\langle xxxxxx \rangle$$

The pattern of connection is known as a Wick contraction. In this simple example, since the six x's are identical, any one of the distinct Wick contractions gives the same value a^{-3} and the final result for $\langle x^6 \rangle$ is just a^{-3} times the number of distinct Wick contractions, namely $5\cdot 3\cdot 1 = 15$. We will soon come to a less trivial example, with distinct x's, in which case distinct Wick contraction gives distinct values.

An important variant is the integral

$$\int_{-\infty}^{+\infty} dx\, e^{-\frac{1}{2}ax^2 + Jx} = \left(\frac{2\pi}{a}\right)^{\frac{1}{2}} e^{J^2/2a} \tag{19}$$

To see this, take the expression in the exponent and "complete the square": $-ax^2/2 + Jx = -(a/2)(x^2 - 2Jx/a) = -(a/2)(x - J/a)^2 + J^2/2a$. The x integral can now be done by shifting $x \to x + J/a$, giving the factor of $(2\pi/a)^{\frac{1}{2}}$. Check that we can also obtain (18) by differentiating with respect to J repeatedly and then setting $J = 0$.

Another important variant is obtained by replacing J by iJ:

$$\int_{-\infty}^{+\infty} dx\, e^{-\frac{1}{2}ax^2 + iJx} = \left(\frac{2\pi}{a}\right)^{\frac{1}{2}} e^{-J^2/2a} \tag{20}$$

To get yet another variant, replace a by $-ia$:

$$\int_{-\infty}^{+\infty} dx\, e^{\frac{1}{2}iax^2+iJx} = \left(\frac{2\pi i}{a}\right)^{\frac{1}{2}} e^{-iJ^2/2a} \tag{21}$$

Let us promote a to a real symmetric N by N matrix A_{ij} and x to a vector x_i $(i, j = 1, \cdots, N)$. Then (19) generalizes to

$$\int_{-\infty}^{+\infty}\int_{-\infty}^{+\infty}\cdots\int_{-\infty}^{+\infty} dx_1 dx_2 \cdots dx_N\, e^{-\frac{1}{2}x\cdot A\cdot x+J\cdot x} = \left(\frac{(2\pi)^N}{\det[A]}\right)^{\frac{1}{2}} e^{\frac{1}{2}J\cdot A^{-1}\cdot J} \tag{22}$$

where $x \cdot A \cdot x = x_i A_{ij} x_j$ and $J \cdot x = J_i x_i$ (with repeated indices summed.)

To derive this important relation, diagonalize A by an orthogonal transformation O so that $A = O^{-1} \cdot D \cdot O$, where D is a diagonal matrix. Call $y_i = O_{ij}x_j$. In other words, we rotate the coordinates in the N-dimensional Euclidean space we are integrating over. The expression in the exponential in the integrand then becomes $-\frac{1}{2}y \cdot D \cdot y + (OJ) \cdot y$. Using $\int_{-\infty}^{+\infty}\cdots\int_{-\infty}^{+\infty} dx_1 \cdots dx_N = \int_{-\infty}^{+\infty}\cdots\int_{-\infty}^{+\infty} dy_1 \cdots dy_N$, we factorize the left-hand side of (22) into a product of N integrals, each of the form $\int_{-\infty}^{+\infty} dy_i e^{-\frac{1}{2}D_{ii}y_i^2+(OJ)_i y_i}$. Plugging into (19) we obtain the right hand side of (22), since $(OJ) \cdot D^{-1} \cdot (OJ) = J \cdot O^{-1}D^{-1}O \cdot J = J \cdot A^{-1} \cdot J$ (where we use the orthogonality of O). (To make sure you got it, try this explicitly for $N = 2$.)

Putting in some i's $(A \to -iA, J \to iJ)$, we find the generalization of (22)

$$\int_{-\infty}^{+\infty}\int_{-\infty}^{+\infty}\cdots\int_{-\infty}^{+\infty} dx_1 dx_2 \cdots dx_N\, e^{(i/2)x\cdot A\cdot x+iJ\cdot x}$$
$$= \left(\frac{(2\pi i)^N}{\det[A]}\right)^{\frac{1}{2}} e^{-(i/2)J\cdot A^{-1}\cdot J} \tag{23}$$

The generalization of (18) is also easy to obtain. Differentiate (22) p times with respect to $J_i, J_j, \cdots J_k$, and J_l, and then set $J = 0$. For example, for $p = 1$ the integrand in (22) becomes $e^{-\frac{1}{2}x\cdot A\cdot x}x_i$ and since the integrand is now odd in x_i the integral vanishes. For $p = 2$ the integrand becomes $e^{-\frac{1}{2}x\cdot A\cdot x}(x_i x_j)$, while on the right hand side we bring down A_{ij}^{-1}. Rearranging and eliminating $\det[A]$ (by setting $J = 0$ in (22)), we obtain

$$\langle x_i x_j \rangle = \frac{\int_{-\infty}^{+\infty}\int_{-\infty}^{+\infty}\cdots\int_{-\infty}^{+\infty} dx_1 dx_2 \cdots dx_N\, e^{-\frac{1}{2}x\cdot A\cdot x}x_i x_j}{\int_{-\infty}^{+\infty}\int_{-\infty}^{+\infty}\cdots\int_{-\infty}^{+\infty} dx_1 dx_2 \cdots dx_N\, e^{-\frac{1}{2}x\cdot A\cdot x}} = A_{ij}^{-1}$$

Just do it. Doing it is easier than explaining how to do it. Then do it for $p = 3$ and 4. You will see immediately how your result generalizes. When the set of indices $i, j, \cdots, k, l$ contains an odd number of elements, $\langle x_i x_j \cdots x_k x_l \rangle$ vanishes trivially. When the set of indices $i, j, \cdots, k, l$ contains an even number of elements, we have

$$\langle x_i x_j \cdots x_k x_l \rangle = \sum_{\text{Wick}} (A^{-1})_{ab} \cdots (A^{-1})_{cd} \tag{24}$$

where we have defined

$$\langle x_i x_j \cdots x_k x_l \rangle$$
$$= \frac{\int_{-\infty}^{+\infty}\int_{-\infty}^{+\infty}\cdots\int_{-\infty}^{+\infty} dx_1 dx_2 \cdots dx_N\, e^{-\frac{1}{2}x\cdot A\cdot x}x_i x_j \cdots x_k x_l}{\int_{-\infty}^{+\infty}\int_{-\infty}^{+\infty}\cdots\int_{-\infty}^{+\infty} dx_1 dx_2 \cdots dx_N\, e^{-\frac{1}{2}x\cdot A\cdot x}} \tag{25}$$

and where the set of indices $\{a, b, \cdots, c, d\}$ represent a permutation of $\{i, j, \cdots, k, l\}$. The sum in (24) is over all such permutations or Wick contractions.

For example,

$$\langle x_i x_j x_k x_l \rangle = (A^{-1})_{ij}(A^{-1})_{kl} + (A^{-1})_{il}(A^{-1})_{jk} + (A^{-1})_{ik}(A^{-1})_{jl} \tag{26}$$

(Recall that A, and thus A^{-1}, is symmetric.) As in the simple case when x does not carry any index, we could connect the x's in $\langle x_i x_j x_k x_l \rangle$ in pairs (Wick contraction) and write a factor $(A^{-1})_{ab}$ if we connect x_a to x_b.

Notice that since $\langle x_i x_j \rangle = (A^{-1})_{ij}$ the right hand side of (24) can also be written in terms of objects like $\langle x_i x_j \rangle$. Thus, $\langle x_i x_j x_k x_l \rangle = \langle x_i x_j \rangle\langle x_k x_l \rangle + \langle x_i x_l \rangle\langle x_j x_k \rangle + \langle x_i x_k \rangle\langle x_j x_l \rangle$.

Please work out $\langle x_i x_j x_k x_l x_m x_n \rangle$; you will become an expert on Wick contractions. Of course, (24) reduces to (18) for $N = 1$.

Perhaps you are like me and do not like to memorize anything, but some of these formulas might be worth memorizing as they appear again and again in theoretical physics (and in this book).

Appendix 3

To do an exponential integral of the form $I = \int_{-\infty}^{+\infty} dq \, e^{-(1/\hbar)f(q)}$ we often have to resort to the steepest-descent approximation, which I will now review for your convenience. In the limit of $\hbar$ small, the integral is dominated by the minimum of $f(q)$. Expanding $f(q) = f(a) + \frac{1}{2} f''(a)(q-a)^2 + O[(q-a)^3]$ and applying (17) we obtain

$$I = e^{-(1/\hbar)f(a)} \left(\frac{2\pi \hbar}{f''(a)} \right)^{\frac{1}{2}} e^{-O(\hbar^{\frac{1}{2}})} \tag{27}$$

For $f(q)$ a function of many variables $q_1, \ldots, q_N$ and with a minimum at $q_j = a_j$, we generalize immediately to

$$I = e^{-(1/\hbar)f(a)} \left(\frac{(2\pi \hbar)^N}{\det f''(a)} \right)^{\frac{1}{2}} e^{-O(\hbar^{\frac{1}{2}})} \tag{28}$$

Here $f''(a)$ denotes the N by N matrix with entries $[f''(a)]_{ij} \equiv (\partial^2 f/\partial q_i \partial q_j)|_{q=a}$. In many situations, we do not even need the factor involving the determinant in (28). If you can derive (28) you are well on your way to becoming a quantum field theorist!

Exercises

I.2.1   Verify (5).

I.2.2   Derive (24).

The mattress in the continuum limit

The path integral representation

$$Z \equiv \langle 0 | e^{-iHT} | 0 \rangle = \int Dq(t) \, e^{i \int_0^T dt [\frac{1}{2} m \dot{q}^2 - V(q)]} \tag{1}$$

(we suppress the factor $\langle 0 | q_f \rangle \langle q_I | 0 \rangle$; we will come back to this issue later in this chapter) which we derived for the quantum mechanics of a single particle, can be generalized almost immediately to the case of N particles with the Hamiltonian

$$H = \sum_a \frac{1}{2m_a} \hat{p}_a^2 + V(\hat{q}_1, \hat{q}_2, \cdots, \hat{q}_N). \tag{2}$$

We simply keep track mentally of the position of the particles q_a with $a = 1, 2, \cdots, N$. Going through the same steps as before, we obtain

$$Z \equiv \langle 0 | e^{-iHT} | 0 \rangle = \int Dq(t) \, e^{iS(q)} \tag{3}$$

with the action

$$S(q) = \int_0^T dt \left(\sum_a \frac{1}{2} m_a \dot{q}_a^2 - V[q_1, q_2, \cdots, q_N] \right).$$

The potential energy $V(q_1, q_2, \ldots, q_N)$ now includes interaction energy between particles, namely terms of the form $v(q_a - q_b)$, as well as the energy due to an external potential, namely terms of the form $w(q_a)$. In particular, let us now write the path integral description of the quantum dynamics of the mattress described in chapter I.1, with the potential

$$V(q_1, q_2, \ldots, q_N) = \sum_{ab} \frac{1}{2} k_{ab} (q_a - q_b)^2 + \cdots$$

We are now just a short hop and skip away from a quantum field theory! Suppose we are only interested in phenomena on length scales much greater than the lattice spacing l (see fig. I.1.1). Mathematically, we take the continuum limit $l \to 0$. In this limit, we can replace the label a on the particles by a two-dimensional position vector $\vec{x}$, and so we write $q(t, \vec{x})$ instead of $q_a(t)$. It is traditional to replace the Latin letter q by the Greek letter φ. The function $\varphi(t, \vec{x})$ is called a field.

The kinetic energy $\sum_a \frac{1}{2} m_a \dot{q}_a^2$ now becomes $\int d^2x \frac{1}{2} \sigma (\partial\varphi/\partial t)^2$. We replace $\sum_a$ by $\int d^2x/l^2$ and denote the mass per unit area m_a/l^2 by σ. We take all the m_a's to be equal; otherwise σ would be a function of $\vec{x}$, the system would be inhomogeneous, and we would have a hard time writing down a Lorentz-invariant action (see later).

We next focus on the first term in V. Assume for simplicity that k_{ab} connect only nearest neighbors on the lattice. For nearest-neighbor pairs $(q_a - q_b)^2 \simeq l^2 (\partial\varphi/\partial x)^2 + \cdots$ in the continuum limit; the derivative is obviously taken in the direction that joins the lattice sites a and b.

Putting it together then, we have

$$S(q) \to S(\varphi) \equiv \int_0^T dt \int d^2x \mathcal{L}(\varphi)$$

$$= \int_0^T dt \int d^2x \frac{1}{2} \left\{ \sigma \left(\frac{\partial\varphi}{\partial t} \right)^2 - \rho \left[\left(\frac{\partial\varphi}{\partial x} \right)^2 + \left(\frac{\partial\varphi}{\partial y} \right)^2 \right] - \tau\varphi^2 - \varsigma\varphi^4 + \cdots \right\} \tag{4}$$

where the parameter ρ is determined by k_{ab} and l. The precise relations do not concern us.

Henceforth in this book, we will take the $T \to \infty$ limit so that we can integrate over all of spacetime in (4).

We can clean up a bit by writing $\rho = \sigma c^2$ and scaling $\varphi \to \varphi/\sqrt{\sigma}$, so that the combination $(\partial\varphi/\partial t)^2 - c^2[(\partial\varphi/\partial x)^2 + (\partial\varphi/\partial y)^2]$ appears in the Lagrangian. The parameter c evidently has the dimension of a velocity and defines the phase velocity of the waves on our mattress.

We started with a mattress for pedagogical reasons. Of course nobody believes that the fields observed in Nature, such as the meson field or the photon field, are actually constructed of point masses tied together with springs. The modern view, which I will call Landau-Ginzburg, is that we start with the desired symmetry, say Lorentz invariance if we want to do particle physics, decide on the fields we want by specifying how they transform under the symmetry (in this case we decided on a scalar field φ), and then write down the action involving no more than two time derivatives (because we don't know how to quantize actions with more than two time derivatives).

We end up with a Lorentz-invariant action (setting $c = 1$)

$$S = \int d^dx \left[\frac{1}{2} (\partial\varphi)^2 - \frac{1}{2} m^2\varphi^2 - \frac{g}{3!}\varphi^3 - \frac{\lambda}{4!}\varphi^4 + \cdots \right] \tag{5}$$

where various numerical factors are put in for later convenience. The relativistic notation $(\partial\varphi)^2 \equiv \partial_\mu\varphi\partial^\mu\varphi = (\partial\varphi/\partial t)^2 - (\partial\varphi/\partial x)^2 - (\partial\varphi/\partial y)^2$ was explained in the note on convention. The dimension of spacetime, d, clearly can be any integer, even though in our mattress model it was actually 3. We often write $d = D + 1$ and speak of a $(D + 1)$-dimensional spacetime.

We see here the power of imposing a symmetry. Lorentz invariance together with the insistence that the Lagrangian involve only at most two powers of $\partial/\partial t$ immediately tells us that the Lagrangian can only have the form[1] $\mathcal{L} = \frac{1}{2}(\partial\varphi)^2 - V(\varphi)$ with V some function of φ. For simplicity, we now restrict V to be a polynomial in φ, although much of the present discussion will not depend on this restriction. We will have a great deal more to say about symmetry later. Here we note that, for example, we could insist that physics is symmetric under $\varphi \to -\varphi$, in which case $V(\varphi)$ would have to be an even polynomial.

Now that you know what a quantum field theory is, you realize why I used the letter q to label the position of the particle in the previous chapter and not the more common $\vec{x}$. In quantum field theory, $\vec{x}$ is a label, not a dynamical variable. The $\vec{x}$ appearing in $\varphi(t, \vec{x})$ corresponds to the label a in $q_a(t)$ in quantum mechanics. The dynamical variable in field theory is not position, but the field φ. The variable $\vec{x}$ simply specifies which field variable we are talking about. I belabor this point because upon first exposure to quantum field theory some students, used to thinking of $\vec{x}$ as a dynamical operator in quantum mechanics, are confused by its role here.

In summary, we have the table

$$
\begin{array}{|c|}
\hline
q \to \varphi \\
\hline
a \to \vec{x} \\
\hline
q_a(t) \to \varphi(t, \vec{x}) = \varphi(x) \\
\hline
\sum_a \to \int d^D x \\
\hline
\end{array}
\tag{6}
$$

Thus we finally have the path integral defining a scalar field theory in $d = (D + 1)$ dimensional spacetime:

$$
Z = \int D\varphi e^{i \int d^d x (\frac{1}{2}(\partial\varphi)^2 - V(\varphi))}
\tag{7}
$$

Note that a $(0 + 1)$-dimensional quantum field theory is just quantum mechanics.

The classical limit

As I have already remarked, the path integral formalism is particularly convenient for taking the classical limit. Remembering that Planck's constant $\hbar$ has the dimension of energy multiplied by time, we see that it appears in the unitary evolution operator $e^{(-i/\hbar)HT}$. Tracing through the derivation of the path integral, we see that we simply divide the overall factor i by $\hbar$ to get

$$
Z = \int D\varphi e^{(i/\hbar) \int d^4 x \mathcal{L}(\varphi)}
\tag{8}
$$

---

[1] Strictly speaking, a term of the form $U(\varphi)(\partial\varphi)^2$ is also possible. In quantum mechanics, a term such as $U(q)(dq/dt)^2$ in the Lagrangian would describe a particle whose mass depends on position. We will not consider such "nasty" terms until much later.

In the limit $\hbar$ much smaller than the relevant action we are considering, we can evaluate the path integral using the stationary phase (or steepest descent) approximation, as I explained in the previous chapter in the context of quantum mechanics. We simply determine the extremum of $\int d^4x \mathcal{L}(\varphi)$. According to the usual Euler-Lagrange variational procedure, this leads to the equation

$$\partial_\mu \frac{\delta \mathcal{L}}{\delta(\partial_\mu \varphi)} - \frac{\delta \mathcal{L}}{\delta \varphi} = 0 \tag{9}$$

We thus recover the classical field equation, exactly as we should, which in our scalar field theory reads

$$(\partial^2 + m^2)\varphi(x) + \frac{g}{2}\varphi(x)^2 + \frac{\lambda}{6}\varphi(x)^3 + \cdots = 0 \tag{10}$$

The vacuum

In the point particle quantum mechanics discussed in chapter I.2 we wrote the path integral for $\langle F| e^{-iHT} |I \rangle$, with some initial and final state, which we can choose at our pleasure. A convenient and particularly natural choice would be to take $|I\rangle = |F\rangle$ to be the ground state. In quantum field theory what should we choose for the initial and final states? A standard choice for the initial and final states is the ground state or the vacuum state of the system, denoted by $|0\rangle$, in which, speaking colloquially, nothing is happening. In other words, we would calculate the quantum transition amplitude from the vacuum to the vacuum, which would enable us to determine the energy of the ground state. But this is not a particularly interesting quantity, because in quantum field theory we would like to measure all energies relative to the vacuum and so, by convention, would set the energy of the vacuum to zero (possibly by having to subtract a constant from the Lagrangian). Incidentally, the vacuum in quantum field theory is a stormy sea of quantum fluctuations, but for this initial pass at quantum field theory, we will not examine it in any detail. We will certainly come back to the vacuum in later chapters.

Disturbing the vacuum

We might enjoy doing something more exciting than watching a boiling sea of quantum fluctuations. We might want to disturb the vacuum. Somewhere in space, at some instant in time, we would like to create a particle, watch it propagate for a while, and then annihilate it somewhere else in space, at some later instant in time. In other words, we want to set up a source and a sink (sometimes referred to collectively as sources) at which particles can be created and annihilated.

To see how to do this, let us go back to the mattress. Bounce up and down on it to create some excitations. Obviously, pushing on the mass labeled by a in the mattress corresponds to adding a term such as $J_a(t)q_a$ to the potential $V(q_1, q_2, \cdots, q_N)$. More generally,

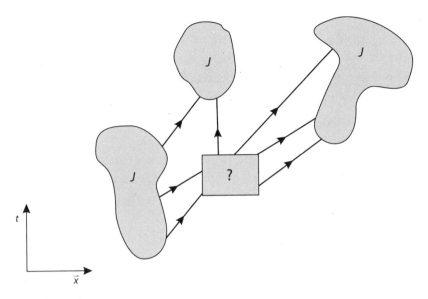

Figure I.3.1

we can add $\sum_a J_a(t)q_a$. When we go to field theory this added term gets promoted to $\int d^D x \, J(x)\varphi(x)$ in the field theory Lagrangian, according to the promotion table (6).

This so-called source function $J(t, \vec{x})$ describes how the mattress is being disturbed. We can choose whatever function we like, corresponding to our freedom to push on the mattress wherever and whenever we like. In particular, $J(x)$ can vanish everywhere in spacetime except in some localized regions.

By bouncing up and down on the mattress we can get wave packets going off here and there (fig. I.3.1). This corresponds precisely to sources (and sinks) for particles. Thus, we really want the path integral

$$Z = \int D\varphi \, e^{i \int d^4 x \left[\frac{1}{2}(\partial\varphi)^2 - V(\varphi) + J(x)\varphi(x)\right]} \tag{11}$$

Free field theory

The functional integral in (11) is impossible to do except when

$$\mathcal{L}(\varphi) = \tfrac{1}{2}[(\partial\varphi)^2 - m^2\varphi^2] \tag{12}$$

The corresponding theory is called the free or Gaussian theory. The equation of motion (9) works out to be $(\partial^2 + m^2)\varphi = 0$, known as the Klein-Gordon equation.[2] Being linear, it can be solved immediately to give $\varphi(\vec{x}, t) = e^{i(\omega t - \vec{k}\cdot\vec{x})}$ with

$$\omega^2 = \vec{k}^2 + m^2 \tag{13}$$

---

[2] The Klein-Gordon equation was actually discovered by Schrödinger before he found the equation that now bears his name. Later, in 1926, it was written down independently by Klein, Gordon, Fock, Kudar, de Donder, and Van Dungen.

In the natural units we are using, $\hbar = 1$ and so frequency ω is the same as energy $\hbar\omega$ and wave vector $\vec{k}$ is the same as momentum $\hbar\vec{k}$. Thus, we recognize (13) as the energy-momentum relation for a particle of mass m, namely the sophisticate's version of the layperson's $E = mc^2$. We expect this field theory to describe a relativistic particle of mass m.

Let us now evaluate (11) in this special case:

$$Z = \int D\varphi e^{i \int d^4x \{ \frac{1}{2}[(\partial\varphi)^2 - m^2\varphi^2] + J\varphi \}} \tag{14}$$

Integrating by parts under the $\int d^4x$ and not worrying about the possible contribution of boundary terms at infinity (we implicitly assume that the fields we are integrating over fall off sufficiently rapidly), we write

$$Z = \int D\varphi e^{i \int d^4x [-\frac{1}{2}\varphi(\partial^2 + m^2)\varphi + J\varphi]} \tag{15}$$

You will encounter functional integrals like this again and again in your study of field theory. The trick is to imagine discretizing spacetime. You don't actually have to do it: Just imagine doing it. Let me sketch how this goes. Replace the function $\varphi(x)$ by the vector $\varphi_i = \varphi(ia)$ with i an integer and a the lattice spacing. (For simplicity, I am writing things as if we were in 1-dimensional spacetime. More generally, just let the index i enumerate the lattice points in some way.) Then differential operators become matrices. For example, $\partial\varphi(ia) \to (1/a)(\varphi_{i+1} - \varphi_i) \equiv \sum_j M_{ij}\varphi_j$, with some appropriate matrix M. Integrals become sums. For example, $\int d^4x J(x)\varphi(x) \to a^4 \sum_i J_i\varphi_i$.

Now, lo and behold, the integral (15) is just the integral we did in (I.2.23)

$$\int_{-\infty}^{+\infty} \int_{-\infty}^{+\infty} \cdots \int_{-\infty}^{+\infty} dq_1 dq_2 \cdots dq_N \, e^{(i/2)q \cdot A \cdot q + iJ \cdot q}$$

$$= \left(\frac{(2\pi i)^N}{\det[A]} \right)^{\frac{1}{2}} e^{-(i/2)J \cdot A^{-1} \cdot J} \tag{16}$$

The role of A in (16) is played in (15) by the differential operator $-(\partial^2 + m^2)$. The defining equation for the inverse, $A \cdot A^{-1} = I$ or $A_{ij}A_{jk}^{-1} = \delta_{ik}$, becomes in the continuum limit

$$-(\partial^2 + m^2)D(x - y) = \delta^{(4)}(x - y) \tag{17}$$

We denote the continuum limit of A_{jk}^{-1} by $D(x - y)$ (which we know must be a function of $x - y$, and not of x and y separately, since no point in spacetime is special). Note that in going from the lattice to the continuum Kronecker is replaced by Dirac. It is very useful to be able to go back and forth mentally between the lattice and the continuum.

Our final result is

$$Z(J) = Ce^{-(i/2)\int\int d^4x d^4y J(x)D(x-y)J(y)} \equiv Ce^{iW(J)} \tag{18}$$

with $D(x)$ determined by solving (17). The overall factor C, which corresponds to the overall factor with the determinant in (16), does not depend on J and, as will become clear in the discussion to follow, is often of no interest to us. I will often omit writing C altogether. Clearly, $C = Z(J = 0)$ so that $W(J)$ is defined by

$$Z(J) \equiv Z(J = 0)e^{iW(J)} \tag{19}$$

Observe that

$$W(J) = -\frac{1}{2} \iint d^4x d^4y J(x) D(x - y) J(y) \tag{20}$$

is a simple quadratic functional of J. In contrast, $Z(J)$ depends on arbitrarily high powers of J. This fact will be of importance in chapter I.7.

Free propagator

The function $D(x)$, known as the propagator, plays an essential role in quantum field theory. As the inverse of a differential operator it is clearly closely related to the Green's function you encountered in a course on electromagnetism.

Physicists are sloppy about mathematical rigor, but even so, they have to be careful once in a while to make sure that what they are doing actually makes sense. For the integral in (15) to converge for large φ we replace $m^2 \to m^2 - i\varepsilon$ so that the integrand contains a factor $e^{-\varepsilon \int d^4x \varphi^2}$, where ε is a positive infinitesimal we will let tend to zero.[3]

We can solve (17) easily by going to momentum space and multiplying together four copies of the representation (I.2.11) of the Dirac delta function

$$\delta^{(4)}(x - y) = \int \frac{d^4k}{(2\pi)^4} e^{ik(x-y)} \tag{21}$$

The solution is

$$D(x - y) = \int \frac{d^4k}{(2\pi)^4} \frac{e^{ik(x-y)}}{k^2 - m^2 + i\varepsilon} \tag{22}$$

which you can check by plugging into (17):

$$-(\partial^2 + m^2) D(x - y) = \int \frac{d^4k}{(2\pi)^4} \frac{k^2 - m^2}{k^2 - m^2 + i\varepsilon} e^{ik(x-y)} = \int \frac{d^4k}{(2\pi)^4} e^{ik(x-y)} = \delta^{(4)}(x - y) \text{ as } \varepsilon \to 0.$$

Note that the so-called $i\varepsilon$ prescription we just mentioned is essential; otherwise the integral giving $D(x)$ would hit a pole. The magnitude of ε is not important as long as it is infinitesimal, but the positive sign of ε is crucial as we will see presently. (More on this in chapter III.8.) Also, note that the sign of k in the exponential does not matter here by the symmetry $k \to -k$.

To evaluate $D(x)$ we first integrate over k^0 by the method of contours. Define $\omega_k \equiv +\sqrt{\vec{k}^2 + m^2}$ with a plus sign. The integrand has two poles in the complex k^0 plane, at $\pm\sqrt{\omega_k^2 - i\varepsilon}$, which in the $\varepsilon \to 0$ limit are equal to $+\omega_k - i\varepsilon$ and $-\omega_k + i\varepsilon$. Thus for ε positive, one pole is in the lower half-plane and the other in the upper half plane, and so as we go along the real k^0 axis from $-\infty$ to $+\infty$ we do not run into the poles. The issue is how to close the integration contour.

For x^0 positive, the factor $e^{ik^0 x^0}$ is exponentially damped for k^0 in the upper half-plane. Hence we should extend the integration contour extending from $-\infty$ to $+\infty$ on the real

---

[3] As is customary, ε is treated as generic, so that ε multiplied by any positive number is still ε.

axis to include the infinite semicircle in the upper half-plane, thus enclosing the pole at $-\omega_k + i\varepsilon$ and giving $-i \int \frac{d^3k}{(2\pi)^3 2\omega_k} e^{-i(\omega_k t - \vec{k}\cdot\vec{x})}$. Again, note that we are free to flip the sign of $\vec{k}$. Also, as is conventional, we use x^0 and t interchangeably. (In view of some reader confusion here in the first edition, I might add that I generally use x^0 with k^0 and t with ω_k; k^0 is a variable that can take on either sign but ω_k is a positive function of $\vec{k}$.)

For x^0 negative, we do the opposite and close the contour in the lower half-plane, thus picking up the pole at $+\omega_k - i\varepsilon$. We now obtain $-i \int (d^3k/(2\pi)^3 2\omega_k) e^{+i(\omega_k t - \vec{k}\cdot\vec{x})}$.

Recall that the Heaviside (we will meet this great and aptly named physicist in chapter IV.4) step function $\theta(t)$ is defined to be equal to 0 for $t < 0$ and equal to 1 for $t > 0$. As for what $\theta(0)$ should be, the answer is that since we are proud physicists and not nitpicking mathematicians we will just wing it when the need arises. The step function allows us to package our two integration results together as

$$D(x) = -i \int \frac{d^3k}{(2\pi)^3 2\omega_k} [e^{-i(\omega_k t - \vec{k}\cdot\vec{x})}\theta(x^0) + e^{i(\omega_k t - \vec{k}\cdot\vec{x})}\theta(-x^0)] \tag{23}$$

Physically, $D(x)$ describes the amplitude for a disturbance in the field to propagate from the origin to x. Lorentz invariance tells us that it is a function of x^2 and the sign of x^0 (since these are the quantities that do not change under a Lorentz transformation). We thus expect drastically different behavior depending on whether x is inside or outside the lightcone defined by $x^2 = (x^0)^2 - \vec{x}^2 = 0$. Without evaluating the d^3k integral we can see roughly how things go. Let us look at some cases.

In the future cone, $x = (t, 0)$ with $t > 0$, $D(x) = -i \int (d^3k/(2\pi)^3 2\omega_k) e^{-i\omega_k t}$ a superposition of plane waves and thus $D(x)$ oscillates. In the past cone, $x = (t, 0)$ with $t < 0$, $D(x) = -i \int (d^3k/(2\pi)^3 2\omega_k) e^{+i\omega_k t}$ oscillates with the opposite phase.

In contrast, for x spacelike rather than timelike, $x^0 = 0$, we have, upon interpreting $\theta(0) = \frac{1}{2}$ (the obvious choice; imagine smoothing out the step function), $D(x) = -i \int (d^3k/(2\pi)^3 2\sqrt{\vec{k}^2 + m^2}) e^{-i\vec{k}\cdot\vec{x}}$. The square root cut starting at $\pm im$ tells us that the characteristic value of $|\vec{k}|$ in the integral is of order m, leading to an exponential decay $\sim e^{-m|\vec{x}|}$, as we would expect. Classically, a particle cannot get outside the lightcone, but a quantum field can "leak" out over a distance of order m^{-1} by the Heisenberg uncertainty principle.

Exercises

I.3.1   Verify that $D(x)$ decays exponentially for spacelike separation.

I.3.2   Work out the propagator $D(x)$ for a free field theory in $(1 + 1)$-dimensional spacetime and study the large x^1 behavior for $x^0 = 0$.

I.3.3   Show that the advanced propagator defined by

$$D_{adv}(x - y) = \int \frac{d^4k}{(2\pi)^4} \frac{e^{ik(x-y)}}{k^2 - m^2 - i \, \text{sgn}(k_0)\varepsilon}$$

(where the sign function is defined by $\mathrm{sgn}(k_0) = +1$ if $k_0 > 0$ and $\mathrm{sgn}(k_0) = -1$ if $k_0 < 0$) is nonzero only if $x^0 > y^0$. In other words, it only propagates into the future. [Hint: both poles of the integrand are now in the upper half of the k_0-plane.] Incidentally, some authors prefer to write $(k_0 - ie)^2 - \vec{k}^2 - m^2$ instead of $k^2 - m^2 - i\,\mathrm{sgn}(k_0)\varepsilon$ in the integrand. Similarly, show that the retarded propagator

$$D_{ret}(x - y) = \int \frac{d^4k}{(2\pi)^4} \frac{e^{ik(x-y)}}{k^2 - m^2 + i\,\mathrm{sgn}(k_0)\varepsilon}$$

propagates into the past.

From field to particle

In the previous chapter we obtained for the free theory

$$W(J) = -\frac{1}{2} \int \int d^4x \, d^4y \, J(x) D(x - y) J(y) \tag{1}$$

which we now write in terms of the Fourier transform $J(k) \equiv \int d^4x \, e^{-ikx} J(x)$:

$$W(J) = -\frac{1}{2} \int \frac{d^4k}{(2\pi)^4} J(k)^* \frac{1}{k^2 - m^2 + i\varepsilon} J(k) \tag{2}$$

[Note that $J(k)^* = J(-k)$ for $J(x)$ real.]

We can jump up and down on the mattress any way we like. In other words, we can choose any $J(x)$ we want, and by exploiting this freedom of choice, we can extract a remarkable amount of physics.

Consider $J(x) = J_1(x) + J_2(x)$, where $J_1(x)$ and $J_2(x)$ are concentrated in two local regions 1 and 2 in spacetime (fig. I.4.1). Then $W(J)$ contains four terms, of the form $J_1^* J_1$, $J_2^* J_2$, $J_1^* J_2$, and $J_2^* J_1$. Let us focus on the last two of these terms, one of which reads

$$W(J) = -\frac{1}{2} \int \frac{d^4k}{(2\pi)^4} J_2(k)^* \frac{1}{k^2 - m^2 + i\varepsilon} J_1(k) \tag{3}$$

We see that $W(J)$ is large only if $J_1(x)$ and $J_2(x)$ overlap significantly in their Fourier transform and if in the region of overlap in momentum space $k^2 - m^2$ almost vanishes. There is a "resonance type" spike at $k^2 = m^2$, that is, if the energy-momentum relation of a particle of mass m is satisfied. (We will use the language of the relativistic physicist, writing "momentum space" for energy-momentum space, and lapse into nonrelativistic language only when the context demands it, such as in "energy-momentum relation.")

We thus interpret the physics contained in our simple field theory as follows: In region 1 in spacetime there exists a source that sends out a "disturbance in the field," which is later absorbed by a sink in region 2 in spacetime. Experimentalists choose to call this

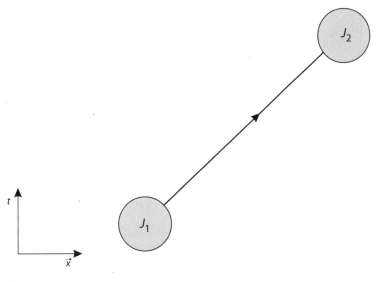

Figure I.4.1

disturbance in the field a particle of mass m. Our expectation based on the equation of motion that the theory contains a particle of mass m is fulfilled.

A bit of jargon: When $k^2 = m^2$, k is said to be on mass shell. Note, however, that in (3) we integrate over all k, including values of k far from the mass shell. For arbitrary k, it is a linguistic convenience to say that a "virtual particle" of momentum k propagates from the source to the sink.

From particle to force

We can now go on to consider other possibilities for $J(x)$ (which we will refer to generically as sources), for example, $J(x) = J_1(x) + J_2(x)$, where $J_a(x) = \delta^{(3)}(\vec{x} - \vec{x}_a)$. In other words, $J(x)$ is a sum of sources that are time-independent infinitely sharp spikes located at $\vec{x}_1$ and $\vec{x}_2$ in space. (If you like more mathematical rigor than is offered here, you are welcome to replace the delta function by lumpy functions peaking at $\vec{x}_a$. You would simply clutter up the formulas without gaining much.) More picturesquely, we are describing two massive lumps sitting at $\vec{x}_1$ and $\vec{x}_2$ on the mattress and not moving at all [no time dependence in $J(x)$].

What do the quantum fluctuations in the field φ, that is, the vibrations in the mattress, do to the two lumps sitting on the mattress? If you expect an attraction between the two lumps, you are quite right.

As before, $W(J)$ contains four terms. We neglect the "self-interaction" term $J_1 J_1$ since this contribution would be present in W regardless of whether J_2 is present or not. We want to study the interaction between the two "massive lumps" represented by J_1 and J_2. Similarly we neglect $J_2 J_2$.

Plugging into (1) and doing the integral over d^3x and d^3y we immediately obtain

$$W(J) = -\iint dx^0 dy^0 \int \frac{dk^0}{2\pi} e^{ik^0(x-y)^0} \int \frac{d^3k}{(2\pi)^3} \frac{e^{i\vec{k}\cdot(\vec{x}_1 - \vec{x}_2)}}{k^2 - m^2 + i\varepsilon} \tag{4}$$

(The factor 2 comes from the two terms $J_2 J_1$ and $J_1 J_2$.) Integrating over y^0 we get a delta function setting k^0 to zero (so that k is certainly not on mass shell, to throw the jargon around a bit). Thus we are left with

$$W(J) = \left(\int dx^0 \right) \int \frac{d^3k}{(2\pi)^3} \frac{e^{i\vec{k}\cdot(\vec{x}_1 - \vec{x}_2)}}{\vec{k}^2 + m^2} \tag{5}$$

Note that the infinitesimal $i\varepsilon$ can be dropped since the denominator $\vec{k}^2 + m^2$ is always positive.

The factor $(\int dx^0)$ should have filled us with fear and trepidation: an integral over time, it seems to be infinite. Fear not! Recall that in the path integral formalism $Z = C e^{iW(J)}$ represents $\langle 0 | e^{-iHT} | 0 \rangle = e^{-iET}$, where E is the energy due to the presence of the two sources acting on each other. The factor $(\int dx^0)$ produces precisely the time interval T. All is well. Setting $iW = -iET$ we obtain from (5)

$$E = -\int \frac{d^3k}{(2\pi)^3} \frac{e^{i\vec{k}\cdot(\vec{x}_1 - \vec{x}_2)}}{\vec{k}^2 + m^2} \tag{6}$$

The integral is evaluated in an appendix. This energy is negative! The presence of two delta function sources, at $\vec{x}_1$ and $\vec{x}_2$, has lowered the energy. (Notice that for the two sources infinitely far apart, we have, as we might expect, $E = 0$: the infinitely rapidly oscillating exponential kills the integral.) In other words, two like objects attract each other by virtue of their coupling to the field φ. We have derived our first physical result in quantum field theory!

We identify E as the potential energy between two static sources. Even without doing the integral, we see by dimensional analysis that the characteristic distance beyond which the integral goes to zero is given by the inverse of the characteristic value of k, which is m. Thus, we expect the attraction between the two sources to decrease rapidly to zero over the distance $1/m$.

The range of the attractive force generated by the field φ is determined inversely by the mass m of the particle described by the field. Got that?

The integral is done in the appendix to this chapter and gives

$$E = -\frac{1}{4\pi r} e^{-mr} \tag{7}$$

The result is as we expected: The potential drops off exponentially over the distance scale $1/m$. Obviously, $dE/dr > 0$: The two massive lumps sitting on the mattress can lower the energy by getting closer to each other.

What we have derived was one of the most celebrated results in twentieth-century physics. Yukawa proposed that the attraction between nucleons in the atomic nucleus is due to their coupling to a field like the φ field described here. The known range of the nuclear force enabled him to predict not only the existence of the particle associated with this field, now called the π meson[1] or the pion, but its mass as well. As you probably know, the pion was eventually discovered with essentially the properties predicted by Yukawa.

Origin of force

That the exchange of a particle can produce a force was one of the most profound conceptual advances in physics. We now associate a particle with each of the known forces: for example, the photon with the electromagnetic force and the graviton with the gravitational force; the former is experimentally well established and while the latter has not yet been detected experimentally hardly anyone doubts its existence. We will discuss the photon and the graviton in the next chapter, but we can already answer a question smart high school students often ask: Why do Newton's gravitational force and Coulomb's electric force both obey the $1/r^2$ law?

We see from (7) that if the mass m of the mediating particle vanishes, the force produced will obey the $1/r^2$ law. If you trace back over our derivation, you will see that this comes from the fact that the Lagrangian density for the simplest field theory involves two powers of the spacetime derivative ∂ (since any term involving one derivative such as $\varphi\,\partial\varphi$ is not Lorentz invariant). Indeed, the power dependence of the potential follows simply from dimensional analysis: $\int d^3k(e^{i\vec{k}\cdot\vec{x}}/k^2) \sim 1/r$.

Connected versus disconnected

We end with a couple of formal remarks of importance to us only in chapter I.7. First, note that we might want to draw a small picture fig. (I.4.2) to represent the integrand $J(x)D(x-y)J(y)$ in $W(J)$: A disturbance propagates from y to x (or vice versa). In fact, this is the beginning of Feynman diagrams! Second, recall that

$$Z(J) = Z(J=0) \sum_{n=0}^{\infty} \frac{[i\,W(J)]^n}{n!}$$

For instance, the $n=2$ term in $Z(J)/Z(J=0)$ is given by

$$\frac{1}{2!}\left(-\frac{i}{2}\right)^2 \iiiint d^4x_1 d^4x_2 d^4x_3 d^4x_4 D(x_1-x_2)$$
$$D(x_3-x_4)J(x_1)J(x_2)J(x_3)J(x_4)$$

The integrand is graphically described in figure I.4.3. The process is said to be disconnected: The propagation from x_1 to x_2 and the propagation from x_3 to x_4 proceed independently. We will come back to the difference between connected and disconnected in chapter I.7.

---

[1] The etymology behind this word is quite interesting (A. Zee, *Fearful Symmetry*: see pp. 169 and 335 to learn, among other things, the French objection and the connection between meson and illusion).

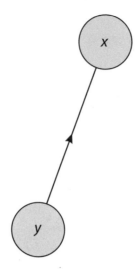

Figure I.4.2

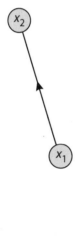

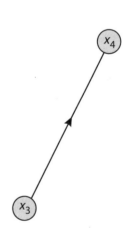

Figure I.4.3

Appendix

Writing $\vec{x} \equiv (\vec{x}_1 - \vec{x}_2)$ and $u \equiv \cos\theta$ with θ the angle between $\vec{k}$ and $\vec{x}$, we evaluate the integral in (6) in spherical coordinates (with $k = |\vec{k}|$ and $r = |\vec{x}|$):

$$I \equiv \frac{1}{(2\pi)^2} \int_0^\infty dk \, k^2 \int_{-1}^{+1} du \, \frac{e^{ikru}}{k^2 + m^2} = \frac{2i}{(2\pi)^2 ir} \int_0^\infty dk \, k \frac{\sin kr}{k^2 + m^2} \tag{8}$$

Since the integrand is even, we can extend the integral and write it as

$$\frac{1}{2} \int_{-\infty}^\infty dk \, k \frac{\sin kr}{k^2 + m^2} = \frac{1}{2i} \int_{-\infty}^\infty dk \, k \frac{1}{k^2 + m^2} e^{ikr}.$$

Since r is positive, we can close the contour in the upper half-plane and pick up the pole at $+im$, obtaining $(1/2i)(2\pi i)(im/2im)e^{-mr} = (\pi/2)e^{-mr}$. Thus, $I = (1/4\pi r)e^{-mr}$.

Exercise

I.4.1   Calculate the analog of the inverse square law in a $(2+1)$-dimensional universe, and more generally in a $(D+1)$-dimensional universe.

I.5 | Coulomb and Newton: Repulsion and Attraction

Why like charges repel

We suggested that quantum field theory can explain both Newton's gravitational force and Coulomb's electric force naturally. Between like objects Newton's force is attractive while Coulomb's force is repulsive. Is quantum field theory "smart enough" to produce this observational fact, one of the most basic in our understanding of the physical universe? You bet!

We will first treat the quantum field theory of the electromagnetic field, known as quantum electrodynamics or QED for short. In order to avoid complications at this stage associated with gauge invariance (about which much more later) I will consider instead the field theory of a massive spin 1 meson, or vector meson. After all, experimentally all we know is an upper bound on the photon mass, which although tiny is not mathematically zero. We can adopt a pragmatic attitude: Calculate with a photon mass m and set $m = 0$ at the end, and if the result does not blow up in our faces, we will presume that it is OK.[1]

Recall Maxwell's Lagrangian for electromagnetism $\mathcal{L} = -\frac{1}{4}F_{\mu\nu}F^{\mu\nu}$, where $F_{\mu\nu} \equiv \partial_\mu A_\nu - \partial_\nu A_\mu$ with $A_\mu(x)$ the vector potential. You can see the reason for the important overall minus sign in the Lagrangian by looking at the coefficient of $(\partial_0 A_i)^2$, which has to be positive, just like the coefficient of $(\partial_0 \varphi)^2$ in the Lagrangian for the scalar field. This says simply that time variation should cost a positive amount of action.

I will now give the photon a small mass by changing the Lagrangian to $\mathcal{L} = -\frac{1}{4}F_{\mu\nu}F^{\mu\nu} + \frac{1}{2}m^2 A_\mu A^\mu + A_\mu J^\mu$. (The mass term is written in analogy to the mass term $m^2\varphi^2$ in the scalar field Lagrangian; we will see shortly that the sign is correct and that this term indeed leads to a photon mass.) I have also added a source $J^\mu(x)$, which in this context is more familiarly known as a current. We will assume that the current is conserved so that $\partial_\mu J^\mu = 0$.

---

[1] When I took a field theory course as a student with Sidney Coleman this was how he treated QED in order to avoid discussing gauge invariance.

Well, you know that the field theory of our vector meson is defined by the path integral $Z = \int DA \, e^{iS(A)} \equiv e^{iW(J)}$ with the action

$$S(A) = \int d^4x \mathcal{L} = \int d^4x \{ \tfrac{1}{2} A_\mu [(\partial^2 + m^2) g^{\mu\nu} - \partial^\mu \partial^\nu] A_\nu + A_\mu J^\mu \} \tag{1}$$

The second equality follows upon integrating by parts [compare (I.3.15)].

By now you have learned that we simply apply (I.3.16). We merely have to find the inverse of the differential operator in the square bracket; in other words, we have to solve

$$[(\partial^2 + m^2) g^{\mu\nu} - \partial^\mu \partial^\nu] D_{\nu\lambda}(x) = \delta^\mu_\lambda \delta^{(4)}(x) \tag{2}$$

As before [compare (I.3.17)] we go to momentum space by defining

$$D_{\nu\lambda}(x) = \int \frac{d^4k}{(2\pi)^4} D_{\nu\lambda}(k) e^{ikx}$$

Plugging in, we find that $[-(k^2 - m^2) g^{\mu\nu} + k^\mu k^\nu] D_{\nu\lambda}(k) = \delta^\mu_\lambda$, giving

$$D_{\nu\lambda}(k) = \frac{-g_{\nu\lambda} + k_\nu k_\lambda / m^2}{k^2 - m^2} \tag{3}$$

This is the photon, or more accurately the massive vector meson, propagator. Thus

$$W(J) = -\frac{1}{2} \int \frac{d^4k}{(2\pi)^4} J^\mu(k)^* \frac{-g_{\mu\nu} + k_\mu k_\nu / m^2}{k^2 - m^2 + i\varepsilon} J^\nu(k) \tag{4}$$

Since current conservation $\partial_\mu J^\mu(x) = 0$ gets translated into momentum space as $k_\mu J^\mu(k) = 0$, we can throw away the $k_\mu k_\nu$ term in the photon propagator. The effective action simplifies to

$$W(J) = \frac{1}{2} \int \frac{d^4k}{(2\pi)^4} J^\mu(k)^* \frac{1}{k^2 - m^2 + i\varepsilon} J_\mu(k) \tag{5}$$

No further computation is needed to obtain a profound result. Just compare this result to (I.4.2). The field theory has produced an extra sign. The potential energy between two lumps of charge density $J^0(x)$ is positive. The electromagnetic force between like charges is repulsive!

We can now safely let the photon mass m go to zero thanks to current conservation. [Note that we could not have done that in (3).] Indeed, referring to (I.4.7) we see that the potential energy between like charges is

$$E = \frac{1}{4\pi r} e^{-mr} \rightarrow \frac{1}{4\pi r} \tag{6}$$

To accommodate positive and negative charges we can simply write $J^\mu = J^\mu_p - J^\mu_n$. We see that a lump with charge density J^0_p is attracted to a lump with charge density J^0_n.

Bypassing Maxwell

Having done electromagnetism in two minutes flat let me now do gravity. Let us move on to the massive spin 2 meson field. In my treatment of the massive spin 1 meson field I took a short cut. Assuming that you are familiar with the Maxwell Lagrangian, I simply added a mass term to it and took off. But I do not feel comfortable assuming that you are equally familiar with the corresponding Lagrangian for the massless spin 2 field (the so-called linearized Einstein Lagrangian, which I will discuss in a later chapter). So here I will follow another strategy.

I invite you to think physically, and together we will arrive at the propagator for a massive spin 2 field. First, we will warm up with the massive spin 1 case.

In fact, start with something even easier: the propagator $D(k) = 1/(k^2 - m^2)$ for a massive spin 0 field. It tells us that the amplitude for the propagation of a spin 0 disturbance blows up when the disturbance is almost a real particle. The residue of the pole is a property of the particle. The propagator for a spin 1 field $D_{\nu\lambda}$ carries a pair of Lorentz indices and in fact we know what it is from (3):

$$D_{\nu\lambda}(k) = \frac{-G_{\nu\lambda}}{k^2 - m^2} \tag{7}$$

where for later convenience we have defined

$$G_{\nu\lambda}(k) \equiv g_{\nu\lambda} - \frac{k_\nu k_\lambda}{m^2} \tag{8}$$

Let us now understand the physics behind $G_{\nu\lambda}$. I expect you to remember the concept of polarization from your course on electromagnetism. A massive spin 1 particle has three degrees of polarization for the obvious reason that in its rest frame its spin vector can point in three different directions. The three polarization vectors $\varepsilon_\mu^{(a)}$ are simply the three unit vectors pointing along the x, y, and z axes, respectively ($a = 1, 2, 3$): $\varepsilon_\mu^{(1)} = (0, 1, 0, 0)$, $\varepsilon_\mu^{(2)} = (0, 0, 1, 0)$, $\varepsilon_\mu^{(3)} = (0, 0, 0, 1)$. In the rest frame $k^\mu = (m, 0, 0, 0)$ and so

$$k^\mu \varepsilon_\mu^{(a)} = 0 \tag{9}$$

Since this is a Lorentz invariant equation, it holds for a moving spin 1 particle as well. Indeed, with a suitable normalization condition this fixes the three polarization vectors $\varepsilon_\mu^{(a)}(k)$ for a particle with momentum k.

The amplitude for a particle with momentum k and polarization a to be created at the source is proportional to $\varepsilon_\lambda^{(a)}(k)$, and the amplitude for it to be absorbed at the sink is proportional to $\varepsilon_\nu^{(a)}(k)$. We multiply the amplitudes together to get the amplitude for propagation from source to sink, and then sum over the three possible polarizations. Now we understand the residue of the pole in the spin 1 propagator $D_{\nu\lambda}(k)$: It represents $\sum_a \varepsilon_\nu^{(a)}(k) \varepsilon_\lambda^{(a)}(k)$. To calculate this quantity, note that by Lorentz invariance it can only be a linear combination of $g_{\nu\lambda}$ and $k_\nu k_\lambda$. The condition $k^\mu \varepsilon_\mu^{(a)} = 0$ fixes it to be proportional to $g_{\nu\lambda} - k_\nu k_\lambda/m^2$. We evaluate the left-hand side for k at rest with $\nu = \lambda = 1$, for instance, and fix the overall and all-crucial sign to be -1. Thus

$$\sum_a \varepsilon_\nu^{(a)}(k)\varepsilon_\lambda^{(a)}(k) = -G_{\nu\lambda}(k) \equiv -\left(g_{\nu\lambda} - \frac{k_\nu k_\lambda}{m^2} \right) \tag{10}$$

We have thus constructed the propagator $D_{\nu\lambda}(k)$ for a massive spin 1 particle, bypassing Maxwell (see appendix 1).

Onward to spin 2! We want to similarly bypass Einstein.

Bypassing Einstein

A massive spin 2 particle has 5 ($2 \cdot 2 + 1 = 5$, remember?) degrees of polarization, characterized by the five polarization tensors $\varepsilon_{\mu\nu}^{(a)}$ ($a = 1, 2, \cdots, 5$) symmetric in the indices μ and ν satisfying

$$k^\mu \varepsilon_{\mu\nu}^{(a)} = 0 \tag{11}$$

and the tracelessness condition

$$g^{\mu\nu} \varepsilon_{\mu\nu}^{(a)} = 0 \tag{12}$$

Let's count as a check. A symmetric Lorentz tensor has $4 \cdot 5/2 = 10$ components. The four conditions in (11) and the single condition in (12) cut the number of components down to $10 - 4 - 1 = 5$, precisely the right number. (Just to throw some jargon around, remember how to construct irreducible group representations? If not, read appendix B.) We fix the normalization of $\varepsilon_{\mu\nu}$ by setting the positive quantity $\sum_a \varepsilon_{12}^{(a)}(k)\varepsilon_{12}^{(a)}(k) = 1$.

So, in analogy with the spin 1 case we now determine $\sum_a \varepsilon_{\mu\nu}^{(a)}(k)\varepsilon_{\lambda\sigma}^{(a)}(k)$. We have to construct this object out of $g_{\mu\nu}$ and k_μ, or equivalently $G_{\mu\nu}$ and k_μ. This quantity must be a linear combination of terms such as $G_{\mu\nu}G_{\lambda\sigma}$, $G_{\mu\nu}k_\lambda k_\sigma$, and so forth. Using (11) and (12) repeatedly (exercise I.5.1) you will easily find that

$$\sum_a \varepsilon_{\mu\nu}^{(a)}(k)\varepsilon_{\lambda\sigma}^{(a)}(k) = (G_{\mu\lambda}G_{\nu\sigma} + G_{\mu\sigma}G_{\nu\lambda}) - \tfrac{2}{3}G_{\mu\nu}G_{\lambda\sigma} \tag{13}$$

The overall sign and proportionality constant are determined by evaluating both sides for k at rest (for $\mu = \lambda = 1$ and $\nu = \sigma = 2$, for instance).

Thus, we have determined the propagator for a massive spin 2 particle

$$D_{\mu\nu,\,\lambda\sigma}(k) = \frac{(G_{\mu\lambda}G_{\nu\sigma} + G_{\mu\sigma}G_{\nu\lambda}) - \tfrac{2}{3}G_{\mu\nu}G_{\lambda\sigma}}{k^2 - m^2} \tag{14}$$

Why we fall

We are now ready to understand one of the fundamental mysteries of the universe: Why masses attract.

Recall from your courses on electromagnetism and special relativity that the energy or mass density out of which mass is composed is part of a stress-energy tensor $T^{\mu\nu}$. For our purposes, in fact, all you need to remember is that $T^{\mu\nu}$ is a symmetric tensor and that the component T^{00} is the energy density. If you don't remember, I will give you a physical explanation in appendix 2.

To couple to the stress-energy tensor, we need a tensor field $\varphi_{\mu\nu}$ symmetric in its two indices. In other words, the Lagrangian of the world should contain a term like $\varphi_{\mu\nu}T^{\mu\nu}$. This is in fact how we know that the graviton, the particle responsible for gravity, has spin 2, just as we know that the photon, the particle responsible for electromagnetism and hence coupled to the current J^μ, has spin 1. In Einstein's theory, which we will discuss in a later chapter, $\varphi_{\mu\nu}$ is of course part of the metric tensor.

Just as we pretended that the photon has a small mass to avoid having to discuss gauge invariance, we will pretend that the graviton has a small mass to avoid having to discuss general coordinate invariance.[2] Aha, we just found the propagator for a massive spin 2 particle. So let's put it to work.

In precise analogy to (4)

$$W(J) = -\frac{1}{2} \int \frac{d^4 k}{(2\pi)^4} J^\mu(k)^* \frac{-g_{\mu\nu} + k_\mu k_\nu/m^2}{k^2 - m^2 + i\varepsilon} J^\nu(k) \tag{15}$$

describing the interaction between two electromagnetic currents, the interaction between two lumps of stress energy is described by

$$W(T) =$$
$$-\frac{1}{2} \int \frac{d^4 k}{(2\pi)^4} T^{\mu\nu}(k)^* \frac{(G_{\mu\lambda}G_{\nu\sigma} + G_{\mu\sigma}G_{\nu\lambda}) - \frac{2}{3}G_{\mu\nu}G_{\lambda\sigma}}{k^2 - m^2 + i\varepsilon} T^{\lambda\sigma}(k) \tag{16}$$

From the conservation of energy and momentum $\partial_\mu T^{\mu\nu}(x) = 0$ and hence $k_\mu T^{\mu\nu}(k) = 0$, we can replace $G_{\mu\nu}$ in (16) by $g_{\mu\nu}$. (Here as is clear from the context $g_{\mu\nu}$ still denotes the flat spacetime metric of Minkowski, rather than the curved metric of Einstein.)

Now comes the punchline. Look at the interaction between two lumps of energy density T^{00}. We have from (16) that

$$W(T) = -\frac{1}{2} \int \frac{d^4 k}{(2\pi)^4} T^{00}(k)^* \frac{1 + 1 - \frac{2}{3}}{k^2 - m^2 + i\varepsilon} T^{00}(k) \tag{17}$$

Comparing with (5) and using the well-known fact that $(1 + 1 - \frac{2}{3}) > 0$, we see that while like charges repel, masses attract. Trumpets, please!

The universe

It is difficult to overstate the importance (not to speak of the beauty) of what we have learned: The exchange of a spin 0 particle produces an attractive force, of a spin 1 particle a repulsive force, and of a spin 2 particle an attractive force, realized in the hadronic strong interaction, the electromagnetic interaction, and the gravitational interaction, respectively. The universal attraction of gravity produces an instability that drives the formation of structure in the early universe.[3] Denser regions become denser yet. The attractive nuclear force mediated by the spin 0 particle eventually ignites the stars. Furthermore, the attractive force between protons and neutrons mediated by the spin 0 particle is able to overcome the repulsive electric force between protons mediated by the spin 1 particle to form a

---

[2] For the moment, I ask you to ignore all subtleties and simply assume that in order to understand gravity it is kosher to let $m \to 0$. I will give a precise discussion of Einstein's theory of gravity in chapter VIII.1.

[3] A good place to read about gravitational instability and the formation of structure in the universe along the line sketched here is in A. Zee, *Einstein's Universe* (formerly known as *An Old Man's Toy*).

variety of nuclei without which the world would certainly be rather boring. The repulsion between likes and hence attraction between opposites generated by the spin 1 particle allow electrically neutral atoms to form.

The world results from a subtle interplay among spin 0, 1, and 2.

In this lightning tour of the universe, we did not mention the weak interaction. In fact, the weak interaction plays a crucial role in keeping stars such as our sun burning at a steady rate.

Time differs from space by a sign

This weaving together of fields, particles, and forces to produce a universe rich with possibilities is so beautiful that it is well worth pausing to examine the underlying physics some more. The expression in (I.4.1) describes the effect of our disturbing the vacuum (or the mattress!) with the source J, calculated to second order. Thus some readers may have recognized that the negative sign in (I.4.6) comes from the elementary quantum mechanical result that in second order perturbation theory the lowest energy state always has its energy pushed downward: for the ground state

all the energy denominators have the same sign.

In essence, this "theorem" follows from the property of 2 by 2 matrices. Let us set the ground state energy to 0 and crudely represent the entirety of the other states by a single state with energy $w > 0$. Then the Hamiltonian including the perturbation v effective to second order is given by

$$H = \begin{pmatrix} w & v \\ v & 0 \end{pmatrix}$$

Since the determinant of H (and hence the product of the two eigenvalues) is manifestly negative, the ground state energy is pushed below 0. [More explicitly, we calculate the eigenvalue ε with the characteristic equation $0 = \varepsilon(\varepsilon - w) - v^2 \approx -(w\varepsilon + v^2)$, and hence $\varepsilon \approx -\frac{v^2}{w}$.] In different fields of physics, this phenomenon is variously known as level repulsion or the seesaw mechanism (see chapter VII.7).

Disturbing the vacuum with a source lowers its energy. Thus it is easy to understand that generically the exchange of a particles leads to an attractive force.

But then why does the exchange of a spin 1 particle produces a repulsion between like objects? The secret lies in the profundity that space differs from time by a sign, namely, that $g_{00} = +1$ while $g_{ii} = -1$ for $i = 1, 2, 3$. In (10), the left-hand side is manifestly positive for $v = \lambda = i$. Taking k to be at rest we understand the minus sign in (10) and hence in (4). Roughly speaking, for spin 2 exchange the sign occurs twice in (16).

Degrees of freedom

Now for a bit of cold water: Logically and mathematically the physics of a particle with mass $m \neq 0$ could well be different from the physics with $m = 0$. Indeed, we know from classical electromagnetism that an electromagnetic wave has 2 polarizations, that is, 2 degrees of freedom. For a massive spin 1 particle we can go to its rest frame, where the rotation group tells us that there are $2 \cdot 1 + 1 = 3$ degrees of freedom. The crucial piece of physics is that we can never bring the massless photon to its rest frame. Mathematically, the rotation group $SO(3)$ degenerates into $SO(2)$, the group of 2-dimensional rotations around the direction of the photon's momentum.

We will see in chapter II.7 that the longitudinal degree of freedom of a massive spin 1 meson decouples as we take the mass to zero. The treatment given here for the interaction between charges (6) is correct. However, in the case of gravity, the $\frac{2}{3}$ in (17) is replaced by 1 in Einstein's theory, as we will see chapter VIII.1. Fortunately, the sign of the interaction given in (17) does not change. Mute the trumpets a bit.

Appendix 1

Pretend that we never heard of the Maxwell Lagrangian. We want to construct a relativistic Lagrangian for a massive spin 1 meson field. Together we will discover Maxwell. Spin 1 means that the field transforms as a vector under the 3-dimensional rotation group. The simplest Lorentz object that contains the 3-dimensional vector is obviously the 4-dimensional vector. Thus, we start with a vector field $A_\mu(x)$.

That the vector field carries mass m means that it satisfies the field equation

$$(\partial^2 + m^2)A_\mu = 0 \tag{18}$$

A spin 1 particle has 3 degrees of freedom [remember, in fancy language, the representation j of the rotation group has dimension $(2j+1)$; here $j = 1$.] On the other hand, the field $A_\mu(x)$ contains 4 components. Thus, we must impose a constraint to cut down the number of degrees of freedom from 4 to 3. The only Lorentz covariant possibility (linear in A_μ) is

$$\partial_\mu A^\mu = 0 \tag{19}$$

It may also be helpful to look at (18) and (19) in momentum space, where they read $(k^2 - m^2)A_\mu(k) = 0$ and $k_\mu A^\mu(k) = 0$. The first equation tells us that $k^2 = m^2$ and the second that if we go to the rest frame $k^\mu = (m, \vec{0})$ then A^0 vanishes, leaving us with 3 nonzero components A^i with $i = 1, 2, 3$.

The remarkable observation is that we can combine (18) and (19) into a single equation, namely

$$(g^{\mu\nu}\partial^2 - \partial^\mu\partial^\nu)A_\nu + m^2 A^\mu = 0 \tag{20}$$

Verify that (20) contains both (18) and (19). Act with ∂_μ on (20). We obtain $m^2\partial_\mu A^\mu = 0$, which implies that $\partial_\mu A^\mu = 0$. (At this step it is crucial that $m \neq 0$ and that we are not talking about the strictly massless photon.) We have thus obtained (19); using (19) in (20) we recover (18).

We can now construct a Lagrangian by multiplying the left-hand side of (20) by $+\frac{1}{2}A_\mu$ (the $\frac{1}{2}$ is conventional but the plus sign is fixed by physics, namely the requirement of positive kinetic energy); thus

$$\mathcal{L} = \tfrac{1}{2}A_\mu[(\partial^2 + m^2)g^{\mu\nu} - \partial^\mu\partial^\nu]A_\nu \tag{21}$$

Integrating by parts, we recognize this as the massive version of the Maxwell Lagrangian. In the limit $m \to 0$ we recover Maxwell.

A word about terminology: Some people insist on calling only $F_{\mu\nu}$ a field and A_μ a potential. Conforming to common usage, we will not make this fine distinction. For us, any dynamical function of spacetime is a field.

Appendix 2: Why does the graviton have spin 2?

First we have to understand why the photon has spin 1. Think physically. Consider a bunch of electrons at rest inside a small box. An observer moving by sees the box Lorentz-Fitzgerald contracted and thus a higher charge density than the observer at rest relative to the box. Thus charge density $J^0(x)$ transforms like the time component of a 4-vector density $J^\mu(x)$. In other words, $J'^0 = J^0/\sqrt{1-v^2}$. The photon couples to $J^\mu(x)$ and has to be described by a 4-vector field $A_\mu(x)$ for the Lorentz indices to match.

What about energy density? The observer at rest relative to the box sees each electron contributing m to the energy enclosed in the box. The moving observer, on the other hand, sees the electrons moving and thus each having an energy $m/\sqrt{1-v^2}$. With the contracted volume and the enhanced energy, the energy density gets enhanced by two factors of $1/\sqrt{1-v^2}$, that is, it transforms like the T^{00} component of a 2-indexed tensor $T^{\mu\nu}$. The graviton couples to $T^{\mu\nu}(x)$ and has to be described by a 2-indexed tensor field $\varphi_{\mu\nu}(x)$ for the Lorentz indices to match.

Exercise

I.5.1 Write down the most general form for $\sum_a \varepsilon^{(a)}_{\mu\nu}(k)\varepsilon^{(a)}_{\lambda\sigma}(k)$ using symmetry repeatedly. For example, it must be invariant under the exchange $\{\mu\nu \leftrightarrow \lambda\sigma\}$. You might end up with something like

$$AG_{\mu\nu}G_{\lambda\sigma} + B(G_{\mu\lambda}G_{\nu\sigma} + G_{\mu\sigma}G_{\nu\lambda}) + C(G_{\mu\nu}k_\lambda k_\sigma + k_\mu k_\nu G_{\lambda\sigma})$$
$$+ D(k_\mu k_\lambda G_{\nu\sigma} + k_\mu k_\sigma G_{\nu\lambda} + k_\nu k_\sigma G_{\mu\lambda} + k_\nu k_\lambda G_{\mu\sigma}) + E k_\mu k_\nu k_\lambda k_\sigma \qquad (22)$$

with various unknown $A, \cdots, E$. Apply $k^\mu \sum_a \varepsilon^{(a)}_{\mu\nu}(k)\varepsilon^{(a)}_{\lambda\sigma}(k) = 0$ and find out what that implies for the constants. Proceeding in this way, derive (13).

I.6 | Inverse Square Law and the Floating 3-Brane

Why inverse square?

In your first encounter with physics, didn't you wonder why an inverse square force law and not, say, an inverse cube law? In chapter I.4 you learned the deep answer. When a massless particle is exchanged between two particles, the potential energy between the two particles goes as

$$V(r) \propto \int d^3k \, e^{i\vec{k}\cdot\vec{x}} \frac{1}{\vec{k}^2} \propto \frac{1}{r} \tag{1}$$

The spin of the exchanged particle controls the overall sign, but the $1/r$ follows just from dimensional analysis, as I remarked earlier. Basically, $V(r)$ is the Fourier transform of the propagator. The $\vec{k}^2$ in the propagator comes from the $(\partial_i \varphi \cdot \partial_i \varphi)$ term in the action, where φ denotes generically the field associated with the massless particle being exchanged, and the $(\partial_i \varphi \cdot \partial_i \varphi)$ form is required by rotational invariance. It couldn't be $\vec{k}$ or $\vec{k}^3$ in (1); $\vec{k}^2$ is the simplest possibility. So you can say that in some sense ultimately the inverse square law comes from rotational invariance!

Physically, the inverse square law goes back to Faraday's flux picture. Consider a sphere of radius r surrounding a charge. The electric flux per unit area going through the sphere varies as $1/4\pi r^2$. This geometric fact is reflected in the factor d^3k in (1).

Brane world

Remarkably, with the tiny bit of quantum field theory I have exposed you to, I can already take you to the frontier of current research, current as of the writing of this book. In string theory, our $(3 + 1)$-dimensional world could well be embedded in a larger universe, the way a $(2 + 1)$-dimensional sheet of paper is embedded in our everyday $(3 + 1)$-dimensional world. We are said to be living on a 3 brane.

So suppose there are n extra dimensions, with coordinates $x^4, x^5, \cdots, x^{n+3}$. Let the characteristic scales associated with these extra coordinates be R. I can't go into the different detailed scenarios describing what R is precisely. For some reason I can't go into either, we are stuck on the 3 brane. In contrast, the graviton is associated intrinsically with the structure of spacetime and so roams throughout the $(n + 3 + 1)$-dimensional universe.

All right, what is the gravitational force law between two particles? It is surely not your grandfather's gravitational force law: We Fourier transform

$$V(r) \propto \int d^{3+n}k \, e^{i\vec{k}\cdot\vec{x}} \frac{1}{\vec{k}^2} \propto \frac{1}{r^{1+n}} \tag{2}$$

to obtain a $1/r^{1+n}$ law.

Doesn't this immediately contradict observation?

Well, no, because Newton's law continues to hold for $r \gg R$. In this regime, the extra coordinates are effectively zero compared to the characteristic length scale r we are interested in. The flux cannot spread far in the direction of the n extra coordinates. Think of the flux being forced to spread in only the three spatial directions we know, just as the electromagnetic field in a wave guide is forced to propagate down the tube. Effectively we are back in $(3 + 1)$-dimensional spacetime and $V(r)$ reverts to a $1/r$ dependence.

The new law of gravity (2) holds only in the opposite regime $r \ll R$. Heuristically, when R is much larger than the separation between the two particles, the flux does not know that the extra coordinates are finite in extent and thinks that it lives in an $(n + 3 + 1)$-dimensional universe.

Because of the weakness of gravity, Newton's force law has not been tested to much accuracy at laboratory distance scales, and so there is plenty of room for theorists to speculate in: R could easily be much larger than the scale of elementary particles and yet much smaller than the scale of everyday phenomena. Incredibly, the universe could have "large extra dimensions"! (The word "large" means large on the scale of particle physics.)

Planck mass

To be quantitative, let us define the Planck mass M_{Pl} by writing Newton's law more rationally as $V(r) = G_N m_1 m_2 (1/r) = (m_1 m_2/M_{Pl}^2)(1/r)$. Numerically, $M_{Pl} \simeq 10^{19}$ Gev. This enormous value obviously reflects the weakness of gravity.

In fundamental units in which $\hbar$ and c are set to unity, gravity defines an intrinsic mass or energy scale much higher than any scale we have yet explored experimentally. Indeed, one of the fundamental mysteries of contemporary particle physics is why this mass scale is so high compared to anything else we know of. I will come back to this so-called hierarchy problem in due time. For the moment, let us ask if this new picture of gravity, new in the waning moments of the last century, can alleviate the hierarchy problem by lowering the intrinsic mass scale of gravity.

Denote the mass scale (the "true scale" of gravity) characteristic of gravity in the $(n + 3 + 1)$-dimensional universe by M_{TG} so that the gravitational potential between two objects of masses m_1 and m_2 separated by a distance $r \ll R$ is given by

$$V(r) = \frac{m_1 m_2}{[M_{TG}]^{2+n}} \frac{1}{r^{1+n}}$$

Note that the dependence on M_{TG} follows from dimensional analysis: two powers to cancel $m_1 m_2$ and n powers to match the n extra powers of $1/r$. For $r \gg R$, as we have argued, the geometric spread of the gravitational flux is cut off by R so that the potential becomes

$$V(r) = \frac{m_1 m_2}{[M_{TG}]^{2+n}} \frac{1}{R^n} \frac{1}{r}$$

Comparing with the observed law $V(r) = (m_1 m_2 / M_{Pl}^2)(1/r)$ we obtain

$$M_{TG}^2 = \frac{M_{Pl}^2}{[M_{TG} R]^n} \tag{3}$$

If $M_{TG} R$ could be made large enough, we have the intriguing possibility that the fundamental scale of gravity M_{TG} may be much lower than what we have always thought.

Accelerators (such as the large Hadron Collider) could offer an exciting verification of this intriguing possibility. If the true scale of gravity M_{TG} lies in an energy range accessible to the accelerator, there may be copious production of gravitons escaping into the higher dimensional universe. Experimentalists would see a massive amount of missing energy.

Exercise

I.6.1   Putting in the numbers, show that the case $n = 1$ is already ruled out.

I.7 | Feynman Diagrams

Feynman brought quantum field theory to the masses.
—J. Schwinger

Anharmonicity in field theory

The free field theory we studied in the last few chapters was easy to solve because the defining path integral (I.3.14) is Gaussian, so we could simply apply (I.2.15). (This corresponds to solving the harmonic oscillator in quantum mechanics.) As I noted in chapter I.3, within the harmonic approximation the vibrational modes on the mattress can be linearly superposed and thus they simply pass through each other. The particles represented by wave packets constructed out of these modes do not interact:[1] hence the term free field theory. To have the modes scatter off each other we have to include anharmonic terms in the Lagrangian so that the equation of motion is no longer linear. For the sake of simplicity let us add only one anharmonic term $-\frac{\lambda}{4!}\varphi^4$ to our free field theory and, recalling (I.3.11), try to evaluate

$$Z(J) = \int D\varphi \, e^{i \int d^4x \{\frac{1}{2}[(\partial\varphi)^2 - m^2\varphi^2] - \frac{\lambda}{4!}\varphi^4 + J\varphi\}} \tag{1}$$

(We suppress the dependence of Z on λ.)

Doing quantum field theory is no sweat, you say, it just amounts to doing the functional integral (1). But the integral is not easy! If you could do it, it would be big news.

Feynman diagrams made easy

As an undergraduate, I heard of these mysterious little pictures called Feynman diagrams and really wanted to learn about them. I am sure that you too have wondered about those

---

[1] A potential source of confusion: Thanks to the propagation of φ, the sources coupled to φ interact, as was seen in chapter I.4, but the particles associated with φ do not interact with each other. This is like saying that charged particles coupled to the photon interact, but (to leading approximation) photons do not interact with each other.

funny diagrams. Well, I want to show you that Feynman diagrams are not such a big deal: Indeed we have already drawn little spacetime pictures in chapters I.3 and I.4 showing how particles can appear, propagate, and disappear.

Feynman diagrams have long posed somewhat of an obstacle for first-time learners of quantum field theory. To derive Feynman diagrams, traditional texts typically adopt the canonical formalism (which I will introduce in the next chapter) instead of the path integral formalism used here. As we will see, in the canonical formalism fields appear as quantum operators. To derive Feynman diagrams, we would have to solve the equation of motion of the field operators perturbatively in λ. A formidable amount of machinery has to be developed.

In the opinion of those who prefer the path integral, the path integral formalism derivation is considerably simpler (naturally!). Nevertheless, the derivation can still get rather involved and the student could easily lose sight of the forest for the trees. There is no getting around the fact that you would have to put in some effort.

I will try to make it as easy as possible for you. I have hit upon the great pedagogical device of letting you discover the Feynman diagrams for yourself. My strategy is to let you tackle two problems of increasing difficulty, what I call the baby problem and the child problem. By the time you get through these, the problem of evaluating (1) will seem much more tractable.

A baby problem

The baby problem is to evaluate the ordinary integral

$$Z(J) = \int_{-\infty}^{+\infty} dq \, e^{-\frac{1}{2}m^2 q^2 - \frac{\lambda}{4!}q^4 + Jq} \tag{2}$$

evidently a much simpler version of (1).

First, a trivial point: we can always scale $q \to q/m$ so that $Z = m^{-1}\mathcal{F}(\frac{\lambda}{m^4}, \frac{J}{m})$, but we won't.

For $\lambda = 0$ this is just one of the Gaussian integrals done in the appendix of chapter I.2. Well, you say, it is easy enough to calculate $Z(J)$ as a series in λ: expand

$$Z(J) = \int_{-\infty}^{+\infty} dq \, e^{-\frac{1}{2}m^2 q^2 + Jq} \left[1 - \frac{\lambda}{4!}q^4 + \frac{1}{2}(\frac{\lambda}{4!})^2 q^8 + \cdots \right]$$

and integrate term by term. You probably even know one of several tricks for computing $\int_{-\infty}^{+\infty} dq \, e^{-\frac{1}{2}m^2 q^2 + Jq} q^{4n}$: you write it as $(\frac{d}{dJ})^{4n} \int_{-\infty}^{+\infty} dq \, e^{-\frac{1}{2}m^2 q^2 + Jq}$ and refer to (I.2.19). So

$$Z(J) = (1 - \frac{\lambda}{4!}(\frac{d}{dJ})^4 + \frac{1}{2}(\frac{\lambda}{4!})^2(\frac{d}{dJ})^8 + \cdots) \int_{-\infty}^{+\infty} dq \, e^{-\frac{1}{2}m^2 q^2 + Jq} \tag{3}$$

$$= e^{-\frac{\lambda}{4!}(\frac{d}{dJ})^4} \int_{-\infty}^{+\infty} dq \, e^{-\frac{1}{2}m^2 q^2 + Jq} = (\frac{2\pi}{m^2})^{\frac{1}{2}} e^{-\frac{\lambda}{4!}(\frac{d}{dJ})^4} e^{\frac{1}{2m^2}J^2} \tag{4}$$

(There are other tricks, such as differentiating $\int_{-\infty}^{+\infty} dq \, e^{-\frac{1}{2}m^2 q^2 + Jq}$ with respect to m^2 repeatedly, but I want to discuss a trick that will also work for field theory.) By expanding

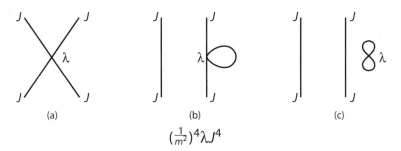

$$\left(\tfrac{1}{m^2}\right)^4 \lambda J^4$$

Figure I.7.1

the two exponentials we can obtain any term in a double series expansion of $Z(J)$ in λ and J. [We will often suppress the overall factor $(2\pi/m^2)^{\frac{1}{2}} = Z(J=0, \lambda=0) \equiv Z(0,0)$ since it will be common to all terms. When we want to be precise, we will define $\tilde{Z} = Z(J)/Z(0,0)$.]

For example, suppose we want the term of order λ and J^4 in $\tilde{Z}$. We extract the order J^8 term in $e^{J^2/2m^2}$, namely, $[1/4!(2m^2)^4]J^8$, replace $e^{-(\lambda/4!)(d/dJ)^4}$ by $-(\lambda/4!)(d/dJ)^4$, and differentiate to get $[8!(-\lambda)/(4!)^3(2m^2)^4]J^4$. Another example: the term of order λ^2 and J^4 is $[12!(-\lambda)^2/(4!)^36!2(2m^2)^6]J^4$. A third example: the term of order λ^2 and J^6 is $\frac{1}{2}(\lambda/4!)^2(d/dJ)^8[1/7!(2m^2)^7]J^{14} = [14!(-\lambda)^2/(4!)^26!7!2(2m^2)^7]J^6$. Finally, the term of order λ and J^0 is $[1/2(2m^2)^2](-\lambda)$.

You can do this as well as I can! Do a few more and you will soon see a pattern. In fact, you will eventually realize that you can associate diagrams with each term and codify some rules. Our four examples are associated with the diagrams in figures I.7.1–I.7.4. You can see, for a reason you will soon understand, that each term can be associated with several diagrams. I leave you to work out the rules carefully to get the numerical factors right (but trust me, the "future of democracy" is not going to depend on them). The rules go something like this: (1) diagrams are made of lines and vertices at which four lines meet; (2) for each vertex assign a factor of $(-\lambda)$; (3) for each line assign $1/m^2$; and (4) for each external end assign J (e.g., figure I.7.3 has seven lines, two vertices, and six ends, giving $\sim [(-\lambda)^2/(m^2)^7]J^6$.) (Did you notice that twice the number of lines is equal to four times the number of vertices plus the number of ends? We will meet relations like that in chapter III.2.)

In addition to the two diagrams shown in figure I.7.3, there are ten diagrams obtained by adding an unconnected straight line to each of the ten diagrams in figure I.7.2. (Do you understand why?)

For obvious reasons, some diagrams (e.g., figure I.7.1a, I.7.3a) are known as tree[2] diagrams and others (e.g., Figs. I.7.1b and I.7.2a) as loop diagrams.

Do as many examples as you need until you feel thoroughly familiar with what is going on, because we are going to do exactly the same thing in quantum field theory. It will look much messier, but only superficially. Be sure you understand how to use diagrams to

---

[2] The Chinese character for tree (A. Zee, *Swallowing Clouds*) is shown in fig. I.7.5. I leave it to you to figure out why this diagram does not appear in our $Z(J)$.

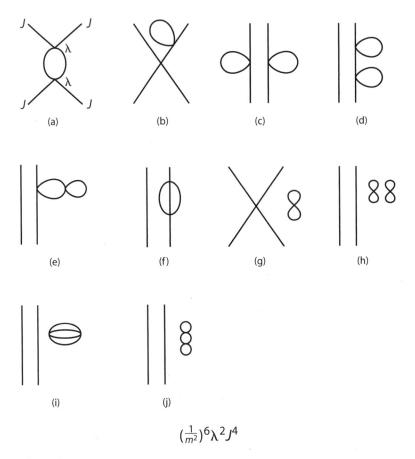

$$(\tfrac{1}{m^2})^6 \lambda^2 J^4$$

Figure I.7.2

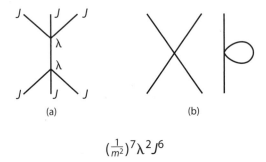

$$(\tfrac{1}{m^2})^7 \lambda^2 J^6$$

Figure I.7.3

Figure I.7.4

Figure I.7.5

represent the double series expansion of $\tilde{Z}(J)$ before reading on. Please. In my experience teaching, students who have not thoroughly understood the expansion of $\tilde{Z}(J)$ have no hope of understanding what we are going to do in the field theory context.

Wick contraction

It is more obvious than obvious that we can expand $Z(J)$ in powers of J, if we please, instead of in powers of λ. As you will see, particle physicists like to classify in power of J. In our baby problem, we can write

$$Z(J) = \sum_{s=0}^{\infty} \frac{1}{s!} J^s \int_{-\infty}^{+\infty} dq \, e^{-\frac{1}{2}m^2q^2 - (\lambda/4!)q^4} q^s \equiv Z(0,0) \sum_{s=0}^{\infty} \frac{1}{s!} J^s G^{(s)} \tag{5}$$

The coefficient $G^{(s)}$, whose analogs are known as "Green's functions" in field theory, can be evaluated as a series in λ with each term determined by Wick contraction (I.2.10). For instance, the $O(\lambda)$ term in $G^{(4)}$ is

$$\frac{-\lambda}{4!Z(0,0)} \int_{-\infty}^{+\infty} dq \, e^{-\frac{1}{2}m^2q^2} q^8 = \frac{-7!!}{4!} \frac{1}{m^8}$$

which of course better be equal[3] to what we obtained above for the λJ^4 term in $\tilde{Z}$. Thus, there are two ways of computing Z: you expand in λ first or you expand in J first.

Connected versus disconnected

You will have noticed that some Feynman diagrams are connected and others are not. Thus, figure I.7.3a is connected while 3b is not. I presaged this at the end of chapter I.4 and in figures I.4.2 and I.4.3. Write

$$Z(J, \lambda) = Z(J = 0, \lambda)e^{W(J,\lambda)} = Z(J = 0, \lambda) \sum_{N=0}^{\infty} \frac{1}{N!} [W(J, \lambda)]^N \tag{6}$$

By definition, $Z(J = 0, \lambda)$ consists of those diagrams with no external source J, such as the one in figure I.7.4. The statement is that W is a sum of connected diagrams while

---

[3] As a check on the laws of arithmetic we verify that indeed $7!!/(4!)^2 = 8!/(4!)^3 2^4$.

Z contains connected as well as disconnected diagrams. Thus, figure I.7.3b consists of two disconnected pieces and comes from the term $(1/2!)[W(J, \lambda)]^2$ in (6), the 2! taking into account that it does not matter which of the two pieces you put "on the left or on the right." Similarly, figure I.7.2i comes from $(1/3!)[W(J, \lambda)]^3$. Thus, it is W that we want to calculate, not Z. If you've had a good course on statistical mechanics, you will recognize that this business of connected graphs versus disconnected graphs is just what underlies the relation between free energy and the partition function.

Propagation: from here to there

All these features of the baby problem are structurally the same as the corresponding features of field theory and we can take over the discussion almost immediately. But before we graduate to field theory, let us consider what I call a child problem, the evaluation of a multiple integral instead of a single integral:

$$Z(J) = \int_{-\infty}^{+\infty} \int_{-\infty}^{+\infty} \cdots \int_{-\infty}^{+\infty} dq_1 dq_2 \cdots dq_N \, e^{-\frac{1}{2}q \cdot A \cdot q - (\lambda/4!)q^4 + J \cdot q} \tag{7}$$

with $q^4 \equiv \sum_i q_i^4$. Generalizing the steps leading to (3) we obtain

$$Z(J) = \left[\frac{(2\pi)^N}{\det[A]} \right]^{\frac{1}{2}} e^{-(\lambda/4!) \sum_i (\partial/\partial J_i)^4} e^{\frac{1}{2} J \cdot A^{-1} \cdot J} \tag{8}$$

Alternatively, just as in (5) we can expand in powers of J

$$Z(J) = \sum_{s=0}^{\infty} \sum_{i_1=1}^{N} \cdots \sum_{i_s=1}^{N} \frac{1}{s!} J_{i_1} \cdots J_{i_s} \int_{-\infty}^{+\infty} \left(\prod_l dq_l \right) e^{-\frac{1}{2}q \cdot A \cdot q - (\lambda/4!)q^4} q_{i_1} \cdots q_{i_s}$$

$$= Z(0, 0) \sum_{s=0}^{\infty} \sum_{i_1=1}^{N} \cdots \sum_{i_s=1}^{N} \frac{1}{s!} J_{i_1} \cdots J_{i_s} G_{i_1 \cdots i_s}^{(s)} \tag{9}$$

which again we can expand in powers of λ and evaluate by Wick contracting.

The one feature the child problem has that the baby problem doesn't is propagation "from here to there". Recall the discussion of the propagator in chapter I.3. Just as in (I.3.16) we can think of the index i as labeling the sites on a lattice. Indeed, in (I.3.16) we had in effect evaluated the "2-point Green's function" $G_{ij}^{(2)}$ to zeroth order in λ (differentiate (I.3.16) with respect to J twice):

$$G_{ij}^{(2)}(\lambda = 0) = \left[\int_{-\infty}^{+\infty} \left(\prod_l dq_l \right) e^{-\frac{1}{2}q \cdot A \cdot q} q_i q_j \right] / Z(0, 0) = (A^{-1})_{ij}$$

(see also the appendix to chapter I.2). The matrix element $(A^{-1})_{ij}$ describes propagation from i to j. In the baby problem, each term in the expansion of $Z(J)$ can be associated with several diagrams but that is no longer true with propagation.

Let us now evaluate the "4-point Green's function" $G^{(4)}_{ijkl}$ to order λ :

$$G^{(4)}_{ijkl} = \int_{-\infty}^{+\infty} \left(\prod_m dq_m \right) e^{-\frac{1}{2}q \cdot A \cdot q} q_i q_j q_k q_l \left[1 - \frac{\lambda}{4!} \sum_n q_n^4 + O(\lambda^2) \right] / Z(0,0)$$

$$= (A^{-1})_{ij}(A^{-1})_{kl} + (A^{-1})_{ik}(A^{-1})_{jl} + (A^{-1})_{il}(A^{-1})_{jk}$$

$$- \lambda \sum_n (A^{-1})_{in}(A^{-1})_{jn}(A^{-1})_{kn}(A^{-1})_{ln} + \cdots + O(\lambda^2) \tag{10}$$

The first three terms describe one excitation propagating from i to j and another propagating from k to l, plus the two possible permutations on this "history." The order λ term tells us that four excitations, propagating from i to n, from j to n, from k to n, and from l to n, meet at n and interact with an amplitude proportional to λ, where n is anywhere on the lattice or mattress. By the way, you also see why it is convenient to define the interaction $(\lambda/4!)\varphi^4$ with a $1/4!$: q_i has a choice of four q_n's to contract with, q_j has three q_n's to contract with, and so on, producing a factor of $4!$ to cancel the $(1/4!)$.

I intentionally did not display in (10) the $O(\lambda)$ terms produced by Wick contracting some of the q_n's with each other. There are two types: (I) Contracting a pair of q_n's produces something like $\lambda(A^{-1})_{ij}(A^{-1})_{kn}(A^{-1})_{ln}(A^{-1})_{nn}$ and (II) contracting the q_n's with each other produces the first three terms in (10) multiplied by $(A^{-1})_{nn}(A^{-1})_{nn}$. We see that (I) and (II) correspond to diagrams b and c in figure I.7.1, respectively. Evidently, the two excitations do not interact with each other. I will come back to (II) later in this chapter.

Perturbative field theory

You should now be ready for field theory!

Indeed, the functional integral in (1) (which I repeat here)

$$Z(J) = \int D\varphi \, e^{i \int d^4 x \{ \frac{1}{2}[(\partial \varphi)^2 - m^2 \varphi^2] - (\lambda/4!)\varphi^4 + J\varphi \}} \tag{11}$$

has the same form as the ordinary integral in (2) and the multiple integral in (7). There is one minor difference: there is no i in (2) and (7), but as I noted in chapter I.2 we can Wick rotate (11) and get rid of the i, but we won't. The significant difference is that J and φ in (11) are functions of a continuous variable x, while J and q in (2) are not functions of anything and in (7) are functions of a discrete variable. Aside from that, everything goes through the same way.

As in (3) and (8) we have

$$Z(J) = e^{-(i/4!)\lambda \int d^4 w [\delta/i\delta J(w)]^4} \int D\varphi e^{i \int d^4 x \{ \frac{1}{2}[(\partial \varphi)^2 - m^2 \varphi^2] + J\varphi \}}$$

$$= Z(0,0) e^{-(i/4!)\lambda \int d^4 w [\delta/i\delta J(w)]^4} e^{-(i/2) \int\int d^4 x d^4 y J(x) D(x-y) J(y)} \tag{12}$$

The structural similarity is total.

The role of $1/m^2$ in (3) and of A^{-1} (8) is now played by the propagator

$$D(x-y) = \int \frac{d^4 k}{(2\pi)^4} \frac{e^{ik \cdot (x-y)}}{k^2 - m^2 + i\varepsilon}$$

Incidentally, if you go back to chapter I.3 you will see that if we were in d-dimensional spacetime, $D(x - y)$ would be given by the same expression with $d^4k/(2\pi)^4$ replaced by $d^dk/(2\pi)^d$. The ordinary integral (2) is like a field theory in 0-dimensional spacetime: if we set $d = 0$, there is no propagating around and $D(x - y)$ collapses to $-1/m^2$. You see that it all makes sense.

We also know that $J(x)$ corresponds to sources and sinks. Thus, if we expand $Z(J)$ as a series in J, the powers of J would indicate the number of particles involved in the process. (Note that in this nomenclature the scattering process $\varphi + \varphi \to \varphi + \varphi$ counts as a 4-particle process: we count the total number of incoming and outgoing particles.) Thus, in particle physics it often makes sense to specify the power of J. Exactly as in the baby and child problems, we can expand in J first:

$$Z(J) = Z(0, 0) \sum_{s=0}^{\infty} \frac{i^s}{s!} \int dx_1 \ldots dx_s J(x_1) \cdots J(x_s) G^{(s)}(x_1, \cdots, x_s)$$

$$= \sum_{s=0}^{\infty} \frac{i^s}{s!} \int dx_1 \ldots dx_s J(x_1) \cdots J(x_s) \int D\varphi \, e^{i \int d^4x \{\frac{1}{2}[(\partial\varphi)^2 - m^2\varphi^2] - (\lambda/4!)\varphi^4\}}$$

$$\varphi(x_1) \cdots \varphi(x_s) \tag{13}$$

In particular, we have the 2-point Green's function

$$G(x_1, x_2) \equiv \frac{1}{Z(0, 0)} \int D\varphi \, e^{i \int d^4x \{\frac{1}{2}[(\partial\varphi)^2 - m^2\varphi^2] - (\lambda/4!)\varphi^4\}} \varphi(x_1)\varphi(x_2) \tag{14}$$

the 4-point Green's function,

$$G(x_1, x_2, x_3, x_4) \equiv \frac{1}{Z(0, 0)} \int D\varphi \, e^{i \int d^4x \{\frac{1}{2}[(\partial\varphi)^2 - m^2\varphi^2] - (\lambda/4!)\varphi^4\}} \varphi(x_1)\varphi(x_2)\varphi(x_3)\varphi(x_4) \tag{15}$$

and so on. [Sometimes $Z(J)$ is called the generating functional as it generates the Green's functions.] Obviously, by translation invariance, $G(x_1, x_2)$ does not depend on x_1 and x_2 separately, but only on $x_1 - x_2$. Similarly, $G(x_1, x_2, x_3, x_4)$ only depends on $x_1 - x_4, x_2 - x_4$, and $x_3 - x_4$. For $\lambda = 0$, $G(x_1, x_2)$ reduces to $iD(x_1 - x_2)$, the propagator introduced in chapter I.3. While $D(x_1 - x_2)$ describes the propagation of a particle between x_1 and x_2 in the absence of interaction, $G(x_1 - x_2)$ describes the propagation of a particle between x_1 and x_2 in the presence of interaction. If you understood our discussion of $G^{(4)}_{ijkl}$, you would know that $G(x_1, x_2, x_3, x_4)$ describes the scattering of particles.

In some sense, there are two ways of doing field theory, what I might call the Schwinger way (12) or the Wick way (13).

Thus, to summarize, Feynman diagrams are just an extremely convenient way of representing the terms in a double series expansion of $Z(J)$ in λ and J.

As I said in the preface, I have no intention of turning you into a whiz at calculating Feynman diagrams. In any case, that can only come with practice. Instead, I tried to give you as clear an account as I can muster of the concept behind this marvellous invention of

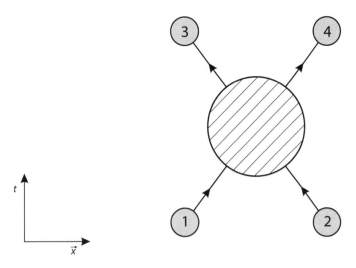

Figure I.7.6

Feynman's, which as Schwinger noted rather bitterly, enables almost anybody to become a field theorist. For the moment, don't worry too much about factors of 4! and 2!

Collision between particles

As I already mentioned, I described in chapter I.4 the strategy of setting up sources and sinks to watch the propagation of a particle (which I will call a meson) associated with the field φ. Let us now set up two sources and two sinks to watch two mesons scatter off each other. The setup is shown in figure I.7.6. The sources localized in regions 1 and 2 both produce a meson, and the two mesons eventually disappear into the sinks localized in regions 3 and 4. It clearly suffices to find in Z a term containing $J(x_1)J(x_2)J(x_3)J(x_4)$. But this is just $G(x_1, x_2, x_3, x_4)$.

Let us be content with first order in λ. Going the Wick way we have to evaluate

$$\frac{1}{Z(0,0)}\left(-\frac{i\lambda}{4!}\right)\int d^4w \int D\varphi\, e^{i\int d^4x\{\frac{1}{2}[(\partial\varphi)^2 - m^2\varphi^2]\}}$$

$$\varphi(x_1)\varphi(x_2)\varphi(x_3)\varphi(x_4)\varphi(w)^4 \tag{16}$$

Just as in (10) we Wick contract and obtain

$$(-i\lambda)\int d^4w\, D(x_1 - w)D(x_2 - w)D(x_3 - w)D(x_4 - w) \tag{17}$$

As a check, let us also derive this the Schwinger way. Replace $e^{-(i/4!)\lambda \int d^4w (\delta/\delta J(w))^4}$ by $-(i/4!)\lambda \int d^4w (\delta/\delta J(w))^4$ and $e^{-(i/2)\iint d^4x d^4y\, J(x)D(x-y)J(y)}$ by

$$\frac{i^4}{4!2^4}\left[\iint d^4x d^4y\, J(x)D(x-y)J(y)\right]^4$$

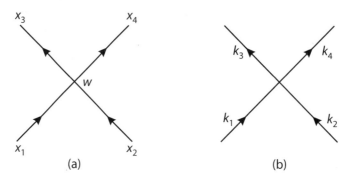

Figure I.7.7

To save writing, it would be sagacious to introduce the abbreviations J_a for $J(x_a)$, $\int_a$ for $\int d^4 x_a$, and D_{ab} for $D(x_a - x_b)$. Dropping overall numerical factors, which I invite you to fill in, we obtain

$$\sim i\lambda \int_w (\frac{\delta}{\delta J_w})^4 \int\int\int\int\int\int\int\int D_{ae}D_{bf}D_{cg}D_{dh}J_aJ_bJ_cJ_dJ_eJ_fJ_gJ_h \tag{18}$$

The four $(\delta/\delta J_w)$'s hit the eight J's in all possible combinations producing many terms, which again I invite you to write out. Two of the three terms are disconnected. The connected term is

$$\sim i\lambda \int_w \int\int\int\int D_{aw}D_{bw}D_{cw}D_{dw}J_aJ_bJ_cJ_d \tag{19}$$

Evidently, this comes from the four $(\delta/\delta J_w)$'s hitting J_e, J_f, J_g, and J_h, thus setting x_e, x_f, x_g, and x_h to w. Compare (19) with $[8!(-\lambda)/(4!)^3(2m^2)^4]J^4$ in the baby problem.

Recall that we embarked on this calculation in order to produce two mesons by sources localized in regions 1 and 2, watch them scatter off each other, and then get rid of them with sinks localized in regions 3 and 4. In other words, we set the source function $J(x)$ equal to a set of delta functions peaked at x_1, x_2, x_3, and x_4. This can be immediately read off from (19): the scattering amplitude is just $-i\lambda \int_w D_{1w}D_{2w}D_{3w}D_{4w}$, exactly as in (17).

The result is very easy to understand (see figure I.7.7a). Two mesons propagate from their birthplaces at x_1 and x_2 to the spacetime point w, with amplitude $D(x_1 - w)D(x_2 - w)$, scatter with amplitude $-i\lambda$, and then propagate from w to their graves at x_3 and x_4, with amplitude $D(w - x_3)D(w - x_4)$ [note that $D(x) = D(-x)$]. The integration over w just says that the interaction point w could have been anywhere. Everything is as in the child problem.

It is really pretty simple once you get it. Still confused? It might help if you think of (12) as some kind of machine $e^{-(i/4!)\lambda \int d^4 w [\delta/i\delta J(w)]^4}$ operating on

$$Z(J, \lambda = 0) = e^{-(i/2)\int\int d^4 x d^4 y J(x)D(x-y)J(y)}$$

When expanded out, $Z(J, \lambda = 0)$ is a bunch of J's thrown here and there in spacetime, with pairs of J's connected by D's. Think of a bunch of strings with the string ends

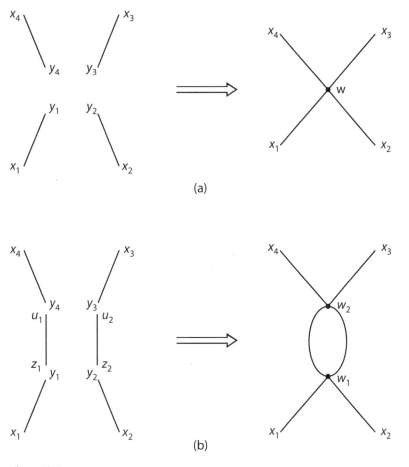

Figure I.7.8

corresponding to the J's. What does the "machine" do? The machine is a sum of terms, for example, the term

$$\sim \lambda^2 \int d^4 w_1 \int d^4 w_2 \left[\frac{\delta}{\delta J(w_1)} \right]^4 \left[\frac{\delta}{\delta J(w_2)} \right]^4$$

When this term operates on a term in $Z(J, \lambda = 0)$ it grabs four string ends and glues them together at the point w_2; then it grabs another 4 string ends and glues them together at the point w_1. The locations w_1 and w_2 are then integrated over. Two examples are shown in figure I.7.8. It is a game you can play with a child! This childish game of gluing four string ends together generates all the Feynman diagrams of our scalar field theory.

Do it once and for all

Now Feynman comes along and says that it is ridiculous to go through this long-winded yakkety-yak every time. Just figure out the rules once and for all.

For example, for the diagram in figure I.7.7a associate the factor $-i\lambda$ with the scattering, the factor $D(x_1 - w)$ with the propagation from x_1 to w, and so forth—conceptually exactly the same as in our baby problem. See, you could have invented Feynman diagrams. (Well, not quite. Maybe not, maybe yes.)

Just as in going from (I.4.1) to (I.4.2), it is easier to pass to momentum space. Indeed, that is how experiments are actually done. A meson with momentum k_1 and a meson with momentum k_2 collide and scatter, emerging with momenta k_3 and k_4 (see figure I.7.7b). Each spacetime propagator gives us

$$D(x_a - w) = \int \frac{d^4 k_a}{(2\pi)^4} \frac{e^{\pm i k_a (x_a - w)}}{k_a^2 - m^2 + i\varepsilon}$$

Note that we have the freedom of associating with the dummy integration variable either a plus or a minus sign in the exponential. Thus in integrating over w in (17) we obtain

$$\int d^4 w \, e^{-i(k_1 + k_2 - k_3 - k_4)w} = (2\pi)^4 \delta^{(4)}(k_1 + k_2 - k_3 - k_4).$$

That the interaction could occur anywhere in spacetime translates into momentum conservation $k_1 + k_2 = k_3 + k_4$. (We put in the appropriate minus signs in two of the D's so that we can think of k_3 and k_4 as outgoing momenta.)

So there are Feynman diagrams in real spacetime and in momentum space. Spacetime Feynman diagrams are literally pictures of what happened. (A trivial remark: the orientation of Feynman diagrams is a matter of idiosyncratic choice. Some people draw them with time running vertically, others with time running horizontally. We follow Feynman in this text.)

The rules

We have thus derived the celebrated momentum space Feynman rules for our scalar field theory:

1. Draw a Feynman diagram of the process (fig. I.7.7b for the example we just discussed).
2. Label each line with a momentum k and associate with it the propagator $i/(k^2 - m^2 + i\varepsilon)$.
3. Associate with each interaction vertex the coupling $-i\lambda$ and $(2\pi)^4 \, \delta^{(4)}(\sum_i k_i - \sum_j k_j)$, forcing the sum of momenta $\sum_i k_i$ coming into the vertex to be equal to the sum of momenta $\sum_j k_j$ going out of the vertex.
4. Momenta associated with internal lines are to be integrated over with the measure $\frac{d^4 k}{(2\pi)^4}$. Incidentally, this corresponds to the summation over intermediate states in ordinary perturbation theory.
5. Finally, there is a rule about some really pesky symmetry factors. They are the analogs of those numerical factors in our baby problem. As a result, some diagrams are to be multiplied by a symmetry factor such as $\frac{1}{2}$. These originate from various combinatorial factors counting the different ways in which the $(\delta/\delta J)$'s can hit the J's in expressions such as (18). I will let you discover a symmetry factor in exercise I.7.2.

We will illustrate by examples what these rules (and the concept of internal lines) mean.

Our first example is just the diagram (fig. I.7.7b) that we have calculated. Applying the rules we obtain

$$-i\lambda(2\pi)^4\delta^{(4)}(k_1 + k_2 - k_3 - k_4) \prod_{a=1}^{4}\left(\frac{i}{k_a^2 - m^2 + i\varepsilon}\right)$$

You would agree that it is silly to drag the factor $\prod_{a=1}^{4}\left(\frac{i}{k_a^2 - m^2 + i\varepsilon}\right)$ around, since it would be common to all diagrams in which two mesons scatter into two mesons. So we append to the Feynman rules an additional rule that we do not associate a propagator with external lines. (This is known in the trade as "amputating the external legs".)

In an actual scattering experiment, the external particles are of course real and on shell, that is, their momenta satisfy $k_a^2 - m^2 = 0$. Thus we better not keep the propagators of the external lines around. Arithmetically, this amounts to multiplying the Green's functions [and what we have calculated thus far are indeed Green's functions, see (16)] by the factor $\prod_a(-i)(k_a^2 - m^2)$ and then set $k_a^2 = m^2$ ("putting the particles on shell" in conversation). At this point, this procedure sounds like formal overkill. We will come back to the rationale behind it at the end of the next chapter.

Also, since there is always an overall factor for momentum conservation we should not drag the overall delta function around either. Thus, we have two more rules:

6. Do not associate a propagator with external lines.
7. A delta function for overall momentum conservation is understood.

Applying these rules we obtain an amplitude which we will denote by $\mathcal{M}$. For example, for the diagram in figure I.7.7b $\mathcal{M} = -i\lambda$.

The birth of particles

As explained in chapter I.1 one of the motivations for constructing quantum field theory was to describe the production of particles. We are now ready to describe how two colliding mesons can produce four mesons. The Feynman diagram in figure I.7.9 (compare fig. I.7.3a) can occur in order λ^2 in perturbation theory. Amputating the external legs, we drop the factor $\prod_{a=1}^{6}[i/k_a^2 - m^2 + i\varepsilon]$ associated with the six external lines, keeping only the propagator associated with the one internal line. For each vertex we put in a factor of $(-i\lambda)$ and a momentum conservation delta function (rule 3). Then we integrate over the internal momentum q (rule 4) to obtain

$$(-i\lambda)^2 \int \frac{d^4q}{(2\pi)^4} \frac{i}{q^2 - m^2 + i\varepsilon}(2\pi)^4\delta^{(4)}(k_1 + k_2 - k_3 - q)(2\pi)^4\delta^{(4)}[q - (k_4 + k_5 + k_6)] \tag{20}$$

The integral over q is a cinch, giving

$$(-i\lambda)^2 \frac{i}{(k_4 + k_5 + k_6)^2 - m^2 + i\varepsilon}(2\pi)^4\delta^{(4)}[k_1 + k_2 - (k_3 + k_4 + k_5 + k_6)] \tag{21}$$

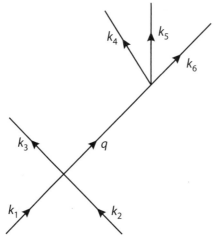

Figure I.7.9

We have already agreed (rule 7) not to drag the overall delta function around. This example teaches us that we didn't have to write down the delta functions and then annihilate (all but one of) them by integrating. In figure I.7.9 we could have simply imposed momentum conservation from the beginning and labeled the internal line as $k_4 + k_5 + k_6$ instead of q.

With some practice you could just write down the amplitude

$$\mathcal{M} = (-i\lambda)^2 \frac{i}{(k_4 + k_5 + k_6)^2 - m^2 + i\varepsilon} \tag{22}$$

directly: just remember, a coupling $(-i\lambda)$ for each vertex and a propagator for each internal (but not external) line. Pretty easy once you get the hang of it. As Schwinger said, the masses could do it.

The cost of not being real

The physics involved is also quite clear: The internal line is associated with a virtual particle whose relativistic 4-momentum $k_4 + k_5 + k_6$ squared is not necessarily equal to m^2, as it would have to be if the particle were real. The farther the momentum of the virtual particle is from the mass shell the smaller the amplitude. You are penalized for not being real.

According to the quantum rules for dealing with identical particles, to obtain the full amplitude we have to symmetrize among the four final momenta. One way of saying it is to note that the line labeled k_3 in figure (I.7.9) could have been labeled k_4, k_5, or k_6, and we have to add all four contributions.

To make sure that you understand the Feynman rules I insist that you go through the path integral calculation to obtain (21) starting with (12) and (13).

I am repeating myself but I think it is worth emphasizing again that there is nothing particularly magical about Feynman diagrams.

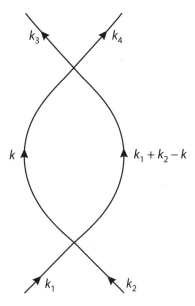

Figure I.7.10

Loops and a first look at divergence

Just as in our baby problem, we have tree diagrams and loop diagrams. So far we have only looked at tree diagrams. Our next example is the loop diagram in figure I.7.10 (compare fig. I.7.2a.) Applying the Feynman rules, we obtain

$$\frac{1}{2}(-i\lambda)^2 \int \frac{d^4k}{(2\pi)^4} \frac{i}{k^2 - m^2 + i\varepsilon} \frac{i}{(k_1 + k_2 - k)^2 - m^2 + i\varepsilon} \tag{23}$$

As above, the physics embodied in (23) is clear: As k ranges over all possible values, the integrand is large only if one or the other or both of the virtual particles associated with the two internal lines are close to being real. Once again, there is a penalty for not being real (see exercise I.7.4).

For large k the integrand goes as $1/k^4$. The integral is infinite! It diverges as $\int d^4k(1/k^4)$. We will come back to this apparent disaster in chapter III.1.

With some practice, you will be able to write down the amplitude by inspection. As another example, consider the three-loop diagram in figure I.7.11 contributing in $O(\lambda^4)$ to meson-meson scattering. First, for each loop pick an internal line and label the momentum it carries, p, q, and r in our example. There is considerable freedom of choice in labeling—your choice may well not agree with mine, but of course the physics should not depend on it. The momenta carried by the other internal lines are then fixed by momentum conservation, as indicated in the figure. Write down a coupling for each vertex, and a propagator for each internal line, and integrate over the internal momenta p, q, and r.

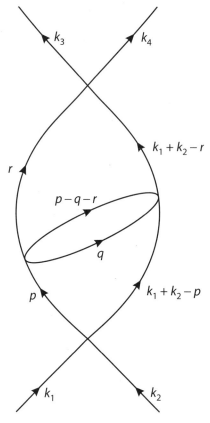

Figure I.7.11

Thus, without worrying about symmetry factors, we have the amplitude

$$(-i\lambda)^4 \int \frac{d^4 p}{(2\pi)^4} \frac{d^4 q}{(2\pi)^4} \frac{d^4 r}{(2\pi)^4} \frac{i}{p^2 - m^2 + i\varepsilon} \frac{i}{(k_1 + k_2 - p)^2 - m^2 + i\varepsilon}$$

$$\frac{i}{q^2 - m^2 + i\varepsilon} \frac{i}{(p - q - r)^2 - m^2 + i\varepsilon} \frac{i}{r^2 - m^2 + i\varepsilon} \frac{i}{(k_1 + k_2 - r)^2 - m^2 + i\varepsilon} \tag{24}$$

Again, this triple integral also diverges: It goes as $\int d^{12} P (1/P^{12})$.

An assurance

When I teach quantum field theory, at this point in the course some students get un-accountably very anxious about Feynman diagrams. I would like to assure the reader that the whole business is really quite simple. Feynman diagrams can be thought of simply as pictures in spacetime of the antics of particles, coming together, colliding and producing other particles, and so on. One student was puzzled that the particles do not move in straight lines. Remember that a quantum particle propagates like a wave; $D(x - y)$ gives us the amplitude for the particle to propagate from x to y. Evidently, it is more convenient to think of particles in momentum space: Fourier told us so. We will see many more examples of Feynman diagrams, and you will soon be well acquainted with them. Another student was concerned about evaluating the integrals in (23) and (24). I haven't taught you how yet, but will eventually. The good news is that in contemporary research on the frontline few theoretical physicists actually have to calculate Feynman diagrams getting all the factors of 2 right. In most cases, understanding the general behavior of the integral is sufficient. But of course, you should always take pride in getting everything right. In chapters II.6, III.6, and III.7 I will calculate Feynman diagrams for you in detail, getting all the factors right so as to be able to compare with experiments.

Vacuum fluctuations

Let us go back to the terms I neglected in (18) and which you are supposed to have figured out. For example, the four $[\delta/\delta J(w)]$'s could have hit J_c, J_d, J_g, and J_h in (18) thus producing something like

$$-i\lambda \int\int\int\int D_{ae}D_{bf}J_aJ_bJ_eJ_f(\int_w D_{ww}D_{ww}).$$

The coefficient of $J(x_1)J(x_2)J(x_3)J(x_4)$ is then $D_{13}D_{24}(-i\lambda\int_w D_{ww}D_{ww})$ plus terms obtained by permuting.

The corresponding physical process is easy to describe in words and in pictures (see figure I.7.12). The source at x_1 produces a particle that propagates freely without any interaction to x_3, where it comes to a peaceful death. The particle produced at x_2 leads a similarly uneventful life before being absorbed at x_4. The two particles did not interact at all. Somewhere off at the point w, which could perfectly well be anywhere in the universe, there was an interaction with amplitude $-i\lambda$. This is known as a vacuum fluctuation:

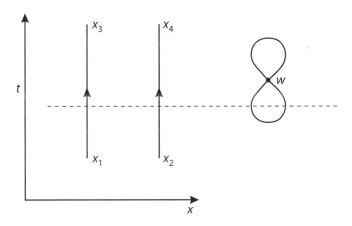

Figure I.7.12

As explained in chapter I.1, quantum mechanics and special relativity inevitably cause particles to pop out of the vacuum, and they could even interact before vanishing again into the vacuum. Look at different time slices (one of which is indicated by the dotted line) in figure I.7.12. In the far past, the universe has no particles. Then it has two particles, then four, then two again, and finally in the far future, none. We will have a lot more to say in chapter VIII.2 about these fluctuations. Note that vacuum fluctuations occur also in our baby and child problems (see, e.g., Figs. I.7.1c, I.7.2g,h,i,j, and so forth).

Two words about history

I believe strongly that any self-respecting physicist should learn about the history of physics, and the history of quantum field theory is among the most fascinating. Unfortunately, I do not have the space to go into it here.[4] The path integral approach to field theory using sources $J(x)$ outlined here is mainly associated with Julian Schwinger, who referred to it as "sorcery" during my graduate student days (so that I could tell people who inquired that I was studying sorcery in graduate school.) In one of the many myths retold around tribal fires by physicists, Richard Feynman came upon his rules in a blinding flash of insight. In 1949 Freeman Dyson showed that the Feynman rules which so mystified people at the Pocono conference a year earlier could actually be obtained from the more formal work of Julian Schwinger and of Shin-Itiro Tomonaga.

Exercises

I.7.1 Work out the amplitude corresponding to figure I.7.11 in (24).

I.7.2 Derive (23) from first principles, that is from (11). It is a bit tedious, but straightforward. You should find a symmetry factor $\frac{1}{2}$.

I.7.3 Draw all the diagrams describing two mesons producing four mesons up to and including order λ^2. Write down the corresponding Feynman amplitudes.

I.7.4 By Lorentz invariance we can always take $k_1 + k_2 = (E, \vec{0})$ in (23). The integral can be studied as a function of E. Show that for both internal particles to become real E must be greater than $2m$. Interpret physically what is happening.

---

[4] An excellent sketch of the history of quantum field theory is given in chapter 1 of *The Quantum Theory of Fields* by S. Weinberg. For a fascinating history of Feynman diagrams, see *Drawing Theories Apart* by D. Kaiser.

$\boxed{\text{I.8}}$ Quantizing Canonically

Quantum electrodynamics is made to appear more difficult than it actually is by the very many equivalent methods by which it may be formulated.

—R. P. Feynman

Always create before we annihilate, not the other way around.

—Anonymous

Complementary formalisms

I adopted the path integral formalism as the quickest way to quantum field theory. But I must also discuss the canonical formalism, not only because historically it was the method used to develop quantum field theory, but also because of its continuing importance. Interestingly, the canonical and the path integral formalisms often appear complementary, in the sense that results difficult to see in one are clear in the other.

Heisenberg and Dirac

Let us begin with a lightning review of Heisenberg's approach to quantum mechanics. Given a classical Lagrangian for a single particle $L = \frac{1}{2}\dot{q}^2 - V(q)$ (we set the mass equal to 1), the canonical momentum is defined as $p \equiv \delta L / \delta \dot{q} = \dot{q}$. The Hamiltonian is then given by $H = p\dot{q} - L = p^2/2 + V(q)$. Heisenberg promoted position $q(t)$ and momentum $p(t)$ to operators by imposing the canonical commutation relation

$$[p, q] = -i \tag{1}$$

Operators evolve in time according to

$$\frac{dp}{dt} = i[H, p] = -V'(q) \tag{2}$$

and

$$\frac{dq}{dt} = i[H, q] = p \tag{3}$$

In other words, operators constructed out of p and q evolve according to $O(t) = e^{iHt}O(0)e^{-iHt}$. In (1) p and q are understood to be taken at the same time. We obtain the operator equation of motion $\ddot{q} = -V'(q)$ by combining (2) and (3).

Following Dirac, we invite ourselves to consider at some instant in time the operator $a \equiv (1/\sqrt{2\omega})(\omega q + ip)$ with some parameter ω. From (1) we have

$$[a, a^{\dagger}] = 1 \tag{4}$$

The operator $a(t)$ evolves according to

$$\frac{da}{dt} = i[H, \frac{1}{\sqrt{2\omega}}(\omega q + ip)] = -i\sqrt{\frac{\omega}{2}}\left(ip + \frac{1}{\omega}V'(q)\right)$$

which can be written in terms of a and $a^{\dagger}$. The ground state $|0\rangle$ is defined as the state such that $a|0\rangle = 0$.

In the special case $V'(q) = \omega^2 q$ we get the particularly simple result $\frac{da}{dt} = -i\omega a$. This is of course the harmonic oscillator $L = \frac{1}{2}\dot{q}^2 - \frac{1}{2}\omega^2 q^2$ and $H = \frac{1}{2}(p^2 + \omega^2 q^2) = \omega(a^{\dagger}a + \frac{1}{2})$.

The generalization to many particles is immediate. Starting with

$$L = \sum_a \frac{1}{2}\dot{q}_a{}^2 - V(q_1, q_2, \cdots, q_N)$$

we have $p_a = \delta L/\delta \dot{q}_a$ and

$$[p_a(t), q_b(t)] = -i\delta_{ab} \tag{5}$$

The generalization to field theory is almost as immediate. In fact, we just use our handy-dandy substitution table (I.3.6) and see that in $D-$dimensional space L generalizes to

$$L = \int d^D x\{\frac{1}{2}(\dot{\varphi}^2 - (\vec{\nabla}\varphi)^2 - m^2\varphi^2) - u(\varphi)\} \tag{6}$$

where we denote the anharmonic term (the interaction term in quantum field theory) by $u(\varphi)$. The canonical momentum density conjugate to the field $\varphi(\vec{x}, t)$ is then

$$\pi(\vec{x}, t) = \frac{\delta L}{\delta \dot{\varphi}(\vec{x}, t)} = \partial_0 \varphi(\vec{x}, t) \tag{7}$$

so that the canonical commutation relation at equal times now reads [using (I.3.6)]

$$[\pi(\vec{x}, t), \varphi(\vec{x}', t)] = [\partial_0\varphi(\vec{x}, t), \varphi(\vec{x}', t)] = -i\delta^{(D)}(\vec{x} - \vec{x}') \tag{8}$$

(and of course also $[\pi(\vec{x}, t), \pi(\vec{x}', t)] = 0$ and $[\varphi(\vec{x}, t), \varphi(\vec{x}', t)] = 0$.) Note that δ_{ab} in (5) gets promoted to $\delta^{(D)}(\vec{x} - \vec{x}')$ in (8). You should check that (8) has the correct dimension. Turning the canonical crank we find the Hamiltonian

$$H = \int d^D x[\pi(\vec{x}, t)\partial_0\varphi(\vec{x}, t) - \mathcal{L}]$$

$$= \int d^D x\{\frac{1}{2}[\pi^2 + (\vec{\nabla}\varphi)^2 + m^2\varphi^2] + u(\varphi)\} \tag{9}$$

For the case $u = 0$, corresponding to the harmonic oscillator, we have a free scalar field theory and can go considerably farther. The field equation reads

$$(\partial^2 + m^2)\varphi = 0 \tag{10}$$

Fourier expanding, we have

$$\varphi(\vec{x}, t) = \int \frac{d^D k}{\sqrt{(2\pi)^D 2\omega_k}} [a(\vec{k}) e^{-i(\omega_k t - \vec{k} \cdot \vec{x})} + a^\dagger(\vec{k}) e^{i(\omega_k t - \vec{k} \cdot \vec{x})}] \tag{11}$$

with $\omega_k = +\sqrt{\vec{k}^2 + m^2}$ so that the field equation (10) is satisfied. The peculiar looking factor $(2\omega_k)^{-\frac{1}{2}}$ is chosen so that the canonical relation

$$[a(\vec{k}), a^\dagger(\vec{k}')] = \delta^{(D)}(\vec{k} - \vec{k}') \tag{12}$$

appropriate for creation and annihilation operators implies the canonical commutation $[\partial_0 \varphi(\vec{x}, t), \varphi(\vec{x}', t)] = -i\delta^{(D)}(\vec{x} - \vec{x}')$ in (8). You should check this but you can see why the factor $(2\omega_k)^{-\frac{1}{2}}$ is needed since in $\partial_0 \varphi$ a factor ω_k is brought down from the exponential.

As in quantum mechanics, the vacuum or ground state $| |0\rangle$ is defined by the condition $a(\vec{k}) |0\rangle = 0$ for all $\vec{k}$ and the single particle state by $|\vec{k}\rangle \equiv a^\dagger(\vec{k}) |0\rangle$. Thus, for example, using (12) we have $\langle 0| \varphi(\vec{x}, t) |\vec{k}\rangle = (1/\sqrt{(2\pi)^D 2\omega_k}) e^{-i(\omega_k t - \vec{k} \cdot \vec{x})}$, which we could think of as the relativistic wave function of a single particle with momentum $\vec{k}$. For later use, we will write this more compactly as $(1/\rho(k)) e^{-ik \cdot x}$, with $\rho(k) \equiv \sqrt{(2\pi)^D 2\omega_k}$ a normalization factor and $k^0 = \omega_k$.

To make contact with the path integral formalism let us calculate $\langle 0| \varphi(\vec{x}, t)\varphi(\vec{0}, 0) |0\rangle$ for $t > 0$. Of the four terms $a^\dagger a^\dagger$, $a^\dagger a$, $aa^\dagger$, and aa in the product of the two fields only $aa^\dagger$ survives, and thus using (12) we obtain $\int [d^D k/(2\pi)^D 2\omega_k] e^{-i(\omega_k t - \vec{k} \cdot \vec{x})}$. In other words, if we define the time-ordered product $T[\varphi(x)\varphi(y)] = \theta(x^0 - y^0)\varphi(x)\varphi(y) + \theta(y^0 - x^0)\varphi(y)\varphi(x)$, we find

$$\langle 0| T[\varphi(\vec{x}, t)\varphi(\vec{0}, 0)] |0\rangle =$$
$$\int \frac{d^D k}{(2\pi)^D 2\omega_k} [\theta(t) e^{-i(\omega_k t - \vec{k} \cdot \vec{x})} + \theta(-t) e^{+i(\omega_k t - \vec{k} \cdot \vec{x})}] \tag{13}$$

Go back to (I.3.23). We discover that $\langle 0| T[\varphi(x)\varphi(0)] |0\rangle = iD(x)$, the propagator for a particle to go from 0 to x we obtained using the path integral formalism. This further justifies the $i\varepsilon$ prescription in (I.3.22). The physical meaning is that we always create before we annihilate, not the other way around. This is a form of causality as formulated in quantum field theory.

A remark: The combination $d^D k/(2\omega_k)$, even though it does not look Lorentz invariant, is in fact a Lorentz invariant measure. To see this, we use (I.2.13) to show that (exercise I.8.1)

$$\int d^{(D+1)} k \delta(k^2 - m^2)\theta(k^0) f(k^0, \vec{k}) = \int \frac{d^D k}{2\omega_k} f(\omega_k, \vec{k}) \tag{14}$$

for any function $f(k)$. Since Lorentz transformations cannot change the sign of k^0, the step function $\theta(k^0)$ is Lorentz invariant. Thus the left-hand side is manifestly Lorentz invariant, and hence the right-hand side must also be Lorentz invariant. This shows that relations such as (13) are Lorentz invariant; they are frame-independent statements.

Scattering amplitude

Now that we have set up the canonical formalism it is instructive to see how the invariant amplitude $\mathcal{M}$ defined in the preceding chapter arises using this alternative formalism. Let us calculate the amplitude $\langle \vec{k}_3 \vec{k}_4 | e^{-iHT} | \vec{k}_1 \vec{k}_2 \rangle = \langle \vec{k}_3 \vec{k}_4 | e^{i \int d^4 x \mathcal{L}(x)} | \vec{k}_1 \vec{k}_2 \rangle$ for meson scattering $\vec{k}_1 + \vec{k}_2 \to \vec{k}_3 + \vec{k}_4$ in order λ with $u(\varphi) = \frac{\lambda}{4!} \varphi^4$. (We have, somewhat sloppily, turned the large transition time T into $\int dx^0$ when going over to the Lagrangian.) Expanding in λ, we obtain $(-i\frac{\lambda}{4!}) \int d^4 x \langle \vec{k}_3 \vec{k}_4 | \varphi^4(x) | \vec{k}_1 \vec{k}_2 \rangle$.

The calculation of the matrix element is not dissimilar to the one we just did for the propagator. There we have the product of two field operators between the vacuum state. Here we have the product of four field operators, all evaluated at the same spacetime point x, sandwiched between two-particle states. There we look for a term of the form $a(\vec{k}) a^\dagger(\vec{k})$. Here, plugging the expansion (11) of the field into the product $\varphi^4(x)$, we look for terms of the form $a^\dagger(\vec{k}_4) a^\dagger(\vec{k}_3) a(\vec{k}_2) a(\vec{k}_1)$, so that we could remove the two incoming particles and produce the two outgoing particles. (To avoid unnecessary complications we assume that all four momenta are different.) We now annihilate and create. The annihilation operator $a(\vec{k}_1)$ could have come from any one of the four φ fields in φ^4, giving a factor of 4, $a(\vec{k}_1)$ could have come from any one of the three remaining φ fields, giving a factor of 3, $a^\dagger(\vec{k}_3)$ could have come from either of the two remaining φ fields, giving a factor of 2, so that we end up with a factor of 4!, which cancels the factor of $\frac{1}{4!}$ included in the definition of λ. (This is of course why, for the sake of convenience, λ is defined the way it is. Recall from the preceding chapter an analogous step.)

As you just learned and as you can see from (11), we obtain a factor of $1/\rho(k) e^{-ik \cdot x}$ for each incoming particle and of $1/\rho(k) e^{+ik \cdot x}$ for each outgoing particle, giving all together

$$\left(\Pi_{\alpha=1}^{4} \frac{1}{\rho(k_\alpha)} \right) \int d^4 x e^{i(k_3 + k_4 - k_1 - k_2) \cdot x} = \left(\Pi_{\alpha=1}^{4} \frac{1}{\rho(k_\alpha)} \right) (2\pi)^4 \delta^4(k_3 + k_4 - k_1 - k_2)$$

It is conventional to refer to $S_{fi} = \langle f | e^{-iHT} | i \rangle$, with some initial and final state, as elements of an "S-matrix" and to define the "transition matrix" T-matrix by $S = I + iT$, that is,

$$S_{fi} = \delta_{fi} + iT_{fi} \tag{15}$$

In general, for initial and final states consisting of scalar particles characterized only by their momenta, we write (using an obvious notation, for example $\sum_i k$ is the sum of the particle momenta in the initial state), invoking momentum conservation:

$$iT_{fi} = (2\pi)^4 \delta^4 \left(\sum_f k - \sum_i k \right) \left(\Pi_\alpha \frac{1}{\rho(k_\alpha)} \right) \mathcal{M}(f \leftarrow i) \tag{16}$$

In our simple example, $iT(\vec{k}_3 \vec{k}_4, \vec{k}_1 \vec{k}_2) = (-i\frac{\lambda}{4!}) \int d^4 x \langle \vec{k}_3 \vec{k}_4 | \varphi^4(x) | \vec{k}_1 \vec{k}_2 \rangle$, and our little calculation showed that $\mathcal{M} = -i\lambda$, precisely as given in the preceding chapter. But this connection between T_{fi} and $\mathcal{M}$, being "merely" kinematical, should hold in general, with the invariant amplitude $\mathcal{M}$ determined by the Feynman rules. I will not give a long boring formal proof, but you should check this assertion by working out some more involved cases, such as the scattering amplitude to order λ^2, recovering (I.7.23), for example.

Thus, quite pleasingly, we see that the invariant amplitude $\mathcal{M}$ determined by the Feynman rules represents the "heart of the matter" with the momentum conservation delta function and normalization factors stripped away.

My pedagogical aim here is merely to make one more contact (we will come across more in later chapters) between the canonical and path integral formalisms, giving the simplest possible example avoiding all subtleties and complications. Those readers into rigor are invited to replace the plane wave states $|\vec{k}_1\vec{k}_2\rangle$ with wave packet states $\int d^3k_1 \int d^3k_2 f_1(\vec{k}_1) f_2(\vec{k}_2) |\vec{k}_1\vec{k}_2\rangle$ for some appropriate functions f_1 and f_2, starting in the far past when the wave packets were far apart, evolving into the far future, so on and so forth, all the while smiling with self-satisfaction. The entire procedure is after all no different from the treatment of scattering[1] in elementary nonrelativistic quantum mechanics.

Complex scalar field

Thus far, we have discussed a hermitean (often called real in a minor abuse of terminology) scalar field. Consider instead (as we will in chapter I.10) a nonhermitean (usually called complex in another minor abuse) scalar field governed by $\mathcal{L} = \partial\varphi^\dagger \partial\varphi - m^2\varphi^\dagger\varphi$.

Again, following Heisenberg, we find the canonical momentum density conjugate to the field $\varphi(\vec{x}, t)$, namely $\pi(\vec{x}, t) = \delta L/[\delta\dot{\varphi}(\vec{x}, t)] = \partial_0\varphi^\dagger(\vec{x}, t)$, so that $[\pi(\vec{x}, t), \varphi(\vec{x}', t)] = [\partial_0\varphi^\dagger(\vec{x}, t), \varphi(\vec{x}', t)] = -i\delta^{(D)}(\vec{x} - \vec{x}')$. Similarly, the canonical momentum density conjugate to the field $\varphi^\dagger(\vec{x}, t)$ is $\partial_0\varphi(\vec{x}, t)$.

Varying $\varphi^\dagger$ we obtain the Euler-Lagrange equation of motion $(\partial^2 + m^2)\varphi = 0$. [Similarly, varying φ we obtain $(\partial^2 + m^2)\varphi^\dagger = 0$.] Once again, we could Fourier expand, but now the nonhermiticity of φ means that we have to replace (11) by

$$\varphi(\vec{x}, t) = \int \frac{d^D k}{\sqrt{(2\pi)^D 2\omega_k}} \left[a(\vec{k})e^{-i(\omega_k t - \vec{k}\cdot\vec{x})} + b^\dagger(\vec{k})e^{i(\omega_k t - \vec{k}\cdot\vec{x})} \right] \tag{17}$$

In (11) hermiticity fixed the second term in the square bracket in terms of the first. Here in contrast, we are forced to introduce two independent sets of creation and annihilation operators $(a, a^\dagger)$ and $(b, b^\dagger)$. You should verify that the canonical commutation relations imply that these indeed behave like creation and annihilation operators.

Consider the current

$$J_\mu = i(\varphi^\dagger \partial_\mu\varphi - \partial_\mu\varphi^\dagger \varphi) \tag{18}$$

Using the equations of motion you should check that $\partial_\mu J^\mu = i(\varphi^\dagger \partial^2\varphi - \partial^2\varphi^\dagger \varphi)$. (This follows immediately from the fact that the equation of motion for $\varphi^\dagger$ is the hermitean

---

[1] For example, M. L. Goldberger and K. M. Watson, *Collision Theory*.

conjugate of the equation of motion for φ.) The current is conserved and the corresponding time-independent charge is given by (verify this!)

$$Q = \int d^D x \, J_0(x) = \int d^D k [a^\dagger(\vec{k}) a(\vec{k}) - b^\dagger(\vec{k}) b(\vec{k})].$$

Thus the particle created by $a^\dagger$ (call it the "particle") and the particle created by $b^\dagger$ (call it the "antiparticle") carry opposite charges. Explicitly, using the commutation relation we have $Q a^\dagger |0\rangle = +a^\dagger |0\rangle$ and $Q b^\dagger |0\rangle = -b^\dagger |0\rangle$.

We conclude that $\varphi^\dagger$ creates a particle and annihilates an antiparticle, that is, it produces one unit of charge. The field φ does the opposite. You should understand this point thoroughly, as we will need it when we come to the Dirac field for the electron and positron.

The energy of the vacuum

As an instructive exercise let us calculate in the free scalar field theory the expectation value $\langle 0| H |0\rangle = \int d^D x \, \frac{1}{2} \langle 0| (\pi^2 + (\vec{\nabla}\varphi)^2 + m^2 \varphi^2) |0\rangle$, which we may loosely refer to as the "energy of the vacuum." It is merely a matter of putting together (7), (11), and (12). Let us focus on the third term in $\langle 0| H |0\rangle$, which in fact we already computed, since

$$\langle 0| \varphi(\vec{x}, t) \varphi(\vec{x}, t) |0\rangle = \langle 0| \varphi(\vec{0}, 0) \varphi(\vec{0}, 0) |0\rangle$$

$$= \lim_{\vec{x}, t \to \vec{0}, 0} \langle 0| \varphi(\vec{x}, t) \varphi(\vec{0}, 0) |0\rangle = \lim_{\vec{x}, t \to \vec{0}, 0} \int \frac{d^D k}{(2\pi)^D 2\omega_k} e^{-i(\omega_k t - \vec{k}\cdot\vec{x})} = \int \frac{d^D k}{(2\pi)^D 2\omega_k}$$

The first equality follows from translation invariance, which also implies that the factor $\int d^D x$ in $\langle 0| H |0\rangle$ can be immediately replaced by V, the volume of space. The calculation of the other two terms proceeds in much the same way: for example, the two factors of $\vec{\nabla}$ in $(\vec{\nabla}\varphi)^2$ just bring down a factor of $\vec{k}^2$. Thus

$$\langle 0| H |0\rangle = V \int \frac{d^D k}{(2\pi)^D 2\omega_k} \frac{1}{2} (\omega_k^2 + \vec{k}^2 + m^2) = V \int \frac{d^D k}{(2\pi)^D} \frac{1}{2} \hbar \omega_k \tag{19}$$

upon restoring $\hbar$.

You should find this result at once gratifying and alarming, gratifying because we recognize it as the zero point energy of the harmonic oscillator integrated over all momentum modes and over all space, and alarming because the integral over $\vec{k}$ clearly diverges. But we should not be alarmed: the energy of any physical configuration, for example the mass of a particle, is to be measured relative to the "energy of the vacuum." We ask for the difference in the energy of the world with and without the particle. In other words, we could simply define the correct Hamiltonian to be $H - \langle 0| H |0\rangle$. We will come back to some of these issues in chapters II.5, III.1, and VIII.2.

Nobody is perfect

In the canonical formalism, time is treated differently from space, and so one might worry about the Lorentz invariance of the resulting field theory. In the standard treatment given in many texts, we would go on from this point and use the Hamiltonian to generate the dynamics, developing a perturbation theory in the interaction $u(\varphi)$. After a number of formal steps, we would manage to derive the Feynman rules, which manifestly define a Lorentz-invariant theory.

Historically, there was a time when people felt that quantum field theory should be defined by its collection of Feynman rules, which gives us a concrete procedure to calculate measurable quantities, such as scattering cross sections. An extreme view along this line held that fields are merely mathematical crutches used to help us arrive at the Feynman rules and should be thrown away at the end of the day.

This view became untenable starting in the 1970s, when it was realized that there is a lot more to quantum field theory than Feynman diagrams. Field theory contains nonperturbative effects invisible by definition to Feynman diagrams. Many of these effects, which we will get to in due time, are more easily seen using the path integral formalism.

As I said, the canonical and the path integral formalism often appear to be complementary, and I will refrain from entering into a discussion about which formalism is superior. In this book, I adopt a pragmatic attitude and use whatever formalism happens to be easier for the problem at hand.

Let me mention, however, some particularly troublesome features in each of the two formalisms. In the canonical formalism fields are quantum operators containing an infinite number of degrees of freedom, and sages once debated such delicate questions as how products of fields are to be defined. On the other hand, in the path integral formalism, plenty of sins can be swept under the rug known as the integration measure (see chapter IV.7).

Appendix 1

It may seem a bit puzzling that in the canonical formalism the propagator has to be defined with time ordering, which we did not need in the path integral formalism. To resolve this apparent puzzle, it suffices to look at quantum mechanics.

Let $A[q(t_1)]$ be a function of q, evaluated at time t_1. What does the path integral $\int Dq(t) \, A[q(t_1)]e^{i \int_0^T dt L(\dot{q},q)}$ represent in the operator language? Well, working backward to (I.2.4) we see that we would slip $A[q(t_1)]$ into the factor $\langle q_{j+1}|e^{-iH\delta t}|q_j\rangle$, where the integer j is determined by the condition that the time t_1 occurs between the times $j\delta t$ and $(j+1)\delta t$. In the resulting factor $\langle q_{j+1}|e^{-iH\delta t}A[q(t_1)]|q_j\rangle$, we could replace the c-number $A[q(t_1)]$ by the operator $A[\hat{q}]$, since $A[\hat{q}]|q_j\rangle = A[q_j]|q_j\rangle \simeq A[q(t_1)]|q_j\rangle$ to the accuracy we are working with. Note that $\hat{q}$ is evidently a Schrödinger operator. Thus, putting in this factor $\langle q_{j+1}|e^{-iH\delta t}A[\hat{q}]|q_j\rangle$ together with all the factors of $\langle q_{i+1}|e^{-iH\delta t}|q_i\rangle$, we find that the integral in question, namely $\int Dq(t) \, A[q(t_1)]e^{i \int_0^T dt L(\dot{q},q)}$, actually represents

$$\langle q_F|e^{-iH(T-t_1)}A[\hat{q}]e^{-iHt_1}|q_I\rangle = \langle q_F|e^{-iHT}A[\hat{q}(t_1)]|q_I\rangle$$

where $\hat{q}(t_1)$ is now evidently a Heisenberg operator. [We have used the standard relation between Heisenberg and Schrödinger operators, namely, $\mathcal{O}_H(t) = e^{iHt}\mathcal{O}_S e^{-iHt}$.] I find this passage back and forth between the Dirac, Schrödinger, and Heisenberg pictures quite instructive, perhaps even amusing.

We are now prepared to ask the more complicated question: what does the path integral $\int Dq(t) \, A[q(t_1)]B[q(t_2)]e^{i \int_0^T dt L(\dot{q},q)}$ represent in the operator language? Here $B[q(t_2)]$ is some other function of q evaluated at time t_2. So we also slip $B[q(t_2)]$ into the appropriate factor in (I.2.4) and replace $B[q(t_2)]$ by $B[\hat{q}]$. But now we see that we have to keep track of whether t_1 or t_2 is the earlier of the two times. If t_2 is earlier, the operator $A[\hat{q}]$ would appear to the left of the operator $B[\hat{q}]$, and if t_1 is earlier, to the right. Explicitly, if t_2 is earlier than t_1, we would end up with the sequence

$$e^{-iH(T-t_1)} A[\hat{q}] e^{-iH(t_1-t_2)} B[\hat{q}] e^{-iHt_2} = e^{-iHT} A[\hat{q}(t_1)] B[\hat{q}(t_2)] \tag{20}$$

upon passing from the Schrödinger to the Heisenberg picture, just as in the simpler situation above. Thus we define the time-ordered product

$$T[A[\hat{q}(t_1)] B[\hat{q}(t_2)]] \equiv \theta(t_1 - t_2) A[\hat{q}(t_1)] B[\hat{q}(t_2)] + \theta(t_2 - t_1) B[\hat{q}(t_2)] A[\hat{q}(t_1)] \tag{21}$$

We just learned that

$$\langle q_F | e^{-iHT} T[A[\hat{q}(t_1)] B[\hat{q}(t_2)]] | q_I \rangle = \int Dq(t) \, A[q(t_1)] B[q(t_2)] e^{i \int_0^T dt L(\dot{q}, q)} \tag{22}$$

The concept of time ordering does not appear on the right-hand side, but is essential on the left-hand side.

Generalizing the discussion here, we see that the Green's functions $G^{(n)}(x_1, x_2, \cdots, x_n)$ introduced in the preceding chapter [see (I.7.13–15)] is given in the canonical formalism by the vacuum expectation value of a time-ordered product of field operators $\langle 0| T\{\varphi(x_1)\varphi(x_2) \cdots \varphi(x_n)\}|0\rangle$. That (13) gives the propagator is a special case of this relationship.

We could also consider $\langle 0| T\{\mathcal{O}_1(x_1)\mathcal{O}_2(x_2) \cdots \mathcal{O}_n(x_n)\}|0\rangle$, the vacuum expectation value of a time-ordered product of various operators $\mathcal{O}_i(x)$ [the current $J_\mu(x)$, for example] made out of the quantum field. Such objects will appear in later chapters [for example, (VII.3.7)].

Appendix 2: Field redefinition

This is perhaps a good place to reveal to the innocent reader that there does not exist an international commission in Brussels mandating what field one is required to use. If we use φ, some other guy is perfectly entitled to use η, assuming that the two fields are related by some invertible function with $\eta = f(\varphi)$. (To be specific, it is often helpful to think of $\eta = \varphi + \alpha\varphi^3$ with some parameter α.) This is known as a field redefinition, an often useful thing to do, as we will see repeatedly.

The S-matrix amplitudes that experimentalists measure are invariant under field redefinition. But this is tautological trivia: the scattering amplitude $\langle \vec{k}_3\vec{k}_4 | e^{-iHT} |\vec{k}_1\vec{k}_2 \rangle$, for example, does not even know about φ and η. The issue is with the formalism we use to calculate the S-matrix.

In the path integral formalism, it is also trivial that we could write $Z(J) = \int D\eta \, e^{i[S(\eta) + \int d^4x J\eta]}$ just as well as $Z(J) = \int D\varphi \, e^{i[S(\varphi) + \int d^4x J\varphi]}$. This result, a mere change of integration variable, was known to Newton and Leibniz. But suppose we write $\tilde{Z}(J) = \int D\varphi \, e^{i[S(\varphi) + \int d^4x J\eta]}$. Now of course any dolt could see that $\tilde{Z}(J) \neq Z(J)$, and a fortiori, the Green's functions (I.7.14,15) obtained by differentiating $\tilde{Z}(J)$ and $Z(J)$ are not equal.

The nontrivial physical statement is that the S-matrix amplitudes obtained from $\tilde{Z}(J)$ and $Z(J)$ are in fact the same. This better be the case, since we are claiming that the path integral formalism provides a way to actual physics. To see how this apparent "miracle" (Green's functions completely different, S-matrix amplitudes the same) occurs, let us think physically. We set up our sources to produce or remove one single field disturbance, as indicated in figure I.4.1. Our friend, who uses $\tilde{Z}(J)$, in contrast, set up his sources to produce or remove $\eta = \varphi + \alpha\varphi^3$ (we specialize for pedagogical clarity), so that once in a while (with a probability determined by α) he is producing three field disturbances instead of one, as shown in figure I.8.1. As a result, while he thinks that he is scattering four mesons, occasionally he is actually scattering six mesons. (Perhaps he should give his accelerator a tune up.)

But to obtain S-matrix amplitudes we are told to multiply the Green's functions by $(k^2 - m^2)$ for each external leg carrying momentum k, and then set k^2 to m^2. When we do this, the diagram in figure I.8.1a survives, since it has a pole that goes like $1/(k^2 - m^2)$ but the extraneous diagram in figure I.8.1b is eliminated. Very simple. One point worth emphasizing is that m here is the actual physical mass of the particle. Let's be precise when we should. Take the single particle state $|\vec{k}\rangle$. Act on it with the Hamiltonian. Then $H |\vec{k}\rangle = \sqrt{\vec{k}^2 + m^2} |\vec{k}\rangle$. The m that appears in the eigenvalue of the Hamiltonian is the actual physical mass. We will come back to the issue of the physical mass in chapter III.3.

In the canonical formalism, the field is an operator, and as we saw just now, the calculation of S-matrix amplitudes involves evaluating products of field operators between physical states. In particular, the matrix elements $\langle \vec{k}| \varphi |0\rangle$ and $\langle 0| \varphi |\vec{k}\rangle$ (related by hermitean conjugation) come in crucially. If we use some other field η,

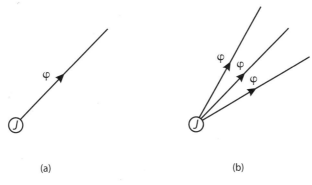

(a) (b)

Figure I.8.1

what matters is merely that $\langle \vec{k}| \eta |0\rangle$ is not zero, in which case we could always write $\langle \vec{k}| \eta |0\rangle = Z^{\frac{1}{2}} \langle \vec{k}| \varphi |0\rangle$ with Z some c-number. We simply divide the scattering amplitude by the appropriate powers of $Z^{\frac{1}{2}}$.

Exercises

I.8.1 Derive (14). Then verify explicitly that $d^D k/(2\omega_k)$ is indeed Lorentz invariant. Some authors prefer to replace $\sqrt{2\omega_k}$ in (11) by $2\omega_k$ when relating the scalar field to the creation and annihilation operators. Show that the operators defined by these authors are Lorentz covariant. Work out their commutation relation.

I.8.2 Calculate $\langle \vec{k}'| H |\vec{k}\rangle$, where $|\vec{k}\rangle = a^\dagger(\vec{k}) |0\rangle$.

I.8.3 For the complex scalar field discussed in the text calculate $\langle 0| T[\varphi(x)\varphi^\dagger(0)] |0\rangle$.

I.8.4 Show that $[Q, \varphi(x)] = -\varphi(x)$.

I.9 Disturbing the Vacuum

Casimir effect

In the preceding chapter, we computed the energy of the vacuum $\langle 0| H |0\rangle$ and obtained the gratifying result that it is given by the zero point energy of the harmonic oscillator integrated over all momentum modes and over space. I explained that the energy of any physical configuration, for example, the mass of a particle, is to be measured relative to this vacuum energy. In effect, we simply subtract off this vacuum energy and define the correct Hamiltonian to be $H - \langle 0| H |0\rangle$.

But what if we disturb the vacuum?

Physically, we could compare the energy of the vacuum before and after we introduce the disturbance, by varying the disturbance for example. Of course, it is not just our textbook scalar field that contributes to the energy of the vacuum. The electromagnetic field, for instance, also undergoes quantum fluctuation and would contribute, with its two polarization degrees of freedom, to the energy density ε of the vacuum the amount $2 \int d^3k/(2\pi)^3 \frac{1}{2}\hbar\omega_k$. In 1948 Casimir had the brilliant insight that we could disturb the vacuum and produce a shift $\Delta\varepsilon$. While ε is not observable, $\Delta\varepsilon$ should be observable since we can control how we disturb the vacuum. In particular, Casimir considered introducing two parallel "perfectly" conducting plates (formally of zero thickness and infinite extent) into the vacuum. The variation of $\Delta\varepsilon$ with the distance d between the plates would lead to a force between the plates, known as the Casimir force. In reality, it is the electromagnetic field that is responsible, not our silly scalar field.

Call the direction perpendicular to the plates the x axis. Because of the boundary conditions the electromagnetic field must satisfy on the conducting plates, the wave vector $\vec{k}$ can only take on the values $(\pi n/d, k_y, k_z)$, with n an integer. Thus the energy per unit area between the plates is changed to $\sum_n \int dk_y dk_z/(2\pi)^2 \sqrt{(\pi n/d)^2 + k_y^2 + k_z^2}$.

To calculate the force, we vary d, but then we would have to worry about how the energy density outside the two plates varies. A clever trick exists for avoiding this worry: we introduce three plates! See figure (I.9.1). We hold the two outer plates fixed and move

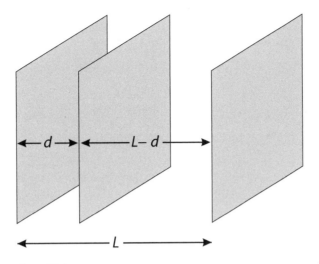

Figure I.9.1

only the inner plate. Now we don't have to worry about the world outside the plates. The separation L between the two outer plates can be taken to be as large as we like.

In the spirit of this book (and my philosophy) of avoiding computational tedium as much as possible, I propose two simplifications: (I) do the calculation for a massless scalar field instead of the electromagnetic field so we won't have to worry about polarization and stuff like that, and (II) retreat to a $(1 + 1)$-dimensional spacetime so we won't have to integrate over k_y and k_z. Readers of quantum field theory texts do not need to watch their authors show off their prowess in doing integrals. As you will see, the calculation is exceedingly instructive and gives us an early taste of the art of extracting finite physical results from seemingly infinite expressions, known as regularization, that we will study in chapters III.1–3.

With this set-up, the energy $E = f(d) + f(L - d)$ with

$$f(d) = \frac{1}{2} \sum_{n=1}^{\infty} \omega_n = \frac{\pi}{2d} \sum_{n=1}^{\infty} n \tag{1}$$

since the modes are given by $\sin(n\pi x/d)$ ($n = 1, \cdots, \infty$) with the corresponding energy $\omega_n = n\pi/d$.

Aagh! What do we do with $\sum_{n=1}^{\infty} n$? None of the ancient Greeks from Zeno on could tell us.

What they should tell us is that we are doing physics, not mathematics! Physical plates cannot keep arbitrarily high frequency waves from leaking out.[1]

To incorporate this piece of all-important physics, we should introduce a factor $e^{-a\omega_n/\pi}$ with a parameter a having the dimension of time (or in our natural units, length) so that modes with $\omega_n \gg \pi/a$ do not contribute: they don't see the plates! The characteristic

---

[1] See footnote 1 in chapter III.1.

frequency π/a parametrizes the high frequency response of the conducting plates. Thus we have

$$f(d) = \frac{\pi}{2d} \sum_{n=1}^{\infty} n e^{-an/d} = -\frac{\pi}{2} \frac{\partial}{\partial a} \sum_{n=1}^{\infty} e^{-an/d} = -\frac{\pi}{2} \frac{\partial}{\partial a} \frac{1}{1 - e^{-a/d}} = \frac{\pi}{2d} \frac{e^{a/d}}{(e^{a/d} - 1)^2}$$

Since we want a^{-1} to be large, we take the limit a small so that

$$f(d) = \frac{\pi d}{2a^2} - \frac{\pi}{24d} + \frac{\pi a^2}{480 d^3} + O(a^4/d^5). \tag{2}$$

Note that $f(d)$ blows up as $a \to 0$, as it should, since we are then back to (1). But the force between two conducting plates shouldn't blow up. Experimentalists might have noticed it by now!

Well, the force is given by

$$F = -\frac{\partial E}{\partial d} = -\{f'(d) - f'(L - d)\} = -\left\{ \left(\frac{1}{2\pi a^2} + \frac{\pi}{24 d^2} + \cdots \right) - (d \to L - d) \right\}$$

$$\xrightarrow[a \to 0]{} -\frac{\pi}{24} \left(\frac{1}{d^2} - \frac{1}{(L - d)^2} \right)$$

$$\xrightarrow[L \gg d]{} -\frac{\pi \hbar c}{24 d^2} \tag{3}$$

Behold, the parameter a we have introduced to make sense of the divergent sum in (1) has disappeared in our final result for the physically measurable force. In the last step, using dimensional analysis we restored $\hbar$ to underline the quantum nature of the force.

The Casimir force between two plates is attractive. Notice that the $1/d^2$ of the force simply follows from dimensional analysis since in natural units force has dimension of an inverse length squared. In a tour de force, experimentalists have measured this tiny force. The fluctuating quantum field is quite real!

To obtain a sensible result we need to regularize in the ultraviolet (namely the high frequency or short time behavior parametrized by a) and in the infrared (namely the long distance cutoff represented by L). Notice how a and L "work" together in (3).

This calculation foreshadows the renormalization of quantum field theories, a topic made mysterious and almost incomprehensible in many older texts. In fact, it is perfectly sensible. We will discuss renormalization in detail in chapters III.1 and III.2, but for now let us review what we just did.

Instead of panicking when faced with the divergent sum in (1), we remind ourselves that we are proud physicists and that physics tells us that the sum should not go all the way to infinity. In a conducting plate, electrons rush about to counteract any applied tangential electric field. But when the incident wave oscillates at sufficiently high frequency, the electrons can't keep up. Thus the idealization of a perfectly conducting plate fails. We regularize (such an ugly term but that's what field theorists use!) the sum in a mathematically convenient way by introducing a damping factor. The single parameter a is supposed to summarize the unknown high frequency physics that causes the electron to fail to keep up. In reality, a^{-1} is related to the plasma frequency of the metal making up the plate.

A priori, the Casimir force between the two plates could end up depending on the parameter a. In that case, the Casimir force would tell us something about the response of a conducting plate to high frequency electric fields, and it would have made for an interesting chapter in a text on solid state physics. Since this is in fact a quantum field theory text, you might have suspected that the Casimir force represents some fundamental physics about fluctuating quantum fields and that a would drop out. That the Casimir force is "universal" makes it unusually interesting. Notice however, as is physically sensible, that the $O(1/d^4)$ correction to the Casimir force does depend on whether the experimentalist used copper or aluminum plates.

We might then wonder whether the leading term $F = -\pi/(24d^2)$ depends on the particular regularization we used. What if we suppress the higher terms in the divergent sum with some other function? We will address this question in the appendix to this chapter.

Amusingly, the 24 in (3) is the same 24 that appears in string theory! (The dimension of spacetime the quantum bosonic string must live in is determined to be $24 + 2 = 26$.) The reader who knows string theory would know what these two cryptic statements are about (summing up the zero modes of the string). Appallingly, in an apparent attempt to make the subject appear even more mysterious than it is, some treatments simply assert that the sum $\sum_{n=1}^{\infty} n$ is by some mathematical sleight-of-hand equal to $-1/12$. Even though it would have allowed us to wormhole from (1) to (3) instantly, this assertion is manifestly absurd. What we did here, however, makes physical sense.

Appendix

Here we address the fundamental issue of whether a physical quantity we extract by cutting off high frequency contributions could depend on how we cut. Let me mention that in recent years the study of Casimir force for actual physical situations has grown into an active area of research, but clearly my aim here is not to give a realistic account of this field, but to study in an easily understood context an issue (as you will see) central to quantum field theory. My hope is that by the time you get to actually regularize a field theory in $(3 + 1)$-dimensional spacetime, you would have amply mastered the essential physics and not have to struggle with the mechanics of regularization.

Let us first generalize a bit the regularization scheme we used and write

$$f(d) = \frac{\pi}{2d} \sum_{n=1}^{\infty} n g\left(\frac{na}{d}\right) = \frac{\pi}{2} \frac{\partial}{\partial a} \sum_{n=1}^{\infty} h\left(\frac{na}{d}\right) \equiv \frac{\pi}{2} \frac{\partial}{\partial a} H\left(\frac{a}{d}\right) \tag{4}$$

Here $g(v) = h'(v)$ is a rapidly decreasing function so that the sums make sense, chosen judiciously to allow ready evaluation of $H(a/d) \equiv \sum_{n=1}^{\infty} h(na/d)$. [In (2), we chose $g(v) = e^{-v}$ and hence $h(v) = -e^{-v}$.] We would like to know how the Casimir force,

$$-F = \frac{\partial f(d)}{\partial d} - (d \to L - d) = \frac{\pi}{2} \frac{\partial^2}{\partial d \partial a} H(\frac{a}{d}) - (d \to L - d) \tag{5}$$

depends on $g(v)$.

Let us try to get as far as we can using physical arguments and dimensional analysis. Expand H as follows: $\pi H(a/d) = \ldots + \gamma_{-2} d^2/a^2 + \gamma_{-1} d/a + \gamma_0 + \gamma_1 a/d + \gamma_2 a^2/d^2 + \cdots$. We might be tempted to just write a Taylor series in a/d, but nothing tells us that $H(a/d)$ might not blow up as $a \to 0$. Indeed, the example in (3) contains a term like d/a, and so we better be cautious.

We will presently argue physically that the series in fact terminate in one direction. The force is given by

$$F = \left(\ldots + \gamma_{-2} \frac{2d}{a^3} + \gamma_{-1} \frac{1}{2a^2} + \gamma_1 \frac{1}{2d^2} + \gamma_2 \frac{2a}{d^3} + \cdots \right) - (d \to L - d) \tag{6}$$

Look at the γ_{-2} term: it contributes to the force a term like $(d - (L - d))/a^3$. But as remarked earlier, the two outer plates could be taken as far apart as we like. The force could not depend on L, and thus on physical grounds γ_{-2} must vanish. Similarly, all γ_{-k} for $k > 2$ must vanish.

Next, we note that the γ_{-1}, although definitely not zero, gets subtracted away since it does not depend on d. (The γ_0 term has already gone away.) At this point, notice that, furthermore, the γ_k terms with $k > 2$ all vanish as $a \to 0$. You could check that all these assertions hold for the $g(v)$ used in the text.

Remarkably, the Casimir force is determined by γ_1 alone: $F = \gamma_1/(2d^2)$. As noted earlier, the force has to be proportional to $1/d^2$. This fact alone shows us that in (6) we only need to keep the γ_1 term. In the text, we found $\gamma_1 = -\pi/12$. In exercise I.9.1 I invite you to go through an amusing calculation obtaining the same value for γ_1 with an entirely different choice of $g(v)$.

This already suggests that the Casimir force is regularization independent, that it tells us more about the vacuum than about metallic conductivity, but still it is highly instructive to study an entire class of damping or regularizing functions to watch how regularization independence emerges. Let us regularize the sum over zero point energies to $f(d) = \frac{1}{2} \sum_{n=1}^{\infty} \omega_n K(\omega_n)$ with

$$K(\omega) = \sum_{\alpha} c_{\alpha} \frac{\Lambda_{\alpha}}{\omega + \Lambda_{\alpha}} \tag{7}$$

Here c_{α} is a bunch of real numbers and Λ_{α} (known as regulators or regulator frequencies) a bunch of high frequencies subject to certain conditions but otherwise chosen entirely at our discretion. For the sum $\sum_{n=1}^{\infty} \omega_n K(\omega_n)$ to converge, we need $K(\omega_n)$ to vanish faster than $1/\omega_n^2$. In fact, for ω much larger than Λ_{α}, $K(\omega) \to \frac{1}{\omega} \sum_{\alpha} c_{\alpha} \Lambda_{\alpha} - \frac{1}{\omega^2} \sum_{\alpha} c_{\alpha} \Lambda_{\alpha}^2 + \cdots$. The requirement that the $1/\omega$ and $1/\omega^2$ terms vanish gives the conditions

$$\sum_{\alpha} c_{\alpha} \Lambda_{\alpha} = 0 \tag{8}$$

and

$$\sum_{\alpha} c_{\alpha} \Lambda_{\alpha}^2 = 0 \tag{9}$$

respectively.

Furthermore, low frequency physics is not to be modified, and so we want $K(\omega) \to 1$ for $\omega << \Lambda_{\alpha}$, thus requiring

$$\sum_{\alpha} c_{\alpha} = 1 \tag{10}$$

At this point, we do not even have to specify the set the index α runs over beyond the fact that the three conditions (8),(9), and (10) require that α must take on at least three values. Note also that some of the c_{α}'s must be negative. Incidentally, we could do with fewer regulators if we are willing to invoke some knowledge of metals, for instance, that $K(\omega) = K(-\omega)$, but that is not the issue here.

We now show that the Casimir force between the two plates does not depend on the choices of c_{α} and Λ_{α}. First, being physicists rather than mathematicians, we freely interchange the two sums in $f(d)$ and write

$$f(d) = \frac{1}{2} \sum_{\alpha} c_{\alpha} \Lambda_{\alpha} \sum_{n} \frac{\omega_n}{\omega_n + \Lambda_{\alpha}} = -\frac{1}{2} \sum_{\alpha} c_{\alpha} \Lambda_{\alpha} \sum_{n} \frac{\Lambda_{\alpha}}{\omega_n + \Lambda_{\alpha}} \tag{11}$$

where, without further ceremony, we have used condition (8). Next, keeping in mind that the sum $\sum_n \Lambda_\alpha / (\omega_n + \Lambda_\alpha)$ is to be put back into (11), we massage it (defining for convenience $b_\alpha = \pi / \Lambda_\alpha$) as follows:

$$\sum_{n=1}^{\infty} \frac{\Lambda}{\omega_n + \Lambda} = \sum_{n=1}^{\infty} \int_0^\infty dt e^{-t(1+n\frac{b}{d})} = \int_0^\infty dt e^{-t} \left(\frac{1}{1 - e^{-\frac{bt}{d}}} - 1 \right)$$
$$= \int_0^\infty dt e^{-t} [\frac{d}{tb} - \frac{1}{2} + \frac{tb}{12d} + O\left(b^3\right)] \tag{12}$$

(To avoid clutter we have temporarily suppressed the index α.) All these manipulations make perfect sense since the entire expression is to be inserted into the sum over α in (11) after we restore the index α. It appears that the result would depend on c_α and λ_α. In fact, mentally restoring and inserting, we see that the $1/b$ term in (12) can be thrown away since

$$\sum_\alpha c_\alpha \Lambda_\alpha / b_\alpha = \pi \sum_\alpha c_\alpha \Lambda_\alpha^2 = 0 \tag{13}$$

[There is in fact a bit of an overkill here since this term corresponds to the γ_{-1} term, which does not appear in the force anyway. Thus, the condition (9) is, strictly speaking, not necessary. We are regularizing not merely the force, but $f(d)$ so that it defines a sensible function.] Similarly, the b^0 term in (12) can be thrown away thanks to (8). Thus, keeping only the b term in (12), we obtain

$$f(d) = -\frac{1}{24d} \int_0^\infty dt e^{-t} t \sum_\alpha c_\alpha \Lambda_\alpha b_\alpha + O\left(\frac{1}{d^3}\right) = -\frac{\pi}{24d} + O\left(\frac{1}{d^3}\right) \tag{14}$$

Indeed, $f(d)$, and a fortiori the Casimir force, do not depend on the c_α's and Λ_α's. To the level of rigor entertained by physicists (but certainly not mathematicians), this amounts to a proof of regularization independence since with enough regulators we could approximate any (reasonable) function $K(\omega)$ that actually describes real conducting plates. Again, as is physically sensible, you could check that the $O(1/d^3)$ term in $f(d)$ does depend on the regularization scheme.

The reason that I did this calculation in detail is that we will encounter this class of regularization, known as Pauli-Villars, in chapter III.1 and especially in the calculation of the anomalous magnetic moment of the electron in chapter III.7, and it is instructive to see how regularization works in a more physical context before dealing with all the complications of relativistic field theory.

Exercises

I.9.1 Choose the damping function $g(v) = 1/(1+v)^2$ instead of the one in the text. Show that this results in the same Casimir force. [Hint: To sum the resulting series, pass to an integral representation $H(\xi) = -\sum_{n=1}^{\infty} 1/(1+n\xi) = -\sum_{n=1}^{\infty} \int_0^\infty dt e^{-(1+n\xi)t} = \int_0^\infty dt e^{-t}/(1 - e^{\xi t})$. Note that the integral blows up logarithmically near the lower limit, as expected.]

I.9.2 Show that with the regularization used in the appendix, the $1/d$ expansion of the force between two conducting plates contains only even powers.

I.9.3 Show off your skill in doing integrals by calculating the Casimir force in $(3 + 1)$-dimensional spacetime. For help, see M. Kardar and R. Golestanian, *Rev. Mod. Phys.* 71: 1233, 1999; J. Feinberg, A. Mann, and M. Revzen, *Ann. Phys.* 288: 103, 2001.

Symmetry, transformation, and invariance

The importance of symmetry in modern physics cannot be overstated.[1]

When a law of physics does not change upon some transformation, that law is said to exhibit a symmetry.

I have already used Lorentz invariance to restrict the form of an action. Lorentz invariance is of course a symmetry of spacetime, but fields can also transform in what is thought of as an internal space. Indeed, we have already seen a simple example of this. I noted in passing in chapter I.3 that we could require the action of a scalar field theory to be invariant under the transformation $\varphi \to -\varphi$ and so exclude terms of odd power in φ, such as φ^3, from the action.

With the φ^3 term included, two mesons could scatter and go into three mesons, for example by the diagrams in (fig. I.10.1). But with this term excluded, you can easily convince yourself that this process is no longer allowed. You will not be able to draw a Feynman diagram with an odd number of external lines. (Think about modifying the integral in our baby problem in chapter I.7 to $\int_{-\infty}^{+\infty} dq\, e^{-\frac{1}{2}m^2 q^2 - g q^3 - \lambda q^4 + Jq}$.) Thus the simple reflection symmetry $\varphi \to -\varphi$ implies that in any scattering process the number of mesons is conserved modulo 2.

Now that we understand one scalar field, let us consider a theory with two scalar fields φ_1 and φ_2 satisfying the reflection symmetry $\varphi_a \to -\varphi_a$ ($a = 1$ or 2):

$$\mathcal{L} = \frac{1}{2}(\partial \varphi_1)^2 - \frac{1}{2}m_1^2 \varphi_1^2 - \frac{\lambda_1}{4}\varphi_1^4 + \frac{1}{2}(\partial \varphi_2)^2 - \frac{1}{2}m_2^2 \varphi_2^2 - \frac{\lambda_2}{4}\varphi_2^4 - \frac{\rho}{2}\varphi_1^2 \varphi_2^2 \tag{1}$$

We have two scalar particles, call them 1 and 2, with mass m_1 and m_2. To lowest order, they scatter in the processes $1 + 1 \to 1 + 1$, $2 + 2 \to 2 + 2$, $1 + 2 \to 1 + 2$, $1 + 1 \to 2 + 2$, and $2 + 2 \to 1 + 1$ (convince yourself). With the five parameters $m_1, m_2, \lambda_1, \lambda_2$, and ρ completely arbitrary, there is no relationship between the two particles.

---

[1] A. Zee, *Fearful Symmetry*.

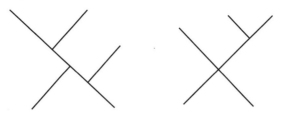

Figure I.10.1

It is almost an article of faith among theoretical physicists, enunciated forcefully by Einstein among others, that the fundamental laws should be orderly and simple, rather than arbitrary and complicated. This orderliness is reflected in the symmetry of the action.

Suppose that $m_1 = m_2$ and $\lambda_1 = \lambda_2$; then the two particles would have the same mass and their interaction, with themselves and with each other, would be the same. The Lagrangian $\mathcal{L}$ becomes invariant under the interchange symmetry $\varphi_1 \longleftrightarrow \varphi_2$.

Next, suppose we further impose the condition $\rho = \lambda_1 = \lambda_2$ so that the Lagrangian becomes

$$\mathcal{L} = \frac{1}{2}\left[(\partial \varphi_1)^2 + (\partial \varphi_2)^2\right] - \frac{1}{2}m^2\left(\varphi_1^2 + \varphi_2^2\right) - \frac{\lambda}{4}\left(\varphi_1^2 + \varphi_2^2\right)^2 \tag{2}$$

It is now invariant under the 2-dimensional rotation $\{\varphi_1(x) \to \cos\theta\, \varphi_1(x) + \sin\theta\, \varphi_2(x)$, $\varphi_2(x) \to -\sin\theta\, \varphi_1(x) + \cos\theta\, \varphi_2(x)\}$ with θ an arbitrary angle. We say that the theory enjoys an "internal" $SO(2)$ symmetry, internal in the sense that the transformation has nothing to do with spacetime. In contrast to the interchange symmetry $\varphi_1 \longleftrightarrow \varphi_2$ the transformation depends on the continuous parameter θ, and the corresponding symmetry is said to be continuous.

We see from this simple example that symmetries exist in hierarchies.

Continuous symmetries

If we stare at the equations of motion $(\partial^2 + m^2)\varphi_a = -\lambda\vec{\varphi}^2\varphi_a$ long enough we see that if we define $J^\mu \equiv i(\varphi_1\partial^\mu\varphi_2 - \varphi_2\partial^\mu\varphi_1)$, then $\partial_\mu J^\mu = i(\varphi_1\partial^2\varphi_2 - \varphi_2\partial^2\varphi_1) = 0$ so that J^μ is a conserved current. The corresponding charge $Q = \int d^D x\, J^0$, just like electric charge, is conserved.

Historically, when Heisenberg noticed that the mass of the newly discovered neutron was almost the same as the mass of a proton, he proposed that if electromagnetism were somehow turned off there would be an internal symmetry transforming a proton into a neutron.

An internal symmetry restricts the form of the theory, just as Lorentz invariance restricts the form of the theory. Generalizing our simple example, we could construct a field theory containing N scalar fields φ_a, with $a = 1, \cdots, N$ such that the theory is invariant under the transformations $\varphi_a \to R_{ab}\varphi_b$ (repeated indices summed), where the matrix R is an element of the rotation group $SO(N)$ (see appendix B for a review of group theory). The fields φ_a transform as a vector $\vec{\varphi} = (\varphi_1, \cdots, \varphi_N)$. We can form only one basic invariant, namely the

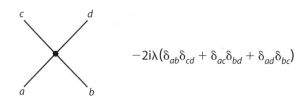

$$-2i\lambda(\delta_{ab}\delta_{cd} + \delta_{ac}\delta_{bd} + \delta_{ad}\delta_{bc})$$

Figure I.10.2

scalar product $\vec{\varphi} \cdot \vec{\varphi} = \varphi_a \varphi_a = \vec{\varphi}^2$ (as always, repeated indices are summed unless otherwise specified). The Lagrangian is thus restricted to have the form

$$\mathcal{L} = \frac{1}{2}\left[(\partial\vec{\varphi})^2 - m^2\vec{\varphi}^2\right] - \frac{\lambda}{4}(\vec{\varphi}^2)^2 \tag{3}$$

The Feynman rules are given in fig. I.10.2. When we draw Feynman diagrams, each line carries an internal index in addition to momentum.

Symmetry manifests itself in physical amplitudes. For example, imagine calculating the propagator $iD_{ab}(x) = \int D\vec{\varphi}\, e^{iS}\varphi_a(x)\varphi_b(0)$. We assume that the measure $D\vec{\varphi}$ is invariant under $SO(N)$. By thinking about how $D_{ab}(x)$ transforms under the symmetry group $SO(N)$ you see easily (exercise I.10.2) that it must be proportional to δ_{ab}. You can check this by drawing a few Feynman diagrams or by considering an ordinary integral $\int d\vec{q}\, e^{-S(q)}q_a q_b$. No matter how complicated a diagram you draw (fig. I.10.3, e.g.) you always get this factor of δ_{ab}. Similarly, scattering amplitudes must also exhibit the symmetry.

Without the $SO(N)$ symmetry, many other terms would be possible (e.g., $\varphi_a \varphi_b \varphi_c \varphi_d$ for some arbitrary choice of a, b, c, and d) in (3).

We can write $R = e^{\theta \cdot T}$ where $\theta \cdot T = \sum_A \theta^A T^A$ is a real antisymmetric matrix. The group $SO(N)$ has $N(N-1)/2$ generators, which we denote by T^A. [Think of the familiar case of $SO(3)$.] Under an infinitesimal transformation (repeated indices summed) $\varphi_a \to R_{ab}\varphi_b \simeq (1 + \theta^A T^A)_{ab}\varphi_b$, or in other words, we have the infinitesimal change $\delta\varphi_a = \theta^A T^A_{ab}\varphi_b$.

Noether's theorem

We now come to one of the most profound observations in theoretical physics, namely Noether's theorem, which states that a conserved current is associated with each generator

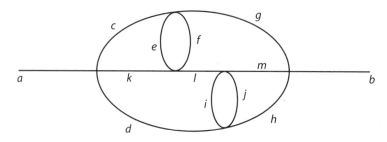

Figure I.10.3

of a continuous symmetry. The appearance of a conserved current for (2) is not an accident.

As is often the case with the most important theorems, the proof of Noether's theorem is astonishingly simple. Denote the fields in our theory generically by φ_a. Since the symmetry is continuous, we can consider an infinitesimal change $\delta\varphi_a$. Since $\mathcal{L}$ does not change, we have

$$
\begin{aligned}
0 = \delta\mathcal{L} &= \frac{\delta\mathcal{L}}{\delta\varphi_a}\delta\varphi_a + \frac{\delta\mathcal{L}}{\delta\partial_\mu\varphi_a}\delta\partial_\mu\varphi_a \\
&= \frac{\delta\mathcal{L}}{\delta\varphi_a}\delta\varphi_a + \frac{\delta\mathcal{L}}{\delta\partial_\mu\varphi_a}\partial_\mu\delta\varphi_a
\end{aligned}
\tag{4}
$$

We would have been stuck at this point, but if we use the equations of motion $\delta\mathcal{L}/\delta\varphi_a = \partial_\mu(\delta\mathcal{L}/\delta\partial_\mu\varphi_a)$ we can combine the two terms and obtain

$$
0 = \partial_\mu\left(\frac{\delta\mathcal{L}}{\delta\partial_\mu\varphi_a}\delta\varphi_a\right)
\tag{5}
$$

If we define

$$
J^\mu \equiv \frac{\delta\mathcal{L}}{\delta\partial_\mu\varphi_a}\delta\varphi_a
\tag{6}
$$

then (5) says that $\partial_\mu J^\mu = 0$. We have found a conserved current! [It is clear from the derivation that the repeated index a is summed in (6)].

Let us immediately illustrate with the simple scalar field theory in (3). Plugging $\delta\varphi_a = \theta^A(T^A)_{ab}\varphi_b$ into (6) and noting that θ^A is arbitrary, we obtain $N(N-1)/2$ conserved currents $J_\mu^A = \partial_\mu\varphi_a(T^A)_{ab}\varphi_b$, one for each generator of the symmetry group $SO(N)$.

In the special case $N = 2$, we can define a complex field $\varphi \equiv (\varphi_1 + i\varphi_2)/\sqrt{2}$. The Lagrangian in (3) can be written as

$$
\mathcal{L} = \partial\varphi^\dagger\partial\varphi - m^2\varphi^\dagger\varphi - \lambda(\varphi^\dagger\varphi)^2,
$$

and is clearly invariant under $\varphi \to e^{i\theta}\varphi$ and $\varphi^\dagger \to e^{-i\theta}\varphi^\dagger$. We find from (6) that $J_\mu = i(\varphi^\dagger\partial_\mu\varphi - \partial_\mu\varphi^\dagger\varphi)$, the current we met already in chapter I.8. Mathematically, this is because the groups $SO(2)$ and $U(1)$ are isomorphic (see appendix B).

For pedagogical clarity I have used the example of scalar fields transforming as a vector under the group $SO(N)$. Obviously, the preceding discussion holds for an arbitrary group G with the fields φ transforming under an arbitrary representation $\mathcal{R}$ of G. The conserved currents are still given by $J_\mu^A = \partial_\mu\varphi_a(T^A)_{ab}\varphi_b$ with T^A the generators evaluated in the representation $\mathcal{R}$. For example, if φ transform as the 5-dimensional representation of $SO(3)$ then T^A is a 5 by 5 matrix.

For physics to be invariant under a group of transformations it is only necessary that the action be invariant. The Lagrangian density $\mathcal{L}$ could very well change by a total divergence: $\delta\mathcal{L} = \partial_\mu K^\mu$, provided that the relevant boundary term could be dropped. Then we would see immediately from (5) that all we have to do to obtain a formula for the conserved current is to modify (6) to $J^\mu \equiv (\delta\mathcal{L}/\delta\partial_\mu\varphi_a)\delta\varphi_a - K^\mu$. As we will see in chapter VIII.4, many supersymmetric field theories are of this type.

Charge as generators

Using the canonical formalism of chapter I.8, we can derive an elegant result for the charge associated with the conserved current

$$Q \equiv \int d^3x \, J^0 = \int d^3x \, \frac{\delta \mathcal{L}}{\delta \partial_0 \varphi_a} \delta \varphi_a$$

Note that Q does not depend on the time at which the integral is evaluated:

$$\frac{dQ}{dt} = \int d^3x \, \partial_0 J^0 = -\int d^3x \, \partial_i J^i = 0 \tag{7}$$

Recognizing that $\delta \mathcal{L}/\delta \partial_0 \varphi_a$ is just the canonical momentum conjugate to the field φ_a, we see that

$$i[Q, \varphi_a] = \delta \varphi_a \tag{8}$$

The charge operator generates the corresponding transformation on the fields. An important special case is for the complex field φ in $SO(2) \simeq U(1)$ theory we discussed; then $[Q, \varphi] = \varphi$ and $e^{i\theta Q} \varphi e^{-i\theta Q} = e^{i\theta} \varphi$.

Exercises

I.10.1 Some authors prefer the following more elaborate formulation of Noether's theorem. Suppose that the action does not change under an infinitesimal transformation $\delta \varphi_a(x) = \theta^A V_a^A$ [with θ^A some parameters labeled by A and V_a^A some function of the fields $\varphi_b(x)$ and possibly also of their first derivatives with respect to x]. It is important to emphasize that when we say the action S does not change we are not allowed to use the equations of motion. After all, the Euler-Lagrange equations of motion follow from demanding that $\delta S = 0$ for any variation $\delta \varphi_a$ subject to certain boundary conditions. Our scalar field theory example nicely illustrates this point, which is confused in some books: $\delta S = 0$ merely because S is constructed using the scalar product of $O(N)$ vectors.

Now let us do something apparently a bit strange. Let us consider the infinitesimal change written above but with the parameters θ^A dependent on x. In other words, we now consider $\delta \varphi_a(x) = \theta^A(x) V_a^A$. Then of course there is no reason for δS to vanish; but, on the other hand, we know that since δS does vanish when θ^A is constant, δS must have the form $\delta S = \int d^4x \, J^\mu(x) \partial_\mu \theta^A(x)$. In practice, this gives us a quick way of reading off the current $J^\mu(x)$; it is just the coefficient of $\partial_\mu \theta^A(x)$ in δS.

Show how all this works for the Lagrangian in (3).

I.10.2 Show that $D_{ab}(x)$ must be proportional to δ_{ab} as stated in the text.

I.10.3 Write the Lagrangian for an $SO(3)$ invariant theory containing a Lorentz scalar field φ transforming in the 5-dimensional representation up to quartic terms. [Hint: It is convenient to write φ as a 3 by 3 symmetric traceless matrix.]

I.10.4 Add a Lorentz scalar field η transforming as a vector under $SO(3)$ to the Lagrangian in exercise I.10.3, maintaining $SO(3)$ invariance. Determine the Noether currents in this theory. Using the equations of motion, check that the currents are conserved.

1.11 Field Theory in Curved Spacetime

General coordinate transformation

In Einstein's theory of gravity, the invariant Minkowskian spacetime interval $ds^2 = \eta_{\mu\nu}dx^\mu dx^\nu = (dt)^2 - (d\vec{x})^2$ is replaced by $ds^2 = g_{\mu\nu}dx^\mu dx^\nu$, where the metric tensor $g_{\mu\nu}(x)$ is a function of the spacetime coordinates x. The guiding principle, known as the principle of general covariance, states that physics, as embodied in the action S, must be invariant under arbitrary coordinate transformations $x \to x'(x)$. More precisely, the principle[1] states that with suitable restrictions the effect of a gravitational field is equivalent to that of a coordinate transformation.

Since

$$ds^2 = g'_{\lambda\sigma}dx'^\lambda dx'^\sigma = g'_{\lambda\sigma}\frac{\partial x'^\lambda}{\partial x^\mu}\frac{\partial x'^\sigma}{\partial x^\nu}dx^\mu dx^\nu = g_{\mu\nu}dx^\mu dx^\nu$$

the metric transforms as

$$g'_{\lambda\sigma}(x')\frac{\partial x'^\lambda}{\partial x^\mu}\frac{\partial x'^\sigma}{\partial x^\nu} = g_{\mu\nu}(x) \tag{1}$$

The inverse of the metric $g^{\mu\nu}$ is defined by $g^{\mu\nu}g_{\nu\rho} = \delta^\mu_\rho$.

A scalar field by its very name does not transform: $\varphi(x) = \varphi'(x')$. The gradient of the scalar field transforms as

$$\partial_\mu\varphi(x) = \frac{\partial x'^\lambda}{\partial x^\mu}\frac{\partial\varphi'(x')}{\partial x'^\lambda} = \frac{\partial x'^\lambda}{\partial x^\mu}\partial'_\lambda\varphi'(x')$$

By definition, a (covariant) vector field transforms as

$$A_\mu(x) = \frac{\partial x'^\lambda}{\partial x^\mu}A'_\lambda(x')$$

[1] For a precise statement of the principle of general covariance, see S. Weinberg, *Gravitation and Cosmology*, p. 92.

so that $\partial_\mu\varphi(x)$ is a vector field. Given two vector fields $A_\mu(x)$ and $B_\nu(x)$, we can contract them with $g^{\mu\nu}(x)$ to form $g^{\mu\nu}(x)A_\mu(x)B_\nu(x)$, which, as you can immediately check, is a scalar. In particular, $g^{\mu\nu}(x)\partial_\mu\varphi(x)\partial_\nu\varphi(x)$ is a scalar. Thus, if we simply replace the Minkowski metric $\eta^{\mu\nu}$ in the Lagrangian $\mathcal{L} = \frac{1}{2}[(\partial\varphi)^2 - m^2\varphi^2] = \frac{1}{2}(\eta^{\mu\nu}\partial_\mu\varphi\partial_\nu\varphi - m^2\varphi^2)$ by the Einstein metric $g^{\mu\nu}$, the Lagrangian is invariant under coordinate transformation.

The action is obtained by integrating the Lagrangian over spacetime. Under a coordinate transformation $d^4x' = d^4x \det(\partial x'/\partial x)$. Taking the determinant of (1), we have

$$g \equiv \det g_{\mu\nu} = \det g'_{\lambda\sigma} \frac{\partial x'^\lambda}{\partial x^\mu} \frac{\partial x'^\sigma}{\partial x^\nu} = g' \left[\det\left(\frac{\partial x'}{\partial x}\right)\right]^2 \tag{2}$$

We see that the combination $d^4x\sqrt{-g} = d^4x'\sqrt{-g'}$ is invariant under coordinate transformation.

Thus, given a quantum field theory we can immediately write down the theory in curved spacetime. All we have to do is promote the Minkowski metric $\eta^{\mu\nu}$ in our Lagrangian to the Einstein metric $g^{\mu\nu}$ and include a factor $\sqrt{-g}$ in the spacetime integration measure.[2] The action S would then be invariant under arbitrary coordinate transformations. For example, the action for a scalar field in curved spacetime is simply

$$S = \int d^4x\sqrt{-g}\frac{1}{2}(g^{\mu\nu}\partial_\mu\varphi\partial_\nu\varphi - m^2\varphi^2) \tag{3}$$

(There is a slight subtlety involving the spin $\frac{1}{2}$ field that we will talk about in part II. We will eventually come to it in chapter VIII.1.)

There is no essential difficulty in quantizing the scalar field in curved spacetime. We simply treat $g_{\mu\nu}$ as given [e.g., the Schwarzschild metric in spherical coordinates: $g_{00} = (1 - 2GM/r)$, $g_{rr} = -(1 - 2GM/r)^{-1}$, $g_{\theta\theta} = -r^2$, and $g_{\phi\phi} = -r^2\sin^2\theta$] and study the path integral $\int D\varphi e^{iS}$, which is still a Gaussian integral and thus do-able. The propagator of the scalar field $D(x, y)$ can be worked out and so on and so forth. (see exercise I.11.1).

At this point, aside from the fact that $g_{\mu\nu}(x)$ carries Lorentz indices while $\varphi(x)$ does not, the metric $g_{\mu\nu}$ looks just like a field and is in fact a classical field. Write the action of the world $S = S_g + S_M$ as the sum of two terms: S_g describing the dynamics of the gravitational field $g_{\mu\nu}$ (which we will come to in chapter VIII.1) and S_M describing the dynamics of all the other fields in the world [the "matter fields," namely φ in our simple example with S_M as given in (3)]. We could quantize gravity by integrating over $g_{\mu\nu}$ as well, thus extending the path integral to $\int Dg D\varphi e^{iS}$.

Easier said than done! As you have surely heard, all attempts to integrate over $g_{\mu\nu}(x)$ have been beset with difficulties, eventually driving theorists to seek solace in string theory. I will explain in due time why Einstein's theory is known as "nonrenormalizable."

---

[2] We also have to replace ordinary derivatives ∂_μ by the covariant derivatives D_μ of general relativity, but acting on a scalar field φ the covariant derivative is just the ordinary derivative $D_\mu\varphi = \partial_\mu\varphi$.

What the graviton listens to

One of the most profound results to come out of Einstein's theory of gravity is a fundamental definition of energy and momentum. What exactly are energy and momentum any way? Energy and momentum are what the graviton listens to. (The graviton is of course the particle associated with the field $g_{\mu\nu}$.)

The stress-energy tensor $T^{\mu\nu}$ is defined as the variation of the matter action S_M with respect to the metric $g_{\mu\nu}$ (holding the coordinates x^μ fixed):

$$T^{\mu\nu}(x) = -\frac{2}{\sqrt{-g}}\frac{\delta S_M}{\delta g_{\mu\nu}(x)} \tag{4}$$

Energy is defined as $E = P^0 = \int d^3x \sqrt{-g}\, T^{00}(x)$ and momentum as $P^i = \int d^3x \sqrt{-g}\, T^{0i}(x)$.

Even if we are not interested in curved spacetime per se, (4) still offers us a simple (and fundamental) way to determine the stress energy of a field theory in flat spacetime. We simply vary around the Minkowski metric $\eta_{\mu\nu}$ by writing $g_{\mu\nu} = \eta_{\mu\nu} + h_{\mu\nu}$ and expand S_M to first order in h. According to (4), we have[3]

$$S_M(h) = S_M(h=0) - \int d^4x \left[\tfrac{1}{2}h_{\mu\nu}T^{\mu\nu} + O(h^2)\right]. \tag{5}$$

The symmetric tensor field $h_{\mu\nu}(x)$ is in fact the graviton field (see chapters I.5 and VIII.1). The stress-energy tensor $T^{\mu\nu}(x)$ is what the graviton field couples to, just as the electromagnetic current $J^\mu(x)$ is what the photon field couples to.

Consider a general $S_M = \int d^4x \sqrt{-g}(A + g^{\mu\nu}B_{\mu\nu} + g^{\mu\nu}g^{\lambda\rho}C_{\mu\nu\lambda\rho} + \cdots)$. Since $-g = 1 + \eta^{\mu\nu}h_{\mu\nu} + O(h^2)$ and $g^{\mu\nu} = \eta^{\mu\nu} - h^{\mu\nu} + O(h^2)$, we find

$$T_{\mu\nu} = 2(B_{\mu\nu} + 2C_{\mu\nu\lambda\rho}\eta^{\lambda\rho} + \cdots) - \eta_{\mu\nu}\mathcal{L} \tag{6}$$

in flat spacetime. Note

$$T \equiv \eta^{\mu\nu}T_{\mu\nu} = -(4A + 2\eta^{\mu\nu}B_{\mu\nu} + 0 \cdot \eta^{\mu\nu}\eta^{\lambda\rho}C_{\mu\nu\lambda\rho} + \cdots) \tag{7}$$

which we have written in a form emphasizing that $C_{\mu\nu\lambda\rho}$ does not contribute to the trace of the stress-energy tensor.

We now show the power of this definition of $T_{\mu\nu}$ by obtaining long-familiar results about the electromagnetic field. Promoting the Lagrangian of the massive spin 1 field to curved spacetime we have[4] $\mathcal{L} = (-\tfrac{1}{4}g^{\mu\nu}g^{\lambda\rho}F_{\mu\lambda}F_{\nu\rho} + \tfrac{1}{2}m^2g^{\mu\nu}A_\mu A_\nu)$ and thus[5] $T_{\mu\nu} = -F_{\mu\lambda}F_\nu{}^\lambda + m^2A_\mu A_\nu - \eta_{\mu\nu}\mathcal{L}$.

---

[3] I use the normal convention in which indices are summed regardless of any symmetry; in other words, $\tfrac{1}{2}h_{\mu\nu}T^{\mu\nu} = \tfrac{1}{2}(h_{01}T^{01} + h_{10}T^{10} + \cdots) = h_{01}T^{01} + \cdots$.

[4] Here we use the fact that the covariant curl is equal to the ordinary curl $D_\mu A_\nu - D_\nu A_\mu = \partial_\mu A_\nu - \partial_\nu A_\mu$ and so $F_{\mu\nu}$ does not involve the metric.

[5] Holding x^μ fixed means that we hold ∂_μ and hence A_μ fixed since A_μ is related to ∂_μ by gauge invariance. We are anticipating (see chapter II.7) here, but you have surely heard of gauge invariance in a nonrelativistic context.

For the electromagnetic field we set $m = 0$. First, $\mathcal{L} = -\frac{1}{4}F_{\mu\nu}F^{\mu\nu} = -\frac{1}{4}(-2F_{0i}^2 + F_{ij}^2) = \frac{1}{2}(\vec{E}^2 - \vec{B}^2)$. Thus, $T_{00} = -F_{0\lambda}F_0{}^\lambda - \frac{1}{2}(\vec{E}^2 - \vec{B}^2) = \frac{1}{2}(\vec{E}^2 + \vec{B}^2)$. That was comforting, to see a result we knew from "childhood." Incidentally, this also makes clear that we can think of $\vec{E}^2$ as kinetic energy and $\vec{B}^2$ as potential energy. Next, $T_{0i} = -F_{0\lambda}F_i{}^\lambda = F_{0j}F_{ij} = \varepsilon_{ijk}E_jB_k = (\vec{E} \times \vec{B})_i$. The Poynting vector has just emerged.

Since the Maxwell Lagrangian $\mathcal{L} = -\frac{1}{4}g^{\mu\nu}g^{\lambda\rho}F_{\mu\lambda}F_{\nu\rho}$ involves only the C term with $C_{\mu\nu\lambda\rho} = -\frac{1}{4}F_{\mu\nu}F_{\lambda\rho}$, we see from (7) that the stress-energy tensor of the electromagnetic field is traceless, an important fact we will need in chapter VIII.1. We can of course check directly that $T = 0$ (exercise I.11.4).[6]

Appendix: A concise introduction to curved spacetime

General relativity is often made to seem more difficult and mysterious than need be. Here I give a concise review of some of its basic elements for later use.

Denote the spacetime coordinates of a point particle by X^μ. To construct its action note that the only coordinate invariant quantity is the "length"[7] of the world line traced out by the particle (fig. I.11.1), namely $\int ds = \int \sqrt{g_{\mu\nu}dX^\mu dX^\nu}$, where $g_{\mu\nu}$ is evaluated at the position of the particle of course. Thus, the action for a point particle must be proportional to

$$\int ds = \int \sqrt{g_{\mu\nu}dX^\mu dX^\nu} = \int \sqrt{g_{\mu\nu}[X(\zeta)]\frac{dX^\mu}{d\zeta}\frac{dX^\nu}{d\zeta}}\,d\zeta$$

where ζ is any parameter that varies monotonically along the world line. The length, being geometric, is manifestly reparametrization invariant, that is, independent of our choice of ζ as long as it is reasonable. This is one of those "more obvious than obvious" facts since $\int \sqrt{g_{\mu\nu}dX^\mu dX^\nu}$ is manifestly independent of ζ. If we insist, we can check the reparametrization invariance of $\int ds$. Obviously the powers of $d\zeta$ match. Explicitly, if we write $\zeta = \zeta(\eta)$, then $dX^\mu/d\zeta = (d\eta/d\zeta)(dX^\mu/d\eta)$ and $d\zeta = (d\zeta/d\eta)d\eta$.

Let us define

$$K \equiv g_{\mu\nu}[X(\zeta)]\frac{dX^\mu}{d\zeta}\frac{dX^\nu}{d\zeta}$$

for ease of writing. Setting the variation of $\int d\zeta \sqrt{K}$ equal to zero, we obtain

$$\int d\zeta \frac{1}{\sqrt{K}}\left(2g_{\mu\nu}\frac{dX^\mu}{d\zeta}\frac{d\delta X^\nu}{d\zeta} + \partial_\lambda g_{\mu\nu}\frac{dX^\mu}{d\zeta}\frac{dX^\nu}{d\zeta}\delta X^\lambda\right) = 0$$

which upon integration by parts (and with $\delta X^\lambda = 0$ at the endpoints as usual) gives the equation of motion

$$\sqrt{K}\frac{d}{d\zeta}\left(\frac{1}{\sqrt{K}}2g_{\mu\lambda}\frac{dX^\mu}{d\zeta}\right) - \partial_\lambda g_{\mu\nu}\frac{dX^\mu}{d\zeta}\frac{dX^\nu}{d\zeta} = 0 \qquad (8)$$

To simplify (8) we exploit our freedom in choosing ζ and set $d\zeta = ds$, so that $K = 1$. We have

$$2g_{\mu\lambda}\frac{d^2X^\mu}{ds^2} + 2\partial_\sigma g_{\mu\lambda}\frac{dX^\sigma}{ds}\frac{dX^\mu}{ds} - \partial_\lambda g_{\mu\nu}\frac{dX^\mu}{ds}\frac{dX^\nu}{ds} = 0$$

---

[6] We see that tracelessness is related to the fact that the electromagnetic field has no mass scale. Pure electromagnetism is said to be scale or dilatation invariant. For more on dilatation invariance see S. Coleman, *Aspects of Symmetry*, p. 67.

[7] We put "length" in quotes because if $g_{\mu\nu}$ had a Euclidean signature then $\int ds$ would indeed be the length and minimizing $\int ds$ would give the shortest path (the geodesic) between the endpoints, but here $g_{\mu\nu}$ has a Minkowskian signature.

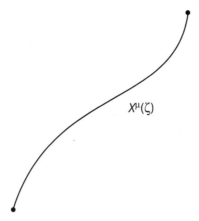

$X^\mu(\zeta)$

Figure I.11.1

which upon multiplication by $g^{\rho\lambda}$ becomes

$$\frac{d^2 X^\rho}{ds^2} + \frac{1}{2}g^{\rho\lambda}(2\partial_\nu g_{\mu\lambda} - \partial_\lambda g_{\mu\nu})\frac{dX^\mu}{ds}\frac{dX^\nu}{ds} = 0$$

that is,

$$\frac{d^2 X^\rho}{ds^2} + \Gamma^\rho_{\mu\nu}[X(s)]\frac{dX^\mu}{ds}\frac{dX^\nu}{ds} = 0 \tag{9}$$

if we define the Riemann-Christoffel symbol by

$$\Gamma^\rho_{\mu\nu} \equiv \tfrac{1}{2}g^{\rho\lambda}(\partial_\mu g_{\nu\lambda} + \partial_\nu g_{\mu\lambda} - \partial_\lambda g_{\mu\nu}) \tag{10}$$

Given the initial position $X^\mu(s_0)$ and velocity $(dX^\mu/ds)(s_0)$ we have four second order differential equations (9) determining the geodesic followed by the particle in curved spacetime. Note that, contrary to the impression given by some texts, unlike (8), (9) is not reparametrization invariant.

To recover Newtonian gravity, three conditions must be met: (1) the particle moves slowly $dX^i/ds \ll dX^0/ds$; (2) the gravitational field is weak, so that the metric is almost Minkowskian $g_{\mu\nu} \simeq \eta_{\mu\nu} + h_{\mu\nu}$; and (3) the gravitational field does not depend on time. Condition (1) means that $d^2 X^\rho/ds^2 + \Gamma^\rho_{00}(dX^0/ds)^2 \simeq 0$, while (2) and (3) imply that $\Gamma^\rho_{00} \simeq -\frac{1}{2}\eta^{\rho\lambda}\partial_\lambda h_{00}$. Thus, (9) reduces to $d^2 X^0/ds^2 \simeq 0$ (which implies that dX^0/ds is a constant) and $d^2 X^i/ds^2 + \frac{1}{2}\partial_i h_{00}(dX^0/ds)^2 \simeq 0$, which since X^0 is proportional to s becomes $d^2 X^i/dt^2 \simeq -\frac{1}{2}\partial_i h_{00}$. Thus, we obtain Newton's equation $\frac{d^2\vec{X}}{dt^2} \simeq -\vec{\nabla}\phi$ if we define the gravitational potential ϕ by $h_{00} = 2\phi$:

$$g_{00} \simeq 1 + 2\phi \tag{11}$$

Referring to the Schwarzschild metric, we see that far from a massive body, $\phi = -GM/r$, as we expect. (Note also that this derivation depends neither on h_{ij} nor on h_{0j}, as long as they are time independent.)

Thus, the action of a point particle is

$$S = -m \int \sqrt{g_{\mu\nu}dX^\mu dX^\nu} = -m \int \sqrt{g_{\mu\nu}[X(\zeta)]\frac{dX^\mu}{d\zeta}\frac{dX^\nu}{d\zeta}}\,d\zeta \tag{12}$$

The m follows from dimensional analysis.

A slick way of deriving S (which also allows us to see the minus sign) is to start with the nonrelativistic action of a particle in a gravitational potential ϕ, namely $S = \int L dt = \int (\frac{1}{2}mv^2 - m - m\phi)dt$. Note that the rest mass

m comes in with a minus sign as it is part of the potential energy in nonrelativistic physics. Now force S into a relativistic form: For small v and ϕ,

$$S = -m \int (1 - \frac{1}{2}v^2 + \phi)dt \simeq -m \int \sqrt{1 - v^2 + 2\phi}\, dt$$

$$= -m \int \sqrt{(1 + 2\phi)(dt)^2 - (d\vec{x})^2}$$

We see[8] that the 2 in (11) comes from the square root in the Lorentz-Fitzgerald quantity $\sqrt{1 - v^2}$.

Now that we have S we can calculate the stress energy of a point particle using (4):

$$T^{\mu\nu}(x) = \frac{m}{\sqrt{-g}} \int d\zeta\, K^{-\frac{1}{2}}\delta^{(4)}[x - X(\zeta)]\frac{dX^\mu}{d\zeta}\frac{dX^\nu}{d\zeta}$$

Setting ζ to s (which we call the proper time τ in this context) we have

$$T^{\mu\nu}(x) = \frac{m}{\sqrt{-g}} \int d\tau\, \delta^{(4)}[x - X(\tau)]\frac{dX^\mu}{d\tau}\frac{dX^\nu}{d\tau}$$

In particular, as we expect the 4-momentum of the particle is given by

$$P^\nu = \int d^3x \sqrt{-g}\, T^{0\nu} = m \int d\tau\, \delta[x^0 - X^0(\tau)]\frac{dX^0}{d\tau}\frac{dX^\nu}{d\tau} = m\frac{dX^\nu}{d\tau}$$

The action (12) given here has two defects: (1) it is difficult to deal with a path integral $\int DX e^{-im \int d\zeta \sqrt{(dX^\mu/d\zeta)(dX_\mu/d\zeta)}}$ involving a square root, and (2) S does not make sense for a massless particle. To remedy these defects, note that classically, S is equivalent to

$$S_{\text{imp}} = -\frac{1}{2} \int d\zeta \left(\frac{1}{\gamma}\frac{dX^\mu}{d\zeta}\frac{dX_\mu}{d\zeta} + \gamma m^2 \right) \tag{13}$$

where $(dX^\mu/d\zeta)(dX_\mu/d\zeta) = g_{\mu\nu}(X)(dX^\mu/d\zeta)(dX^\nu/d\zeta)$. Varying with respect to $\gamma(\zeta)$ we obtain $m^2\gamma^2 = (dX^\mu/d\zeta)(dX_\mu/d\zeta)$. Eliminating γ in S_{imp} we recover S.

The path integral $\int DX e^{iS_{\text{imp}}}$ has a standard quadratic form.[9] Quantum mechanics of a relativistic point particle is best formulated in terms of S_{imp}, not S. Furthermore, for $m = 0$, $S_{\text{imp}} = -\frac{1}{2} \int d\zeta [\gamma^{-1}(dX^\mu/d\zeta)(dX_\mu/d\zeta)]$ makes perfect sense. Note that varying with respect to γ now gives the well-known fact that for a massless particle $g_{\mu\nu}(X)dX^\mu dX^\nu = 0$.

The action S_{imp} will provide the starting point for our discussion on string theory in chapter VIII.5.

Exercises

I.11.1   Integrate by parts to obtain for the scalar field action

$$S = -\int d^4x \sqrt{-g}\frac{1}{2}\varphi(\frac{1}{\sqrt{-g}}\partial_\mu\sqrt{-g}g^{\mu\nu}\partial_\nu + m^2)\varphi$$

---

[8] We should not conclude from this that $g_{ij} = \delta_{ij}$. The point is that to leading order in v/c, our particle is sensitive only to g_{00}, as we have just shown. Indeed, restoring c in the Schwarzschild metric we have

$$ds^2 = (1 - \frac{2GM}{c^2r})c^2dt^2 - (1 - \frac{2GM}{c^2r})^{-1}dr^2 - r^2d\theta^2 - r^2\sin^2\theta d\phi^2$$

$$\rightarrow c^2dt^2 - d\vec{x}^2 - \frac{2GM}{r}dt^2 + O(1/c^2)$$

[9] One technical problem, which we will address in chapter III.4, is that in the integral over $X(\zeta)$ apparently different functions $X(\zeta)$ may in fact be the same physically, related by a reparametrization.

and write the equation of motion for φ in curved spacetime. Discuss the propagator of the scalar field $D(x, y)$ (which is of course no longer translation invariant, i.e., it is no longer a function of $x - y$).

I.11.2 Use (4) to find E for a scalar field theory in flat spacetime. Show that the result agrees with what you would obtain using the canonical formalism of chapter I.8.

I.11.3 Show that in flat spacetime P^μ as derived here from the stress energy tensor $T^{\mu\nu}$ when interpreted as an operator in the canonical formalism satisfies $[P^\mu, \varphi(x)] = -i\partial^\mu\varphi(x)$, and thus does exactly what you expect the energy and momentum operators to do, namely to be conjugate to time and space and hence represented by $-i\partial^\mu$.

I.11.4 Show that for the Maxwell field $T_{ij} = -(E_i E_j + B_i B_j) + \frac{1}{2}\delta_{ij}(\vec{E}^2 + \vec{B}^2)$ and hence $T = 0$.

What have you learned so far?

Now that we have reached the end of part I, let us take stock of what you have learned. Quantum field theory is not that difficult; it just consists of doing one great big integral

$$Z(J) = \int D\varphi \, e^{i \int d^{D+1}x[\frac{1}{2}(\partial\varphi)^2 - \frac{1}{2}m^2\varphi^2 - \lambda\varphi^4 + J\varphi]} \qquad (1)$$

By repeatedly functionally differentiating $Z(J)$ and then setting $J = 0$ we obtain

$$\int D\varphi \, \varphi(x_1)\varphi(x_2) \cdots \varphi(x_n) e^{i \int d^{D+1}x[\frac{1}{2}(\partial\varphi)^2 - \frac{1}{2}m^2\varphi^2 - \lambda\varphi^4]} \qquad (2)$$

which tells us about the amplitude for n particles associated with the field φ to come into and go out of existence at the spacetime points $x_1, x_2, \cdots, x_n$, interacting with each other in between. Birth and death, with some kind of life in between.

Ah, if we could only do the integral in (1)! But we can't. So one way of going about it is to evaluate the integral as a series in λ :

$$\sum_{k=0}^{\infty} \frac{(-i\lambda)^k}{k!} \int D\varphi \, \varphi(x_1)\varphi(x_2) \cdots \varphi(x_n)[\int d^{D+1}y\varphi(y)^4]^k e^{i \int d^{D+1}x[\frac{1}{2}(\partial\varphi)^2 - \frac{1}{2}m^2\varphi^2]}$$

$$(3)$$

To keep track of the terms in the series we draw little diagrams.

Quantum field theorists try to dream up ways to evaluate (1), and failing that, they invent tricks and methods for extracting the physics they are interested in, by hook and by crook, without actually evaluating (1).

To see that quantum field theory is a straightforward generalization of quantum mechanics, look at how (1) reduces appropriately. We have written the theory in $(D + 1)$-dimensional spacetime, that is, D spatial dimensions and 1 temporal dimension. Consider (1) in $(0 + 1)$-dimensional spacetime, that is, no space; it becomes

$$Z(J) = \int D\varphi \, e^{i \int dt[\frac{1}{2}(\frac{d\varphi}{dt})^2 - \frac{1}{2}m^2\varphi^2 - \lambda\varphi^4 + J\varphi]} \qquad (4)$$

where we now denote the spacetime coordinate x just by time t. We recognize this as the quantum mechanics of an anharmonic oscillator with the position of the mass point tied to the spring denoted by φ and with an external force J pushing on the oscillator.

In the quantum field theory (1), each term in the action makes physical sense: The first two terms generalize the harmonic oscillator to include spatial variations, the third term the anharmonicity, and the last term an external probe. You can think of a quantum field theory as an infinite collection of anharmonic oscillators, one at each point in space.

We have here a scalar field φ. In previous and future chapters, the notion of field was and will be generalized ever so slightly: The field can transform according to a nontrivial representation of the Lorentz group. We have already encountered fields transforming as a vector and a tensor and will presently encounter a field transforming as a spinor. Lorentz invariance and whatever other symmetries we have constrain the form of the action. The integral will look more complicated but the approach is exactly as outlined here.

That's just about all there is to quantum field theory.

Part II | Dirac and the Spinor

The Dirac Equation

Staring into a fire

According to a physics legend, apparently even true, Dirac was staring into a fire one evening in 1928 when he realized that he wanted, for reasons that are no longer relevant, a relativistic wave equation linear in spacetime derivatives $\partial_\mu \equiv \partial/\partial x^\mu$. At that time, the Klein-Gordon equation $(\partial^2 + m^2)\varphi = 0$, which describes a free particle of mass m and quadratic in spacetime derivatives, was already well known. This is in fact the equation of motion of the scalar field theory we studied earlier.

At first sight, what Dirac wanted does not make sense. The equation is supposed to have the form "some linear combination of ∂_μ acting on some field ψ is equal to some constant times the field." Denote the linear combination by $c^\mu \partial_\mu$. If the c^μ's are four ordinary numbers, then the four-vector c^μ defines some direction and the equation cannot be Lorentz invariant.

Nevertheless, let us follow Dirac and write, using modern notation,

$$(i\gamma^\mu \partial_\mu - m)\psi = 0 \tag{1}$$

At this point, the four quantities $i\gamma^\mu$ are just the coefficients of ∂_μ and m is just a constant. We have already argued that γ^μ cannot simply be four numbers. Well, let us see what these objects have to be in order for this equation to contain the correct physics.

Acting on (1) with $(i\gamma^\mu \partial_\mu + m)$, Dirac obtained $-(\gamma^\mu \gamma^\nu \partial_\mu \partial_\nu + m^2)\psi = 0$. It is traditional to define, in addition to the commutator $[A, B] = AB - BA$ familiar from quantum mechanics, the anticommutator $\{A, B\} = AB + BA$. Since derivatives commute, $\gamma^\mu \gamma^\nu \partial_\mu \partial_\nu = \frac{1}{2}\{\gamma^\mu, \gamma^\nu\}\partial_\mu \partial_\nu$, and we have $(\frac{1}{2}\{\gamma^\mu, \gamma^\nu\}\partial_\mu \partial_\nu + m^2)\psi = 0$. In a moment of inspiration Dirac realized that if

$$\{\gamma^\mu, \gamma^\nu\} = 2\eta^{\mu\nu} \tag{2}$$

with $\eta^{\mu\nu}$ the Minkowski metric he would obtain $(\partial^2 + m^2)\psi = 0$, which describes a particle of mass m, and thus (1) would also describe a particle of mass m.

Since $\eta^{\mu\nu}$ is a diagonal matrix with diagonal elements $\eta^{00} = 1$ and $\eta^{jj} = -1$, (2) says that $(\gamma^0)^2 = 1$, $(\gamma^j)^2 = -1$, and $\gamma^\mu\gamma^\nu = -\gamma^\nu\gamma^\mu$ for $\mu \neq \nu$. This last statement, that the coefficients γ^μ anticommute with each other, implies that they indeed cannot be ordinary numbers. Dirac's thought would make sense if we could find four such objects.

Clifford algebra

A set of objects γ^μ (clearly d of them in d-dimensional spacetime) satisfying the relation (2) is said to form a Clifford algebra. I will develop the mathematics of Clifford algebra later. Suffice it for you to check here that the following 4 by 4 matrices satisfy (2):

$$\gamma^0 = \begin{pmatrix} I & 0 \\ 0 & -I \end{pmatrix} = I \otimes \tau_3 \tag{3}$$

$$\gamma^i = \begin{pmatrix} 0 & \sigma^i \\ -\sigma^i & 0 \end{pmatrix} = \sigma^i \otimes i\tau_2 \tag{4}$$

Here σ and τ denote the standard Pauli matrices. For historical reasons the four matrices γ^μ are known as gamma matrices—not a very imaginative name! (Our convention is such that whether an index on a Pauli matrix is upper or lower has no significance. On the other hand, we define $\gamma_\mu \equiv \eta_{\mu\nu}\gamma^\nu$ and it does matter whether the index on a gamma matrix is upper or lower; it is to be treated just like the index on any Lorentz vector. This convention is useful because then $\gamma^\mu\partial_\mu = \gamma_\mu\partial^\mu$.)

The direct product notation is convenient for computation: For example, $\gamma^i\gamma^j = (\sigma^i \otimes i\tau_2)(\sigma^j \otimes i\tau_2) = (\sigma^i\sigma^j \otimes i^2\tau_2\tau_2) = -(\sigma^i\sigma^j \otimes I)$ and thus $\{\gamma^i, \gamma^j\} = -\{\sigma^i, \sigma^j\} \otimes I = -2\delta^{ij}$ as desired.

You can convince yourself that the γ^μ's cannot be smaller than 4 by 4 matrices. The mathematics forces the Dirac spinor ψ to have 4 components! The physical content of the Dirac equation (1) is most transparent if we transform to momentum space: we plug $\psi(x) = \int [d^4 p/(2\pi)^4] e^{-ipx} \psi(p)$ into (1) and obtain

$$(\gamma^\mu p_\mu - m)\psi(p) = 0 \tag{5}$$

Since (5) is Lorentz invariant, as we will show below, we can examine its physical content in any frame, in particular the rest frame $p^\mu = (m, \vec{0})$, in which it becomes

$$(\gamma^0 - 1)\psi = 0 \tag{6}$$

As $(\gamma^0 - 1)^2 = -2(\gamma^0 - 1)$ we recognize $(\gamma^0 - 1)$ as a projection operator up to a trivial normalization. Indeed, using the explicit form in (3), we see that there is nothing mysterious to Dirac's equation: When written out, (6) reads

$$\begin{pmatrix} 0 & 0 \\ 0 & I \end{pmatrix} \psi = 0$$

thus telling us that 2 of the 4 components in ψ are zero.

This makes perfect sense since we know that the electron has 2 physical degrees of freedom, not 4. Viewed in this light, the mysterious Dirac equation is no more and no less than a projection that gets rid of the unwanted degrees of freedom. Compare our discussion of the equation of motion of a massive spin 1 particle (chapter I.5). There also, 1 of the 4 components of A^μ is projected out. Indeed, the Klein-Gordon equation $(\partial^2 + m^2)\varphi(x) = 0$ just projects out those Fourier components $\varphi(k)$ not satisfying the mass shell condition $k^2 = m^2$. Our discussion provides a unified view of the equations of motion in relativistic physics: They just project out the unphysical components.

A convenient notation introduced by Feynman, $\rlap{/}{a} \equiv \gamma^\mu a_\mu$ for any 4-vector a_μ, is now standard. The Dirac equation then reads $(i\rlap{/}{\partial} - m)\psi = 0$.

Cousins of the gamma matrices

Under a Lorentz transformation $x'^\nu = \Lambda^\nu{}_\mu x^\mu$, the 4 components of the vector field A_μ transform like, well, a vector. How do the 4 components of ψ transform? Surely not in the same way as A_μ since even under rotation ψ and A_μ transform quite differently: one as spin $\frac{1}{2}$ and the other as spin 1. Let us write $\psi(x) \to \psi'(x') \equiv S(\Lambda)\psi(x)$ and try to determine the 4 by 4 matrix $S(\Lambda)$.

It is a good idea to first sort out (and name) the 16 linearly independent 4 by 4 matrices. We already know five of them: the identity matrix and the γ^μ's. The strategy is simply to multiply the γ^μ's together, thus generating more 4 by 4 matrices until we get all 16. Since the square of a gamma matrix γ^μ is equal to ± 1 and the γ^μ's anticommute with each other, we have to consider only $\gamma^\mu\gamma^\nu$, $\gamma^\mu\gamma^\nu\gamma^\lambda$, and $\gamma^\mu\gamma^\nu\gamma^\lambda\gamma^\rho$ with μ, ν, λ, and ρ all different from one another. Thus, the only product of four gamma matrices that we have to consider is

$$\gamma^5 \equiv i\gamma^0\gamma^1\gamma^2\gamma^3 \tag{7}$$

This combination is so important that it has its own name! (The peculiar name comes about because in some old-fashioned notation the time coordinate was called x^4 with a corresponding γ^4.) We have

$$\gamma^5 = i(I \otimes \tau_3)(\sigma^1 \otimes i\tau_2)(\sigma^2 \otimes i\tau_2)(\sigma^3 \otimes i\tau_2) = i^4(I \otimes \tau_3)(\sigma^1\sigma^2\sigma^3 \otimes \tau_2)$$

and so

$$\gamma^5 = I \otimes \tau_1 = \begin{pmatrix} 0 & I \\ I & 0 \end{pmatrix} \tag{8}$$

With the factor of i included, γ^5 is manifestly hermitean. An important property is that γ^5 anticommutes with the γ^μ's:

$$\{\gamma^5, \gamma^\mu\} = 0 \tag{9}$$

Continuing, we see that the products of three gamma matrices, all different, can be written as $\gamma^\mu \gamma^5$ (e.g., $\gamma^1 \gamma^2 \gamma^3 = -i\gamma^0 \gamma^5$). Finally, using (2) we can write the product of two gamma matrices as $\gamma^\mu \gamma^\nu = \eta^{\mu\nu} - i\sigma^{\mu\nu}$, where

$$\sigma^{\mu\nu} \equiv \frac{i}{2}[\gamma^\mu, \gamma^\nu] \tag{10}$$

There are $4 \cdot 3/2 = 6$ of these $\sigma^{\mu\nu}$ matrices.

Count them, we got all 16. The set of 16 matrices $\{1, \gamma^\mu, \sigma^{\mu\nu}, \gamma^\mu \gamma^5, \gamma^5\}$ forms a complete basis of the space of all 4 by 4 matrices, that is, any 4 by 4 matrix can be written as a linear combination of these 16 matrices.

It is instructive to write out $\sigma^{\mu\nu}$ explicitly in the representation (3) and (4):

$$\sigma^{0i} = i \begin{pmatrix} 0 & \sigma^i \\ \sigma^i & 0 \end{pmatrix} \tag{11}$$

$$\sigma^{ij} = \varepsilon^{ijk} \begin{pmatrix} \sigma^k & 0 \\ 0 & \sigma^k \end{pmatrix} \tag{12}$$

We see that σ^{ij} are just the Pauli matrices doubly stacked, for example,

$$\sigma^{12} = \begin{pmatrix} \sigma^3 & 0 \\ 0 & \sigma^3 \end{pmatrix}$$

Lorentz transformation

Recall from a course on quantum mechanics that a general rotation can be written as $e^{i\vec{\theta} \cdot \vec{J}}$ with $\vec{J}$ the 3 generators of rotation and $\vec{\theta}$ 3 rotation parameters. Recall also that the Lorentz group contains boosts in addition to rotations, with $\vec{K}$ denoting the 3 generators of boosts. Recall from a course on electromagnetism that the 6 generators $\{\vec{J}, \vec{K}\}$ transform under the Lorentz group as the components of an antisymmetric tensor just like the electromagnetic field $F_{\mu\nu}$ and thus can be denoted by $J_{\mu\nu}$. I will discuss these matters in more detail in chapter II.3. For the moment, suffice it to note that with this notation we can write a Lorentz transformation as $\Lambda = e^{-\frac{i}{2}\omega_{\mu\nu}J^{\mu\nu}}$, with J^{ij} generating rotations, J^{0i} generating boosts, and the antisymmetric tensor $\omega_{\mu\nu} = -\omega_{\nu\mu}$ with its $6 = 4 \cdot 3/2$ components corresponding to the 3 rotation and 3 boost parameters.

Given the preceding discussion and the fact that there are six matrices $\sigma^{\mu\nu}$, we suspect that up to an overall numerical factor the $\sigma^{\mu\nu}$'s must represent the 6 generators $J^{\mu\nu}$ of the Lorentz group acting on a spinor. In fact, our suspicion is confirmed by thinking about what a rotation $e^{-\frac{i}{2}\omega_{ij}J^{ij}}$ does. Referring to (12) we see that if J^{ij} is represented by $\frac{1}{2}\sigma^{ij}$ this would correspond exactly to how a spin $\frac{1}{2}$ particle transforms in quantum mechanics. More precisely, separate the 4 components of the Dirac spinor into 2 sets of 2 components:

$$\psi = \begin{pmatrix} \phi \\ \chi \end{pmatrix} \tag{13}$$

From (12) we see that under a rotation around the 3rd axis, $\phi \to e^{-i\omega_{12}\frac{1}{2}\sigma^3}\phi$ and $\chi \to e^{-i\omega_{12}\frac{1}{2}\sigma^3}\chi$. It is gratifying to see that ϕ and χ transform like 2-component Pauli spinors.

We have thus figured out that a Lorentz transformation Λ acting on ψ is represented by $S(\Lambda) = e^{-(i/4)\omega_{\mu\nu}\sigma^{\mu\nu}}$, and so, acting on ψ, the generators $J^{\mu\nu}$ are indeed represented by $\frac{1}{2}\sigma^{\mu\nu}$. Therefore we would expect that if $\psi(x)$ satisfies the Dirac equation (1) then $\psi'(x') \equiv S(\Lambda)\psi(x)$ would satisfy the Dirac equation in the primed frame,

$$(i\gamma^\mu \partial'_\mu - m)\psi'(x') = 0 \tag{14}$$

where $\partial'_\mu \equiv \partial/\partial x'^\mu$. To show this, calculate $[\sigma^{\mu\nu}, \gamma^\lambda] = 2i(\gamma^\mu \eta^{\nu\lambda} - \gamma^\nu \eta^{\mu\lambda})$ and hence for ω infinitesimal $S\gamma^\lambda S^{-1} = \gamma^\lambda - (i/4)\omega_{\mu\nu}[\sigma^{\mu\nu}, \gamma^\lambda] = \gamma^\lambda + \gamma^\mu \omega_\mu{}^\lambda$. Building up a finite Lorentz transformation by compounding infinitesimal transformations (just as in the standard discussion of the rotation group in quantum mechanics), we have $S\gamma^\lambda S^{-1} = \Lambda^\lambda{}_\mu \gamma^\mu$.

Dirac bilinears

The Clifford algebra tells us that $(\gamma^0)^2 = +1$ and $(\gamma^i)^2 = -1$; hence the necessity for the i in (4). One consequence of the i is that γ^0 is hermitean while γ^i is antihermitean, a fact conveniently expressed as

$$(\gamma^\mu)^\dagger = \gamma^0 \gamma^\mu \gamma^0 \tag{15}$$

Thus, contrary to what you might think, the bilinear $\psi^\dagger \gamma^\mu \psi$ is not hermitean; rather, $\bar{\psi}\gamma^\mu\psi$ is hermitean with $\bar{\psi} \equiv \psi^\dagger \gamma^0$. The necessity for introducing $\bar{\psi}$ in addition to $\psi^\dagger$ in relativistic physics is traced back to the $(+,-,-,-)$ signature of the Minkowski metric.

It follows that $(\sigma^{\mu\nu})^\dagger = \gamma^0 \sigma^{\mu\nu} \gamma^0$. Hence, $S(\Lambda)^\dagger = \gamma^0 e^{(i/4)\omega_{\mu\nu}\sigma^{\mu\nu}} \gamma^0$, (which incidentally, clearly shows that S is not unitary, a fact we knew since σ_{0i} is not hermitean), and so

$$\bar{\psi}'(x') = \psi(x)^\dagger S(\Lambda)^\dagger \gamma^0 = \bar{\psi}(x)e^{+(i/4)\omega_{\mu\nu}\sigma^{\mu\nu}}. \tag{16}$$

We have

$$\bar{\psi}'(x')\psi'(x') = \bar{\psi}(x)e^{+(i/4)\omega_{\mu\nu}\sigma^{\mu\nu}}e^{-(i/4)\omega_{\mu\nu}\sigma^{\mu\nu}}\psi(x) = \bar{\psi}(x)\psi(x)$$

You are probably used to writing $\psi^\dagger \psi$ in nonrelativistic physics. In relativistic physics you have to get used to writing $\bar{\psi}\psi$. It is $\bar{\psi}\psi$, not $\psi^\dagger\psi$, that transforms as a Lorentz scalar.

There are obviously 16 Dirac bilinears $\bar{\psi}\Gamma\psi$ that we can form, corresponding to the 16 linearly independent Γ's. You can now work out how various fermion bilinears transform (exercise II.1.1). The notation is rather nice: Various objects transform the way it looks like they should transform. We simply look at the Lorentz indices they carry. Thus, $\bar{\psi}(x)\gamma^\mu\psi(x)$ transforms as a Lorentz vector.

Parity

An important discrete symmetry in physics is that of parity or reflection in a mirror[1]

$$x^\mu \to x'^\mu = (x^0, -\vec{x}) \tag{17}$$

Multiply the Dirac equation (1) by γ^0: $\gamma^0(i\gamma^\mu \partial_\mu - m)\psi(x) = 0 = (i\gamma^\mu \partial'_\mu - m)\gamma^0\psi(x)$, where $\partial'_\mu \equiv \partial/\partial x'^\mu$. Thus,

$$\psi'(x') \equiv \eta\gamma^0\psi(x) \tag{18}$$

satisfies the Dirac equation in the space-reflected world (where η is an arbitrary phase that we can set to 1).

Note, for example, $\bar{\psi}'(x')\psi'(x') = \bar{\psi}(x)\psi(x)$ but $\bar{\psi}'(x')\gamma^5\psi'(x') = \bar{\psi}(x)\gamma^0\gamma^5\gamma^0\psi(x) = -\bar{\psi}(x)\gamma^5\psi(x)$. Under a Lorentz transformation $\bar{\psi}(x)\gamma^5\psi(x)$ and $\bar{\psi}(x)\psi(x)$ transform in the same way but under space reflection they transform in an opposite way; in other words, while $\bar{\psi}(x)\psi(x)$ transforms as a scalar, $\bar{\psi}(x)\gamma^5\psi(x)$ transforms as a pseudoscalar.

You are now ready to do the all-important exercises in this chapter.

The Dirac Lagrangian

An interesting question: What Lagrangian would give Dirac's equation? The answer is

$$\mathcal{L} = \bar{\psi}(i\slashed{\partial} - m)\psi \tag{19}$$

Since ψ is complex we can vary ψ and $\bar{\psi}$ independently to obtain the Euler-Lagrange equation of motion. Thus, $\partial_\mu(\delta\mathcal{L}/\delta\partial_\mu\psi) - \delta\mathcal{L}/\delta\psi = 0$ gives $\partial_\mu(i\bar{\psi}\gamma^\mu) + m\bar{\psi} = 0$, which upon hermitean conjugation and multiplication by γ^0 gives the Dirac equation (1). The other variational equation $\partial_\mu(\delta\mathcal{L}/\delta\partial_\mu\bar{\psi}) - \delta\mathcal{L}/\delta\bar{\psi} = 0$ gives the Dirac equation even more directly. (If you are disturbed by the asymmetric treatment of ψ and $\bar{\psi}$, you can always integrate by parts in the action, have ∂_μ act on $\bar{\psi}$ in the Lagrangian, and then average the two forms of the Lagrangian. The action $S = \int d^4x \mathcal{L}$ treats ψ and $\bar{\psi}$ symmetrically.)

Slow and fast electrons

Given a set of gamma matrices it is straightforward to solve the Dirac equation

$$(\slashed{p} - m)\psi(p) = 0 \tag{20}$$

for $\psi(p)$: It is a simple matrix equation (see exercise II.1.3).

---

[1] Rotations consist of all linear transformations $x^i \to R^{ij}x^j$ such that det $R = +1$. Those transformations with det $R = -1$ are composed of parity followed by a rotation. In $(3 + 1)$-dimensional spacetime, parity can be defined as reversing one of the spatial coordinates or all three spatial coordinates. The two operations are related by a rotation. Note that in odd dimensional spacetime, parity is not the same as space inversion, in which all spatial coordinates are reversed (see exercise II.1.12).

Note that if somebody uses the gamma matrices γ^μ, you are free to use instead $\gamma'^\mu = W^{-1}\gamma^\mu W$ with W any 4 by 4 matrix with an inverse. Obviously, γ'^μ also satisfy the Clifford algebra. This freedom of choice corresponds to a simple change of basis. Physics cannot depend on the choice of basis, but which basis is the most convenient depends on the physics.

For example, suppose we want to study a slowly moving electron. Let us use the basis defined by (3) and (4), and the 2-component decomposition of ψ (13). Since (6) tells us that $\chi(p) = 0$ for an electron at rest, we expect $\chi(p)$ to be much smaller than $\phi(p)$ for a slowly moving electron.

In contrast, for momentum much larger than the mass, we can approximate (20) by $\not{p}\psi(p) = 0$. Multiplying on the left by γ^5, we see that if $\psi(p)$ is a solution then $\gamma^5\psi(p)$ is also a solution since γ^5 anticommutes with γ^μ. Since $(\gamma^5)^2 = 1$, we can form two projection operators $P_L \equiv \frac{1}{2}(1 - \gamma^5)$ and $P_R \equiv \frac{1}{2}(1 + \gamma^5)$, satisfying $P_L^2 = P_L$, $P_R^2 = P_R$, and $P_L P_R = 0$. It is extremely useful to introduce the two combinations $\psi_L = \frac{1}{2}(1 - \gamma^5)\psi$ and $\psi_R = \frac{1}{2}(1 + \gamma^5)\psi$. Note that $\gamma^5\psi_L = -\psi_L$ and $\gamma^5\psi_R = +\psi_R$. Physically, a relativistic electron has two degrees of freedom known as helicities: it can spin either clockwise or anticlockwise around the direction of motion. I leave it to you as an exercise to show that ψ_L and ψ_R correspond to precisely these two possibilities. The subscripts L and R indicate left and right handed. Thus, for fast moving electrons, a basis known as the Weyl basis, designed so that γ^5, rather than γ^0, is diagonal, is more convenient. Instead of (3), we choose

$$\gamma^0 = \begin{pmatrix} 0 & I \\ I & 0 \end{pmatrix} = I \otimes \tau_1 \tag{21}$$

We keep γ^i as in (4). This defines the Weyl basis. We now calculate

$$\gamma^5 \equiv i\gamma^0\gamma^1\gamma^2\gamma^3 = i(I \otimes \tau_1)(\sigma^1\sigma^2\sigma^3 \otimes i^3\tau_2) = -(I \otimes \tau_3) = \begin{pmatrix} -I & 0 \\ 0 & I \end{pmatrix} \tag{22}$$

which is indeed diagonal as desired. The decomposition into left and right handed fields is of course defined regardless of what basis we feel like using, but in the Weyl basis we have the nice feature that ψ_L has two upper components and ψ_R has two lower components. The spinors ψ_L and ψ_R are known as Weyl spinors.

Note that in going from the Dirac to the Weyl basis γ^0 and γ^5 trade places (up to a sign):

$$\text{Dirac} : \gamma^0 \text{diagonal}; \quad \text{Weyl} : \gamma^5 \text{diagonal}. \tag{23}$$

Physics dictates which basis to use: We prefer to have γ^0 diagonal when we deal with slowly moving spin $\frac{1}{2}$ particles, while we prefer to have γ^5 diagonal when we deal with fast moving spin $\frac{1}{2}$ particles .

I note in passing that if we define $\sigma^\mu \equiv (I, \vec{\sigma})$ and $\bar{\sigma}^\mu \equiv (I, -\vec{\sigma})$ we can write

$$\gamma^\mu = \begin{pmatrix} 0 & \sigma^\mu \\ \bar{\sigma}^\mu & 0 \end{pmatrix}$$

more compactly in the Weyl basis. (We develop this further in appendix E.)

Chirality or handedness

Regardless of whether a Dirac field $\psi(x)$ is massive or massless, it is enormously useful to decompose ψ into left and right handed fields $\psi(x) = \psi_L(x) + \psi_R(x) \equiv \frac{1}{2}(1 - \gamma^5)\psi(x) + \frac{1}{2}(1 + \gamma^5)\psi(x)$. As an exercise, show that you can write the Dirac Lagrangian as

$$\mathcal{L} = \bar{\psi}(i\partial\!\!\!/ - m)\psi = \bar{\psi}_L i \partial\!\!\!/ \psi_L + \bar{\psi}_R i \partial\!\!\!/ \psi_R - m(\bar{\psi}_L \psi_R + \bar{\psi}_R \psi_L) \tag{24}$$

The kinetic energy connects left to left and right to right, while the mass term connects left to right and right to left.

The transformation $\psi \to e^{i\theta}\psi$ leaves the Lagrangian $\mathcal{L}$ invariant. Applying Noether's theorem, we obtain the conserved current associated with this symmetry $J^\mu = \bar{\psi}\gamma^\mu\psi$. Projecting into left and right handed fields we see that they transform the same way: $\psi_L \to e^{i\theta}\psi_L$ and $\psi_R \to e^{i\theta}\psi_R$.

If $m = 0$, $\mathcal{L}$ enjoys an additional symmetry, known as a chiral symmetry, under which $\psi \to e^{i\phi\gamma^5}\psi$. Noether's theorem tells us that the axial current $J^{5\mu} \equiv \bar{\psi}\gamma^\mu\gamma^5\psi$ is conserved. The left and right handed fields transform in opposite ways: $\psi_L \to e^{-i\phi}\psi_L$ and $\psi_R \to e^{i\phi}\psi_R$. These points are particularly obvious when $\mathcal{L}$ is written in terms of ψ_L and ψ_R, as in (24).

In 1956 Lee and Yang proposed that the weak interaction does not preserve parity. It was eventually realized (with these four words I brush over a beautiful chapter in the history of particle physics; I urge you to read about it!) that the weak interaction Lagrangian has the generic form

$$\mathcal{L} = G\bar{\psi}_{1L}\gamma^\mu\psi_{2L}\bar{\psi}_{3L}\gamma_\mu\psi_{4L} \tag{25}$$

where $\psi_{1,2,3,4}$ denotes four Dirac fields and G the Fermi coupling constant. This Lagrangian clearly violates parity: Under a spatial reflection, left handed fields are transformed into right handed fields and vice versa.

Incidentally, henceforth when I say a Lagrangian has a certain form, I will usually indicate only one or more of the relevant terms in the Lagrangian, as in (25). The other terms in the Lagrangian, such as $\bar{\psi}_1(i\partial\!\!\!/ - m_1)\psi_1$, are understood. If the term is not hermitean, then it is understood that we also add its hermitean conjugate.

Interactions

As we saw in (25) given the classification of bilinears in the spinor field you worked out in an exercise it is easy to introduce interactions. As another example, we can couple a scalar field φ to the Dirac field by adding the term $g\varphi\bar{\psi}\psi$ (with g some coupling constant) to the Lagrangian $\mathcal{L} = \bar{\psi}(i\partial\!\!\!/ - m)\psi$ (and of course also adding the Lagrangian for φ). Similarly, we can couple a vector field A_μ by adding the term $eA_\mu\bar{\psi}\gamma^\mu\psi$. We note that in this case we can introduce the covariant derivative $D_\mu = \partial_\mu - ieA_\mu$ and write

$\mathcal{L} = \bar{\psi}(i\not{\partial} - m)\psi + eA_\mu \bar{\psi}\gamma^\mu \psi = \bar{\psi}(i\gamma^\mu D_\mu - m)\psi$. Thus, the Lagrangian for a Dirac field interacting with a vector field of mass μ reads

$$\mathcal{L} = \bar{\psi}(i\gamma^\mu D_\mu - m)\psi - \tfrac{1}{4}F_{\mu\nu}F^{\mu\nu} - \tfrac{1}{2}\mu^2 A_\mu A^\mu \tag{26}$$

If the mass μ vanishes, this is the Lagrangian for quantum electrodynamics. Varying with respect to $\bar{\psi}$, we obtain the Dirac equation in the presence of an electromagnetic field:

$$[i\gamma^\mu(\partial_\mu - ieA_\mu) - m]\psi = 0 \tag{27}$$

Charge conjugation and antimatter

With coupling to the electromagnetic field, we have the concept of charge and hence of charge conjugation. Let us try to flip the charge e. Take the complex conjugate of (27): $[-i\gamma^{\mu*}(\partial_\mu + ieA_\mu) - m]\psi^* = 0$. Complex conjugating (2) we see that the $-\gamma^{\mu*}$ also satisfy the Clifford algebra and thus must be the γ^μ matrices expressed in a different basis, that is, there exists a matrix $C\gamma^0$ (the notation with an explicit factor of γ^0 is standard; see below) such that $-\gamma^{\mu*} = (C\gamma^0)^{-1}\gamma^\mu(C\gamma^0)$. Plugging in, we find that

$$[i\gamma^\mu(\partial_\mu + ieA_\mu) - m]\psi_c = 0 \tag{28}$$

where we have defined $\psi_c \equiv C\gamma^0\psi^*$. Thus, if ψ is the field of the electron, then ψ_c is the field of a particle with a charge opposite to that of the electron but with the same mass, namely the positron.

The discovery of antimatter was one of the most momentous in twentieth-century physics. We will discuss antimatter in more detail in the next chapter.

It may be instructive to look at the specific form of the charge conjugation matrix C. We can write the defining equation for C as $C\gamma^0\gamma^{\mu*}\gamma^0 C^{-1} = -\gamma^\mu$. Complex conjugating the equation $(\gamma^\mu)^\dagger = \gamma^0\gamma^\mu\gamma^0$, we obtain $(\gamma^\mu)^T = \gamma^0\gamma^{\mu*}\gamma^0$ if γ^0 is real. Thus,

$$(\gamma^\mu)^T = -C^{-1}\gamma^\mu C \tag{29}$$

which explains why C is defined with a γ^0 attached.

In both the Dirac and the Weyl bases γ^2 is the only imaginary gamma matrix. Then the defining equation for C just says that $C\gamma^0$ commutes with γ^2 but anticommutes with the other three γ matrices. So evidently $C = \gamma^2\gamma^0$ [up to an arbitrary phase not fixed by (29)] and indeed $\gamma^2\gamma^{\mu*}\gamma^2 = \gamma^\mu$. Note that we have the simple (and satisfying) relation

$$\psi_c = \gamma^2\psi^* \tag{30}$$

You can easily convince yourself (exercise II.1.9) that the charge conjugate of a left handed field is right handed and vice versa. As we will see later, this fact turns out to be crucial in the construction of grand unified theory. Experimentally, it is known that the neutrino is left handed. Thus, we can now predict that the antineutrino is right handed.

Furthermore, ψ_c transforms as a spinor. Let's check: Under a Lorentz transformation $\psi \to e^{-(i/4)\omega_{\mu\nu}\sigma^{\mu\nu}}\psi$, complex conjugating we have $\psi^* \to e^{+(i/4)\omega_{\mu\nu}(\sigma^{\mu\nu})^*}\psi^*$; hence $\psi_c \to \gamma^2 e^{+(i/4)\omega_{\mu\nu}(\sigma^{\mu\nu})^*}\psi^* = e^{-(i/4)\omega_{\mu\nu}\sigma^{\mu\nu}}\psi_c$. [Recall from (10) that $\sigma^{\mu\nu}$ is defined with an explicit i.]

Note that $C^T = \gamma^0\gamma^2 = -C$ in both the Dirac and the Weyl bases.

Majorana neutrino

Since ψ_c transforms as a spinor, Majorana[2] noted that Lorentz invariance allows not only the Dirac equation $i\slashed{\partial}\psi = m\psi$ but also the Majorana equation

$$i\slashed{\partial}\psi = m\psi_c \tag{31}$$

Complex conjugating this equation and multiplying by γ^2, we have $-\gamma^2 i\gamma^{\mu*}\partial_\mu\psi^* = \gamma^2 m(-\gamma^2)\psi$, that is, $i\slashed{\partial}\psi_c = m\psi$. Thus, $-\partial^2\psi = i\slashed{\partial}(i\slashed{\partial}\psi) = i\slashed{\partial}m\psi_c = m^2\psi$. As we anticipated, m is indeed the mass, known as a Majorana mass, of the particle associated with ψ.

The Majorana equation (31) can be obtained from the Lagrangian[3]

$$\mathcal{L} = \bar{\psi}i\slashed{\partial}\psi - \tfrac{1}{2}m(\psi^T C\psi + \bar{\psi}C\bar{\psi}^T) \tag{32}$$

upon varying $\bar{\psi}$.

Since ψ and ψ_c carry opposite charge, the Majorana equation, unlike the Dirac equation, can only be applied to electrically neutral fields. However, as ψ_c is right handed if ψ is left handed, the Majorana equation, again unlike the Dirac equation, preserves handedness. Thus, the Majorana equation is almost tailor made for the neutrino.

From its conception the neutrino was assumed to be massless, but couple of years ago experimentalists established that it has a small but nonvanishing mass. As of this writing, it is not known whether the neutrino mass is Dirac or Majorana. We will see in chapter VII.7 that a Majorana mass for the neutrino arises naturally in the $SO(10)$ grand unified theory.

Finally, there is the possibility that $\psi = \psi_c$, in which case ψ is known as a Majorana spinor.

Time reversal

Finally, we come to time reversal,[4] which as you probably know, is much more confusing to discuss than parity and charge conjugation. In a famous 1932 paper Wigner showed that

---

[2] Ettore Majorana had a brilliant but tragically short career. In his early thirties, he disappeared off the coast of Sicily during a boat trip. The precise cause of his death remains a mystery. See F. Guerra and N. Robotti, *Ettore Majorana: Aspects of His Scientific and Academic Activity*.

[3] Upon recalling that C is antisymmetric, you may have worried that $\psi^T C\psi = C_{\alpha\beta}\psi_\alpha\psi_\beta$ vanishes. In future chapters we will learn that ψ has to be treated as anticommuting "Grassmannian numbers."

[4] Incidentally, I do not feel that we completely understand the implications of time-reversal invariance. See A. Zee, "Night thoughts on consciousness and time reversal," in: *Art and Symmetry in Experimental Physics*: pp. 246–249 .

time reversal is represented by an antiunitary operator. Since this peculiar feature already appears in nonrelativistic quantum physics, it is in some sense not the responsibility of a book on relativistic quantum field theory to explain time reversal as an antiunitary operator. Nevertheless, let me try to be as clear as possible. I adopt the approach of "letting the physics, namely the equations, lead us."

Take the Schrödinger equation $i(\partial/\partial t)\Psi(t) = H\Psi(t)$ (and for definiteness, think of $H = -(1/2m)\nabla^2 + V(\vec{x})$, just simple one particle nonrelativistic quantum mechanics.) We suppress the dependence of Ψ on $\vec{x}$. Consider the transformation $t \to t' = -t$. We want to find a $\Psi'(t')$ such that $i(\partial/\partial t')\Psi'(t') = H\Psi'(t')$. Write $\Psi'(t') = T\Psi(t)$, where T is some operator to be determined (up to some arbitrary phase factor η). Plugging in, we have $i[\partial/\partial(-t)]T\Psi(t) = HT\Psi(t)$. Multiply by T^{-1}, and we obtain $T^{-1}(-i)T(\partial/\partial t)\Psi(t) = T^{-1}HT\Psi(t)$. Since H does not involve time in any way, we want $T^{-1}H = HT^{-1}$. Then $T^{-1}(-i)T(\partial/\partial t)\Psi(t) = H\Psi(t)$. We are forced to conclude, as Wigner was, that

$$T^{-1}(-i)T = i \tag{33}$$

Speaking colloquially, we can say that in quantum physics time goes with an i and so flipping time means flipping i as well.

Let $T = UK$, where K complex conjugates everything to its right. Then $T^{-1} = KU^{-1}$ and (33) holds if $U^{-1}iU = i$, that is, if U^{-1} is just an ordinary (unitary) operator that does nothing to i. We will determine U as we go along. The presence of K makes T "antiunitary."

We check that this works for a spinless particle in a plane wave state $\Psi(t) = e^{i(\vec{k}\cdot\vec{x}-Et)}$. Plugging in, we have $\Psi'(t') = T\Psi(t) = UK\Psi(t) = U\Psi^*(t) = Ue^{-i(\vec{k}\cdot\vec{x}-Et)}$; since Ψ has only one component, U is just a phase factor[5] η that we can choose to be 1. Rewriting, we have $\Psi'(t) = e^{-i(\vec{k}\cdot\vec{x}+Et)} = e^{i(-\vec{k}\cdot\vec{x}-Et)}$. Indeed, Ψ' describes a plane wave moving in the opposite direction. Crucially, $\Psi'(t) \propto e^{-iEt}$ and thus has positive energy as it should. Note that acting on a spinless particle $T^2 = UKUK = UU^*K^2 = +1$.

Next consider a spin $\frac{1}{2}$ nonrelativistic electron. Acting with T on the spin-up state $\begin{pmatrix} 1 \\ 0 \end{pmatrix}$ we want to obtain the spin-down state $\begin{pmatrix} 0 \\ 1 \end{pmatrix}$. Thus, we need a nontrivial matrix $U = \eta\sigma_2$ to flip the spin:

$$T\begin{pmatrix} 1 \\ 0 \end{pmatrix} = U\begin{pmatrix} 1 \\ 0 \end{pmatrix} = i\eta\begin{pmatrix} 0 \\ 1 \end{pmatrix}$$

Similarly, T acting on the spin-down state produces the spin-up state. Note that acting on a spin $\frac{1}{2}$ particle

$$T^2 = \eta\sigma_2 K\eta\sigma_2 K = \eta\sigma_2\eta^*\sigma_2^* KK = -1$$

This is the origin of Kramer's degeneracy: In a system with an odd number of electrons in an electric field, no matter how complicated, each energy level is twofold degenerate. The proof is very simple: Since the system is time reversal invariant, Ψ and $T\Psi$ have the same energy. Suppose they actually represent the same state. Then $T\Psi = e^{i\alpha}\Psi$, but then

---

[5] It is a phase factor rather than an arbitrary complex number because we require that $|\Psi'|^2 = |\Psi|^2$.

$T^2 \Psi = T(T\Psi) = T e^{i\alpha} \Psi = e^{-i\alpha} T \Psi = \Psi \neq -\Psi$. So Ψ and $T\Psi$ must represent two distinct states.

All of this is beautiful stuff, which as I noted earlier you could and should have learned in a decent course on quantum mechanics. My responsibility here is to show you how it works for the Dirac equation. Multiplying (1) by γ^0 from the left, we have $i(\partial/\partial t)\psi(t) = H\psi(t)$ with $H = -i\gamma^0\gamma^i\partial_i + \gamma^0 m$. Once again, we want $i(\partial/\partial t')\psi'(t') = H\psi'(t')$ with $\psi'(t') = T\psi(t)$ and T some operator to be determined. The discussion above carries over if $T^{-1}HT = H$, that is, $KU^{-1}HUK = H$. Thus, we require $KU^{-1}\gamma^0 UK = \gamma^0$ and $KU^{-1}(i\gamma^0\gamma^i)UK = i\gamma^0\gamma^i$. Multiplying by K on the left and on the right, we see that we have to solve for a U such that $U^{-1}\gamma^0 U = \gamma^{0*}$ and $U^{-1}\gamma^i U = -\gamma^{i*}$. We now restrict ourselves to the Dirac and Weyl bases, in both of which γ^2 is the only imaginary guy. Okay, what flips γ^1 and γ^3 but not γ^0 and γ^2? Well, $U = \eta\gamma^1\gamma^3$ (with η an arbitrary phase factor) works:

$$\psi'(t') = \eta\gamma^1\gamma^3 K \psi(t) \tag{34}$$

Since the γ^i's are the same in both the Dirac and the Weyl bases, in either we have from (4)

$$U = \eta(\sigma^1 \otimes i\tau_2)(\sigma^3 \otimes i\tau_2) = \eta i\sigma^2 \otimes 1$$

As we expect, acting on the 2-component spinors contained in ψ, the time reversal operator T involves multiplying by $i\sigma^2$. Note also that as in the nonrelativistic case $T^2\psi = -\psi$.

It may not have escaped your notice that γ^0 appears in the parity operator (18), γ^2 in charge conjugation (30), and $\gamma^1\gamma^3$ in time reversal (34). If we change a Dirac particle to its antiparticle and flip spacetime, γ^5 appears.

CPT theorem

There exists a profound theorem stating that any local Lorentz invariant field theory must be invariant under[6] $\mathcal{CPT}$, the combined action of charge conjugation, parity, and time reversal. The pedestrian proof consists simply of checking that any Lorentz invariant local interaction you can write down [such as (25)], while it may break charge conjugation, parity, or time reversal separately, respects $\mathcal{CPT}$. The more fundamental proof involves considerable formal machinery that I will not develop here. You are urged to read about the phenomenological study of charge conjugation, parity, time reversal, and $\mathcal{CPT}$, surely one of the most fascinating chapters in the history of physics.[7]

---

[6] A rather pedantic point, but potentially confusing to some students, is that I distinguish carefully between the action of charge conjugation $\mathcal{C}$ and the matrix C: Charge conjugation $\mathcal{C}$ involves taking the complex conjugate of ψ and then scrambling the components with $C\gamma^0$. Similarly, I distinguish between the operation of time reversal $\mathcal{T}$ and the matrix T.

[7] See, e.g., J. J. Sakurai, *Invariance Principles and Elementary Particles* and E. D. Commins, *Weak Interactions*.

Two stories

I end this chapter with two of my favorite physics stories—one short and one long.

Paul Dirac was notoriously a man of few words. Dick Feynman told the story that when he first met Dirac at a conference, Dirac said after a long silence, "I have an equation; do you have one too?"

Enrico Fermi did not usually take notes, but during the 1948 Pocono conference (see chapter I.7) he took voluminous notes during Julian Schwinger's lecture. When he got back to Chicago, he assembled a group consisting of two professors, Edward Teller and Gregory Wentzel, and four graduate students, Geoff Chew, Murph Goldberger, Marshall Rosenbluth, and Chen-Ning Yang (all to become major figures later). The group met in Fermi's office several times a week, a couple of hours each time, to try to figure out what Schwinger had done. After 6 weeks, everyone was exhausted. Then someone asked, "Didn't Feynman also speak?" The three professors, who had attended the conference, said yes. But when pressed, not Fermi, nor Teller, nor Wentzel could recall what Feynman had said. All they remembered was his strange notation: p with a funny slash through it.[8]

Exercises

II.1.1 Show that the following bilinears in the spinor field $\bar{\psi}\psi$, $\bar{\psi}\gamma^\mu\psi$, $\bar{\psi}\sigma^{\mu\nu}\psi$, $\bar{\psi}\gamma^\mu\gamma^5\psi$, and $\bar{\psi}\gamma^5\psi$ transform under the Lorentz group and parity as a scalar, a vector, a tensor, a pseudovector or axial vector, and a pseudoscalar, respectively. [Hint: For example, $\bar{\psi}\gamma^\mu\gamma^5\psi \to \bar{\psi}[1+(i/4)\omega\sigma]\gamma^\mu\gamma^5[1-(i/4)\omega\sigma]\psi$ under an infinitesimal Lorentz transformation and $\to \bar{\psi}\gamma^0\gamma^\mu\gamma^5\gamma^0\psi$ under parity. Work out these transformation laws and show that they define an axial vector.]

II.1.2 Write all the bilinears in the preceding exercise in terms of ψ_L and ψ_R.

II.1.3 Solve $(\not{p} - m)\psi(p) = 0$ explicitly (by rotational invariance it suffices to solve it for $\vec{p}$ along the 3rd direction, say). Verify that indeed χ is much smaller than ϕ for a slowly moving electron. What happens for a fast moving electron?

II.1.4 Exploiting the fact that χ is much smaller than ϕ for a slowly moving electron, find the approximate equation satisfied by ϕ.

II.1.5 For a relativistic electron moving along the z-axis, perform a rotation around the z-axis. In other words, study the effect of $e^{-(i/4)\omega\sigma^{12}}$ on $\psi(p)$ and verify the assertion in the text regarding ψ_L and ψ_R.

II.1.6 Solve the massless Dirac equation.

II.1.7 Show explicitly that (25) violates parity.

II.1.8 The defining equation for C evidently fixes C only up to an overall constant. Show that this constant is fixed by requiring $(\psi_c)_c = \psi$.

---

[8] C. N. Yang, Lecture at the Schwinger Memorial Session of the American Physical Society meeting in Washington D. C., 1995.

II.1.9   Show that the charge conjugate of a left handed field is right handed and vice versa.

II.1.10   Show that $\psi C \psi$ is a Lorentz scalar.

II.1.11   Work out the Dirac equation in $(1 + 1)$-dimensional spacetime.

II.1.12   Work out the Dirac equation in $(2 + 1)$-dimensional spacetime. Show that the apparently innocuous mass term violates parity and time reversal. [Hint: The three γ^μ's are just the three Pauli matrices with appropriate factors of i.]

II.2 Quantizing the Dirac Field

Anticommutation

We will use the canonical formalism of chapter I.8 to quantize the Dirac field.

Long and careful study of atomic spectroscopy revealed that the wave function of two electrons had to be antisymmetric upon exchange of their quantum numbers. It follows that we cannot put two electrons into the same energy level so that they will have the same quantum numbers. In 1928 Jordan and Wigner showed how this requirement of an antisymmetric wave function can be formalized by having the creation and annihilation operators for electrons satisfy anticommutation rather than commutation relations as in (I.8.12).

Let us start out with a state with no electron $|0\rangle$ and denote by $b_\alpha^\dagger$ the operator creating an electron with the quantum numbers α. In other words, the state $b_\alpha^\dagger |0\rangle$ is the state with an electron having the quantum numbers α. Now suppose we want to have another electron with the quantum numbers β, so we construct the state $b_\beta^\dagger b_\alpha^\dagger |0\rangle$. For this to be antisymmetric upon interchanging α and β we must have

$$\{b_\alpha^\dagger, b_\beta^\dagger\} \equiv b_\alpha^\dagger b_\beta^\dagger + b_\beta^\dagger b_\alpha^\dagger = 0 \tag{1}$$

Upon hermitean conjugation, we have $\{b_\alpha, b_\beta\} = 0$. In particular, $b_\alpha^\dagger b_\alpha^\dagger = 0$, so that we cannot create two electrons with the same quantum numbers.

To this anticommutation relation we add

$$\{b_\alpha, b_\beta^\dagger\} = \delta_{\alpha\beta} \tag{2}$$

One way of arguing for this is to say that we would like the number operator to be $N = \sum_\alpha b_\alpha^\dagger b_\alpha$, just as in the bosonic case. Show with one line of algebra that $[AB, C] = A[B, C] + [A, C]B$ or $[AB, C] = A\{B, C\} - \{A, C\}B$. (A heuristic way of remembering the minus sign in the anticommuting case is that we have to move C past B in order for C to do its anticommuting with A.) For the desired number operator to work we need $[\sum_\alpha b_\alpha^\dagger b_\alpha, b_\beta^\dagger] = +b_\beta^\dagger$ (so that as usual $N |0\rangle = 0$, and $N b_\beta^\dagger |0\rangle = b_\beta^\dagger |0\rangle$) and so we have (2).

The Dirac field

Let us now turn to the free Dirac Lagrangian

$$\mathcal{L} = \bar{\psi}(i\slashed{\partial} - m)\psi \tag{3}$$

The momentum conjugate to ψ is $\pi_\alpha = \delta\mathcal{L}/\delta\partial_t\psi_\alpha = i\psi_\alpha^\dagger$. We anticipate that the correct canonical procedure requires imposing the anticommutation relation:

$$\{\psi_\alpha(\vec{x}, t), \psi_\beta^\dagger(\vec{0}, t)\} = \delta^{(3)}(\vec{x})\delta_{\alpha\beta} \tag{4}$$

We will derive this below.

The Dirac field satisfies

$$(i\slashed{\partial} - m)\psi = 0 \tag{5}$$

Plugging in plane waves $u(p, s)e^{-ipx}$ and $v(p, s)e^{ipx}$ for ψ, we have

$$(\slashed{p} - m)u(p, s) = 0 \tag{6}$$

and

$$(\slashed{p} + m)v(p, s) = 0 \tag{7}$$

The index $s = \pm 1$ reminds us that each of these two equations has two solutions, spin up and spin down. Evidently, under a Lorentz transformation the two spinors u and v transform in the same way as ψ. Thus, if we define $\bar{u} \equiv u^\dagger\gamma^0$ and $\bar{v} \equiv v^\dagger\gamma^0$, then $\bar{u}u$ and $\bar{v}v$ are Lorentz scalars.

This subject is full of "peculiar" signs and so I will proceed very carefully and show you how every sign makes sense.

First, since (6) and (7) are linear we have to fix the normalization of u and v. Since $\bar{u}(p, s)u(p, s)$ and $\bar{v}(p, s)v(p, s)$ are Lorentz scalars, the normalization condition we impose on them in the rest frame will hold in any frame.

Our strategy is to do things in the rest frame using a particular basis and then invoke Lorentz invariance and basis independence. In the rest frame, (6) and (7) reduce to $(\gamma^0 - 1)u = 0$ and $(\gamma^0 + 1)v = 0$. In particular, in the Dirac basis $\gamma^0 = \begin{pmatrix} I & 0 \\ 0 & -I \end{pmatrix}$, so the two independent spinors u (labeled by spin $s = \pm 1$) have the form

$$\begin{pmatrix} 1 \\ 0 \\ 0 \\ 0 \end{pmatrix} \text{ and } \begin{pmatrix} 0 \\ 1 \\ 0 \\ 0 \end{pmatrix}$$

while the two independent spinors v have the form

$$\begin{pmatrix} 0 \\ 0 \\ 1 \\ 0 \end{pmatrix} \text{ and } \begin{pmatrix} 0 \\ 0 \\ 0 \\ 1 \end{pmatrix}$$

The normalization conditions we have implicitly chosen are then $\bar{u}(p, s)u(p, s) = 1$ and $\bar{v}(p, s)v(p, s) = -1$. Note the minus sign thrust upon us. Clearly, we also have the orthogonality condition $\bar{u}v = 0$ and $\bar{v}u = 0$. Lorentz invariance and basis independence then tell us that these four relations hold in general.

Furthermore, in the rest frame

$$\sum_s u_\alpha(p, s)\bar{u}_\beta(p, s) = \begin{pmatrix} I & 0 \\ 0 & 0 \end{pmatrix}_{\alpha\beta} = \frac{1}{2}(\gamma^0 + 1)_{\alpha\beta}$$

and

$$\sum_s v_\alpha(p, s)\bar{v}_\beta(p, s) = \begin{pmatrix} 0 & 0 \\ 0 & -I \end{pmatrix}_{\alpha\beta} = \frac{1}{2}(\gamma^0 - 1)_{\alpha\beta}$$

Thus, in general

$$\sum_s u_\alpha(p, s)\bar{u}_\beta(p, s) = \left(\frac{\not{p} + m}{2m}\right)_{\alpha\beta} \tag{8}$$

and

$$\sum_s v_\alpha(p, s)\bar{v}_\beta(p, s) = \left(\frac{\not{p} - m}{2m}\right)_{\alpha\beta} \tag{9}$$

Another way of deriving (8) is to note that the left hand side is a 4 by 4 matrix (it is like a column vector multiplied by a row vector on the right) and so must be a linear combination of the sixteen 4 by 4 matrices we listed in chapter II.1. Argue that γ^5 and $\gamma^\mu\gamma^5$ are ruled out by parity and that $\sigma^{\mu\nu}$ is ruled out by Lorentz invariance and the fact that only one Lorentz vector, namely p^μ, is available. Hence the right hand side must be a linear combination of $\not{p}$ and m. Fix the relative coefficient by acting with $\not{p} - m$ from the left. The normalization is fixed by setting $\alpha = \beta$ and summing over α. Similarly for (9). In particular, setting $\alpha = \beta$ and summing over α, we recover $\bar{v}(p, s)v(p, s) = -1$.

We are now ready to promote $\psi(x)$ to an operator. In analogy with (I.8.11) we expand the field in plane waves[1]

$$\psi_\alpha(x) =$$
$$\int \frac{d^3 p}{(2\pi)^{\frac{3}{2}}(E_p/m)^{\frac{1}{2}}} \sum_s [b(p, s)u_\alpha(p, s)e^{-ipx} + d^\dagger(p, s)v_\alpha(p, s)e^{ipx}] \tag{10}$$

(Here $E_p = p_0 = +\sqrt{\vec{p}^2 + m^2}$ and $px = p_\mu x^\mu$.) The normalization factor $(E_p/m)^{\frac{1}{2}}$ is slightly different from that in (I.8.11) for reasons we will see. Otherwise, the rationale

---

[1] The notation is standard. See e.g., J. A. Bjorken and S. D. Drell, *Relativistic Quantum Mechanics*.

for (10) is essentially the same as in (I.8.11). We integrate over momentum $\vec{p}$, sum over spin s, expand in plane waves, and give names to the coefficients in the expansion. Because ψ is complex, we have, similar to the complex scalar field in chapter I.8, a b operator and a $d^\dagger$ operator.

Just as in chapter I.8, the operators b and $d^\dagger$ must carry the same charge. Thus, if b annihilates an electron with charge $e = -|e|$, $d^\dagger$ must remove charge e; that is, it creates a positron with charge $-e = |e|$.

A word on notation: in (10) $b(p, s)$, $d^\dagger(p, s)$, $u(p, s)$, and $v(p, s)$ are written as functions of the 4-momentum p but strictly speaking they are functions of $\vec{p}$ only, with p^0 always understood to be $+\sqrt{\vec{p}^2 + m^2}$.

Thus let $b^\dagger(p, s)$ and $b(p, s)$ be the creation and annihilation operators for an electron of momentum p and spin s. Our introductory discussion indicates that we should impose

$$\{b(p, s), b^\dagger(p', s')\} = \delta^{(3)}(\vec{p} - \vec{p}')\delta_{ss'} \tag{11}$$

$$\{b(p, s), b(p', s')\} = 0 \tag{12}$$

$$\{b^\dagger(p, s), b^\dagger(p', s')\} = 0 \tag{13}$$

There is a corresponding set of relations for $d^\dagger(p, s)$ and $d(p, s)$ the creation and annihilation operators for a positron, for instance,

$$\{d(p, s), d^\dagger(p', s')\} = \delta^{(3)}(\vec{p} - \vec{p}')\delta_{ss'} \tag{14}$$

We now have to show that we indeed obtain (4). Write

$$\bar{\psi}(0) = \int \frac{d^3 p'}{(2\pi)^{\frac{3}{2}}(E_{p'}/m)^{\frac{1}{2}}} \sum_{s'} [b^\dagger(p', s')\bar{u}(p', s') + d(p', s')\bar{v}(p', s')]$$

Nothing to do but to plow ahead:

$$\{\psi(\vec{x}, 0), \bar{\psi}(0)\}$$
$$= \int \frac{d^3 p}{(2\pi)^3 (E_p/m)} \sum_s [u(p, s)\bar{u}(p, s)e^{i\vec{p}\cdot\vec{x}} + v(p, s)\bar{v}(p, s)e^{-i\vec{p}\cdot\vec{x}}]$$

if we take b and $b^\dagger$ to anticommute with d and $d^\dagger$. Using (8) and (9) we obtain

$$\{\psi(\vec{x}, 0), \bar{\psi}(0)\} = \int \frac{d^3 p}{(2\pi)^3 (2E_p)} [(\not{p} + m)e^{i\vec{p}\cdot\vec{x}} + (\not{p} - m)e^{-i\vec{p}\cdot\vec{x}}]$$
$$= \int \frac{d^3 p}{(2\pi)^3 (2E_p)} 2p^0 \gamma^0 e^{-i\vec{p}\cdot\vec{x}} = \gamma^0 \delta^{(3)}(\vec{x})$$

which is just (4) slightly disguised.

Similarly, writing schematically, we have $\{\psi, \psi\} = 0$ and $\{\psi^\dagger, \psi^\dagger\} = 0$.

We are of course free to normalize the spinors u and v however we like. One alternative normalization is to define u and v as the u and v given here multiplied by $(2m)^{\frac{1}{2}}$, thus changing (8) and (9) to $\sum_s u\bar{u} = \not{p} + m$ and $\sum_s v\bar{v} = \not{p} - m$. Multiplying the numerator and denominator in (10) by $(2m)^{\frac{1}{2}}$, we see that the normalization factor $(E_p/m)^{\frac{1}{2}}$ is changed to $(2E_p)^{\frac{1}{2}}$ [thus making it the same as the normalization factor for the scalar field in (I.8.11)].

This alternative normalization (let us call it "any mass normalization") is particularly convenient when we deal with massless spin-$\frac{1}{2}$ particles: we could set $m = 0$ everywhere without ever encountering m in the denominator as in (8) and (9).

The advantage of the normalization used here (let us call it "rest normalization") is that the spinors assume simple forms in the rest frame, as we have just seen. This would prove to be advantageous when we calculate the magnetic moment of the electron in chapter III.6, for example. Of course, multiplying and dividing here and there by $(2m)^{\frac{1}{2}}$ is a trivial operation, and there is not much sense in arguing over the relative advantages of one normalization over another.

In chapter II.6 we will calculate electron scattering at energies high compared to the mass m, so that effectively we could set $m = 0$. Actually, even then, "rest normalization" has the slight advantage of providing a (rather weak) check on the calculation. We set $m = 0$ everywhere we can, such as in the numerator of (8) and (9), but not where we can't, such as in the denominator. Then m must cancel out in physical quantities such as the differential scattering cross section.

Energy of the vacuum

An important exercise at this point is to calculate the Hamiltonian starting with the Hamiltonian density

$$\mathcal{H} = \pi \frac{\partial \psi}{\partial t} - \mathcal{L} = \bar{\psi}(i\vec{\gamma} \cdot \vec{\partial} + m)\psi \tag{15}$$

Inserting (10) into this expression and integrating, we have

$$H = \int d^3x \mathcal{H} = \int d^3x \bar{\psi}(i\vec{\gamma} \cdot \vec{\partial} + m)\psi = \int d^3x \bar{\psi} i\gamma^0 \frac{\partial \psi}{\partial t} \tag{16}$$

which works out to be

$$H = \int d^3p \sum_s E_p[b^\dagger(p, s)b(p, s) - d(p, s)d^\dagger(p, s)] \tag{17}$$

We can see the all important minus sign in (17) schematically: In (16) $\bar{\psi}$ gives a factor $\sim (b^\dagger + d)$, while $\partial/\partial t$ acting on ψ brings down a relative minus sign giving $\sim (b - d^\dagger)$, thus giving us $\sim (b^\dagger + d)(b - d^\dagger) \sim b^\dagger b - dd^\dagger$ (orthogonality between spinors $\bar{v}u = 0$ kills the cross terms).

To bring the second term in (17) into the right order, we anticommute $-d(p, s)d^\dagger(p, s)$ $= d^\dagger(p, s)d(p, s) - \delta^{(3)}(\vec{0})$ so that

$$H = \int d^3p \sum_s E_p[b^\dagger(p, s)b(p, s) + d^\dagger(p, s)d(p, s)]$$

$$- \delta^{(3)}(\vec{0}) \int d^3p \sum_s E_p \tag{18}$$

The first two terms tell us that each electron and each positron of momentum p and spin s has exactly the same energy E_p, as it should. But what about the last term? That $\delta^{(3)}(\vec{0})$ should fill us with fear and loathing.

It is OK: Noting that $\delta^{(3)}(\vec{p}) = [1/(2\pi)^3] \int d^3x \, e^{i\vec{p}\vec{x}}$, we see that $\delta^{(3)}(\vec{0}) = [1/(2\pi)^3] \int d^3x$ (we encounter the same maneuver in exercise I.8.2) and so the last term contributes to H

$$E_0 = -\frac{1}{h^3} \int d^3x \int d^3p \sum_s 2(\tfrac{1}{2}E_p) \tag{19}$$

(since in natural units we have $\hbar = 1$ and hence $h = 2\pi$). We have an energy $-\frac{1}{2}E_p$ in each unit-size phase-space cell $(1/h^3)d^3x \, d^3p$ in the sense of statistical mechanics, for each spin and for the electron and positron separately (hence the factor of 2). This infinite additive term E_0 is precisely the analog of the zero point energy $\frac{1}{2}\hbar\omega$ of the harmonic oscillator you encountered in your quantum mechanics course. But it comes in with a minus sign!

The sign is bizarre and peculiar! Each mode of the Dirac field contributes $-\frac{1}{2}\hbar\omega$ to the vacuum energy. In contrast, each mode of a scalar field contributes $\frac{1}{2}\hbar\omega$ as we saw in chapter I.8. This fact is of crucial importance in the development of supersymmetry, which we will discuss in chapter VIII.4.

Fermion propagating through spacetime

In analogy with (I.8.14), the propagator for the electron is given by $iS_{\alpha\beta}(x) \equiv \langle 0| T \psi_\alpha(x)\bar{\psi}_\beta(0) |0\rangle$, where the argument of $\bar{\psi}$ has been set to 0 by translation invariance. As we will see, the anticommuting character of ψ requires us to define the time-ordered product with a minus sign, namely

$$T\psi(x)\bar{\psi}(0) \equiv \theta(x^0)\psi(x)\bar{\psi}(0) - \theta(-x^0)\bar{\psi}(0)\psi(x) \tag{20}$$

Referring to (10), we obtain for $x^0 > 0$,

$$iS(x) = \langle 0| \psi(x)\bar{\psi}(0) |0\rangle = \int \frac{d^3p}{(2\pi)^3(E_p/m)} \sum_s u(p,s)\bar{u}(p,s)e^{-ipx}$$

$$= \int \frac{d^3p}{(2\pi)^3(E_p/m)} \frac{\not{p}+m}{2m} e^{-ipx}$$

For $x^0 < 0$, we have to be a bit careful about the spinorial indices:

$$iS_{\alpha\beta}(x) = -\langle 0| \bar{\psi}_\beta(0)\psi_\alpha(x) |0\rangle$$

$$= -\int \frac{d^3p}{(2\pi)^3(E_p/m)} \sum_s \bar{v}_\beta(p,s)v_\alpha(p,s)e^{-ipx}$$

$$= -\int \frac{d^3p}{(2\pi)^3(E_p/m)} (\frac{\not{p}-m}{2m})_{\alpha\beta} e^{-ipx}$$

using the identity (9).

Putting things together we obtain

$$iS(x) = \int \frac{d^3p}{(2\pi)^3(E_p/m)} \left[\theta(x^0)\frac{\not{p}+m}{2m}e^{-ipx} - \theta(-x^0)\frac{\not{p}-m}{2m}e^{+ipx} \right] \tag{21}$$

We will now show that this fermion propagator can be written more elegantly as a 4-dimensional integral:

$$iS(x) = i \int \frac{d^4 p}{(2\pi)^4} e^{-ip\cdot x} \frac{\not{p}+m}{p^2-m^2+i\varepsilon} = \int \frac{d^4 p}{(2\pi)^4} e^{-ip\cdot x} \frac{i}{\not{p}-m+i\varepsilon} \tag{22}$$

To show that (22) is indeed equivalent to (21) we go through essentially the same steps as after (I.8.14). In the complex p^0 plane the integrand has poles at $p^0 = \pm\sqrt{\vec{p}^2+m^2-i\varepsilon} \simeq \pm(E_p - i\varepsilon)$. For $x^0 > 0$ the factor $e^{-ip^0 x^0}$ tells us to close the contour in the lower half-plane. We go around the pole at $+(E_p - i\varepsilon)$ clockwise and obtain

$$iS(x) = (-i)i \int \frac{d^3 p}{(2\pi)^3} e^{-ip\cdot x} \frac{\not{p}+m}{2E_p}$$

producing the first term in (21). For $x^0 < 0$ we are now told to close the contour in the upper half-plane and thus we go around the pole at $-(E_p - i\varepsilon)$ anticlockwise. We obtain

$$iS(x) = i^2 \int \frac{d^3 p}{(2\pi)^3} e^{+iE_p x^0 + i\vec{p}\cdot\vec{x}} \frac{1}{-2E_p} (-E_p \gamma^0 - \vec{p}\vec{\gamma} + m)$$

and flipping $\vec{p}$ we have

$$iS(x) = -\int \frac{d^3 p}{(2\pi)^3} e^{ip\cdot x} \frac{1}{2E_p} (E_p \gamma^0 - \vec{p}\vec{\gamma} - m) = -\int \frac{d^3 p}{(2\pi)^3} e^{ip\cdot x} \frac{1}{2E_p} (\not{p} - m)$$

precisely the second term in (21) with the minus sign and all. Thus, we must define the time-ordered product with the minus sign as in (20).

After all these steps, we see that in momentum space the fermion propagator has the elegant form

$$iS(p) = \frac{i}{\not{p}-m+i\varepsilon} \tag{23}$$

This makes perfect sense: $S(p)$ comes out to be the inverse of the Dirac operator $\not{p} - m$, just as the scalar boson propagator $D(k) = 1/(k^2-m^2+i\varepsilon)$ is the inverse of the Klein-Gordon operator $k^2 - m^2$.

Poetic but confusing metaphors

In closing this chapter let me ask you some rhetorical questions. Did I speak of an electron going backward in time? Did I mumble something about a sea of negative energy electrons? This metaphorical language, when used by brilliant minds, the likes of Dirac and Feynman, was evocative and inspirational, but unfortunately confused generations of physics students and physicists. The presentation given here is in the modern spirit, which seeks to avoid these potentially confusing metaphors.

Exercises

II.2.1   Use Noether's theorem to derive the conserved current $J^\mu = \bar{\psi}\gamma^\mu\psi$. Calculate $[Q, \psi]$, thus showing that b and $d^\dagger$ must carry the same charge.

II.2.2   Quantize the Dirac field in a box of volume of V and show that the vacuum energy E_0 is indeed proportional to V. [Hint: The integral over momentum $\int d^3 p$ is replaced by a sum over discrete values of the momentum.]

The Lorentz algebra

In chapter II.1 we followed Dirac's brilliantly idiosyncratic way of deriving his equation. We develop here a more logical and mathematical theory of the Dirac spinor. A deeper understanding of the Dirac spinor not only gives us a certain satisfaction, but is also indispensable, as we will see later, in studying supersymmetry, one of the foundational concepts of superstring theory; and of course, most of the fundamental particles such as the electron and the quarks carry spin $\frac{1}{2}$ and are described by spinor fields.

Let us begin by reminding ourselves how the rotation group works. The three generators J_i ($i = 1, 2, 3$ or x, y, z) of the rotation group satisfy the commutation relation

$$[J_i, J_j] = i\epsilon_{ijk}J_k \tag{1}$$

When acting on the spacetime coordinates, written as a column vector

$$\begin{pmatrix} x^0 \\ x^1 \\ x^2 \\ x^3 \end{pmatrix}$$

the generators of rotations are represented by the hermitean matrices

$$J_1 = \begin{pmatrix} 0 & 0 & 0 & 0 \\ 0 & 0 & 0 & 0 \\ 0 & 0 & 0 & -i \\ 0 & 0 & i & 0 \end{pmatrix} \tag{2}$$

with J_2 and J_3 obtained by cyclic permutations. You should verify by laboriously multiplying these three matrices that (1) is satisfied. Note that the signs of J_i are fixed by the commutation relation (1).

Now add the Lorentz boosts. A boost in the $x \equiv x^1$ direction transforms the spacetime coordinates:

$$t' = (\cosh \varphi) t + (\sinh \varphi) x; \qquad x' = (\sinh \varphi) t + (\cosh \varphi) x \qquad (3)$$

or for infinitesimal φ

$$t' = t + \varphi x; \qquad x' = x + \varphi t \qquad (4)$$

In other words, the infinitesimal generator of a Lorentz boost in the x direction is represented by the hermitean matrix ($x^0 \equiv t$ as usual)

$$i K_1 = \begin{pmatrix} 0 & 1 & 0 & 0 \\ 1 & 0 & 0 & 0 \\ 0 & 0 & 0 & 0 \\ 0 & 0 & 0 & 0 \end{pmatrix} \qquad (5)$$

Similarly,

$$i K_2 = \begin{pmatrix} 0 & 0 & 1 & 0 \\ 0 & 0 & 0 & 0 \\ 1 & 0 & 0 & 0 \\ 0 & 0 & 0 & 0 \end{pmatrix} \qquad (6)$$

I leave it to you to write down K_3. Note that K_i is defined to be antihermitean.

Check that $[J_i, K_j] = i\epsilon_{ijk}K_k$. To see that this implies that the boost generators K_i transform as a 3-vector $\vec{K}$ under rotation, as you would expect, apply a rotation through an infinitesimal angle θ around the 3-axis. Then (you might wish to review the material in appendix B at this point) $K_1 \to e^{i\theta J_3}K_1 e^{-i\theta J_3} = K_1 + i\theta[J_3, K_1] + O(\theta^2) = K_1 + i\theta(i K_2) + O(\theta^2) = \cos\theta\, K_1 - \sin\theta\, K_2$ to the order indicated.

You are now about to do one of the most significant calculations in the history of twentieth century physics. By brute force compute $[K_1, K_2]$, evidently an antisymmetric matrix. You will discover that it is equal to $-i J_3$. Two Lorentz boosts produce a rotation! (You might recall from your course on electromagnetism that this mathematical fact is responsible for the physics of the Thomas precession.)

Mathematically, the generators of the Lorentz group satisfy the following algebra [known to the cognoscenti as $SO(3, 1)$]:

$$[J_i, J_j] = i\epsilon_{ijk}J_k \qquad (7)$$

$$[J_i, K_j] = i\epsilon_{ijk}K_k \qquad (8)$$

$$[K_i, K_j] = -i\epsilon_{ijk}J_k \qquad (9)$$

Note the all-important minus sign!

How do we study this algebra? The crucial observation is that the algebra falls apart into two pieces if we form the combinations $J_{\pm i} \equiv \frac{1}{2}(J_i \pm i K_i)$. You should check that

$$[J_{+i}, J_{+j}] = i\epsilon_{ijk}J_{+k} \tag{10}$$

$$[J_{-i}, J_{-j}] = i\epsilon_{ijk}J_{-k} \tag{11}$$

and most remarkably

$$[J_{+i}, J_{-j}] = 0 \tag{12}$$

This last commutation relation tells us that J_+ and J_- form two separate $SU(2)$ algebras. (For more, see appendix B.)

From algebra to representation

This means that you can simply use what you have already learned about angular momentum in elementary quantum mechanics and the representation of $SU(2)$ to determine all the representations of $SO(3, 1)$. As you know, the representations of $SU(2)$ are labeled by $j = 0, \frac{1}{2}, 1, \frac{3}{2}, \cdots$. We can think of each representation as consisting of $(2j + 1)$ objects ψ_m with $m = -j, -j + 1, \cdots, j - 1, j$ that transform into each other under $SU(2)$. It follows immediately that the representations of $SO(3, 1)$ are labeled by (j^+, j^-) with j^+ and j^- each taking on the values $0, \frac{1}{2}, 1, \frac{3}{2}, \ldots$. Each representation consists of $(2j^+ + 1)(2j^- + 1)$ objects $\psi_{m^+ m^-}$ with $m^+ = -j^+, -j^+ + 1, \ldots, j^+ - 1, j^+$ and $m^- = -j^-, -j^- + 1, \ldots, j^- - 1, j^-$.

Thus, the representations of $SO(3, 1)$ are $(0, 0)$, $(\frac{1}{2}, 0)$, $(0, \frac{1}{2})$, $(1, 0)$, $(0, 1)$, $(\frac{1}{2}, \frac{1}{2})$, and so on, in order of increasing dimension. We recognize the 1-dimensional representation $(0, 0)$ as clearly the trivial one, the Lorentz scalar. By counting dimensions, we expect that the 4-dimensional representation $(\frac{1}{2}, \frac{1}{2})$ has to be the Lorentz vector, the defining representation of the Lorentz group (see exercise II.3.1).

Spinor representations

What about the representation $(\frac{1}{2}, 0)$? Let us write the two objects as ψ_α with $\alpha = 1, 2$. Well, what does the notation $(\frac{1}{2}, 0)$ mean? It says that $J_{+i} = \frac{1}{2}(J_i + i K_i)$ acting on ψ_α is represented by $\frac{1}{2}\sigma_i$ while $J_{-i} = \frac{1}{2}(J_i - i K_i)$ acting on ψ_α is represented by 0. By adding and subtracting we find that

$$J_i = \frac{1}{2}\sigma_i \tag{13}$$

and

$$i K_i = \frac{1}{2}\sigma_i \tag{14}$$

where the equal sign means "represented by" in this context. (By convention we do not distinguish between upper and lower indices on the 3-dimensional quantities J_i, K_i, and σ_i.) Note again that K_i is anti-hermitean.

Similarly, let us denote the two objects in $(0, \frac{1}{2})$ by the peculiar symbol $\bar{\chi}^{\dot{\alpha}}$. I should emphasize the trivial but potentially confusing point that unlike the bar used in chapter II.1, the bar on $\bar{\chi}^{\dot{\alpha}}$ is a typographical element: Think of the symbol $\bar{\chi}$ as a letter in the Hittite alphabet if you like. Similarly, the symbol $\dot{\alpha}$ bears no relation to α: we do not obtain $\dot{\alpha}$ by operating on α in any way. The rather strange notation is known informally as "dotted and undotted" and more formally as the van der Waerden notation—a bit excessive for our rather modest purposes at this point but I introduce it because it is the notation used in supersymmetric physics and superstring theory. (Incidentally, Dirac allegedly said that he wished he had invented the dotted and undotted notation.) Repeating the same steps as above, you will find that on the representation $(0, \frac{1}{2})$ we have $J_i = \frac{1}{2}\sigma_i$ and $iK_i = -\frac{1}{2}\sigma_i$. The minus sign is crucial.

The 2-component spinors ψ_α and $\bar{\chi}^{\dot{\alpha}}$ are called Weyl spinors and furnish perfectly good representations of the Lorentz group. Why then does the Dirac spinor have 4 components?

The reason is parity. Under parity, $\vec{x} \to -\vec{x}$ and $\vec{p} \to -\vec{p}$, and thus $\vec{J} \to \vec{J}$ and $\vec{K} \to -\vec{K}$, and so $\vec{J}_+ \leftrightarrow \vec{J}_-$. In other words, under parity the representations $(\frac{1}{2}, 0) \leftrightarrow (0, \frac{1}{2})$. Therefore, to describe the electron we must use both of these 2-dimensional representations, or in mathematical notation, the 4-dimensional reducible representation $(\frac{1}{2}, 0) \oplus (0, \frac{1}{2})$.

We thus stack two Weyl spinors together to form a Dirac spinor

$$\Psi = \begin{pmatrix} \psi_\alpha \\ \bar{\chi}^{\dot{\alpha}} \end{pmatrix} \tag{15}$$

The spinor $\Psi(p)$ is of course a function of 4-momentum p [and by implication also $\psi_\alpha(p)$ and $\bar{\chi}^{\dot{\alpha}}(p)$] but we will suppress the p dependence for the time being. Referring to (13) and (14) we see that acting on Ψ the generators of rotation

$$\vec{J} = \begin{pmatrix} \frac{1}{2}\vec{\sigma} & 0 \\ 0 & \frac{1}{2}\vec{\sigma} \end{pmatrix}$$

where once again the equality means "represented by," and the generators of boost

$$i\vec{K} = \begin{pmatrix} \frac{1}{2}\vec{\sigma} & 0 \\ 0 & -\frac{1}{2}\vec{\sigma} \end{pmatrix}$$

Note once again the all-important minus sign.

Parity forces us to have a 4-component spinor but we know on the other hand that the electron has only two physical degrees of freedom. Let us go to the rest frame. We must project out two of the components contained in $\Psi(p_r)$ with the rest momentum $p_r \equiv (m, \vec{0})$. With the benefit of hindsight, we write the projection operator as $\mathcal{P} = \frac{1}{2}(1 - \gamma^0)$. You are probably guessing from the notation that γ^0 will turn out to be one of the gamma matrices, but at this point, logically γ^0 is just some 4 by 4 matrix. The condition $\mathcal{P}^2 = \mathcal{P}$ implies that $(\gamma^0)^2 = 1$ so that the eigenvalues of γ^0 are ± 1. Since $\psi_\alpha \leftrightarrow \bar{\chi}^{\dot{\alpha}}$ under parity we naturally guess that ψ_α and $\bar{\chi}^{\dot{\alpha}}$ correspond to the left and right handed fields of chapter II.1.

We cannot simply use the projection to set for example $\bar{\chi}^{\dot{\alpha}}$ to 0. Parity means that we should treat ψ_α and $\bar{\chi}^{\dot{\alpha}}$ on the same footing. We choose

$$\gamma^0 = \begin{pmatrix} 0 & 1 \\ 1 & 0 \end{pmatrix},$$

or more explicitly

$$\begin{pmatrix} 0 & 0 & 1 & 0 \\ 0 & 0 & 0 & 1 \\ 1 & 0 & 0 & 0 \\ 0 & 1 & 0 & 0 \end{pmatrix}$$

(Different choices of γ^0 correspond to the different basis choices discussed in chapter II.1.) In other words, in the rest frame $\psi_\alpha - \bar{\chi}^{\dot{\alpha}} = 0$. The projection to two degrees of freedom can be written as

$$(\gamma^0 - 1)\Psi(p_r) = 0 \tag{16}$$

Indeed, we recognize this as just the Weyl basis introduced in chapter II.1.

The Dirac equation

We have derived the Dirac equation, a bit in disguise!

Since our derivation is based on a step-by-step study of the spinor representation of the Lorentz group, we know how to obtain the equation satisfied by $\Psi(p)$ for any p: We simply boost. Writing $\Psi(p) = e^{-i\vec{\varphi}\vec{K}}\Psi(p_r)$, we have $(e^{-i\vec{\varphi}\vec{K}}\gamma^0 e^{i\vec{\varphi}\vec{K}} - 1)\Psi(p) = 0$. Introducing the notation $\gamma^\mu p_\mu / m \equiv e^{-i\vec{\varphi}\vec{K}}\gamma^0 e^{i\vec{\varphi}\vec{K}}$, we obtain the Dirac equation

$$(\gamma^\mu p_\mu - m)\Psi(p) = 0 \tag{17}$$

You can work out the details as an exercise.

The derivation here represents the deep group theoretic way of looking at the Dirac equation: It is a projection boosted into an arbitrary frame.

Note that this is an example of the power of symmetry, which pervades modern physics and this book: Our knowledge of how the electron field transforms under the rotation group, namely that it has spin $\frac{1}{2}$, allows us to know how it transforms under the Lorentz group. Symmetry rules!

In appendix E we will develop the dotted and undotted notation further for later use in the chapter on supersymmetry.

In light of your deeper group theoretic understanding it is a good idea to reread chapter II.1 and compare it with this chapter.

Exercises

II.3.1 Show by explicit computation that $(\frac{1}{2}, \frac{1}{2})$ is indeed the Lorentz vector.

II.3.2 Work out how the six objects contained in the $(1, 0)$ and $(0, 1)$ transform under the Lorentz group. Recall from your course on electromagnetism how the electric and magnetic fields $\vec{E}$ and $\vec{B}$ transform. Conclude that the electromagnetic field in fact transforms as $(1, 0) \oplus (0, 1)$. Show that it is parity that once again forces us to use a reducible representation.

II.3.3 Show that

$$e^{-i\vec{\varphi}\vec{K}}\gamma^0 e^{i\vec{\varphi}\vec{K}} = \begin{pmatrix} 0 & e^{-\vec{\varphi}\vec{\sigma}} \\ e^{\vec{\varphi}\vec{\sigma}} & 0 \end{pmatrix}$$

and

$$e^{\vec{\varphi}\vec{\sigma}} = \cosh\varphi + \vec{\sigma} \cdot \hat{\varphi}\sinh\varphi$$

with the unit vector $\hat{\varphi} \equiv \vec{\varphi}/\varphi$. Identifying $\vec{p} = m\hat{\varphi}\sinh\varphi$, derive the Dirac equation. Show that

$$\gamma^i = \begin{pmatrix} 0 & \sigma_i \\ -\sigma_i & 0 \end{pmatrix}$$

II.3.4 Show that a spin $\frac{3}{2}$ particle can be described by a vector-spinor $\Psi_{\alpha\mu}$, namely a Dirac spinor carrying a Lorentz index. Find the corresponding equations of motion, known as the Rarita-Schwinger equations. [Hint: The object $\Psi_{\alpha\mu}$ has 16 components, which we need to cut down to $2 \cdot \frac{3}{2} + 1 = 4$ components.]

II.4 | Spin-Statistics Connection

> There is no one fact in the physical world which has
> a greater impact on the way things are, than the Pauli
> exclusion principle.[1]

Degrees of intellectual incompleteness

In a course on nonrelativistic quantum mechanics you learned about the Pauli exclusion principle[2] and its later generalization stating that particles with half integer spins, such as electrons, obey Fermi-Dirac statistics and want to stay apart, while in contrast particles with integer spins, such as photons or pairs of electrons, obey Bose-Einstein statistics and love to stick together. From the microscopic structure of atoms to the macroscopic structure of neutron stars, a dazzling wealth of physical phenomena would be incomprehensible without this spin-statistics rule. Many elements of condensed matter physics, for instance, band structure, Fermi liquid theory, superfluidity, superconductivity, quantum Hall effect, and so on and so forth, are consequences of this rule.

Quantum statistics, one of the most subtle concepts in physics, rests on the fact that in the quantum world, all elementary particles and hence all atoms, are absolutely identical to, and thus indistinguishable from, one other.[3] It should be recognized as a triumph of quantum field theory that it is able to explain absolute identity and indistinguishability easily and naturally. Every electron in the universe is an excitation in one and the same electron field ψ. Otherwise, one might be able to imagine that the electrons we now

---

[1] I. Duck and E. C. G. Sudarshan, *Pauli and the Spin-Statistics Theorem*, p. 21.

[2] While a student in Cambridge, E. C. Stoner came to within a hair of stating the exclusion principle. Pauli himself in his famous paper (*Zeit. f. Physik* 31: 765, 1925) only claimed to "summarize and generalize Stoner's idea." However, later in his Nobel Prize lecture Pauli was characteristically ungenerous toward Stoner's contribution. A detailed and fascinating history of the spin and statistics connection may be found in Duck and Sudarshan, op. cit.

[3] Early in life, I read in one of George Gamow's popular physics books that he could not explain quantum statistics—all he could manage for Fermi statistics was an analogy, invoking Greta Garbo's famous remark "I vont to be alone."—and that one would have to go to school to learn about it. Perhaps this spurs me, later in life, to write popular physics books also. See A. Zee, *Einstein's Universe*, p. x.

know came off an assembly line somewhere in the early universe and could all be slightly different owing to some negligence in the manufacturing process.

While the spin-statistics rule has such a profound impact in quantum mechanics, its explanation had to wait for the development of relativistic quantum field theory. Imagine a civilization that for some reason developed quantum mechanics but has yet to discover special relativity. Physicists in this civilization eventually realize that they have to invent some rule to account for the phenomena mentioned above, none of which involves motion fast compared to the speed of light. Physics would have been intellectually unsatisfying and incomplete.

One interesting criterion in comparing different areas of physics is their degree of intellectual incompleteness.

Certainly, in physics we often accept a rule that cannot be explained until we move to the next level. For instance, in much of physics, we take as a given the fact that the charge of the proton and the charge of electron are exactly equal and opposite. Quantum electrodynamics by itself is not capable of explaining this striking fact either. This fact, charge quantization, can only be deduced by embedding quantum electrodynamics into a larger structure, such as a grand unified theory, as we will see in chapter VII.6. (In chapter IV.4 we will learn that the existence of magnetic monopoles implies charge quantization, but monopoles do not exist in pure quantum electrodynamics.)

Thus, the explanation of the spin-statistics connection, by Fierz and by Pauli in the late 1930s, and by Lüders and Zumino and by Burgoyne in the late 1950s, ranks as one of the great triumphs of relativistic quantum field theory. I do not have the space to give a general and rigorous proof[4] here. I will merely sketch what goes terribly wrong if we violate the spin-statistics connection.

The price of perversity

A basic quantum principle states that if two observables commute then they are simultaneously diagonalizable and hence observable. A basic relativistic principle states that if two spacetime points are spacelike with respect to each other then no signal can propagate between them, and hence the measurement of an observable at one of the points cannot influence the measurement of another observable at the other point.

Consider the charge density $J_0 = i(\varphi^\dagger \partial_0 \varphi - \partial_0 \varphi^\dagger \varphi)$ in a charged scalar field theory. According to the two fundamental principles just enunciated, $J_0(\vec{x}, t = 0)$ and $J_0(\vec{y}, t = 0)$ should commute for $\vec{x} \neq \vec{y}$. In calculating the commutator of $J_0(\vec{x}, t = 0)$ with $J_0(\vec{y}, t = 0)$, we simply use the fact that $\varphi(\vec{x}, t = 0)$ and $\partial_0 \varphi(\vec{x}, t = 0)$ commute with $\varphi(\vec{y}, t = 0)$ and $\partial_0 \varphi(\vec{y}, t = 0)$, so we just move the field at $\vec{x}$ steadily past the field at $\vec{y}$. The commutator vanishes almost trivially.

---

[4] See I. Duck and E. C. G. Sudarshan, *Pauli and the Spin-Statistics Theorem*, and R. F. Streater and A. S. Wightman, *PCT, Spin Statistics, and All That*.

Now suppose we are perverse and quantize the creation and annihilation operators in the expansion (I.8.11)

$$\varphi(\vec{x}, t = 0) = \int \frac{d^D k}{\sqrt{(2\pi)^D 2\omega_k}} [a(\vec{k}) e^{i\vec{k}\cdot\vec{x}} + a^\dagger(\vec{k}) e^{-i\vec{k}\cdot\vec{x}}] \tag{1}$$

according to anticommutation rules

$$\{a(\vec{k}), a^\dagger(\vec{q})\} = \delta^{(D)}(\vec{k} - \vec{q})$$

and

$$\{a(\vec{k}), a(\vec{q})\} = 0 = \{a^\dagger(\vec{k}), a^\dagger(\vec{q})\}$$

instead of the correct commutation rules.

What is the price of perversity?

Now when we try to move $J_0(\vec{y}, t = 0)$ past $J_0(\vec{x}, t = 0)$, we have to move the field at $\vec{y}$ past the field at $\vec{x}$ using the anticommutator

$$\{\varphi(\vec{x}, t = 0), \varphi(\vec{y}, t = 0)\}$$
$$= \iint \frac{d^D k}{\sqrt{(2\pi)^D 2\omega_k}} \frac{d^D q}{\sqrt{(2\pi)^D 2\omega_q}} \{[a(\vec{k}) e^{i\vec{k}\cdot\vec{x}} + a^\dagger(\vec{k}) e^{-i\vec{k}\cdot\vec{x}}], [a(\vec{q}) e^{i\vec{q}\cdot\vec{y}} + a^\dagger(\vec{q}) e^{-i\vec{q}\cdot\vec{y}}]\}$$
$$= \int \frac{d^D k}{(2\pi)^D 2\omega_k} (e^{i\vec{k}\cdot(\vec{x}-\vec{y})} + e^{-i\vec{k}\cdot(\vec{x}-\vec{y})}) \tag{2}$$

You see the problem? In a normal scalar-field theory that obeys the spin-statistics connection, we would have computed the commutator, and then in the last expression in (2) we would have gotten $(e^{i\vec{k}\cdot(\vec{x}-\vec{y})} - e^{-i\vec{k}\cdot(\vec{x}-\vec{y})})$ instead of $(e^{i\vec{k}\cdot(\vec{x}-\vec{y})} + e^{-i\vec{k}\cdot(\vec{x}-\vec{y})})$. The integral

$$\int \frac{d^D k}{(2\pi)^D 2\omega_k} (e^{i\vec{k}\cdot(\vec{x}-\vec{y})} - e^{-i\vec{k}\cdot(\vec{x}-\vec{y})})$$

would obviously vanish and all would be well. With the plus sign, we get in (2) a nonvanishing piece of junk. A disaster if we quantize the scalar field as anticommuting! A spin 0 field has to be commuting. Thus, relativity and quantum physics join hands to force the spin-statistics connection.

It is sometimes said that because of electromagnetism you do not sink through the floor and because of gravity you do not float to the ceiling, and you would be sinking or floating in total darkness were it not for the weak interaction, which regulates stellar burning. Without the spin-statistics connection, electrons would not obey Pauli exclusion. Matter would just collapse.[5]

Exercise

II.4.1 Show that we would also get into trouble if we quantize the Dirac field with commutation instead of anticommutation rules. Calculate the commutator $[J^0(\vec{x}, 0), J^0(0)]$.

---

[5] The proof of the stability of matter, given by Dyson and Lenard, depends crucially on Pauli exclusion.

Vacuum Energy, Grassmann Integrals, and Feynman Diagrams for Fermions

The vacuum is a boiling sea of nothingness, full of sound
and fury, signifying a great deal.

—Anonymous

Fermions are weird

I developed the quantum field theory of a scalar field $\varphi(x)$ first in the path integral formalism and then in the canonical formalism. In contrast, I have thus far developed the quantum field theory of the free spin $\frac{1}{2}$ field $\psi(x)$ only in the canonical formalism. We learned that the spin-statistics connection forces the field operator $\psi(x)$ to satisfy anticommutation relations. This immediately suggests something of a mystery in writing down the path integral for the spinor field ψ. In the path integral formalism $\psi(x)$ is not an operator but merely an integration variable. How do we express the fact that its operator counterpart in the canonical formalism anticommutes?

We presumably cannot represent ψ as a commuting variable, as we did φ. Indeed, we will discover that in the path integral formalism ψ is to be treated not as an ordinary complex number but as a novel kind of mathematical entity known as a Grassmann number.

If you thought about it, you would realize that some novel mathematical structure is needed. In chapter I.3 we promoted the coordinates of point particles $q_i(t)$ in quantum mechanics to the notion of a scalar field $\varphi(\vec{x}, t)$. But you already know from quantum mechanics that a spin $\frac{1}{2}$ particle has the peculiar property that its wave function turns into minus itself when rotated through 2π. Unlike particle coordinates, half integral spin is not an intuitive concept.

Vacuum energy

To motivate the introduction of Grassmann-valued fields I will discuss the notion of vacuum energy. The reason for this apparently strange strategy will become clear shortly.

Quantum field theory was first developed to describe the scattering of photons and electrons, and later the scattering of particles. Recall that in chapter I.7 while studying the scattering of particles we encountered diagrams describing vacuum fluctuations, which we simply neglected (see fig. I.7.12). Quite naturally, particle physicists considered these fluctuations to be of no importance. Experimentally, we scatter particles off each other. Who cares about fluctuations in the vacuum somewhere else? It was only in the early 1970s that physicists fully appreciated the importance of vacuum fluctuations. We will come back to the importance of the vacuum[1] in a later chapter.

In chapters I.8 and II.2 we calculated the vacuum energy of a free scalar field and of a free spinor field using the canonical formalism. To motivate the use of Grassmann numbers to formulate the path integral for the spinor field I will adopt the following strategy. First, I use the path integral formalism to obtain the result we already have for the free scalar field using the canonical formalism. Then we will see that in order to produce the result we already have for the free spinor field we must modify the path integral.

By definition, vacuum fluctuations occur even when there are no sources to produce particles. Thus, let us consider the generating functional of a free scalar field theory in the absence of sources:[2]

$$ Z = \int D\varphi e^{i \int d^4x \frac{1}{2}[(\partial\varphi)^2 - m^2\varphi^2]} = C \left(\frac{1}{\det[\partial^2 + m^2]} \right)^{\frac{1}{2}} = Ce^{-\frac{1}{2}\text{Tr} \log(\partial^2 + m^2)} \tag{1} $$

For the first equality we used (I.2.15) and absorbed inessential factors into the constant C. In the second equality we used the important identity

$$ \det M = e^{\text{Tr} \log M} \tag{2} $$

which you encountered in exercise I.11.2.

Recall that $Z = \langle 0| e^{-iHT} |0 \rangle$ (with $T \to \infty$ understood so that we integrate over all of spacetime in (1)), which in this case is just e^{-iET} with E the energy of the vacuum. Evaluating the trace in (1)

$$ \text{Tr}\mathcal{O} = \int d^4x \langle x| \mathcal{O} |x \rangle $$

$$ = \int d^4x \int \frac{d^4k}{(2\pi)^4} \int \frac{d^4q}{(2\pi)^4} \langle x|k \rangle \langle k| \mathcal{O} |q \rangle \langle q|x \rangle $$

we obtain

$$ iET = \frac{1}{2} VT \int \frac{d^4k}{(2\pi)^4} \log(k^2 - m^2 + i\varepsilon) + A $$

where A is an infinite constant corresponding to the multiplicative factor C in (1). Recall that in the derivation of the path integral we had lots of divergent multiplicative factors; this is where they can come in. The presence of A is a good thing here since it solves a problem you might have noticed: The argument of the log is not dimensionless. Let us define m' by writing

---

[1] Indeed, we have already discussed one way to observe the effects of vacuum fluctuations in chapter I.9.
[2] Strictly speaking, to render the expressions here well defined we should replace m^2 by $m^2 - i\varepsilon$ as discussed earlier.

$$A = -\frac{1}{2}VT \int \frac{d^4k}{(2\pi)^4} \log(k^2 - m'^2 + i\varepsilon)$$

In other words, we do not calculate the vacuum energy as such, but only the difference between it and the vacuum energy we would have had if the particle had mass m' instead of m. The arbitrarily long time T cancels out and E is proportional to the volume of space V, as might be expected. Thus, the (difference in) vacuum energy density is

$$\frac{E}{V} = -\frac{i}{2} \int \frac{d^4k}{(2\pi)^4} \log\left[\frac{k^2 - m^2 + i\varepsilon}{k^2 - m'^2 + i\varepsilon}\right] \tag{3}$$

$$= -\frac{i}{2} \int \frac{d^3k}{(2\pi)^3} \int \frac{d\omega}{2\pi} \log\left[\frac{\omega^2 - \omega_k^2 + i\varepsilon}{\omega^2 - \omega_k'^2 + i\varepsilon}\right]$$

where $\omega_k' \equiv +\sqrt{\vec{k}^2 + m'^2}$. We treat the (convergent) integral over ω by integrating by parts:

$$\int \frac{d\omega}{2\pi} \frac{d\omega}{d\omega} \log\left[\frac{\omega^2 - \omega_k^2 + i\varepsilon}{\omega^2 - \omega_k'^2 + i\varepsilon}\right] = -2 \int \frac{d\omega}{2\pi} \omega \left[\frac{\omega}{\omega^2 - \omega_k^2 + i\varepsilon} - (\omega_k \rightarrow \omega_k')\right]$$

$$= -i2\omega_k^2 \left(\frac{1}{-2\omega_k}\right) - (\omega_k \rightarrow \omega_k')$$

$$= +i(\omega_k - \omega_k') \tag{4}$$

Indeed, restoring $\hbar$ we get the result we want:

$$\frac{E}{V} = \int \frac{d^3k}{(2\pi)^3} \left(\frac{1}{2}\hbar\omega_k - \frac{1}{2}\hbar\omega_k'\right) \tag{5}$$

We had to go through a few arithmetical steps to obtain this result, but the important point is that using the path integral formalism we have managed to obtain a result previously obtained using the canonical formalism.

A peculiar sign for fermions

Our goal is to figure out the path integral for the spinor field. Recall from chapter II.2 that the vacuum energy of the spinor field comes out to have the opposite sign to the vacuum energy of the scalar field, a sign that surely ranks among the "top ten" signs of theoretical physics. How are we to get it using the path integral?

As explained in chapter I.3, the origin of (1) lies in the simple Gaussian integration formula

$$\int_{-\infty}^{+\infty} dx e^{-\frac{1}{2}ax^2} = \sqrt{\frac{2\pi}{a}} = \sqrt{2\pi} e^{-\frac{1}{2}\log a}$$

Roughly speaking, we have to find a new type of integral so that the analog of the Gaussian integral would go something like $e^{+\frac{1}{2}\log a}$.

Grassmann math

It turns out that the mathematics we need was invented long ago by Grassmann. Let us postulate a new kind of number, called the Grassmann or anticommuting number, such that if η and ξ are Grassmann numbers, then $\eta\xi = -\xi\eta$. In particular, $\eta^2 = 0$. Heuristically, this mirrors the anticommutation relation satisfied by the spinor field. Grassmann assumed that any function of η can be expanded in a Taylor series. Since $\eta^2 = 0$, the most general function of η is $f(\eta) = a + b\eta$, with a and b two ordinary numbers.

How do we define integration over η? Grassmann noted that an essential property of ordinary integrals is that we can shift the dummy integration variable: $\int_{-\infty}^{+\infty} dx f(x+c) = \int_{-\infty}^{+\infty} dx f(x)$. Thus, we should also insist that the Grassmann integral obey the rule $\int d\eta f(\eta + \xi) = \int d\eta f(\eta)$, where ξ is an arbitrary Grassmann number. Plugging into the most general function given above, we find that $\int d\eta b\xi = 0$. Since ξ is arbitrary this can only hold if we define $\int d\eta b = 0$ for any ordinary number b, and in particular $\int d\eta \equiv \int d\eta 1 = 0$

Since given three Grassmann numbers χ, η, and ξ, we have $\chi(\eta\xi) = (\eta\xi)\chi$, that is, the product $(\eta\xi)$ commutes with any Grassmann number χ, we feel that the product of two anticommuting numbers should be an ordinary number. Thus, the integral $\int d\eta\eta$ is just an ordinary number that we can simply take to be 1: This fixes the normalization of $d\eta$. Thus Grassmann integration is extraordinarily simple, being defined by two rules:

$$\int d\eta = 0 \tag{6}$$

and

$$\int d\eta\eta = 1 \tag{7}$$

With these two rules we can integrate any function of η :

$$\int d\eta f(\eta) = \int d\eta(a + b\eta) = b \tag{8}$$

if b is an ordinary number so that $f(\eta)$ is Grassmannian, and

$$\int d\eta f(\eta) = \int d\eta(a + b\eta) = -b \tag{9}$$

if b is Grassmannian so that $f(\eta)$ is an ordinary number. Note that the concept of a range of integration does not exist for Grassmann integration. It is much easier to master Grassmann integration than ordinary integration!

Let η and $\bar{\eta}$ be two independent Grassmann numbers and a an ordinary number. Then the Grassmannian analog of the Gaussian integral gives

$$\int d\eta \int d\bar{\eta} e^{\bar{\eta}a\eta} = \int d\eta \int d\bar{\eta}(1 + \bar{\eta}a\eta) = \int d\eta\eta a = a = e^{+\log a} \tag{10}$$

Precisely what we had wanted!

We can generalize immediately: Let $\eta = (\eta_1, \eta_2, \cdots, \eta_N)$ be N Grassmann numbers, and similarly for $\bar{\eta}$; we then have

$$\int d\eta \int d\bar{\eta} e^{\bar{\eta}A\eta} = \det A \qquad (11)$$

for $A = \{A_{ij}\}$ an antisymmetric N by N matrix. (Note that contrary to the bosonic case, the inverse of A need not exist.) We can further generalize to a functional integral.

As we will see shortly, we now have all the mathematics we need.

Grassmann path integral

In analogy with the generating functional for the scalar field

$$Z = \int D\varphi e^{iS(\varphi)} = \int D\varphi e^{i\int d^4x \frac{1}{2}[(\partial\varphi)^2 - (m^2 - i\varepsilon)\varphi^2]}$$

we would naturally write the generating functional for the spinor field as

$$Z = \int D\psi D\bar{\psi} e^{iS(\psi,\bar{\psi})} = \int D\psi \int D\bar{\psi} e^{i\int d^4x \bar{\psi}(i\partial\!\!\!/ - m + i\varepsilon)\psi}$$

Treating the integration variables ψ and $\bar{\psi}$ as Grassmann-valued Dirac spinors, we immediately obtain

$$Z = \int D\psi \int D\bar{\psi} e^{i\int d^4x \bar{\psi}(i\partial\!\!\!/ - m + i\varepsilon)\psi} = C' \det(i\partial\!\!\!/ - m + i\varepsilon)$$

$$= C' e^{\text{tr}\log(i\partial\!\!\!/ - m + i\varepsilon)} \qquad (12)$$

where C' is some multiplicative constant. Using the cyclic property of the trace, we note that (here m is understood to be $m - i\varepsilon$)

$$\text{tr}\log(i\partial\!\!\!/ - m) = \text{tr}\log\gamma^5(i\partial\!\!\!/ - m)\gamma^5 = \text{tr}\log(-i\partial\!\!\!/ - m)$$

$$= \frac{1}{2}[\text{tr}\log(i\partial\!\!\!/ - m) + \text{tr}\log(-i\partial\!\!\!/ - m)]$$

$$= \frac{1}{2}\text{tr}\log(\partial^2 + m^2). \qquad (13)$$

Thus, $Z = C' e^{\frac{1}{2}\text{tr}\log(\partial^2 + m^2 - i\varepsilon)}$ [compare with (1)!].

We see that we get the same vacuum energy we obtained in chapter II.2 using the canonical formalism if we remember that the trace operation here contains a factor of 4 compared to the trace operation in (1), since $(i\partial\!\!\!/ - m)$ is a 4 by 4 matrix.

Heuristically, we can now see the necessity for Grassmann variables. If we were to treat ψ and $\bar{\psi}$ as complex numbers in (12), we would obtain something like $(1/\det[i\partial\!\!\!/ - m]) = e^{-\text{tr}\log(i\partial\!\!\!/ - m)}$ and so have the wrong sign for the vacuum energy. We want the determinant to come out in the numerator rather than in the denominator.

Dirac propagator

Now that we have learned that the Dirac field is to be quantized by a Grassmann path integral we can introduce Grassmannian spinor sources η and $\bar{\eta}$:

$$Z(\eta, \bar{\eta}) = \int D\psi D\bar{\psi} e^{i\int d^4x[\bar{\psi}(i\partial\!\!\!/ - m)\psi + \bar{\eta}\psi + \bar{\psi}\eta]} \qquad (14)$$

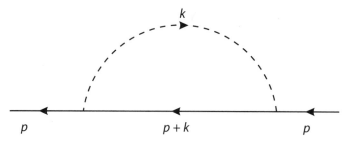

Figure II.5.1

and proceed pretty much as before. Completing the square just as in the case of the scalar field, we have

$$\bar{\psi} K \psi + \bar{\eta}\psi + \bar{\psi}\eta = (\bar{\psi} + \bar{\eta}K^{-1})K(\psi + K^{-1}\eta) - \bar{\eta}K^{-1}\eta \tag{15}$$

and thus

$$Z(\eta, \bar{\eta}) = C'' e^{-i\bar{\eta}(i\slashed{\partial}-m)^{-1}\eta} \tag{16}$$

The propagator $S(x)$ for the Dirac field is the inverse of the operator $(i\slashed{\partial} - m)$: in other words, $S(x)$ is determined by

$$(i\slashed{\partial} - m)S(x) = \delta^{(4)}(x) \tag{17}$$

As you can verify, the solution is

$$iS(x) = \int \frac{d^4 p}{(2\pi)^4} \frac{ie^{-ipx}}{\slashed{p} - m + i\varepsilon} \tag{18}$$

in agreement with (II.2.22).

Feynman rules for fermions

We can now derive the Feynman rules for fermions in the same way that we derived the Feynman rules for a scalar field. For example, consider the theory of a scalar field interacting with a Dirac field

$$\mathcal{L} = \bar{\psi}(i\gamma^\mu \partial_\mu - m)\psi + \tfrac{1}{2}[(\partial\varphi)^2 - \mu^2\varphi^2] - \lambda\varphi^4 + f\varphi\bar{\psi}\psi \tag{19}$$

The generating functional

$$Z(\eta, \bar{\eta}, J) = \int D\psi\, D\bar{\psi}\, D\varphi\, e^{iS(\psi, \bar{\psi}, \varphi) + i\int d^4 x (J\varphi + \bar{\eta}\psi + \bar{\psi}\eta)} \tag{20}$$

can be evaluated as a double series in the couplings λ and f. The Feynman rules (not repeating the rules involving only the boson) are as follows:

1. Draw a diagram with straight lines for the fermion and dotted lines for the boson, and label each line with a momentum, for example, as in figure II.5.1.

2. Associate with each fermion line the propagator

$$\frac{i}{\not{p} - m + i\varepsilon} = i\frac{\not{p} + m}{p^2 - m^2 + i\varepsilon} \tag{21}$$

3. Associate with each interaction vertex the coupling factor if and the factor $(2\pi)^4\delta^{(4)}(\sum_{\text{in}} p - \sum_{\text{out}} p)$ expressing momentum conservation (the two sums are taken over the incoming and the outgoing momenta, respectively).

4. Momenta associated with internal lines are to be integrated over with the measure $\int [d^4 p/(2\pi)^4]$.

5. External lines are to be amputated. For an incoming fermion line write $u(p, s)$ and for an outgoing fermion line $\bar{u}(p', s')$. The sources and sinks have to recognize the spin polarization of the fermion being produced and absorbed. [For antifermions, we would have $\bar{v}(p, s)$ and $v(p', s')$. You can see from (II.2.10) that an outgoing antifermion is associated with v rather than $\bar{v}$.]

6. A factor of (-1) is to be associated with each closed fermion line. The spinor index carried by the fermion should be summed over, thus leading to a trace for each closed fermion line. [For an example, see (II.7.7–9).]

Note that rule 6 is unique to fermions, and is needed to account for their negative contribution to the vacuum energy. The Feynman diagram corresponding to vacuum fluctuation has no external line. I will discuss these points in detail in chapter IV.3.

For the theory of a massive vector field interacting with a Dirac field mentioned in chapter II.1

$$\mathcal{L} = \bar{\psi}[i\gamma^\mu(\partial_\mu - ieA_\mu) - m]\psi - \tfrac{1}{4}F_{\mu\nu}F^{\mu\nu} + \tfrac{1}{2}\mu^2 A_\mu A^\mu \tag{22}$$

the rules differ from above as follows. The vector boson propagator is given by

$$\frac{i}{k^2 - \mu^2}\left(\frac{k_\mu k_\nu}{\mu^2} - g_{\mu\nu}\right) \tag{23}$$

and thus each vector boson line is associated not only with a momentum, but also with indices μ and ν. The vertex (figure II.5.2) is associated with $ie\gamma^\mu$.

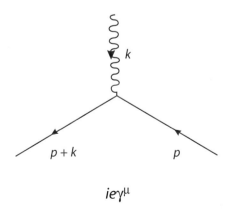

$$ie\gamma^\mu$$

Figure II.5.2

If the vector boson line in figure II.5.2 is external and on shell, we have to specify its polarization. As discussed in chapter I.5, a massive vector boson has three degrees of polarizations described by the polarization vector $\varepsilon_\mu^{(a)}$ for $a = 1, 2, 3$. The amplitude for emitting or absorbing a vector boson with polarization a is $ie\gamma^\mu\varepsilon_\mu^{(a)} = ie\ \rlap{/}{\varepsilon}^{(a)}$.

In Schwinger's sorcery, the source for producing a vector boson $J_\mu(x)$, in contrast to the source for producing a scalar meson $J(x)$, carries a Lorentz index. Work in momentum space. Current conservation $k^\mu J_\mu(k) = 0$ implies that we can decompose $J_\mu(k) = \sum_{a=1}^3 J^{(a)}(k)\varepsilon_\mu^{(a)}(k)$. The clever experimentalist sets up her machine, that is, chooses the functions $J^{(a)}(k)$, so as to produce a vector boson of the desired momentum k and polarization a. Current conservation requires $k^\mu\varepsilon_\mu^{(a)}(k) = 0$. For $k^\mu = (\omega(k), 0, 0, k)$, we could choose

$$\varepsilon_\mu^{(1)}(k) = (0, 1, 0, 0), \quad \varepsilon_\mu^{(2)}(k) = (0, 0, 1, 0), \quad \varepsilon_\mu^{(3)}(k) = (-k, 0, 0, \omega(k))/m \tag{24}$$

In the canonical formalism, we have in analogy with the expansion of the scalar field φ in (I.8.11)

$$A_\mu(\vec{x}, t) = \int \frac{d^D k}{\sqrt{(2\pi)^D 2\omega_k}} \sum_{a=1}^3 \{a^{(a)}(\vec{k})\varepsilon_\mu^{(a)}(k)e^{-i(\omega_k t - \vec{k}\cdot\vec{x})} + a^{(a)\dagger}(\vec{k})\varepsilon_\mu^{(a)*}(k)e^{i(\omega_k t - \vec{k}\cdot\vec{x})}\} \tag{25}$$

(I trust you not to confuse the letter a used to denote annihilation and used to label polarization.) The point is that in contrast to φ, A_μ carries a Lorentz index, which the creation and annihilation operators have to "know about" (through the polarization label.) It is instructive to compare with the expansion of the fermion field ψ in (II.2.10): the spinor index α on ψ is carried in the expansion by the spinors $u(p, s)$ and $v(p, s)$. In each case, an index (μ in the case of the vector and α in the case of the spinor) known to the Lorentz group is "traded" for a label specifying the spin polarization (a and s respectively.)

A minor technicality: notice that I have complex conjugated the polarization vector associated with the creation operator $a^{(a)\dagger}(\vec{k})$ in (25) even though the polarization vectors in (24) are real. This is because experimentalists sometimes enjoy using circularly polarized photons with polarization vectors $\varepsilon_\mu^{(1)}(k) = (0, 1, i, 0)/\sqrt{2}$, $\varepsilon_\mu^{(2)}(k) = (0, 1, -i, 0)/\sqrt{2}$.

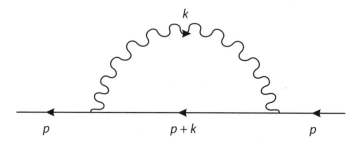

Figure II.5.3

Exercises

II.5.1 Write down the Feynman amplitude for the diagram in figure II.5.1 for the scalar theory (19). The answer is given in chapter III.3.

II.5.2 Applying the Feynman rules for the vector theory (22) show that the amplitude for the diagram in figure II.5.3 is given by

$$(ie)^2 i^2 \int \frac{d^4k}{(2\pi)^4} \frac{1}{k^2 - \mu^2} \left(\frac{k_\mu k_\nu}{\mu^2} - g_{\mu\nu} \right) \bar{u}(p) \gamma^\nu \frac{\not{p} + \not{k} + m}{(p+k)^2 - m^2} \gamma^\mu u(p) \tag{26}$$

II.6 Electron Scattering and Gauge Invariance

Electron-proton scattering

We will now finally calculate a physical process that experimentalists can go out and measure. Consider scattering an electron off a proton. (For the moment let us ignore the strong interaction that the proton also participates in. We will learn in chapter III.6 how to take this fact into account. Here we pretend that the proton, just like the electron, is a structureless spin-$\frac{1}{2}$ fermion obeying the Dirac equation.) To order e^2 the relevant Feynman diagram is given in figure II.6.1 in which the electron and the proton exchange a photon.

But wait, from chapter I.5 we only know how to write down the propagator $i D_{\mu\nu} = i\left(\frac{k_\mu k_\nu}{\mu^2} - \eta_{\mu\nu}\right)/(k^2 - \mu^2)$ for a hypothetical massive photon. (Trivial notational change: the mass of the photon is now called μ, since m is reserved for the mass of the electron and M for the mass of the proton.) In that chapter I outlined our philosophy: we will plunge ahead and calculate with a nonzero μ and hope that at the end we can set μ to zero. Indeed, when we calculated the potential energy between two external charges, we find that we can let $\mu \to 0$ without any signs of trouble [see (I.5.6)]. In this chapter and the next, we would like to see whether this will always be the case.

Applying the Feynman rules, we obtain the amplitude for the diagram in figure II.6.1 (with $k = P - p$ the momentum transfer in the scattering)

$$\mathcal{M}(P, P_N) = (-ie)(ie)\frac{i}{(P - p)^2 - \mu^2}\left(\frac{k_\mu k_\nu}{\mu^2} - \eta_{\mu\nu}\right)\bar{u}(P)\gamma^\mu u(p)\bar{u}(P_N)\gamma^\nu u(p_N) \tag{1}$$

We have suppressed the spin labels and used the subscript N (for nucleon) to refer to the proton.

Now notice that

$$k_\mu \bar{u}(P)\gamma^\mu u(p) = (P - p)_\mu \bar{u}(P)\gamma^\mu u(p) = \bar{u}(P)(\slashed{P} - \slashed{p})u(p) = \bar{u}(P)(m - m)u(p) = 0 \tag{2}$$

by virtue of the equations of motion satisfied by $\bar{u}(P)$ and $u(p)$. Similarly, $k^\mu \bar{u}(P_N)\gamma_\mu u(p_N) = 0$.

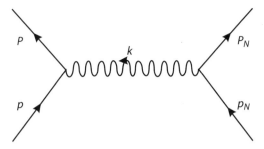

Figure II.6.1

This important observation implies that the $k_\mu k_\nu / \mu^2$ term in the photon propagator does not enter. Thus

$$\mathcal{M}(P, P_N) = -ie^2 \frac{1}{(P-p)^2 - \mu^2} \bar{u}(P)\gamma^\mu u(p) \bar{u}(P_N)\gamma_\mu u(p_N) \tag{3}$$

and we can now set the photon mass μ to zero with impunity and replace $(P-p)^2 - \mu^2$ in the denominator by $(P-p)^2$.

Note that the identity that allows us to set μ to zero is just the momentum space version of electromagnetic current conservation $\partial_\mu J^\mu = \partial_\mu(\bar{\psi}\gamma^\mu\psi) = 0$. You would notice that this calculation is intimately related to the one we did in going from (I.5.4) to (I.5.5), with $\bar{u}(P)\gamma^\mu u(p)$ playing the role of $J^\mu(k)$.

Potential scattering

That the proton mass M is so much larger than the electron mass m allows us to make a useful approximation familiar from elementary physics. In the limit M/m tending to infinity, the proton hardly moves, and we could use, for the proton, the spinors for a particle at rest given in chapter II.2, so that $\bar{u}(P_N)\gamma_0 u(p_N) \approx 1$ and $\bar{u}(P_N)\gamma_i u(p_N) \approx 0$. Thus

$$\mathcal{M} = \frac{-ie^2}{k^2}\bar{u}(P)\gamma^0 u(p) \tag{4}$$

We recognize that we are scattering the electron in the Coulomb potential generated by the proton. Work out the (familiar) kinematics: $p = (E, 0, 0, |\vec{p}|)$ and $P = (E, 0, |\vec{p}|\sin\theta, |\vec{p}|\cos\theta)$. We see that $k = P - p$ is purely spacelike and $k^2 = -\vec{k}^2 = -4|\vec{p}|^2 \sin^2(\theta/2)$. Recall from (I.4.7) that

$$\int d^3x\, e^{i\vec{k}\cdot\vec{x}} \left(-\frac{e}{4\pi r}\right) = -\frac{e}{\vec{k}^2} \tag{5}$$

We represent potential scattering by the Feynman diagram in figure II.6.2: the proton has disappeared and been replaced by a cross, which supplies the virtual photon the electron interacts with. It is in this sense that you could think of the Coulomb potential picturesquely as a swarm of virtual photons.

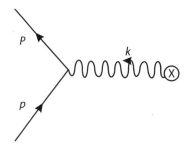

Figure II.6.2

Once again, it is instructive to use the canonical formalism to derive this expression for $\mathcal{M}$ for potential scattering. We want the transition amplitude $\langle P, S | e^{-iHT} | p, s \rangle$ with the single electron state $| p, s \rangle \equiv b^\dagger(p, s) | 0 \rangle$. The term in the Lagrangian describing the electron interacting with the external c-number potential $A_\mu(x)$ is given in (II.5.22) and thus to leading order we have the transition amplitude $ie \int d^4x \langle P, S | \bar{\psi}(x) \gamma^\mu \psi(x) | p, s \rangle A_\mu(x)$. Using (II.2.10–11) we evaluate this as

$$ie \int d^4x (1/\rho(P))(1/\rho(p))(\bar{u}(P, S)\gamma^\mu u(p, s))e^{i(P-p)x} A_\mu(x)$$

Here $\rho(p)$ denotes the fermion normalization factor $\sqrt{(2\pi)^3 E_p/m}$ in (II.2.10). Given that the Coulomb potential has only a time component and does not depend on time, we see that integration over time gives us an energy conservation delta function, and integration over space the Fourier transform of the potential, as in (5). Thus the above becomes $(1/\rho(P))(1/\rho(p))(2\pi)\delta(E_P - E_p)(\frac{-ie^2}{k^2})\bar{u}(P, S)\gamma^0 u(p, s)$. Satisfyingly, we have recovered the Feynman amplitude up to normalization factors and an energy conservation delta function, just as in (I.8.16) except for the substitution of boson for fermion normalization factors. Notice that we have energy conservation but not 3-momentum conservation, a fact we understand perfectly well when we dribble a basketball ball, for example.

Electron-electron scattering

Next, we graduate to two electrons scattering off each other: $e^-(p_1) + e^-(p_2) \to e^-(P_1) + e^-(P_2)$. Here we have a new piece of physics: the two electrons are identical. A profound tenet of quantum physics states that we cannot distinguish between the two outgoing electrons. Now there are two Feynman diagrams (see fig. II.6.3) to order e^2, obtained by interchanging the two outgoing electrons. The electron carrying momentum P_1 could have "come from" the incoming electron carrying momentum p_1 or the incoming electron carrying momentum p_2.

We have for figure II.6.3a the amplitude

$$A(P_1, P_2) = (ie^2/(P_1 - p_1)^2)\bar{u}(P_1)\gamma^\mu u(p_1)\bar{u}(P_2)\gamma_\mu u(p_2)$$

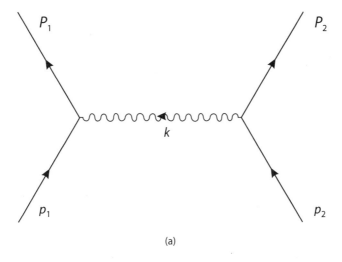

(a)

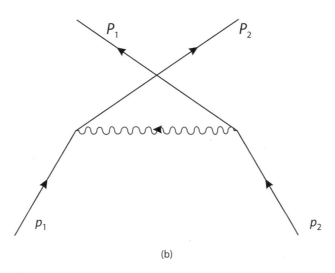

(b)

Figure II.6.3

as before. We have only indicated the dependence of A on the final momenta, suppressing the other dependence. By Fermi statistics, the amplitude for the diagram in figure II.6.3 is then $-A(P_2, P_1)$. Thus the invariant amplitude for two electrons of momentum p_1 and p_2 to scatter into two electrons with momentum P_1 and P_2 is

$$\mathcal{M} = A(P_1, P_2) - A(P_2, P_1) \tag{6}$$

To obtain the cross section we have to square the amplitude

$$|\mathcal{M}|^2 = [|A(P_1, P_2)|^2 + (P_1 \leftrightarrow P_2)] - 2\,\mathrm{Re}\,A(P_2, P_1)^* A(P_1, P_2) \tag{7}$$

At this point we have to do a fair amount of arithmetic, but keep in mind that there is nothing conceptually intricate in what follows. First, we have to learn to complex conjugate spinor amplitudes. Using (II.1.15), note that in general $(\bar{u}(p')\gamma_\mu \cdots \gamma_\nu u(p))^* = u(p)^\dagger \gamma_\nu^\dagger \cdots \gamma_\mu^\dagger \gamma^0 u(p') = \bar{u}(p)\gamma_\nu \cdots \gamma_\mu u(p')$. Here $\gamma_\mu \cdots \gamma_\nu$ represents a product of any number of γ matrices. Complex conjugation reverses the order of the product and interchanges the two spinors. Thus we have

$$|A(P_1, P_2)|^2 = \frac{e^4}{k^4}[\bar{u}(P_1)\gamma^\mu u(p_1)\bar{u}(p_1)\gamma^\nu u(P_1)][\bar{u}(P_2)\gamma_\mu u(p_2)\bar{u}(p_2)\gamma_\nu u(P_2)] \tag{8}$$

which factorizes with one factor involving spinors carrying momentum with subscript 1 and another factor involving spinors carrying momentum with subscript 2. In contrast, the interference term $A(P_2, P_1)^* A(P_1, P_2)$ does not factorize.

In the simplest experiments, the initial electrons are unpolarized, and the polarization of the outgoing electrons is not measured. We average over initial spins and sum over final spins using (II.2.8):

$$\sum_s u(p, s)\bar{u}(p, s) = \frac{\not{p} + m}{2m} \tag{9}$$

In averaging and summing $|A(P_1, P_2)|^2$ we encounter the object (displaying the spin labels explicitly)

$$\tau^{\mu\nu}(P_1, p_1) \equiv \frac{1}{2}\sum\sum \bar{u}(P_1, S)\gamma^\mu u(p_1, s)\bar{u}(p_1, s)\gamma^\nu u(P_1, S) \tag{10}$$

$$= \frac{1}{2(2m)^2} tr(\not{P}_1 + m)\gamma^\mu (\not{p}_1 + m)\gamma^\nu \tag{11}$$

which is to be multiplied by $\tau_{\mu\nu}(P_2, p_2)$.

Well, we, or rather you, have to develop some technology for evaluating the trace of products of gamma matrices. The key observation is that the square of a gamma matrix is either $+1$ or -1, and different gamma matrices anticommute. Clearly, the trace of a product of an odd number of gamma matrices vanish. Furthermore, since there are only four different gamma matrices, the trace of a product of six gamma matrices can always be reduced to the trace of a product of four gamma matrices, since there are always pairs of gamma matrices that are equal and can be brought together by anticommuting. Similarly for the trace of a product of an even higher number of gamma matrices.

Hence $\tau^{\mu\nu}(P_1, p_1) = \frac{1}{2(2m)^2}(\text{tr}(\not{P}_1\gamma^\mu \not{p}_1\gamma^\nu) + m^2\text{tr}(\gamma^\mu\gamma^\nu))$. Writing $\text{tr}(\not{P}_1\gamma^\mu \not{p}_1\gamma^\nu) = P_{1\rho}p_{1\lambda}\text{tr}(\gamma^\rho\gamma^\mu\gamma^\lambda\gamma^\nu)$ and using the expressions for the trace of a product of an even number of gamma matrices listed in appendix D, we obtain $\tau^{\mu\nu}(P_1, p_1) = \frac{1}{2(2m)^2}4(P_1^\mu p_1^\nu - \eta^{\mu\nu}P_1 \cdot p_1 + P_1^\nu p_1^\mu + m^2\eta^{\mu\nu})$.

In averaging and summing $A(P_2, P_1)^* A(P_1, P_2)$ we encounter the more involved object

$$\kappa \equiv \frac{1}{2^2}\sum\sum\sum\sum \bar{u}(P_1)\gamma^\mu u(p_1)\bar{u}(P_2)\gamma_\mu u(p_2)\bar{u}(p_1)\gamma^\nu u(P_2)\bar{u}(p_2)\gamma_\nu u(P_1) \tag{12}$$

where for simplicity of notation we have suppressed the spin labels. Applying (9) we can write κ as a single trace. The evaluation of κ is quite tedious, since it involves traces of products of up to eight gamma matrices.

We will be content to study electron-electron scattering in the relativistic limit in which m may be neglected compared to the momenta. As explained in chapter II.2, while we are using the "rest normalization" for spinors we can nevertheless set m to 0 wherever possible. Then

$$\kappa = \frac{1}{4(2m)^4}\operatorname{tr}(\not{P}_1\gamma^\mu \not{p}_1\gamma^\nu \not{P}_2\gamma_\mu \not{p}_2\gamma_\nu) \tag{13}$$

Applying the identities in appendix D to (13) we obtain $\operatorname{tr}(\not{P}_1\gamma^\mu \not{p}_1\gamma^\nu \not{P}_2\gamma_\mu \not{p}_2\gamma_\nu) = -2\operatorname{tr}(\not{P}_1\gamma^\mu \not{p}_1 \not{p}_2\gamma_\mu \not{P}_2) = -32 p_1 \cdot p_2 P_1 \cdot P_2$.

In the same limit $\tau^{\mu\nu}(P_1, p_1) = \frac{2}{(2m)^2}(P_1^\mu p_1^\nu + P_1^\nu p_1^\mu - \eta^{\mu\nu}P_1 \cdot p_1)$ and thus

$$\tau^{\mu\nu}(P_1, p_1)\tau_{\mu\nu}(P_2, p_2) = \frac{4}{(2m)^4}(P_1^\mu p_1^\nu + P_1^\nu p_1^\mu - \eta^{\mu\nu}P_1 \cdot p_1)(2P_{2\mu}p_{2\nu} - \eta_{\mu\nu}P_2 \cdot p_2) \tag{14}$$

$$= \frac{4 \cdot 2}{(2m)^4}(p_1 \cdot p_2 P_1 \cdot P_2 + p_1 \cdot P_2 p_2 \cdot P_1) \tag{15}$$

An amusing story to break up this tedious calculation: Murph Goldberger, who was a graduate student at the University of Chicago after working on the Manhattan Project during the war and whom I mentioned in chapter II.1 regarding the Feynman slash, told me that Enrico Fermi marvelled at this method of taking a trace that young people were using to sum over spin-$\frac{1}{2}$ polarizations. Fermi and others in the older generation had simply memorized the specific form of the spinors in the Dirac basis (which you know from doing exercise II.1.3) and consequently the expressions for $\bar{u}(P, S)\gamma^\mu u(p, s)$. They simply multiplied these expressions together and added up the different possibilities. Fermi was skeptical of the fancy schmancy method the young Turks were using and challenged Murph to a race on the blackboard. Of course, with his lightning speed, Fermi won. To me, it is amazing, living in the age of string theory, that another generation once regarded the trace as fancy math. I confessed that I was even a bit doubtful of this story until I looked at Feynman's book *Quantum Electrodynamics*, but guess what, Feynman indeed constructed, on page 100 in the edition I own, a table showing the result for the amplitude squared for various spin polarizations. Some pages later, he mentioned that polarizations could also be summed using the spur (the original German word for trace). Another amusing aside: spur is cognate with the English word spoor, meaning animal droppings, and hence also meaning track, trail, and trace. All right, back to work!

While it is not the purpose of this book to teach you to calculate cross sections for a living, it is character building to occasionally push calculations to the bitter end. Here is a good place to introduce some useful relativistic kinematics. In calculating the cross section for the scattering process $p_1 + p_2 \rightarrow P_1 + P_2$ (with the masses of the four particles all different in general) we typically encounter Lorentz invariants such as $p_1 \cdot P_2$. A priori, you might think there are six such invariants, but in fact, you know that there are only physical variables, the incident energy E and the scattering angle θ. The cleanest way to organize these invariants is to introduce what are called Mandelstam variables:

$$s \equiv (p_1 + p_2)^2 = (P_1 + P_2)^2 \tag{16}$$

$$t \equiv (P_1 - p_1)^2 = (P_2 - p_2)^2 \tag{17}$$

$$u \equiv (P_2 - p_1)^2 = (P_1 - p_2)^2 \tag{18}$$

You know that there must be an identity reducing the three variables s, t, and u to two. Show that (with an obvious notation)

$$s + t + u = m_1^2 + m_2^2 + M_1^2 + M_2^2 \tag{19}$$

For our calculation here, we specialize to the center of mass frame in the relativistic limit $p_1 = E(1, 0, 0, 1)$, $p_2 = E(1, 0, 0, -1)$, $P_1 = E(1, \sin\theta, 0, \cos\theta)$, and $P_2 = E(1, -\sin\theta, 0, -\cos\theta)$. Hence

$$p_1 \cdot p_2 = P_1 \cdot P_2 = 2E^2 = \frac{1}{2}s \tag{20}$$

$$p_1 \cdot P_1 = p_2 \cdot P_2 = 2E^2 \sin^2\frac{\theta}{2} = -\frac{1}{2}t \tag{21}$$

and

$$p_1 \cdot P_2 = p_2 \cdot P_1 = 2E^2 \cos^2\frac{\theta}{2} = -\frac{1}{2}u \tag{22}$$

Also, in this limit $(P_1 - p_1)^4 = (-2p_1 \cdot P_1)^2 = 16E^4 \sin^4(\theta/2) = t^4$. Putting it together, we obtain $\frac{1}{4}\sum\sum\sum\sum|\mathcal{M}|^2 = (e^4/4m^4)f(\theta)$, where

$$\begin{aligned}
f(\theta) &= \frac{s^2 + u^2}{t^2} + \frac{2s^2}{tu} + \frac{s^2 + t^2}{u^2} \\
&= \frac{s^4 + t^4 + u^4}{t^2 u^2} \\
&= \left(\frac{1 + \cos^4(\theta/2)}{\sin^4(\theta/2)} + \frac{2}{\sin^2(\theta/2)\cos^2(\theta/2)} + \frac{1 + \sin^4(\theta/2)}{\cos^4(\theta/2)}\right) \tag{23} \\
&= 2\left(\frac{1}{\sin^4(\theta/2)} + 1 + \frac{1}{\cos^4(\theta/2)}\right) \tag{24}
\end{aligned}$$

The physical origin of each of the terms in (23) [before we simplify with trigonometric identities] to get to (24) is clear. The first term strongly favors forward scattering due to the photon propagator $\sim 1/k^2$ blowing up at $k \sim 0$. The third term is required by the indistinguishability of the two outgoing electrons: the scattering must be symmetric under $\theta \to \pi - \theta$, since experimentalists can't tell whether a particular incoming electron has scattered forward or backward. The second term is the most interesting of all: it comes from quantum interference. If we had mistakenly thought that electrons are bosons and taken the plus sign in (6), the second term in $f(\theta)$ would come with a minus sign. This makes a big difference: for instance, $f(\pi/2)$ would be $5 - 8 + 5 = 2$ instead of $5 + 8 + 5 = 18$.

Since the conversion of a squared probability amplitude to a cross section is conceptually the same as in nonrelativistic quantum mechanics (divide by the incoming flux, etc.), I will relegate the derivation to an appendix and let you go the last few steps and obtain the differential cross section as an exercise:

$$\frac{d\sigma}{d\Omega} = \frac{\alpha^2}{8E^2}f(\theta) \tag{25}$$

with the fine structure constant $\alpha \equiv e^2/4\pi \approx 1/137$.

An amazing subject

When you think about it, theoretical physics is truly an amazing business. After the appropriate equipments are assembled and high energy electrons are scattered off each other, experimentalists indeed would find the differential cross section given in (25). There is almost something magical about it.

Appendix: Decay rate and cross section

To make contact with experiments, we have to convert transition amplitudes into the scattering cross sections and decay rates that experimentalists actually measure. I assume that you are already familiar with the physical concepts behind these measurements from a course on nonrelativistic quantum mechanics, and thus here we focus more on those aspects specific to quantum field theory.

To be able to count states, we adopt an expedient probably already familiar to you from quantum statistical mechanics, namely that we enclose our system in a box, say a cube with length L on each side with L much larger than the characteristic size of our system. With periodic boundary conditions, the allowed plane wave states $e^{i\vec{p}\cdot\vec{x}}$ carry momentum

$$\vec{p} = \frac{2\pi}{L}(n_x, n_y, n_z) \tag{26}$$

where the n_i's are three integers. The allowed values of momentum form a lattice of points in momentum space with spacing $2\pi/L$ between points. Experimentalists measure momentum with finite resolution, small but much larger than $2\pi/L$. Thus, an infinitesimal volume d^3p in momentum space contains $d^3p/(2\pi/L)^3 = V d^3p/(2\pi)^3$ states with $V = L^3$ the volume of the box. We obtain the correspondence

$$\int \frac{d^3p}{(2\pi)^3} f(p) \leftrightarrow \frac{1}{V} \sum_p f(p) \tag{27}$$

In the sum the values of p ranges over the discrete values in (26). The correspondence (27) between continuum normalization and the discrete box normalization implies that

$$\delta^{(3)}(\vec{p} - \vec{p}') \leftrightarrow \frac{V}{(2\pi)^3} \delta_{\vec{p}\vec{p}'} \tag{28}$$

with the Kronecker delta $\delta_{\vec{p}\vec{p}'}$ equal to 1 if $\vec{p} = \vec{p}'$ and 0 otherwise. One way of remembering these correspondences is simply by dimensional matching.

Let us now look at the expansion (I.8.17) of a complex scalar field

$$\varphi(\vec{x}, t) = \int \frac{d^3k}{\sqrt{(2\pi)^3 2\omega_k}} [a(\vec{k})e^{-i(\omega_k t - \vec{k}\cdot\vec{x})} + b^\dagger(\vec{k})e^{i(\omega_k t - \vec{k}\cdot\vec{x})}] \tag{29}$$

in terms of creation and annihilation operators. Henceforth, in order not to clutter up the page, I will abuse notation slightly, for example, dropping the arrows on vectors when there is no risk of confusion. Going over to the box normalization, we replace the commutation relation $[a(k), a^\dagger(k')] = \delta^{(3)}(\vec{k} - \vec{k}')$ by

$$[a(k), a^\dagger(k')] = \frac{V}{(2\pi)^3} \delta_{\vec{k}\vec{k}'} \tag{30}$$

We now normalize the creation and annihilation operators by

$$a(k) = \left(\frac{V}{(2\pi)^3}\right)^{\frac{1}{2}} \tilde{a}(k) \tag{31}$$

so that

$$[\tilde{a}(k), \tilde{a}^\dagger(k')] = \delta_{k,k'} \tag{32}$$

Thus the state $|\vec{k}\rangle \equiv \tilde{a}^\dagger(\vec{k}) |0\rangle$ is properly normalized: $\langle \vec{k} | \vec{k} \rangle = 1$. Using (27) and (31), we end up with

$$\varphi(x) = \frac{1}{V^{\frac{1}{2}}} \sum_k \frac{1}{\sqrt{2\omega_k}} (\tilde{a} e^{-ikx} + \tilde{b}^\dagger e^{ikx}) \tag{33}$$

We specified a complex, rather than a real, scalar field because then, as you showed in exercise I.8.4, a conserved current can be defined with the corresponding charge $Q = \int d^3x J^0 = \int d^3k (a^\dagger(k)a(k) - b^\dagger(k)b(k)) \rightarrow \sum_k (\tilde{a}^\dagger(k)\tilde{a}(k) - \tilde{b}^\dagger(k)\tilde{b}(k))$. It follows immediately that $\langle \vec{k} | Q | \vec{k} \rangle = 1$, so that for the state $|\vec{k}\rangle$ we have one particle in the box.

To derive the formula for the decay rate, we focus, for the sake of pedagogical clarity, on a toy Lagrangian $\mathcal{L} = g(\eta^\dagger \xi^\dagger \varphi + \text{h.c.})$ describing the decay $\varphi \rightarrow \eta + \xi$ of a meson into two other mesons. (As usual, we display only the part of the Lagrangian that is of immediate interest. In other words, we suppress the stuff you have long since mastered: $\mathcal{L} = \partial \varphi^\dagger \partial \varphi - m_\varphi \varphi^\dagger \varphi + \cdots$ and all the rest.)

The transition amplitude $\langle \vec{p}, \vec{q} | e^{-iHT} | \vec{k} \rangle$ is given to lowest order by $\mathcal{A} = i \langle \vec{p}, \vec{q} | \int d^4x (g\eta^\dagger(x)\xi^\dagger(x)\varphi(x)) | \vec{k} \rangle$. Here we use the states we "carefully" normalized above, namely the ones created by the various "analogs" of $\tilde{a}^\dagger$. (Just as in quantum mechanics, strictly, we should use wave packets instead of plane wave states. I assume that you have gone through that at least once.) Plugging in the various "analogs" of (33), we have

$$\mathcal{A} = ig(\frac{1}{V^{\frac{1}{2}}})^3 \sum_{p'} \sum_{q'} \sum_{k'} \frac{1}{\sqrt{2\omega_{p'} 2\omega_{q'} 2\omega_{k'}}} \int d^4x e^{i(p'+q'-k')} \langle \vec{p}, \vec{q} | \tilde{a}^\dagger(p')\tilde{a}^\dagger(q')\tilde{a}(k') | \vec{k} \rangle$$

$$= ig \frac{1}{V^{\frac{3}{2}}} \frac{1}{\sqrt{2\omega_p 2\omega_q 2\omega_k}} (2\pi)^4 \delta^{(4)}(p+q-k) \tag{34}$$

Here we have committed various minor transgressions against notational consistency. For example, since the three particles φ, η, and ξ have different masses, the symbol ω represents, depending on context, different functions of its subscript (thus $\omega_p = \sqrt{\vec{p}^2 + m_\eta^2}$, and so forth). Similarly, $\tilde{a}(k')$ should really be written as $\tilde{a}_\varphi(k')$, and so forth. Also, we confound 3- and 4-momenta. I would like to think that these all fall under the category of what the Catholic church used to call venial sins. In any case, you know full well what I am talking about.

Next, we square the transition amplitude $\mathcal{A}$ to find the transition probability. You might be worried, because it appears that we will have to square the Dirac delta function. But fear not, we have enclosed ourselves in a box. Furthermore, we are in reality calculating $\langle \vec{p}, \vec{q} | e^{-iHT} | \vec{k} \rangle$, the amplitude for the state $|\vec{k}\rangle$ to become the state $|\vec{p}, \vec{q}\rangle$ after a large but finite time T. Thus we could in all comfort write

$$[(2\pi)^4 \delta^{(4)}(p+q-k)]^2 = (2\pi)^4 \delta^{(4)}(p+q-k) \int d^4x e^{i(p+q-k)x}$$

$$= (2\pi)^4 \delta^{(4)}(p+q-k) \int d^4x = (2\pi)^4 \delta^{(4)}(p+q-k)VT \tag{35}$$

Thus the transition probability per unit time, aka the transition rate, is equal to

$$\frac{|\mathcal{A}|^2}{T} = \frac{V}{V^3} \left(\frac{1}{2\omega_p 2\omega_q 2\omega_k} \right) (2\pi)^4 \delta^{(4)}(p+q-k)g^2 \tag{36}$$

Recall that there are $V d^3p/(2\pi)^3$ states in the volume d^3p in momentum space. Hence, multiplying the number of final states $(V d^3p/(2\pi)^3)(V d^3q/(2\pi)^3)$ by the transition rate $|\mathcal{A}|^2/T$, we obtain the differential decay rate of a meson into two mesons carrying off momenta in some specified range d^3p and d^3q:

$$d\Gamma = \frac{1}{2\omega_k} \frac{V}{V^3} \left(V \frac{d^3p}{(2\pi)^3 2\omega_p} \right) \left(V \frac{d^3q}{(2\pi)^3 2\omega_q} \right) (2\pi)^4 \delta^{(4)}(p+q-k)g^2 \tag{37}$$

Yes sir, indeed, the factors of V cancel, as they should.

To obtain the total decay rate Γ we integrate over d^3p and d^3q. Notice the factor $1/2\omega_k$: the decay rate for a moving particle is smaller than that of a resting particle by a factor m/ω_k. We have derived time dilation, as we had better.

We are now ready to generalize to the decay of a particle carrying momentum P into n particles carrying momenta $k_1, \cdots, k_n$. For definiteness, we suppose that these are all Bose particles. First, we draw all the relevant Feynman diagrams and compute the invariant amplitude $\mathcal{M}$. (In our toy example, $\mathcal{M} = ig$.) Second, the transition probability contains a factor $1/V^{n+1}$, one factor of $1/V$ for each particle, but when we squared the momentum conservation delta function we also obtained a factor of VT, which converts the transition probability into a transition rate and knocks off one power of V, leaving the factor $1/V^n$. Next, when we sum over final states, we have a factor $V d^3k_i/((2\pi)^3 2\omega_{k_i})$ for each particle in the final state. Thus the factors of V indeed cancel.

The differential decay rate of a boson of mass M in its rest frame is thus given by

$$d\Gamma = \frac{1}{2M} \frac{d^3k_1}{(2\pi)^3 2\omega(k_1)} \cdots \frac{d^3k_n}{(2\pi)^3 2\omega(k_n)} (2\pi)^4 \delta^{(4)} \left(P - \sum_{i=1}^{n} k_i \right) |\mathcal{M}|^2 \tag{38}$$

At this point, we recall that, as explained in chapter II.2, in the expansion of a fermion field into creation and annihilation operators [see (II.2.10)], we have a choice of two commonly used normalizations, trivially related by a factor $(2m)^{\frac{1}{2}}$. If you choose to use the "rest normalization" so that spinors come out nice in the rest frame, then the field expansion contains the normalization factor $(E_p/m)^{\frac{1}{2}}$ instead of the factor $(2\omega_k)^{\frac{1}{2}}$ for a Bose field [see (I.8.11)]. This entails the trivial replacement, for each fermion, of the factor $2\omega(k) = 2\sqrt{\vec{k}^2 + m^2}$ by $E(p)/m = \sqrt{\vec{p}^2 + m^2}/m$. In particular, for the decay rate of a fermion the factor $1/2M$ should be removed. If you choose the "any mass renormalization," you have to remember to normalize the spinors appearing in $\mathcal{M}$ correctly, but you need not touch the phase space factors derived here.

We next turn to scattering cross sections. As I already said, the basic concepts involved should already be familiar to you from nonrelativistic quantum mechanics. Nevertheless, it may be helpful to review the basic notions involved. For the sake of definiteness, consider some happy experimentalist sending a beam of hapless electrons crashing into a stationary proton. The flux of the beam is defined as the number of electrons crossing an imagined unit area per unit time and is thus given by $F = nv$, where n and v denote the density and velocity of the electrons in the beam. The measured event rate divided by the flux of the beam is defined to be the cross section σ, which has the dimension of an area and could be thought of as the effective size of the proton as seen by the electrons.

It may be more helpful to go to the rest frame of the electrons, in which the proton is plowing through the cloud of electrons like a bulldozer. In time Δt the proton moves through a distance $v\Delta t$ and thus sweeps through a volume $\sigma v\Delta t$, which contains $n\sigma v\Delta t$ electrons. Dividing this by Δt gives us the event rate $nv\sigma$.

To measure the differential cross section, the experimentalist sets up, typically in the lab frame in which the target particle is at rest, a detector spanning a solid angle $d\Omega = \sin\theta d\theta d\phi$ and counts the number of events per unit time.

All of this is familiar stuff. Now we could essentially take over our calculation of the differential decay rate almost in its entirety to calculate the differential cross section for the process $p_1 + p_2 \to k_1 + k_2 + \cdots + k_n$. With two particles in the initial state we now have a factor of $(1/V)^{n+2}$ in the transition probability. But as before, the square of the momentum conservation delta function produces one power of V and counting the momentum final states gives a factor V^n, so that we are left with a factor of $1/V$. You might be worried about this remaining factor of $1/V$, but recall that we still have to divide by the flux, given by $|\vec{v}_1 - \vec{v}_2|n$. Since we have normalized to one particle in the box the density n is $1/V$. Once again, all factors of V cancel, as they must.

The procedure is thus to draw all relevant diagrams to the order desired and calculate the Feynman amplitude $\mathcal{M}$ for the process $p_1 + p_2 \to k_1 + k_2 + \cdots + k_n$. Then the differential cross section is given by (again assuming all particles to be bosons)

$$d\sigma = \frac{1}{|\vec{v}_1 - \vec{v}_2| 2\omega(p_1) 2\omega(p_2)} \frac{d^3k_1}{(2\pi)^3 2\omega(k_1)} \cdots \frac{d^3k_n}{(2\pi)^3 2\omega(k_n)} (2\pi)^4 \delta^{(4)} \left(p_1 + p_2 - \sum_{i=1}^{n} k_i \right) |\mathcal{M}|^2$$

$$\tag{39}$$

We are implicitly working in a collinear frame in which the velocities of the incoming particles, $\vec{v}_1$ and $\vec{v}_2$, point in opposite directions. This class of frames includes the familiar center of mass frame and the lab frame (in which $\vec{v}_2 = 0$). In a collinear frame, $p_1 = E_1(1, 0, 0, v_1)$ and $p_2 = E_2(1, 0, 0, v_2)$, and a simple calculation shows that $((p_1 p_2)^2 - m_1^2 m_2^2) = (E_1 E_2 (v_1 - v_2))^2$. We could write the factor $|\vec{v}_1 - \vec{v}_2| E_1 E_2$ in $d\sigma$ in the more invariant-looking form $((p_1 p_2)^2 - m_1^2 m_2^2)^{\frac{1}{2}}$, thus showing explicitly that the differential cross section is invariant under Lorentz boosts in the direction of the beam, as physically must be the case.

An often encountered case involves two particles scattering into two particles in the center of mass frame. Let us do the phase space integral $\int (d^3k_1/2\omega_1)(d^3k_2/2\omega_2)\delta^{(4)}(P - k_1 - k_2)$ here for easy reference. We will do it in two different ways for your edification.

We could immediately integrate over d^3k_2 thus knocking out the 3-dimensional momentum conservation delta function $\delta^3(\vec{k}_1 + \vec{k}_2)$. Writing $d^3k_1 = k_1^2 dk_1 d\Omega$, we integrate over the remaining energy conservation delta function $\delta(\sqrt{k_1^2 + m_1^2} + \sqrt{k_1^2 + m_2^2} - E_{\text{total}})$. Using (I.2.13), we find that the integral over k_1 gives $k_1\omega_1\omega_2/E_{\text{total}}$, where $\omega_1 \equiv \sqrt{k_1^2 + m_1^2}$ and $\omega_2 \equiv \sqrt{k_1^2 + m_2^2}$, with k_1 determined by $\sqrt{k_1^2 + m_1^2} + \sqrt{k_1^2 + m_2^2} = E_{\text{total}}$. Thus we obtain

$$\int \frac{d^3k_1}{2\omega_1} \frac{d^3k_2}{2\omega_2} \delta^{(4)}(P - k_1 - k_2) = \frac{k_1}{4E_{\text{total}}} \int d\Omega \tag{40}$$

Once again, if you use the "rest normalization" for fermions, remember to make the replacement as explained above for the decay rate. The factor of $\frac{1}{4}$ should be replaced by $m_f/2$ for one fermion and one boson, and by $m_1 m_2$ for two fermions.

Alternatively, we use (I.8.14) and regressing, write $d^3k_2 = \int d^4k_2 \theta(k_2^0)\delta(k_2^2 - m_2^2)2\omega_2$. Integrate over d^4k_2 and knock out the 4-dimensional delta function, leaving us with $\int 2\omega_2 dk_1 k_1^2 d\Omega/(2\omega_1 2\omega_2)\delta((P - k_1)^2 - m_2^2)$. The argument of the delta function is $E_{\text{total}}^2 - 2E_{\text{total}}k_1 + m_1^2 - m_2^2$, and thus integrating over k_1 we get a factor of $2E_{\text{total}}$ in the denominator, giving a result in agreement with (40).

For the record, you could work out the kinematics and obtain

$$k_1 = \sqrt{(E_{\text{total}}^2 - (m_1 + m_2)^2)(E_{\text{total}}^2 - (m_1 - m_2)^2)}/2E_{\text{total}}$$

Evidently, this phase space integral also applies to the decay into two particles in the rest frame of the parent particle, in which case we replace E_{total} by M. In particular, for our toy example, we have

$$\Gamma = \frac{g^2}{16\pi M^3}\sqrt{(M^2 - (m + \mu)^2)(M^2 - (m - \mu)^2)} \tag{41}$$

The differential cross section for two-into-two scattering in the center of mass frame is given by

$$\frac{d\sigma}{d\Omega} = \frac{1}{(2\pi)^2|\vec{v}_1 - \vec{v}_2|2\omega(p_1)2\omega(p_2)} \frac{k_1}{E_{\text{total}}} F|\mathcal{M}|^2 \tag{42}$$

In particular, in the text we calculated electron-electron scattering in the relativistic limit. As shown there, we can write $|\mathcal{M}|^2 = |\widehat{\mathcal{M}}|^2/(2m)^4$ in terms of some reduced invariant amplitude $\widehat{\mathcal{M}}$. The factor $1/(2m)^4$ transforms the factors $2\omega(p)$ into $2E$. Things simplify enormously, with $|\vec{v}_1 - \vec{v}_2| = 2$ and $k_1 = \frac{1}{2}E_{\text{total}}$, so that finally

$$\frac{d\sigma}{d\Omega} = \frac{1}{2^4(4\pi)^2 E^2}|\widehat{\mathcal{M}}|^2 \tag{43}$$

Last, we come to the statistical factor S that must be included in calculating the total decay rate and the total cross section to avoid over-counting if there are identical particles in the final state. The factor S has nothing to do with quantum field theory per se and should already be familiar to you from nonrelativistic quantum mechanics. The rule is that if there are n_i identical particles of type i in the final state, the total decay rate or the total cross section must be multiplied by $S = \Pi_i 1/n_i!$ to account for indistinguishability.

To see the necessity for this factor, it suffices to think about the simplest case of two identical Bose particles. To be specific, consider electron-positron annihilation into two photons (which we will study in chapter II.8). For simplicity, average and sum over all spin polarizations. Let us calculate $d\sigma/d\Omega$ according to (43) above. This is the probability that a photon will check into a detector set up at angles θ and ϕ relative to the beam direction. If the detector clicks, then we know that the other photon emerged at an angle $\pi - \theta$ relative to the beam direction. Thus the total cross section should be

$$\sigma = \frac{1}{2}\int d\Omega \frac{d\sigma}{d\Omega} = \frac{1}{2}\int_0^\pi d\theta \frac{d\sigma}{d\theta} \tag{44}$$

(The second equality is for all the elementary cases we will encounter in which $d\sigma$ does not depend on the azimuthal angle ϕ.) In other words, to avoid double counting, we should divide by 2 if we integrate over the full angular range of θ.

More formally, we argue as follows. In quantum mechanics, a set of states $|\alpha\rangle$ is complete if $1 = \sum_\alpha |\alpha\rangle\langle\alpha|$ ("decomposition of 1"). Acting with this on $|\beta\rangle$ we see that these states must be normalized according to $\langle\alpha|\beta\rangle = \delta_{\alpha\beta}$.

Now consider the state

$$|k_1, k_2\rangle \equiv \frac{1}{\sqrt{2}}\tilde{a}^\dagger(k_1)\tilde{a}^\dagger(k_2)|0\rangle = |k_2, k_1\rangle \qquad (45)$$

containing two identical bosons. By repeatedly using the commutation relation (32), we compute $\langle q_1, q_2 | k_1, k_2 \rangle = \langle 0 | \tilde{a}(q_1)\tilde{a}(q_2)\tilde{a}^\dagger(k_1)\tilde{a}^\dagger(k_2)|0\rangle = \frac{1}{2}(\delta_{q_1 k_1}\delta_{q_2 k_2} + \delta_{q_2 k_1}\delta_{q_1 k_2})$. Thus $\sum_{q_1}\sum_{q_2}|q_1, q_2\rangle\langle q_1, q_2 | k_1, k_2\rangle = \sum_{q_1}\sum_{q_2}|q_1, q_2\rangle\frac{1}{2}(\delta_{q_1 k_1}\delta_{q_2 k_2} + \delta_{q_2 k_1}\delta_{q_1 k_2}) = \frac{1}{2}(|k_1, k_2\rangle + |k_2, k_1\rangle) = |k_1, k_2\rangle$. Thus the states $|k_1, k_2\rangle$ are normalized properly. In the sum over states, we have $1 = \cdots + \sum_{q_1}\sum_{q_2}|q_1, q_2\rangle\langle q_1, q_2| + \cdots$.

In other words, if we are to sum over q_1 and q_2 independently, then we must normalize our states as in (45) with the factor of $1/\sqrt{2}$. But then this factor would appear multiplying $\mathcal{M}$. In calculating the total decay rate or the total cross section, we are effectively summing over a complete set of final states. In summary, we have two options: either we treat the integration over $d^3k_1 d^3k_2$ as independent in which case we have to multiply the integral by $\frac{1}{2}$, or we integrate over only half of phase space.

We readily generalize from this factor of $\frac{1}{2}$ to the statistical factor S.

In closing, let me mention two interesting pieces of physics.

To calculate the cross section σ, we have to divide by the flux, and hence σ is proportional to $1/|\vec{v}_1 - \vec{v}_2|$. For exothermal processes, such as electron-positron annihilation into photons or slow neutron capture, σ could become huge as the relative velocity $v_{\mathrm{rel}} \to 0$. Fermi exploited this fact to great advantage in studying nuclear fission. Note that although the cross section, which has dimension of an area, formally goes to infinity, the reaction rate (the number of reactions per unit time) remains finite.

Positronium decay into photons is an example of a bound state decaying in finite time. In positronium, the positron and electron are not approaching each other in plane wave states, as we assumed in our cross section calculation. Rather, the probability (per unit volume) that the positron finds itself near the electron is given by $|\psi(0)|^2$ according to elementary quantum mechanics, with $\psi(x)$ the bound state wave function for whatever state of positronium we are interested in. In other words, $|\psi(0)|^2$ gives the volume density of positrons near the electron. Since $v\sigma$ is a volume divided by time, the decay rate is given by $\Gamma = v\sigma|\psi(0)|^2$.

Exercises

II.6.1 Show that the differential cross section for a relativistic electron scattering in a Coulomb potential is given by

$$\frac{d\sigma}{d\Omega} = \frac{\alpha^2}{4\vec{p}^2 v^2 \sin^4(\theta/2)}(1 - v^2\sin^2(\theta/2)).$$

Known as the Mott cross section, it reduces to the Rutherford cross section you derived in a course on quantum mechanics in the limit the electron velocity $v \to 0$.

II.6.2 To order e^2 the amplitude for positron scattering off a proton is just minus the amplitude (3) for electron scattering off a proton. Thus, somewhat counterintuitively, the differential cross sections for positron scattering off a proton and for electron scattering off a proton are the same to this order. Show that to the next order this is no longer true.

II.6.3 Show that the trace of a product of odd number of gamma matrices vanishes.

II.6.4 Prove the identity $s + t + u = \sum_a m_a^2$.

II.6.5 Verify the differential cross section for relativistic electron electron scattering given in (25).

II.6.6 For those who relish long calculations, determine the differential cross section for electron-electron scattering without taking the relativistic limit.

II.6.7 Show that the decay rate for one boson of mass M into two bosons of masses m and μ is given by

$$\Gamma = \frac{|\mathcal{M}|^2}{16\pi M^3}\sqrt{(M^2 - (m + \mu)^2)(M^2 - (m - \mu)^2)}$$

II.7 | Diagrammatic Proof of Gauge Invariance

Gauge invariance

Conceptually, rather than calculate cross sections, we have the more important task of proving that we can indeed set the photon mass μ equal to zero with impunity in calculating any physical process. With $\mu = 0$, the Lagrangian given in chapter II.1 becomes the Lagrangian for quantum electrodynamics:

$$\mathcal{L} = \bar{\psi}[i\gamma^{\mu}(\partial_{\mu} - ieA_{\mu}) - m]\psi - \tfrac{1}{4}F_{\mu\nu}F^{\mu\nu} \tag{1}$$

We are now ready for one of the most important observations in the history of theoretical physics. Behold, the Lagrangian is left invariant by the gauge transformation

$$\psi(x) \to e^{i\Lambda(x)}\psi(x) \tag{2}$$

and

$$A_{\mu}(x) \to A_{\mu}(x) + \frac{1}{ie}e^{-i\Lambda(x)}\partial_{\mu}e^{i\Lambda(x)} = A_{\mu}(x) + \frac{1}{e}\partial_{\mu}\Lambda(x) \tag{3}$$

which implies

$$F_{\mu\nu}(x) \to F_{\mu\nu}(x) \tag{4}$$

You are of course already familiar with (3) and the invariance of $F_{\mu\nu}$ from classical electromagnetism.

In contemporary theoretical physics, gauge invariance[1] is regarded as fundamental and all important, as we will see later. The modern philosophy is to look at (1) as a consequence of (2) and (3). If we want to construct a gauge invariant relativistic field theory involving a spin $\tfrac{1}{2}$ and a spin 1 field, then we are forced to quantum electrodynamics.

---

[1] The discovery of gauge invariance was one of the most arduous in the history of physics. Read J. D. Jackson and L. B. Okun, "Historical roots of gauge invariance," *Rev. Mod. Phys.* 73, 2001 and learn about the sad story of a great physicist whose misfortune in life was that his name differed from that of another physicist by only one letter.

You will notice that in (3) I have carefully given two equivalent forms. While it is simpler, and commonly done in most textbooks, to write the second form, we should also keep the first form in mind. Note that $\Lambda(x)$ and $\Lambda(x) + 2\pi$ give exactly the same transformation. Mathematically speaking, the quantities $e^{i\Lambda(x)}$ and $\partial_\mu \Lambda(x)$ are well defined, but $\Lambda(x)$ is not.

After these apparently formal but actually physically important remarks, we are ready to work on the proof. I will let you give the general proof, but I will show you the way by working through some representative examples.

Recall that the propagator for the hypothetical massive photon is $i D_{\mu\nu} = i(k_\mu k_\nu / \mu^2 - g_{\mu\nu})/(k^2 - \mu^2)$. We can set the μ^2 in the denominator equal to zero without further ado and write the photon propagator effectively as $i D_{\mu\nu} = i(k_\mu k_\nu / \mu^2 - g_{\mu\nu})/k^2$. The dangerous term is $k_\mu k_\nu / \mu^2$. We want to show that it goes away.

A specific example

First consider electron-electron scattering to order e^4. Of the many diagrams, focus on the two in figure II.7.1.a. The Feynman amplitude is then

$$\bar{u}(p') \left(\gamma^\lambda \frac{1}{\not{p} + \not{k} - m} \gamma^\mu + \gamma^\mu \frac{1}{\not{p}' - \not{k} - m} \gamma^\lambda \right) u(p) \frac{i}{k^2} \left(\frac{k_\mu k^\nu}{\mu^2} - \delta_\mu^\nu \right) \Gamma_{\lambda\nu} \tag{5}$$

where $\Gamma_{\lambda\nu}$ is some factor whose detailed structure does not concern us. For the specific case shown in figure II.7.1a we can of course write out $\Gamma_{\lambda\nu}$ explicitly if we want. Note the plus sign here from interchanging the two photons since photons obey Bose statistics.

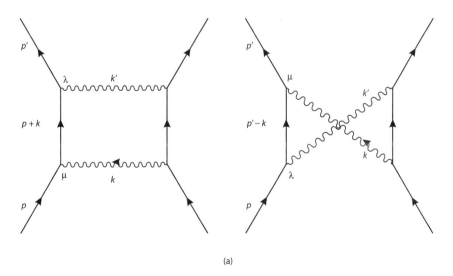

(a)

Figure II.7.1

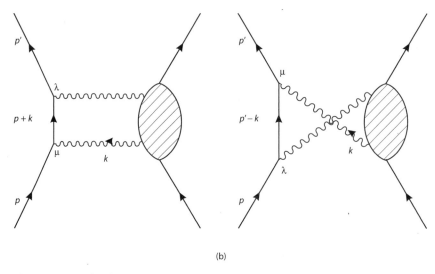

(b)

Figure II.7.1 (*continued*)

Focus on the dangerous term. Contracting the $\bar{u}(p')(\cdots)u(p)$ factor in (5) with k_μ we have

$$\bar{u}(p') \left(\gamma^\lambda \frac{1}{\not{p}+\not{k}-m} \not{k} + \not{k} \frac{1}{\not{p}'-\not{k}-m} \gamma^\lambda \right) u(p) \tag{6}$$

The trick is to write the $\not{k}$ in the numerator of the first term as $(\not{p}+\not{k}-m)-(\not{p}-m)$, and in the numerator of the second term as $(\not{p}'-m)-(\not{p}'-\not{k}-m)$. Using $(\not{p}-m)u(p)=0$ and $\bar{u}(p')(\not{p}'-m)=0$, we see that the expression in (6) vanishes. This proves the theorem in this simple example. But since the explicit form of $\Gamma_{\lambda\nu}$ did not enter, the proof would have gone through even if figure II.7.1a were replaced by the more general figure II.7.1b, where arbitrarily complicated processes could be going on under the shaded blob.

Indeed we can generalize to figure II.7.1c. Apart from the photon carrying momentum k that we are focusing on, there are already n photons attached to the electron line. These n photons are just "spectators" in the proof in the same way that the photon carrying momentum k' in figure II.7.1a never came into the proof that (6) vanishes. The photon we are focusing on can attach to the electron line in $n+1$ different places. You can now extend the proof as an exercise.

Photon landing on an internal line

In the example we just considered, the photon line in question lands on an external electron line. The fact that the line is "capped at the two ends" by $\bar{u}(p')$ and $u(p)$ is crucial in the proof. What if the photon line in question lands on an internal line?

An example is shown in figure II.7.2, contributing to electron-electron scattering in order e^8. The figure contains three distinct diagrams. The electron "on the left" emits

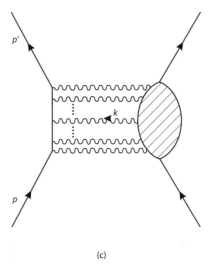

(c)

Figure II.7.1 (*continued*)

three photons, which attach to an internal electron loop. The electron "on the right" emits a photon with momentum k, which can attach to the loop in three distinct ways.

Since what we care about is whether the $k_\mu k_\rho / \mu^2$ piece in the photon propagator $i(k_\mu k_\rho / \mu^2 - g_{\mu\rho})/k^2$ goes away or not, we can for our purposes replace that photon propagator by k_μ. To save writing slightly, we define $p_1 = p + q_1$ and $p_2 = p_1 + q_2$ (see the momentum labels in figure II.7.2): Let's focus on the relevant part of the three diagrams, referring to them as A, B, and C.

$$A = \int \frac{d^4 p}{(2\pi)^4} \operatorname{tr} \left(\gamma^\nu \frac{1}{\not{p}_2 + \not{k} - m} \gamma^\sigma \frac{1}{\not{p}_1 + \not{k} - m} \gamma^\lambda \frac{1}{\not{p} + \not{k} - m} \not{k} \frac{1}{\not{p} - m} \right) \tag{7}$$

$$B = \int \frac{d^4 p}{(2\pi)^4} \operatorname{tr} \left(\gamma^\nu \frac{1}{\not{p}_2 + \not{k} - m} \gamma^\sigma \frac{1}{\not{p}_1 + \not{k} - m} \not{k} \frac{1}{\not{p}_1 - m} \gamma^\lambda \frac{1}{\not{p} - m} \right) \tag{8}$$

and

$$C = \int \frac{d^4 p}{(2\pi)^4} \operatorname{tr} \left(\gamma^\nu \frac{1}{\not{p}_2 + \not{k} - m} \not{k} \frac{1}{\not{p}_2 - m} \gamma^\sigma \frac{1}{\not{p}_1 - m} \gamma^\lambda \frac{1}{\not{p} - m} \right) \tag{9}$$

This looks like an unholy mess, but it really isn't. We use the same trick we used before. In C write $\not{k} = (\not{p}_2 + \not{k} - m) - (\not{p}_2 - m)$, so that

$$C = \int \frac{d^4 p}{(2\pi)^4} \left[\operatorname{tr} \left(\gamma^\nu \frac{1}{\not{p}_2 - m} \gamma^\sigma \frac{1}{\not{p}_1 - m} \gamma^\lambda \frac{1}{\not{p} - m} \right) \right.$$
$$\left. - \operatorname{tr} \left(\gamma^\nu \frac{1}{\not{p}_2 + \not{k} - m} \gamma^\sigma \frac{1}{\not{p}_1 - m} \gamma^\lambda \frac{1}{\not{p} - m} \right) \right] \tag{10}$$

In B write $\not{k} = (\not{p}_1 + \not{k} - m) - (\not{p}_1 - m)$, so that

$$B = \int \frac{d^4 p}{(2\pi)^4} \left[\operatorname{tr}(\gamma^\nu \frac{1}{\not{p}_2 + \not{k} - m} \gamma^\sigma \frac{1}{\not{p}_1 - m} \gamma^\lambda \frac{1}{\not{p} - m}) \right.$$
$$\left. - \operatorname{tr} \left(\gamma^\nu \frac{1}{\not{p}_2 + \not{k} - m} \gamma^\sigma \frac{1}{\not{p}_1 + \not{k} - m} \gamma^\lambda \frac{1}{\not{p} - m} \right) \right] \tag{11}$$

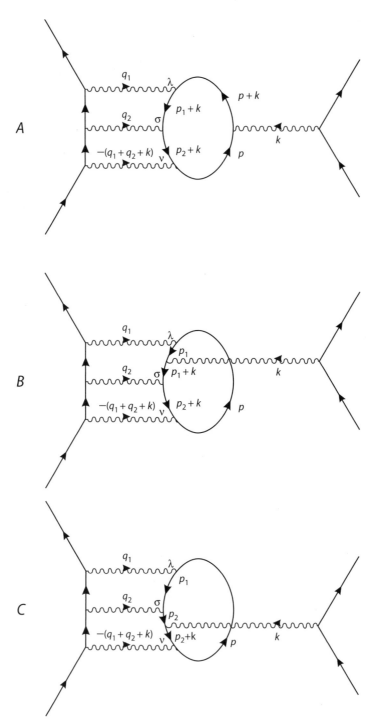

Figure II.7.2

Finally, in A write $\not{k} = (\not{p} + \not{k} - m) - (\not{p} - m)$

$$A = \int \frac{d^4 p}{(2\pi)^4} \left[\text{tr} \left(\gamma^\nu \frac{1}{\not{p}_2 + \not{k} - m} \gamma^\sigma \frac{1}{\not{p}_1 + \not{k} - m} \gamma^\lambda \frac{1}{\not{p} - m} \right) \right.$$
$$\left. - \text{tr} \left(\gamma^\nu \frac{1}{\not{p}_2 + \not{k} - m} \gamma^\sigma \frac{1}{\not{p}_1 + \not{k} - m} \gamma^\lambda \frac{1}{\not{p} + \not{k} - m} \right) \right] \tag{12}$$

Now you see what is happening. When we add the three diagrams together terms cancel in pairs, leaving us with

$$A + B + C = \int \frac{d^4 p}{(2\pi)^4} \left[\text{tr} \left(\gamma^\nu \frac{1}{\not{p}_2 - m} \gamma^\sigma \frac{1}{\not{p}_1 - m} \gamma^\lambda \frac{1}{\not{p} - m} \right) \right.$$
$$\left. - \text{tr} \left(\gamma^\nu \frac{1}{\not{p}_2 + \not{k} - m} \gamma^\sigma \frac{1}{\not{p}_1 + \not{k} - m} \gamma^\lambda \frac{1}{\not{p} + \not{k} - m} \right) \right] \tag{13}$$

If we shift (see exercise II.7.2) the dummy integration variable $p \to p - k$ in the second term, we see that the two terms cancel. Indeed, the $k_\mu k_\rho / \mu^2$ piece in the photon propagator goes away and we can set $\mu = 0$.

I will leave the general proof to you. We have done it for one particular process. Try it for some other process. You will see how it goes.

Ward-Takahashi identity

Let's summarize. Given any physical amplitude $T^{\mu\cdots}(k, \cdots)$ with external electrons on shell [this is jargon for saying that all necessary factors $u(p)$ and $\bar{u}(p)$ are included in $T^{\mu\cdots}(k, \cdots)$] describing a process with a photon carrying momentum k coming out of, or going into, a vertex labeled by the Lorentz index μ, we have

$$k_\mu T^{\mu\cdots}(k, \cdots) = 0 \tag{14}$$

This is sometimes known as a Ward-Takahashi identity.

The bottom line is that we can write $i D_{\mu\nu} = -i g^{\mu\nu}/k^2$ for the photon propagator. Since we can discard the $k_\mu k_\nu / \mu^2$ term in the photon propagator $i(k_\mu k_\nu / \mu^2 - g_{\mu\nu})/k^2$ we can also add in a $k_\mu k_\nu / k^2$ term with an arbitrary coefficient. Thus, for the photon propagator we can use

$$i D_{\mu\nu} = \frac{i}{k^2} \left[(1 - \xi) \frac{k_\mu k_\nu}{k^2} - g_{\mu\nu} \right] \tag{15}$$

where we can choose the number ξ to simplify our calculation as much as possible. Evidently, the choice of ξ amounts to a choice of gauge for the electromagnetic field. In particular, the choice $\xi = 1$ is known as the Feynman gauge, and the choice $\xi = 0$ is known as the Landau gauge. If you find an especially nice choice, you can have a gauge named after you as well! For fairly simple calculations, it is often advisable to calculate with an arbitrary ξ. The fact that the end result must not depend on ξ provides a useful check on the arithmetic.

This completes the derivation of the Feynman rules for quantum electrodynamics: They are the same rules as those given in chapter II.5 for the massive vector boson theory except for the photon propagator given in (15).

We have given here a diagrammatic proof of the gauge invariance of quantum electrodynamics. We will worry later (in chapter IV.7) about the possibility that the shift of integration momentum used in the proof may not be allowed in some cases.

The longitudinal mode

We now come back to the worry we had in chapter I.5. Consider a massive spin 1 meson moving along the z−direction. The 3 polarization vectors are fixed by the condition $k^\lambda \varepsilon_\lambda = 0$ with $k^\lambda = (\omega, 0, 0, k)$ (recall chapter I.5) and the normalization $\varepsilon^\lambda \varepsilon_\lambda = -1$, so that $\varepsilon_\lambda^{(1)} = (0, 1, 0, 0)$, $\varepsilon_\lambda^{(2)} = (0, 0, 1, 0)$, $\varepsilon_\lambda^{(3)} = (-k, 0, 0, \omega)/\mu$. Note that as $\mu \to 0$, the longitudinal polarization vector $\varepsilon_\lambda^{(3)}$ becomes proportional to $k_\lambda = (\omega, 0, 0, -k)$. The amplitude for emitting a meson with a longitudinal polarization in the process described by (14) is given by $\varepsilon_\lambda^{(3)} T^{\lambda \cdots} = (-k T^{0 \cdots} + \omega T^{3 \cdots})/\mu = (-k T^{0 \cdots} + \sqrt{k^2 + \mu^2} T^{3 \cdots})/\mu \simeq (-k T^{0 \cdots} + (k + \frac{\mu^2}{2k}) T^{3 \cdots})/\mu$ (for $\mu \ll k$), namely $-(k_\lambda T^{\lambda \cdots}/\mu) + \frac{\mu}{2k} T^{3 \cdots}$ with $k_\lambda = (k, 0, 0, -k)$. Upon using (14) we see that the amplitude $\varepsilon_\lambda^{(3)} T^{\lambda \cdots} \to \frac{\mu}{2k} T^{3 \cdots} \to 0$ as $\mu \to 0$.

The longitudinal mode of the photon does not exist because it decouples from all physical processes.

Here is an apparent paradox. Mr. Boltzmann tells us that in thermal equilibrium each degree of freedom is associated with $\frac{1}{2} T$. Thus, by measuring some thermal property (such as the specific heat) of a box of photon gas to an accuracy of $2/3$ an experimentalist could tell if the photon is truly massless rather than have a mass of a zillionth of an electron volt.

The resolution is of course that as the coupling of the longitudinal mode vanishes as $\mu \to 0$ the time it takes for the longitudinal mode to come to thermal equilibrium goes to infinity. Our crafty experimentalist would have to be very patient.

Emission and absorption of photons

According to chapter II.5, the amplitude for emitting or absorbing an external on-shell photon with momentum k and polarization a ($a = 1, 2$) is given by $\varepsilon_\mu^{(a)}(k) T^{\mu \cdots}(k, \cdots)$. Thanks to (14), we are free to vary the polarization vector

$$\varepsilon_\mu^{(a)}(k) \to \varepsilon_\mu^{(a)}(k) + \lambda k_\mu \tag{16}$$

for arbitrary λ. You should recognize (16) as the momentum space version of (3). As we will see in the next chapter, by a judicious choice of $\varepsilon_\mu^{(a)}(k)$, we can simplify a given calculation considerably. In one choice, known as the "transverse gauge," the 4-vectors $\varepsilon_\mu^{(a)}(k) = (0, \vec{\varepsilon}(k))$ for $a = 1, 2$ do not have time components. (For a photon moving in the z-direction, this is just the choice specified in the preceding section.)

Exercises

II.7.1   Extend the proof to cover figure II.7.1c. [Hint: To get oriented, note that figure II.7.1b corresponds to $n = 1$.]

II.7.2   You might have worried whether the shift of integration variable is allowed. Rationalizing the denominators in the first integral

$$\int \frac{d^4 p}{(2\pi)^4} \, \mathrm{tr}(\gamma^\nu \frac{1}{\not{p}_2 - m} \gamma^\sigma \frac{1}{\not{p}_1 - m} \gamma^\lambda \frac{1}{\not{p} - m})$$

in (13) and imagining doing the trace, you can convince yourself that this integral is only logarithmically divergent and hence that the shift is allowed. This issue will come up again in chapter IV.7 and we are anticipating a bit here.

Photon scattering on an electron

We now apply what we just learned to calculating the amplitude for the Compton scattering of a photon on an electron, namely, the process $\gamma(k) + e(p) \to \gamma(k') + e(p')$. First step: draw the Feynman diagrams, and notice that there are two, as indicated in figure II.8.1. The electron can either absorb the photon carrying momentum k first or emit the photon carrying momentum k' first. Think back to the spacetime stories we talked about in chapter I.7. The plot of our biopic here is boringly simple: the electron comes along, absorbs and then emits a photon, or emits and absorbs a photon, and then continues on its merry way. Because this is a quantum movie, the two alternate plots are shown superposed.

So, apply the Feynman rules (chapter II.5) to get (just to make the writing a bit easier, we take the polarization vectors ε and ε' to be real)

$$\mathcal{M} = A(\varepsilon', k'; \varepsilon, k) + (\varepsilon' \leftrightarrow \varepsilon, k' \leftrightarrow -k) \tag{1}$$

where

$$
\begin{aligned}
A(\varepsilon', k'; \varepsilon, k) &= (-ie)^2 \bar{u}(p') \, \not{\varepsilon}' \, \frac{i}{\not{p} + \not{k} - m} \, \not{\varepsilon} u(p) \\
&= \frac{i(-ie)^2}{2pk} \bar{u}(p') \, \not{\varepsilon}' (\not{p} + \not{k} + m) \, \not{\varepsilon} u(p)
\end{aligned}
\tag{2}
$$

In either case, absorb first or emit first, the electron is penalized for not being real, by the factor of $1/((p+k)^2 - m^2) = 1/(2pk)$ in one case, and $1/((p-k')^2 - m^2) = -1/(2pk')$ in the other.

At this point, to obtain the differential cross section, you just have take a deep breath and calculate away. I will show you, however, that we could simplify the calculation considerably by a clever choice of polarization vectors and of the frame of reference. (The calculation is still a big mess, though!) For a change, we will be macho guys and not average and sum over the photon polarizations.

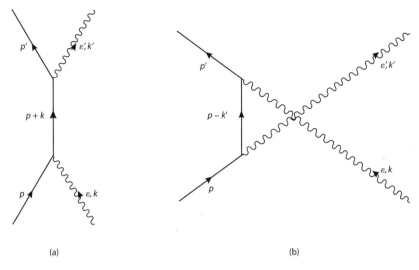

(a)

(b)

Figure II.8.1

In any case, we have $\varepsilon k = 0$ and $\varepsilon' k' = 0$. Now choose the transverse gauge introduced in the preceding chapter, so that ε and ε' have zero time components. Then calculate in the lab frame. Since $p = (m, 0, 0, 0)$, we have the additional relations

$$\varepsilon p = 0 \tag{3}$$

and

$$\varepsilon' p = 0 \tag{4}$$

Why is this a shrewd choice? Recall that $\not{a}\not{b} = 2ab - \not{b}\not{a}$. Thus, we could move $\not{p}$ past $\not{\varepsilon}$ or $\not{\varepsilon}'$ at the rather small cost of flipping a sign. Notice in (1) that $(\not{p} + \not{k} + m)\not{\varepsilon}u(p) = \not{\varepsilon}(-\not{p} - \not{k} + m)u(p) = -\not{\varepsilon}\not{k}u(p)$ (where we have used $\varepsilon k = 0$.) Thus

$$A(\varepsilon', k'; \varepsilon, k) = ie^2 \bar{u}(p') \frac{\not{\varepsilon}' \not{\varepsilon} \not{k}}{2pk} u(p) \tag{5}$$

To obtain the differential cross section, we need $|\mathcal{M}|^2$. We will wimp out a bit and suppose, just as in chapter II.6, that the initial electron is unpolarized and the polarization of the final electron is not measured. Then averaging over initial polarization and summing over final polarization we have [applying II.2.8]

$$\frac{1}{2}\Sigma\Sigma|A(\varepsilon', k'; \varepsilon, k)|^2 = \frac{e^4}{2(2m)^2(2pk)^2} \text{tr}(\not{p}' + m) \not{\varepsilon}' \not{\varepsilon} \not{k}(\not{p} + m) \not{k} \not{\varepsilon} \not{\varepsilon}' \tag{6}$$

In evaluating the trace, keep in mind that the trace of an odd number of gamma matrices vanishes. The term proportional to m^2 contains $\not{k}\not{k} = k^2 = 0$ and hence vanishes. We are left with $\text{tr}(\not{p}' \not{\varepsilon}' \not{\varepsilon} \not{k} \not{p} \not{k} \not{\varepsilon} \not{\varepsilon}') = 2kp \, \text{tr}(\not{p}' \not{\varepsilon}' \not{\varepsilon} \not{k} \not{\varepsilon} \not{\varepsilon}') = -2kp \, \text{tr}(\not{p}' \not{\varepsilon}' \not{\varepsilon} \not{\varepsilon} \not{k} \not{\varepsilon}') = 2kp \, \text{tr}(\not{p}' \not{\varepsilon}' \not{k} \not{\varepsilon}') = 8kp[2(k\varepsilon')^2 + k'p]$.

Work through the steps as indicated and the strategy should be clear. We anticommute judiciously to exploit the "zero relations" $\varepsilon p = 0$, $\varepsilon' p = 0$, $\varepsilon k = 0$, and $\varepsilon' k' = 0$ and the normalization conditions $\not\varepsilon\not\varepsilon = \varepsilon^2 = -1$ and $\not\varepsilon'\not\varepsilon' = \varepsilon'^2 = -1$ as much as possible.

We obtain

$$\frac{1}{2}\Sigma\Sigma|A(\varepsilon', k'; \varepsilon, k)|^2 = \frac{e^4}{2(2m)^2(2pk)^2}8kp[2(k\varepsilon')^2 + k'p] \tag{7}$$

The other term

$$\frac{1}{2}\Sigma\Sigma|A(\varepsilon, -k; \varepsilon', k')|^2 = \frac{e^4}{2(2m)^2(2pk)^2}8(-k'p)[2(k'\varepsilon)^2 - kp]$$

follows immediately by inspecting figure II.8.1 and interchanging $(\varepsilon' \leftrightarrow \varepsilon, k' \leftrightarrow -k)$.

Just as in chapter II.6, the interference term

$$\frac{1}{2}\Sigma\Sigma A(\varepsilon, -k; \varepsilon', k')^* A(\varepsilon', k'; \varepsilon, k) = \frac{e^4}{2(2m)^2(2pk)(-2pk')}\mathrm{tr}(\not p' + m)\not\varepsilon'\not\varepsilon\not k(\not p + m)\not k'\not\varepsilon'\not\varepsilon \tag{8}$$

is the most tedious to evaluate. Call the trace T. Clearly, it would be best to eliminate $p' = p + k - k'$, since we could "do more" with $\not p$ than $\not p'$. Divide and conquer: write $T = P + Q_1 + Q_2$. First, massage $P \equiv \mathrm{tr}(\not p + m)\not\varepsilon'\not\varepsilon\not k(\not p + m)\not k'\not\varepsilon'\not\varepsilon = m^2\mathrm{tr}\,\not\varepsilon'\not\varepsilon\not k\not k'\not\varepsilon'\not\varepsilon + \mathrm{tr}\,\not p\not\varepsilon'\not\varepsilon\not k\not p\not k'\not\varepsilon'\not\varepsilon$. In the second term, we could sail the first $\not p$ past the $\not\varepsilon$ and $\not\varepsilon'$ (ah, so nice to work in the rest frame for this problem!) to find the combination $\not p\not k\not p = 2kp\,\not p - m^2\,\not k$. The m^2 term gives a contribution that cancels the first term in P, leaving us with $P = 2kp\,\mathrm{tr}\,\not\varepsilon'\not\varepsilon\not p\not k'\not\varepsilon'\not\varepsilon = 2kp\,\mathrm{tr}\,\not p\not k'\not\varepsilon'(2\varepsilon'\varepsilon - \not\varepsilon'\not\varepsilon)\not\varepsilon = 8(kp)(k'p)[2(\varepsilon\varepsilon')^2 - 1]$. Similarly, $Q_1 = \mathrm{tr}\,\not k\not\varepsilon'\not\varepsilon\not k\not p\not k'\not\varepsilon'\not\varepsilon = -2k\varepsilon'\,\mathrm{tr}\,\not k\not p\not k'\not\varepsilon = -8(\varepsilon'k)^2k'p$ and $Q_2 = -\mathrm{tr}\,\not k'\not\varepsilon'\not\varepsilon\not k\not p\not k'\not\varepsilon'\not\varepsilon = 8(\varepsilon k')^2kp'$.

Putting it all together and writing $kp' = k'p = m\omega'$ and $k'p' = kp = m\omega$, we find

$$\frac{1}{2}\Sigma\Sigma|\mathcal{M}|^2 = \frac{e^4}{(2m)^2}\left[\frac{\omega'}{\omega} + \frac{\omega}{\omega'} + 4(\varepsilon\varepsilon')^2 - 2\right] \tag{9}$$

We calculate the differential cross section as in chapter II.6 with some minor differences since we are in the lab frame, obtaining

$$d\sigma = \frac{m}{(2\pi)^2 2\omega}\left[\int \frac{d^3k'}{2\omega'}\frac{d^3p'}{E_{p'}}\delta^{(4)}(k' + p' - k - p)\right]\frac{1}{2}\Sigma\Sigma|\mathcal{M}|^2 \tag{10}$$

As described in the appendix to chapter II.6, we could use (I.8.14) and write $\int \frac{d^3p'}{E_{p'}}(\cdots) = \int d^4p'\theta(p'^0)\delta(p'^2 - m^2)(\cdots)$. Doing the integral over d^4p' to knock out the 4-dimensional delta function, we are left with a delta function enforcing the mass shell condition $0 = p'^2 - m^2 = (p + k - k')^2 - m^2 = 2p(k - k') - 2kk' = 2m(\omega - \omega') - 2\omega\omega'(1 - \cos\theta)$, with θ the scattering angle of the photon. Thus, the frequency of the outgoing photon and of the incoming photon are related by

$$\omega' = \frac{\omega}{1 + \frac{2\omega}{m}\sin^2\frac{\theta}{2}} \tag{11}$$

giving the frequency shift that won Arthur Compton the Nobel Prize. You realize of course that this formula, though profound at the time, is "merely" relativistic kinematics and has nothing to do with quantum field theory per se.

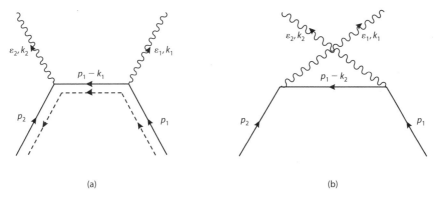

Figure II.8.2

What quantum field theory gives us is the Klein-Nishina formula (1929)

$$\frac{d\sigma}{d\Omega} = \frac{1}{(2m)^2} \left(\frac{e^2}{4\pi}\right)^2 \left(\frac{\omega'}{\omega}\right)^2 \left[\frac{\omega'}{\omega} + \frac{\omega}{\omega'} + 4(\varepsilon\varepsilon')^2 - 2\right]. \tag{12}$$

You ought to be impressed by the year.

Electron-positron annihilation

Here and in chapter II.6 we calculated the cross sections for some interesting scattering processes. At the end of that chapter we marvelled at the magic of theoretical physics. Even more magical is the annihilation of matter and antimatter, a process that occurs only in relativistic quantum field theory. Specifically, an electron and a positron meet and annihilate each other, giving rise to two photons: $e^-(p_1) + e^+(p_2) \to \gamma(\varepsilon_1, k_1) + \gamma(\varepsilon_2, k_2)$. (Annihilating into one physical, that is, on-shell, photon is kinematically impossible.) This process, often featured in science fiction, is unknown in nonrelativistic quantum mechanics. Without quantum field theory, you would be clueless on how to calculate, say, the angular distribution of the outgoing photons.

But having come this far, you simply apply the Feynman rules to the diagrams in figure II.8.2, which describe the process to order e^2. We find the amplitude $\mathcal{M} = A(k_1, \varepsilon_1; k_2, \varepsilon_2) + A(k_2, \varepsilon_2; k_1, \varepsilon_1)$ (Bose statistics for the two photons!), where

$$A(k_1, \varepsilon_1; k_2, \varepsilon_2) = (ie)(-ie)\bar{v}(p_2) \, \slashed{\varepsilon}_2 \frac{i}{\slashed{p}_1 - \slashed{k}_1 - m} \, \slashed{\varepsilon}_1 u(p_1) \tag{13}$$

Students of quantum field theory are sometimes confused that while the incoming electron goes with the spinor u, the incoming positron goes with $\bar{v}$, and not with v. You could check this by inspecting the hermitean conjugate of (II.2.10). Even simpler, note that $\bar{v}(\cdots)u$ [with $(\cdots)$ a bunch of gamma matrices contracted with various momenta] transforms correctly under the Lorentz group, while $v(\cdots)u$ does not (and does not even make sense, since they are both column spinors.) Or note that the annihilation operator d for the positron is associated with $\bar{v}$, not v.

I want to emphasize that the positron carries momentum $p_2 = (+\sqrt{\vec{p}_2^2 + m^2}, \vec{p}_2)$ on its way to that fatal rendezvous with the electron. Its energy $p_2^0 = +\sqrt{\vec{p}_2^2 + m^2}$ is manifestly positive. Nor is any physical particle traveling backward in time. The honest experimentalist who arranged for the positron to be produced wouldn't have it otherwise. Remember my rant at the end of chapter II.2?

In figure II.8.2a I have labeled the various lines with arrows indicating momentum flow. The external particles are physical and there would have been serious legal issues if their energies were not positive. There is no such restriction on the virtual particle being exchanged, though. Which way we draw the arrow on the virtual particle is purely up to us. We could reverse the arrow, and then the momentum label would become $p_2 - k_2 = k_1 - p_1$: the time component of this "composite" 4-vector can be either positive or negative.

To make the point totally clear, we could also label the lines by dotted arrows showing the flow of (electron) charge. Indeed, on the positron line, momentum and (electron) charge flow in opposite directions.

Crossing

I now invite you to discover something interesting by staring at the expression in (13) for a while.

Got it? Does it remind you of some other amplitude?

No? How about looking at the amplitude for Compton scattering in (2)?

Notice that the two amplitudes could be turned into each other (up to an irrelevant sign) by the exchange

$$p \leftrightarrow p_1, k \leftrightarrow -k_1, p' \leftrightarrow -p_2, k' \leftrightarrow k_2, \varepsilon \leftrightarrow \varepsilon_1, \varepsilon' \leftrightarrow \varepsilon_2, u(p) \leftrightarrow u(p_1), u(p') \leftrightarrow v(p_2) \tag{14}$$

This is known as crossing. Diagrammatically, we are effectively turning the diagrams in figures II.8.1 and II.8.2 into each other by $90°$ rotations. Crossing expresses in precise terms what people who like to mumble something about negative energy traveling backward in time have in mind.

Once again, it is advantageous to work in the electron rest frame and in the transverse gauge, so that we have $\varepsilon_1 p_1 = 0$ and $\varepsilon_2 p_1 = 0$ as well as $\varepsilon_1 k_1 = 0$ and $\varepsilon_2 k_1 = 0$. Averaging over the electron and positron polarizations we obtain

$$\frac{d\sigma}{d\Omega} = \frac{\alpha^2}{8m} \left(\frac{\omega_1}{|\vec{p}|}\right) \left[\frac{\omega'}{\omega} + \frac{\omega}{\omega'} - 4(\varepsilon\varepsilon')^2 + 2\right] \tag{15}$$

with $\omega_1 = m(m + E)/(m + E - p\cos\theta)$, $\omega_2 = (E - m - p\cos\theta)\omega_1/m$, and $p = |\vec{p}|$ and E the positron momentum and energy, respectively.

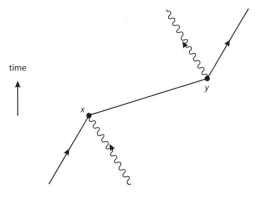

Figure II.8.3

Special relativity and quantum mechanics require antimatter

The formalism in chapter II.2 makes it totally clear that antimatter is obligatory. For us to be able to add the operators b and $d^\dagger$ in (II.2.10) they must carry the same electric charge, and thus b and d carry opposite charge. No room for argument there. Still, it would be comforting to have a physical argument that special relativity and quantum mechanics mandate antimatter.

Compton scattering offers a context for constructing a nice heuristic argument. Think of the process in spacetime. We have redrawn figure II.8.1a in figure II.8.3: the electron is hit by the photon at the point x, propagates to the point y, and emits a photon. We have assumed implicitly that $(y^0 - x^0) > 0$, since we don't know what propagating backward in time means. (If the reader knows how to build a time machine, let me know.) But special relativity tells us that another observer moving by (along the 1-direction say) would see the time difference $(y'^0 - x'^0) = \cosh\varphi(y^0 - x^0) - \sinh\varphi(y^1 - x^1)$, which could be negative for large enough boost parameter φ, provided that $(y^1 - x^1) > (y^0 - x^0)$, that is, if the separation between the two spacetime points x and y were spacelike. Then this observer would see the field disturbance propagating from y to x. Since we see negative electric charge propagating from x to y, the other observer must see positive electric charge propagating from y to x. Without special relativity, as in nonrelativistic quantum mechanics, we simply write down the Schödinger equation for the electron and that is that. Special relativity allows different observers to see different time ordering and hence opposite charges flowing toward the future.

Exercises

II.8.1 Show that averaging and summing over photon polarizations amounts to replacing the square bracket in (9) by $2[\frac{\omega'}{\omega} + \frac{\omega}{\omega'} - \sin^2\theta]$. [Hint: We are working in the transverse gauge.]

II.8.2 Repeat the calculation of Compton scattering for circularly polarized photons.

Part III | Renormalization and Gauge Invariance

III.1 | Cutting Off Our Ignorance

Who is afraid of infinities? Not I, I just cut them off.
—Anonymous

An apparent sleight of hand

The pioneers of quantum field theory were enormously puzzled by the divergent integrals that they often encountered in their calculations, and they spent much of the 1930s and 1940s struggling with these infinities. Many leading lights of the day, driven to desperation, advocated abandoning quantum field theory altogether. Eventually, a so-called renormalization procedure was developed whereby the infinities were argued away and finite physical results were obtained. But for many years, well into the late 1960s and even the 1970s many physicists looked upon renormalization theory suspiciously as a sleight of hand. Jokes circulated that in quantum field theory infinity is equal to zero and that under the rug in a field theorist's office had been swept many infinities.

Eventually, starting in the 1970s a better understanding of quantum field theory was developed through the efforts of Ken Wilson and many others. Field theorists gradually came to realize that there is no problem of divergences in quantum field theory at all. We now understand quantum field theory as an effective low energy theory in a sense I will explain briefly here and in more detail in chapter VIII.3.

Field theory blowing up

We have to see an infinity before we can talk about how to deal with infinities. Well, we saw one in chapter I.7. Recall that the order λ^2 correction (I.7.23) to the meson-meson scattering amplitude diverges. With $K \equiv k_1 + k_2$, we have

$$\mathcal{M} = \tfrac{1}{2}(-i\lambda)^2 i^2 \int \frac{d^4k}{(2\pi)^4} \frac{1}{k^2 - m^2 + i\varepsilon} \frac{1}{(K-k)^2 - m^2 + i\varepsilon} \tag{1}$$

As I remarked back in chapter I.7, even without doing any calculations we can see the problem that confounded the pioneers of quantum field theory. The integrand goes as $1/k^4$

for large k and thus the integral diverges logarithmically as $\int d^4k/k^4$. (The ordinary integral $\int^\infty dr \, r^n$ diverges linearly for $n = 0$, quadratically for $n = 1$, and so on, and $\int^\infty dr/r$ diverges logarithmically.) Since this divergence is associated with large values of k it is known as an ultraviolet divergence.

To see how to deal with this apparent infinity, we have to distinguish between two conceptually separate issues, associated with the terrible names "regularization" and "renormalization" for historical reasons.

Parametrization of ignorance

Suppose we are studying quantum electrodynamics instead of this artificial φ^4 theory. It would be utterly unreasonable to insist that the theory of an electron interacting with a photon would hold to arbitrarily high energies. At the very least, with increasingly higher energies other particles come in, and eventually electrodynamics becomes merely part of a larger electroweak theory. Indeed, these days it is thought that as we go to higher and higher energies the whole edifice of quantum field theory will ultimately turn out to be an approximation to a theory whose identity we don't yet know, but probably a string theory according to some physicists.

The modern view is that quantum field theory should be regarded as an effective low energy theory, valid up to some energy (or momentum in a Lorentz invariant theory) scale Λ. We can imagine living in a universe described by our toy φ^4 theory. As physicists in this universe explore physics to higher and higher momentum scales they will eventually discover that their universe is a mattress constructed out of mass points and springs. The scale Λ is roughly the inverse of the lattice spacing.

When I teach quantum field theory, I like to write "Ignorance is no shame" on the blackboard for emphasis when I get to this point. Every physical theory should have a domain of validity beyond which we are ignorant of the physics. Indeed were this not true physics would not have been able to progress. It is a good thing that Feynman, Schwinger, Tomonaga, and others who developed quantum electrodynamics did not have to know about the charm quark for example.

I emphasize that Λ should be thought of as physical, parametrizing our threshold of ignorance, and not as a mathematical construct.[1] Indeed, physically sensible quantum field theories should all come with an implicit Λ. If anyone tries to sell you a field theory claiming that it holds up to arbitrarily high energies, you should check to see if he sold used cars for a living. (As I wrote this, a colleague who is an editor of *Physical Review Letters* told me that he worked as a garbage collector during high school vacations, adding jokingly that this experience prepared him well for his present position.)

---

[1] We saw a particularly vivid example of this in chapter I.8. When we define a conducting plate as a surface on which a tangential electric field vanishes, we are ignorant of the physics of the electrons rushing about to counter any such imposed field. At extremely high frequencies, the electrons can't rush about fast enough and new physics comes in, namely that high frequency modes do not see the plates. In calculating the Casimir force we parametrize our ignorance with $a \sim \Lambda^{-1}$.

Figure III.1.1

Thus, in evaluating (1) we should integrate only up to Λ, known as a cutoff. We literally cut off the momentum integration (fig. III.1.1).[2] The integral is said to have been "regularized."

Since my philosophy in this book is to emphasize the conceptual rather than the computational, I will not actually do the integral but merely note that it is equal to $2iC \log(\Lambda^2/K^2)$ where C is some numerical constant that you can compute if you want (see appendix 1 to this chapter). For the sake of simplicity I also assumed that $m^2 << K^2$ so that we could neglect m^2 in the integrand. It is convenient to use the kinematic variables $s \equiv K^2 = (k_1 + k_2)^2$, $t \equiv (k_1 - k_3)^2$, and $u \equiv (k_1 - k_4)^2$ introduced in chapter II.6. (Writing out the k_j's explicitly in the center-of-mass frame, you see that s, t, and u are related to rather mundane quantities such as the center-of-mass energy and the scattering angle.) After all this, the meson-meson scattering amplitude reads

$$\mathcal{M} = -i\lambda + iC\lambda^2 [\log\left(\frac{\Lambda^2}{s}\right) + \log\left(\frac{\Lambda^2}{t}\right) + \log\left(\frac{\Lambda^2}{u}\right)] + O(\lambda^3) \qquad (2)$$

This much is easy enough to understand. After regularization, we speak of cutoff-dependent quantities instead of divergent quantities, and $\mathcal{M}$ depends logarithmically on the cutoff.

---

[2] A. Zee, *Einstein's Universe*, p. 204. Cartooning schools apparently teach that physicists in general, and quantum field theorists in particular, all wear lab coats.

What is actually measured

Now that we have dealt with regularization, let us turn to renormalization, a terrible word because it somehow implies we are doing normalization again when in fact we haven't yet.

The key here is to imagine what we would tell an experimentalist about to measure meson-meson scattering. We tell her (or him if you insist) that we need a cutoff Λ and she is not bothered at all; to an experimentalist it makes perfect sense that any given theory has a finite domain of validity.

Our calculation is supposed to tell her how the scattering will depend on the center-of-mass energy and the scattering angle. So we show her the expression in (2). She points to λ and exclaims, "What in the world is that?"

We answer, "The coupling constant," but she says, "What do you mean, coupling constant, it's just a Greek letter!"

A confused student, Confusio, who has been listening in, pipes up, "Why the fuss? I have been studying physics for years and years, and the teachers have shown us lots of equations with Latin and Greek letters, for example, Hooke's law $F = -kx$, and nobody gets upset about k being just a Latin letter."

Smart Experimentalist: "But that is because if you give me a spring I can go out and measure k. That's the whole point! Mr. Egghead Theorist here has to tell me how to measure this λ."

Woah, that is a darn smart experimentalist. We now have to think more carefully what a coupling constant really means. Think about α, the coupling constant of quantum electrodynamics. Well, it is the coefficient of $1/r$ in Coulomb's law. Fine, Monsieur Coulomb measured it using metallic balls or something. But a modern experimentalist could just as well have measured α by scattering an electron at such and such an energy and at such and such a scattering angle off a proton. We explain all this to our experimentalist friend.

SE, nodding, agrees: "Oh yes, recently my colleague so and so measured the coupling for meson-meson interaction by scattering one meson off another at such and such an energy and at such and such a scattering angle, which correspond to your variables s, t, and u having values s_0, t_0, and u_0. But what does the coupling constant my colleague measured, let us call it λ_P, with the subscript meaning "physical," have to do with your theoretical λ, which, as far as I am concerned, is just a Greek letter in something you call a Lagrangian!"

Confusio, "Hey, if she's going to worry about small lambda, I am going to worry about big lambda. How do I know how big the domain of validity is?"

SE: "Confusio, you are not as dumb as you look! Mr. Egghead Theorist, if I use your formula (2), what is the precise value of Λ that I am supposed to plug in? Does it depend on your mood, Mr. Theorist? If you wake up feeling optimistic, do you use 2Λ instead of Λ? And if your girl friend left you, you use $\frac{1}{2}\Lambda$?"

We assert, "Ha, we know the answer to that one. Look at (2): $\mathcal{M}$ is supposed to be an actual scattering amplitude and should not depend on Λ. If someone wants to change Λ

we just shift λ in such a way so that $\mathcal{M}$ does not change. In fact, a couple lines of arithmetic will show you precisely what $d\lambda/d\Lambda$ has to be (see exercise III.1.3)."

SE: "Okay, so λ is secretly a function of Λ. Your notation is lousy."

We admit, "Exactly, this bad notation has confused generations of physicists."

SE: "I am still waiting to hear how the λ_P my experimental colleague measured is related to your λ."

We say, "Aha, that's easy. Just look at (2), which is repeated here for clarity and for your reading convenience:

$$\mathcal{M} = -i\lambda + iC\lambda^2 \left[\log\left(\frac{\Lambda^2}{s}\right) + \log\left(\frac{\Lambda^2}{t}\right) + \log\left(\frac{\Lambda^2}{u}\right) \right] + O(\lambda^3) \tag{3}$$

According to our theory, λ_P is given by

$$-i\lambda_P = -i\lambda + iC\lambda^2 \left[\log\left(\frac{\Lambda^2}{s_0}\right) + \log\left(\frac{\Lambda^2}{t_0}\right) + \log\left(\frac{\Lambda^2}{u_0}\right) \right] + O(\lambda^3) \tag{4}$$

To show you clearly what is involved, let us denote the sum of logarithms in the square bracket in (3) and in (4) by L and by L_0, respectively, so that we can write (3) and (4) more compactly as

$$\mathcal{M} = -i\lambda + iC\lambda^2 L + O(\lambda^3) \tag{5}$$

and

$$-i\lambda_P = -i\lambda + iC\lambda^2 L_0 + O(\lambda^3) \tag{6}$$

That is how λ_P and λ are related."

SE: "If you give me the scattering amplitude expressed in terms of the physical coupling λ_P then it's of use to me, but it's not of use in terms of λ. I understand what λ_P is, but not λ."

We answer: "Fine, it just takes two lines of algebra to eliminate λ in favor of λ_P. Big deal. Solving (6) for λ gives

$$-i\lambda = -i\lambda_P - iC\lambda^2 L_0 + O(\lambda^3) = -i\lambda_P - iC\lambda_P^2 L_0 + O(\lambda_P^3) \tag{7}$$

The second equality is allowed to the order of approximation indicated. Now plug this into (5)

$$\mathcal{M} = -i\lambda + iC\lambda^2 L + O(\lambda^3) = -i\lambda_P - iC\lambda_P^2 L_0 + iC\lambda_P^2 L + O(\lambda_P^3) \tag{8}$$

Please check that all manipulations are legitimate up to the order of approximation indicated."

The "miracle"

Lo and behold! The miracle of renormalization!

Now in the scattering amplitude $\mathcal{M}$ we have the combination $L - L_0 = [\log(s_0/s) + \log(t_0/t) + \log(u_0/u)]$. In other words, the scattering amplitude comes out as

$$\mathcal{M} = -i\lambda_P + iC\lambda_P^2 \left[\log\left(\frac{s_0}{s}\right) + \log\left(\frac{t_0}{t}\right) + \log\left(\frac{u_0}{u}\right) \right] + O(\lambda_P^3) \tag{9}$$

We announce triumphantly to our experimentalist friend that when the scattering amplitude is expressed in terms of the physical coupling constant λ_P as she had wanted, the cutoff Λ disappear completely!

The answer should always be in terms of physically measurable quantities

The lesson here is that we should express physical quantities not in terms of "fictitious" theoretical quantities such as λ, but in terms of physically measurable quantities such as λ_P.

By the way, in the literature, λ_P is often denoted by λ_R and for historical reasons called the "renormalized coupling constant." I think that the physics of "renormalization" is much clearer with the alternative term "physical coupling constant," hence the subscript P. We never did have a "normalized coupling constant."

Suddenly Confusio pipes up again; we have almost forgotten him!

Confusio: "You started out with an $\mathcal{M}$ in (2) with two unphysical quantities λ and Λ, and their "unphysicalness" sort of cancel each other out."

SE: "Yeah, it is reminiscent of what distinguishes the good theorists from the bad ones. The good ones always make an even number of sign errors, and the bad ones always make an odd number."

Integrating over only the slow modes

In the path integral formulation, the scattering amplitude $\mathcal{M}$ discussed here is obtained by evaluating the integral (chapter I.7)

$$\int D\varphi \; \varphi(x_1)\varphi(x_2)\varphi(x_3)\varphi(x_4)e^{i\int d^dx\{\frac{1}{2}[(\partial\varphi)^2-m^2\varphi^2]-\frac{\lambda}{4!}\varphi^4\}}$$

The regularization used here corresponds roughly to restricting ourselves, in the integral $\int D\varphi$, to integrating over only those field configurations $\varphi(x)$ whose Fourier transform $\varphi(k)$ vanishes for $k \gtrsim \Lambda$. In other words, the fields corresponding to the internal lines in the Feynman diagrams in fig. (I.7.10) are not allowed to fluctuate too energetically. We will come back to this path integral formulation later when we discuss the renormalization group.

Alternative lifestyles

I might also mention that there are a number of alternative ways of regularizing Feynman diagrams, each with advantages and disadvantages that make them suitable for some calculations but not others. The regularization used here, known as Pauli-Villars, has the advantage of being physically transparent. Another often used regularization is known as dimensional regularization. We pretend that we are calculating in d-dimensional space-time. After the Feynman integral has been beaten down to a suitable form, we do an analytic continuation in d and set $d = 4$ at the end of the day. The cutoff dependences of various integrals now show up as poles as we let $d \to 4$. Just as the cutoff Λ disappears when the scattering amplitude is expressed in terms of the physical coupling constant λ_P, in dimensional regularization the scattering amplitude expressed in terms of λ_P is free of poles. While dimensional regularization proves to be useful in certain contexts as I will note in a later chapter, it is considerably more abstract and formal than Pauli-Villars regularization. Each to his or her own taste when it comes to regularizing.

Since the emphasis in this book is on the conceptual rather than the computational, I won't discuss other regularization schemes but will merely sketch how Pauli-Villars and dimensional regularizations work in two appendixes to this chapter.

Appendix 1: Pauli-Villars regularization

The important message of this chapter is the conceptual point that when physical amplitudes are expressed in term of physical coupling constants the cutoff dependence disappears. The actual calculation of the Feynman integral is unimportant. But I will show you how to do the integral just in case you would like to do Feynman integrals for a living.

Let us start with the convergent integral

$$\int \frac{d^4k}{(2\pi)^4} \frac{1}{(k^2 - c^2 + i\varepsilon)^3} = \frac{-i}{32\pi^2 c^2} \tag{10}$$

The dependence on c^2 follows from dimensional analysis. The overall factor is calculated in appendix D.

Applying the identity (D.15)

$$\frac{1}{xy} = \int_0^1 d\alpha \frac{1}{[\alpha x + (1 - \alpha)y]^2} \tag{11}$$

to (1) we have

$$\mathcal{M} = \frac{1}{2}(-i\lambda)^2 i^2 \int \frac{d^4k}{(2\pi)^4} \int_0^1 d\alpha \frac{1}{D}$$

with

$$D = [\alpha(K - k)^2 + (1 - \alpha)k^2 - m^2 + i\varepsilon]^2 = [(k - \alpha K)^2 + \alpha(1 - \alpha)K^2 - m^2 + i\varepsilon]^2$$

Shift the integration variable $k \to k + \alpha K$ and we meet the integral $\int [d^4k/(2\pi)^4] [1/(k^2 - c^2 + i\varepsilon)^2]$, where $c^2 = m^2 - \alpha(1 - \alpha)k^2$. Pauli-Villars proposed replacing it by

$$\int \frac{d^4k}{(2\pi)^4} \left[\frac{1}{(k^2 - c^2 + i\varepsilon)^2} - \frac{1}{(k^2 - \Lambda^2 + i\varepsilon)^2} \right] \tag{12}$$

with $\Lambda^2 \gg c^2$. For k much smaller than Λ the added second term in the integrand is of order Λ^{-4} and is negligible compared to the first term since Λ is much larger than c. For k much larger than Λ, the two terms almost cancel and the integrand vanishes rapidly with increasing k, effectively cutting off the integral.

Upon differentiating (12) with respect to c^2 and using (10) we deduce that (12) must be equal to $(i/16\pi^2) \log(\Lambda^2/c^2)$. Thus, the integral

$$\int^\Lambda \frac{d^4k}{(2\pi)^4} \frac{1}{(k^2 - c^2 + i\varepsilon)^2} = \frac{i}{16\pi^2} \log\left(\frac{\Lambda^2}{c^2}\right) \tag{13}$$

is indeed logarithmically dependent on the cutoff, as anticipated in the text.

For what it is worth, we obtain

$$\mathcal{M} = \frac{i\lambda^2}{32\pi^2} \int_0^1 d\alpha \, \log\left(\frac{\Lambda^2}{m^2 - \alpha(1-\alpha)K^2 - i\varepsilon}\right) \tag{14}$$

Appendix 2: Dimensional regularization

The basic idea behind dimensional regularization is very simple. When we reach $I = \int [d^4k/(2\pi)^4][1/(k^2 - c^2 + i\varepsilon)^2]$ we rotate to Euclidean space and generalize to d dimensions (see appendix D):

$$I(d) = i \int \frac{d_E^d k}{(2\pi)^d} \frac{1}{(k^2 + c^2)^2} = i \left[\frac{2\pi^{d/2}}{\Gamma(d/2)}\right] \frac{1}{(2\pi)^d} \int_0^\infty dk \, k^{d-1} \frac{1}{(k^2 + c^2)^2}$$

As I said, I don't want to get bogged down in computation in this book, but we've got to do what we've got to do. Changing the integration variable by setting $k^2 + c^2 = c^2/x$ we find

$$\int_0^\infty dk \, k^{d-1} \frac{1}{(k^2 + c^2)^2} = \tfrac{1}{2} c^{d-4} \int_0^1 dx (1-x)^{d/2-1} x^{1-d/2},$$

which we are supposed to recognize as the integral representation of the beta function. After the dust settles, we obtain

$$i \int \frac{d_E^d k}{(2\pi)^d} \frac{1}{(k^2 + c^2)^2} = i \frac{1}{(4\pi)^{d/2}} \Gamma\left(\frac{4-d}{2}\right) c^{d-4} \tag{15}$$

As $d \to 4$, the right-hand side becomes

$$i \frac{1}{(4\pi)^2} \left[\frac{2}{4-d} - \log c^2 + \log(4\pi) - \gamma + O(d-4)\right]$$

where $\gamma = 0.577\cdots$ denotes the Euler-Mascheroni constant.

Comparing with (13) we see that $\log \Lambda^2$ in Pauli-Villars regularization has been effectively replaced by the pole $2/(4-d)$. As noted in the text, when physical quantities are expressed in terms of physical coupling constants, all such poles cancel.

Exercises

III.1.1 Work through the manipulations leading to (9) without referring to the text.

III.1.2 Regard (1) as an analytic function of K^2. Show that it has a cut extending from $4m^2$ to infinity. [Hint: If you can't extract this result directly from (1) look at (14). An extensive discussion of this exercise will be given in chapter III.8.]

III.1.3 Change Λ to $e^\varepsilon \Lambda$. Show that for $\mathcal{M}$ not to change, to the order indicated λ must change by $\delta\lambda = 6\varepsilon C\lambda^2 + O(\lambda^3)$, that is,

$$\Lambda \frac{d\lambda}{d\Lambda} = 6C\lambda^2 + O(\lambda^3)$$

Old view versus new view

We learned that if we were to write the meson meson scattering amplitude in terms of a physically measured coupling constant λ_P, the dependence on the cutoff Λ would disappear (at least to order λ_P^2). Were we lucky or what?

Well, it turns out that there are quantum field theories in which this would happen and that there are quantum field theories in which this would not happen, which gives us a binary classification of quantum field theories. Again, for historical reasons, the former are known as "renormalizable theories" and are considered "nice." The latter are known as "nonrenormalizable theories," evoking fear and loathing in theoretical physicists.

Actually, with the new view of field theories as effective low energy theories to some underlying theory, physicists now look upon nonrenormalizable theories in a much more sympathetic light than a generation ago. I hope to make all these remarks clear in this and a later chapter.

High school dimensional analysis

Let us begin with some high school dimensional analysis. In natural units in which $\hbar = 1$ and $c = 1$, length and time have the same dimension, the inverse of the dimension of mass (and of energy and momentum). Particle physicists tend to count dimension in terms of mass as they are used to thinking of energy scales. Condensed matter physicists, on the other hand, usually speak of length scales. Thus, a given field operator has (equal and) opposite dimensions in particle physics and in condensed matter physics. We will use the convention of the particle physicists.

Since the action $S \equiv \int d^4x \mathcal{L}$ appears in the path integral as e^{iS}, it is clearly dimensionless, thus implying that the Lagrangian (Lagrangian density, strictly speaking) $\mathcal{L}$ has the same dimension as the 4th power of a mass. We will use the notation $[\mathcal{L}] = 4$ to indicate that $\mathcal{L}$ has dimension 4. In this notation $[x] = -1$ and $[\partial] = 1$. Consider the scalar field theory $\mathcal{L} = \frac{1}{2}[(\partial\varphi)^2 - m^2\varphi^2] - \lambda\varphi^4$. For the term $(\partial\varphi)^2$ to have dimension 4, we see that $[\varphi] = 1$ (since $2(1 + [\varphi]) = 4$). This then implies that $[\lambda] = 0$, that is, the coupling λ is dimensionless. The rule is simply that for each term in $\mathcal{L}$, the dimensions of the various pieces, including the coupling constant and mass, have to add up to 4 (thus, e.g., $[\lambda] + 4[\varphi] = 4$).

How about the fermion field ψ? Applying this rule to the Lagrangian $\mathcal{L} = \bar{\psi}i\gamma^\mu\partial_\mu\psi + \cdots$ we see that $[\psi] = \frac{3}{2}$. (Henceforth we will suppress the $\cdots$; it is understood that we are looking at a piece of the Lagrangian. Furthermore, since we are doing dimensional analysis we will often suppress various irrelevant factors, such as numerical factors and the gamma matrices in the Fermi interaction that we will come to presently.) Looking at the coupling $f\varphi\bar{\psi}\psi$ we see that the Yukawa coupling f is dimensionless. In contrast, in the theory of the weak interaction with $\mathcal{L} = G\bar{\psi}\psi\bar{\psi}\psi$ we see that the Fermi coupling G has dimension -2 (since $-2 + 4(\frac{3}{2}) = 4$; got that?).

From the Maxwell Lagrangian $-\frac{1}{4}F_{\mu\nu}F^{\mu\nu}$ we see that $[A_\mu] = 1$ and hence A_μ has the same dimension as ∂_μ: The vector field has the same dimension as the scalar field. The electromagnetic coupling $eA_\mu\bar{\psi}\gamma^\mu\psi$ tells us that e is dimensionless, which we can also deduce from Coulomb's law written in natural units $V(r) = \alpha/r$, with the fine structure constant $\alpha = e^2/4\pi$.

Scattering amplitude blows up

We are now ready for a heuristic argument regarding the nonrenormalizability of a theory. Consider Fermi's theory of the weak interaction. Imagine calculating the amplitude $\mathcal{M}$ for a four-fermion interaction, say neutrino-neutrino scattering at an energy much smaller than Λ. In lowest order, $\mathcal{M} \sim G$. Let us try to write down the amplitude to the next order: $\mathcal{M} \sim G + G^2(?)$, where we will try to guess what (?) is. Since all masses and energies are by definition small compared to the cutoff Λ, we can simply set them equal to zero. Since $[G] = -2$, by high school dimensional analysis the unknown factor (?) must have dimension $+2$. The only possibility for (?) is Λ^2. Hence, the amplitude to the next order must have the form $\mathcal{M} \sim G + G^2\Lambda^2$. We can also check this conclusion by looking at the Feynman diagram in figure III.2.1: Indeed it goes as $G^2\int^\Lambda d^4p\,(1/p)(1/p) \sim G^2\Lambda^2$.

Without a cutoff on the theory, or equivalently with $\Lambda = \infty$, theorists realized that the theory was sick: Infinity was the predicted value for a physical quantity. Fermi's weak interaction theory was said to be nonrenormalizable. Furthermore, if we try to calculate to higher order, each power of G is accompanied by another factor of Λ^2.

In desperation, some theorists advocated abandoning quantum field theory altogether. Others expended an enormous amount of effort trying to "cure" weak interaction theory. For instance, one approach was to speculate that the series (with coefficients suppressed) $\mathcal{M} \sim G[1 + G\Lambda^2 + (G\Lambda^2)^2 + (G\Lambda^2)^4 + \cdots]$ summed to $Gf(G\Lambda^2)$, where the unknown function f might have the property that $f(\infty)$ was finite. In hindsight, we now know that this is not a fruitful approach.

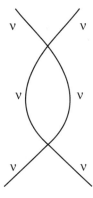

Figure III.2.1

Instead, what happened was that toward the late 1960s S. Glashow, A. Salam, and S. Weinberg, building on the efforts of many others, succeeded in constructing an electroweak theory unifying the electromagnetic and weak interactions, as I will discuss in chapter VII.2. Fermi's weak interaction theory emerges within electroweak theory as a low energy effective theory.

Fermi's theory cried out

In modern terms, we think of the cutoff Λ as really being there and we hear the cutoff dependence of the four-fermion interaction amplitude $\mathcal{M} \sim G + G^2 \Lambda^2$ as the sound of the theory crying out that something dramatic has to happen at the energy scale $\Lambda \sim (1/G)^{\frac{1}{2}}$. The second term in the perturbation series becomes comparable to the first, so at the very least perturbation theory fails.

Here is another way of making the same point. Suppose that we don't know anything about cutoff and all that. With G having mass dimension -2, just by high school dimensional analysis we see that the neutrino-neutrino scattering amplitude at center-of-mass energy E has to go as $\mathcal{M} \sim G + G^2 E^2 + \cdots$. When E reaches the scale $\sim (1/G)^{\frac{1}{2}}$ the amplitude reaches order unity and some new physics must take over just because the cross section is going to violate the unitarity bound from basic quantum mechanics. (Remember phase shift and all that?)

In fact, what that something is goes back to Yukawa, who at the same time that he suggested the meson theory for the nuclear forces also suggested that an intermediate vector boson could account for the Fermi theory of the weak interaction. (In the 1930s the distinction between the strong and the weak interactions was far from clear.) Schematically, consider a theory of a vector boson of mass M coupled to a fermion field via a dimensionless coupling constant g:

$$\mathcal{L} = \bar{\psi}(i\gamma^\mu \partial_\mu - m)\psi - \tfrac{1}{4}F_{\mu\nu}F^{\mu\nu} + M^2 A_\mu A^\mu + g A_\mu \bar{\psi}\gamma^\mu \psi \tag{1}$$

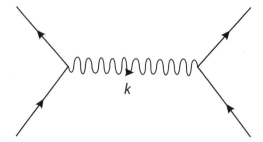

Figure III.2.2

Let's calculate fermion-fermion scattering. The Feynman diagram in figure III.2.2 generates an amplitude $(-ig)^2(\bar{u}\gamma^\mu u)[i/(k^2 - M^2 + i\varepsilon)](\bar{u}\gamma_\mu u)$, which when the momentum transfer k is much less than M becomes $i(g^2/M^2)(\bar{u}\gamma^\mu u)(\bar{u}\gamma_\mu u)$. But this is just as if the fermions are interacting via a Fermi theory of the form $G(\bar{\psi}\gamma^\mu\psi)(\bar{\psi}\gamma_\mu\psi)$ with $G = g^2/M^2$.

If we blithely calculate with the low energy effective theory $G(\bar{\psi}\gamma^\mu\psi)(\bar{\psi}\gamma_\mu\psi)$, it cries out that it is going to fail. Yes sir indeed, at the energy scale $(1/G)^{\frac{1}{2}} = M/g$, the vector boson is produced. New physics appears.

I find it sobering and extremely appealing that theories in physics have the ability to announce their own eventual failure and hence their domains of validity, in contrast to theories in some other areas of human thought.

Einstein's theory is now crying out

The theory of gravity is also notoriously nonrenormalizable. Simply comparing Newton's law $V(r) = G_N M_1 M_2/r$ with Coulomb's $V(r) = \alpha/r$ we see that Newton's gravitational constant G_N has mass dimension -2. No more need be said. We come to the same morose conclusion that the theory of gravity, just like Fermi's theory of weak interaction, is nonrenormalizable. To repeat the argument, if we calculate graviton-graviton scattering at energy E, we encounter the series $\sim [1 + G_N E^2 + (G_N E^2)^2 + \cdots]$.

Just as in our discussion of the Fermi theory, the nonrenormalizability of quantum gravity tells us that at the Planck energy scale $(1/G_N)^{\frac{1}{2}} \equiv M_{\text{Planck}} \sim 10^{19} m_{\text{proton}}$ new physics must appear. Fermi's theory cried out, and the new physics turned out to be the electroweak theory. Einstein's theory is now crying out. Will the new physics turn out to be string theory?[1]

Exercise

III.2.1   Consider the d-dimensional scalar field theory $S = \int d^d x (\frac{1}{2}(\partial\varphi)^2 + \frac{1}{2}m^2\varphi^2 + \lambda\varphi^4 + \cdots + \lambda_n\varphi^n + \cdots)$. Show that $[\varphi] = (d-2)/2$ and $[\lambda_n] = n(2-d)/2 + d$. Note that φ is dimensionless for $d = 2$.

[1] J. Polchinski, *String Theory*.

III.3 | Counterterms and Physical Perturbation Theory

Renormalizability

The heuristic argument of the previous chapter indicates that theories whose coupling has negative mass dimension are nonrenormalizable. What about theories with dimensionless couplings, such as quantum electrodynamics and the φ^4 theory? As a matter of fact, both of these theories have been proved to be renormalizable. But it is much more difficult to prove that a theory is renormalizable than to prove that it is nonrenormalizable. Indeed, the proof that nonabelian gauge theory (about which more later) is renormalizable took the efforts of many eminent physicists, culminating in the work of 't Hooft, Veltman, B. Lee, Zinn-Justin, and many others.

Consider again the simple φ^4 theory. First, a trivial remark: The physical coupling constant λ_P is a function of s_0, t_0, and u_0 [see (III.1.4)]. For theoretical purposes it is much less cumbersome to set s_0, t_0, and u_0 equal to μ^2 and thus use, instead of (III.1.4), the simpler definition

$$-i\lambda_P = -i\lambda + 3iC\lambda^2 \log\left(\frac{\Lambda^2}{\mu^2}\right) + O(\lambda^3) \tag{1}$$

This is purely for theoretical convenience.[1]

We saw that to order λ^2 the meson-meson scattering amplitude when expressed in terms of the physical coupling λ_P is independent of the cutoff Λ. How do we prove that this is true to all orders in λ? Dimensional analysis only tells us that to any order in λ the dependence of the meson scattering amplitude on the cutoff must be a sum of terms going as $[\log(\Lambda/\mu)]^p$ with some power p.

The meson-meson scattering amplitude is certainly not the only quantity that depends on the cutoff. Consider the inverse of the φ propagator to order λ^2 as shown in figure III.3.1.

---

[1] In fact, the kinematic point $s_0 = t_0 = u_0 = \mu^2$ cannot be reached experimentally, but that's of concern to theorists.

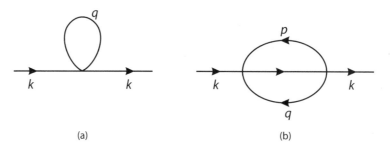

(a)   (b)

Figure III.3.1

The Feynman diagram in figure III.3.1a gives something like

$$-i\lambda \int^{\Lambda} \left[\frac{d^4q}{(2\pi)^4} \right] \left[\frac{i}{q^2 - m^2 + i\varepsilon} \right]$$

The precise value does not concern us; we merely note that it depends quadratically on the cutoff Λ but not on k^2. The diagram in figure III.3.1b involves a double integral

$$I(k, m, \Lambda; \lambda) \equiv (-i\lambda)^2 \int^{\Lambda} \int^{\Lambda} \frac{d^4p}{(2\pi)^4} \frac{d^4q}{(2\pi)^4}$$

$$\frac{i}{p^2 - m^2 + i\varepsilon} \frac{i}{q^2 - m^2 + i\varepsilon} \frac{i}{(p+q+k)^2 - m^2 + i\varepsilon} \tag{2}$$

Counting powers of p and q we see that the integral $\sim \int (d^8 P / P^6)$ and so I depends quadratically on the cutoff Λ.

By Lorentz invariance I is a function of k^2, which we can expand in a series $D + Ek^2 + Fk^4 + \cdots$. The quantity D is just I with the external momentum k set equal to zero and so depends quadratically on the cutoff Λ. Next, we can obtain E by differentiating I with respect to k twice and then setting k equal to zero. This clearly decreases the powers of p and q in the integrand by 2 and so E depends only logarithmically on the cutoff Λ. Similarly, we can obtain F by differentiating I with respect to k four times and then setting k equal to zero. This decreases the powers of p and q in the integrand by 4 and thus F is given by an integral that goes as $\sim \int d^8 P / P^{10}$ for large P. The integral is convergent and hence cutoff independent. We can clearly repeat the argument ad infinitum. Thus F and the terms in $(\cdots)$ are cutoff independent as the cutoff goes to infinity and we don't have to worry about them.

Putting it altogether, we have the inverse propagator $k^2 - m^2 + a + bk^2$ up to $O(k^2)$ with a and b, respectively, quadratically and logarithmically cutoff dependent. The propagator is changed to

$$\frac{1}{k^2 - m^2} \to \frac{1}{(1+b)k^2 - (m^2 - a)} \tag{3}$$

The pole in k^2 is shifted to $m_P^2 \equiv m^2 + \delta m^2 \equiv (m^2 - a)(1+b)^{-1}$, which we identify as the physical mass. This shift is known as mass renormalization. Physically, it is quite reasonable that quantum fluctuations will shift the mass.

What about the fact that the residue of the pole in the propagator is no longer 1 but $(1 + b)^{-1}$?

To understand this shift in the residue, recall that we blithely normalized the field φ so that $\mathcal{L} = \frac{1}{2}(\partial\varphi)^2 + \cdots$. That the coefficient of k^2 in the lowest order inverse propagator $k^2 - m^2$ is equal to 1 reflects the fact that the coefficient of $\frac{1}{2}(\partial\varphi)^2$ in $\mathcal{L}$ is equal to 1. There is certainly no guarantee that with higher order corrections included the coefficient of $\frac{1}{2}(\partial\varphi)^2$ in an effective $\mathcal{L}$ will stay at 1. Indeed, we see that it is shifted to $(1 + b)$. For historical reasons, this is known as "wave function renormalization" even though there is no wave function anywhere in sight. A more modern term would be field renormalization. (The word renormalization makes some sense in this case, as we did normalize the field without thinking too much about it.)

Incidentally, it is much easier to say "logarithmic divergent" than to say "logarithmically dependent on the cutoff Λ," so we will often slip into this more historical and less accurate jargon and use the word divergent. In φ^4 theory, the wave function renormalization and the coupling renormalization are logarithmically divergent, while the mass renormalization is quadratically divergent.

Bare versus physical perturbation theory

What we have been doing thus far is known as bare perturbation theory. We should have put the subscript 0 on what we have been calling φ, m, and λ. The field φ_0 is known as the bare field, and m_0 and λ_0 are known as the bare mass and bare coupling, respectively. I did not put on the subscript 0 way back in part I because I did not want to clutter up the notation before you, the student, even knew what a field was.

Seen in this light, using bare perturbation theory seems like a really stupid thing to do, and it is. Shouldn't we start out with a zeroth order theory already written in terms of the physical mass m_P and physical coupling λ_P that experimentalists actually measure, and perturb around that theory? Yes, indeed, and this way of calculating is known as renormalized or dressed perturbation theory, or as I prefer to call it, physical perturbation theory.

We write

$$\mathcal{L} = \frac{1}{2}[(\partial\varphi)^2 - m_P^2\varphi^2] - \frac{\lambda_P}{4!}\varphi^4 + A(\partial\varphi)^2 + B\varphi^2 + C\varphi^4 \tag{4}$$

(A word on notation: The pedantic would probably want to put a subscript P on the field φ, but let us clutter up the notation as little as possible.) Physical perturbation theory works as follows. The Feynman rules are as before, but with the crucial difference that for the coupling we use λ_P and for the propagator we write $i/(k^2 - m_P^2 + i\varepsilon)$ with the physical mass already in place. The last three terms in (4) are known as counterterms. The coefficients A, B, and C are determined iteratively (see later) as we go to higher and higher order in perturbation theory. They are represented as crosses in Feynman diagrams, as indicated in figure III.3.2, with the corresponding Feynman rules. All momentum integrals are cut off.

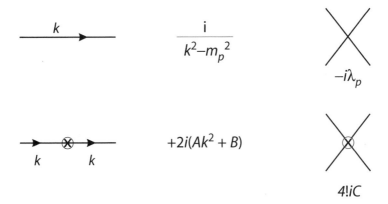

Figure III.3.2

Let me now explain how A, B, and C are determined iteratively. Suppose we have determined them to order λ_P^N. Call their values to this order A_N, B_N, and C_N. Draw all the diagrams that appear in order λ_P^{N+1}. We determine A_{N+1}, B_{N+1}, and C_{N+1} by requiring that the propagator calculated to the order λ_P^{N+1} has a pole at m_P with a residue equal to 1, and that the meson-meson scattering amplitude evaluated at some specified values of the kinematic variables has the value $-i\lambda_P$. In other words, the counterterms are fixed by the condition that m_P and λ_P are what we say they are. Of course, A, B, and C will be cutoff dependent. Note that there are precisely three conditions to determine the three unknowns A_{N+1}, B_{N+1}, and C_{N+1}.

Explained in this way, you can see that it is almost obvious that physical perturbation theory works, that is, it works in the sense that all the physical quantities that we calculate will be cutoff independent. Imagine, for example, that you labor long and hard to calculate the meson-meson scattering amplitude to order λ_P^{17}. It would contain some cutoff dependent and some cutoff independent terms. Then you simply add a contribution given by C_{17} and adjust C_{17} to cancel the cutoff dependent terms.

But ah, you start to worry. You say, "What if I calculate the amplitude for two mesons to go into four mesons, that is, diagrams with six external legs? If I get a cutoff dependent answer, then I am up the creek, as there is no counterterm of the form $D\varphi^6$ in (4) to soak up the cutoff dependence." Very astute of you, but this worry is covered by the following power counting theorem.

Degree of divergence

Consider a diagram with B_E external φ lines. First, a definition: A diagram is said to have a superficial degree of divergence D if it diverges as Λ^D. (A logarithmic divergence $\log \Lambda$ counts as $D = 0$.) The theorem says that D is given by

$$D = 4 - B_E \tag{5}$$

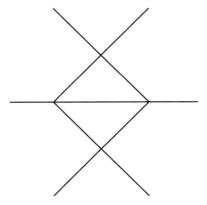

Figure III.3.3

I will give a proof later, but I will first illustrate what (5) means. For the inverse propagator, which has $B_E = 2$, we are told that $D = 2$. Indeed, we encountered a quadratic divergence. For the meson-meson scattering amplitude, $B_E = 4$, and so $D = 0$, and indeed, it is logarithmically divergent.

According to the theorem, if you calculate a diagram with six external legs (what is technically sometimes known as the six-point function), $B_E = 6$ and $D = -2$. The theorem says that your diagram is convergent or cutoff independent (i.e., the cutoff dependence disappears as Λ^{-2}). You didn't have to worry. You should draw a few diagrams to check this point. Diagrams with more external legs are even more convergent.

The proof of the theorem follows from simple power counting. In addition to B_E and D, let us define B_I as the number of internal lines, V as the number of vertices, and L as the number of loops. (It is helpful to focus on a specific diagram such as the one in figure III.3.3 with $B_E = 6$, $D = -2$, $B_I = 5$, $V = 4$, and $L = 2$.)

The number of loops is just the number of $\int [d^4 k/(2\pi)^4]$ we have to do. Each internal line carries with it a momentum to be integrated over, so we seem to have B_I integrals to do. But the actual number of integrals to be done is of course decreased by the momentum conservation delta functions associated with the vertices, one to each vertex. There are thus V delta functions, but one of them is associated with overall momentum conservation of the entire diagram. Thus, the number of loops is

$$L = B_I - (V - 1) \tag{6}$$

[If you have trouble following this argument, you should work things out for the diagram in figure III.3.3 for which this equation reads $2 = 5 - (4 - 1)$.]

For each vertex there are four lines coming out (or going in, depending on how you look at it). Each external line comes out of (or goes into) one vertex. Each internal line connects two vertices. Thus,

$$4V = B_E + 2B_I \tag{7}$$

(For figure III.3.3 this reads $4 \cdot 4 = 6 + 2 \cdot 5$.)

Finally, for each loop there is a $\int d^4k$ while for each internal line there is a $i/(k^2 - m^2 + i\varepsilon)$, bringing the powers of momentum down by 2. Hence,

$$D = 4L - 2B_I \tag{8}$$

(For figure III.3.3 this reads $-2 = 4 \cdot 2 - 2 \cdot 5$.)

Putting (6), (7), and (8) together, we obtain the theorem (5).[2] As you can plainly see, this formalizes the power counting we have been doing all along.

Degree of divergence with fermions

To test if you understood the reasoning leading to (5), consider the Yukawa theory we met in (II.5.19). (We suppress the counterterms for typographical clarity.)

$$\mathcal{L} = \bar{\psi}(i\gamma^\mu \partial_\mu - m_P)\psi + \tfrac{1}{2}[(\partial\varphi)^2 - \mu_P^2 \varphi^2] - \lambda_P \varphi^4 + f_P \varphi \bar{\psi} \psi \tag{9}$$

Now we have to count F_I and F_E, the number of internal and external fermion lines, respectively, and keep track of V_f and V_λ, the number of vertices with the coupling f and λ, respectively. We have five equations altogether. For instance, (7) splits into two equations because we now have to count fermion lines as well as boson lines. For example, we now have

$$V_f + 4V_\lambda = B_E + 2B_I \tag{10}$$

We (that is, you) finally obtain

$$D = 4 - B_E - \tfrac{3}{2}F_E \tag{11}$$

So the divergent amplitudes, that is, those classes of diagrams with $D \geq 0$, have $(B_E, F_E) = (0, 2)$, $(2, 0)$, $(1, 2)$, and $(4, 0)$. We see that these correspond to precisely the six terms in the Lagrangian (9), and thus we need six counterterms.

Note that this counting of superficial powers of divergence shows that all terms with mass dimension ≤ 4 are generated. For example, suppose that in writing down the Lagrangian (9) we forgot to include the $\lambda_P \varphi^4$ term. The theory would demand that we include this term: We have to introduce it as a counterterm [the term with $(B_E, F_E) = (4, 0)$ in the list above].

A common feature of (5) and (11) is that they both depend only on the number of external lines and not on the number of vertices V. Thus, for a given number of external lines, no matter to what order of perturbation theory we go, the superficial degree of divergence remains the same. Further thought reveals that we are merely formalizing the dimension-counting argument of the preceding chapter. [Recall that the mass dimension of a Bose field $[\varphi]$ is 1 and of a Fermi field $[\psi]$ is $\tfrac{3}{2}$. Hence the coefficients 1 and $\tfrac{3}{2}$ in (11).]

---

[2] The superficial degree of divergence measures the divergence of the Feynman diagram as all internal momenta are scaled uniformly by $k \to ak$ with a tending to infinity. In a more rigorous treatment we have to worry about the momenta in some subdiagram (a piece of the full diagram) going to infinity with other momenta held fixed.

Our discussion hardly amounts to a rigorous proof that theories such as the Yukawa theory are renormalizable. If you demand rigor, you should consult the many field theory tomes on renormalization theory, and I do mean tomes, in which such arcane topics as overlapping divergences and Zimmerman's forest formula are discussed in exhaustive and exhausting detail.

Nonrenormalizable field theories

It is instructive to see how nonrenormalizable theories reveal their unpleasant personalities when viewed in the context of this discussion. Consider the Fermi theory of the weak interaction written in a simplified form:

$$\mathcal{L} = \bar{\psi}(i\gamma^\mu \partial_\mu - m_P)\psi + G(\bar{\psi}\psi)^2.$$

The analogs of (6), (7), and (8) now read $L = F_I - (V - 1)$, $4V = F_E + 2F_I$, and $D = 4L - F_I$. Solving for the superficial degree of divergence in terms of the external number of fermion lines, we find

$$D = 4 - \tfrac{3}{2}F_E + 2V \tag{12}$$

Compared to the corresponding equations for renormalizable theories (5) and (11), D now depends on V. Thus if we calculate fermion-fermion scattering ($F_E = 4$), for example, the divergence gets worse and worse as we go to higher and higher order in the perturbation series. This confirms the discussion of the previous chapter. But the really bad news is that for any F_E, we would start running into divergent diagrams when V gets sufficiently large, so we would have to include an unending stream of counterterms $(\bar{\psi}\psi)^3$, $(\bar{\psi}\psi)^4$, $(\bar{\psi}\psi)^5, \cdots$, each with an arbitrary coupling constant to be determined by an experimental measurement. The theory is severely limited in predictive power.

At one time nonrenormalizable theories were considered hopeless, but they are accepted in the modern view based on the effective field theory approach, which I will discuss in chapter VIII.3.

Dependence on dimension

The superficial degree of divergence clearly depends on the dimension d of spacetime since each loop is associated with $\int d^d k$. For example, consider the Fermi interaction $G(\bar{\psi}\psi)^2$ in $(1 + 1)$-dimensional spacetime. Of the three equations that went into (12) one is changed to $D = 2L - F_I$, giving

$$D = 2 - \tfrac{1}{2}F_E \tag{13}$$

the analog of (12) for 2-dimensional spacetime. In contrast to (12), V no longer enters, and the only superficial diagrams have $F_E = 2$ and 4, which we can cancel by the appropriate counterterms. The Fermi interaction is renormalizable in $(1 + 1)$-dimensional spacetime. I will come back to it in chapter VII.4.

The Weisskopf phenomenon

I conclude by pointing out that the mass correction to a Bose field and to a Fermi field diverge differently. Since this phenomenon was first discovered by Weisskopf, I refer to it as the Weisskopf phenomenon. To see this go back to (11) and observe that B_E and F_E contribute differently to the superficial degree of divergence D. For $B_E = 2$, $F_E = 0$, we have $D = 2$, and thus the mass correction to a Bose field diverges quadratically, as we have already seen explicitly [the quantity a in (3)]. But for $F_E = 2$, $B_E = 0$, we have $D = 1$ and it looks like the fermion mass is linearly divergent. Actually, in 4-dimensional field theory we cannot possibly get a linear dependence on the cutoff. To see this, it is easiest to look at, as an example, the Feynman integral you wrote down for exercise II.5.1, for the diagram in figure II.5.1:

$$(if)^2 i^2 \int \frac{d^4 k}{(2\pi)^4} \frac{1}{k^2 - \mu^2} \frac{\not{p} + \not{k} + m}{(p+k)^2 - m^2} \equiv A(p^2) \not{p} + B(p^2) \tag{14}$$

where I define the two unknown functions $A(p^2)$ and $B(p^2)$ for convenience. (For the purpose of this discussion it doesn't matter whether we are doing bare or physical perturbation theory. If the latter, then I have suppressed the subscript P for the sake of notational clarity.)

Look at the integrand for large k. You see that the integral goes as $\int^\Lambda d^4 k (\not{k}/k^4)$ and looks linearly divergent, but by reflection symmetry $k \to -k$ the integral to leading order vanishes. The integral in (14) is merely logarithmically divergent. The superficial degree of divergence D often gives an exaggerated estimate of how bad the divergence can be (hence the adjective "superficial"). In fact, staring at (14) we can prove more, for instance, that $B(p^2)$ must be proportional to m. As an exercise you can show, using the Feynman rules given in chapter II.5, that the same conclusion holds in quantum electrodynamics.

For a boson, quantum fluctuations give $\delta\mu \propto \Lambda^2/\mu$, while for a fermion such as the electron, quantum correction to its mass $\delta m \propto m \log(\Lambda/m)$ is much more benign. It is interesting to note that in the early twentieth century physicists thought of the electron as a ball of charge of radius a. The electrostatic energy of such a ball, of the order e^2/a, was identified as the electron mass. Interpreting $1/a$ as Λ, we could say that in classical physics the electron mass is proportional to Λ and diverges linearly. Thus, one way of stating the Weisskopf phenomenon is that "bosons behave worse than a classical charge, but fermions behave better."

As Weisskopf explained in 1939, the difference in the degree of the divergence can be understood heuristically in terms of quantum statistics. The "bad" behavior of bosons has to do with their gregariousness. A fermion would push away the virtual fermions fluctuating in the vacuum, thus creating a cavity in the vacuum charge distribution surrounding it. Hence its self-energy is less singular than would be the case were quantum statistics not taken into account. A boson does the opposite.

The "bad" behavior of bosons will come back to haunt us later.

Power of $\hbar$ counts the number of loops

This is a convenient place to make a useful observation, though unrelated to divergences and cut-off dependence. Suppose we restore Planck's unit of action $\hbar$. In the path integral, the integrand becomes $e^{iS/\hbar}$ (recall chapter I.2) so that effectively $\mathcal{L} \to \mathcal{L}/\hbar$. Consider $\mathcal{L} = -\frac{1}{2}\varphi(\partial^2 + m^2)\varphi - \frac{\lambda}{4!}\varphi^4$, just to be definite. The coupling $\lambda \to \lambda/\hbar$, so that each vertex is now associated with a factor of $1/\hbar$. Recall that the propagator is essentially the inverse of the operator $(\partial^2 + m^2)$, and so in momentum space $1/(k^2 - m^2) \to \hbar/(k^2 - m^2)$. Thus the powers of $\hbar$ are given by the number of internal lines minus the number of vertices $P = B_I - V = L - 1$, where we used (6). You can check that this holds in general and not just for φ^4 theory.

This observation shows that organizing Feynman diagrams by the number of loops amounts to an expansion in Planck's constant (sometimes called a semi-classical expansion), with the tree diagrams providing the leading term. We will come across this again in chapter IV.3.

Exercises

III.3.1 Show that in $(1+1)$-dimensional spacetime the Dirac field ψ has mass dimension $\frac{1}{2}$, and hence the Fermi coupling is dimensionless.

III.3.2 Derive (11) and (13).

III.3.3 Show that $B(p^2)$ in (14) vanishes when we set $m = 0$. Show that the same behavior holds in quantum electrodynamics.

III.3.4 We showed that the specific contribution (14) to δm is logarithmically divergent. Convince yourself that this is actually true to any finite order in perturbation theory.

III.3.5 Show that the result $P = L - 1$ holds for all the theories we have studied.

III.4 | Gauge Invariance: A Photon Can Find No Rest

When the central identity blows up

I explained in chapter I.7 that the path integral for a generic field theory can be formally evaluated in what deserves to be called the Central Identity of Quantum Field Theory:

$$\int D\varphi \, e^{-\frac{1}{2}\varphi \cdot K \cdot \varphi - V(\varphi) + J \cdot \varphi} = e^{-V(\delta/\delta J)} e^{\frac{1}{2} J \cdot K^{-1} \cdot J} \tag{1}$$

For any field theory we can always gather up all the fields, put them into one giant column vector, and call the vector φ. We then single out the term quadratic in φ write it as $\frac{1}{2}\varphi \cdot K \cdot \varphi$, and call the rest $V(\varphi)$. I am using a compact notation in which spacetime coordinates and any indices on the field, including Lorentz indices, are included in the indices of the formal matrix K. We will often use (1) with $V = 0$:

$$\int D\varphi \, e^{-\frac{1}{2}\varphi \cdot K \cdot \varphi + J \cdot \varphi} = e^{\frac{1}{2} J \cdot K^{-1} \cdot J} \tag{2}$$

But what if K does not have an inverse?

This is not an esoteric phenomenon that occurs in some pathological field theory, but in one of the most basic actions of physics, the Maxwell action

$$S(A) = \int d^4x \, \mathcal{L} = \int d^4x \left[\frac{1}{2} A_\mu (\partial^2 g^{\mu\nu} - \partial^\mu \partial^\nu) A_\nu + A_\mu J^\mu \right]. \tag{3}$$

The formal matrix K in (2) is proportional to the differential operator $(\partial^2 g^{\mu\nu} - \partial^\mu \partial^\nu) \equiv Q^{\mu\nu}$. A matrix does not have an inverse if some of its eigenvalues are zero, that is, if when acting on some vector, the matrix annihilates that vector. Well, observe that $Q^{\mu\nu}$ annihilates vectors of the form $\partial_\nu \Lambda(x) : Q^{\mu\nu} \partial_\nu \Lambda(x) = 0$. Thus $Q^{\mu\nu}$ has no inverse.

There is absolutely nothing mysterious about this phenomenon; we have already encountered it in classical physics. Indeed, when we first learned the concepts of electricity, we were told that only the "voltage drop" between two points has physical meaning. At a more sophisticated level, we learned that we can always add any constant (or indeed any function of time) to the electrostatic potential (which is of course just "voltage") since by definition its gradient is the electric field. At an even more sophisticated level, we see that solving Maxwell's equation (which of course comes from just extremizing the action) amounts to finding the inverse Q^{-1}. [In the notation I am using here Maxwell's equation $\partial_\mu F^{\mu\nu} = J^\nu$ is written as $Q_{\mu\nu} A^\nu = J_\mu$, and the solution is $A^\nu = (Q^{-1})^{\nu\mu} J_\mu$.]

Well, Q^{-1} does not exist! What do we do? We learned that we must impose an additional constraint on the gauge potential A^μ, known as "fixing a gauge."

A mundane nonmystery

To emphasize the rather mundane nature of this gauge fixing problem (which some older texts tend to make into something rather mysterious and almost hopelessly difficult to understand), consider just an ordinary integral $\int_{-\infty}^{+\infty} dA e^{-A \cdot K \cdot A}$, with $A = (a, b)$ a 2-component vector and $K = \begin{pmatrix} 1 & 0 \\ 0 & 0 \end{pmatrix}$, a matrix without an inverse. Of course you realize what the problem is: We have $\int_{-\infty}^{+\infty} \int_{-\infty}^{+\infty} da \, db \, e^{-a^2}$ and the integral over b does not exist. To define the integral we insert into it a delta function $\delta(b - \xi)$. The integral becomes defined and actually does not depend on the arbitrary number ξ. More generally, we can insert $\delta[f(b)]$ with f some function of our choice. In the context of an ordinary integral, this procedure is of course ludicrous overkill, but we will use the analog of this procedure in what follows. In this baby problem, we could regard the variable b, and thus the integral over it, as "redundant." As we will see, gauge invariance is also a redundancy in our description of massless spin 1 particles.

A massless spin 1 field is intrinsically different from a massive spin 1 field—that's the crux of the problem. The photon has only two polarization degrees of freedom. (You already learned in classical electrodynamics that an electromagnetic wave has two transverse degrees of freedom.) This is the true physical origin of gauge invariance.

In this sense, gauge invariance is, strictly speaking, not a "real" symmetry but merely a reflection of the fact that we used a redundant description: a Lorentz vector field to describe two physical degrees of freedom.

Restricting the functional integral

I will now discuss the method for dealing with this redundancy invented by Faddeev and Popov. As you will see presently, it is the analog of the method we used in our baby problem above. Even in the context of electromagnetism this method is a bit of overkill, but it will prove to be essential for nonabelian gauge theories (as we will see in chapter VII.1) and for gravity. I will describe the method using a completely general and somewhat abstract language. In the next section, I will then apply the discussion here to a specific example. If you have some trouble with this section, you might find it helpful to go back and forth between the two sections.

Suppose we have to do the integral $I \equiv \int DA e^{iS(A)}$; this can be an ordinary integral or a path integral. Suppose that under the transformation $A \to A_g$ the integrand and the measure do not change, that is, $S(A) = S(A_g)$ and $DA = DA_g$. The transformations obviously form a group, since if we transform again with g', the integrand and the measure do not change under the combined effect of g and g' and $A_g \to (A_g)_{g'} = A_{gg'}$. We would like to write the integral I in the form $I = (\int Dg)J$, with J independent of g. In other words, we want to factor out the redundant integration over g. Note that Dg is the invariant measure over the group of transformations and $\int Dg$ is the volume of the group. Be aware of the compactness of the notation in the case of a path integral: A and g are both functions of the spacetime coordinates x.

I want to emphasize that this hardly represents anything profound or mysterious. If you have to do the integral $I = \int dx\, dy\, e^{iS(x,y)}$ with $S(x, y)$ some function of $x^2 + y^2$, you know perfectly well to go to polar coordinates $I = (\int d\theta)J = (2\pi)J$, where $J = \int dr\, r e^{iS(r)}$ is an integral over the radial coordinate r only. The factor 2π is precisely the volume of the group of rotations in 2 dimensions.

Faddeev and Popov showed how to do this "going over to polar coordinates" in a unified and elegant way. Following them, we first write the numeral "one" as $1 = \Delta(A) \int Dg\delta[f(A_g)]$, an equality that merely defines $\Delta(A)$. Here f is some function of our choice and $\Delta(A)$, known as the Faddeev-Popov determinant, of course depends on f. Next, note that $[\Delta(A_{g'})]^{-1} = \int Dg\delta[f(A_{g'g})] = \int Dg''\delta[f(A_{g''})] = [\Delta(A)]^{-1}$, where the second equality follows upon defining $g'' = g'g$ and noting that $Dg'' = Dg$. In other words, we showed that $\Delta(A) = \Delta(A_g)$: the Faddeev-Popov determinant is gauge invariant. We now insert 1 into the integral I we have to do:

$$I = \int DA e^{iS(A)}$$

$$= \int DA e^{iS(A)} \Delta(A) \int Dg\delta[f(A_g)]$$

$$= \int Dg \int DA e^{iS(A)} \Delta(A)\delta[f(A_g)] \tag{4}$$

As physicists and not mathematicians, we have merrily interchanged the order of integration.

At the physicist's level of rigor, we are always allowed to change integration variables until proven guilty. So let us change A to $A_{g^{-1}}$; then

$$I = \left(\int Dg\right) \int DA e^{iS(A)} \Delta(A)\delta[f(A)] \tag{5}$$

where we have used the fact that DA, $S(A)$, and $\Delta(A)$ are all invariant under $A \to A_{g^{-1}}$.

That's it. We've done it. The group integration $(\int Dg)$ has been factored out.

The volume of a compact group is finite, but in gauge theories there is a separate group at every point in spacetime, and hence $(\int Dg)$ is an infinite factor. (This also explains why there is no gauge fixing problem in theories with global symmetries introduced in chapter I.10.) Fortunately, in the path integral Z for field theory we do not care about overall factors in Z, as was explained in chapter I.3, and thus the factor $(\int Dg)$ can simply be thrown away.

Fixing the electromagnetic gauge

Let us now apply the Faddeev-Popov method to electromagnetism. The transformation leaving the action invariant is of course $A_\mu \to A_\mu - \partial_\mu \Lambda$, so g in the present context is denoted by Λ and $A_g \equiv A_\mu - \partial_\mu \Lambda$. Note also that since the integral I we started with is independent of f it is still independent of f in spite of its appearance in (5). Choose $f(A) = \partial A - \sigma$, where σ is a function of x. In particular, I is independent of σ and so we can integrate I with an arbitrary functional of σ, in particular, the functional $e^{-(i/2\xi)\int d^4 x \sigma(x)^2}$.

We now turn the crank. First, we calculate

$$[\Delta(A)]^{-1} \equiv \int Dg \, \delta[f(A_g)] = \int D\Lambda \, \delta(\partial A - \partial^2 \Lambda - \sigma) \tag{6}$$

Next we note that in (5) $\Delta(A)$ appears multiplied by $\delta[f(A)]$ and so in evaluating $[\Delta(A)]^{-1}$ in (6) we can effectively set $f(A) = \partial A - \sigma$ to zero. Thus from (6) we have $\Delta(A)$ "=" $[\int D\Lambda \, \delta(\partial^2 \Lambda)]^{-1}$. But this object does not even depend on A, so we can throw it away. Thus, up to irrelevant overall factors that could be thrown away I is just $\int DA e^{iS(A)} \delta(\partial A - \sigma)$.

Integrating over $\sigma(x)$ as we said we were going to do, we finally obtain

$$Z = \int D\sigma \, e^{-(i/2\xi)\int d^4 x \sigma(x)^2} \int DA e^{iS(A)} \delta(\partial A - \sigma)$$

$$= \int DA e^{iS(A) - (i/2\xi)\int d^4 x(\partial A)^2} \tag{7}$$

Nifty trick by Faddeev and Popov, eh?

Thus, $S(A)$ in (3) is effectively replaced by

$$S_{\text{eff}}(A) = S(A) - \frac{1}{2\xi} \int d^4 x (\partial A)^2$$

$$= \int d^4 x \left\{ \frac{1}{2} A_\mu \left[\partial^2 g^{\mu\nu} - \left(1 - \frac{1}{\xi}\right) \partial^\mu \partial^\nu \right] A_\nu + A_\mu J^\mu \right\} \tag{8}$$

and $Q^{\mu\nu}$ by $Q_{\text{eff}}^{\mu\nu} = \partial^2 g^{\mu\nu} - (1 - 1/\xi)\partial^\mu \partial^\nu$ or in momentum space $Q_{\text{eff}}^{\mu\nu} = -k^2 g^{\mu\nu} + (1 - 1/\xi)k^\mu k^\nu$, which does have an inverse. Indeed, you can check that

$$Q_{\text{eff}}^{\mu\nu} \left[-g_{\nu\lambda} + (1 - \xi)\frac{k_\nu k_\lambda}{k^2} \right] \frac{1}{k^2} = \delta_\lambda^\mu$$

Thus, the photon propagator can be chosen to be

$$\frac{(-i)}{k^2} \left[g_{\nu\lambda} - (1 - \xi)\frac{k_\nu k_\lambda}{k^2} \right] \tag{9}$$

in agreement with the conclusion in chapter II.7.

While the Faddeev-Popov argument is a lot slicker, many physicists still prefer the explicit Feynman argument given in chapter II.7. I do. When we deal with the Yang-Mills theory and the Einstein theory, however, the Faddeev-Popov method is indispensable, as I have already noted.

A photon can find no rest

Let us understand the physics behind the necessity for imposing by hand a (gauge fixing) constraint in gauge theories. In chapter I.5 we sidestepped this whole issue of fixing the gauge by treating the massive vector meson instead of the photon. In effect, we changed $Q^{\mu\nu}$ to $(\partial^2 + m^2)g^{\mu\nu} - \partial^\mu\partial^\nu$, which does have an inverse (in fact we even found the inverse explicitly). We then showed that we could set the mass m to 0 in physical calculations.

There is, however, a huge and intrinsic difference between massive and massless particles. Consider a massive particle moving along. We can always boost it to its rest frame, or in more mathematical terms, we can always Lorentz transform the momentum of a massive particle to the reference momentum $q^\mu = m(1, 0, 0, 0)$. (As is the case elsewhere in this book, if there is no risk of confusion, we write column vectors as row vectors for typographical convenience.) To study the spin degrees of freedom, we should evidently sit in the rest frame of the particle and study how its states respond to rotation. The fancy pants way of saying this is we should study how the states of the particle transform under that particular subgroup of the Lorentz group (known as the little group) consisting of those Lorentz transformations Λ that leave q^μ invariant, namely $\Lambda^\mu_\nu q^\nu = q^\mu$. For $q^\mu = m(1, 0, 0, 0)$, the little group is obviously the rotation group $SO(3)$. We then apply what we learned in nonrelativistic quantum mechanics and conclude that a spin j particle has $(2j + 1)$ spin states (or polarizations in classical physics), as already noted back in chapter I.5.

But if the particle is massless, we can no longer find a Lorentz boost that would bring us to its rest frame. A photon can find no rest!

For a massless particle, the best we can do is to transform the particle's momentum to the reference momentum $q^\mu = \omega(1, 0, 0, 1)$ for some arbitrarily chosen ω. Again, this is just a fancy way of saying that we can always call the direction of motion the third axis. What is the little group that leaves q^μ invariant? Obviously, rotations around the third axis, forming the group $O(2)$, leave q^μ invariant. The spin states of a massless particle of any spin around its direction of motion are known as helicity states, as was already mentioned in chapter II.1. For a particle of spin j, the helicities $\pm j$ are transformed into each other by parity and time reversal, and thus both helicities must be present if the interactions the particle participates in respect these discrete symmetries, as is the case with the photon and the graviton.[1] In particular, the photon, as we have seen repeatedly, has only two polarization degrees of freedom, instead of three, since we no longer have the full rotation group $SO(3)$.

---

[1] But not with the neutrino.

(You already learned in classical electrodynamics that an electromagnetic wave has two transverse degrees of freedom.) For more on this, see appendix B.

In this sense, gauge invariance is strictly speaking not a "real" symmetry but merely a reflection of the fact that we used a redundant description: we used a vector field A_μ with its four degrees of freedom to describe two physical degrees of freedom. This is the true physical origin of gauge invariance.

The condition $\Lambda^\mu_\nu q^\nu = q^\mu$ should leave us with a 3-parameter subgroup. To find the other transformations, it suffices to look in the neighborhood of the identity, that is, at Lorentz transformations of the form $\Lambda(\alpha, \beta) = I + \alpha A + \beta B + \cdots$. By inspection, we see that

$$
A = \begin{pmatrix} 0 & 1 & 0 & 0 \\ 1 & 0 & 0 & -1 \\ 0 & 0 & 0 & 0 \\ 0 & 1 & 0 & 0 \end{pmatrix} = i(K_1 + J_2), \qquad B = \begin{pmatrix} 0 & 0 & 1 & 0 \\ 0 & 0 & 0 & 0 \\ 1 & 0 & 0 & -1 \\ 0 & 0 & 1 & 0 \end{pmatrix} = i(K_2 - J_1) \tag{10}
$$

where we used the notation for the generators of the Lorentz group from chapter II.3. Note that A and B are to a large extent determined by the fact that J and K are symmetric and antisymmetric, respectively.

By direct computation or by invoking the celebrated minus sign in (II.3.9), we find that $[A, B] = 0$. Also, $[J_3, A] = B$ and $[J_3, B] = -A$ so that, as expected, (A, B) form a 2-component vector under $O(2)$ rotations around the third axis. (For those who must know, the generators A, B, and J_3 generate the group $ISO(2)$, the invariance group of the Euclidean 2-plane, consisting of two translations and one rotation.)

The preceding paragraph establishing the little group for massless particles applies for any spin, including zero. Now specialize to a spin 1 massless particle with the two polarization vectors $\epsilon^\pm(q) = (1/\sqrt{2})(0, 1, \pm i, 0)$. The polarization vectors are defined by how they transform under rotation $e^{i\phi J_3}$. So it is natural to ask how $\epsilon^\pm(q)$ transform under $\Lambda(\alpha, \beta)$. Inspecting (10), we see that

$$
\epsilon^\pm(q) \to \epsilon^\pm(q) + \frac{1}{\sqrt{2}}(\alpha \pm i\beta)q \tag{11}
$$

We recognize (11) as a gauge transformation (as was explained in chapter II.7). For a massless spin 1 particle, the gauge transformation is contained in the Lorentz transformations!

Suppose we construct the corresponding spin 1 field as in chapter II.5 [and in analogy to (I.8.11) and (II.2.10)]:

$$
A_\mu(x) = \int \frac{d^3k}{\sqrt{(2\pi)^3 2\omega_k}} \sum_{\alpha=1,2} [a^{(\alpha)}(\vec{k})\varepsilon^{(\alpha)}_\mu(k)e^{-i(\omega_k t - \vec{k}\cdot\vec{x})} + a^{\dagger(\alpha)}(\vec{k})\varepsilon^{*(\alpha)}_\mu(k)e^{i(\omega_k t - \vec{k}\cdot\vec{x})}] \tag{12}
$$

with $\omega_k = |\vec{k}|$. The polarization vectors $\varepsilon^{(\alpha)}_\mu(k)$ are of coursed determined by the condition $k^\mu \varepsilon^{(\alpha)}_\mu(k) = 0$, which we could easily satisfy by defining $\varepsilon^{(\alpha)}(k) = \Lambda(q \to k)\varepsilon^{(\alpha)}(q)$, where $\Lambda(q \to k)$ denotes a Lorentz transformation that brings the reference momentum q to k.

Note that the $\varepsilon^{(\alpha)}(k)$ thus constructed has a vanishing time component. (To see this, first boost $\varepsilon^{(\alpha)}_\mu(q)$ along the third axis and then rotate, for example.) Hence, $k^\mu \varepsilon^{(\alpha)}_\mu(k) = -\vec{k} \cdot \vec{\varepsilon}^{(\alpha)}(k) = 0$. These properties of $\varepsilon^{(\alpha)}_\mu(k)$ translate into $A_0(x) = 0$ and $\vec{\nabla} \cdot \vec{A}(x) = 0$.

These two constraints cut the four degrees of freedom contained in $A_\mu(x)$ down to two and fix what is known as the Coulomb or radiation gauge.[2]

Given the enormous importance of gauge invariance, it might be instructive to review the logic underlying the "poor man's approach" to gauge invariance (which, as I mentioned in chapter I.5, I learned from Coleman) adopted in this book for pedagogical reasons. You could have fun faking Feynman's mannerism and accent, saying, "Aw shucks, all that fancy talk about little groups! Who needs it? Those experimentalists won't ever be able to prove that the photon mass is mathematically zero anyway."

So start, as in chapter I.5, with the two equations needed for describing a spin 1 massive particle:

$$(\partial^2 + m^2)A_\mu = 0 \tag{13}$$

and

$$\partial_\mu A^\mu = 0 \tag{14}$$

Equation (14) is needed to cut the number of degrees of freedom contained in A^μ down from four to three.

Lo and behold, (13) and (14) are equivalent to the single equation

$$\partial^\mu(\partial_\mu A_\nu - \partial_\nu A_\mu) + m^2 A_\nu = 0 \tag{15}$$

Obviously, (13) and (14) together imply (15). To verify that (15) implies (13) and (14), we act with ∂^ν on (15) and obtain

$$m^2 \partial A = 0 \tag{16}$$

which for $m \neq 0$ requires $\partial A = 0$, namely (14). Plugging this into (15) we obtain (13).

Having packaged two equations into one, we note that we can derive this single equation (15) by varying the Lagrangian

$$\mathcal{L} = -\frac{1}{4}F_{\mu\nu}F^{\mu\nu} + \frac{1}{2}m^2 A^2 \tag{17}$$

with $F_{\mu\nu} \equiv \partial_\mu A_\nu - \partial_\nu A_\mu$.

Next, suppose we include a source J_μ for this particle by changing the Lagrangian to

$$\mathcal{L} = -\frac{1}{4}F_{\mu\nu}F^{\mu\nu} + \frac{1}{2}m^2 A^2 + A_\mu J^\mu \tag{18}$$

with the resulting equation of motion

$$\partial^\mu(\partial_\mu A_\nu - \partial_\nu A_\mu) + m^2 A_\nu = -J_\nu \tag{19}$$

But now observe that when we act with ∂^ν on (19) we obtain

$$m^2 \partial A = -\partial J \tag{20}$$

---

[2] For a much more detailed and leisurely discussion, see S. Weinberg, *Quantum Theory of Fields,* pp. 69–74 and 246–255.

We recover (14) only if $\partial_\mu J^\mu = 0$, that is, if the source producing the particle, commonly know as the current, is conserved.

Put more vividly, suppose the experimentalists who constructed the accelerator (or whatever) to produce the spin 1 particle messed up and failed to insure that $\partial_\mu J^\mu = 0$; then $\partial A \neq 0$ and a spin 0 excitation would also be produced. To make sure that the beam of spin 1 particles is not contaminated with spin 0 particles, the accelerator builders must assure us that the source J_μ in the Lagrangian (18) is indeed conserved.

Now, if we want to study massless spin 1 particles, we simply set $m = 0$ in (18). The "poor man" ends up (just like the "rich man") using the Lagrangian

$$\mathcal{L} = -\frac{1}{4} F_{\mu\nu} F^{\mu\nu} + A_\mu J^\mu \tag{21}$$

to describe the photon. Lo and behold (as we exclaimed in chapter II.7), $\mathcal{L}$ is left invariant by the gauge transformation $A_\mu \to A_\mu - \partial_\mu \Lambda$ for any $\Lambda(x)$. (As was also explained in that chapter, the third polarization decouples in the limit $m \to 0$.) The "poor man" has thus discovered gauge invariance!

However, as I warned in chapter I.5, depending on his or her personality, the poor man could also wake up in the middle of the night worrying that physics might be discontinuous in the limit $m \to 0$. Thus the little group discussion is needed to remove that nightmare. But then a "real" physicist in the Feynman mode could always counter that for any physical measurement everything must be okay as long as the duration of the experiment is short compared to the characteristic time $1/m$. More on this issue in chapter VIII.1.

A reflection on gauge symmetry

As we will see later and as you might have heard, much of the world beyond electromagnetism is also described by gauge theories. But as we saw here, gauge theories are also deeply disturbing and unsatisfying in some sense: They are built on a redundancy of description. The electromagnetic gauge transformation $A_\mu \to A_\mu - \partial_\mu \Lambda$ is not truly a symmetry stating that two physical states have the same properties. Rather, it tells us that the two gauge potentials A_μ and $A_\mu - \partial_\mu \Lambda$ describe the same physical state. In your orderly study of physics, the first place where A_μ becomes indispensable is the Schrödinger equation, as I will explain in chapter IV.4. Within classical physics, you got along perfectly well with just $\vec{E}$ and $\vec{B}$. Some physicists are looking for a formulation of quantum electrodynamics without using A_μ, but so far have failed to turn up an attractive alternative to what we have. It is conceivable that a truly deep advance in theoretical physics would involve writing down quantum electrodynamics without writing A_μ.

Slower in its maturity

Quantum field theory at its birth was relativistic. Later in its maturity, it found applications in condensed matter physics. We will have a lot more to say about the role of quantum field theory in condensed matter, but for now, we have the more modest goal of learning how to take the nonrelativistic limit of a quantum field theory.

The Lorentz invariant scalar field theory

$$\mathcal{L} = (\partial \Phi^\dagger)(\partial \Phi) - m^2 \Phi^\dagger \Phi - \lambda (\Phi^\dagger \Phi)^2 \tag{1}$$

(with $\lambda > 0$ as always) describes a bunch of interacting bosons. It should certainly contain the physics of slowly moving bosons. For clarity consider first the relativistic Klein-Gordon equation

$$(\partial^2 + m^2)\Phi = 0 \tag{2}$$

for a free scalar field. A mode with energy $E = m + \varepsilon$ would oscillate in time as $\Phi \propto e^{-iEt}$. In the nonrelativistic limit, the kinetic energy ε is much smaller than the rest mass m. It makes sense to write $\Phi(\vec{x}, t) = e^{-imt}\varphi(\vec{x}, t)$, with the field φ oscillating in time much more slowly than e^{-imt}. Plugging into (2) and using the identity $(\partial/\partial t)e^{-imt}(\cdots) = e^{-imt}(-im + \partial/\partial t)(\cdots)$ twice, we obtain $(-im + \partial/\partial t)^2 \varphi - \vec{\nabla}^2 \varphi + m^2 \varphi = 0$. Dropping the term $(\partial^2/\partial t^2)\varphi$ as small compared to $-2im(\partial/\partial t)\varphi$, we find Schrödinger's equation, as we had better:

$$i\frac{\partial}{\partial t}\varphi = -\frac{\vec{\nabla}^2}{2m}\varphi \tag{3}$$

By the way, the Klein-Gordon equation was actually discovered before Schrödinger's equation.

Having absorbed this, you can now easily take the nonrelativistic limit of a quantum field theory. Simply plug

$$\Phi(\vec{x}, t) = \frac{1}{\sqrt{2m}} e^{-imt} \varphi(\vec{x}, t) \tag{4}$$

into (1) . (The factor $1/\sqrt{2m}$ is for later convenience.) For example,

$$\frac{\partial \Phi^\dagger}{\partial t} \frac{\partial \Phi}{\partial t} - m^2 \Phi^\dagger \Phi \rightarrow \frac{1}{2m} \left\{ \left[\left(im + \frac{\partial}{\partial t} \right) \varphi^\dagger \right] \left[\left(-im + \frac{\partial}{\partial t} \right) \varphi \right] - m^2 \varphi^\dagger \varphi \right\}$$

$$\simeq \frac{1}{2} i \left(\varphi^\dagger \frac{\partial \varphi}{\partial t} - \frac{\partial \varphi^\dagger}{\partial t} \varphi \right) \tag{5}$$

After an integration by parts we arrive at

$$\mathcal{L} = i\varphi^\dagger \partial_0 \varphi - \frac{1}{2m} \partial_i \varphi^\dagger \partial_i \varphi - g^2 (\varphi^\dagger \varphi)^2 \tag{6}$$

where $g^2 = \lambda/4m^2$.

As we saw in chapter I.10 the theory (1) enjoys a conserved Noether current $J_\mu = i(\Phi^\dagger \partial_\mu \Phi - \partial_\mu \Phi^\dagger \Phi)$. The density J_0 reduces to $\varphi^\dagger \varphi$, precisely as you would expect, while J_i reduces to $(i/2m)(\varphi^\dagger \partial_i \varphi - \partial_i \varphi^\dagger \varphi)$. When you first took a course in quantum mechanics, didn't you wonder why the density $\rho \equiv \varphi^\dagger \varphi$ and the current $J_i = (i/2m)(\varphi^\dagger \partial_i \varphi - \partial_i \varphi^\dagger \varphi)$ look so different? As to be expected, various expressions inevitably become uglier when reduced from a more symmetric to a less symmetric theory.

Number is conjugate to phase angle

Let me point out some differences between the relativistic and nonrelativistic case.

The most striking is that the relativistic theory is quadratic in time derivative, while the nonrelativistic theory is linear in time derivative. Thus, in the nonrelativistic theory the momentum density conjugate to the field φ, namely $\delta\mathcal{L}/\delta\partial_0\varphi$, is just $i\varphi^\dagger$, so that $[\varphi^\dagger(\vec{x}, t), \varphi(\vec{x}', t)] = -\delta^{(D)}(\vec{x} - \vec{x}')$. In condensed matter physics it is often illuminating to write $\varphi = \sqrt{\rho} e^{i\theta}$ so that

$$\mathcal{L} = \frac{i}{2} \partial_0 \rho - \rho \partial_0 \theta - \frac{1}{2m} \left[\rho(\partial_i \theta)^2 + \frac{1}{4\rho}(\partial_i \rho)^2 \right] - g^2 \rho^2 \tag{7}$$

The first term is a total divergence. The second term tells us something of great importance[1] in condensed matter physics: in the canonical formalism (chapter I.8), the momentum density conjugate to the phase field $\theta(x)$ is $\delta\mathcal{L}/\delta\partial_0\theta = -\rho$ and thus Heisenberg tells us that

$$[\rho(\vec{x}, t), \theta(\vec{x}', t)] = i\delta^{(D)}(\vec{x} - \vec{x}') \tag{8}$$

---

[1] See P. Anderson, *Basic Notions of Condensed Matter Physics*, p. 235.

Integrating and defining $N \equiv \int d^D x \rho(\vec{x}, t) =$ the total number of bosons, we find one of the most important relations in condensed matter physics

$$[N, \theta] = i \tag{9}$$

Number is conjugate to phase angle, just as momentum is conjugate to position. Marvel at the elegance of this! You would learn in a condensed matter course that this fundamental relation underlies the physics of the Josephson junction.

You may know that a system of bosons with a "hard core" repulsion between them is a superfluid at zero temperature. In particular, Bogoliubov showed that the system contains an elementary excitation obeying a linear dispersion relation.[2] I will discuss superfluidity in chapter V.1.

In the path integral formalism, going from the complex field $\varphi = \varphi_1 + i\varphi_2$ to ρ and θ amounts to a change of integration variables, as I remarked back in chapter I.8. In the canonical formalism, since one deals with operators, one has to tread with somewhat more finesse.

The sign of repulsion

In the nonrelativistic theory (7) it is clear that the bosons repel each other: Piling particles into a high density region would cost you an energy density $g^2 \rho^2$. But it is less clear in the relativistic theory that $\lambda(\Phi^\dagger \Phi)^2$ with λ positive corresponds to repulsion. I outline one method in exercise III.5.3, but here let's just take a flying heuristic guess. The Hamiltonian (density) involves the negative of the Lagrangian and hence goes as $\lambda(\Phi^\dagger \Phi)^2$ for large Φ and would thus be unbounded below for $\lambda < 0$. We know physically that a free Bose gas tends to condense and clump, and with an attractive interaction it surely might want to collapse. We naturally guess that $\lambda > 0$ corresponds to repulsion.

I next give you a more foolproof method. Using the central identity of quantum field theory we can rewrite the path integral for the theory in (1) as

$$Z = \int D\Phi D\sigma e^{i \int d^4 x [(\partial \Phi^\dagger)(\partial \Phi) - m^2 \Phi^\dagger \Phi + 2\sigma \Phi^\dagger \Phi + (1/\lambda)\sigma^2]} \tag{10}$$

Condensed matter physicists call the transformation from (1) to the Lagrangian $\mathcal{L} = (\partial \Phi^\dagger)(\partial \Phi) - m^2 \Phi^\dagger \Phi + 2\sigma \Phi^\dagger \Phi + (1/\lambda)\sigma^2$ the Hubbard-Stratonovich transformation. In field theory, a field that does not have kinetic energy, such as σ, is known as an auxiliary field and can be integrated out in the path integral. When we come to the superfield formalism in chapter VIII.4, auxiliary fields will play an important role.

Indeed, you might recall from chapter III.2 how a theory with an intermediate vector boson could generate Fermi's theory of the weak interaction. The same physics is involved here: The theory (10) in which the Φ field is coupled to an "intermediate σ boson" can generate the theory (1).

---

[2] For example, L.D. Landau and E. M. Lifschitz, *Statisical Physics*, p. 238.

If σ were a "normal scalar field" of the type we have studied, that is, if the terms quadratic in σ in the Lagrangian had the form $\frac{1}{2}(\partial\sigma)^2 - \frac{1}{2}M^2\sigma^2$, then its propagator would be $i/(k^2 - M^2 + i\varepsilon)$. The scattering amplitude between two Φ bosons would be proportional to this propagator. We learned in chapter I.4 that the exchange of a scalar field leads to an attractive force.

But σ is not a normal field as evidenced by the fact that the Lagrangian contains only the quadratic term $+(1/\lambda)\sigma^2$. Thus its propagator is simply $i/(1/\lambda) = i\lambda$, which (for $\lambda > 0$) has a sign opposite to the normal propagator evaluated at low-momentum transfer $i/(k^2 - M^2 + i\varepsilon) \simeq -i/M^2$. We conclude that σ exchange leads to a repulsive force.

Incidentally, this argument also shows that the repulsion is infinitely short ranged, like a delta function interaction. Normally, as we learned in chapter I.4 the range is determined by the interplay between the k^2 and the M^2 terms. Here the situation is as if the M^2 term is infinitely large. We can also argue that the interaction $\lambda(\Phi^\dagger\Phi)^2$ involves creating two bosons and then annihilating them both at the same spacetime point.

Finite density

One final point of physics that people trained as particle physicists do not always remember: Condensed matter physicists are not interested in empty space, but want to have a finite density $\bar{\rho}$ of bosons around. We learned in statistical mechanics to add a chemical potential term $\mu\varphi^\dagger\varphi$ to the Lagrangian (6). Up to an irrelevant (in this context!) additive constant, we can rewrite the resulting Lagrangian as

$$\mathcal{L} = i\varphi^\dagger\partial_0\varphi - \frac{1}{2m}\partial_i\varphi^\dagger\partial_i\varphi - g^2(\varphi^\dagger\varphi - \bar{\rho})^2 \tag{11}$$

Amusingly, mass appears in different places in relativistic and nonrelativistic field theories. To proceed further, I have to develop the concept of spontaneous symmetry breaking. Thus, adios for now. We will come back to superfluidity in due time.

Exercises

III.5.1   Obtain the Klein-Gordon equation for a particle in an electrostatic potential (such as that of the nucleus) by the gauge principle of replacing $(\partial/\partial t)$ in (2) by $\partial/\partial t - ieA_0$. Show that in the nonrelativistic limit this reduces to the Schrödinger's equation for a particle in an external potential.

III.5.2   Take the nonrelativistic limit of the Dirac Lagrangian.

III.5.3   Given a field theory we can compute the scattering amplitude of two particles in the nonrelativistic limit. We then postulate an interaction potential $U(\vec{x})$ between the two particles and use nonrelativistic quantum mechanics to calculate the scattering amplitude, for example in Born approximation. Comparing the two scattering amplitudes we can determine $U(\vec{x})$. Derive the Yukawa and the Coulomb potentials this way. The application of this method to the $\lambda(\Phi^\dagger\Phi)^2$ interaction is slightly problematic since the delta function interaction is a bit singular, but it should be all right for determining whether the force is repulsive or attractive.

III.6 | The Magnetic Moment of the Electron

Dirac's triumph

I said in the preface that the emphasis in this book is not on computation, but how can I not tell you about the greatest triumph of quantum field theory?

After Dirac wrote down his equation, the next step was to study how the electron interacts with the electromagnetic field. According to the gauge principle already used to write the Schrödinger's equation in an electromagnetic field, to obtain the Dirac equation for an electron in an external electromagnetic field we merely have to replace the ordinary derivative ∂_μ by the covariant derivative $D_\mu = \partial_\mu - ieA_\mu$:

$$(i\gamma^\mu D_\mu - m)\psi = 0 \tag{1}$$

Recall (II.1.27).

Acting on this equation with $(i\gamma^\mu D_\mu + m)$, we obtain $-(\gamma^\mu \gamma^\nu D_\mu D_\nu + m^2)\psi = 0$. We have $\gamma^\mu \gamma^\nu D_\mu D_\nu = \frac{1}{2}(\{\gamma^\mu, \gamma^\nu\} + [\gamma^\mu, \gamma^\nu])D_\mu D_\nu = D_\mu D^\mu - i\sigma^{\mu\nu} D_\mu D_\nu$ and $i\sigma^{\mu\nu} D_\mu D_\nu = (i/2)\sigma^{\mu\nu}[D_\mu, D_\nu] = (e/2)\sigma^{\mu\nu} F_{\mu\nu}$. Thus

$$\left(D_\mu D^\mu - \frac{e}{2}\sigma^{\mu\nu} F_{\mu\nu} + m^2\right)\psi = 0 \tag{2}$$

Now consider a weak constant magnetic field pointing in the 3rd direction for definiteness, weak so that we can ignore the $(A_i)^2$ term in $(D_i)^2$. By gauge invariance, we can choose $A_0 = 0$, $A_1 = -\frac{1}{2}Bx^2$, and $A_2 = \frac{1}{2}Bx^1$ (so that $F_{12} = \partial_1 A_2 - \partial_2 A_1 = B$). As we will see, this is one calculation in which we really have to keep track of factors of 2. Then

$$(D_i)^2 = (\partial_i)^2 - ie(\partial_i A_i + A_i \partial_i) + O(A_i^2)$$
$$= (\partial_i)^2 - 2\frac{ie}{2}B(x^1\partial_2 - x^2\partial_1) + O(A_i^2)$$
$$= \vec{\nabla}^2 - e\vec{B}\cdot \vec{x}\times\vec{p} + O(A_i^2) \tag{3}$$

Note that we used $\partial_i A_i + A_i \partial_i = (\partial_i A_i) + 2A_i \partial_i = 2A_i \partial_i$, where in $(\partial_i A_i)$ the partial derivative acts only on A_i. You may have recognized $\vec{L} \equiv \vec{x} \times \vec{p}$ as the orbital angular momentum operator. Thus, the orbital angular momentum generates an orbital magnetic moment that interacts with the magnetic field.

This calculation makes good physical sense. If we were studying the interaction of a charged scalar field Φ with an external electromagnetic field we would start with

$$(D_\mu D^\mu + m^2)\Phi = 0 \tag{4}$$

obtained by replacing the ordinary derivative in the Klein-Gordon equation by covariant derivatives. We would then go through the same calculation as in (3). Comparing (4) with (2) we see that the spin of the electron contributes the additional term $(e/2)\sigma^{\mu\nu}F_{\mu\nu}$.

As in chapter II.1 we write $\psi = \begin{pmatrix} \phi \\ \chi \end{pmatrix}$ in the Dirac basis and focus on ϕ since in the nonrelativistic limit it dominates χ. Recall that in that basis $\sigma^{ij} = \varepsilon^{ijk}\begin{pmatrix} \sigma^k & 0 \\ 0 & \sigma^k \end{pmatrix}$. Thus $(e/2)\sigma^{\mu\nu}F_{\mu\nu}$ acting on ϕ is effectively equal to $(e/2)\sigma^3(F_{12} - F_{21}) = (e/2)2\sigma^3 B = 2e\vec{B} \cdot \vec{S}$ since $\vec{S} = (\vec{\sigma}/2)$. Make sure you understand all the factors of 2! Meanwhile, according to what I told you in chapter II.1, we should write $\phi = e^{-imt}\Psi$, where Ψ oscillates much more slowly than e^{-imt} so that $(\partial_0^2 + m^2)e^{-imt}\Psi \simeq e^{-imt}[-2im(\partial/\partial t)\Psi]$. Putting it all together, we have

$$\left[-2im\frac{\partial}{\partial t} - \vec{\nabla}^2 - e\vec{B} \cdot (\vec{L} + 2\vec{S})\right]\Psi = 0 \tag{5}$$

There you have it! As if by magic, Dirac's equation tells us that a unit of spin angular momentum interacts with a magnetic field twice as much as a unit of orbital angular momentum, an observational fact that had puzzled physicists deeply at the time. The calculation leading to (5) is justly celebrated as one of the greatest in the history of physics.

The story is that Dirac did not do this calculation until a day after he discovered his equation, so sure was he that the equation had to be right. Another version is that he dreaded the possibility that the magnetic moment would come out wrong and that Nature would not take advantage of his beautiful equation.

Another way of seeing that the Dirac equation contains a magnetic moment is by the Gordon decomposition, the proof of which is given in an exercise:

$$\bar{u}(p')\gamma^\mu u(p) = \bar{u}(p')\left[\frac{(p'+p)^\mu}{2m} + \frac{i\sigma^{\mu\nu}(p'-p)_\nu}{2m}\right]u(p) \tag{6}$$

Looking at the interaction with an electromagnetic field $\bar{u}(p')\gamma^\mu u(p)A_\mu(p'-p)$, we see that the first term in (6) only depends on the momentum $(p'+p)^\mu$ and would have been there even if we were treating the interaction of a charged scalar particle with the electromagnetic field to first order. The second term involves spin and gives the magnetic moment. One way of saying this is that $\bar{u}(p')\gamma^\mu u(p)$ contains a magnetic moment component.

The anomalous magnetic moment

With improvements in experimental techniques, it became clear by the late 1940's that the magnetic moment of the electron was larger than the value calculated by Dirac by a factor of 1.00118 ± 0.00003. The challenge to any theory of quantum electrodynamics was to calculate this so-called anomalous magnetic moment. As you probably know, Schwinger's spectacular success in meeting this challenge established the correctness of relativistic quantum field theory, at least in dealing with electromagnetic phenomena, beyond any doubt.

Before we plunge into the calculation, note that Lorentz invariance and current conservation tell us (see exercise III.6.3) that the matrix element of the electromagnetic current must have the form (here $|p, s\rangle$ denotes a state with an electron of momentum p and polarization s)

$$\langle p', s' | J^\mu(0) | p, s \rangle = \bar{u}(p', s') \left[\gamma^\mu F_1(q^2) + \frac{i\sigma^{\mu\nu} q_\nu}{2m} F_2(q^2) \right] u(p, s) \tag{7}$$

where $q \equiv (p' - p)$. The functions $F_1(q^2)$ and $F_2(q^2)$, about which Lorentz invariance can tell us nothing, are known as form factors. To leading order in momentum transfer q, (7) becomes

$$\bar{u}(p', s') \left\{ \frac{(p' + p)^\mu}{2m} F_1(0) + \frac{i\sigma^{\mu\nu} q_\nu}{2m} [F_1(0) + F_2(0)] \right\} u(p, s)$$

by the Gordon decomposition. The coefficient of the first term is the electric charge observed by experimentalists and is by definition equal to 1. (To see this, think of potential scattering, for example. See chapter II.6.) Thus $F_1(0) = 1$. The magnetic moment of the electron is shifted from the Dirac value by a factor $1 + F_2(0)$.

Schwinger's triumph

Let us now calculate $F_2(0)$ to order $\alpha = e^2/4\pi$. First draw all the relevant Feynman diagrams to this order (fig. III.6.1). Except for figure 1b, all the Feynman diagrams are clearly proportional to $\bar{u}(p', s')\gamma^\mu u(p, s)$ and thus contribute to $F_1(q^2)$, which we don't care about. Happy are we! We only have to calculate one Feynman diagram.

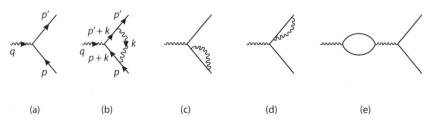

(a)   (b)   (c)   (d)   (e)

Figure III.6.1

It is convenient to normalize the contribution of figure 1b by comparing it to the lowest order contribution of figure 1a and write the sum of the two contributions as $\bar{u}(\gamma^\mu + \Gamma^\mu)u$. Applying the Feynman rules, we find

$$\Gamma^\mu = \int \frac{d^4k}{(2\pi)^4} \frac{-i}{k^2} \left(ie\gamma^\nu \frac{i}{\not{p}' + \not{k} - m} \gamma^\mu \frac{i}{\not{p} + \not{k} - m} ie\gamma_\nu \right) \tag{8}$$

I will now go through the calculation in some detail not only because it is important, but also because we will be using a variety of neat tricks springing from the brilliant minds of Schwinger and Feynman. You should verify all the steps of course.

Simplifying somewhat we obtain $\Gamma^\mu = -ie^2 \int [d^4k/(2\pi)^4](N^\mu/D)$, where

$$N^\mu = \gamma^\nu(\not{p}' + \not{k} + m)\gamma^\mu(\not{p} + \not{k} + m)\gamma_\nu \tag{9}$$

and

$$\frac{1}{D} = \frac{1}{(p'+k)^2 - m^2} \frac{1}{(p+k)^2 - m^2} \frac{1}{k^2} = 2 \int d\alpha \, d\beta \frac{1}{D}. \tag{10}$$

We have used the identity (D.16). The integral is evaluated over the triangle in the $(\alpha$-$\beta)$ plane bounded by $\alpha = 0$, $\beta = 0$, and $\alpha + \beta = 1$, and

$$\mathcal{D} = [k^2 + 2k(\alpha p' + \beta p)]^3 = [l^2 - (\alpha + \beta)^2 m^2]^3 + O(q^2) \tag{11}$$

where we completed a square by defining $k = l - (\alpha p' + \beta p)$. The momentum integration is now over d^4l.

Our strategy is to massage N^μ into a form consisting of a linear combination of γ^μ, p^μ, and p'^μ. Invoking the Gordon decomposition (6) we can write (7) as

$$\bar{u} \left\{ \gamma^\mu[F_1(q^2) + F_2(q^2)] - \frac{1}{2m}(p'+p)^\mu F_2(q^2) \right\} u$$

Thus, to extract $F_2(0)$ we can throw away without ceremony any term proportional to γ^μ that we encounter while massaging N^μ. So, let's proceed.

Eliminating k in favor of l in (9) we obtain

$$N^\mu = \gamma^\nu[\not{l} + \not{P}' + m]\gamma^\mu[\not{l} + \not{P} + m]\gamma_\nu \tag{12}$$

where $P'^\mu \equiv (1-\alpha)p'^\mu - \beta p^\mu$ and $P^\mu \equiv (1-\beta)p^\mu - \alpha p'^\mu$. I will use the identities in appendix D repeatedly, without alerting you every time I use one. It is convenient to organize the terms in N^μ by powers of m. (Here I give up writing in complete grammatical sentences.)

1. The m^2 term: a γ^μ term, throw away.
2. The m terms: organize by powers of l. The term linear in l integrates to 0 by symmetry. Thus, we are left with the term independent of l:

$$m(\gamma^\nu \not{P}' \gamma^\mu \gamma_\nu + \gamma^\nu \gamma^\mu \not{P} \gamma_\nu) = 4m[(1-2\alpha)p'^\mu + (1-2\beta)p^\mu]$$
$$\rightarrow 4m(1-\alpha-\beta)(p'+p)^\mu \tag{13}$$

In the last step I used a handy trick; since $\mathcal{D}$ is symmetric under $\alpha \longleftrightarrow \beta$, we can symmetrize the terms we get in N^μ.

3. Finally, the most complicated m^0 term. The term quadratic in l: note that we can effectively replace $l^\sigma l^\tau$ inside $\int d^4l/(2\pi)^4$ by $\frac{1}{4}\eta^{\sigma\tau}l^2$ by Lorentz invariance (this step is possible because we have shifted the integration variable so that $\mathcal{D}$ is a Lorentz invariant function of l^2.) Thus, the term quadratic in l gives rise to a γ^μ term. Throw it away. Again we throw away the term linear in l, leaving [use (D.6) here!]

$$\gamma^\nu \not{P}'\gamma^\mu \not{P}\gamma_\nu = -2\not{P}\gamma^\mu \not{P}'$$
$$\rightarrow -2[(1-\beta)\not{p} - \alpha m]\gamma^\mu[(1-\alpha)\not{p}' - \beta m] \tag{14}$$

where in the last step we remembered that Γ^μ is to be sandwiched between $\bar{u}(p')$ and $u(p)$. Again, it is convenient to organize the terms in (14) by powers of m. With the various tricks we have already used, we find that the m^2 term can be thrown away, the m term gives $2m(p'+p)^\mu[\alpha(1-\alpha) + \beta(1-\beta)]$, and the m^0 term gives $2m(p'+p)^\mu[-2(1-\alpha)(1-\beta)]$. Putting it altogether, we find that $N^\mu \rightarrow 2m(p'+p)^\mu(\alpha + \beta)(1 - \alpha - \beta)$

We can now do the integral $\int [d^4l/(2\pi)^4](1/\mathcal{D})$ using (D.11). Finally, we obtain

$$\Gamma^\mu = -2ie^2 \int d\alpha\, d\beta (\frac{-i}{32\pi^2}) \frac{1}{(\alpha+\beta)^2 m^2} N^\mu$$
$$= -\frac{e^2}{8\pi^2} \frac{1}{2m}(p'+p)^\mu \tag{15}$$

and thus, trumpets please:

$$F_2(0) = \frac{e^2}{8\pi^2} = \frac{\alpha}{2\pi} \tag{16}$$

Schwinger's announcement of this result in 1948 had an electrifying impact on the theoretical physics community.

I gave you in this chapter not one, but two, of the great triumphs of twentieth century physics, although admittedly the first is not a result of field theory per se.

Exercises

III.6.1 Evaluate $\bar{u}(p')(\not{p}'\gamma^\mu + \gamma^\mu \not{p})u(p)$ in two different ways and thus prove Gordon decomposition.

III.6.2 Check that (7) is consistent with current conservation. [Hint: By translation invariance (we suppress the spin variable)

$$\langle p'| J^\mu(x) |p\rangle = \langle p'| J^\mu(0) |p\rangle e^{i(p'-p)x}$$

and hence

$$\langle p'| \partial_\mu J^\mu(x) |p\rangle = i(p'-p)_\mu \langle p'| J^\mu(0) |p\rangle e^{i(p'-p)x}$$

Thus current conservation implies that $q^\mu \langle p'| J^\mu(0) |p\rangle = 0$.]

III.6.3 By Lorentz invariance the right hand side of (7) has to be a vector. The only possibilities are $\bar{u}\gamma^\mu u$, $(p+p')^\mu \bar{u}u$, and $(p-p')^\mu \bar{u}u$. The last term is ruled out because it would not be consistent with current conservation. Show that the form given in (7) is in fact the most general allowed.

III.6.4 In chapter II.6, when discussing electron-proton scattering, we ignored the strong interaction that the proton participates in. Argue that the effects of the strong interaction could be included phenomenologically by replacing the vertex $\bar{u}(P, S)\gamma^\mu u(p, s)$ in (II.6.1) by

$$\langle P, S| J^\mu(0) |p, s\rangle = \bar{u}(P, S) \left[\gamma^\mu F_1(q^2) + \frac{i\sigma^{\mu\nu} q_\nu}{2m} F_2(q^2) \right] u(p, s) \tag{17}$$

Careful measurements of electron-proton scattering, thus determining the two proton form factors $F_1(q^2)$ and $F_2(q^2)$, earned R. Hofstadter the 1961 Nobel Prize. While we could account for the general behavior of these two form factors, we are still unable to calculate them from first principles (in contrast to the corresponding form factors for the electron.) See chapters IV.2 and VII.3.

A photon can fluctuate into an electron and a positron

One early triumph of quantum electrodynamics is the understanding of how quantum fluctuations affect the way the photon propagates. A photon can always metamorphose into an electron and a positron that, after a short time mandated by the uncertainty principle, annihilate each other becoming a photon again. The process, which keeps on repeating itself, is depicted in figure III.7.1.

Quantum fluctuations are not limited to what we just described. The electron and positron can interact by exchanging a photon, which in turn can change into an electron and a positron, and so on and so forth. The full process is shown in figure III.7.2, where the shaded object, denoted by $i\Pi_{\mu\nu}(q)$ and known as the vacuum polarization tensor, is given by an infinite number of Feynman diagrams, as shown in figure III.7.3. Figure III.7.1 is obtained from figure III.7.2 by approximating $i\Pi_{\mu\nu}(q)$ by its lowest order diagram.

It is convenient to rewrite the Lagrangian $\mathcal{L} = \bar{\psi}[i\gamma^\mu(\partial_\mu - ieA_\mu) - m]\psi - \frac{1}{4}F_{\mu\nu}F^{\mu\nu}$ by letting $A \to (1/e)A$, which we are always allowed to do, so that

$$\mathcal{L} = \bar{\psi}[i\gamma^\mu(\partial_\mu - iA_\mu) - m]\psi - \frac{1}{4e^2}F_{\mu\nu}F^{\mu\nu} \tag{1}$$

Note that the gauge transformation leaving $\mathcal{L}$ invariant is given by $\psi \to e^{i\alpha}\psi$ and $A_\mu \to A_\mu + \partial_\mu\alpha$. The photon propagator (chapter III.4), obtained roughly speaking by inverting $(1/4e^2)F_{\mu\nu}F^{\mu\nu}$, is now proportional to e^2:

$$iD_{\mu\nu}(q) = \frac{-ie^2}{q^2}\left[g_{\mu\nu} - (1-\xi)\frac{q_\mu q_\nu}{q^2}\right] \tag{2}$$

Every time a photon is exchanged, the amplitude gets a factor of e^2. This is just a trivial but convenient change and does not affect the physics in the slightest. For example, in the Feynman diagram we calculated in chapter II.6 for electron-electron scattering, the factor e^2 can be thought of as being associated with the photon propagator rather than as coming from the interaction vertices. In this interpretation e^2 measures the ease with which the

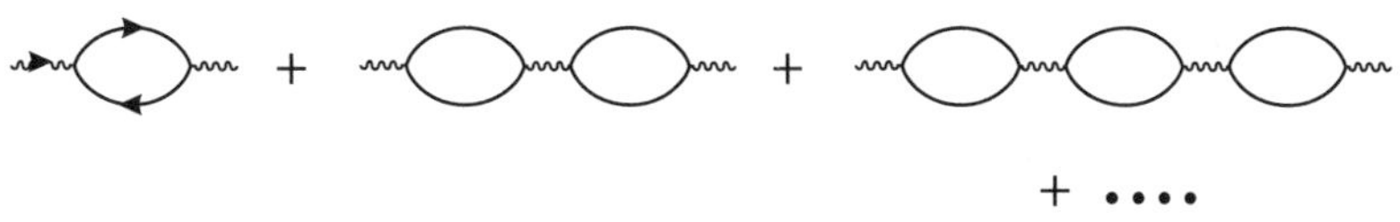

Figure III.7.1

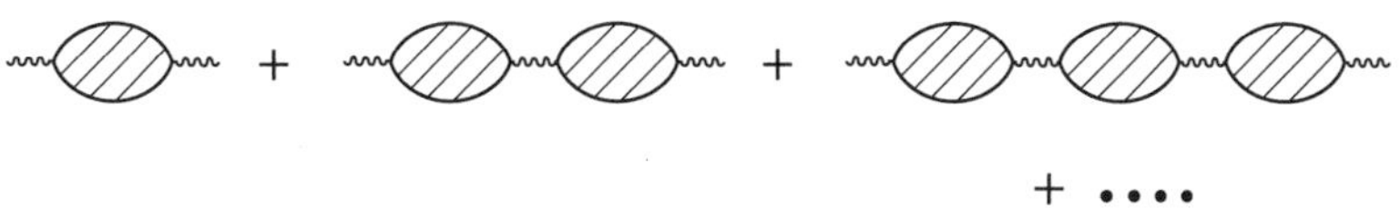

Figure III.7.2

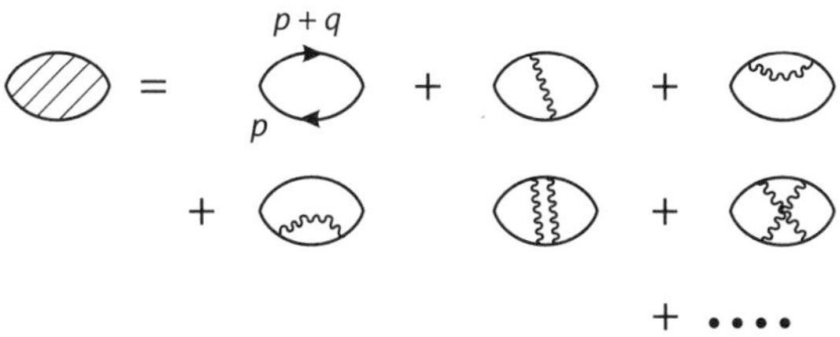

Figure III.7.3

photon propagates through spacetime. The smaller e^2, the more action it takes to have the photon propagate, and the harder for the photon to propagate, the weaker the effect of electromagnetism.

The diagrammatic proof of gauge invariance given in chapter II.7 implies that $q^\mu \Pi_{\mu\nu}(q) = 0$. Together with Lorentz invariance, this requires that

$$\Pi_{\mu\nu}(q) = (q_\mu q_\nu - g_{\mu\nu}q^2)\Pi(q^2) \tag{3}$$

The physical or renormalized photon propagator as shown in figure III.7.2 is then given by the geometric series

$$
\begin{aligned}
i D^P_{\mu\nu}(q) &= i D_{\mu\nu}(q) + i D_{\mu\lambda}(q) i \Pi^{\lambda\rho}(q) i D_{\rho\nu}(q) \\
&\quad + i D_{\mu\lambda}(q) i \Pi^{\lambda\rho}(q) i D_{\rho\sigma}(q) i \Pi^{\sigma\kappa}(q) i D_{\kappa\nu}(q) + \cdots \\
&= \frac{-ie^2}{q^2} g_{\mu\nu}\{1 - e^2\Pi(q^2) + [e^2\Pi(q^2)]^2 + \cdots\} + q_\mu q_\nu \text{ term} \\
&= \frac{-ie^2}{q^2} g_{\mu\nu} \frac{1}{1 + e^2\Pi(q^2)} + q_\mu q_\nu \text{ term}
\end{aligned}
\tag{4}
$$

Because of (3) the $(1 - \xi)(q_\mu q_\lambda/q^2)$ part of $D_{\mu\lambda}(q)$ is annihilated when it encounters $\Pi_{\lambda\rho}(q)$. Thus, in $i D^P_{\mu\nu}(q)$ the gauge parameter ξ enters only into the $q_\mu q_\nu$ term and drops out in physical amplitudes, as explained in chapter II.7.

The residue of the pole in $i D^P_{\mu\nu}(q)$ is the physical or renormalized charge squared:

$$e^2_R = e^2 \frac{1}{1 + e^2 \Pi(0)} \tag{5}$$

Respect for gauge invariance

In order to determine e_R in terms of e, let us calculate to lowest order

$$i \Pi^{\mu\nu}(q) = (-) \int \frac{d^4 p}{(2\pi)^4} \, \text{tr} \left(i\gamma^\nu \frac{i}{\not{p} + \not{q} - m} i\gamma^\mu \frac{i}{\not{p} - m} \right) \tag{6}$$

For large p the integrand goes as $1/p^2$ with a subleading term going as m^2/p^4 causing the integral to have a quadratically divergent and a logarithmically divergent piece. (You see, it is easy to slip into bad language.) Not a conceptual problem at all, as I explained in chapter III.1. We simply regularize. But now there is a delicate point: Since gauge invariance plays a crucial role, we must make sure that our regularization respects gauge invariance.

In the Pauli-Villars regularization (III.1.13) we replace (6) by

$$i \Pi^{\mu\nu}(q) = (-) \int \frac{d^4 p}{(2\pi)^4} \left[\text{tr} \left(i\gamma^\nu \frac{i}{\not{p} + \not{q} - m} i\gamma^\mu \frac{i}{\not{p} - m} \right) \right.$$
$$\left. - \sum_a c_a \, \text{tr} \left(i\gamma^\nu \frac{i}{\not{p} + \not{q} - m_a} i\gamma^\mu \frac{i}{\not{p} - m_a} \right) \right] \tag{7}$$

Now the integrand goes as $(1 - \sum_a c_a)(1/p^2)$ with a subleading term going as $(m^2 - \sum_a c_a m^2_a)(1/p^4)$, and thus the integral would converge if we choose c_a and m_a such that

$$\sum_a c_a = 1 \tag{8}$$

and

$$\sum_a c_a m^2_a = m^2 \tag{9}$$

Clearly, we have to introduce at least two regulator masses. We are confessing to ignorance of the physics above the mass scale m_a. The integral in (7) is effectively cut off when the momentum p exceeds m_a.

Does a bell ring for you? It should, as this discussion conceptually parallels that in the appendix to chapter I.9.

The gauge invariant form (3) we expect to get actually suggests that we need fewer regulator terms than we think. Imagine expanding (6) in powers of q. Since

$$\Pi_{\mu\nu}(q) = (q_\mu q_\nu - g_{\mu\nu} q^2)[\Pi(0) + \cdots]$$

we are only interested in terms of $O(q^2)$ and higher in the Feynman integral. If we expand the integrand in (6), we see that the term of $O(q^2)$ goes as $1/p^4$ for large p, thus giving a logarithmically divergent (speaking bad language again!) contribution. (Incidentally, you may recall that this sort of argument was also used in chapter III.3.) It seems that we need only one regulator. This argument is not rigorous because we have not proved that $\Pi(q^2)$ has a power series expansion in q^2, but instead of worrying about it let us proceed with the calculation.

Once the integral is convergent, the proof of gauge invariance given in chapter II.7 now goes through. Let us recall briefly how the proof went. In computing $q^\mu \Pi_{\mu\nu}(q)$ we use the identity

$$\frac{1}{\not p + \not q - m} \not q \frac{1}{\not p - m} = \frac{1}{\not p - m} - \frac{1}{\not p + \not q - m}$$

to split the integrand into two pieces that cancel upon shifting the integration variable $p \to p + q$. Recall from exercise (II.7.2) that we were concerned that in some cases the shift may not be allowed, but it is allowed if the integral is sufficiently convergent, as is indeed the case now that we have regularized. In any event, the proof is in the eating of the pudding, and we will see by explicit calculation that $\Pi_{\mu\nu}(q)$ indeed has the form in (3).

Having learned various computational tricks in the previous chapter you are now ready to tackle the calculation. I will help by walking you through it. In order not to clutter up the page I will suppress the regulator terms in (7) in the intermediate steps and restore them toward the end. After a few steps you should obtain

$$i\Pi_{\mu\nu}(q) = -\int \frac{d^4p}{(2\pi)^4} \frac{N_{\mu\nu}}{D}$$

where $N_{\mu\nu} = \text{tr}[\gamma_\nu(\not p + \not q + m)\gamma_\mu(\not p + m)]$ and

$$\frac{1}{D} = \int_0^1 d\alpha \frac{1}{\mathcal{D}}$$

with $\mathcal{D} = [l^2 + \alpha(1-\alpha)q^2 - m^2 + i\varepsilon]^2$, where $l = p + \alpha q$. Eliminating p in favor of l and beating on $N_{\mu\nu}$ you will find that $N_{\mu\nu}$ is effectively equal to

$$-4\left(\frac{1}{2}g_{\mu\nu}l^2 + \alpha(1-\alpha)(2q_\mu q_\nu - g_{\mu\nu}q^2) - m^2 g_{\mu\nu}\right)$$

Integrate over l using (D.12) and (D.13) and, writing the contribution from the regulators explicitly, obtain

$$\Pi_{\mu\nu}(q) = -\frac{1}{4\pi^2} \int_0^1 d\alpha \left[F_{\mu\nu}(m) - \sum_a c_a F_{\mu\nu}(m_a) \right] \tag{10}$$

where

$$F_{\mu\nu}(m)$$
$$= \frac{1}{2}g_{\mu\nu}\left\{\Lambda^2 - 2[m^2 - \alpha(1-\alpha)q^2]\log\frac{\Lambda^2}{m^2 - \alpha(1-\alpha)q^2} + m^2 - \alpha(1-\alpha)q^2\right\}$$
$$- [\alpha(1-\alpha)(2q_\mu q_\nu - g_{\mu\nu}q^2) - m^2 g_{\mu\nu}]\left[\log\frac{\Lambda^2}{m^2 - \alpha(1-\alpha)q^2} - 1\right] \tag{11}$$

Remember, you are doing the calculation; I am just pointing the way. In appendix D, Λ was introduced to give meaning to various divergent integrals. Since our integral is convergent, we should not need Λ, and indeed, it is gratifying to see that in (10) Λ drops out thanks to the conditions (8) and (9). Some other terms drop out as well, and we end up with

$$\Pi_{\mu\nu}(q) = -\frac{1}{2\pi^2}(q_\mu q_\nu - g_{\mu\nu}q^2) \int_0^1 d\alpha\, \alpha(1-\alpha)$$

$$\{\log[m^2 - \alpha(1-\alpha)q^2] - \sum_a c_a \log[m_a^2 - \alpha(1-\alpha)q^2]\} \tag{12}$$

Lo and behold! The vacuum polarization tensor indeed has the form $\Pi_{\mu\nu}(q) = (q_\mu q_\nu - g_{\mu\nu}q^2)\Pi(q^2)$. Our regularization scheme does respect gauge invariance.

For $q^2 \ll m_a^2$ (the kinematic regime we are interested in had better be much lower than our threshold of ignorance) we simply define $\log M^2 \equiv \sum_a c_a \log m_a^2$ in (12) and obtain

$$\Pi(q^2) = \frac{1}{2\pi^2} \int_0^1 d\alpha\, \alpha(1-\alpha) \log \frac{M^2}{m^2 - \alpha(1-\alpha)q^2} \tag{13}$$

Note that our heuristic argument is indeed correct. In the end, effectively we need only one regulator, but in the intermediate steps we needed two. Actually, this bickering over the number of regulators is beside the point.

In chapter III.1 I mentioned dimensional regularization as an alternative to Pauli-Villars regularization. Historically, dimensional regularization was invented to preserve gauge invariance in nonabelian gauge theories (which I will discuss in a later chapter). It is instructive to calculate Π using dimensional regularization (exercise III.7.1).

Electric charge

Physically, we end up with a result for $\Pi(q^2)$ containing a parameter M^2 expressing our threshold of ignorance. We conclude that

$$e_R^2 = e^2 \frac{1}{1 + (e^2/12\pi^2)\log(M^2/m^2)} \simeq e^2 \left(1 - \frac{e^2}{12\pi^2}\log\frac{M^2}{m^2}\right) \tag{14}$$

Quantum fluctuations effectively diminish the charge. I will explain the physical origin of this effect in a later chapter on renormalization group flow.

You might argue that physically charge is measured by how strongly one electron scatters off another electron. To order e^4, in addition to the diagrams in chapter II.7, we also have, among others, the diagrams shown in figure III.7.4a,b,c. We have computed 4a, but what about 4b and 4c? In many texts, it is shown that contributions of III.7.4b and III.7.4c to charge renormalization cancel. The advantage of using the Lagrangian in (1) is that this fact becomes self-evident: Charge is a measure of how the photon propagates.

To belabor a more or less self-evident point let us imagine doing physical or renormalized perturbation theory as explained in chapter III.3. The Lagrangian is written in terms of physical or renormalized fields (and as before we drop the subscript P on the fields)

$$\mathcal{L} = \bar{\psi}(i\gamma^\mu(\partial_\mu - iA_\mu) - m_P)\psi - \frac{1}{4e_P^2}F_{\mu\nu}F^{\mu\nu}$$

$$+ A\bar{\psi}i\gamma^\mu(\partial_\mu - iA_\mu)\psi + B\bar{\psi}\psi - CF_{\mu\nu}F^{\mu\nu} \tag{15}$$

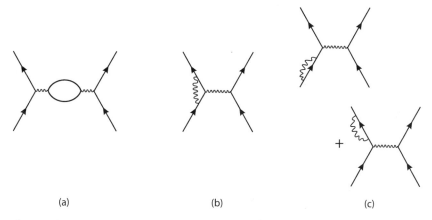

(a)   (b)   (c)

Figure III.7.4

where the coefficients of the counterterms A, B, and C are determined iteratively. The point is that gauge invariance guarantees that $\bar{\psi} i \gamma^\mu \partial_\mu \psi$ and $\bar{\psi} \gamma^\mu A_\mu \psi$ always occur in the combination $\bar{\psi} i \gamma^\mu (\partial_\mu - i A_\mu) \psi$: The strength of the coupling of A_μ to $\bar{\psi} \gamma^\mu \psi$ cannot change. What can change is the ease with which the photon propagates through spacetime.

This statement has profound physical implications. Experimentally, it is known to a high degree of accuracy that the charges of the electron and the proton are opposite and exactly equal. If the charges were not exactly equal, there would be a residual electrostatic force between macroscopic objects. Suppose we discovered a principle that tells us that the bare charges of the electron and the proton are exactly equal (indeed, as we will see, in grand unification theories, this fact follows from group theory). How do we know that quantum fluctuations would not make the charges slightly unequal? After all, the proton participates in the strong interaction and the electron does not and thus many more diagrams would contribute to the long range electromagnetic scattering between two protons. The discussion here makes clear that this equality will be maintained for the obvious reason that charge renormalization has to do with the photon. In the end, it is all due to gauge invariance.

Modifying the Coulomb potential

We have focused on charge renormalization, which is determined completely by $\Pi(0)$, but in (13) we obtained the complete function $\Pi(q^2)$, which tells us how the q dependence of the photon propagator is modified. According to the discussion in chapter I.5, the Coulomb potential is just the Fourier transform of the photon propagator (see also exercise III.5.3). Thus, the Coulomb interaction is modified from the venerable $1/r$ law at a distance scale of the order of $(2m)^{-1}$, namely the inverse of the characteristic value of q in $\Pi(q^2)$. This modification was experimentally verified as part of the Lamb shift in atomic spectroscopy, another great triumph of quantum electrodynamics.

Exercises

III.7.1 Calculate $\Pi_{\mu\nu}(q)$ using dimensional regularization. The procedure is to start with (6), evaluate the trace in $N_{\mu\nu}$, shift the integration momentum from p to l, and so forth, proceeding exactly as in the text, until you have to integrate over the loop momentum l. At that point you "pretend" that you are living in d-dimensional spacetime, so that the term like $l_\mu l_\nu$ in $N_{\mu\nu}$, for example, is to be effectively replaced by $(1/d)g_{\mu\nu}l^2$. The integration is to be performed using (III.1.15) and various generalizations thereof. Show that the form (3) automatically emerges when you continue to $d = 4$.

III.7.2 Study the modified Coulomb's law as determined by the Fourier integral $\int d^3q \{1/\vec{q}^{\,2}[1 + e^2\Pi(\vec{q}^{\,2})]\}e^{i\vec{q}\cdot\vec{x}}$.

III.8 | Becoming Imaginary and Conserving Probability

When Feynman amplitudes go imaginary

Let us admire the polarized vacuum, viz (III.7.13):

$$\Pi(q^2) = \frac{1}{2\pi^2} \int_0^1 d\alpha\, \alpha(1-\alpha) \log \frac{\Lambda^2}{m^2 - \alpha(1-\alpha)q^2 - i\varepsilon} \tag{1}$$

Dear reader, you have come a long way in quantum field theory, to be able to calculate such an amazing effect. Quantum fluctuations alter the way a photon propagates!

For a spacelike photon, with q^2 negative, Π is real and positive for momentum small compared to the threshold of our ignorance Λ. For a timelike photon, we see that if $q^2 > 0$ is large enough, the argument of the logarithm may go negative, and thus Π becomes complex. As you know, the logarithmic function $\log z$ could be defined in the complex z plane with a cut that can be taken conventionally to go along the negative real axis, so that for w real and positive, $\log(-w \pm i\varepsilon) = \log(w) \pm i\pi$ [since in polar coordinates $\log(\rho e^{i\theta}) = \log(\rho) + i\theta$].

We now invite ourselves to define a function in the complex plane: $\Pi(z) \equiv 1/(2\pi^2) \int_0^1 d\alpha\, \alpha(1-\alpha) \log \Lambda^2/(m^2 - \alpha(1-\alpha)z)$. The integrand has a cut on the positive real z axis extending from $z = m^2/(\alpha(1-\alpha))$ to infinity (fig. III.8.1). Since the maximum value of $\alpha(1-\alpha)$ in the integration range is $\frac{1}{4}$, $\Pi(z)$ is an analytic function in the complex z plane with a cut along the real axis starting at $z_c = 4m^2$. The integral over α smears all those cuts of the integrand into one single cut.

For timelike photons with large enough q^2, a mathematician might be paralyzed wondering which side of the cut to go to, but we as physicists know, as per the $i\varepsilon$ prescription from chapter I.3, that we should approach the cut from above, namely that we should take $\Pi(q^2 + i\varepsilon)$ (with ε, as always, a positive infinitesimal) as the physical value. Ultimately, causality tells us which side of the cut we should be on.

That the imaginary part of Π starts at $\sqrt{q^2} > 2m$ provides a strong hint of the physics behind amplitudes going complex. We began the preceding chapter talking about how

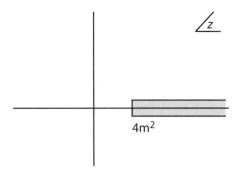

Figure III.8.1

a photon merrily propagating along could always metamorphose into a virtual electron-positron pair that, after a short time dictated by the uncertainly principle, annihilate each other to become a photon again. For $\sqrt{q^2} > 2m$ the pair is no longer condemned to be virtual and to fluctuate out of existence almost immediately. The pair has enough energy to get real. (If you did the exercises religiously, you would recognize that these points were already developed in exercises I.7.4 and III.1.2.)

Physically, we could argue more forcefully as follows. Imagine a gauge boson of mass M coupling to electrons just like the photon. (Indeed, in this book we started out supposing that the photon has a mass.) The vacuum polarization diagram then provides a one-loop correction to the vector boson propagator. For $M > 2m$, the vector boson becomes unstable against decay into an electron-positron pair. At the same time, Π acquires an imaginary part. You might suspect that $\text{Im}\Pi$ might have something to do with the decay rate. We will verify these suspicions later and show that, hey, your physical intuition is pretty good.

When we ended the preceding chapter talking about the modifications to the Coulomb potential, we thought of a spacelike virtual photon being exchanged between two charges as in electron-electron or electron-proton scattering (chapter II.6). Use crossing (chapter II.8) to map electron-electron scattering into electron-positron scattering. The vacuum polarization diagram then appears (fig. III.8.2) as a correction to electron-positron scattering. One function Π covers two different physical situations.

Incidentally, the title of this section should, strictly speaking, have the word "complex," but it is more dramatic to say "When Feynman amplitudes go imaginary," if only to echo certain movie titles.

Dispersion relations and high frequency behavior

> One of the most remarkable discoveries in elementary particle physics has been that of the existence of the complex plane.
>
> —J. Schwinger

Considering that amplitudes are calculated in quantum field theory as integrals over products of propagators, it is more or less clear that amplitudes are analytic functions

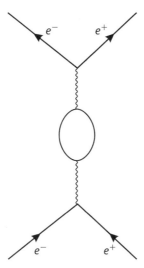

Figure III.8.2

of the external kinematic variables. Another example is the scattering amplitude $\mathcal{M}$ in chapter III.1: it is manifestly an analytic function of s, t, and u with various cuts. From the late 1950s until the early 1960s, considerable effort was devoted to studying analyticity in quantum field theory, resulting in a vast literature.

Here we merely touch upon some elementary aspects. Let us start with an embarrassingly simple baby example: $f(z) = \int_0^1 d\alpha\, 1/(z - \alpha) = \log((z - 1)/z)$. The integrand has a pole at $z = \alpha$, which got smeared by the integral over α into a cut stretching from 0 to 1. At the level of physicist rigor, we may think of a cut as a lot of poles mashed together and a pole as an infinitesimally short cut.

We will mostly encounter real analytic functions, namely functions satisfying $f^*(z) = f(z^*)$ (such as $\log z$). Furthermore, we focus on functions that have cuts along the real axis, as exemplified by $\Pi(z)$. For the class of analytic functions specified here, the discontinuity of the function across the cut is given by $\operatorname{disc} f(x) \equiv f(x + i\varepsilon) - f(x - i\varepsilon) = f(x + i\varepsilon) - f(x + i\varepsilon)^* = 2i \operatorname{Im} f(x + i\varepsilon)$. Define, for σ real, $\rho(\sigma) = \operatorname{Im} f(\sigma + i\varepsilon)$. Using Cauchy's theorem with a contour C that goes around the cut as indicated in figure III.8.3, we could write

$$f(z) = \oint_C \frac{dz'}{2\pi i} \frac{f(z')}{z' - z}. \tag{2}$$

Assuming that $f(z)$ vanishes faster than $1/z$ as $z \to \infty$, we can drop the contribution from infinity and write

$$f(z) = \frac{1}{\pi} \int d\sigma \frac{\rho(\sigma)}{\sigma - z} \tag{3}$$

where the integral ranges over the cut. Note that we can check this equation using the identity (I.2.14):

$$\operatorname{Im} f(x + i\varepsilon) = \frac{1}{\pi} \int d\sigma \rho(\sigma) \operatorname{Im} \frac{1}{\sigma - x - i\varepsilon} = \frac{1}{\pi} \int d\sigma \rho(\sigma) \pi \delta(\sigma - x)$$

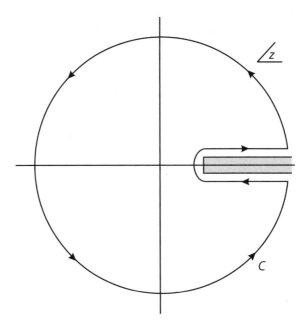

Figure III.8.3

This relation tells us that knowing the imaginary part of f along the cut allows us to construct f in its entirety, including a fortiori its real part on and away from the cut. Relations of this type, known collectively as dispersion relations, go back at least to the work of Kramers and Krönig on optics and are enormously useful in many areas of physics. We will use it in, for example, chapter VII.4.

We implicitly assumed that the integral over σ converges. If not, we can always (formally) subtract $f(0) = \frac{1}{\pi} \int d\sigma \rho(\sigma)/\sigma$ from $f(z)$ as given above and write

$$f(z) = f(0) + \frac{z}{\pi} \int d\sigma \frac{\rho(\sigma)}{\sigma(\sigma - z)} \tag{4}$$

The integral over σ now enjoys an additional factor of $1/\sigma$ and hence is more convergent. In this case, to reconstruct $f(z)$, we need, in addition to knowledge of the imaginary part of f on the cut, an unknown constant $f(0)$. Evidently, we could repeat this process until we obtain a convergent integral.

A bell rings, and you, the astute reader, see the connection with the renormalization procedure of introducing counterterms. In the dispersion weltanschauung, divergent Feynman integrals correspond to integrals over σ that do not converge. Once again, divergent integrals do not bend real physicists out of shape: we simply admit to ignorance of the high σ regime.

During the height of the dispersion program, it was jokingly said that particle theorists either group or disperse, depending on whether you like group theory or complex analysis better.

Imaginary part of Feynman integrals

Going back to the calculation of vacuum polarization in the preceding chapter, we see that the numerator $N_{\mu\nu}$, which comes from the spin of the photon and of the electron, is irrelevant in determining the analytic structure of the Feynman diagram. It is the denominator D that counts. Thus, to get at a conceptual understanding of analyticity in quantum field theory, we could dispense with spins and study the analog of vacuum polarization in the scalar field theory with the interaction term $\mathcal{L} = g(\eta^\dagger \xi^\dagger \varphi + \text{h.c.})$, introduced in the appendix to chapter II.6. The φ propagator is corrected by the analog of the diagrams in figure III.7.1 to [compare with (4)]

$$
\begin{aligned}
iD^P(q) &= \frac{i}{q^2 - M^2 + i\epsilon} + \frac{i}{q^2 - M^2 + i\epsilon} i\Pi(q^2) \frac{i}{q^2 - M^2 + i\epsilon} + \cdots \\
&= \frac{i}{q^2 - M^2 + \Pi(q^2) + i\epsilon}
\end{aligned}
\tag{5}
$$

To order g^2 we have

$$
i\Pi(q^2) = i^4 g^2 \int \frac{d^4 k}{(2\pi)^4} \frac{1}{k^2 - \mu^2 + i\varepsilon} \frac{1}{(q-k)^2 - m^2 + i\varepsilon}
\tag{6}
$$

As in the preceding chapter we need to regulate the integral, but we will leave that implicit.

Having practiced with the spinful calculation of the preceding chapter, you can now whiz through this spinless calculation and obtain

$$
\Pi(z) = \frac{g^2}{16\pi^2} \int_0^1 d\alpha \, \log \frac{\Lambda^2}{\alpha m^2 + (1-\alpha)\mu^2 - \alpha(1-\alpha)z}
\tag{7}
$$

with Λ some cutoff. (Please do whiz and not imagine that you could whiz.) We use the same Greek letter Π and allow the two particles in the loop to have different masses, in contrast to the situation in quantum electrodynamics.

As before, for z real and negative, the argument of the log is real and positive, and Π is real. By the same token, for z real and positive enough, the argument of the log becomes negative for some value of α, and $\Pi(z)$ goes complex. Indeed,

$$
\begin{aligned}
\text{Im}\,\Pi(\sigma + i\varepsilon) &= -\frac{g^2}{16\pi^2} \int_0^1 d\alpha(-\pi)\theta[\alpha(1-\alpha)\sigma - \alpha m^2 - (1-\alpha)\mu^2] \\
&= \frac{g^2}{16\pi} \int_{\alpha_-}^{\alpha_+} d\alpha \\
&= \frac{g^2}{16\pi\sigma} \sqrt{(\sigma - (m+\mu)^2)(\sigma - (m-\mu)^2)}
\end{aligned}
\tag{8}
$$

with $\alpha_\pm$ the two roots of the quadratic equation obtained by setting the argument of the step function to zero.

Decay and distintegration

At this point you might already be flipping back to the expression given in chapter II.6 for the decay rate of a particle. Earlier we entertained the suspicion that the imaginary part of $\Pi(z)$ corresponds to decay. To confirm our suspicion, let us first go back to elementary quantum mechanics. The higher energy levels in a hydrogen atom, say, are, strictly speaking, not eigenstates of the Hamiltonian: an electron in a higher energy level will, in a finite time, emit a photon and jump to a lower energy level. Phenomenologically, however, the level could be assigned a complex energy $E - i\frac{1}{2}\Gamma$. The probability of staying in this level then goes with time like $|\psi(t)|^2 \propto |e^{-i(E-i\frac{1}{2}\Gamma)t}|^2 = e^{-\Gamma t}$. (Note that in elementary quantum mechanics, the Coulomb and radiation components of the electromagnetic field are treated separately: the former is included in the Schrödinger equation but not the latter. One of the aims of quantum field theory is to remedy this artificial split.)

We now go back to (5) and field theory: note that $\Pi(q^2)$ effectively shifts $M^2 \to M^2 - \Pi(q^2)$. Recall from (III.3.3) that we have counter terms available to, well, counter two cutoff-dependent pieces of $\Pi(q^2)$. But we have nothing to counter the imaginary part of $\Pi(q^2)$ with, and so it better be cutoff independent. Indeed it is! The cutoff only appears in the real part in (7).

We conclude that the effect of Π going imaginary is to shift the mass of the φ meson by a cutoff-independent amount from M to $\sqrt{M^2 - i\operatorname{Im}\Pi(M^2)} \approx M - i\operatorname{Im}\Pi(M^2)/(2M)$. Note that to order g^2 it suffices to evaluate Π at the unshifted mass squared M^2, since the shift in mass is itself of order g^2. Thus $\Gamma = \operatorname{Im}\Pi(M^2)/M$ gives the decay rate, as we suspected. We obtain (g has dimension of mass and so the dimension is correct)

$$\Gamma = \frac{g^2}{16\pi M^3}\sqrt{[M^2 - (m+\mu)^2][M^2 - (m-\mu)^2]} \tag{9}$$

precisely what we had in (II.6.7). You and I could both take a bow for getting all the factors exactly right!

Note that both the treatment given in elementary quantum mechanics and here are in the spirit of treating the decay as a small perturbation. As the width becomes large, at some point it no longer makes good sense to talk of the field associated with the particle φ.

Taking the imaginary part directly

We ought to be able to take the imaginary part of the Feynman integral in (6) directly, rather than having to first calculate it as an integral over the Feynman parameter α. I will now show you how to do this using a trick. For clarity and convenience, change notation from (6), label the momentum carried by the two internal lines in figure III.8.4a separately, and restore the momentum conservation delta function, so that

$$i\Pi(q) = (ig)^2 i^2 \int \frac{d^4 k_\eta}{(2\pi)^4} \frac{d^4 k_\xi}{(2\pi)^4} (2\pi)^4 \delta^4(k_\eta + k_\xi - q) \left\{ \frac{1}{k_\eta^2 - m_\eta^2 + i\epsilon} \frac{1}{k_\xi^2 - m_\xi^2 + i\epsilon} \right\} \tag{10}$$

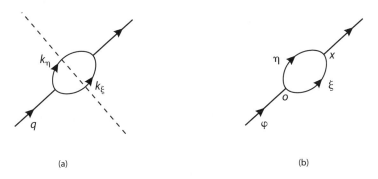

(a)   (b)

Figure III.8.4

Write the propagator as $1/(k^2 - m^2 + i\epsilon) = \mathcal{P}(1/(k^2 - m^2)) - i\pi\delta(k^2 - m^2)$ and, noting an explicit overall factor of i, take the real part of the curly bracket above, thus obtaining

$$\text{Im}\,\Pi(q) = -g^2 \int d\Phi(\mathcal{P}_\eta \mathcal{P}_\xi - \Delta_\eta \Delta_\xi) \tag{11}$$

For the sake of compactness, we have introduced the notation

$$d\Phi = \frac{d^4 k_\eta}{(2\pi)^4}\frac{d^4 k_\xi}{(2\pi)^4}(2\pi)^4\delta^4(k_\eta + k_\xi - q), \quad \mathcal{P}_\eta = \mathcal{P}\frac{1}{k_\eta^2 - m_\eta^2}, \quad \Delta_\eta = \pi\delta(k_\eta^2 - m_\eta^2)$$

and so on.

We welcome the product of two delta functions; they are what we want, restricting the two particles η and ξ on shell. But yuck, what do we do with the product of the two principal values? They don't correspond to anything too physical that we know of.

To get rid of the two principal values, we use a trick.[1] First, we regress and recall that we started out with Feynman diagrams as spacetime diagrams (for example, fig. I.7.6) of the process under study. Here (fig. III.8.4b) a φ excitation turns into an η and a ξ with amplitude ig at some spacetime point, which by translation invariance we could take to be the origin; the η and the ξ excitations propagate to some point x with amplitude $iD_\eta(x)$ and $iD_\xi(x)$, respectively, and then recombine into φ with amplitude ig (note: not $-ig$). Fourier transforming this product of spacetime amplitudes gives

$$i\Pi(q) = (ig)^2 i^2 \int dx\, e^{-iqx} D_\eta(x) D_\xi(x) \tag{12}$$

To see that this is indeed the same as (10), all you have to do is to plug in the expression (I.3.22) for $D_\eta(x)$ and $D_\xi(x)$.

Incidentally, while many "professors of Feynman diagrams" think almost exclusively in momentum space, Feynman titled his 1949 paper "Space-Time Approach to Quantum Electrodynamics," and on occasions it is useful to think of the spacetime roots of a given Feynman diagram. Now is one of those occasions.

---

[1] C. Itzkyson and J.-B. Zuber, *Quantum Field Theory*, p. 367.

Next, go back to exercise I.3.3 and recall that the advanced propagator $D_{adv}(x)$ and retarded propagator $D_{ret}(x)$ vanish for $x^0 < 0$ and $x^0 > 0$, respectively, and thus the product $D_{adv}(k)D_{ret}(k)$ manifestly vanishes for all x. Also recall that the advanced and retarded propagators $D_{adv}(k)$ and $D_{ret}(k)$ differ from the Feynman propagator $D(k)$ by simply, but crucially, having their poles in different half-planes in the complex k^0 plane. Thus

$$0 = -ig^2 \int dx\, e^{-iqx} D_{\eta,adv}(x) D_{\xi,ret}(x)$$

$$= \int \frac{d^4k_\eta}{(2\pi)^4} \frac{d^4k_\xi}{(2\pi)^4} (2\pi)^4 \delta^4(k_\eta + k_\xi - q) \frac{1}{k_\eta^2 - m_\eta^2 - i\sigma_\eta\epsilon} \frac{1}{k_\xi^2 - m_\xi^2 + i\sigma_\xi\epsilon} \tag{13}$$

where we used the shorthand $\sigma_\eta = \text{sgn}(k_\eta^0)$ and $\sigma_\xi = \text{sgn}(k_\xi^0)$. (The sign function is defined by $\text{sgn}(x) = \pm 1$ according to whether $x > 0$ or < 0.) Taking the imaginary part of 0, we obtain [compare (11)]

$$0 = -g^2 \int d\Phi (\mathcal{P}_\eta \mathcal{P}_\xi + \sigma_\eta \sigma_\xi \Delta_\eta \Delta_\xi) \tag{14}$$

Subtracting (14) from (11) to get rid of the rather unpleasant term $\mathcal{P}_\eta \mathcal{P}_\xi$, we find finally

$$\text{Im}\,\Pi(q) = +g^2 \int d\Phi (1 + \sigma_\eta \sigma_\xi)(\Delta_\eta \Delta_\xi)$$

$$= g^2 \pi^2 \int \frac{d^4k_\eta}{(2\pi)^4} \frac{d^4k_\xi}{(2\pi)^4} (2\pi)^4 \delta^4(k_\eta + k_\xi - q) \theta(k_\eta^0)\delta(k_\eta^2 - m_\eta^2)\theta(k_\xi^0)\delta(k_\xi^2 - m_\xi^2)(1 + \sigma_\eta\sigma_\xi) \tag{15}$$

Thus $\Pi(q)$ develops an imaginary part only when the three delta functions can be satisfied simultaneously.

To see what these three conditions imply, we can, since $\Pi(q)$ is a function of q^2, go to a frame in which $q = (Q, \vec{0})$ with $Q > 0$ with no loss of generality. Since $k_\eta^0 + k_\xi^0 = Q > 0$ and since $(1 + \sigma_\eta\sigma_\xi)$ vanishes unless k_η^0 and k_ξ^0 have the same sign, k_η^0 and k_ξ^0 must be both positive if $\text{Im}\,\Pi(q)$ is to be nonzero, but that is already mandated by the two step functions. Furthermore, we need to solve the conservation of energy condition $Q = \sqrt{\vec{k}^2 + m_\eta^2} + \sqrt{\vec{k}^2 + m_\xi^2}$ for some 3-vector $\vec{k}$. This is possible only if $Q > m_\eta + m_\xi$, in which case, using the identity (I.8.14)

$$\theta(k^0)\delta(k^2 - \mu^2) = \theta(k^0)\frac{\delta(k^0 - \varepsilon_k)}{2\varepsilon_k}, \tag{16}$$

we obtain

$$\text{Im}\,\Pi(q) = \frac{1}{2}g^2 \int \frac{d^3k_\eta}{(2\pi)^3 2\omega_\eta} \frac{d^3k_\xi}{(2\pi)^3 2\omega_\xi} (2\pi)^4 \delta^4(k_\eta + k_\xi - q) \tag{17}$$

We see that (and as we will see more generally) $\text{Im}\,\Pi(q)$ works out to be a finite integral over delta functions. Indeed, no counter term is needed.

Some readers might feel that this trick of invoking the advanced and retarded propagators is perhaps a bit "too tricky." For them, I will show a more brute force method in appendix 1.

Unitarity and the Cutkosky cutting rule

The simple example we just went through in detail illustrates what is known as the Cutkosky cutting rule, which states that to calculate the imaginary part of a Feynman amplitude we first "cut" through a diagram (as indicated by the dotted line in figure II.8.4a). For each internal line cut, replace the propagator $1/(k^2 - m^2 + i\epsilon)$ by $\delta(k^2 - m^2)$; in other words, put the virtual excitation propagating through the cut onto the mass shell. Thus, in our example, we could jump from (10) to (15) directly. This validates our intuition that Feynman amplitudes go imaginary when virtual particles can "get real."

For a precise statement of the cutting rule, see below. (The Cutkosky cut is not be confused with the Cauchy cut in the complex plane, of course.)

The Cutkosky cutting rule in fact follows in all generality from unitarity. A basic postulate of quantum mechanics is that the time evolution operator e^{-iHT} is unitary and hence preserves probability. Recall from chapter I.8 that it is convenient to split off from the S matrix $S_{fi} = \langle f | e^{-iHT} | i \rangle$ the piece corresponding to "nothing is happening": $S = I + iT$. Unitarity $S^\dagger S = I$ then implies $2 \operatorname{Im} T = i(T^\dagger - T) = T^\dagger T$. Sandwiching this between initial and final states and inserting a complete set of intermediate states ($1 = \sum_n |n\rangle\langle n|$) we have

$$2 \operatorname{Im} T_{fi} = \sum_n T_{fn}^\dagger T_{ni} \tag{18}$$

which some readers might recognize as a generalization of the optical theorem from elementary quantum mechanics.

It is convenient to introduce $\mathcal{F} = -i\mathcal{M}$. (We are merely taking out an explicit factor of i in $\mathcal{M}$: in our simple example, $\mathcal{M}$ corresponds to $i\Pi$, $\mathcal{F}$ to Π.) Then the relation (I.8.16) between T and $\mathcal{M}$ becomes $T_{fi} = (2\pi)^4 \delta^{(4)}(P_{fi})(\Pi_{fi} 1/\rho)\mathcal{F}(f \leftarrow i)$, where for the sake of compactness we have introduced some obvious notations [thus $(\Pi_{fi} \frac{1}{\rho})$ denotes the product of the normalization factors $1/\rho$ (see chapter I.8), one for each of the particle in the state i and in the state f, and P_{fi} the sum of the momenta in f minus the sum of the momenta in i.]

With this notation, the left-hand side of the generalized optical theorem becomes $2\operatorname{Im}T_{fi} = 2(2\pi)^4 \delta^{(4)}(P_{fi})(\Pi_{fi} 1/\rho)\operatorname{Im} \mathcal{F}(f \leftarrow i)$ and the right-hand side

$$\sum_n T_{fn}^\dagger T_{ni} = \sum_n (2\pi)^4 \delta^{(4)}(P_{fn})(2\pi)^4 \delta^{(4)}(P_{ni}) \left(\Pi_{fn} \frac{1}{\rho}\right) \left(\Pi_{ni} \frac{1}{\rho}\right) (\mathcal{F}(n \leftarrow f))^* \mathcal{F}(n \leftarrow i)$$

The product of two delta functions $\delta^{(4)}(P_{fn})\delta^{(4)}(P_{ni}) = \delta^{(4)}(P_{fi})\delta^{(4)}(P_{ni})$, and thus we could cancel off $\delta^{(4)}(P_{fi})$. Also $(\Pi_{fn} 1/\rho)(\Pi_{ni} 1/\rho)/(\Pi_{fi} 1/\rho) = (\Pi_n 1/\rho^2)$, and we happily recover the more familiar factor ρ^2 [namely $(2\pi)^3 2\omega$ for bosons]. Thus finally, the generalized optical theorem tells that

$$2\operatorname{Im}\mathcal{F}(f \leftarrow i) = \sum_n (2\pi)^4 \delta^{(4)}(P_{ni}) \left(\Pi_n \frac{1}{\rho^2}\right) (\mathcal{F}(n \leftarrow f))^* \mathcal{F}(n \leftarrow i), \tag{19}$$

namely, that the imaginary part of the Feynman amplitude $\mathcal{F}(f \leftarrow i)$ is given by a sum of $(\mathcal{F}(n \leftarrow f))^* \mathcal{F}(n \leftarrow i)$ over intermediate states $|n\rangle$. The particles in the intermediate state are of course physical and on shell. We are to sum over all possible states $|n\rangle$ allowed by quantum numbers and by the kinematics.

According to Cutkosky, given a Feynman diagram, to obtain its imaginary part, we simply cut through it in the several different ways allowed, corresponding to the different possible intermediate state $|n\rangle$. Particles in the state $|n\rangle$ are manifestly real, not virtual.

Note that unitarity and hence the optical theorem are nonlinear in the transition amplitude. This has proved to be enormously useful in actual computation. Suppose we are perturbing in some coupling g and we know $\mathcal{F}(n \leftarrow i)$ and $\mathcal{F}(n \leftarrow f)$ to order g^N. The optical theorem gives us $\operatorname{Im} \mathcal{F}(f \leftarrow i)$ to order g^{N+1}, and we could then construct $\mathcal{F}(f \leftarrow i)$ to order g^{N+1} using a dispersion relation.

The application of the Cutkosky rule to the vacuum polarization function discussed in this chapter is particularly simple: there is only one possible way of cutting the Feynman diagram for Π. Here the initial and final states $|i\rangle$ and $|f\rangle$ both consist of a single φ meson, while the intermediate state $|n>$ consists of an η and a ξ meson.

Referring to the appendix to chapter II.6, we recall that $\sum_n$ corresponds to $\int \frac{d^3 k_\eta}{(2\pi)^3} \frac{d^3 k_\xi}{(2\pi)^3}$. Thus the optical theorem as stated in (19) says that

$$\operatorname{Im} \Pi(q) = \frac{1}{2} g^2 \int \frac{d^3 k_\eta}{(2\pi)^3 2\omega_\eta} \frac{d^3 k_\xi}{(2\pi)^3 2\omega_\xi} (2\pi)^4 \delta^4(k_\eta + k_\xi - q) \tag{20}$$

precisely what we obtained in (17). You and I could take another bow, since we even get the factor of 2 correctly (as we must!).

We should also say, in concluding this chapter, "Vive Cauchy!"

Appendix 1: Taking the imaginary part by brute force

For those readers who like brute force, we will extract the imaginary part of

$$\Pi = -ig^2 \int \frac{d^4 k}{(2\pi)^4} \frac{1}{k^2 - \mu^2 + i\varepsilon} \frac{1}{(q-k)^2 - m^2 + i\varepsilon} \tag{21}$$

by a more straightforward method, as promised in the text. Since Π depends only on q^2, we have the luxury of setting $q = (M, \vec{0})$. We already know that in the complex M^2 plane, Π has a cut on the axis starting at $M^2 = (m + \mu)^2$. Let us verify this by brute force.

We could restrict ourselves to $M > 0$. Factorizing, we find that the denominator of the integrand is a product of four factors, $k^0 - (\varepsilon_k - i\varepsilon)$, $k^0 + (\varepsilon_k - i\varepsilon)$, $k^0 - (M + E_k - i\varepsilon)$, and $k^0 - (M - E_k + i\varepsilon)$, and thus the integrand has four poles in the complex k^0 plane. (Evidently, $\varepsilon_k = \sqrt{\vec{k}^2 + \mu^2}$, $E_k = \sqrt{\vec{k}^2 + m^2}$, and if you are perplexed over the difference between ε_k and ε, then you are hopelessly confused.) We now integrate over k^0, choosing to close the contour in the lower half plane and going around picking up poles. Picking up the pole at $\varepsilon_k - i\varepsilon$, we obtain $\Pi_1 = -g^2 \int (d^3 k / (2\pi)^3)(1/(2\varepsilon_k(\varepsilon_k - M - E_k)(\varepsilon_k - M + E_k)))$. Picking up the pole at $M + E_k - i\varepsilon$, we obtain $\Pi_2 = -g^2 \int (d^3 k / (2\pi)^3)(1/((M + E_k - \varepsilon_k)(M + E_k + \varepsilon_k)(2E_k)))$. We now regard $\Pi = \Pi_1 + \Pi_2$ as a function of M:

$$\Pi = -g^2 \int \frac{d^3 k}{(2\pi)^3} \frac{1}{(M + E_k - \varepsilon_k)} \left[\frac{1}{2\varepsilon_k(M - \varepsilon_k - E_k + i\varepsilon)} + \frac{1}{2E_k(M + E_k + \varepsilon_k - i\varepsilon)} \right] \tag{22}$$

In spite of appearances, there is no pole at $M \approx \varepsilon_k - E_k$. (Since this pole would lead to a cut at $\mu - m$, there better not be!) For $M > 0$ we only care about the pole at $M \approx \varepsilon_k + E_k = \sqrt{\vec{k}^2 + \mu^2} + \sqrt{\vec{k}^2 + m^2}$. When we integrate over $\vec{k}$, this pole gets smeared into a cut starting at $m + \mu$. So far so good.

To calculate the discontinuity across the cut, we use the identity (16) once again Restoring the $i\varepsilon$'s and throwing away the term we don't care about, we have effectively

$$
\Pi = -g^2 \int \frac{d^3k}{(2\pi)^3} \frac{1}{2\varepsilon_k(\varepsilon_k - M - E_k + i\varepsilon)(\varepsilon_k - M + E_k - i\varepsilon)}
$$

$$
= -2\pi g^2 \int \frac{d^4k}{(2\pi)^4} \theta(k^0)\delta(k^2 - \mu^2) \frac{1}{(M - \varepsilon_k + E_k - i\varepsilon)(M - (\varepsilon_k + E_k) + i\varepsilon)}
$$

The discontinuity of Π across the cut just specified is determined by applying (I.2.13) to the factor $1/(M - (\varepsilon_k + E_k) + i\varepsilon)$, giving $\mathrm{Im}\Pi = 2\pi^2 g^2 \int (d^4k/(2\pi)^4)\theta(k^0)\delta(k^2 - \mu^2)\delta(M - (\varepsilon_k + E_k))/2E_k$. Use the identity (16) again in the form

$$
\theta(q^0 - k^0)\theta((q - k)^2 - m^2) = \theta(q^0 - k^0)\frac{\theta((q^0 - k^0) - E_k)}{2E_k} \tag{23}
$$

and we obtain

$$
\mathrm{Im}\Pi = 2\pi^2 g^2 \int \frac{d^4k}{(2\pi)^4} \theta(k^0)\delta(k^2 - \mu^2)\theta(q^0 - k^0)\delta((q - k)^2 - m^2) \tag{24}
$$

Remarkably, as Cutkosky taught us, to obtain the imaginary part we simply replace the propagators in (21) by delta functions.

Appendix 2: A dispersion representation for the two-point amplitude

I would like to give you a bit more flavor of the dispersion program once active and now being revived (see part N). Consider the two-point amplitude $i\mathcal{D}(x) \equiv \langle 0| T(\mathcal{O}(x)\mathcal{O}(0)) |0\rangle$, with $\mathcal{O}(x)$ some operator in the canonical formalism. For example, for $\mathcal{O}(x)$ equal to the field $\varphi(x)$, $\mathcal{D}(x)$ would be the propagator. In chapter I.8, we were able to evaluate $\mathcal{D}(x)$ for a free field theory, because then we could solve the field equation of motion and expand $\varphi(x)$ in terms of creation and annihilation operators. But what can we do in a fully interacting field theory? There is no hope of solving the operator field equation of motion.

The goal of the dispersion program of the 1950s and 1960s is to say as much as possible about $\mathcal{D}(x)$ based on general considerations such as analyticity.

OK, so first write $i\mathcal{D}(x) = \theta(x^0)\langle 0| e^{iPx}\mathcal{O}(0)e^{-iPx}\mathcal{O}(0) |0\rangle + \theta(-x^0)\langle 0| \mathcal{O}(0)e^{iPx}\mathcal{O}(0)e^{-iPx} |0\rangle$, where we used spacetime translation $\mathcal{O}(x) = e^{iP \cdot x}\mathcal{O}(0)e^{-iP \cdot x}$. By the way, if you are not totally sure of this relation, differentiate it to obtain $\partial_\mu \mathcal{O}(x) = i[P_\mu, \mathcal{O}(x)]$ which you should recognize as the relativistic version of the usual Heisenberg equations (I.8.2, 3). Now insert $1 = \sum_n |n\rangle\langle n|$, with $|n\rangle$ a complete set of intermediate states, to obtain $\langle 0| e^{iPx}\mathcal{O}(0)e^{-iPx}\mathcal{O}(0) |0\rangle = \langle 0| \mathcal{O}(0)e^{-iPx}\mathcal{O}(0) |0\rangle = \sum_n \langle 0| \mathcal{O}(0) |n\rangle\langle n| e^{-iPx}\mathcal{O}(0) |0\rangle = \sum_n e^{-iP_n x}|\mathcal{O}_{0n}|^2$, where we used $P^\mu |0\rangle = 0$ and $P^\mu |n\rangle = P_n^\mu |n\rangle$ and defined $\mathcal{O}_{0n} \equiv \langle 0| \mathcal{O}(0) |n\rangle$. Next, use the integral representations for the step function $\theta(t) = -i \int (d\omega/2\pi)e^{i\omega t}/(\omega - i\varepsilon)$ and $\theta(-t) = i \int (d\omega/2\pi)e^{i\omega t}/(\omega + i\varepsilon)$. Again, if you are not sure of this, simply differentiate $\frac{d}{dt}\theta(t) = -i \frac{d}{dt} \int (d\omega/2\pi)e^{i\omega t}/(\omega - i\varepsilon) = \int (d\omega/2\pi)e^{i\omega t}$, which you recognize from (I.2.12) as indeed the integral representation of the delta function $\delta(t) = \frac{d}{dt}\theta(t)$. In other words, the representation used here is the integral of the representation in (I.2.12).

Putting it all together, we obtain

$$
i\mathcal{D}(q) = \int d^4x e^{iq \cdot x} i\mathcal{D}(x) = -i(2\pi)^3 \sum_n |\mathcal{O}_{0n}|^2 \left\{ \frac{\delta^{(3)}(\vec{q} - \vec{P}_n)}{P_n^0 - q^0 - i\varepsilon} + \frac{\delta^{(3)}(\vec{q} + \vec{P}_n)}{P_n^0 + q^0 - i\varepsilon} \right\}. \tag{25}
$$

The integral over d^3x produced the 3-dimensional delta function, while the integral over $dx^0 = dt$ picked up the denominator in the integral representation for the step function.

Now take the imaginary part using $\mathrm{Im}1/(P_n^0 - q^0 - i\varepsilon) = \pi\delta(q^0 - P_n^0)$. We thus obtain

$$
\mathrm{Im}(i \int d^4x e^{iqx}\langle 0| T(\mathcal{O}(x)\mathcal{O}(0)) |0\rangle) = \pi(2\pi)^3 \sum_n |\mathcal{O}_{0n}|^2 (\delta^{(4)}(q - P_n) + \delta^{(4)}(q + P_n)) \tag{26}
$$

with the more pleasing 4-dimensional delta function. Note that for $q^0 > 0$ the term involving $\delta^{(4)}(q + P_n)$ drops out, since the energies of physical states must be positive.

What have we accomplished? Even though we are totally incapable of calculating $\mathcal{D}(q)$, we have managed to represent its imaginary part in terms of physical quantities that are measurable in principle, namely the absolute square $|\mathcal{O}_{0n}|^2$ of the matrix element of $\mathcal{O}(0)$ between the vacuum state and the state $|n\rangle$. For example, if $\mathcal{O}(x)$ is the meson field $\varphi(x)$ in a φ^4 theory, the state $|n\rangle$ would consist of the single-meson state, the three-meson state, and so on. The general hope during the dispersion era was that by keeping a few states we could obtain a decent approximation to $\mathcal{D}(q)$. Note that the result does not depend on perturbing in some coupling constant.

The contribution of the single meson state $|\vec{k}\rangle$ has a particularly simple form, as you might expect. With our normalization of single-particle states (as in chapter I.8), Lorentz invariance implies $\langle \vec{k}| \mathcal{O}(0) |0\rangle = Z^{\frac{1}{2}}/\sqrt{(2\pi)^3 2\omega_k}$, with $\omega_k = \sqrt{\vec{k}^2 + m^2}$ and $Z^{\frac{1}{2}}$ an unknown constant, measuring the "strength" with which $\mathcal{O}$ is capable of producing the single meson from the vacuum. Putting this into (25) and recognizing that the sum over single-meson states is now given by $\int d^3k \, |\vec{k}\rangle\langle\vec{k}|$ [with the normalization $\langle \vec{k}'|\vec{k}\rangle = \delta^3(\vec{k}' - \vec{k})$], we find that the single-meson contribution to $i\mathcal{D}(q)$ is given by

$$-i(2\pi)^3 \int d^3k \frac{Z}{(2\pi)^3 2\omega_k} \left\{ \frac{\delta^{(3)}(\vec{q} - \vec{k})}{\omega_k - q^0 - i\varepsilon} + (q \to -q) \right\} = -i \frac{Z}{2\omega_q} \left\{ \frac{1}{\omega_q - q^0 - i\varepsilon} + (q \to -q) \right\}$$

$$= \frac{iZ}{q^2 - m^2 + i\varepsilon} \tag{27}$$

This is a very satisfying result: even though we cannot calculate $i\mathcal{D}(q)$, we know that it has a pole at a position determined by the meson mass with a residue that depends on how $\mathcal{O}$ is normalized.

As a check, we can also easily calculate the contribution of the single-meson state to $-\text{Im}\mathcal{D}(q)$. Plugging into (26), we find, for $q^0 > 0$, $\pi Z \int (d^3k/2\omega_k)\delta^4(q - k) = (\pi Z/2\omega_q)\delta(q^0 - \omega_q) = \pi Z\delta(q^2 - m^2)$, where we used (I.8.14, 16) in the last step.

Given our experience with the vacuum polarization function, we would expect $\mathcal{D}(q)$ (which by Lorentz invariance is a function of q^2) to have a cut starting at $q^2 = (3m)^2$. To verify this, simply look at (26) and choose $\vec{q} = 0$. The contribution of the three-meson state occurs at $\sqrt{q^2} = q^0 = P^0_{"3"} = \sqrt{\vec{k}_1^2 + m^2} + \sqrt{\vec{k}_2^2 + m^2} + \sqrt{\vec{k}_3^2 + m^2} \geq 3m$. The sum over states is now a triple integration over $\vec{k}_1$, $\vec{k}_2$, and $\vec{k}_3$, subject to the constraint $\vec{k}_1 + \vec{k}_2 + \vec{k}_3 = 0$. Knowing the imaginary part of $\mathcal{D}(q)$ we can now write a dispersion relation of the kind in (3).

Finally, if you stare at (26) long enough (see exercise III.8.3) you will discover the relation

$$\text{Im}(i \int d^4x \, e^{iqx} \langle 0| T(\mathcal{O}(x)\mathcal{O}(0)) |0\rangle) = \frac{1}{2} \int d^4x \, e^{iqx} \langle 0| [\mathcal{O}(x), \mathcal{O}(0)] |0\rangle \tag{28}$$

The discussion here is relevant to the discussion of field redefinition in chapter I.8. Suppose our friend uses $\eta = Z^{\frac{1}{2}}\varphi + \alpha\varphi^3$ instead of φ; then the present discussion shows that his propagator $\int d^4x \, e^{iqx} \langle 0| T(\eta(x)\eta(0)) |0\rangle$ still has a pole at $q^2 - m^2$. The important point is that physics fixes the pole to be at the same location.

Here we have taken $\mathcal{O}$ to be a Lorentz scalar. In applications (see chapter VII.3) the role of $\mathcal{O}$ is often played by the electromagnetic current $J^\mu(x)$ (treated as an operator). The same discussion holds except that we have to keep track of some Lorentz indices. Indeed, we recognize that the vacuum polarization function $\Pi_{\mu\nu}$ then corresponds to the function $\mathcal{D}$ in this discussion.

Exercises

III.8.1   Evaluate the imaginary part of the vacuum polarization function, and by explicit calculation verify that it is related to the decay rate of a vector particle into an electron and a positron.

III.8.2   Suppose we add a term $g\varphi^3$ to our scalar φ^4 theory. Show that to order g^4 there is a "box diagram" contributing to meson scattering $p_1 + p_2 \to p_3 + p_4$ with the amplitude

$$\mathcal{I} = g^4 \int \frac{d^4k}{(2\pi)^4} \frac{1}{(k^2 - m^2 - i\varepsilon)((k + p_2)^2 - m^2 - i\varepsilon)((k - p_1)^2 - m^2 - i\varepsilon)((k + p_2 - p_3)^2 - m^2 - i\varepsilon)}$$

Calculate the integral explicitly as a function of $s = (p_1 + p_2)^2$ and $t = (p_3 - p_2)^2$. Study the analyticity property of $\mathcal{I}$ as a function of s for fixed t. Evaluate the discontinuity of $\mathcal{I}$ across the cut and verify Cutkosky's cutting rule. Check that the optical theorem works. What about the analyticity property of $\mathcal{I}$ as a function of t for fixed s? And as a function of $u = (p_3 - p_1)^2$?

III.8.3 Prove (28). [Hint: Do unto $\int d^4x \, e^{iqx} \langle 0| \, [\mathcal{O}(x), \mathcal{O}(0)] \, |0\rangle$ what we did to $\int d^4x \, e^{iqx} \langle 0| \, |T(\mathcal{O}(x)\mathcal{O}(0)) \, |0\rangle$, namely, insert $1 = \sum_n |n\rangle\langle n|$ (with $|n\rangle$ a complete set of states) between $\mathcal{O}(x)$ and $\mathcal{O}(0)$ in the commutator. Now we don't have to bother with representing the step function.]

Part IV | Symmetry and Symmetry Breaking

IV.1 | Symmetry Breaking

A symmetric world would be dull

While we would like to believe that the fundamental laws of Nature are symmetric, a completely symmetric world would be rather dull, and as a matter of fact, the real world is not perfectly symmetric. More precisely, we want the Lagrangian, but not the world described by the Lagrangian, to be symmetric. Indeed, a central theme of modern physics is the study of how symmetries of the Lagrangian can be broken. We will see in subsequent chapters that our present understanding of the fundamental laws is built upon an understanding of symmetry breaking.

Consider the Lagrangian studied in chapter I.10:

$$\mathcal{L} = \frac{1}{2}\left[(\partial\vec{\varphi})^2 - \mu^2\vec{\varphi}^2\right] - \frac{\lambda}{4}(\vec{\varphi}^2)^2 \tag{1}$$

where $\vec{\varphi} = (\varphi_1, \varphi_2, \cdots, \varphi_N)$. This Lagrangian exhibits an $O(N)$ symmetry under which $\vec{\varphi}$ transforms as an N-component vector.

We can easily add terms that do not respect the symmetry. For instance, add terms such as φ_1^2, φ_1^4 and $\varphi_1^2\vec{\varphi}^2$ and break the $O(N)$ symmetry down to $O(N-1)$, under which $\varphi_2, \ldots, \varphi_N$ rotate as an $(N-1)$-component vector. This way of breaking the symmetry, "by hand" as it were, is known as explicit breaking.

We can break the symmetry in stages. Obviously, if we want to, we can break it down to $O(N-M)$ by hand, for any $M < N$.

Note that in this example, with the terms we added, the reflection symmetry $\varphi_a \rightarrow -\varphi_a$ (any a) still holds. It is easy enough to break this symmetry as well, by adding a term such as φ_a^3, for example.

Breaking the symmetry by hand is not very interesting. Indeed, we might as well start with a nonsymmetric Lagrangian in the first place.

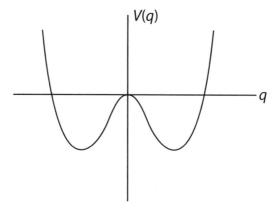

Figure IV.1.1

Spontaneous symmetry breaking

A more subtle and interesting way is to let the system "break the symmetry itself," a phenomenon known as spontaneous symmetry breaking. I will explain by way of an example. Let us flip the sign of the $\vec{\varphi}^2$ term in (1) and write

$$\mathcal{L} = \frac{1}{2}\left[(\partial\vec{\varphi})^2 + \mu^2\vec{\varphi}^2\right] - \frac{\lambda}{4}(\vec{\varphi}^2)^2 \tag{2}$$

Naively, we would conclude that for small λ the field φ creates a particle of mass $\sqrt{-\mu^2} = i\mu$. Something is obviously wrong.

The essential physics is analogous to what would happen if we give the spring constant in an anharmonic oscillator the wrong sign and write $L = \frac{1}{2}(\dot{q}^2 + kq^2) - (\lambda/4)q^4$. We all know what to do in classical mechanics. The potential energy $V(q) = -\frac{1}{2}kq^2 + (\lambda/4)q^4$ [known as the double-well potential (figure IV.1.1)] has two minima at $q = \pm v$, where $v \equiv (k/\lambda)^{\frac{1}{2}}$. At low energies, we choose either one of the two minima and study small oscillations around that minimum. Committing to one or the other of the two minima breaks the reflection symmetry $q \rightarrow -q$ of the system.

In quantum mechanics, however, the particle can tunnel between the two minima, the tunneling barrier being $V(0) - V(\pm v)$. The probability of being in one or the other of the two minima must be equal, thus respecting the reflection symmetry $q \rightarrow -q$ of the Hamiltonian. In particular, the ground state wave function $\psi(q) = \psi(-q)$ is even.

Let us try to extend the same reasoning to quantum field theory. For a generic scalar field Lagrangian $\mathcal{L} = \frac{1}{2}(\partial_0\varphi)^2 - \frac{1}{2}(\partial_i\varphi)^2 - V(\varphi)$ we again have to find the minimum of the potential energy $\int d^D x [\frac{1}{2}(\partial_i\varphi)^2 + V(\varphi)]$, where D is the dimension of space. Clearly, any spatial variation in φ only increases the energy, and so we set $\varphi(x)$ to equal a spacetime independent quantity φ and look for the minimum of $V(\varphi)$. In particular, for the example in (2), we have

$$V(\varphi) = -\frac{1}{2}\mu^2\vec{\varphi}^2 + \frac{\lambda}{4}(\vec{\varphi}^2)^2 \tag{3}$$

As we will see, the $N = 1$ case is dramatically different from the $N \geq 2$ cases.

Difference between quantum mechanics and quantum field theory

Study the $N = 1$ case first. The potential $V(\varphi)$ looks exactly the same as the potential in figure IV.1.1 with the horizontal axis relabeled as φ. There are two minima at $\varphi = \pm v = \pm(\mu^2/\lambda)^{\frac{1}{2}}$.

But some thought reveals a crucial difference between quantum field theory and quantum mechanics. The tunneling barrier is now $[V(0) - V(\pm v)] \int d^D x$ (where D denotes the dimension of space) and hence infinite (or more precisely, extensive with the volume of the system)! Tunneling is shut down, and the ground state wave function is concentrated around either $+v$ or $-v$. We have to commit to one or the other of the two possibilities for the ground state and build perturbation theory around it. It does not matter which one we choose: The physics is equivalent. But by making a choice, we break the reflection symmetry $\varphi \to -\varphi$ of the Lagrangian.

The reflection symmetry is broken spontaneously! We did not put symmetry breaking terms into the Lagrangian by hand but yet the reflection symmetry is broken.[1]

Let's choose the ground state at $+v$ and write $\varphi = v + \varphi'$. Expanding in φ' we find after a bit of arithmetic that

$$\mathcal{L} = \frac{\mu^4}{4\lambda} + \frac{1}{2}(\partial\varphi')^2 - \mu^2\varphi'^2 - O(\varphi'^3) \tag{4}$$

The physical particle created by the shifted field φ' has mass $\sqrt{2}\mu$. The physical mass squared has to come out positive since, after all, it is just $-V''(\varphi)|_{\varphi=v}$, as you can see after a moment's thought.

Similarly, you would recognize that the first term in (4) is just $-V(\varphi)|_{\varphi=v}$. If we are only interested in the scattering of the mesons associated with φ' this term does not enter at all. Indeed, we are always free to add an arbitrary constant to $\mathcal{L}$ to begin with. We had quite arbitrarily set $V(\varphi = 0)$ equal to 0. The same situation appears in quantum mechanics: In the discussion of the harmonic oscillator the zero point energy $\frac{1}{2}\hbar\omega$ is not observable; only transitions between energy levels are physical. We will return to this point in chapter VIII.2.

Yet another way of looking at (2) is that quantum field theory amounts to doing the Euclidean functional integral

$$Z = \int D\varphi e^{-\int d^d x\{\frac{1}{2}[(\partial\varphi)^2 - \mu^2\varphi^2] + \frac{\lambda}{4}(\varphi^2)^2\}}$$

and perturbation theory just corresponds to studying the small oscillations around a minimum of the Euclidean action. Normally, with μ^2 positive, we expand around the minimum $\varphi = 0$. With μ^2 negative, $\varphi = 0$ is a local maximum and not a minimum.

In quantum field theory what is called the ground state is also known as the vacuum, since it is literally the state in which the field is "at rest," with no particles present. Here we

---

[1] An insignificant technical aside for the nitpickers: Strictly speaking, in field theory the ground state wave function should be called a wave functional, since $\Psi[\varphi(\vec{x})]$ is a functional of the function $\varphi(\vec{x})$.

have two physically equivalent vacua from which we are to choose one. The value assumed by φ in the ground state, either v or $-v$ in our example, is known as the vacuum expectation value of φ. The field φ is said to have acquired a vacuum expectation value.

Continuous symmetry

Let us now turn to (2) with $N \geq 2$. The potential (3) is shown in figure IV.1.2 for $N = 2$. The shape of the potential has been variously compared to the bottom of a punted wine bottle or a Mexican hat. The potential is minimized at $\vec{\varphi}^2 = \mu^2/\lambda$. Something interesting is going on: We have an infinite number of vacua characterized by the direction of $\vec{\varphi}$ in that vacuum. Because of the $O(2)$ symmetry of the Lagrangian they are all physically equivalent. The result had better not depend on our choice. So let us choose $\vec{\varphi}$ to point in the 1 direction, that is, $\varphi_1 = v \equiv +\sqrt{\mu^2/\lambda}$ and $\varphi_2 = 0$.

Now consider fluctuations around this field configuration, in other words, write $\varphi_1 = v + \varphi_1'$ and $\varphi_2 = \varphi_2'$, plug into (2) for $N = 2$, and expand $\mathcal{L}$ out. I invite you to do the arithmetic. You should find (after dropping the primes on the fields; why clutter the notation, right?)

$$\mathcal{L} = \frac{\mu^4}{4\lambda} + \frac{1}{2}\left[(\partial\varphi_1)^2 + (\partial\varphi_2)^2\right] - \mu^2\varphi_1^2 + O(\varphi^3) \tag{5}$$

The constant term is exactly as in (4), and just like the field φ' in (4), the field φ_1 has mass $\sqrt{2}\mu$. But now note the remarkable feature of (5): the absence of a φ_2^2 term. The field φ_2 is massless!

Emergence of massless boson

That φ_2 comes out massless is not an accident. I will now explain that the masslessness is a general and exact phenomenon.

Referring back to figure IV.1.2 we can easily understand the particle spectrum. Excitation in the φ_1 field corresponds to fluctuation in the radial direction, "climbing the wall" so to speak, while excitation in the φ_2 field corresponds to fluctuation in the angular direction, "rolling along the gutter" so to speak. It costs no energy for a marble to roll along the minima of the potential energy, going from one minimum to another. Another way of saying this is to picture a long wavelength excitation of the form $\varphi_2 = a \sin(\omega t - \vec{k}\vec{x})$ with a small. In a region of length scale small compared to $|\vec{k}|^{-1}$, the field φ_2 is essentially constant and thus the field $\vec{\varphi}$ is just rotated slightly away from the 1 direction, which by the $O(2)$ symmetry is equivalent to the vacuum. It is only when we look at regions of length scale large compared to $|\vec{k}|^{-1}$ that we realize that the excitation costs energy. Thus, as $|\vec{k}| \to 0$, we expect the energy of the excitation to vanish.

We now understand the crucial difference between the $N = 1$ and the $N = 2$ cases: In the former we have a reflection symmetry, which is discrete, while in the latter we have an $O(2)$ symmetry, which is continuous.

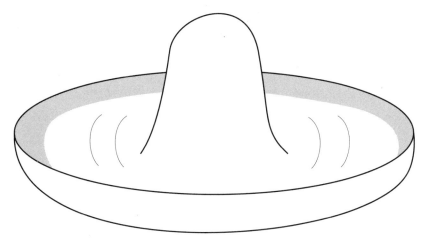

Figure IV.1.2

We have worked out the $N = 2$ case in detail. You should now be able to generalize our discussion to arbitrary $N \geq 2$ (see exercise IV.1.1).

Meanwhile, it is worth looking at $N = 2$ from another point of view. Many field theories can be written in more than one form and it is important to know them under different guises. Construct the complex field $\varphi = (1/\sqrt{2})(\varphi_1 + i\varphi_2)$; we have $\varphi^\dagger\varphi = \frac{1}{2}(\varphi_1^2 + \varphi_2^2)$ and so can write (2) as

$$\mathcal{L} = \partial\varphi^\dagger\partial\varphi + \mu^2\varphi^\dagger\varphi - \lambda(\varphi^\dagger\varphi)^2 \tag{6}$$

which is manifestly invariant under the $U(1)$ transformation $\varphi \to e^{i\alpha}\varphi$ (recall chapter I.10). You may recognize that this amounts to saying that the groups $O(2)$ and $U(1)$ are locally isomorphic. Just as we can write a vector in Cartesian or polar coordinates we are free to parametrize the field by $\varphi(x) = \rho(x)e^{i\theta(x)}$ (as in chapter III.5) so that $\partial_\mu\varphi = (\partial_\mu\rho + i\rho\partial_\mu\theta)e^{i\theta}$. We obtain $\mathcal{L} = \rho^2(\partial\theta)^2 + (\partial\rho)^2 + \mu^2\rho^2 - \lambda\rho^4$. Spontaneous symmetry breaking means setting $\rho = v + \chi$ with $v = +\sqrt{\mu^2/2\lambda}$, whereupon

$$\mathcal{L} = v^2(\partial\theta)^2 + \left[(\partial\chi)^2 - 2\mu^2\chi^2 - 4\sqrt{\frac{\mu^2\lambda}{2}}\chi^3 - \lambda\chi^4\right] + \left(\sqrt{\frac{2\mu^2}{\lambda}}\chi + \chi^2\right)(\partial\theta)^2 \tag{7}$$

We recognize the phase $\theta(x)$ as the massless field. We have arranged the terms in the Lagrangian in three groups: the kinetic energy of the massless field θ, the kinetic and potential energy of the massive field χ, and the interaction between θ and χ. (The additive constant in (5) has been dropped to minimize clutter.)

Goldstone's theorem

We will now prove Goldstone's theorem, which states that whenever a continuous symmetry is spontaneously broken, massless fields, known as Nambu[2]-Goldstone bosons, emerge.

Recall that associated with every continuous symmetry is a conserved charge Q. That Q generates a symmetry is stated as

$$[H, Q] = 0 \tag{8}$$

Let the vacuum (or ground state in quantum mechanics) be denoted by $|0\rangle$. By adding an appropriate constant to the Hamiltonian $H \to H + c$ we can always write $H|0\rangle = 0$. Normally, the vacuum is invariant under the symmetry transformation, $e^{i\theta Q}|0\rangle = |0\rangle$, or in other words $Q|0\rangle = 0$.

But suppose the symmetry is spontaneously broken, so that the vacuum is not invariant under the symmetry transformation; in other words, $Q|0\rangle \neq 0$. Consider the state $Q|0\rangle$. What is its energy? Well,

$$HQ|0\rangle = [H, Q]|0\rangle = 0 \tag{9}$$

[The first equality follows from $H|0\rangle = 0$ and the second from (8).] Thus, we have found another state $Q|0\rangle$ with the same energy as $|0\rangle$.

Note that the proof makes no reference to either relativity or fields. You can also see that it merely formalizes the picture of the marble rolling along the gutter.

In quantum field theory, we have local currents, and so

$$Q = \int d^D x\, J^0(\vec{x}, t)$$

where D denotes the dimension of space and conservation of Q says that the integral can be evaluated at any time. Consider the state

$$|s\rangle = \int d^D x\, e^{-i\vec{k}\vec{x}} J^0(\vec{x}, t)|0\rangle$$

which has[3] spatial momentum $\vec{k}$. As $\vec{k}$ goes to zero it goes over to $Q|0\rangle$, which as we learned in (9) has zero energy. Thus, as the momentum of the state $|s\rangle$ goes to zero, its energy goes to zero. In a relativistic theory, this means precisely that $|s\rangle$ describes a massless particle.

---

[2] Y. Nambu, quite deservedly, received the 2008 physics Nobel Prize for his profound contribution to our understanding of spontaneous symmetry breaking.

[3] Acting on it with P^i (exercise I.11.3) and using $P^i|0\rangle = 0$, we have

$$P^i|s\rangle = \int d^D x\, e^{-i\vec{k}\vec{x}}[P^i, J^0(\vec{x}, t)]|0\rangle = -i \int d^D e^{-i\vec{k}\cdot\vec{x}} \partial^i J^0(\vec{x}, t)|0\rangle = k^i|s\rangle$$

upon integrating by parts.

The proof makes clear that the theorem practically exudes generality: It applies to any spontaneously broken continuous symmetry.

Counting Nambu-Goldstone bosons

From our proof, we see that the number of Nambu-Goldstone bosons is clearly equal to the number of conserved charges that do not leave the vacuum invariant, that is, do not annihilate $|0\rangle$. For each such charge Q^α, we can construct a zero-energy state $Q^\alpha |0\rangle$.

In our example, we have only one current $J_\mu = i(\varphi_1 \partial_\mu \varphi_2 - \varphi_2 \partial_\mu \varphi_1)$ and hence one Nambu-Goldstone boson. In general, if the Lagrangian is left invariant by a symmetry group G with $n(G)$ generators, but the vacuum is left invariant by only a subgroup H of G with $n(H)$ generators, then there are $n(G) - n(H)$ Nambu-Goldstone bosons. If you want to show off your mastery of mathematical jargon you can say that the Nambu-Goldstone bosons live in the coset space G/H.

Ferromagnet and spin wave

The generality of the proof suggests that the usefulness of Goldstone's theorem is not restricted to particle physics. In fact, it originated in condensed matter physics, the classic example there being the ferromagnet. The Hamiltonian, being composed of just the interaction of nonrelativistic electrons with the ions in the solid, is of course invariant under the rotation group $SO(3)$, but the magnetization $\vec{M}$ picks out a direction, and the ferromagnet is left invariant only under the subgroup $SO(2)$ consisting of rotations about the axis defined by $\vec{M}$. The Nambu-Goldstone theorem is easy to visualize physically. Consider a "spin wave" in which the local magnetization $\vec{M}(\vec{x})$ varies slowly from point to point. A physicist living in a region small compared to the wavelength does not even realize that he or she is no longer in the "vacuum." Thus, the frequency of the wave must go to zero as the wavelength goes to infinity. This is of course exactly the same heuristic argument given earlier. Note that quantum mechanics is needed only to translate the wave vector $\vec{k}$ into momentum and the frequency ω into energy. I will come back to magnets and spin wave in chapters V.3 and VI.5.

Quantum fluctuations and the dimension of spacetime

Our discussion of spontaneous symmetry breaking is essentially classical. What happens when quantum fluctuations are included? I will address this question in detail in chapter IV.3, but for now let us go back to (5). In the ground state, $\varphi_1 = v$ and $\varphi_2 = 0$. Recall that in the mattress model of a scalar field theory the mass term comes from the springs holding the mattress to its equilibrium position. The term $-\mu^2 \varphi_1'^2$ (note the prime) in (5) tells us that it costs action for φ'_1 to wander away from its ground state value $\varphi'_1 = 0$. But now we are worried: φ_2 is massless. Can it wander away from its ground state value? To answer this question let us calculate the mean square fluctuation

$$
\begin{aligned}
\langle(\varphi_2(0))^2\rangle &= \frac{1}{Z} \int D\varphi e^{iS(\varphi)} [\varphi_2(0)]^2 \\
&= \lim_{x \to 0} \frac{1}{Z} \int D\varphi e^{iS(\varphi)} \varphi_2(x)\varphi_2(0) \\
&= \lim_{x \to 0} \int \frac{d^d k}{(2\pi)^d} \frac{e^{ikx}}{k^2}
\end{aligned}
\tag{10}
$$

(We recognized the functional integral that defines the propagator; recall chapter I.7.) The upper limit of the integral in (10) is cut off at some Λ (which would correspond to the inverse of the lattice spacing when applying these ideas to a ferromagnet) and so as explained in chapter III.1 (and as you will see in chapter VIII.3) we are not particularly worried about the ultraviolet divergence for large k. But we do have to worry about a possible infrared divergence for small k. (Note that for a massive field $1/k^2$ in (10) would have been replaced by $1/(k^2 + \mu^2)$ and there would be no infrared divergence.)

We see that there is no infrared divergence for $d > 2$. Our picture of spontaneously breaking a continuous symmetry is valid in our $(3 + 1)$-dimensional world.

However, for $d \leq 2$ the mean square fluctuation of φ_2 comes out infinite, so our naive picture is totally off. We have arrived at the Coleman-Mermin-Wagner theorem (proved independently by a particle theorist and two condensed matter theorists), which states that spontaneous breaking of a continuous symmetry is impossible for $d = 2$. Note that while our discussion is given for $O(2)$ symmetry the conclusion applies to any continuous symmetry since the argument depends only on the presence of Nambu-Goldstone fields.

In our examples, symmetry is spontaneously broken by a scalar field φ, but nothing says that the field φ must be elementary. In many condensed matter systems, superconductors, for example, symmetries are spontaneously broken, but we know that the system consists of electrons and atomic nuclei. The field φ is generated dynamically, for example as a bound state of two electrons in superconductors. More on this in chapter V.4. The spontaneous breaking of a symmetry by a dynamically generated field is sometimes referred to as dynamical symmetry breaking.[4]

Exercises

IV.1.1 Show explicitly that there are $N - 1$ Nambu-Goldstone bosons in the $G = O(N)$ example (2).

IV.1.2 Construct the analog of (2) with N complex scalar fields and invariant under $SU(N)$. Count the number of Nambu-Goldstone bosons when one of the scalar fields acquires a vacuum expectation value.

---

[4] This chapter is dedicated to the memory of the late Jorge Swieca.

IV.2 | The Pion as a Nambu-Goldstone Boson

Crisis for field theory

After the spectacular triumphs of quantum field theory in the electromagnetic interaction, physicists in the 1950s and 1960s were naturally eager to apply it to the strong and weak interactions. As we have already seen, field theory when applied to the weak interaction appeared not to be renormalizable. As for the strong interaction, field theory appeared totally untenable for other reasons. For one thing, as the number of experimentally observed hadrons (namely strongly interacting particles) proliferated, it became clear that were we to associate a field with each hadron the resulting field theory would be quite a mess, with numerous arbitrary coupling constants. But even if we were to restrict ourselves to nucleons and pions, the known coupling constant of the interaction between pions and nucleons is a large number. (Hence the term strong interaction in the first place!) The perturbative approach that worked so spectacularly well in quantum electrodynamics was doomed to failure.

Many eminent physicists at the time advocated abandoning quantum field theory altogether, and at certain graduate schools, quantum field theory was even dropped from the curriculum. It was not until the early 1970s that quantum field theory made a triumphant comeback. A field theory for the strong interaction was formulated, not in terms of hadrons, but in terms of quarks and gluons. I will get to that in chapter VII.3.

Pion weak decay

To understand the crisis facing field theory, let us go back in time and imagine what a field theorist might be trying to do in the late 1950s. Since this is not a book on particle physics, I will merely sketch the relevant facts. You are urged to consult one of the texts on the subject.[1] By that time, many semileptonic decays such as $n \to p + e^- + \bar{\nu}$, $\pi^- \to e^- + \bar{\nu}$,

---

[1] See, e.g., E. Commins and P. H. Bucksbaum, *Weak Interactions of Leptons and Quarks.*

and $\pi^- \to \pi^0 + e^- + \bar{\nu}$ had been measured. Neutron β decay $n \to p + e^- + \bar{\nu}$ was of course the process for which Fermi invented his theory, which by that time had assumed the form $\mathcal{L} = G[\bar{e}\gamma^\mu(1 - \gamma_5)\nu][\bar{p}\gamma_\mu(1 - \gamma_5)n]$, where n is a neutron field annihilating a neutron, p a proton field annihilating a proton, ν a neutrino field annihilating a neutrino (or creating an antineutrino as in β decay), and e an electron field annihilating an electron.

It became clear that to write down a field for each hadron and a Lagrangian for each decay process, as theorists were in fact doing for a while, was a losing battle. Instead, we should write

$$\mathcal{L} = G[\bar{e}\gamma^\mu(1 - \gamma_5)\nu](J_\mu - J_{5\mu}) \tag{1}$$

with J_μ and $J_{5\mu}$ two currents transforming as a Lorentz vector, and axial vector respectively. We think of J_μ and $J_{5\mu}$ as quantum operators in a canonical formulation of field theory. Our task would then be to calculate the matrix elements between hadron states, $\langle p| (J_\mu - J_{5\mu}) |n\rangle$, $\langle 0| (J_\mu - J_{5\mu}) |\pi^-\rangle$, $\langle \pi^0| (J_\mu - J_{5\mu}) |\pi^-\rangle$, and so on, corresponding to the three decay processes I listed above. (I should make clear that although I am talking about weak decays, the calculation of these matrix elements is a problem in the strong interaction. In other words, in understanding these decays, we have to treat the strong interaction to all orders in the strong coupling, but it suffices to treat the weak interaction to lowest order in the weak coupling G.) Actually, there is a precedent for the attitude we are adopting here. To account for nuclear β decay $(Z, A) \to (Z + 1, A) + e^- + \bar{\nu}$, Fermi certainly did not write a separate Lagrangian for each nucleus. Rather, it was the task of the nuclear theorist to calculate the matrix element $\langle Z + 1, A| [\bar{p}\gamma_\mu(1 - \gamma_5)n] |Z, A\rangle$. Similarly, it is the task of the strong interaction theorist to calculate matrix elements such as $\langle p| (J_\mu - J_{5\mu}) |n\rangle$.

For the story I am telling, let me focus on trying to calculate the matrix element of the axial vector current J_5^μ between a neutron and a proton. Here we make a trivial change in notation: We no longer indicate that we have a neutron in the initial state and a proton in the final state, but instead we specify the momentum p of the neutron and the momentum p' of the proton. Incidentally, in (1) the fields and the currents are of course all functions of the spacetime coordinates x. Thus, we want to calculate $\langle p'| J_5^\mu(x) |p\rangle$, but by translation invariance this is equal to $\langle p'| J_5^\mu(0) |p\rangle e^{-i(p'-p)\cdot x}$. Henceforth, we simply calculate $\langle p'| J_5^\mu(0) |p\rangle$ and suppress the 0. Note that spin labels have already been suppressed.

Lorentz invariance and parity can take us some distance: They imply that[2]

$$\langle p'| J_5^\mu |p\rangle = \bar{u}(p')[\gamma^\mu\gamma^5 F(q^2) + q^\mu\gamma^5 G(q^2)]u(p) \tag{2}$$

with $q \equiv p' - p$ [compare with (III.6.7)]. But Lorentz invariance and parity can only take us so far: We know nothing about the "form factors" $F(q^2)$ and $G(q^2)$.

---

[2] Another possible term of the form $(p' + p)^\mu\gamma^5$ can be shown to vanish by charge conjugation and isospin symmetries.

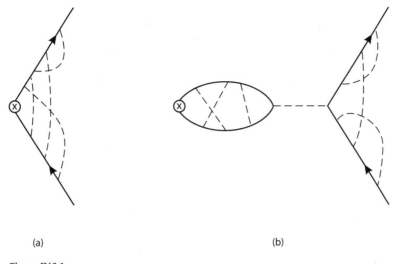

(a)   (b)

Figure IV.2.1

Similarly, for the matrix element $\langle 0| J_5^\mu |\pi^- \rangle$ Lorentz invariance tells us that

$$\langle 0| J_5^\mu |k\rangle = f k^\mu \tag{3}$$

I have again labeled the initial state by the momentum k of the pion. The right-hand side of (3) has to be a vector but since k is the only vector available it has to be proportional to k. Just like $F(q^2)$ and $G(q^2)$, the constant f is a strong interaction quantity that we don't know how to calculate. On the other hand, $F(q^2)$, $G(q^2)$, and f can all be measured experimentally. For instance, the rate for the decay $\pi^- \rightarrow e^- + \bar{\nu}$ clearly depends on f^2.

Too many diagrams

Let us look over the shoulder of a field theorist trying to calculate $\langle p'| J_5^\mu |p\rangle$ and $\langle 0| J_5^\mu |k\rangle$ in (2) in the late 1950s. He would draw Feynman diagrams such as the ones in figures IV.2.1 and IV.2.2 and soon realize that it would be hopeless. Because of the strong coupling, he would have to calculate an infinite number of diagrams, even if the strong interaction were described by a field theory, a notion already rejected by many luminaries of the time.

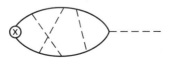

Figure IV.2.2

In telling the story of the breakthrough I am not going to follow the absolutely fascinating history of the subject, full of total confusion and blind alleys. Instead, with the benefit of hindsight, I am going to tell the story using what I regard as the best pedagogical approach.

The pion is very light

The breakthrough originated in the observation that the mass of the π^- at 139 Mev was considerably less than the mass of the proton at 938 Mev. For a long time this was simply taken as a fact not in any particular need of an explanation. But eventually some theorists wondered why one hadron should be so much lighter than another.

Finally, some theorists took the bold step of imagining an "ideal world" in which the π^- is massless. The idea was that this ideal world would be a good approximation of our world, to an accuracy of about 15% ($\sim 139/938$).

Do you remember one circumstance in which a massless spinless particle would emerge naturally? Yes, spontaneous symmetry breaking! In one of the blinding insights that have characterized the history of particle physics, some theorists proposed that the π mesons are the Nambu-Goldstone bosons of some spontaneous broken symmetry.

Indeed, let's multiply (3) by k_μ :

$$k_\mu \langle 0| J_5^\mu |k\rangle = f k^2 = f m_\pi^2 \tag{4}$$

which is equal to zero in the ideal world. Recall from our earlier discussion on translation invariance that

$$\langle 0| J_5^\mu (x) |k\rangle = \langle 0| J_5^\mu (0) |k\rangle e^{-ik\cdot x}$$

and hence

$$\langle 0| \partial_\mu J_5^\mu (x) |k\rangle = -ik_\mu \langle 0| J_5^\mu (0) |k\rangle e^{-ik\cdot x}$$

Thus, if the axial current is conserved, $\partial_\mu J_5^\mu (x) = 0$, in the ideal world, $k_\mu \langle 0| J_5^\mu |k\rangle = 0$ and (4) would indeed imply $m_\pi^2 = 0$.

The ideal world we are discussing enjoys a symmetry known as the chiral symmetry of the strong interaction. The symmetry is spontaneously broken in the ground state we inhabit, with the π meson as the Nambu-Goldstone boson. The Noether current associated with this symmetry is the conserved J_5^μ.

In fact, you should recognize that the manipulation here is closely related to the proof of the Nambu-Goldstone theorem given in chapter IV.1.

Goldberger-Treiman relation

Now comes the punchline. Multiply (2) by $(p' - p)_\mu$. By the same translation invariance argument we just used,

$$(p' - p)_\mu \langle p'| J_5^\mu (0) |p\rangle = i \langle p'| \partial_\mu J_5^\mu (x) |p\rangle e^{i(p'-p)\cdot x}$$

and hence vanishes if $\partial_\mu J_5^\mu = 0$. On the other hand, multiplying the right-hand side of (2) by $(p' - p)_\mu$ we obtain $\bar{u}(p')[(\not{p}' - \not{p})\gamma^5 F(q^2) + q^2\gamma^5 G(q^2)]u(p)$. Using the Dirac equation (do it!) we conclude that

$$0 = 2m_N F(q^2) + q^2 G(q^2) \tag{5}$$

with m_N the nucleon mass.

The form factors $F(q^2)$ and $G(q^2)$ are each determined by an infinite number of Feynman diagrams we have no hope of calculating, but yet we have managed to relate them! This represents a common strategy in many areas of physics: When faced with various quantities we don't know how to calculate, we can nevertheless try to relate them.

We can go farther by letting $q \to 0$ in (5). Referring to (2) we see that $F(0)$ is measured experimentally in $n \to p + e^- + \bar{\nu}$ (the momentum transfer is negligible on the scale of the strong interaction). But oops, we seem to have a problem: We predict the nucleon mass $m_N = 0$!

In fact, we are saved by examining figure IV.2.1b: There are an infinite number of diagrams exhibiting a pole due to none other than the massless π meson, which you can see gives

$$f q^\mu \frac{1}{q^2} g_{\pi NN} \bar{u}(p')\gamma^5 u(p) \tag{6}$$

When the π propagator joins onto the nucleon line, an infinite number of diagrams summed together gives the experimentally measured pion-nucleon coupling constant $g_{\pi NN}$. Thus, referring to (2), we see that for $q \sim 0$ the form factor $G(q^2) \sim f(1/q^2)g_{\pi NN}$. Plugging into (5), we obtain the celebrated Goldberger-Treiman relation

$$2m_N F(0) + f g_{\pi NN} = 0 \tag{7}$$

relating four experimentally measured quantities. As might be expected, it holds with about a 15% error, consistent with our not living in a world with an exactly massless π meson.

Toward a theory of the strong interaction

The art of relating infinite sets of Feynman diagrams without calculating them, and it is an art form involving a great deal of cleverness, was developed into a subject called dispersion relations and S-matrix theory, which we mentioned briefly in chapter III.8. Our present understanding of the strong interaction was built on this foundation. You could see from this example that an important component of dispersion relations was the study of the analyticity properties of Feynman diagrams as described in chapter III.8. The essence of the Goldberger-Treiman argument is separating the infinite number of diagrams into those with a pole in the complex q^2-plane and those without a pole (but with a cut.)

The discovery that the strong interaction contains a spontaneously broken symmetry provided a crucial clue to the underlying theory of the strong interaction and ultimately led to the concepts of quarks and gluons.

A note for the historian of science: Whether theoretical physicists regard a quantity as small or large depends (obviously) on the cultural and mental framework they grew up in. Treiman once told me that the notion of setting 138 Mev to zero, when the energy released per nucleon in nuclear fission is of order 10 Mev, struck the generation that grew up with the atomic bomb (as Treiman did—he was with the armed forces in the Pacific) as surely the height of absurdity. Now of course a new generation of young string theorists is perfectly comfortable in regarding anything less than the Planck energy 10^{19} Gev as essentially zero.

IV.3 | Effective Potential

Quantum fluctuations and symmetry breaking

The important phenomenon of spontaneous symmetry breaking was based on minimizing the classical potential energy $V(\varphi)$ of a quantum field theory. It is natural to wonder how quantum fluctuations would change this picture.

To motivate the discussion, consider once again (III.3.3)

$$\mathcal{L} = \frac{1}{2}(\partial\varphi)^2 - \frac{1}{2}\mu^2\varphi^2 - \frac{1}{4!}\lambda\varphi^4 + A(\partial\varphi)^2 + B\varphi^2 + C\varphi^4 \tag{1}$$

(Speaking of quantum fluctuations, we have to include counterterms as indicated.) What have you learned about this theory? For $\mu^2 > 0$, the action is extremized at $\varphi = 0$, and quantizing the small fluctuations around $\varphi = 0$ we obtain scalar particles that scatter off each other. For $\mu^2 < 0$, the action is extremized at some $\varphi_{\min}$, and the discrete symmetry $\varphi \to -\varphi$ is spontaneously broken, as you learned in chapter IV.1. What happens when $\mu = 0$? To break or not to break, that is the question.

A quick guess is that quantum fluctuations would break the symmetry. The $\mu = 0$ theory is posed on the edge of symmetry breaking, and quantum fluctuations ought to push it over the brink. Think of a classical pencil perfectly balanced on its tip. Then "switch on" quantum mechanics.

Wisdom of the son-in-law

Let us follow Schwinger and Jona-Lasinio and develop the formalism that enables us to answer this question. Consider a scalar field theory defined by

$$Z = e^{iW(J)} = \int D\varphi\, e^{i[S(\varphi) + J\varphi]} \tag{2}$$

[with the convenient shorthand $J\varphi = \int d^4x\, J(x)\varphi(x)$]. If we can do the functional integral, we obtain the generating functional $W(J)$. As explained in chapter I.7, by differentiating

W with respect to the source $J(x)$ repeatedly, we can obtain any Green's function and hence any scattering amplitude we want. In particular,

$$\varphi_c(x) \equiv \frac{\delta W}{\delta J(x)} = \frac{1}{Z} \int D\varphi e^{i[S(\varphi) + J\varphi]} \varphi(x) \tag{3}$$

The subscript c is used traditionally to remind us (see appendix 2 in chapter I.8) that in a canonical formalism $\varphi_c(x)$ is the expectation value $\langle 0| \hat{\varphi} |0 \rangle$ of the quantum operator $\hat{\varphi}$. It is certainly not to be confused with the integration dummy variable φ in (3). The relation (3) determines $\varphi_c(x)$ as a functional of J.

Given a functional W of J we can perform a Legendre transform to obtain a functional Γ of φ_c. Legendre transform is just the fancy term for the simple relation

$$\Gamma(\varphi_c) = W(J) - \int d^4x J(x) \varphi_c(x) \tag{4}$$

The relation is simple, but be careful about what it says: It defines a functional of $\varphi_c(x)$ through the implicit dependence of J on φ_c. On the right-hand side of (4) J is to be eliminated in favor of φ_c by solving (3). We expand the functional $\Gamma(\varphi_c)$ in the form

$$\Gamma(\varphi_c) = \int d^4x [-V_{\text{eff}}(\varphi_c) + Z(\varphi_c)(\partial \varphi_c)^2 + \cdots] \tag{5}$$

where $(\cdots)$ indicates terms with higher and higher powers of ∂. We will soon see the wisdom of the notation $V_{\text{eff}}(\varphi_c)$.

The point of the Legendre transform is that the functional derivative of Γ is nice and simple:

$$\frac{\delta \Gamma(\varphi_c)}{\delta \varphi_c(y)} = \int d^4x \frac{\delta J(x)}{\delta \varphi_c(y)} \frac{\delta W(J)}{\delta J(x)} - \int d^4x \frac{\delta J(x)}{\delta \varphi_c(y)} \varphi_c(x) - J(y)$$
$$= -J(y) \tag{6}$$

a relation we can think of as the "dual" of $\delta W(J)/\delta J(x) = \varphi_c(x)$.

If you vaguely feel that you have seen this sort of manipulation before in your physics eduction, you are quite right! It was in a course on thermodynamics, where you learned about the Legendre transform relating the free energy to the energy: $F = E - TS$ with F a function of the temperature T and E a function of the entropy S. Thus J and φ are "conjugate" pairs just like T and S (or even more clearly magnetic field H and magnetization M). Convince yourself that this is far more than a mere coincidence.

For J and φ_c independent of x we see from (5) that the condition (6) reduces to

$$V'_{\text{eff}}(\varphi_c) = J \tag{7}$$

This relation makes clear what the effective potential $V_{\text{eff}}(\varphi_c)$ is good for. Let's ask what happens when there is no external source J. The answer is immediate: (7) tells us that

$$V'_{\text{eff}}(\varphi_c) = 0, \tag{8}$$

In other words, the vacuum expectation value of $\hat{\varphi}$ in the absence of an external source is determined by minimizing $V_{\text{eff}}(\varphi_c)$.

First order in quantum fluctuations

All of these formal manipulations are not worth much if we cannot evaluate $W(J)$. In fact, in most cases we can only evaluate $e^{iW(J)} = \int D\varphi e^{i[S(\varphi)+J\varphi]}$ in the steepest descent approximation (see chapter I.2). Let us turn the crank and find the steepest descent "point" $\varphi_s(x)$, namely the solution of (henceforth I will drop the subscript c as there is little risk of confusion)

$$\left. \frac{\delta[S(\varphi) + \int d^4 y J(y)\varphi(y)]}{\delta\varphi(x)} \right|_{\varphi_s} = 0 \qquad (9)$$

or more explicitly,

$$\partial^2 \varphi_s(x) + V'[\varphi_s(x)] = J(x) \qquad (10)$$

Write the dummy integration variable in (2) as $\varphi = \varphi_s + \widetilde{\varphi}$ and expand to quadratic order in $\widetilde{\varphi}$ to obtain

$$
\begin{aligned}
Z = e^{(i/\hbar)W(J)} &= \int D\varphi e^{(i/\hbar)[S(\varphi)+J\varphi]} \\
&\simeq e^{(i/\hbar)[S(\varphi_s)+J\varphi_s]} \int D\widetilde{\varphi} e^{(i/\hbar)\int d^4 x \frac{1}{2}[(\partial\widetilde{\varphi})^2 - V''(\varphi_s)\widetilde{\varphi}^2]} \\
&= e^{(i/\hbar)[S(\varphi_s)+J\varphi_s] - \frac{1}{2}\operatorname{tr}\log[\partial^2 + V''(\varphi_s)]}
\end{aligned}
\qquad (11)
$$

We have used (II.5.2) to represent the determinant we get upon integrating over $\widetilde{\varphi}$. Note that I have put back Planck's constant $\hbar$. Here φ_s, as a solution of (10), is to be regarded as a function of J.

Now that we have determined

$$W(J) = [S(\varphi_s) + J\varphi_s] + \frac{i\hbar}{2} \operatorname{tr}\log[\partial^2 + V''(\varphi_s)] + O(\hbar^2)$$

it is straightforward to Legendre transform. I will go painfully slowly here:

$$\varphi = \frac{\delta W}{\delta J} = \frac{\delta[S(\varphi_s) + J\varphi_s]}{\delta\varphi_s}\frac{\delta\varphi_s}{\delta J} + \varphi_s + O(\hbar) = \varphi_s + O(\hbar)$$

To leading order in $\hbar$, φ (namely the object formerly known as φ_c) is equal to φ_s. Thus, from (4) we obtain

$$\Gamma(\varphi) = S(\varphi) + \frac{i\hbar}{2} \operatorname{tr}\log[\partial^2 + V''(\varphi)] + O(\hbar^2) \qquad (12)$$

Nice though this formula looks, in practice it is impossible to evaluate the trace for arbitrary $\varphi(x)$: We have to find all the eigenvalues of the operator $\partial^2 + V''(\varphi)$, take their log, and sum. Our task simplifies drastically if we are content with studying $\Gamma(\varphi)$ for

φ independent of x, in which case $V''(\varphi)$ is a constant and the operator $\partial^2 + V''(\varphi)$ is translation invariant and easily treated in momentum space:

$$
\begin{aligned}
\text{tr} \log[\partial^2 + V''(\varphi)] &= \int d^4x \, \langle x| \log[\partial^2 + V''(\varphi)]|x\rangle \\
&= \int d^4x \int \frac{d^4k}{(2\pi)^4} \langle x|k\rangle\langle k| \log[\partial^2 + V''(\varphi)]|k\rangle\langle k|x\rangle \\
&= \int d^4x \int \frac{d^4k}{(2\pi)^4} \log[-k^2 + V''(\varphi)]
\end{aligned}
\tag{13}
$$

Referring to (5), we obtain

$$
V_{\text{eff}}(\varphi) = V(\varphi) - \frac{i\hbar}{2} \int \frac{d^4k}{(2\pi)^4} \log\left[\frac{k^2 - V''(\varphi)}{k^2}\right] + O(\hbar^2)
\tag{14}
$$

known as the Coleman-Weinberg effective potential. What we computed is the order $\hbar$ correction to the classical potential $V(\varphi)$. Note that we have added a φ independent constant to make the argument of the logarithm dimensionless.

We can give a nice physical interpretation of (14). Let the universe be suffused with the scalar field $\varphi(x)$ taking on the value φ, a background field so to speak. For $V(\varphi) = \frac{1}{2}\mu^2\varphi^2 + (1/4!)\lambda\varphi^4$, we have $V''(\varphi) = \mu^2 + \frac{1}{2}\lambda\varphi^2 \equiv \mu(\varphi)^2$, which, as the notation $\mu(\varphi)^2$ suggests, we recognize as the φ-dependent effective mass squared of a scalar particle propagating in the background field φ. The mass squared μ^2 in the Lagrangian is corrected by a term $\frac{1}{2}\lambda\varphi^2$ due to the interaction of the particle with the background field φ. Now we see clearly what (14) tells us: The first term $V(\varphi)$ is the classical energy density contained in the background φ, while the second term is the vacuum energy density of a scalar field with mass squared equal to $V''(\varphi)$ [see (II.5.3) and exercise IV.3.4].

Your renormalization theory at work

The integral in (14) is quadratically divergent, or more correctly, quadratically dependent on the cutoff. But no sweat, we were instructed to introduce three counterterms (of which only two are relevant here since φ is independent of x). Thus, we actually have

$$
V_{\text{eff}}(\varphi) = V(\varphi) + \frac{\hbar}{2} \int \frac{d^4k_E}{(2\pi)^4} \log\left[\frac{k_E^2 + V''(\varphi)}{k_E^2}\right] + B\varphi^2 + C\varphi^4 + O(\hbar^2)
\tag{15}
$$

where we have Wick rotated to a Euclidean integral (see appendix D). Using (D.9) and integrating up to $k_E^2 = \Lambda^2$, we obtain (suppressing $\hbar$)

$$
V_{\text{eff}}(\varphi) = V(\varphi) + \frac{\Lambda^2}{32\pi^2} V''(\varphi) - \frac{[V''(\varphi)]^2}{64\pi^2} \log \frac{e^{\frac{1}{2}}\Lambda^2}{V''(\varphi)} + B\varphi^2 + C\varphi^4
\tag{16}
$$

As expected, since the integrand in (15) goes as $1/k_E^2$ for large k_E^2 the integral depends quadratically and logarithmically on the cutoff Λ^2.

Watch renormalization theory at work! Since V is a quartic polynomial in φ, $V''(\varphi)$ is a quadratic polynomial and $[V''(\varphi)]^2$ a quartic polynomial. Thus, we have just enough counterterms $B\varphi^2 + C\varphi^4$ to absorb the cutoff dependence. This is a particularly transparent example of how the method of adding counterterms works.

To see how bad things can happen in a nonrenormalizable theory, suppose in contrast that V is a polynomial of degree 6 in φ. Then we are allowed to have three counterterms $B\varphi^2 + C\varphi^4 + D\varphi^6$, but that is not enough since $[V''(\varphi)]^2$ is now a polynomial of degree 8. This means that we should have started out with V a polynomial of degree 8, but then $[V''(\varphi)]^2$ would be a polynomial of degree 12. Clearly, the process escalates into an infinite-degree polynomial. We see the hallmark of a nonrenormalizable theory: its insatiable appetite for counterterms.

Imposing renormalization conditions

Waking up from the nightmare of an infinite number of counterterms chasing us, let us go back to the sweetly renormalizable φ^4 theory. In chapter III.3 we fix the counterterms by imposing conditions on various scattering amplitudes. Here we would have to fix the coefficients B and C by imposing two conditions on $V_{\text{eff}}(\varphi)$ at appropriate values of φ. We are working in field space, so to speak, rather than momentum space, but the conceptual framework is the same.

We could proceed with the general quartic polynomial $V(\varphi)$, but instead let us try to answer the motivating question of this chapter: What happens when $\mu = 0$, that is, when $V(\varphi) = (1/4!)\lambda\varphi^4$? The arithmetic is also simpler.

Evaluating (16) we get

$$V_{\text{eff}}(\varphi) = \left(\frac{\Lambda^2}{64\pi^2}\lambda + B\right)\varphi^2 + \left(\frac{1}{4!}\lambda + \frac{\lambda^2}{(16\pi)^2}\log\frac{\varphi^2}{\Lambda^2} + C\right)\varphi^4 + O(\lambda^3)$$

(after absorbing some φ-independent constants into C). We see explicitly that the Λ dependence can be absorbed into B and C.

We started out with a purely quartic $V(\varphi)$. Quantum fluctuations generate a quadratically divergent φ^2 term that we can cancel with the B counterterm. What does $\mu = 0$ mean? It means that $(d^2V/d\varphi^2)|_{\varphi=0}$ vanishes. To say that we have a $\mu = 0$ theory means that we have to maintain a vanishing renormalized mass squared, defined here as the coefficient of φ^2. Thus, we impose our first condition

$$\left.\frac{d^2V_{\text{eff}}}{d\varphi^2}\right|_{\varphi=0} = 0 \tag{17}$$

This is a somewhat long-winded way of saying that we want $B = -(\Lambda^2/64\pi^2)\lambda$ to this order.

Similarly, we might think that the second condition would be to set $(d^4V_{\text{eff}}/d\varphi^4)|_{\varphi=0}$ equal to some coupling, but differentiating the $\varphi^4 \log\varphi$ term in V_{eff} four times we are going to get a term like $\log\varphi$, which is not defined at $\varphi = 0$. We are forced to impose our condition on $d^4 V_{\text{eff}}/d\varphi^4$ not at $\varphi = 0$ but at φ equal to some arbitrarily chosen mass M. (Recall that φ has the dimension of mass.) Thus, the second condition reads

$$\left. \frac{d^4 V_{\text{eff}}}{d\varphi^4} \right|_{\varphi = M} = \lambda(M) \tag{18}$$

where $\lambda(M)$ is a coupling manifestly dependent on M.

Plugging

$$V_{\text{eff}}(\varphi) = (\frac{1}{4!}\lambda + \frac{\lambda^2}{(16\pi)^2} \log \frac{\varphi^2}{\Lambda^2} + C)\varphi^4 + O(\lambda^3)$$

into (18) we see that $\lambda(M)$ is equal to λ plus $O(\lambda^2)$ corrections, among which is a term like $\lambda^2 \log M$. We can get a clean relation by differentiating $\lambda(M)$:

$$M \frac{d\lambda(M)}{dM} = \frac{3}{16\pi^2}\lambda^2 + O(\lambda^3)$$

$$= \frac{3}{16\pi^2}\lambda(M)^2 + O[\lambda(M)^3] \tag{19}$$

where the second equality is correct to the order indicated. This interesting relation tells us how the coupling $\lambda(M)$ depends on the mass scale M at which it is defined. Recall exercise III.1.3. We will come back to this relation in chapter VI.7 on the renormalization group.

Meanwhile, let us press on. Using (18) to determine C and plugging it into V_{eff} we obtain

$$V_{\text{eff}}(\varphi) = \frac{1}{4!}\lambda(M)\varphi^4 + \frac{\lambda(M)^2}{(16\pi)^2}\varphi^4 \left(\log \frac{\varphi^2}{M^2} - \frac{25}{6} \right) + O[\lambda(M)^3] \tag{20}$$

You are no longer surprised, I suppose, that C and the cutoff Λ have both disappeared. That's a renormalizable theory for you!

The fact that V_{eff} does not depend on the arbitrarily chosen M, namely, $M(dV_{\text{eff}}/dM) = 0$, reproduces (19) to the order indicated.

Breaking by quantum fluctuations

Now we can answer the motivating question: To break or not to break?

Quantum fluctuations generate a correction to the potential of the form $+\varphi^4 \log \varphi^2$, but $\log \varphi^2$ is whopping big and negative for small φ! The $O(\hbar)$ correction overwhelms the classical $O(\hbar^0)$ potential $+\varphi^4$ near $\varphi = 0$. Quantum fluctuations break the discrete symmetry $\varphi \to -\varphi$.

It is easy enough to determine the minima $\pm\varphi_{\text{min}}$ of $V_{\text{eff}}(\varphi)$ (which you should plot as a function of φ to get a feeling for). But closer inspection shows us that we cannot take the precise value of φ_{min} seriously; V_{eff} has the form $\lambda\varphi^4(1 + \lambda \log \varphi + \cdots)$ suggesting that the expansion parameter is actually $\lambda \log \varphi$ rather than λ. [Try to convince yourself that $(\cdots)$ starts with $(\lambda \log \varphi)^2$.] The minima φ_{min} of V_{eff} clearly occurs when the expansion parameter is of order unity. In an exercise in chapter IV.7 you will see a clever way of getting around this problem.

Fermions

In (11) φ_s plays the role of an external field while $\tilde{\varphi}$ corresponds to a quantum field we integrate over. The role of $\tilde{\varphi}$ can also be played by a fermion field ψ. Consider adding $\bar{\psi}(i\slashed{\partial} - m - f\varphi)\psi$ to the Lagrangian. In the path integral

$$Z = \int D\varphi\, D\bar{\psi}\, D\psi\, e^{i\int d^4x[\frac{1}{2}(\partial\varphi)^2 - V(\varphi) + \bar{\psi}(i\slashed{\partial} - m - f\varphi)\psi]} \tag{21}$$

we can always choose to integrate over ψ first, obtaining

$$Z = \int D\varphi\, e^{i\int d^4x[\frac{1}{2}(\partial\varphi)^2 - V(\varphi)] + \text{tr}\log(i\slashed{\partial} - m - f\varphi)} \tag{22}$$

Repeating the steps in (13) we find that the fermion field contributes

$$V_F(\varphi) = +i\int \frac{d^4p}{(2\pi)^4}\, \text{tr}\log \frac{\slashed{p} - m - f\varphi}{\slashed{p}} \tag{23}$$

to $V_{\text{eff}}(\varphi)$. (The trace in (23) is taken over the gamma matrices.) Again from chapter II.5, we see that physically $V_F(\varphi)$ represents the vacuum energy of a fermion with the effective mass $m(\varphi) \equiv m + f\varphi$.

We can massage the trace of the logarithm using $\text{tr}\log M = \log\det M$ (II.5.12) and cyclically permuting factors in a determinant):

$$\text{tr}\log(\slashed{p} - a) = \text{tr}\log \gamma^5(\slashed{p} - a)\gamma^5 = \text{tr}\log(-\slashed{p} - a)$$
$$= \tfrac{1}{2}\text{tr}(\log(\slashed{p} - a) + \log(\slashed{p} + a)) + \tfrac{1}{2}\text{tr}\log(-1)$$
$$= \tfrac{1}{2}\text{tr}\log(-1)(p^2 - a^2). \tag{24}$$

Hence,

$$\text{tr}\log \frac{(\slashed{p} - a)}{\slashed{p}} = \frac{1}{2}\text{tr}\log \frac{p^2 - a^2}{p^2} = 2\log \frac{p^2 - a^2}{p^2} \tag{25}$$

and so

$$V_F(\varphi) = 2i\int \frac{d^4p}{(2\pi)^4}\, \log \frac{p^2 - m(\varphi)^2}{p^2} \tag{26}$$

Contrast the overall sign with the sign in (14): the difference in sign between fermionic and bosonic loops was explained in chapter II.5.

Thus, in the end the effective potential generated by the quantum fluctuations has a pleasing interpretation: It is just the energy density due to the fluctuating energy, entirely analogous to the zero point energy of the harmonic oscillator, of quantum fields living in the background φ (see exercise IV.3.5).

Exercises

IV.3.1   Consider the effective potential in $(0+1)$-dimensional spacetime:

$$V_{\text{eff}}(\varphi) = V(\varphi) + \frac{\hbar}{2}\int \frac{dk_E}{(2\pi)}\, \log \frac{k_E^2 + V''(\varphi)}{k_E^2} + O(\hbar^2)$$

No counterterm is needed since the integral is perfectly convergent. But $(0 + 1)$-dimensional field theory is just quantum mechanics. Evaluate the integral and show that V_{eff} is in complete accord with your knowledge of quantum mechanics.

IV.3.2 Study V_{eff} in $(1 + 1)$−dimensional spacetime.

IV.3.3 Consider a massless fermion field ψ coupled to a scalar field φ by $f\varphi\bar{\psi}\psi$ in $(1 + 1)$-dimensional spacetime. Show that

$$V_F = \frac{1}{2\pi}(f\varphi)^2 \log \frac{\varphi^2}{M^2} \tag{27}$$

after a suitable counterterm has been added. This result is important in condensed matter physics, as we will see in chapter V.5 on the Peierls instability.

IV.3.4 Understand (14) using Feynman diagrams. Show that V_{eff} is generated by an infinite number of diagrams. [Hint: Expand the logarithm in (14) as a series in $V''(\varphi)/k^2$ and try to associate a Feynman diagram with each term in the series.]

IV.3.5 Consider the electrodynamics of a complex scalar field

$$\mathcal{L} = -\tfrac{1}{4}F_{\mu\nu}F^{\mu\nu} + \left[(\partial^\mu + ieA^\mu)\varphi^\dagger\right]\left[(\partial_\mu - ieA_\mu)\varphi\right]$$
$$+ \mu^2\varphi^\dagger\varphi - \lambda(\varphi^\dagger\varphi)^2 \tag{28}$$

In a universe suffused with the scalar field $\varphi(x)$ taking on the value φ independent of x as in the text, the Lagrangian will contain a term $(e^2\varphi^\dagger\varphi)A_\mu A^\mu$ so that the effective mass squared of the photon field becomes $M(\varphi)^2 \equiv e^2\varphi^\dagger\varphi$. Show that its contribution to $V_{\text{eff}}(\varphi)$ has the form

$$\int \frac{d^4k}{(2\pi)^4} \log \frac{k^2 - M(\varphi)^2}{k^2} \tag{29}$$

Compare with (14) and (26). [Hint: Use the Landau gauge to simplify the calculation.] If you need help, I strongly urge you to read S. Coleman and E. Weinberg, *Phys. Rev.* D7: 1883, 1973, a paragon of clarity in exposition.

IV.4 | Magnetic Monopole

Quantum mechanics and magnetic monopoles

Curiously enough, while electric charges are commonplace nobody has ever seen a magnetic charge or monopole. Within classical physics we can perfectly well modify one of Maxwell's equations to $\vec{\nabla} \cdot \vec{B} = \rho_M$, with ρ_M denoting the density of magnetic monopoles. The only price we have to pay is that the magnetic field $\vec{B}$ can no longer be represented as $\vec{B} = \vec{\nabla} \times \vec{A}$ since otherwise $\vec{\nabla} \cdot \vec{B} = \vec{\nabla} \cdot \vec{\nabla} \times \vec{A} = \varepsilon_{ijk}\partial_i\partial_j A_k = 0$ identically. Newton and Leibniz told us that derivatives commute with each other.

So what, you say. Indeed, who cares that $\vec{B}$ cannot be written as $\vec{\nabla} \times \vec{A}$? The vector potential $\vec{A}$ was introduced into physics only as a mathematical crutch, and indeed that is still how students are often taught in a course on classical electromagnetism. As the distinguished nineteenth-century physicist Heaviside thundered, "Physics should be purged of such rubbish as the scalar and vector potentials; only the fields $\vec{E}$ and $\vec{B}$ are physical."

With the advent of quantum mechanics, however, Heaviside was proved to be quite wrong. Recall, for example, the nonrelativistic Schrödinger equation for a charged particle in an electromagnetic field:

$$\left[-\frac{1}{2m}(\vec{\nabla} - ie\vec{A})^2 + e\phi \right] \psi = E\psi \tag{1}$$

Charged particles couple directly to the vector and scalar potentials $\vec{A}$ and ϕ, which are thus seen as being more fundamental, in some sense, than the electromagnetic fields $\vec{E}$ and $\vec{B}$, as I alluded to in chapter III.4. Quantum physics demands the vector potential.

Dirac noted brilliantly that these remarks imply an intrinsic conflict between quantum mechanics and the concept of magnetic monopoles. Upon closer analysis, he found that quantum mechanics does not actually forbid the existence of magnetic monopoles. It allows magnetic monopoles, but only those carrying a specific amount of magnetic charge.

Differential forms

For the following discussion and for the next chapter on Yang-Mills theory, it is highly convenient to use the language of differential forms. Fear not, we will need only a few elementary concepts. Let x^μ be D real variables (thus, the index μ takes on D values) and A_μ (not necessarily the electromagnetic gauge potential in this purely mathematical section) be D functions of the x's. In our applications, x^μ represent coordinates and, as we will see, differential forms have natural geometric interpretations.

We call the object $A \equiv A_\mu dx^\mu$ a 1-form. The differentials dx^μ are treated following Newton and Leibniz. If we change coordinates $x \to x'$, then as usual $dx^\mu = (\partial x^\mu/\partial x'^\nu)dx'^\nu$ so that $A \equiv A_\mu dx^\mu = A_\mu(\partial x^\mu/\partial x'^\nu)dx'^\nu \equiv A'_\nu dx'^\nu$. This reproduces the standard transformation law of vectors under coordinate transformation $A'_\nu = A_\mu(\partial x^\mu/\partial x'^\nu)$. As an example, consider $A = \cos\theta \, d\varphi$. Regarding θ and φ as angular coordinates on a 2-sphere (namely the surface of a 3-ball), we have $A_\theta = 0$ and $A_\varphi = \cos\theta$. Similarly, we define a p-form as $H = (1/p!)H_{\mu_1\mu_2\cdots\mu_p}dx^{\mu_1}dx^{\mu_2}\cdots dx^{\mu_p}$. (Repeated indices are summed, as always.) The "degenerate" example is that of a 0-form, call it Λ, which is just a scalar function of the coordinates x^μ. An example of a 2-form is $F = (1/2!)F_{\mu\nu}dx^\mu dx^\nu$.

We now face the question of how to think about products of differentials. In an elementary course on calculus we learned that $dx\, dy$ represents the area of an infinitesimal rectangle with length dx and width dy. At that level, we more or less automatically regard $dy\, dx$ as the same as $dx\, dy$. The order of writing the differentials does not matter. However, think about making a coordinate transformation so that $x = x(x', y')$ and $y = y(x', y')$ are now functions of the new coordinates x' and y'. Now look at

$$dx\, dy = \left(\frac{\partial x}{\partial x'}dx' + \frac{\partial x}{\partial y'}dy'\right)\left(\frac{\partial y}{\partial x'}dx' + \frac{\partial y}{\partial y'}dy'\right) \tag{2}$$

Note that the coefficient of $dx'dy'$ is $(\partial x/\partial x')(\partial y/\partial y')$ and that the coefficient of $dy'dx'$ is $(\partial x/\partial y')(\partial y/\partial x')$. We see that it is much better if we regard the differentials dx^μ as anticommuting objects [what mathematicians would call Grassmann variables (recall chapter (II.5)] so that $dy'dx' = -dx'dy'$ and $dx'dx' = 0 = dy'dy'$. Then (2) simplifies neatly to

$$dx\, dy = \left(\frac{\partial x}{\partial x'}\frac{\partial y}{\partial y'} - \frac{\partial x}{\partial y'}\frac{\partial y}{\partial x'}\right)dx'dy' \equiv J(x, y; x', y')dx'dy' \tag{3}$$

We obtain the correct Jacobian $J(x, y; x', y')$ for transforming the area element $dx\, dy$ to the area element $dx'dy'$.

In many texts, $dx\, dy$ is written as $dx \wedge dy$. We will omit the wedge—no reason to clutter up the page.

This little exercise tells us that we should define $dx^\mu dx^\nu = -dx^\nu dx^\mu$ and regard the area element $dx^\mu dx^\nu$ as directional. The area elements $dx^\mu dx^\nu$ and $dx^\nu dx^\mu$ have the same magnitude but point in opposite directions.

We now define a differential operation d to act on any form. Acting on a p-form H, it gives by definition

$$dH = \frac{1}{p!} \partial_\nu H_{\mu_1\mu_2\cdots\mu_p} dx^\nu dx^{\mu_1} dx^{\mu_2} \cdots dx^{\mu_p}$$

Thus, $d\Lambda = \partial_\nu \Lambda dx^\nu$ and

$$dA = \partial_\nu A_\mu dx^\nu dx^\mu = \tfrac{1}{2}(\partial_\nu A_\mu - \partial_\mu A_\nu) dx^\nu dx^\mu$$

In the last step, we used $dx^\mu dx^\nu = -dx^\nu dx^\mu$.

We see that this mathematical formalism is almost tailor made to describe electromagnetism. If we call $A \equiv A_\mu dx^\mu$ the potential 1-form and think of A_μ as the electromagnetic potential, then $F = dA$ is in fact the field 2-form. If we write F out in terms of its components $F = (1/2!) F_{\mu\nu} dx^\mu dx^\nu$, then $F_{\mu\nu}$ is indeed equal to the electromagnetic field $(\partial_\mu A_\nu - \partial_\nu A_\mu)$.

Note that x^μ is not a form, and dx^μ is not d acting on a form.

If you like, you can think of differential forms as "merely" an elegantly compact notation. The point is to think of physical objects such as A and F as entities, without having to commit to any particular coordinate system. This is particularly convenient when one has to deal with objects more complicated than A and F, for example in string theory. By using differential forms, we avoid drowning in a sea of indices.

An important identity is

$$dd = 0 \tag{4}$$

which says that acting with d on any form twice gives zero. Verify this as an exercise. In particular $dF = ddA = 0$. If you write this out in components you will recognize it as a standard identity (the "Bianchi identity") in electromagnetism.

Closed is not necessarily globally exact

It is convenient here to introduce some jargon. A p-form α is said to be closed if $d\alpha = 0$. It is said to be exact if there exists a $(p-1)$-form β such that $\alpha = d\beta$.

Talking the talk, we say that (4) tells us that exact forms are closed.

Is the converse of (4) true? Kind of. The Poincaré lemma states that a closed form is locally exact. In other words, if $dH = 0$ with H some p-form, then locally

$$H = dK \tag{5}$$

for some $(p-1)$-form K. However, it may or may not be the case that $H = dK$ globally, that is, everywhere. Actually, whether you know it or not, you are already familiar with the Poincaré lemma. For example, surely you learned somewhere that if the curl of a vector field vanishes, the vector field is locally the gradient of some scalar field.

Forms are ready made to be integrated over. For example, given the 2-form $F = (1/2!) F_{\mu\nu} dx^\mu dx^\nu$, we can write $\int_M F$ for any 2-manifold M. Note that the measure is already included and there is no need to specify a coordinate choice. Again, whether you know it or not, you are already familiar with the important theorem

$$\int_M dH = \int_{\partial M} H \tag{6}$$

with H a p-form and ∂M the boundary of a $(p + 1)$-dimensional manifold M.

Dirac quantization of magnetic charge

After this dose of mathematics, we are ready to do some physics. Consider a sphere surrounding a magnetic monopole with magnetic charge g. Then the electromagnetic field 2-form is given by $F = (g/4\pi)d\cos\theta \, d\varphi$. This is almost a definition of what we mean by a magnetic monopole (see exercise IV.4.3.) In particular, calculate the magnetic flux by integrating F over the sphere S^2

$$\int_{S^2} F = g \tag{7}$$

As I have already noted, the area element is automatically included. Indeed, you might have recognized $d\cos\theta \, d\varphi = -\sin\theta \, d\theta \, d\varphi$ as precisely the area element on a unit sphere. Note that in "ordinary notation" (7) implies the magnetic field $\vec{B} = (g/4\pi r^2)\hat{r}$, with $\hat{r}$ the unit vector in the radial direction.

I will now give a rather mathematical, but rigorous, derivation, originally developed by Wu and Yang, of Dirac's quantization of the magnetic charge g.

First, let us recall how gauge invariance works, from, for example, (II.7.3). Under a transformation of the electron field $\psi(x) \rightarrow e^{i\Lambda(x)}\psi(x)$, the electromagnetic gauge potential changes by

$$A_\mu(x) \rightarrow A_\mu(x) + \frac{1}{ie}e^{-i\Lambda(x)}\partial_\mu e^{i\Lambda(x)}$$

or in the language of forms,

$$A \rightarrow A + \frac{1}{ie}e^{-i\Lambda}de^{i\Lambda} \tag{8}$$

Differentiating, we can of course write

$$A_\mu(x) \rightarrow A_\mu(x) + \frac{1}{e}\partial_\mu\Lambda(x)$$

as is commonly done. The form given in (8) reminds us that gauge transformation is defined as multiplication by a phase factor $e^{i\Lambda(x)}$, so that $\Lambda(x)$ and $\Lambda(x) + 2\pi$ describe exactly the same transformation.

In quantum mechanics A is physical, pace Heaviside, and so we should ask what A would give rise to $F = (g/4\pi) \, d\cos\theta \, d\varphi$. Easy, you say; clearly $A = (g/4\pi)\cos\theta \, d\varphi$. (In checking this by calculating dA, remember that $dd = 0$.)

But not so fast; your mathematician friend says that $d\varphi$ is not defined at the north and south poles. Put his objection into everyday language: If you are standing on the north pole, what is your longitude? So strictly speaking it is forbidden to write $A = (g/4\pi)\cos\theta \, d\varphi$.

But, you are smart enough to counter, then what about $A_N = (g/4\pi)(\cos\theta - 1)\,d\varphi$, eh? When you act with d on A_N you obtain the desired F; the added piece $(g/4\pi)(-1)\,d\varphi$ gets annihilated by d thanks once again to the identity (4). At the north pole, $\cos\theta = 1$, A_N vanishes, and is thus perfectly well defined.

OK, but your mathematician friend points out that your A_N is not defined at the south pole, where it is equal to $(g/4\pi)(-2)d\varphi$.

Right, you respond, I anticipated that by adding the subscript N. I am now also forced to define $A_S = (g/4\pi)(\cos\theta + 1)\,d\varphi$. Note that d acting on A_S again gives the desired F. But now A_S is defined everywhere except at the north pole.

In mathematical jargon, we say that the gauge potential A is defined locally, but not globally. The gauge potential A_N is defined on a "coordinate patch" covering the northern hemisphere and extending past the equator as far south as we want as long as we do not include the south pole. Similarly, A_S is defined on a "coordinate patch" covering the southern hemisphere and extending past the equator as far north as we want as long as we do not include the north pole.

But what happens where the two coordinate patches overlap, for example, along the equator. The gauge potentials A_N and A_S are not the same:

$$A_S - A_N = 2\frac{g}{4\pi}d\varphi \tag{9}$$

Now what? Aha, but this is a gauge theory: If A_S and A_N are related by a gauge transformation, then all is well. Thus, referring to (8) we require that $2(g/4\pi)\,d\varphi = (1/ie)e^{-i\Lambda}de^{i\Lambda}$ for some phase function $e^{i\Lambda}$. By inspection we have $e^{i\Lambda} = e^{i2(eg/4\pi)\varphi}$.

But $\varphi = 0$ and $\varphi = 2\pi$ describe exactly the same point. In order for $e^{i\Lambda}$ to make sense, we must have $e^{i2(eg/4\pi)(2\pi)} = e^{i2(eg/4\pi)(0)} = 1$; in other words, $e^{ieg} = 1$, or

$$g = \frac{2\pi}{e}n \tag{10}$$

where n denotes an integer. This is Dirac's famous discovery that the magnetic charge on a magnetic monopole is quantized in units of $2\pi/e$. A "dual" way of putting this is that if the monopole exists then electric charge is quantized in units of $2\pi/g$.

Note that the whole point is that F is locally but not globally exact; otherwise by (6) the magnetic charge $g = \int_{S^2} F$ would be zero.

I show you this rigorous mathematical derivation partly to cut through a lot of the confusion typical of the derivations in elementary texts and partly because this type of argument is used repeatedly in more advanced areas of physics, such as string theory.

Electromagnetic duality

That a duality may exist between electric and magnetic fields has tantalized theoretical physicists for a century and a half. By the way, if you read Maxwell, you will discover that he often talked about magnetic charges. You can check that Maxwell's equations are invariant under the elegant transformation $(\vec{E} + i\vec{B}) \to e^{i\theta}(\vec{E} + i\vec{B})$ if magnetic charges exist.

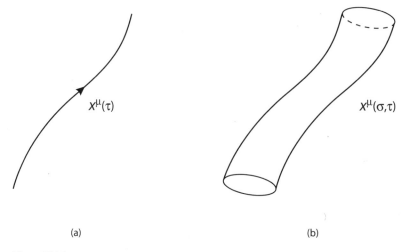

(a)   (b)

Figure IV.4.1

One intriguing feature of (10) is that if e is small, then g is large, and vice versa. What would magnetic charges look like if they exist? They wouldn't look any different from electric charges: They too interact with a $1/r$ potential, with likes repelling and opposites attracting. In principle, we could have perfectly formulated electromagnetism in terms of magnetic charges, with magnetic and electric fields exchanging their roles, but the theory would be strongly coupled, with the coupling g rather than e.

Theoretical physicists are interested in duality because it allows them a glimpse into field theories in the strongly coupled regime. Under duality, a weakly coupled field theory is mapped into a strongly coupled field theory. This is exactly the reason why the discovery some years ago that certain string theories are dual to others caused such enormous excitement in the string theory community: We get to know how string theories behave in the strongly coupled regime. More on duality in chapter VI.3.

Forms and geometry

The geometric character of differential forms is further clarified by thinking about the electromagnetic current of a charged particle tracing out the world line $X^\mu(\tau)$ in D-dimensional spacetime (see figure IV.4.1a):

$$J^\mu(x) = \int d\tau \frac{dX^\mu}{d\tau} \delta^{(D)}[x - X(\tau)] \tag{11}$$

The interpretation of this elementary formula from electromagnetism is clear: $dX^\mu/d\tau$ is the 4-velocity at a given value of the parameter τ ("proper time") and the delta function ensures that the current at x vanishes unless the particle passes through x. Note that $J^\mu(x)$ is invariant under the reparametrization $\tau \to \tau'(\tau)$.

The generalization to an extended object is more or less obvious. Consider a string. It traces out a world sheet $X^\mu(\tau, \sigma)$ in spacetime (see figure 1b), where σ is a parameter telling us where we are along the length of the string. [For example, for a closed string, σ is conventionally taken to range between 0 and 2π with $X^\mu(\tau, 0) = X^\mu(\tau, 2\pi)$.] The current associated with the string is evidently given by

$$J^{\mu\nu}(x) = \int d\tau d\sigma \, \det \begin{pmatrix} \partial_\tau X^\mu & \partial_\tau X^\nu \\ \partial_\sigma X^\mu & \partial_\sigma X^\nu \end{pmatrix} \delta^{(D)}[x - X(\tau, \sigma)] \tag{12}$$

where $\partial_\tau \equiv \partial/\partial\tau$ and so forth. The determinant is forced on us by the requirement of invariance under reparametrization $\tau \to \tau'(\tau, \sigma), \sigma \to \sigma'(\tau, \sigma)$. It follows that $J^{\mu\nu}$ is an antisymmetric tensor. Hence, the analog of the electromagnetic potential A_μ coupling to the current J^μ is an antisymmetric tensor field $B_{\mu\nu}$ coupling to the current $J^{\mu\nu}$. Thus, string theory contains a 2-form potential $B = \frac{1}{2} B_{\mu\nu} dx^\mu dx^\nu$ and the corresponding 3-form field $H = dB$. In fact, string theory typically contains numerous p-forms.

Aharonov-Bohm effect

The reality of the gauge potential A was brought home forcefully in 1959 by Aharonov and Bohm. Consider a magnetic field B confined to a region Ω as illustrated in figure IV.4.2. The quantum physics of an electron is described by solving the Schrödinger equation (1). In Feynman's path integral formalism the amplitude associated with a path P is modified by a multiplicative factor $e^{ie \int_P \vec{A} \cdot d\vec{x}}$, where the line integral is evaluated along the path P. Thus, in the path integral calculation of the probability for an electron to propagate from a to b (fig. IV.4.2), there will be interference between the contributions from path 1 and path 2 of the form

$$\left(e^{ie \int_{P_1} \vec{A} \cdot d\vec{x}}\right)\left(e^{ie \int_{P_2} \vec{A} \cdot d\vec{x}}\right)^* = \left(e^{ie \oint \vec{A} \cdot d\vec{x}}\right)$$

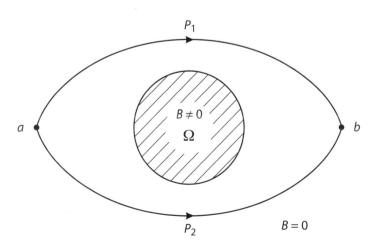

Figure IV.4.2

but $\oint \vec{A} \cdot d\vec{x} = \int \vec{B} \cdot d\vec{S}$ is precisely the flux enclosed by the closed curve $(P_1 - P_2)$, namely the curve going from a to b along P_1 and then returning from b to a along $(-P_2)$ since complex conjugation in effect reverses the direction of the path P_2. Remarkably, the electron feels the effect of the magnetic field even though it never wanders into a region with a magnetic field present.

When the Aharonov-Bohm paper was first published, no less an authority than Niels Bohr was deeply disturbed. The effect has since been conclusively demonstrated in a series of beautiful experiments by Tonomura and collaborators.

Coleman once told of a gedanken prank that connects the Aharonov-Bohm effect to Dirac quantization of magnetic charge. Let us fabricate an extremely thin solenoid so that it is essentially invisible and thread it into the lab of an unsuspecting experimentalist, perhaps our friend from chapter III.1. We turn on a current and generate a magnetic field through the solenoid. When the experimentalist suddenly sees the magnetic flux coming out of apparently nowhere, she gets so excited that she starts planning to go to Stockholm.

What is the condition that prevents the experimentalist from discovering the prank? A careful experimentalist might start scattering electrons around to see if she can detect a solenoid. The condition that she does not see an Aharonov-Bohm effect and thus does not discover the prank is precisely that the flux going through the solenoid is an integer times $2\pi/e$. This implies that the apparent magnetic monopole has precisely the magnetic charge predicted by Dirac!

Exercises

IV.4.1   Prove $dd = 0$.

IV.4.2   Show by writing out the components explicitly that $dF = 0$ expresses something that you are familiar with but disguised in a compact notation.

IV.4.3   Consider $F = (g/4\pi)\, d\cos\theta\, d\varphi$. By transforming to Cartesian coordinates show that this describes a magnetic field pointing outward along the radial direction.

IV.4.4   Restore the factors of $\hbar$ and c in Dirac's quantization condition.

IV.4.5   Write down the reparametrization-invariant current $J^{\mu\nu\lambda}$ of a membrane.

IV.4.6   Let $g(x)$ be the element of a group G. The 1-form $v = g\,dg^{\dagger}$ is known as the Cartan-Maurer form. Then tr v^N is trivially closed on an N-dimensional manifold since it is already an N-form. Consider $Q = \int_{S^N} \mathrm{tr}\, v^N$ with S^N the N-dimensional sphere. Discuss the topological meaning of Q. These considerations will become important later when we discuss topology in field theory in chapter V.7. [Hint: Study the case $N = 3$ and $G = SU(2)$.]

IV.5 | Nonabelian Gauge Theory

> Most such ideas are eventually discarded or shelved. But some persist and may become obsessions. Occasionally an obsession does finally turn out to be something good.
>
> —C. N. Yang talking about an idea that he first had as a student and that he kept coming back to year after year.[1]

Local transformation

It was quite a nice little idea.

To explain the idea Yang was talking about, recall our discussion of symmetry in chapter I.10. For the sake of definiteness let $\varphi(x) = \{\varphi_1(x), \varphi_2(x), \cdots, \varphi_N(x)\}$ be an N-component complex scalar field transforming as $\varphi(x) \to U\varphi(x)$, with U an element of $SU(N)$. Since $\varphi^\dagger \to \varphi^\dagger U^\dagger$ and $U^\dagger U = 1$, we have $\varphi^\dagger \varphi \to \varphi^\dagger \varphi$ and $\partial \varphi^\dagger \partial \varphi \to \partial \varphi^\dagger \partial \varphi$. The invariance of the Lagrangian $\mathcal{L} = \partial \varphi^\dagger \partial \varphi - V(\varphi^\dagger \varphi)$ under $SU(N)$ is obvious for any polynomial V.

In the theoretical physics community there are many more people who can answer well-posed questions than there are people who can pose the truly important questions. The latter type of physicist can invariably also do much of what the former type can do, but the reverse is certainly not true.

In 1954 C.N. Yang and R. Mills asked what will happen if the transformation varies from place to place in spacetime, or in other words, if $U = U(x)$ is a function of x.

Clearly, $\varphi^\dagger \varphi$ is still invariant. But in contrast $\partial \varphi^\dagger \partial \varphi$ is no longer invariant. Indeed,

$$\partial_\mu \varphi \to \partial_\mu (U\varphi) = U\partial_\mu \varphi + (\partial_\mu U)\varphi = U[\partial_\mu \varphi + (U^\dagger \partial_\mu U)\varphi]$$

To cancel the unwanted term $(U^\dagger \partial_\mu U)\varphi$, we generalize the ordinary derivative ∂_μ to a covariant derivative D_μ, which when acting on φ, gives

$$D_\mu \varphi(x) = \partial_\mu \varphi(x) - iA_\mu(x)\varphi(x) \tag{1}$$

The field A_μ is called a gauge potential in direct analogy with electromagnetism.

---

[1] C. N. Yang, *Selected Papers 1945–1980 with Commentary*, p. 19.

How must A_μ transform, so that $D_\mu \varphi(x) \to U(x) D_\mu \varphi(x)$? In other words, we would like $D_\mu \varphi(x)$ to transform the way $\partial_\mu \varphi(x)$ transformed when U did not depend on x. If so, then $[D_\mu \varphi(x)]^\dagger D_\mu \varphi(x) \to [D_\mu \varphi(x)]^\dagger D_\mu \varphi(x)$ and can be used as an invariant kinetic energy term for the field φ.

Working backward, we see that $D_\mu \varphi(x) \to U(x) D_\mu \varphi(x)$ if (and it goes without saying that you should be checking this)

$$A_\mu \to U A_\mu U^\dagger - i(\partial_\mu U) U^\dagger = U A_\mu U^\dagger + i U \partial_\mu U^\dagger \tag{2}$$

(The equality follows from $U U^\dagger = 1$.) We refer to A_μ as the nonabelian gauge potential and to (2) as a nonabelian gauge transformation.

Let us now make a series of simple observations.

1. Clearly, A_μ have to be N by N matrices. Work out the transformation law for $A_\mu^\dagger$ using (2) and show that the condition $A_\mu - A_\mu^\dagger = 0$ is preserved by the gauge transformation. Thus, it is consistent to take A_μ to be hermitean. Specifically, you should work out what this means for the group $SU(2)$ so that $U = e^{i\theta \cdot \tau/2}$ where $\theta \cdot \tau = \theta^a \tau^a$, with τ^a the familiar Pauli matrices.

2. Writing $U = e^{i\theta \cdot T}$ with T^a the generators of $SU(N)$, we have

$$A_\mu \to A_\mu + i\theta^a [T^a, A_\mu] + \partial_\mu \theta^a T^a \tag{3}$$

under an infinitesimal transformation $U \simeq 1 + i\theta \cdot T$. For most purposes, the infinitesimal form (3) suffices.

3. Taking the trace of (3) we see that the trace of A_μ does not transform and so we can take A_μ to be traceless as well as hermitean. This means that we can always write $A_\mu = A_\mu^a T^a$ and thus decompose the matrix field A_μ into component fields A_μ^a. There are as many A_μ^a's as there are generators in the group [3 for $SU(2)$, 8 for $SU(3)$, and so forth.]

4. You are reminded in appendix B that the Lie algebra of the group is defined by $[T^a, T^b] = if^{abc} T^c$, where the numbers f^{abc} are called structure constants. For example, $f^{abc} = \varepsilon^{abc}$ for $SU(2)$. Thus, (3) can be written as

$$A_\mu^a \to A_\mu^a - f^{abc} \theta^b A_\mu^c + \partial_\mu \theta^a \tag{4}$$

Note that if θ does not depend on x, the A_μ^a's transform as the adjoint representation of the group.

5. If $U(x) = e^{i\theta(x)}$ is just an element of the abelian group $U(1)$, all these expressions simplify and A_μ is just the abelian gauge potential familiar from electromagnetism, with (2) the usual abelian gauge transformation. Hence, A_μ is known as the nonabelian gauge potential.

A transformation U that depends on the spacetime coordinates x is known as a gauge transformation or local transformation. A Lagrangian $\mathcal{L}$ invariant under a gauge transformation is said to be gauge invariant.

Construction of the field strength

We can now immediately write a gauge invariant Lagrangian, namely

$$\mathcal{L} = (D_\mu \varphi)^\dagger (D_\mu \varphi) - V(\varphi^\dagger \varphi) \tag{5}$$

but the gauge potential A_μ does not yet have dynamics of its own. In the familiar example of $U(1)$ gauge invariance, we have written the coupling of the electromagnetic potential A_μ to the matter field φ, but we have yet to write the Maxwell term $-\frac{1}{4} F_{\mu\nu} F^{\mu\nu}$ in the Lagrangian. Our first task is to construct a field strength $F_{\mu\nu}$ out of A_μ. How do we do that? Yang and Mills apparently did it by trial and error. As an exercise you might also want to try that before reading on.

At this point the language of differential forms introduced in chapter IV.4 proves to be of use. It is convenient to absorb a factor of $-i$ by defining $A_\mu^M \equiv -i A_\mu^P$, where A_μ^P denotes the gauge potential we have been using all along. Until further notice, when we write A_μ we mean A_μ^M. Referring to (1) we see that the covariant derivative has the cleaner form $D_\mu = \partial_\mu + A_\mu$. (Incidentally, the superscripts M and P indicate the potential appearing in the mathematical and physical literature, respectively.) As before, let us introduce $A = A_\mu dx^\mu$, now a matrix 1-form, that is, a form that also happens to be a matrix in the defining representation of the Lie algebra [e.g., an N by N traceless hermitean matrix for $SU(N)$.] Note that

$$A^2 = A_\mu A_\nu dx^\mu dx^\nu = \tfrac{1}{2}[A_\mu, A_\nu] dx^\mu dx^\nu$$

is not zero for a nonabelian gauge potential. (Obviously, there is no such object in electromagnetism.)

Our task is to construct a 2-form $F = \frac{1}{2} F_{\mu\nu} dx^\mu dx^\nu$ out of the 1-form A. We adopt a direct approach. Out of A we can construct only two possible 2-forms: dA and A^2. So F must be a linear combination of the two.

In the notation we are using the transformation law (2) reads

$$A \to UAU^\dagger + U dU^\dagger \tag{6}$$

with U a 0-form (and so $dU^\dagger = \partial_\mu U^\dagger dx^\mu$.) Applying d to (6) we have

$$dA \to U dAU^\dagger + dU AU^\dagger - U A dU^\dagger + dU dU^\dagger \tag{7}$$

Note the minus sign in the third term, from moving the 1-form d past the 1-form A. On the other hand, squaring (6) we have

$$A^2 \to U A^2 U^\dagger + U A dU^\dagger + U dU^\dagger U AU^\dagger + U dU^\dagger U dU^\dagger \tag{8}$$

Applying d to $UU^\dagger = 1$ we have $U dU^\dagger = -dU U^\dagger$. Thus, we can rewrite (8) as

$$A^2 \to U A^2 U^\dagger + U A dU^\dagger - dU AU^\dagger - dU dU^\dagger \tag{9}$$

Lo and behold! If we add (7) and (9), six terms knock each other off, leaving us with something nice and clean:

$$dA + A^2 \rightarrow U(dA + A^2)U^\dagger \tag{10}$$

The mathematical structure thus led Yang and Mills to define the field strength

$$F = dA + A^2 \tag{11}$$

Unlike A, the field strength 2-form F transforms homogeneously (10):

$$F \rightarrow UFU^\dagger \tag{12}$$

In the abelian case A^2 vanishes and F reduces to the usual electromagnetic form. In the nonabelian case, F is not gauge invariant, but gauge covariant.

Of course, you can also construct $F^a_{\mu\nu}$ without using differential forms. As an exercise you should do it starting with (4). The exercise will make you appreciate differential forms! At the very least, we can regard differential forms as an elegantly compact notation that suppresses the indices a and μ in (4). At the same time, the fact that (11) emerges so smoothly clearly indicates a profound underlying mathematical structure. Indeed, there is a one-to-one translation between the physicist's language of gauge theory and the mathematician's language of fiber bundles.

Let me show you another route to (11). In analogy to d, define $D = d + A$, understood as an operator acting on a form to its right. Let us calculate

$$D^2 = (d + A)(d + A) = d^2 + dA + Ad + A^2$$

The first term vanishes, the second can be written as $dA = (dA) - Ad$; the parenthesis emphasizes that d acts only on A. Thus,

$$D^2 = (dA) + A^2 = F \tag{13}$$

Pretty slick? I leave it as an exercise for you to show that D^2 transforms homogeneously and hence so does F.

Elegant though differential forms are, in physics it is often desirable to write more explicit formulas. We can write (11) out as

$$F = (\partial_\mu A_\nu + A_\mu A_\nu)dx^\mu dx^\nu = \tfrac{1}{2}(\partial_\mu A_\nu - \partial_\nu A_\mu + [A_\mu, A_\nu])dx^\mu dx^\nu \tag{14}$$

With the definition $F \equiv \tfrac{1}{2}F_{\mu\nu}dx^\mu dx^\nu$ we have

$$F_{\mu\nu} = \partial_\mu A_\nu - \partial_\nu A_\mu + [A_\mu, A_\nu] \tag{15}$$

At this point, we might also want to switch back to physicist's notation. Recall that A_μ in (15) is actually $A_\mu^M \equiv -iA_\mu^P$ and so by analogy define $F_{\mu\nu}^M = -iF_{\mu\nu}^P$. Thus,

$$F_{\mu\nu} = \partial_\mu A_\nu - \partial_\nu A_\mu - i[A_\mu, A_\nu] \tag{16}$$

where, until further notice, A_μ stands for A_μ^P. (One way to see the necessity for the i in (16) is to remember that physicists like to take A_μ to be a hermitean matrix and the commutator of two hermitean matrices is antihermitean.)

As long as we are being explicit we might as well go all the way and exhibit the group indices as well as the Lorentz indices. We already wrote $A_\mu = A_\mu^a T^a$ and so we naturally write $F_{\mu\nu} = F_{\mu\nu}^a T^a$. Then (16) becomes

$$F_{\mu\nu}^a = \partial_\mu A_\nu^a - \partial_\nu A_\mu^a + f^{abc} A_\mu^b A_\nu^c \tag{17}$$

I mention in passing that for $SU(2)$ A and F transform as vectors and the structure constant f^{abc} is just ε^{abc}, so the vector notation $\vec{F}_{\mu\nu} = \partial_\mu \vec{A}_\nu - \partial_\nu \vec{A}_\mu + \vec{A}_\mu \times \vec{A}_\nu$ is often used.

The Yang-Mills Lagrangian

Given that F transforms homogeneously (12) we can immediately write down the analog of the Maxwell Lagrangian, namely the Yang-Mills Lagrangian

$$\mathcal{L} = -\frac{1}{2g^2} \text{ tr } F_{\mu\nu} F^{\mu\nu} \tag{18}$$

We are normalizing T^a by tr $T^a T^b = \frac{1}{2}\delta^{ab}$ so that $\mathcal{L} = -(1/4g^2) F_{\mu\nu}^a F^{a\mu\nu}$. The theory described by this Lagrangian is known as pure Yang-Mills theory or nonabelian gauge theory.

Apart from the quadratic term $(\partial_\mu A_\nu^a - \partial_\nu A_\mu^a)^2$, the Lagrangian $\mathcal{L} = -(1/4g^2) F_{\mu\nu}^a F^{a\mu\nu}$ also contains a cubic term $f^{abc} A^{b\mu} A^{c\nu} (\partial_\mu A_\nu^a - \partial_\nu A_\mu^a)$ and a quartic term $(f^{abc} A_\mu^b A_\nu^c)^2$. As in electromagnetism the quadratic term describes the propagation of a massless vector boson carrying an internal index a, known as the nonabelian gauge boson or the Yang-Mills boson. The cubic and quartic terms are not present in electromagnetism and describe the self-interaction of the nonabelian gauge boson. The corresponding Feynman rules are given in figure IV.5.1a, 1b, and c.

The physics behind this self-interaction of the Yang-Mills bosons is not hard to understand. The photon couples to charged fields but is not charged itself. Just as the charge

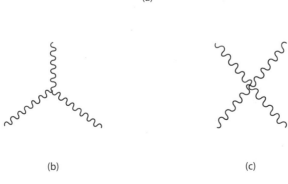

(a)

(b)   (c)

Figure IV.5.1

of a field tells us how the field transforms under the $U(1)$ gauge group, the analog of the charge of a field in a nonabelian gauge theory is the representation the field belongs to. The Yang-Mills bosons couple to all fields transforming nontrivially under the gauge group. But the Yang-Mills bosons themselves transform nontrivially: In fact, as we have noted, they transform under the adjoint representation. Thus, they must couple to themselves.

Pure Maxwell theory is free and so essentially trivial. It contains a noninteracting photon. In contrast, pure Yang-Mills theory contains self-interaction and is highly nontrivial. Note that the structure coefficients f^{abc} are completely fixed by group theory, and thus in contrast to a scalar field theory, the cubic and quartic self-interactions of the gauge bosons, including their relative strengths, are totally fixed by symmetry. If any 4-dimensional field theory can be solved exactly, pure Yang-Mills theory may be it, but in spite of the enormous amount of theoretical work devoted to it, it remains unsolved (see chapters VII.3 and VII.4).

't Hooft's double-line formalism

While it is convenient to use the component fields A^a_μ for many purposes, the matrix field $A_\mu = A^a_\mu T^a$ embodies the mathematical structure of nonabelian gauge theory more elegantly. The propagator for the components of the matrix field in a $U(N)$ gauge theory has the form

$$\langle 0| T A_\mu(x)^i_j A_\nu(0)^k_l |0\rangle$$
$$= \langle 0| T A^a_\mu(x) A^b_\nu(0) |0\rangle (T^a)^i_j (T^b)^k_l \tag{19}$$
$$\propto \delta^{ab} (T^a)^i_j (T^b)^k_l \propto \delta^i_l \delta^k_j$$

[We have gone from an $SU(N)$ to a $U(N)$ theory for the sake of simplicity. The generators of $SU(N)$ satisfy a traceless condition $\mathrm{Tr} T^a = 0$, as a result of which we would have to subtract $\frac{1}{N} \delta^i_j \delta^k_l$ from the right-hand side.] The matrix structure $A^i_{\mu j}$ naturally suggests that we, following 't Hooft, introduce a double-line formalism, in which the gauge potential is described by two lines, each associated with one of the two indices i and j. We choose the convention that the upper index flows into the diagram, while the lower index flows out of the diagram. The propagator in (19) is represented in figure IV.5.2a. The double-line formalism allows us to reproduce the index structure $\delta^i_l \delta^k_j$ naturally. The cubic and quartic couplings are represented in figure IV.5.2b and c.

The constant g introduced in (18) is known as the Yang-Mills coupling constant. We can always write the quadratic term in (18) in the convention commonly used in electromagnetism by a trivial rescaling $A \to gA$. After this rescaling, the cubic and quartic couplings of the Yang-Mills boson go as g and g^2, respectively. The covariant derivative in (1) becomes $D_\mu \varphi = \partial_\mu \varphi - ig A_\mu \varphi$, showing that g also measures the coupling of the Yang-Mills boson to matter. The convention we used, however, brings out the mathematical structure more clearly. As written in (18), g^2 measures the ease with which the Yang-Mills boson can propagate. Recall that in chapter III.7 we also found this way of defining the coupling as

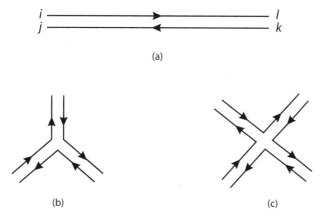

(a)

(b)   (c)

Figure IV.5.2

a measure of propagation useful in electromagnetism. We will see in chapter VIII.1 that Newton's coupling appears in the same way in the Einstein-Hilbert action for gravity.

The θ term

Besides tr $F_{\mu\nu}F^{\mu\nu}$, we can also form the dimension-4 term $\varepsilon^{\mu\nu\lambda\rho}$ tr $F_{\mu\nu}F_{\lambda\rho}$. Clearly, this term violates time reversal invariance T and parity P since it involves one time index and three space indices. We will see later that the strong interaction is described by a nonabelian gauge theory, with the Lagrangian containing the so-called θ term $(\theta/32\pi^2)\varepsilon^{\mu\nu\lambda\rho}$ tr $F_{\mu\nu}F_{\lambda\rho}$. As you will show in exercise IV.5.3 this term is a total divergence and does not contribute to the equation of motion. Nevertheless, it induces an electric dipole moment for the neutron. The experimental upper bound on the electric dipole moment for the neutron translates into an upper bound on θ of the order 10^{-9}. I will not go into how particle physicists resolve the problem of making sure that θ is small enough or vanishes outright.

Coupling to matter fields

We took the scalar field φ to transform in the fundamental representation of the group. In general, φ can transform in an arbitrary representation $\mathcal{R}$ of the gauge group G. We merely have to write the covariant derivative more generally as

$$D_\mu\varphi = (\partial_\mu - iA_\mu^a T_{(\mathcal{R})}^a)\varphi \tag{20}$$

where $T_{(\mathcal{R})}^a$ represents the ath generator in the representation $\mathcal{R}$ (see exercise IV.5.1).

Clearly, the prescription to turn a globally symmetric theory into a locally symmetric theory is to replace the ordinary derivative ∂_μ acting on any field, boson or fermion, belonging to the representation $\mathcal{R}$ by the covariant derivative $D_\mu = (\partial_\mu - i A_\mu^a T_{(\mathcal{R})}^a)$. Thus, the coupling of the nonabelian gauge potential to a fermion field is given by

$$\mathcal{L} = \bar{\psi}(i\gamma^\mu D_\mu - m)\psi = \bar{\psi}(i\gamma^\mu \partial_\mu + \gamma^\mu A_\mu^a T_{(\mathcal{R})}^a - m)\psi. \tag{21}$$

Fields listen to the Yang-Mills gauge bosons according to the representation $\mathcal{R}$ that they belong to, and those that belong to the trivial identity representation do not hear the call of the gauge bosons. In the special case of a $U(1)$ gauge theory, also known as electromagnetism, $\mathcal{R}$ corresponds to the electric charge of the field. Those fields that transform trivially under $U(1)$ are electrically neutral.

Appendix

Let me show you another context, somewhat surprising at first sight, in which the Yang-Mills structure pops up.[2] Consider Schrödinger's equation

$$i\frac{\partial}{\partial t}\Psi(t) = H(t)\Psi(t) \tag{22}$$

with a time dependent Hamiltonian $H(t)$. The setup is completely general: For instance, we could be talking about spin states in a magnetic field or about a single particle nonrelativistic Hamiltonian with the wave function $\Psi(\vec{x}, t)$. We suppress the dependence of H and Ψ on variables other than time t.

First, solve the eigenvalue problem of $H(t)$. Suppose that because of symmetry or some other reason the spectrum of $H(t)$ contains an n-fold degeneracy, in other words, there exist n distinct solutions of the equations $H(t)\psi_a(t) = E(t)\psi_a(t)$, with $a = 1, \cdots, n$. Note that $E(t)$ can vary with time and that we are assuming that the degeneracy persists with time, that is, the degeneracy does not occur "accidentally" at one instant in time. We can always replace $H(t)$ by $H(t) - E(t)$ so that henceforth we have $H(t)\psi_a(t) = 0$. Also, the states can be chosen to be orthogonal so that $\langle\psi_b(t)|\psi_a(t)\rangle = \delta_{ba}$. (For notational reasons it is convenient to jump back and forth between the Schrödinger and the Dirac notation. To make it absolutely clear, we have

$$\langle\psi_b(t)|\psi_a(t)\rangle = \int d\vec{x}\, \psi_b^*(\vec{x}, t)\psi_a(\vec{x}, t)$$

if we are talking about single particle quantum mechanics.)

Let us now study (22) in the adiabatic limit, that is, we assume that the time scale over which $H(t)$ varies is much longer than $1/\Delta E$, where ΔE denotes the energy gap separating the states $\psi_a(t)$ from neighboring states. In that case, if $\Psi(t)$ starts out in the subspace spanned by $\{\psi_a(t)\}$ it will stay in that subspace and we can write $\Psi(t) = \sum_a c_a(t)\psi_a(t)$. Plugging this into (22) we obtain immediately $\sum_a[(dc_a/dt)\psi_a(t) + c_a(t)(\partial\psi_a/\partial t)] = 0$. Taking the scalar product with $\psi_b(t)$, we obtain

$$\frac{dc_b}{dt} = -\sum_a A_{ba}c_a \tag{23}$$

with the n by n matrix

$$A_{ba}(t) \equiv i\langle\psi_b(t)|\frac{\partial\psi_a}{\partial t}\rangle \tag{24}$$

Now suppose somebody else decides to use a different basis, $\psi_a'(t) = U_{ac}^*(t)\psi_c(t)$, related to ours by a unitary transformation. (The complex conjugate on the unitary matrix U is just a notational choice so that our final equation will come out looking the same as a celebrated equation in the text; see below.) I have also passed

---

[2] F. Wilczek and A. Zee, "Appearance of Gauge Structure in Simple Dynamical Systems," *Phys. Rev. Lett.* 52:2111, 1984.

to the repeated indices summed notation. Differentiate to obtain $(\partial \psi'_a/\partial t) = U^*_{ac}(t)(\partial \psi_c/\partial t) + (dU^*_{ac}/dt)\psi_c(t)$. Contracting this with $\psi'^*_b(t) = U_{bd}(t)\psi^*_d(t)$ and multiplying by i, we find

$$A' = UAU^\dagger + iU\frac{\partial U^\dagger}{\partial t} \tag{25}$$

Suppose the Hamiltonian $H(t)$ depends on d parameters $\lambda^1, \cdots, \lambda^d$. We vary the parameters, thus tracing out a path defined by $\{\lambda^\mu(t), \mu = 1, \cdots, d\}$ in the d-dimensional parameter space. For example, for a spin Hamiltonian, $\{\lambda^\mu\}$ could represent an external magnetic field. Now (23) becomes

$$\frac{dc_b}{dt} = -\sum_a (A_\mu)_{ba} c_a \frac{d\lambda^\mu}{dt} \tag{26}$$

if we define $(A_\mu)_{ba} \equiv i\langle\psi_b|\partial_\mu\psi_a\rangle$, where $\partial_\mu \equiv \partial/\partial\lambda^\mu$, and (25) generalizes to

$$A'_\mu = UA_\mu U^\dagger + iU\partial_\mu U^\dagger \tag{27}$$

We have recovered (IV.5.2). Lo and behold, a Yang-Mills gauge potential A_μ has popped up in front of our very eyes!

The "transport" equation (26) can be formally solved by writing $c(\lambda) = Pe^{-\int A_\mu d\lambda^\mu}$, where the line integral is over a path connecting an initial point in the parameter space to some final point λ and P denotes a path ordering operation. We break the path into infinitesimal segments and multiply together the noncommuting contribution $e^{-A_\mu \Delta\lambda^\mu}$ from each segment, ordered along the path. In particular, if the path is a closed curve, by the time we return to the initial values of the parameters, the wave function will have acquired a matrix phase factor, known as the nonabelian Berry's phase. This discussion is clearly intimately related to the discussion of the Aharonov-Bohm phase in the preceding chapter.

To see this nonabelian phase, all we have to do is to find some quantum system with degeneracy in its spectrum and vary some external parameter such as a magnetic field.[3] In their paper, Yang and Mills spoke of the degeneracy of the proton and neutron under isospin in an idealized world and imagined transporting a proton from one point in the universe to another. That a proton at one point can be interpreted as a neutron at another necessitates the introduction of a nonabelian gauge potential. I find it amusing that this imagined transport can now be realized analogously in the laboratory.

You will realize that the discussion here parallels the discussion in the text leading up to (IV.5.2). The spacetime dependent symmetry transformation corresponds to a parameter dependent change of basis. When I discuss gravity in chapter VIII.1 it will become clear that moving the basis $\{\psi_a\}$ around in the parameter space is the precise analog of parallel transporting a local coordinate frame in differential geometry and general relativity. We will also encounter the quantity $Pe^{-\int A_\mu d\lambda^\mu}$ again in chapter VII.1 in the guise of a Wilson loop.

Exercises

IV.5.1   Write down the Lagrangian of an $SU(2)$ gauge theory with a scalar field in the $I = 2$ representation.

IV.5.2   Prove the Bianchi identity $DF \equiv dF + [A, F] = 0$. Write this out explicitly with indices and show that in the abelian case it reduces to half of Maxwell's equations.

IV.5.3   In 4-dimensions $\varepsilon^{\mu\nu\lambda\rho} \, \mathrm{tr} \, F_{\mu\nu}F_{\lambda\rho}$ can be written as $\mathrm{tr} \, F^2$. Show that $d \, \mathrm{tr} \, F^2 = 0$ in any dimensions.

IV.5.4   Invoking the Poincaré lemma (IV.4.5) and the result of exercise IV.5.3 show that $\mathrm{tr} \, F^2 = d \, \mathrm{tr}(AdA + \frac{2}{3}A^3)$. Write this last equation out explicitly with indices. Identify these quantities in the case of electromagnetism.

---

[3] A. Zee, "On the Non-Abelian Gauge Structure in Nuclear Quadrupole Resonance," *Phys. Rev.* A38:1, 1988. The proposed experiment was later done by A. Pines.

IV.5.5 For a challenge show that tr F^n, which appears in higher dimensional theories such as string theory, are all total divergences. In other words, there exists a $(2n - 1)$-form $\omega_{2n-1}(A)$ such that tr $F^n = d\omega_{2n-1}(A)$. [Hint: A compact representation of the form $\omega_{2n-1}(A) = \int_0^1 dt\, f_{2n-1}(t, A)$ exists.] Work out $\omega_5(A)$ explicitly and try to generalize knowing ω_3 and ω_5. Determine the $(2n - 1)$-form $f_{2n-1}(t, A)$. For help, see B. Zumino et al., *Nucl. Phys.* B239:477, 1984.

IV.5.6 Write down the Lagrangian of an $SU(3)$ gauge theory with a fermion field in the fundamental or defining triplet representation.

IV.6 | The Anderson-Higgs Mechanism

The gauge potential eats the Nambu-Goldstone boson

As I noted earlier the ability to ask good questions is of crucial importance in physics. Here is an excellent question: How does spontaneous symmetry breaking manifest itself in gauge theories?

Going back to chapter IV.1, we gauge the $U(1)$ theory in (IV.1.6) by replacing $\partial_\mu \varphi$ with $D_\mu \varphi = (\partial_\mu - ieA_\mu)\varphi$ so that

$$\mathcal{L} = -\tfrac{1}{4} F_{\mu\nu} F^{\mu\nu} + (D\varphi)^\dagger D\varphi + \mu^2 \varphi^\dagger \varphi - \lambda(\varphi^\dagger \varphi)^2 \tag{1}$$

Now when we go to polar coordinates $\varphi = \rho e^{i\theta}$ we have $D_\mu \varphi = [\partial_\mu \rho + i\rho(\partial_\mu \theta - eA_\mu)]e^{i\theta}$ and thus

$$\mathcal{L} = -\tfrac{1}{4} F_{\mu\nu} F^{\mu\nu} + \rho^2(\partial_\mu \theta - eA_\mu)^2 + (\partial\rho)^2 + \mu^2 \rho^2 - \lambda\rho^4 \tag{2}$$

(Compare this with $\mathcal{L} = \rho^2(\partial_\mu \theta)^2 + (\partial\rho)^2 + \mu^2 \rho^2 - \lambda\rho^4$ in the absence of the gauge field.) Under a gauge transformation $\varphi \to e^{i\alpha}\varphi$ (so that $\theta \to \theta + \alpha$) and $eA_\mu \to eA_\mu + \partial_\mu \alpha$, and thus the combination $B_\mu \equiv A_\mu - (1/e)\partial_\mu \theta$ is gauge invariant. The first two terms in $\mathcal{L}$ thus become $-\tfrac{1}{4} F_{\mu\nu} F^{\mu\nu} + e^2 \rho^2 B_\mu^2$. Note that $F_{\mu\nu} = \partial_\mu A_\nu - \partial_\nu A_\mu = \partial_\mu B_\nu - \partial_\nu B_\mu$ has the same form in terms of the potential B_μ.

Upon spontaneous symmetry breaking, we write $\rho = (1/\sqrt{2})(v + \chi)$, with $v = \sqrt{\mu^2/\lambda}$. Hence

$$\mathcal{L} = -\frac{1}{4} F_{\mu\nu} F^{\mu\nu} + \frac{1}{2} M^2 B_\mu^2 + e^2 v\chi B_\mu^2 + \frac{1}{2} e^2 \chi^2 B_\mu^2$$

$$+ \frac{1}{2}(\partial\chi)^2 - \mu^2 \chi^2 - \sqrt{\lambda}\mu\chi^3 - \frac{\lambda}{4}\chi^4 + \frac{\mu^4}{4\lambda} \tag{3}$$

The theory now consists of a vector field B_μ with mass

$$M = ev \qquad (4)$$

interacting with a scalar field χ with mass $\sqrt{2}\mu$. The phase field θ, which would have been the Nambu-Goldstone boson in the ungauged theory, has disappeared. We say that the gauge field A_μ has eaten the Nambu-Goldstone boson; it has gained weight and changed its name to B_μ.

Recall that a massless gauge field has only 2 degrees of freedom, while a massive gauge field has 3 degrees of freedom. A massless gauge field has to eat a Nambu-Goldstone boson in order to have the requisite number of degrees of freedom. The Nambu-Goldstone boson becomes the longitudinal degree of freedom of the massive gauge field. We do not lose any degrees of freedom, as we had better not.

This phenomenon of a massless gauge field becoming massive by eating a Nambu-Goldstone boson was discovered by numerous particle physicists[1] and is known as the Higgs mechanism. People variously call φ, or more restrictively χ, the Higgs field. The same phenomenon was discovered in the context of condensed matter physics by Landau, Ginzburg, and Anderson, and is known as the Anderson mechanism.

Let us give a slightly more involved example, an $O(3)$ gauge theory with a Higgs field φ^a ($a = 1, 2, 3$) transforming in the vector representation. The Lagrangian contains the kinetic energy term $\frac{1}{2}(D_\mu \varphi^a)^2$, with $D_\mu \varphi^a = \partial_\mu \varphi^a + g \varepsilon^{abc} A_\mu^b \varphi^c$ as indicated in (IV.5.20). Upon spontaneous symmetry breaking, $\vec{\varphi}$ acquires a vacuum expectation value which without loss of generality we can choose to point in the 3-direction, so that $\langle \varphi^a \rangle = v \delta^{a3}$. We set $\varphi^3 = v$ and see that

$$\tfrac{1}{2}(D_\mu \varphi^a)^2 \to \tfrac{1}{2}(gv)^2 (A_\mu^1 A^{\mu 1} + A_\mu^2 A^{\mu 2}) \qquad (5)$$

The gauge potential A_μ^1 and A_μ^2 acquires mass gv [compare with (4)] while A_μ^3 remains massless.

A more elaborate example is that of an $SU(5)$ gauge theory with φ transforming as the 24-dimensional adjoint representation. (See appendix B for the necessary group theory.) The field φ is a 5 by 5 hermitean traceless matrix. Since the adjoint representation transforms as $\varphi \to \varphi + i\theta^a [T^a, \varphi]$, we have $D_\mu \varphi = \partial_\mu \varphi - ig A_\mu^a [T^a, \varphi]$ with $a = 1, \cdots, 24$ running over the 24 generators of $SU(5)$. By a symmetry transformation the vacuum expectation value of φ can be taken to be diagonal $\langle \varphi_j^i \rangle = v_j \delta_j^i$ ($i, j = 1, \cdots, 5$), with $\sum_j v_j = 0$. (This is the analog of our choosing $\langle \vec{\varphi} \rangle$ to point in the 3-direction in the preceding example.) We have in the Lagrangian

$$\mathrm{tr}(D_\mu \varphi)(D^\mu \varphi) \to g^2 \, \mathrm{tr}[T^a, \langle \varphi \rangle][\langle \varphi \rangle, T^b] A_\mu^a A^{\mu b} \qquad (6)$$

The gauge boson masses squared are given by the eigenvalues of the 24 by 24 matrix $g^2 \, \mathrm{tr}[T^a, \langle \varphi \rangle][\langle \varphi \rangle, T^b]$, which we can compute laboriously for any given $\langle \varphi \rangle$.

---

[1] Including P. Higgs, F. Englert, R. Brout, G. Guralnik, C. Hagen, and T. Kibble.

It is easy to see, however, which gauge bosons remain massless. As a specific example (which will be of interest to us in chapter VII.6), suppose

$$\langle \varphi \rangle = v \begin{pmatrix} 2 & 0 & 0 & 0 & 0 \\ 0 & 2 & 0 & 0 & 0 \\ 0 & 0 & 2 & 0 & 0 \\ 0 & 0 & 0 & -3 & 0 \\ 0 & 0 & 0 & 0 & -3 \end{pmatrix} \tag{7}$$

Which generators T^a commute with $\langle \varphi \rangle$? Clearly, generators of the form $\begin{pmatrix} A & 0 \\ 0 & 0 \end{pmatrix}$ and of the form $\begin{pmatrix} 0 & 0 \\ 0 & B \end{pmatrix}$. Here A represents 3 by 3 hermitean traceless matrices (of which there are $3^2 - 1 = 8$, the so-called Gell-Mann matrices) and B represents 2 by 2 hermitean traceless matrices (of which there are $2^2 - 1 = 3$, namely the Pauli matrices). Furthermore, the generator

$$\begin{pmatrix} 2 & 0 & 0 & 0 & 0 \\ 0 & 2 & 0 & 0 & 0 \\ 0 & 0 & 2 & 0 & 0 \\ 0 & 0 & 0 & -3 & 0 \\ 0 & 0 & 0 & 0 & -3 \end{pmatrix} \tag{8}$$

being proportional to $\langle \varphi \rangle$, obviously commutes with $\langle \varphi \rangle$. Clearly, these generators generate $SU(3)$, $SU(2)$, and $U(1)$, respectively. Thus, in the 24 by 24 mass-squared matrix $g^2 \operatorname{tr}[T^a, \langle \varphi \rangle][\langle \varphi \rangle, T^b]$ there are blocks of submatrices that vanish, namely, an 8 by 8 block, a 3 by 3 block, and a 1 by 1 block. We have $8 + 3 + 1 = 12$ massless gauge bosons. The remaining $24 - 12 = 12$ gauge bosons acquire mass.

Counting massless gauge bosons

In general, consider a theory with the global symmetry group G spontaneously broken to a subgroup H. As we learned in chapter IV.1, $n(G) - n(H)$ Nambu-Goldstone bosons appear. Now suppose the symmetry group G is gauged. We start with $n(G)$ massless gauge bosons, one for each generator. Upon spontaneous symmetry breaking, the $n(G) - n(H)$ Nambu-Goldstone bosons are eaten by $n(G) - n(H)$ gauge bosons, leaving $n(H)$ massless gauge bosons, exactly the right number since the gauge bosons associated with the surviving gauge group H should remain massless.

In our simple example, $G = U(1)$, $H = $ nothing: $n(G) = 1$ and $n(H) = 0$. In our second example, $G = O(3)$, $H = O(2) \simeq U(1) : n(G) = 3$ and $n(H) = 1$, and so we end up with one massless gauge boson. In the third example, $G = SU(5)$, $H = SU(3) \otimes SU(2) \otimes U(1)$ so that $n(G) = 24$ and $n(H) = 12$. Further examples and generalizations are worked out in the exercises.

Gauge boson mass spectrum

It is easy enough to work out the mass spectrum explicitly. The covariant derivative of a Higgs field is $D_\mu \varphi = \partial_\mu \varphi + g A_\mu^a T^a \varphi$, where g is the gauge coupling, T^a are the generators of the group G when acting on φ, and A_μ^a the gauge potential corresponding to the ath generator. Upon spontaneous symmetry breaking we replace φ by its vacuum expectation value $\langle \varphi \rangle = v$. Hence $D_\mu \varphi$ is replaced by $g A_\mu^a T^a v$. The kinetic term $\frac{1}{2}(D^\mu \varphi \cdot D_\mu \varphi)$ [here $(\cdot)$ denotes the scalar product in the group G] in the Lagrangian thus becomes

$$\tfrac{1}{2} g^2 (T^a v \cdot T^b v) A^{\mu a} A_\mu^b \equiv \tfrac{1}{2} A^{\mu a} (\mu^2)^{ab} A_\mu^b$$

where we have introduced the mass-squared matrix

$$(\mu^2)^{ab} = g^2 (T^a v \cdot T^b v) \tag{9}$$

for the gauge bosons. [You will recognize (9) as the generalization of (4); also compare (5) and (6).] We diagonalize $(\mu^2)^{ab}$ to obtain the masses of the gauge bosons. The eigenvectors tell us which linear combinations of A_μ^a correspond to mass eigenstates.

Note that μ^2 is an $n(G)$ by $n(G)$ matrix with $n(H)$ zero eigenvalues, whose existence can also be seen explicitly. Let T^c be a generator of H. The statement that H remains unbroken by the vacuum expectation value v means that the symmetry transformation generated by T^c leaves v invariant; in other words, $T^c v = 0$, and hence the gauge boson associated with T^c remains massless, as it should. All these points are particularly evident in the $SU(5)$ example we worked out.

Feynman rules in spontaneously broken gauge theories

It is easy enough to derive the Feynman rules for spontaneously broken gauge theories. Take, for example, (3). As usual, we look at the terms quadratic in the fields, Fourier transform, and invert. We see that the gauge boson propagator is given by

$$\frac{-i}{k^2 - M^2 + i\varepsilon} (g_{\mu\nu} - \frac{k_\mu k_\nu}{M^2}) \tag{10}$$

and the χ propagator by

$$\frac{i}{k^2 - 2\mu^2 + i\varepsilon} \tag{11}$$

I leave it to you to work out the rules for the interaction vertices.

As I said in another context, field theories often exist in several equivalent forms. Take the $U(1)$ theory in (1) and instead of polar coordinates go to Cartesian coordinates $\varphi = (1/\sqrt{2})(\varphi_1 + i\varphi_2)$ so that

$$D_\mu \varphi = \partial_\mu \varphi - ie A_\mu \varphi = \frac{1}{\sqrt{2}}[(\partial_\mu \varphi_1 + e A_\mu \varphi_2) + i(\partial_\mu \varphi_2 - e A_\mu \varphi_1)]$$

Then (1) becomes

$$\mathcal{L} = -\tfrac{1}{4}F_{\mu\nu}F^{\mu\nu} + \tfrac{1}{2}[(\partial_\mu\varphi_1 + eA_\mu\varphi_2)^2 + (\partial_\mu\varphi_2 - eA_\mu\varphi_1)^2] \tag{12}$$
$$+ \tfrac{1}{2}\mu^2(\varphi_1^2 + \varphi_2^2) - \tfrac{1}{4}\lambda(\varphi_1^2 + \varphi_2^2)^2$$

Spontaneous symmetry breaking means setting $\varphi_1 \to v + \varphi_1'$ with $v = \sqrt{\mu^2/\lambda}$.

The physical content of (12) and (3) should be the same. Indeed, expand the Lagrangian (12) to quadratic order in the fields:

$$\mathcal{L} = \tfrac{\mu^4}{4\lambda} - \tfrac{1}{4}F_{\mu\nu}F^{\mu\nu} + \tfrac{1}{2}M^2A_\mu^2 - MA_\mu\partial^\mu\varphi_2 + \tfrac{1}{2}[(\partial_\mu\varphi_1')^2 - 2\mu^2\varphi_1'^2]$$
$$+ \tfrac{1}{2}(\partial_\mu\varphi_2)^2 + \cdots \tag{13}$$

The spectrum, a gauge boson A with mass $M = ev$ and a scalar boson φ_1' with mass $\sqrt{2}\mu$, is identical to the spectrum in (3). (The particles there were named B and χ.)

But oops, you may have noticed something strange: the term $-MA_\mu\partial^\mu\varphi_2$ which mixes the fields A_μ and φ_2. Besides, why is φ_2 still hanging around? Isn't he supposed to have been eaten? What to do?

We can of course diagonalize but it is more convenient to get rid of this mixing term. Referring to the Fadeev-Popov quantization of gauge theories discussed in chapter III.4 we note that the gauge fixing term generates a term to be added to $\mathcal{L}$. We can cancel the undesirable mixing term by choosing the gauge function to be $f(A) = \partial A + \xi ev\varphi_2 - \sigma$. Going through the steps, we obtain the effective Lagrangian $\mathcal{L}_{\mathrm{eff}} = \mathcal{L} - (1/2\xi)(\partial A + \xi M\varphi_2)^2$ [compare with (III.4.7)]. The undesirable cross term $-MA_\mu\partial^\mu\varphi_2$ in $\mathcal{L}$ is now canceled upon integration by parts. In $\mathcal{L}_{\mathrm{eff}}$ the terms quadratic in A now read $-\tfrac{1}{4}F_{\mu\nu}F^{\mu\nu} + \tfrac{1}{2}M^2A_\mu^2 - (1/2\xi)(\partial A)^2$ while the terms quadratic in φ_2 read $\tfrac{1}{2}[(\partial_\mu\varphi_2)^2 - \xi M^2\varphi_2^2]$, immediately giving us the gauge boson propagator

$$\frac{-i}{k^2 - M^2 + i\varepsilon}\left[g_{\mu\nu} - (1-\xi)\frac{k_\mu k_\nu}{k^2 - \xi M^2 + i\varepsilon}\right] \tag{14}$$

and the φ_2 propagator

$$\frac{i}{k^2 - \xi M^2 + i\varepsilon} \tag{15}$$

This one-parameter class of gauge choices is known as the R_ξ gauge. Note that the would-be Goldstone field φ_2 remains in the Lagrangian, but the very fact that its mass depends on the gauge parameter ξ brands it as unphysical. In any physical process, the ξ dependence in the φ_2 and A propagators must cancel out so as to leave physical amplitudes ξ independent. In exercise IV.6.9 you will verify that this is indeed the case in a simple example.

Different gauges have different advantages

You might wonder why we would bother with the R_ξ gauge. Why not just use the equivalent formulation of the theory in (3), known as the unitary gauge, in which the gauge boson propagator (10) looks much simpler than (14) and in which we don't have to deal with the unphysical φ_2 field? The reason is that the R_ξ gauge and the unitary gauge complement each other. In the R_ξ gauge, the gauge boson propagator (14) goes as $1/k^2$ for large k and so renormalizability can be proved rather easily. On the other hand, in the unitary gauge all fields are physical (hence the name "unitary") but the gauge boson propagator (10) apparently goes as $k_\mu k_\nu/k^2$ for large k; to prove renormalizability we must show that the $k_\mu k_\nu$ piece of the propagator does not contribute. Using both gauges, we can easily prove that the theory is both renormalizable and unitary. By the way, note that in the limit $\xi \to \infty$ (14) goes over to (10) and φ_2 disappear, at least formally.

In practical calculations, there are typically many diagrams to evaluate. In the R_ξ gauge, the parameter ξ darn well better disappears when we add everything up to form the physical mass shell amplitude. The R_ξ gauge is attractive precisely because this requirement provides a powerful check on practical calculations.

I remarked earlier that strictly speaking, gauge invariance is not so much a symmetry as the reflection of a redundancy in the degrees of freedom used. (The photon has only 2 degrees of freedom but we use a field A_μ with 4 components.) A purist would insist, in the same vein, that there is no such thing as spontaneously breaking a gauge symmetry. To understand this remark, note that spontaneous breaking amounts to setting $\rho \equiv |\varphi|$ to v and θ to 0 in (2). The statement $|\varphi| = v$ is perfectly $U(1)$ invariant: It defines a circle in φ space. By picking out the point $\theta = 0$ on the circle in a globally symmetric theory we break the symmetry. In contrast, in a gauge theory, we can use the gauge freedom to fix $\theta = 0$ everywhere in spacetime. Hence the purists. I will refrain from such hair-splitting in this book and continue to use the convenient language of symmetry breaking even in a gauge theory.

Exercises

IV.6.1 Consider an $SU(5)$ gauge theory with a Higgs field φ transforming as the 5-dimensional representation: φ^i, $i = 1, 2, \cdots, 5$. Show that a vacuum expectation value of φ breaks $SU(5)$ to $SU(4)$. Now add another Higgs field φ', also transforming as the 5-dimensional representation. Show that the symmetry can either remain at $SU(4)$ or be broken to $SU(3)$.

IV.6.2 In general, there may be several Higgs fields belonging to various representations labeled by α. Show that the mass squared matrix for the gauge bosons generalize immediately to $(\mu^2)^{ab} = \sum_\alpha g^2(T_\alpha^a v_\alpha \cdot T_\alpha^b v_\alpha)$, where v_α is the vacuum expectation value of φ_α and T_α^a is the ath generator represented on φ_α. Combine the situations described in exercises IV.6.1 and IV.6.2 and work out the mass spectrum of the gauge bosons.

IV.6.3 The gauge group G does not have to be simple; it could be of the form $G_1 \otimes G_2 \otimes \cdots \otimes G_k$, with coupling constants $g_1, g_2, \cdots, g_k$. Consider, for example, the case $G = SU(2) \otimes U(1)$ and a Higgs field φ transforming like the doublet under $SU(2)$ and like a field with charge $\frac{1}{2}$ under $U(1)$, so that $D_\mu \varphi = \partial_\mu \varphi - i[g A_\mu^a(\tau^a/2) + g'B_\mu \frac{1}{2}]\varphi$. Let $\langle\varphi\rangle = \begin{pmatrix} 0 \\ v \end{pmatrix}$. Determine which linear combinations of the gauge bosons A_μ^a and B_μ acquire mass.

IV.6.4 In chapter IV.5 you worked out an $SU(2)$ gauge theory with a scalar field φ in the $I = 2$ representation. Write down the most general quartic potential $V(\varphi)$ and study the possible symmetry breaking pattern.

IV.6.5 Complete the derivation of the Feynman rules for the theory in (3) and compute the amplitude for the physical process $\chi + \chi \to B + B$.

IV.6.6 Derive (14). [Hint: The procedure is exactly the same as that used to obtain (III.4.9).] Write $\mathcal{L} = \frac{1}{2} A_\mu Q^{\mu\nu} A_\nu$ with $Q^{\mu\nu} = (\partial^2 + M^2) g^{\mu\nu} - [1 - (1/\xi)] \partial^\mu \partial^\nu$ or in momentum space $Q^{\mu\nu} = -(k^2 - M^2) g^{\mu\nu} + [1 - (1/\xi)] k^\mu k^\nu$. The propagator is the inverse of $Q^{\mu\nu}$.

IV.6.7 Work out the $(\cdots)$ in (13) and the Feynman rules for the various interaction vertices.

IV.6.8 Using the Feynman rules derived in exercise IV.6.7 calculate the amplitude for the physical process $\varphi_1' + \varphi_1' \to A + A$ and show that the dependence on ξ cancels out. Compare with the result in exercise IV.6.5. [Hint: There are two diagrams, one with A exchange and the other with φ_2 exchange.]

IV.6.9 Consider the theory defined in (12) with $\mu = 0$. Using the result of exercise IV.3.5 show that

$$V_{\text{eff}}(\varphi) = \frac{1}{4} \lambda \varphi^4 + \frac{1}{64\pi^2} (10\lambda^2 + 3e^4) \varphi^4 \left(\log \frac{\varphi^2}{M^2} - \frac{25}{6} \right) + \cdots \tag{16}$$

where $\varphi^2 = \varphi_1^2 + \varphi_2^2$. This potential has a minimum away from $\varphi = 0$ and thus the gauge symmetry is spontaneously broken by quantum fluctuations. In chapter IV.3 we did not have the e^4 term and argued that the minimum we got there was not to be trusted. But here we can balance the $\lambda \varphi^4$ against $e^4 \varphi^4 \log(\varphi^2/M^2)$ for λ of the same order of magnitude as e^4. The minimum can be trusted. Show that the spectrum of this theory consists of a massive scalar boson and a massive vector boson, with

$$\frac{m^2(\text{scalar})}{m^2(\text{vector})} = \frac{3}{2\pi} \frac{e^2}{4\pi} \tag{17}$$

For help, see S. Coleman and E. Weinberg, *Phys. Rev.* D7: 1888, 1973.

IV.7 | Chiral Anomaly

Classical versus quantum symmetry

I have emphasized the importance of asking good questions. Here is another good one: Is a symmetry of classical physics necessarily a symmetry of quantum physics?

We have a symmetry of classical physics if a transformation $\varphi \to \varphi + \delta\varphi$ leaves the action $S(\varphi)$ invariant. We have a symmetry of quantum physics if the transformation leaves the path integral $\int D\varphi e^{iS(\varphi)}$ invariant.

When our question is phrased in this path integral language, the answer seems obvious: Not necessarily. Indeed, the measure $D\varphi$ may or may not be invariant.

Yet historically, field theorists took as almost self-evident the notion that any symmetry of classical physics is necessarily a symmetry of quantum physics, and indeed, almost all the symmetries they encountered in the early days of field theory had the property of being symmetries of both classical and quantum physics. For instance, we certainly expect quantum mechanics to be rotational invariant. It would be very odd indeed if quantum fluctuations were to favor a particular direction.

You have to appreciate the frame of mind that field theorists operated in to understand their shock when they discovered in the late 1960s that quantum fluctuations can indeed break classical symmetries. Indeed, they were so shocked as to give this phenomenon the rather misleading name "anomaly," as if it were some kind of sickness of field theory. With the benefits of hindsight, we now understand the anomaly as being no less conceptually innocuous as the elementary fact that when we change integration variables in an integral we better not forget the Jacobian.

With the passing of time, field theorists have developed many different ways of looking at the all important subject of anomaly. They are all instructive and shed different lights on how the anomaly comes about. For this introductory text I choose to show the existence of anomaly by an explicit Feynman diagram calculation. The diagram method is certainly more laborious and less slick than other methods, but the advantage is that you will see a classical symmetry vanishing in front of your very eyes! No smooth formal argument for us.

The lesser of two evils

Consider the theory of a single massless fermion $\mathcal{L} = \bar{\psi} i \gamma^\mu \partial_\mu \psi$. You can hardly ask for a simpler theory! Recall from chapter II.1 that $\mathcal{L}$ is manifestly invariant under the separate transformations $\psi \to e^{i\theta} \psi$ and $\psi \to e^{i\theta\gamma^5} \psi$, corresponding to the conserved vector current $J^\mu = \bar{\psi} \gamma^\mu \psi$ and the conserved axial current $J_5^\mu = \bar{\psi} \gamma^\mu \gamma^5 \psi$ respectively. You should verify that $\partial_\mu J^\mu = 0$ and $\partial_\mu J_5^\mu = 0$ follow immediately from the classical equation of motion $i \gamma^\mu \partial_\mu \psi = 0$.

Let us now calculate the amplitude for a spacetime history in which a fermion-anti-fermion pair is created at x_1 and another such pair is created at x_2 by the vector current, with the fermion from one pair annihilating the antifermion from the other pair and the remaining fermion-antifermion pair being subsequently annihilated by the axial current. This is a long-winded way of describing the amplitude $\langle 0 | T J_5^\lambda(0) J^\mu(x_1) J^\nu(x_2) | 0 \rangle$ in words, but I want to make sure that you know what I am talking about. Feynman tells us that the Fourier transform of this amplitude is given by the two "triangle" diagrams in figure IV.7.1a and b.

$$\Delta^{\lambda\mu\nu}(k_1, k_2) = (-1) i^3 \int \frac{d^4 p}{(2\pi)^4}$$

$$\text{tr} \left(\gamma^\lambda \gamma^5 \frac{1}{\not{p} - \not{q}} \gamma^\nu \frac{1}{\not{p} - \not{k}_1} \gamma^\mu \frac{1}{\not{p}} + \gamma^\lambda \gamma^5 \frac{1}{\not{p} - \not{q}} \gamma^\mu \frac{1}{\not{p} - \not{k}_2} \gamma^\nu \frac{1}{\not{p}} \right) \tag{1}$$

with $q = k_1 + k_2$. Note that the two terms are required by Bose statistics. The overall factor of (-1) comes from the closed fermion loop.

Classically, we have two symmetries implying $\partial_\mu J^\mu = 0$ and $\partial_\mu J_5^\mu = 0$. In the quantum theory, if $\partial_\mu J^\mu = 0$ continues to hold, then we should have $k_{1\mu} \Delta^{\lambda\mu\nu} = 0$ and $k_{2\nu} \Delta^{\lambda\mu\nu} = 0$, and if $\partial_\mu J_5^\mu = 0$ continues to hold, then $q_\lambda \Delta^{\lambda\mu\nu} = 0$. Now that we have things all set up, we merely have to calculate $\Delta^{\lambda\mu\nu}$ to see if the two symmetries hold up under quantum fluctuations. No big deal.

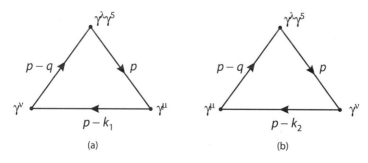

(a)   (b)

Figure IV.7.1

Before we blindly calculate, however, let us ask ourselves how sad we would be if either of the two currents J^μ and J_5^μ fails to be conserved. Well, we would be very upset if the vector current is not conserved. The corresponding charge $Q = \int d^3x J^0$ counts the number of fermions. We wouldn't want our fermions to disappear into thin air or pop out of nowhere. Furthermore, it may please us to couple the photon to the fermion field ψ. In that case, you would recall from chapter II.7 that we need $\partial_\mu J^\mu = 0$ to prove gauge invariance and hence show that the photon has only two degrees of polarization. More explicitly, imagine a photon line coming into the vertex labeled by μ in figure IV.7.1a and b with propagator $(i/k_1^2)[\xi(k_{1\mu}k_{1\rho}/k_1^2) - g_{\mu\rho}]$. The gauge dependent term $\xi(k_{1\mu}k_{1\rho}/k_1^2)$ would not go away if the vector current is not conserved, that is, if $k_{1\mu}\Delta^{\lambda\mu\nu}$ fails to vanish.

On the other hand, quite frankly, just between us friends, we won't get too upset if quantum fluctuation violates axial current conservation. Who cares if the axial charge $Q^5 = \int d^3x J_5^0$ is not constant in time?

Shifting integration variable

So, do $k_{1\mu}\Delta^{\lambda\mu\nu}$ and $k_{2\nu}\Delta^{\lambda\mu\nu}$ vanish? We will look over Professor Confusio's shoulders as he calculates $k_{1\mu}\Delta^{\lambda\mu\nu}$. (We are now in the 1960s, long after the development of renormalization theory as described in chapter III.1 and Confusio has managed to get a tenure track assistant professorship.) He hits $\Delta^{\lambda\mu\nu}$ as written in (1) with $k_{1\mu}$ and using what he learned in chapter II.7 writes $\not{k}_1$ in the first term as $\not{p} - (\not{p} - \not{k}_1)$ and in the second term as $(\not{p} - \not{k}_2) - (\not{p} - \not{q})$, thus obtaining

$$k_{1\mu}\Delta^{\lambda\mu\nu}(k_1, k_2)$$

$$= i \int \frac{d^4p}{(2\pi)^4} \operatorname{tr}(\gamma^\lambda\gamma^5 \frac{1}{\not{p} - \not{q}} \gamma^\nu \frac{1}{\not{p} - \not{k}_1} - \gamma^\lambda\gamma^5 \frac{1}{\not{p} - \not{k}_2} \gamma^\nu \frac{1}{\not{p}}) \tag{2}$$

Just as in chapter II.7, Confusio recognizes that in the integrand the first term is just the second term with the shift of the integration variable $p \to p - k_1$. The two terms cancel and Professor Confusio publishes a paper saying $k_{1\mu}\Delta^{\lambda\mu\nu} = 0$, as we all expect.

Remember back in chapter II.7 I said we were going to worry later about whether it is legitimate to shift integration variables. Now is the time to worry!

You could have asked your calculus teacher long ago when it is legitimate to shift integration variables. When is $\int_{-\infty}^{+\infty} dp f(p + a)$ equal to $\int_{-\infty}^{+\infty} dp f(p)$? The difference between these two integrals is

$$\int_{-\infty}^{+\infty} dp(a\frac{d}{dp}f(p) + \cdots) = a(f(+\infty) - f(-\infty)) + \cdots$$

Clearly, if $f(+\infty)$ and $f(-\infty)$ are two different constants, then it is not okay to shift. But if the integral $\int_{-\infty}^{+\infty} dp f(p)$ is convergent, or even logarithmically divergent, it is certainly okay. It was okay in chapter II.7 but definitely not here in (2)!

As usual, we rotate the Feynman integrand to Euclidean space. Generalizing our observation above to d-dimensional Euclidean space, we have

$$\int d_E^d p[f(p+a) - f(p)] = \int d_E^d p[a^\mu \partial_\mu f(p) + \cdots]$$

which by Gauss's theorem is given by a surface integral over an infinitely large sphere enclosing all of Euclidean spacetime and hence equal to

$$\lim_{P \to \infty} a^\mu \left(\frac{P_\mu}{P}\right) f(P) S_{d-1}(P)$$

where $S_{d-1}(P)$ is the area of a $(d-1)$-dimensional sphere (see appendix D) and where an average over the surface of the sphere is understood. (Recall from our experience evaluating Feynman diagrams that the average of $P^\mu P^\nu / P^2$ is equal to $\frac{1}{4}\eta^{\mu\nu}$ by a symmetry argument, with the normalization $\frac{1}{4}$ fixed by contracting with $\eta_{\mu\nu}$.) Rotating back, we have for a 4-dimensional Minkowskian integral

$$\int d^4 p[f(p+a) - f(p)] = \lim_{P \to \infty} ia^\mu \left(\frac{P_\mu}{P}\right) f(P)(2\pi^2 P^3) \tag{3}$$

Note the i from Wick rotating back.

Applying (3) with

$$f(p) = \text{tr}\left(\gamma^\lambda \gamma^5 \frac{1}{\not{p} - \not{k}_2} \gamma^\nu \frac{1}{\not{p}}\right) = \frac{\text{tr}[\gamma^5(\not{p} - \not{k}_2)\gamma^\nu \not{p} \gamma^\lambda]}{(p - k_2)^2 p^2} = \frac{4i\varepsilon^{\tau\nu\sigma\lambda} k_{2\tau} p_\sigma}{(p - k_2)^2 p^2}$$

we obtain

$$k_{1\mu} \Delta^{\lambda\mu\nu} = \frac{i}{(2\pi)^4} \lim_{P \to \infty} i(-k_1)^\mu \frac{P_\mu}{P} \frac{4i\varepsilon^{\tau\nu\sigma\lambda} k_{2\tau} P_\sigma}{P^4} 2\pi^2 P^3 = \frac{i}{8\pi^2} \varepsilon^{\lambda\nu\tau\sigma} k_{1\tau} k_{2\sigma}$$

Contrary to what Confusio said, $k_{1\mu} \Delta^{\lambda\mu\nu} \neq 0$.

As I have already said, this would be a disaster. Fermion number is not conserved and matter would be disintegrating all around us! What is the way out?

In fact, we are only marginally smarter than Professor Confusio. We did not notice that the integral defining $\Delta^{\lambda\mu\nu}$ in (1) is linearly divergent and is thus not well defined.

Oops, even before we worry about calculating $k_{1\mu} \Delta^{\lambda\mu\nu}$ and $k_{2\nu} \Delta^{\lambda\mu\nu}$ we better worry about whether or not $\Delta^{\lambda\mu\nu}$ depends on the physicist doing the calculation. In other words, suppose another physicist chooses[1] to shift the integration variable p in the linearly divergent integral in (1) by an arbitrary 4-vector a and define

$$\Delta^{\lambda\mu\nu}(a, k_1, k_2)$$
$$= (-1)i^3 \int \frac{d^4 p}{(2\pi)^4} \text{tr}(\gamma^\lambda \gamma^5 \frac{1}{\not{p} + \not{a} - \not{k}} \gamma^\nu \frac{1}{\not{p} + \not{a} - \not{k}_1} \gamma^\mu \frac{1}{\not{p} + \not{a}})$$
$$+ \{\mu, k_1 \leftrightarrow \nu, k_2\} \tag{4}$$

There can be as many results for the Feynman diagrams in figure IV.7.1a and b as there are physicists! That would be the end of physics, or at least quantum field theory, for sure.

---

[1] This is the freedom of choice in labeling internal momenta mentioned in chapter I.7.

Well, whose result should we declare to be correct?

The only sensible answer is that we trust the person who chooses an a such that $k_{1\mu}\Delta^{\lambda\mu\nu}(a, k_1, k_2)$ and $k_{2\nu}\Delta^{\lambda\mu\nu}(a, k_1, k_2)$ vanish, so that the photon will have the right number of degrees of freedom should we introduce a photon into the theory.

Let us compute $\Delta^{\lambda\mu\nu}(a, k_1, k_2) - \Delta^{\lambda\mu\nu}(k_1, k_2)$ by applying (3) to $f(p) = \mathrm{tr}(\gamma^\lambda\gamma^5\frac{1}{\not p - \not q}\gamma^\nu\frac{1}{\not p - \not k_1}\gamma^\mu\frac{1}{\not p})$. Noting that

$$f(P) = \lim_{P \to \infty} \frac{\mathrm{tr}(\gamma^\lambda\gamma^5\,\not P\gamma^\nu\,\not P\gamma^\mu\,\not P)}{P^6}$$

$$= \frac{2P^\mu \mathrm{tr}(\gamma^\lambda\gamma^5\,\not P\gamma^\nu\,\not P) - P^2\,\mathrm{tr}(\gamma^\lambda\gamma^5\,\not P\gamma^\nu\gamma^\mu)}{P^6} = \frac{+4i\,P^2 P_\sigma \varepsilon^{\sigma\nu\mu\lambda}}{P^6}$$

we see that

$$\Delta^{\lambda\mu\nu}(a, k_1, k_2) - \Delta^{\lambda\mu\nu}(k_1, k_2) = \frac{4i}{8\pi^2}\lim_{P \to \infty} a^\omega \frac{P_\omega P_\sigma}{P^2}\varepsilon^{\sigma\nu\mu\lambda} + \{\mu, k_1 \leftrightarrow \nu, k_2\}$$

$$= \frac{i}{8\pi^2}\varepsilon^{\sigma\nu\mu\lambda}a_\sigma + \{\mu, k_1 \leftrightarrow \nu, k_2\} \tag{5}$$

There are two independent momenta k_1 and k_2 in the problem, so we can take $a = \alpha(k_1 + k_2) + \beta(k_1 - k_2)$. Plugging into (5), we obtain

$$\Delta^{\lambda\mu\nu}(a, k_1, k_2) = \Delta^{\lambda\mu\nu}(k_1, k_2) + \frac{i\beta}{4\pi^2}\varepsilon^{\lambda\mu\nu\sigma}(k_1 - k_2)_\sigma \tag{6}$$

Note that α drops out.

As expected, $\Delta^{\lambda\mu\nu}(a, k_1, k_2)$ depends on β, and hence on a. Our unshakable desire to have a conserved vector current, that is, $k_{1\mu}\Delta^{\lambda\mu\nu}(a, k_1, k_2) = 0$, now fixes the parameter β upon recalling

$$k_{1\mu}\Delta^{\lambda\mu\nu}(k_1, k_2) = \frac{i}{8\pi^2}\varepsilon^{\lambda\nu\tau\sigma}k_{1\tau}k_{2\sigma}$$

Hence, we must choose to deal with $\Delta^{\lambda\mu\nu}(a, k_1, k_2)$ with $\beta = -\frac{1}{2}$.

One way of viewing all this is to say that the Feynman rules do not suffice in determining $\langle 0| T J_5^\lambda(0)J^\mu(x_1)J^\nu(x_2)|0\rangle$. They have to be supplemented by vector current conservation. The amplitude $\langle 0| T J_5^\lambda(0)J^\mu(x_1)J^\nu(x_2)|0\rangle$ is defined by $\Delta^{\lambda\mu\nu}(a, k_1, k_2)$ with $\beta = -\frac{1}{2}$.

Quantum fluctuation violates axial current conservation

Now we come to the punchline of the story. We insisted that the vector current be conserved. Is the axial current also conserved?

To answer this question, we merely have to compute

$$q_\lambda \Delta^{\lambda\mu\nu}(a, k_1, k_2) = q_\lambda \Delta^{\lambda\mu\nu}(k_1, k_2) + \frac{i}{4\pi^2}\varepsilon^{\mu\nu\lambda\sigma}k_{1\lambda}k_{2\sigma} \tag{7}$$

By now, you know how to do this:

$$q_\lambda \Delta^{\lambda\mu\nu}(k_1, k_2) = i \int \frac{d^4 p}{(2\pi)^4} \, \text{tr} \left(\gamma^5 \frac{1}{\not{p} - \not{q}} \gamma^\nu \frac{1}{\not{p} - \not{k_1}} \gamma^\mu \right.$$

$$\left. - \gamma^5 \frac{1}{\not{p} - \not{k_2}} \gamma^\nu \frac{1}{\not{p}} \gamma^\mu \right) + \{\mu, k_1 \leftrightarrow \nu, k_2\}$$

$$= \frac{i}{4\pi^2} \varepsilon^{\mu\nu\lambda\sigma} k_{1\lambda} k_{2\sigma} \tag{8}$$

Indeed, you recognize that the integration has already been done in (2). We finally obtain

$$q_\lambda \Delta^{\lambda\mu\nu}(a, k_1, k_2) = \frac{i}{2\pi^2} \varepsilon^{\mu\nu\lambda\sigma} k_{1\lambda} k_{2\sigma} \tag{9}$$

The axial current is not conserved!

In summary, in the simple theory $\mathcal{L} = \bar{\psi} i \gamma^\mu \partial_\mu \psi$ while the vector and axial currents are both conserved classically, quantum fluctuation destroys axial current conservation. This phenomenon is known variously as the anomaly, the axial anomaly, or the chiral anomaly.

Consequences of the anomaly

As I said, the anomaly is an extraordinarily rich subject. I will content myself with a series of remarks, the details of which you should work out as exercises.

1. Suppose we gauge our simple theory $\mathcal{L} = \bar{\psi} i \gamma^\mu (\partial_\mu - ie A_\mu) \psi$ and speak of A_μ as the photon field. Then in figure IV.7.1 we can think of two photon lines coming out of the vertices labeled μ and ν. Our central result (9) can then be written elegantly as two operator equations:

   CLASSICAL PHYSICS: $\partial_\mu J_5^\mu = 0$ $\qquad\qquad$ (10)

   QUANTUM PHYSICS: $\partial_\mu J_5^\mu = \frac{e^2}{(4\pi)^2} \varepsilon^{\mu\nu\lambda\sigma} F_{\mu\nu} F_{\lambda\sigma}$ $\qquad$ (11)

   The divergence of the axial current $\partial_\mu J_5^\mu$ is not zero, but is an operator capable of producing two photons.

2. Applying the same type of argument as in chapter IV.2 we can calculate the rate of the decay $\pi^0 \to \gamma + \gamma$. Indeed, historically people used the erroneous result (10) to deduce that this experimentally observed decay cannot occur! See exercise IV.7.2. The resolution of this apparent paradox led to the correct result (11).

3. Writing the Lagrangian in terms of left and right handed fields ψ_R and ψ_L and introducing the left and right handed currents $J_R^\mu \equiv \bar{\psi}_R \gamma^\mu \psi_R$ and $J_L^\mu \equiv \bar{\psi}_L \gamma^\mu \psi_L$, we can repackage the anomaly as

   $$\partial_\mu J_R^\mu = \frac{1}{2} \frac{e^2}{(4\pi)^2} \varepsilon^{\mu\nu\lambda\sigma} F_{\mu\nu} F_{\lambda\sigma}$$

   and

   $$\partial_\mu J_L^\mu = -\frac{1}{2} \frac{e^2}{(4\pi)^2} \varepsilon^{\mu\nu\lambda\sigma} F_{\mu\nu} F_{\lambda\sigma} \tag{12}$$

(Hence the name chiral!) We can think of left handed and right handed fermions running around the loop in figure IV.7.1, contributing oppositely to the anomaly.

4. Consider the theory $\mathcal{L} = \bar{\psi}(i\gamma^\mu \partial_\mu - m)\psi$. Then invariance under the transformation $\psi \to e^{i\theta\gamma^5}\psi$ is spoiled by the mass term. Classically, $\partial_\mu J_5^\mu = 2m\bar{\psi}i\gamma^5\psi$: The axial current is explicitly not conserved. The anomaly now says that quantum fluctuation produces an additional term. In the theory $\mathcal{L} = \bar{\psi}[i\gamma^\mu(\partial_\mu - ieA_\mu) - m]\psi$, we have

$$\partial_\mu J_5^\mu = 2m\bar{\psi}i\gamma^5\psi + \frac{e^2}{(4\pi)^2}\varepsilon^{\mu\nu\lambda\sigma}F_{\mu\nu}F_{\lambda\sigma} \tag{13}$$

5. Recall that in chapter III.7 we introduced Pauli-Villars regulators to calculate vacuum polarization. We subtract from the integrand what the integrand would have been if the electron mass were replaced by some regulator mass. The analog of electron mass in (1) is in fact 0 and so we subtract from the integrand what the integrand would have been if 0 were replaced by a regulator mass M. In other words, we now define

$$\Delta^{\lambda\mu\nu}(k_1, k_2) = (-1)i^3 \int \frac{d^4p}{(2\pi)^4} \, \mathrm{tr}\left(\gamma^\lambda\gamma^5 \frac{1}{\slashed{p} - \slashed{q}}\gamma^\nu \frac{1}{\slashed{p} - \slashed{k}_1}\gamma^\mu \frac{1}{\slashed{p}}\right.$$
$$\left. -\gamma^\lambda\gamma^5 \frac{1}{\slashed{p} - \slashed{q} - M}\gamma^\nu \frac{1}{\slashed{p} - \slashed{k}_1 - M}\gamma^\mu \frac{1}{\slashed{p} - M}\right)$$
$$+ \{\mu, k_1 \leftrightarrow \nu, k_2\}. \tag{14}$$

Note that as $p \to \infty$ the integrand now vanishes faster than $1/p^3$. This is in accordance with the philosophy of regularization outlined in chapters III.1 and III.7: For $p \ll M$, the threshold of ignorance, the integrand is unchanged. But for $p \gg M$, the integrand is cut off. Now the integral in (14) is superficially logarithmically divergent and we can shift the integration variable p at will.

So how does the chiral anomaly arise? By including the regulator mass M we have broken axial current conservation explicitly. The anomaly is the statement that this breaking persists even when we let M tend to infinity. It is extremely instructive (see exercise IV.7.4) to work this out.

6. Consider the nonabelian theory $\mathcal{L} = \bar{\psi}i\gamma^\mu(\partial_\mu - igA_\mu^a T^a)\psi$. We merely have to include in the Feynman amplitude a factor of T^a at the vertex labeled by μ and a factor of T^b at the vertex labeled by ν. Everything goes through as before except that in summing over all the different fermions that run around the loop we obtain a factor $\mathrm{tr}\, T^a T^b$. Thus, we see instantly that in a nonabelian gauge theory

$$\partial_\mu J_5^\mu = \frac{g^2}{(4\pi)^2}\varepsilon^{\mu\nu\lambda\sigma}\, \mathrm{tr}\, F_{\mu\nu}F_{\lambda\sigma} \tag{15}$$

where $F_{\mu\nu} = F_{\mu\nu}^a T^a$ is the matrix field strength defined in chapter IV.5. Nonabelian symmetry tells us something remarkable: The object $\varepsilon^{\mu\nu\lambda\sigma}\, \mathrm{tr}\, F_{\mu\nu}F_{\lambda\sigma}$ contains not only a term quadratic in A, but also terms cubic and quartic in A, and hence there is also a chiral anomaly with three and four gauge bosons coming in, as indicated in figure IV.7.2a and b. Some people refer to the anomaly produced in figures IV.7.1 and IV.7.2 as the triangle, square, and pentagon anomaly. Historically, after the triangle anomaly was discovered, there was a controversy as to whether the square and pentagon anomaly existed. The nonabelian

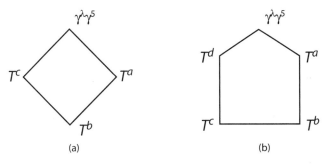

Figure IV.7.2

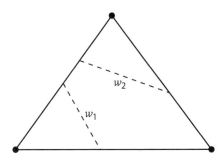

Figure IV.7.3

symmetry argument given here makes things totally obvious, but at the time people calculated Feynman diagrams explicitly and, as we just saw, there are subtleties lying in wait for the unwary.

7. We will see in chapter V.7 that the anomaly has deep connections to topology.

8. We computed the chiral anomaly in the free theory $\mathcal{L} = \bar{\psi}(i\gamma^\mu\partial_\mu - m)\psi$. Suppose we couple the fermion to a scalar field by adding $f\varphi\bar{\psi}\psi$ or to the electromagnetic field for that matter. Now we have to calculate higher order diagrams such as the three-loop diagram in figure IV.7.3. You would expect that the right-hand side of (9) would be multiplied by $1 + h(f, e, \cdots)$, where h is some unknown function of all the couplings in the theory.

Surprise! Adler and Bardeen proved that $h = 0$. This apparently miraculous fact, known as the nonrenormalization of the anomaly, can be understood heuristically as follows. Before we integrate over the momenta of the scalar propagators in figure IV.7.3 (labeled by w_1 and w_2) the Feynman integrand has seven fermion propagators and thus is more than sufficiently convergent that we can shift integration variables with impunity. Thus, before we integrate over w_1 and w_2 all the appropriate Ward identities are satisfied, for instance, $q_\lambda \Delta^{\lambda\mu\nu}_{3\ loops}(k_1, k_2; w_1, w_2) = 0$. You can easily complete the proof. You will give a proof[2] based on topology in exercise V.7.13.

---

[2] For a simple proof not involving topology, see J. Collins, *Renormalization*, p. 352.

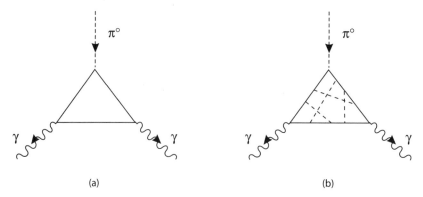

(a)   (b)

Figure IV.7.4

9. The preceding point was of great importance in the history of particle physics as it led directly to the notion of color, as we will discuss in chapter VII.3. The nonrenormalization of the anomaly allowed the decay amplitude for $\pi^0 \to \gamma + \gamma$ to be calculated with confidence in the late 1960s. In the quark model of the time, the amplitude is given by an infinite number of Feynman diagrams, as indicated in figure IV.7.4 (with a quark running around the fermion loop), but the nonrenormalization of the anomaly tells us that only figure IV.7.4a contributes. In other words, the amplitude does not depend on the details of the strong interaction. That it came out a factor of 3 too small suggested that quarks come in 3 copies, as we will see in chapter VII.3.

10. It is natural to speculate as to whether quarks and leptons are composites of yet more fundamental fermions known as preons. The nonrenormalization of the chiral anomaly provides a powerful tool for this sort of theoretical speculation. No matter how complicated the relevant interactions might be, as long as they are described by field theory as we know it, the anomaly at the preon level must be the same as the anomaly at the quark-lepton level. This so-called anomaly matching condition[3] severely constrains the possible preon theories.

11. Historically, field theorists were deeply suspicious of the path integral, preferring the canonical approach. When the chiral anomaly was discovered, some people even argued that the existence of the anomaly proved that the path integral was wrong. Look, these people said, the path integral

$$\int D\bar{\psi} D\psi \; e^{i \int d^4x \, \bar{\psi} i \gamma^\mu (\partial_\mu - iA_\mu)\psi} \tag{16}$$

is too stupid to tell us that it is not invariant under the chiral transformation $\psi \to e^{i\theta\gamma^5}\psi$. Fujikawa resolved the controversy by showing that the path integral did know about the anomaly: Under the chiral transformation the measure $D\bar{\psi} D\psi$ changes by a Jacobian. Recall that this was how I motivated this chapter: The action may be invariant but not the path integral.

[3] G. 't Hooft, in: G. 't Hooft et al., eds., *Recent Developments in Gauge Theories*; A. Zee, *Phys. Lett.* 95B:290, 1980.

Exercises

IV.7.1 Derive (11) from (9). The momentum factors $k_{1\lambda}$ and $k_{2\sigma}$ in (9) become the two derivatives in $F_{\mu\nu}F_{\lambda\sigma}$ in (11).

IV.7.2 Following the reasoning in chapter IV.2 and using the erroneous (10) show that the decay amplitude for the decay $\pi^0 \to \gamma + \gamma$ would vanish in the ideal world in which the π^0 is massless. Since the π^0 does decay and since our world is close to the ideal world, this provided the first indication historically that (10) cannot possibly be valid.

IV.7.3 Repeat all the calculations in the text for the theory $\mathcal{L} = \bar{\psi}(i\gamma^\mu \partial_\mu - m)\psi$.

IV.7.4 Take the Pauli-Villars regulated $\Delta^{\lambda\mu\nu}(k_1, k_2)$ and contract it with q_λ. The analog of the trick in chapter II.7 is to write $\dslash{q}\gamma^5$ in the second term as $[2M + (\dslash{p} - M) - (\dslash{p} - \dslash{q} + M)]\gamma^5$. Now you can freely shift integration variables. Show that

$$q_\lambda \Delta^{\lambda\mu\nu}(k_1, k_2) = -2M\,\Delta^{\mu\nu}(k_1, k_2) \tag{17}$$

where

$$\Delta^{\mu\nu}(k_1, k_2) \equiv (-1)i^3 \int \frac{d^4p}{(2\pi)^4}$$

$$\text{tr}\left(\gamma^5 \frac{1}{\dslash{p} - \dslash{q} - M}\gamma^\nu \frac{1}{\dslash{p} - \dslash{k}_1 - M}\gamma^\mu \frac{1}{\dslash{p} - M}\right) + \{\mu, k_1 \leftrightarrow \nu, k_2\}$$

Evaluate $\Delta^{\mu\nu}$ and show that $\Delta^{\mu\nu}$ goes as $1/M$ in the limit $M \to \infty$ and so the right hand side of (17) goes to a finite limit. The anomaly is what the regulator leaves behind as it disappears from the low energy spectrum: It is like the smile of the Cheshire cat. [We can actually argue that $\Delta^{\mu\nu}$ goes as $1/M$ without doing a detailed calculation. By Lorentz invariance and because of the presence of γ^5, $\Delta^{\mu\nu}$ must be proportional to $\varepsilon^{\mu\nu\lambda\rho}k_{1\lambda}k_{2\rho}$, but by dimensional analysis, $\Delta^{\mu\nu}$ must be some constant times $\varepsilon^{\mu\nu\lambda\rho}k_{1\lambda}k_{2\rho}/M$. You might ask why we can't use something like $1/(k_1^2)^{\frac{1}{2}}$ instead of $1/M$ to make the dimension come out right. The answer is that from your experience in evaluating Feynman diagrams in $(3+1)$-dimensional spacetime you can never get a factor like $1/(k_1^2)^{\frac{1}{2}}$.]

IV.7.5 There are literally N ways of deriving the anomaly. Here is another. Evaluate

$$\Delta^{\lambda\mu\nu}(k_1, k_2) = (-1)i^3 \int \frac{d^4p}{(2\pi)^4}$$

$$\text{tr}\left(\gamma^\lambda\gamma^5 \frac{1}{\dslash{p} - \dslash{q} - m}\gamma^\nu \frac{1}{\dslash{p} - \dslash{k}_1 - m}\gamma^\mu \frac{1}{\dslash{p} - m}\right) + \{\mu, k_1 \leftrightarrow \nu, k_2\}$$

in the massive fermion case not by brute force but by first using Lorentz invariance to write

$$\Delta^{\lambda\mu\nu}(k_1, k_2) = \varepsilon^{\lambda\mu\nu\sigma}k_{1\sigma}A_1 + \cdots + \varepsilon^{\mu\nu\sigma\tau}k_{1\sigma}k_{2\tau}k_2^\lambda A_8$$

where $A_i \equiv A_i(k_1^2, k_2^2, q^2)$ are eight functions of the three Lorentz scalars in the problem. You are supposed to fill in the dots. By counting powers as in chapters III.3 and III.7 show that two of these functions are given by superficially logarithmically divergent integrals while the other six are given by perfectly convergent integrals. Next, impose Bose statistics and vector current conservation $k_{1\mu}\Delta^{\lambda\mu\nu} = 0 = k_{2\nu}\Delta^{\lambda\mu\nu}$ to show that we can avoid calculating the superficially logarithmically divergent integrals. Compute the convergent integrals and then evaluate $q_\lambda \Delta^{\lambda\mu\nu}(k_1, k_2)$.

IV.7.6 Discuss the anomaly by studying the amplitude

$$\langle 0 | T J_5^\lambda(0) J_5^\mu(x_1) J_5^\nu(x_2) | 0 \rangle$$

given in lowest orders by triangle diagrams with axial currents at each vertex. [Hint: Call the momentum space amplitude $\Delta_5^{\lambda\mu\nu}(k_1, k_2)$.] Show by using $(\gamma^5)^2 = 1$ and Bose symmetry that

$$\Delta_5^{\lambda\mu\nu}(k_1, k_2) = \tfrac{1}{3}[\Delta^{\lambda\mu\nu}(a, k_1, k_2) + \Delta^{\mu\nu\lambda}(a, k_2, -q) + \Delta^{\nu\lambda\mu}(a, -q, k_1)]$$

Now use (9) to evaluate $q_\lambda \Delta_5^{\lambda\mu\nu}(k_1, k_2)$.

IV.7.7 Define the fermionic measure $D\psi$ in (16) carefully by going to Euclidean space. Calculate the Jacobian upon a chiral transformation and derive the anomaly. [Hint: For help, see K. Fujikawa, *Phys. Rev. Lett.* 42: 1195, 1979.]

IV.7.8 Compute the pentagon anomaly by Feynman diagrams in order to check remark 6 in the text. In other words, determine the coefficient c in $\partial_\mu J_5^\mu = \cdots + c\varepsilon^{\mu\nu\lambda\sigma}$ tr $A_\mu A_\nu A_\lambda A_\sigma$.

Part V | Field Theory and Collective Phenomena

I mentioned in the introduction that one of the more intellectually satisfying developments in the last two or three decades has been the increasingly important role played by field theoretic methods in condensed matter physics. This is a rich and diverse subject; in this and subsequent chapters I can barely describe the tip of the iceberg and will have to content myself with a few selected topics.

Historically, field theory was introduced into condensed matter physics in a rather direct and straightforward fashion. The nonrelativistic electrons in a condensed matter system can be described by a field ψ, along the lines discussed in chapter III.5. Field theoretic Lagrangians may then be written down, Feynman diagrams and rules developed, and so on and so forth. This is done in a number of specialized texts. What we present here is to a large extent the more modern view of an effective field theoretic description of a condensed matter system, valid at low energy and momentum. One of the fascinations of condensed matter physics is that due to highly nontrivial many body effects the low energy degrees of freedom might be totally different from the electrons we started out with. A particularly striking example (to be discussed in chapter VI.2) is the quantum Hall system, in which the low energy effective degree of freedom carries fractional charge and statistics.

Another advantage of devoting a considerable portion of a field theory book to condensed matter physics is that historically and pedagogically it is much easier to understand the renormalization group in condensed matter physics than in particle physics.

I will defiantly not stick to a legalistic separation between condensed matter and particle physics. Some of the topics treated in Parts V and VI actually belong to particle physics. And of course I cannot be responsible for explaining condensed matter physics, any more than I could be responsible for explaining particle physics in chapter IV.2.

Repulsive bosons

Consider a finite density $\bar{\rho}$ of nonrelativistic bosons interacting with a short ranged repulsion. Return to (III.5.11):

$$\mathcal{L} = i\varphi^\dagger \partial_0 \varphi - \frac{1}{2m} \partial_i \varphi^\dagger \partial_i \varphi - g^2 (\varphi^\dagger \varphi - \bar{\rho})^2 \tag{1}$$

The last term is exactly the Mexican well potential of chapter IV.1, forcing the magnitude of φ to be close to $\sqrt{\bar{\rho}}$, thus suggesting that we use polar variables $\varphi \equiv \sqrt{\rho} e^{i\theta}$ as we did in (III.5.7). Plugging in and dropping the total derivative $(i/2)\partial_0 \rho$, we obtain

$$\mathcal{L} = -\rho \partial_0 \theta - \frac{1}{2m} \left[\frac{1}{4\rho} (\partial_i \rho)^2 + \rho (\partial_i \theta)^2 \right] - g^2 (\rho - \bar{\rho})^2 \tag{2}$$

Spontaneous symmetry breaking

As in chapter IV.1 write $\sqrt{\rho} = \sqrt{\bar{\rho}} + h$ (the vacuum expectation value of φ is $\sqrt{\bar{\rho}}$), assume $h \ll \sqrt{\bar{\rho}}$, and expand[1]:

$$\mathcal{L} = -2\sqrt{\bar{\rho}} h \partial_0 \theta - \frac{\bar{\rho}}{2m} (\partial_i \theta)^2 - \frac{1}{2m} (\partial_i h)^2 - 4g^2 \bar{\rho} h^2 + \cdots \tag{3}$$

Picking out the terms up to quadratic in h in (3) we use the "central identity of quantum field theory" (see appendix A) to integrate out h, obtaining

$$\mathcal{L} = \bar{\rho} \partial_0 \theta \frac{1}{4g^2 \bar{\rho} - (1/2m)\partial_i^2} \partial_0 \theta - \frac{\bar{\rho}}{2m} (\partial_i \theta)^2 + \cdots$$

$$= \frac{1}{4g^2} (\partial_0 \theta)^2 - \frac{\bar{\rho}}{2m} (\partial_i \theta)^2 + \cdots \tag{4}$$

---

[1] Note that we have dropped the (potentially interesting) term $-\bar{\rho} \partial_0 \theta$ because it is a total divergence.

In the second equality we assumed that we are looking at processes with wave number k small compared to $\sqrt{8g^2\bar{\rho}m}$ so that $(1/2m)\partial_i^2$ is negligible compared to $4g^2\bar{\rho}$. Thus, we see that there exists in this fluid of bosons a gapless mode (often referred to as the phonon) with the dispersion

$$\omega^2 = \frac{2g^2\bar{\rho}}{m}\vec{k}^2 \tag{5}$$

The learned among you will have realized that we have obtained Bogoliubov's classic result without ever doing a Bogoliubov rotation.[2]

Let me briefly remind you of Landau's idealized argument[3] that a linearly dispersing mode (that is, ω is linear in k) implies superfluidity. Consider a mass M of fluid flowing down a tube with velocity v. It could lose momentum and slow down to velocity v' by creating an excitation of momentum k: $Mv = Mv' + \hbar k$. This is only possible with sufficient energy to spare if $\frac{1}{2}Mv^2 \geq \frac{1}{2}Mv'^2 + \hbar\omega(k)$. Eliminating v' we obtain for M macroscopic $v \geq \omega/k$. For a linearly dispersing mode this gives a critical velocity $v_c \equiv \omega/k$ below which the fluid cannot lose momentum and is hence super. [Thus, from (5) the idealized $v_c = g\sqrt{2\bar{\rho}/m}$.]

Suitably scaling the distance variable, we can summarize the low energy physics of superfluidity in the compact Lagrangian

$$\mathcal{L} = \frac{1}{4g^2}(\partial_\mu\theta)^2 \tag{6}$$

which we recognize as the massless version of the scalar field theory we studied in part I, but with the important proviso that θ is a phase angle field, that is, $\theta(x)$ and $\theta(x) + 2\pi$ are really the same. This gapless mode is evidently the Nambu-Goldstone boson associated with the spontaneous breaking of the global $U(1)$ symmetry $\varphi \to e^{i\alpha}\varphi$.

Linearly dispersing gapless mode

The physics here becomes particularly clear if we think about a gas of free bosons. We can give a momentum $\hbar\vec{k}$ to any given boson at the cost of only $(\hbar\vec{k})^2/2m$ in energy. There exist many low energy excitations in a free boson system. But as soon as a short ranged repulsion is turned on between the bosons, a boson moving with momentum $\vec{k}$ would affect all the other bosons. A density wave is set up as a result, with energy proportional to k as we have shown in (5). The gapless mode has gone from quadratically dispersing to linearly dispersing. There are far fewer low energy excitations. Specifically, recall that the density of states is given by $N(E) \propto k^{D-1}(dk/dE)$. For example, for $D = 2$ the density of states goes from $N(E) \propto$ constant (in the presence of quadratically dispersing modes) to $N(E) \propto E$ (in the presence of linearly dispersing modes) at low energies.

---

[2] L. D. Landau and E. M. Lifshitz, *Statistical Physics*, p. 238.
[3] Ibid., p. 192.

As was emphasized by Feynman[4] among others, the physics of superfluidity lies not in the presence of gapless excitations, but in the paucity of gapless excitations. (After all, the Fermi liquid has a continuum of gapless modes.) There are too few modes that the superfluid can lose energy and momentum to.

Relativistic versus nonrelativistic

This is a good place to discuss one subtle difference between spontaneous symmetry breaking in relativistic and nonrelativistic theories. Consider the relativistic theory studied in chapter IV.1: $\mathcal{L} = (\partial \Phi^\dagger)(\partial \Phi) - \lambda(\Phi^\dagger \Phi - v^2)^2$. It is often convenient to take the $\lambda \to \infty$ limit holding v fixed. In the language used in chapter IV.1 "climbing the wall" costs infinitely more energy than "rolling along the gutter." The resulting theory is defined by

$$\mathcal{L} = (\partial \Phi^\dagger)(\partial \Phi) \tag{7}$$

with the constraint $\Phi^\dagger \Phi = v^2$. This is known as a nonlinear σ model, about which much more in chapter VI.4.

The existence of a Nambu-Goldstone boson is particularly easy to see in the nonlinear σ model. The constraint is solved by $\Phi = v e^{i\theta}$, which when plugged into $\mathcal{L}$ gives $\mathcal{L} = v^2(\partial \theta)^2$. There it is: the Nambu-Goldstone boson θ.

Let's repeat this in the nonrelativistic domain. Take the limit $g^2 \to \infty$ with $\bar{\rho}$ held fixed so that (1) becomes

$$\mathcal{L} = i\varphi^\dagger \partial_0 \varphi - \frac{1}{2m} \partial_i \varphi^\dagger \partial_i \varphi \tag{8}$$

with the constraint $\varphi^\dagger \varphi = \bar{\rho}$. But now if we plug the solution of the constraint $\varphi = \sqrt{\bar{\rho}} e^{i\theta}$ into $\mathcal{L}$ (and drop the total derivative $-\bar{\rho} \partial_0 \theta$), we get $\mathcal{L} = -(\bar{\rho}/2m)(\partial_i \theta)^2$ with the equation of motion $\partial_i^2 \theta = 0$. Oops, what is this? It's not even a propagating degree of freedom? Where is the Nambu-Goldstone boson?

Knowing what I already told you, you are not going to be puzzled by this apparent paradox[5] for long, but believe me, I have stumped quite a few excellent relativistic minds with this one. The Nambu-Goldstone boson is still there, but as we can see from (5) its propagation velocity ω/k scales to infinity as g and thus it disappears from the spectrum for any nonzero $\vec{k}$.

Why is it that we are allowed to go to this "nonlinear" limit in the relativistic case? Because we have Lorentz invariance! The velocity of a linearly dispersing mode, if such a mode exists, is guaranteed to be equal to 1.

---

[4] R. P. Feynman, *Statistical Mechanics*.

[5] This apparent paradox was discussed by A. Zee, "From Semionics to Topological Fluids" in O. J. P. Ébolic et al., eds., *Particle Physics*, p. 415.

Exercises

V.1.1 Verify that the approximation used to reach (3) is consistent.

V.1.2 To confine the superfluid in an external potential $W(\vec{x})$ we would add the term $-W(\vec{x})\varphi^{\dagger}(\vec{x}, t)\varphi(\vec{x}, t)$ to (1). Derive the corresponding equation of motion for φ. The equation, known as the Gross-Pitaevski equation, has been much studied in recent years in connection with the Bose-Einstein condensate.

Euclid, Boltzmann, Hawking, and Field Theory at Finite Temperature

Statistical mechanics and Euclidean field theory

I mentioned in chapter I.2 that to define the path integral more rigorously we should perform a Wick rotation $t = -it_E$. The scalar field theory, instead of being defined by the Minkowskian path integral

$$Z = \int D\varphi e^{(i/\hbar) \int d^d x [\frac{1}{2}(\partial\varphi)^2 - V(\varphi)]} \tag{1}$$

is then defined by the Euclidean functional integral

$$Z = \int D\varphi e^{-(1/\hbar) \int d^d_E x [\frac{1}{2}(\partial\varphi)^2 + V(\varphi)]} = \int D\varphi e^{-(1/\hbar)\mathcal{E}(\varphi)} \tag{2}$$

where $d^d x = -i d^d_E x$, with $d^d_E x \equiv dt_E d^{(d-1)} x$. In (1) $(\partial\varphi)^2 = (\partial\varphi/\partial t)^2 - (\vec{\nabla}\varphi)^2$, while in (2) $(\partial\varphi)^2 = (\partial\varphi/\partial t_E)^2 + (\vec{\nabla}\varphi)^2$: The notation is a tad confusing but I am trying not to introduce too many extraneous symbols. You may or may not find it helpful to think of $(\vec{\nabla}\varphi)^2 + V(\varphi)$ as one unit, untouched by Wick rotation. I have introduced $\mathcal{E}(\varphi) \equiv \int d^d_E x [\frac{1}{2}(\partial\varphi)^2 + V(\varphi)]$, which may naturally be regarded as a static energy functional of the field $\varphi(x)$. Thus, given a configuration $\varphi(x)$ in d-dimensional space, the more it varies, the less likely it is to contribute to the Euclidean functional integral Z.

The Euclidean functional integral (2) may remind you of statistical mechanics. Indeed, Herr Boltzmann taught us that in thermal equilibrium at temperature $T = 1/\beta$, the probability for a configuration to occur in a classical system or the probability for a state to occur in a quantum system is just the Boltzmann factor $e^{-\beta E}$ suitably normalized, where E is to be interpreted as the energy of the configuration in a classical system or as the energy eigenvalue of the state in a quantum system. In particular, recall the classical statistical mechanics of an N-particle system for which

$$E(p, q) = \sum_i \frac{1}{2m} p_i^2 + V(q_1, q_2, \cdots, q_N)$$

The partition function is given (up to some overall constant) by

$$Z = \prod_i \int dp_i dq_i e^{-\beta E(p,q)}$$

After doing the integrals over p we are left with the (reduced) partition function

$$Z = \prod_i \int dq_i e^{-\beta V(q_1, q_2, \cdots, q_N)}$$

Promoting this to a field theory as in chapter I.3, letting $i \to x$ and $q_i \to \varphi(x)$ as before, we see that the partition function of a classical field theory with the static energy functional $\mathcal{E}(\varphi)$ has precisely the form in (2), upon identifying the symbol $\hbar$ as the temperature $T = 1/\beta$. Thus,

$$\boxed{\begin{array}{l} \text{Euclidean quantum field theory in } d\text{-dimensional spacetime} \\ \sim \text{Classical statistical mechanics in } d\text{-dimensional space} \end{array}} \tag{3}$$

Functional integral representation of the quantum partition function

More interestingly, we move on to quantum statistical mechanics. The integration over phase space $\{p, q\}$ is replaced by a trace, that is, a sum over states: Thus the partition function of a quantum mechanical system (say of a single particle to be definite) with the Hamiltonian H is given by

$$Z = \mathrm{tr}\ e^{-\beta H} = \sum_n \langle n| e^{-\beta H} |n\rangle$$

In chapter I.2 we worked out the integral representation of $\langle F| e^{-iHT} |I\rangle$. (You should not confuse the time T with the temperature T of course.) Suppose we want an integral representation of the partition function. No need to do any more work! We simply replace the time T by $-i\beta$, set $|I\rangle = |F\rangle = |n\rangle$ and sum over $|n\rangle$ to obtain

$$Z = \mathrm{tr}\ e^{-\beta H} = \int_{\mathrm{PBC}} Dq e^{-\int_0^\beta d\tau L(q)} \tag{4}$$

Tracing the steps from (I.2.3) to (I.2.5) you can verify that here $L(q) = \frac{1}{2}(dq/d\tau)^2 + V(q)$ is precisely the Lagrangian corresponding to H in the Euclidean time τ. The integral over τ runs from 0 to β. The trace operation sets the initial and final states equal and so the functional integral should be done over all paths $q(\tau)$ with the boundary condition $q(0) = q(\beta)$. The subscript PBC reminds us of this all important periodic boundary condition.

The extension to field theory is immediate. If H is the Hamiltonian of a quantum field theory in D-dimensional space [and hence $d = (D + 1)$-dimensional spacetime], then the partition function (4) is

$$Z = \mathrm{tr}\ e^{-\beta H} = \int_{\mathrm{PBC}} D\varphi e^{-\int_0^\beta d\tau \int d^D x \mathcal{L}(\varphi)} \tag{5}$$

with the integral evaluated over all paths $\varphi(\vec{x}, \tau)$ such that

$$\varphi(\vec{x}, 0) = \varphi(\vec{x}, \beta) \tag{6}$$

(Here φ represents all the Bose fields in the theory.)

A remarkable result indeed! To study a field theory at finite temperature all we have to do is rotate it to Euclidean space and impose the boundary condition (6). Thus,

> Euclidean quantum field theory in $(D + 1)$-dimensional
>
> spacetime, $0 \leq \tau < \beta$ $\qquad\qquad\qquad$ (7)
>
> $\sim$ Quantum statistical mechanics in D-dimensional space

In the zero temperature limit $\beta \to \infty$ we recover from (5) the standard Wick-rotated quantum field theory over an infinite spacetime, as we should.

Surely you would hit it big with mystical types if you were to tell them that temperature is equivalent to cyclic imaginary time. At the arithmetic level this connection comes merely from the fact that the central objects in quantum physics e^{-iHT} and in thermal physics $e^{-\beta H}$ are formally related by analytic continuation. Some physicists, myself included, feel that there may be something profound here that we have not quite understood.

Finite temperature Feynman diagrams

If we so desire, we can develop the finite temperature perturbation theory of (5), working out the Feynman rules and so forth. Everything goes through as before with one major difference stemming from the condition (6) $\varphi(\vec{x}, \tau = 0) = \varphi(\vec{x}, \tau = \beta)$. Clearly, when we Fourier transform with the factor $e^{i\omega\tau}$, the Euclidean frequency ω can take on only discrete values $\omega_n \equiv (2\pi/\beta)n$, with n an integer. The propagator of the scalar field becomes $1/(k_4^2 + \vec{k}^2) \to 1/(\omega_n^2 + \vec{k}^2)$. Thus, to evaluate the partition function, we simply write the relevant Euclidean Feynman diagrams and instead of integrating over frequency we sum over a discrete set of frequencies $\omega_n = (2\pi T)n$, $n = -\infty, \cdots, +\infty$. In other words, after you beat a Feynman integral down to the form $\int d_E^d k F(k_E^2)$, all you have to do is replace it by $2\pi T \sum_n \int d^D k F[(2\pi T)^2 n^2 + \vec{k}^2]$.

It is instructive to see what happens in the high-temperature $T \to \infty$ limit. In summing over ω_n, the $n = 0$ term dominates since the combination $(2\pi T)^2 n^2 + \vec{k}^2$ occurs in the denominator. Hence, the diagrams are evaluated effectively in D-dimensional space. We lose a dimension! Thus,

> Euclidean quantum field theory in D-dimensional spacetime
>
> $\sim$ High-temperature quantum statistical mechanics in $\qquad$ (8)
>
> D-dimensional space

This is just the statement that at high-temperature quantum statistical mechanics goes classical [compare (3)].

An important application of quantum field theory at finite temperature is to cosmology: The early universe may be described as a soup of elementary particles at some high temperature.

Hawking radiation

Hawking radiation from black holes is surely the most striking prediction of gravitational physics of the last few decades. The notion of black holes goes all the way back to Michell and Laplace, who noted that the escape velocity from a sufficiently massive object may exceed the speed of light. Classically, things fall into black holes and that's that. But with quantum physics a black hole can in fact radiate like a black body at a characteristic temperature T.

Remarkably, with what little we learned in chapter I.11 and here, we can actually determine the black hole temperature. I hasten to add that a systematic development would be quite involved and fraught with subtleties; indeed, entire books are devoted to this subject. However, what we need to do is more or less clear. Starting with chapter I.11, we would have to develop quantum field theory (for instance, that of a scalar field φ) in curved spacetime, in particular in the presence of a black hole, and ask what a vacuum state (i.e., a state devoid of φ quanta) in the far past evolves into in the far future. We would find a state filled with a thermal distribution of φ quanta. We will not do this here.

In hindsight, people have given numerous heuristic arguments for Hawking radiation. Here is one. Let us look at the Schwarzschild solution (see chapter I.11)

$$ds^2 = \left(1 - \frac{2GM}{r}\right) dt^2 - \left(1 - \frac{2GM}{r}\right)^{-1} dr^2 - r^2 d\theta^2 - r^2 \sin^2 \theta \, d\phi^2 \tag{9}$$

At the horizon $r = 2GM$, the coefficients of dt^2 and dr^2 change sign, indicating that time and space, and hence energy and momentum, are interchanged. Clearly, something strange must occur. With quantum fluctuations, particle and antiparticle pairs are always popping in and out of the vacuum, but normally, as we had discussed earlier, the uncertainty principle limits the amount of time Δt the pairs can exist to $\sim 1/\Delta E$. Near the black hole horizon, the situation is different. A pair can fluctuate out of the vacuum right at the horizon, with the particle just outside the horizon and the antiparticle just inside; heuristically the Heisenberg restriction on Δt may be evaded since what is meant by energy changes as we cross the horizon. The antiparticle falls in while the particle escapes to spatial infinity. Of course, a hand-waving argument like this has to be backed up by detailed calculations.

If black holes do indeed radiate at a definite temperature T, and that is far from obvious a priori, we can estimate T easily by dimensional analysis. From (9) we see that only the combination GM, which evidently has the dimension of a length, can come in. Since T has the dimension of mass, that is, length inverse, we can only have $T \propto 1/GM$.

To determine T precisely, we resort to a rather slick argument. I warn you from the outset that the argument will be slick and should be taken with a grain of salt. It is only meant to whet your appetite for a more correct treatment.

Imagine quantizing a scalar field theory in the Schwarzschild metric, along the line described in chapter I.11. If upon Wick rotation the field "feels" that time is periodic with period β, then according to what we have learned in this chapter the quanta of the scalar field would think that they are living in a heat bath with temperature $T = 1/\beta$.

Setting $t \to -i\tau$, we rotate the metric to

$$ds^2 = -\left[\left(1 - \frac{2GM}{r}\right)d\tau^2 + \left(1 - \frac{2GM}{r}\right)^{-1}dr^2 + r^2 d\theta^2 + r^2 \sin^2\theta d\phi^2\right] \tag{10}$$

In the region just outside the horizon $r \gtrsim 2GM$, we perform the general coordinate transformation $(\tau, r) \to (\alpha, R)$ so that the first two terms in ds^2 become $R^2 d\alpha^2 + dR^2$, namely the length element squared of flat 2-dimensional Euclidean space in polar coordinates.

To leading order, we can write the Schwarzchild factor $(1 - 2GM/r)$ as $(r - 2GM)/(2GM) \equiv \gamma^2 R^2$ with the constant γ to be determined. Then the second term becomes $dr^2/(\gamma^2 R^2) = (4GM)^2 \gamma^2 dR^2$, and thus we set $\gamma = 1/(4GM)$ to get the desired dR^2. The first two terms in $-ds^2$ are then given by $R^2(d\tau/(4GM))^2 + dR^2$. Thus the Euclidean time is related to the polar angle by $\tau = 4GM\alpha$ and so has a period of $8\pi GM = \beta$. We obtain thus the Hawking temperature

$$T = \frac{1}{8\pi GM} = \frac{\hbar c^3}{8\pi GM} \tag{11}$$

Restoring $\hbar$ by dimensional analysis, we see that Hawking radiation is indeed a quantum effect.

It is interesting to note that the Wick rotated geometry just outside the horizon is given by the direct product of a plane with a 2-sphere of radius $2GM$, although, this observation is not needed for the calculation we just did.

Exercises

V.2.1 Study the free field theory $\mathcal{L} = \frac{1}{2}(\partial\varphi)^2 - \frac{1}{2}m^2\varphi^2$ at finite temperature and derive the Bose-Einstein distribution.

V.2.2 It probably does not surprise you that for fermionic fields the periodic boundary condition (6) is replaced by an antiperiodic boundary condition $\psi(\vec{x}, 0) = -\psi(\vec{x}, \beta)$ in order to reproduce the results of chapter II.5. Prove this by looking at the simplest fermionic functional integral. [Hint: The clearest exposition of this satisfying fact may be found in appendix A of R. Dashen, B. Hasslacher, and A. Neveu, *Phys. Rev.* D12: 2443, 1975.]

V.2.3 It is interesting to consider quantum field theory at finite density, as may occur in dense astrophysical objects or in heavy ion collisions. (In the previous chapter we studied a system of bosons at finite density and zero temperature.) In statistical mechanics we learned to go from the partition function to the grand partition function $Z = \text{tr } e^{-\beta(H-\mu N)}$, where a chemical potential μ is introduced for every conserved particle number N. For example, for noninteracting relativistic fermions, the Lagrangian is modified to $\mathcal{L} = \bar{\psi}(i\not{\partial} - m)\psi + \mu\bar{\psi}\gamma^0\psi$. Note that finite density, as well as finite temperature, breaks Lorentz invariance. Develop the subject of quantum field theory at finite density as far as you can.

V.3 | Landau-Ginzburg Theory of Critical Phenomena

The emergence of nonanalyticity

Historically, the notion of spontaneous symmetry breaking, originating in the work of Landau and Ginzburg on second-order phase transitions, came into particle physics from condensed matter physics.

Consider a ferromagnetic material in thermal equilibrium at temperature T. The magnetization $\vec{M}(x)$ is defined as the average of the atomic magnetic moments taken over a region of a size much larger than the length scale characteristic of the relevant microscopic physics. (In this chapter, we are discussing a nonrelativistic theory and x denotes the spatial coordinates only.) We know that at low temperatures, rotational invariance is spontaneously broken and that the material exhibits a bulk magnetization pointing in some direction. As the temperature is raised past some critical temperature T_c the bulk magnetization suddenly disappears. We understand that with increased thermal agitation the atomic magnetic moments point in increasingly random directions, canceling each other out. More precisely, it was found experimentally that just below T_c the magnetization $|\vec{M}|$ vanishes as $\sim (T_c - T)^\beta$, where the so-called critical exponent $\beta \simeq 0.37$.

This sudden change is known as a second order phase transition, an example of a critical phenomenon. Historically, critical phenomena presented a challenge to theoretical physicists. In principle, we are to compute the partition function $Z = \text{tr } e^{-\mathcal{H}/T}$ with the microscopic Hamiltonian $\mathcal{H}$, but Z is apparently smooth in T except possibly at $T = 0$. Some physicists went as far as saying that nonanalytic behavior such as $(T_c - T)^\beta$ is impossible and that within experimental error $|\vec{M}|$ actually vanishes as a smooth function of T. Part of the importance of Onsager's famous exact solution in 1944 of the 2-dimensional Ising model is that it settled this question definitively. The secret is that an infinite sum of terms each of which may be analytic in some variable need not be analytic in that variable. The trace in tr $e^{-\mathcal{H}/T}$ sums over an infinite number of terms.

Arguing from symmetry

In most situations, it is essentially impossible to calculate Z starting with the microscopic Hamiltonian. Landau and Ginzburg had the brilliant insight that the form of the free energy G as a function of $\vec{M}$ for a system with volume V could be argued from general principles. First, for $\vec{M}$ constant in x, we have by rotational invariance

$$G = V[a\vec{M}^2 + b(\vec{M}^2)^2 + \cdots]$$ (1)

where $a, b, \cdots$ are unknown (but expected to be smooth) functions of T. Landau and Ginzburg supposed that a vanishes at some temperature T_c. Unless there is some special reason, we would expect that for T near T_c we have $a = a_1(T - T_c) + \cdots$ [rather than, say, $a = a_2(T - T_c)^2 + \cdots$]. But you already learned in chapter IV.1 what would happen. For $T > T_c$, G is minimized at $\vec{M} = 0$, but as T drops below T_c, new minima suddenly develop at $|\vec{M}| = \sqrt{(-a/2b)} \sim (T_c - T)^{\frac{1}{2}}$. Rotational symmetry is spontaneously broken, and the mysterious nonanalytic behavior pops out easily.

To include the possibility of $\vec{M}$ varying in space, Landau and Ginzburg argued that G must have the form

$$G = \int d^3x\{\partial_i\vec{M}\partial_i\vec{M} + a\vec{M}^2 + b(\vec{M}^2)^2 + \cdots\}$$ (2)

where the coefficient of the $(\partial_i\vec{M})^2$ term has been set to 1 by rescaling $\vec{M}$. You would recognize (2) as the Euclidean version of the scalar field theory we have been studying. By dimensional analysis we see that $1/\sqrt{a}$ sets the length scale. More precisely, for $T > T_c$, let us turn on a perturbing external magnetic field $\vec{H}(x)$ by adding the term $-\vec{H} \cdot \vec{M}$. Assuming $\vec{M}$ small and minimizing G we obtain $(-\partial^2 + a)\vec{M} \simeq \vec{H}$, with the solution

$$\vec{M}(x) = \int d^3y \int \frac{d^3k}{(2\pi)^3} \frac{e^{i\vec{k}\cdot(\vec{x}-\vec{y})}}{\vec{k}^2 + a}\vec{H}(y)$$

$$= \int d^3y \frac{1}{4\pi|\vec{x}-\vec{y}|}e^{-\sqrt{a}|\vec{x}-\vec{y}|}\vec{H}(y)$$ (3)

[Recall that we did the integral in (I.4.7)—admire the unity of physics!]

It is standard to define a correlation function $< \vec{M}(x)\vec{M}(0) >$ by asking what the magnetization $\vec{M}(x)$ will be if we use a magnetic field sharply localized at the origin to create a magnetization $\vec{M}(0)$ there. We expect the correlation function to die off as $e^{-|\vec{x}|/\xi}$ over some correlation length ξ that goes to infinity as T approaches T_c from above. The critical exponent ν is traditionally defined by $\xi \sim 1/(T - T_c)^\nu$.

In Landau-Ginzburg theory, also known as mean field theory, we obtain $\xi = 1/\sqrt{a}$ and hence $\nu = \frac{1}{2}$.

The important point is not how well the predicted critical exponents such as β and ν agree with experiment but how easily they emerge from Landau-Ginzburg theory. The theory provides a starting point for a complete theory of critical phenomena, which was eventually developed by Kadanoff, Fisher, Wilson, and others using the renormalization group (to be discussed in chapter VI.8).

The story goes that Landau had a logarithmic scale with which he ranked theoretical physicists, with Einstein on top, and that after working out Landau-Ginzburg theory he moved himself up by half a notch.

Exercise

V.3.1   Another important critical exponent γ is defined by saying that the susceptibility $\chi \equiv (\partial M/\partial H)|_{H=0}$ diverges $\sim 1/|T - T_c|^\gamma$ as T approaches T_c. Determine γ in Landau-Ginzburg theory. [Hint: Instructively, there are two ways of doing it: (a) Add $-\vec{H} \cdot \vec{M}$ to (1) for $\vec{M}$ and $\vec{H}$ constant in space and solve for $\vec{M}(\vec{H})$. (b) Calculate the susceptibility function $\chi_{ij}(x - y) \equiv [\partial M_i(x)/\partial H_j(y)]|_{H=0}$ and integrate over space.]

Pairing and condensation

When certain materials are cooled below a certain critical temperature T_c, they suddenly become superconducting. Historically, physicists had long suspected that the superconducting transition, just like the superfluid transition, has something to do with Bose-Einstein condensation. But electrons are fermions, not bosons, and thus they first have to pair into bosons, which then condense. We now know that this general picture is substantially correct: Electrons form Cooper pairs, whose condensation is responsible for superconductivity.

With brilliant insight, Landau and Ginzburg realized that without having to know the detailed mechanism driving the pairing of electrons into bosons, they could understand a great deal about superconductivity by studying the field $\varphi(x)$ associated with these condensing bosons. In analogy with the ferromagnetic transition in which the magnetization $\vec{M}(x)$ in a ferromagnet suddenly changes from zero to a nonzero value when the temperature drops below some critical temperature, they proposed that $\varphi(x)$ becomes nonzero for temperatures below T_c. (In this chapter x denotes spatial coordinates only.) In statistical physics, quantities such as $\vec{M}(x)$ and $\varphi(x)$ that change through a phase transition are known as order parameters.

The field $\varphi(x)$ carries two units of electric charge and is therefore complex. The discussion now unfolds much as in chapter V.3 except that $\partial_i \varphi$ should be replaced by $D_i \varphi \equiv (\partial_i - i2eA_i)\varphi$ since φ is charged. Following Landau and Ginzburg and including the energy of the external magnetic field, we write the free energy as

$$\mathcal{F} = \frac{1}{4}F_{ij}^2 + |D_i\varphi|^2 + a|\varphi|^2 + \frac{b}{2}|\varphi|^4 + \cdots \tag{1}$$

which is clearly invariant under the $U(1)$ gauge transformation $\varphi \to e^{i2e\Lambda}\varphi$ and $A_i \to A_i + \partial_i \Lambda$. As before, setting the coefficient of $|D_i\varphi|^2$ equal to 1 just amounts to a normalization choice for φ.

The similarity between (1) and (IV.6.1) should be evident.

Meissner effect

A hallmark of superconductivity is the Meissner effect, in which an external magnetic field $\vec{B}$ permeating the material is expelled from it as the temperature drops below T_c. This indicates that a constant magnetic field inside the material is not favored energetically. The effective laws of electromagnetism in the material must somehow change at T_c. Normally, a constant magnetic field would cost an energy of the order $\sim \vec{B}^2 V$, where V is the volume of the material. Suppose that the energy density is changed from the standard $\vec{B}^2$ to $\vec{A}^2$ (where as usual $\vec{\nabla} \times \vec{A} = \vec{B}$). For a constant magnetic field $\vec{B}$, $\vec{A}$ grows as the distance and hence the total energy would grow faster than V. After the material goes superconducting, we have to pay an unacceptably large amount of extra energy to maintain the constant magnetic field and so it is more favorable to expel the magnetic field.

Note that a term like $\vec{A}^2$ in the effective energy density preserves rotational and translational invariance but violates electromagnetic gauge invariance. But we already know how to break gauge invariance from chapter IV.6. Indeed, the $U(1)$ gauge theory described there and the theory of superconductivity described here are essentially the same, related by a Wick rotation.

As in chapter V.3 we suppose that for temperature $T \simeq T_c$, $a \simeq a_1(T - T_c)$ while b remains positive. The free energy $\mathcal{F}$ is minimized by $\varphi = 0$ above T_c, and by $|\varphi| = \sqrt{-a/b} \equiv v$ below T_c. All this is old hat to you, who have learned that upon symmetry breaking in a gauge theory the gauge field gains a mass. We simply read off from (1) that

$$\mathcal{F} = \tfrac{1}{4}F_{ij}^2 + (2ev)^2 A_i^2 + \cdots \tag{2}$$

which is precisely what we need to explain the Meissner effect.

London penetration length and coherence length

Physically, the magnetic field does not drop precipitously from some nonzero value outside the superconductor to zero inside, but drops over some characteristic length scale, called the London penetration length. The magnetic field leaks into the superconductor a bit over a length scale l, determined by the competition between the energy in the magnetic field $F_{ij}^2 \sim (\partial A)^2 \sim A^2/l^2$ and the Meissner term $(2ev)^2 A^2$ in (2). Thus, Landau and Ginzburg obtained the London penetration length $l_L \sim (1/ev) = (1/e)\sqrt{b/-a}$.

Similarly, the characteristic length scale over which the order parameter φ varies is known as the coherence length l_φ, which can be estimated by balancing the second and third terms in (1), roughly $(\partial \varphi)^2 \sim \varphi^2/l_\varphi^2$ and $a\varphi^2$ against each other, giving a coherence length of order $l_\varphi \sim 1/\sqrt{-a}$.

Putting things together, we have

$$\frac{l_L}{l_\varphi} \sim \frac{\sqrt{b}}{e} \tag{3}$$

You might recognize from chapter IV.6 that this is just the ratio of the mass of the scalar field to the mass of the vector field.

As I remarked earlier, the concept of spontaneous symmetry breaking went from condensed matter physics to particle physics. After hearing a talk at the University of Chicago on the Bardeen-Cooper-Schrieffer theory of superconductivity by the young Schrieffer, Nambu played an influential role in bringing spontaneous symmetry breaking to the particle physics community.

Exercises

V.4.1   Vary (1) to obtain the equation for A and determine the London penetration length more carefully.

V.4.2   Determine the coherence length more carefully.

Noninteracting hopping electrons

The appearance of the Dirac equation and a relativistic field theory in a solid would be surprising indeed, but yes, it is possible.

Consider the Hamiltonian

$$H = -t \sum_j (c_{j+1}^\dagger c_j + c_j^\dagger c_{j+1}) \tag{1}$$

describing noninteracting electrons hopping on a 1-dimensional lattice (figure V.5.1). Here c_j annihilates an electron on site j. Thus, the first term describes an electron hopping from site j to site $j + 1$ with amplitude t. We have suppressed the spin labels. This is just about the simplest solid state model; a good place to read about it is in Feynman's "Freshman lectures."

Fourier transforming $c_j = \sum_k e^{ikaj} c(k)$ (where a is the spacing between sites), we immediately find the energy spectrum $\varepsilon(k) = -2t \cos ka$ (fig. V.5.2). Imposing a periodic boundary condition on a lattice with N sites, we have $k = (2\pi/Na)n$ with n an integer from $-\frac{1}{2}N$ to $\frac{1}{2}N$. As $N \to \infty$, k becomes a continuous rather than a discrete variable. As usual, the Brillouin zone is defined by $-\pi/a < k \leq \pi/a$.

There is absolutely nothing relativistic about any of this. Indeed, at the bottom of the spectrum the energy (up to an irrelevant additive constant) goes as $\varepsilon(k) \simeq 2t\frac{1}{2}(ka)^2 \equiv k^2/2m_{\text{eff}}$. The electron disperses nonrelativistically with an effective mass m_{eff}.

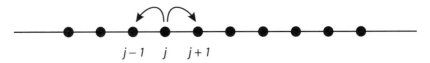

$j-1 \quad j \quad j+1$

Figure V.5.1

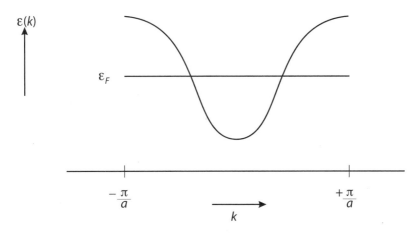

Figure V.5.2

But now let us fill the system with electrons up to some Fermi energy ε_F (see fig. V.5.2). Focus on an electron near the Fermi surface and measure its energy from ε_F and momentum from $+k_F$. Suppose we are interested in electrons with energy small compared to ε_F, that is, $E \equiv \varepsilon - \varepsilon_F \ll \varepsilon_F$, and momentum small compared to k_F, that is, $p \equiv k - k_F \ll k_F$. These electrons obey a linear energy-momentum dispersion $E = v_F p$ with the Fermi velocity $v_F = (\partial \varepsilon / \partial k)|_{k=k_F}$. We will call the field associated with these electrons ψ_R, where the subscript indicates that they are "right moving." It satisfies the equation of motion $(\partial / \partial t + v_F \partial / \partial x)\psi_R = 0$.

Similarly, the electrons with momentum around $-k_F$ obey the dispersion $E = -v_F p$. We will call the field associated with these electrons ψ_L with L for "left moving," satisfying $(\partial / \partial t - v_F \partial / \partial x)\psi_L = 0$.

Emergence of the Dirac equation

The Lagrangian summarizing all this is simply

$$\mathcal{L} = i\psi_R^\dagger \left(\frac{\partial}{\partial t} + v_F \frac{\partial}{\partial x} \right) \psi_R + i\psi_L^\dagger \left(\frac{\partial}{\partial t} - v_F \frac{\partial}{\partial x} \right) \psi_L \tag{2}$$

Introducing a 2-component field $\psi = \begin{pmatrix} \psi_L \\ \psi_R \end{pmatrix}$, $\bar{\psi} \equiv \psi^\dagger \gamma^0 \equiv \psi^\dagger \sigma_2$, and choosing units so that $v_F = 1$, we may write $\mathcal{L}$ more compactly as

$$\mathcal{L} = i\psi^\dagger \left(\frac{\partial}{\partial t} - \sigma_3 \frac{\partial}{\partial x} \right) \psi = \bar{\psi} i \gamma^\mu \partial_\mu \psi \tag{3}$$

with $\gamma^0 = \sigma_2$ and $\gamma^1 = i\sigma_1$ satisfying the Clifford algebra $\{\gamma^\mu, \gamma^\nu\} = 2g^{\mu\nu}$.

Amazingly enough, the $(1+1)$-dimensional Dirac Lagrangian emerges in a totally nonrelativistic situation!

An instability

I will now go on to discuss an important phenomenon known as Peierls's instability. I will necessarily have to be a bit sketchy. I don't have to tell you again that this is not a text on solid state physics, but in any case you will not find it difficult to fill in the gaps.

Peierls considered a distortion in the lattice, with the ion at site j displaced from its equilibrium position by $\cos[q(ja)]$. (Shades of our mattress from chapter I.1!) A lattice distortion with wave vector $q = 2k_F$ will connect electrons with momentum k_F with electrons with momentum $-k_F$. In other words, it connects right moving ones with left moving electrons, or in our field theoretic language ψ_R with ψ_L. Since the right moving electrons and the left moving electrons on the surface of the Fermi sea have the same energy (namely ε_F, duh!) we have the always interesting situation of degenerate perturbation theory: $\begin{pmatrix} \varepsilon_F & 0 \\ 0 & \varepsilon_F \end{pmatrix} + \begin{pmatrix} 0 & \delta \\ \delta & 0 \end{pmatrix}$ with eigenvalues $\varepsilon_F \pm \delta$. A gap opens at the surface of the Fermi sea. Here δ represents the perturbation. Thus, Peierls concluded that the spectrum changes drastically and the system is unstable under a perturbation with wave vector $2k_F$.

A particularly interesting situation occurs when the system is half filled with electrons (so that the density is one electron per site—recall that electrons have up and down spin). In other words, $k_F = \pi/2a$ and thus $2k_F = \pi/a$. A lattice distortion of the form shown in figure V.5.3 has precisely this wave vector. Peierls showed that a half-filled system would want to distort the lattice in this way, doubling the unit cell. It is instructive to see how this physical phenomenon emerges in a field theoretic formulation.

Denote the displacement of the ion at site j by d_j. In the continuum limit, we should be able to replace d_j by a scalar field. Show that a perturbation connecting ψ_R and ψ_L couples to $\bar{\psi}\psi$ and $\bar{\psi}\gamma^5\psi$, and that a linear combination $\bar{\psi}\psi$ and $\bar{\psi}\gamma^5\psi$ can always be rotated to $\bar{\psi}\psi$ by a chiral transformation (see exercise V.5.1.) Thus, we extend (3) to

$$\mathcal{L} = \bar{\psi}i\gamma^\mu\partial_\mu\psi + \tfrac{1}{2}[(\partial_t\varphi)^2 - v^2(\partial_x\varphi)^2] - \tfrac{1}{2}\mu^2\varphi^2 + g\varphi\bar{\psi}\psi + \cdots \qquad (4)$$

Remember that you worked out the effective potential $V_{\text{eff}}(\varphi)$ of this $(1+1)$-dimensional field theory in exercise IV.3.2: $V_{\text{eff}}(\varphi)$ goes as $\varphi^2 \log \varphi^2$ for small φ, which overwhelms the $\tfrac{1}{2}\mu^2\varphi^2$ term. Thus, the symmetry $\varphi \to -\varphi$ is dynamically broken. The field φ acquires a vacuum expectation value and ψ becomes massive. In other words, the electron spectrum develops a gap.

Figure V.5.3

Exercise

V.5.1   Parallel to the discussion in chapter II.1 you can see easily that the space of 2 by 2 matrices is spanned by the four matrices I, γ^μ, and $\gamma^5 \equiv \gamma^0\gamma^1 = \sigma_3$. (Note the peculiar but standard notation of γ^5.) Convince yourself that $\frac{1}{2}(I \pm \gamma^5)$ projects out right- and left handed fields just as in $(3+1)$-dimensional spacetime. Show that in the bilinear $\bar{\psi}\gamma^\mu\psi$ left handed fields are connected to left handed fields and right handed to right handed and that in the scalar $\bar{\psi}\psi$ and the pseudoscalar $\bar{\psi}\gamma^5\psi$ right handed is connected to left handed and vice versa. Finally, note that under the transformation $\psi \to e^{i\theta\gamma^5}\psi$ the scalar and the pseudoscalar rotate into each other. Check that this transformation leaves the massless Dirac Lagrangian (3) invariant.

V.6 | Solitons

Breaking the shackles of Feynman diagrams

When I teach quantum field theory I like to tell the students that by the mid-1970s field theorists were breaking the shackles of Feynman diagrams. A bit melodramatic, yes, but by that time Feynman diagrams, because of their spectacular successes in quantum electrodynamics, were dominating the thinking of many field theorists, perhaps to excess. As a student I was even told that Feynman diagrams define quantum field theory, that quantum fields were merely the "slices of venison"[1] used to derive the Feynman rules, and should be discarded once the rules were obtained. The prevailing view was that it barely made sense to write down $\varphi(x)$. This view was forever shattered with the discovery of topological solitons, as we will now discuss.

Small oscillations versus lumps

Consider once again our favorite toy model $\mathcal{L} = \frac{1}{2}(\partial\varphi)^2 - V(\varphi)$ with the infamous double-well potential $V(\varphi) = (\lambda/4)(\varphi^2 - v^2)^2$ in $(1 + 1)$-dimensional spacetime. In chapter IV.1 we learned that of the two vacua $\varphi = \pm v$ we are to pick one and study small oscillations around it. So, pick one and write $\varphi = v + \chi$, expand $\mathcal{L}$ in χ, and study the dynamics of the χ meson with mass $\mu = (\lambda v^2)^{\frac{1}{2}}$. Physics then consists of suitably quantized waves oscillating about the vacuum v.

But that is not the whole story. We can also have a time independent field configuration with $\varphi(x)$ (in this and the next chapter x will denote only space unless it is clearly meant to be otherwise from the context) taking on the value $-v$ as $x \to -\infty$ and $+v$ as $x \to +\infty$, and changing from $-v$ to $+v$ around some point x_0 over some length scale l as shown in

---

[1] Gell-Mann used to speak about how pheasant meat is cooked in France between two slices of venison which are then discarded. He forcefully advocated a program to extract and study the algebraic structure of quantum field theories which are then discarded.

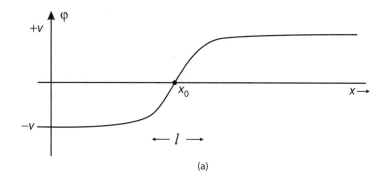

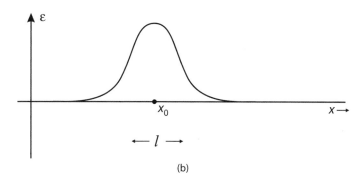

Figure V.6.1

figure V.6.1a. [Note that if we consider the Euclidean version of the field theory, identify the time coordinate as the y coordinate, and think of $\varphi(x, y)$ as the magnetization (as in chapter V.3), then the configuration here describes a "domain wall" in a 2-dimensional magnetic system.]

Think about the energy per unit length

$$\varepsilon(x) = \frac{1}{2}\left(\frac{d\varphi}{dx}\right)^2 + \frac{\lambda}{4}(\varphi^2 - v^2)^2 \tag{1}$$

for this configuration, which I plot in figure V.6.1b. Far away from x_0 we are in one of the two vacua and there is no energy density. Near x_0, the two terms in $\varepsilon(x)$ both contribute to the energy or mass $M = \int dx\, \varepsilon(x)$: the "spatial variation" (in a slight abuse of terminology often called the "kinetic energy") term $\int dx\, \frac{1}{2}(d\varphi/dx)^2 \sim l(v/l)^2 \sim v^2/l$, and the "potential energy" term $\int dx\, \lambda(\varphi^2 - v^2)^2 \sim l\lambda v^4$. To minimize the total energy the spatial variation term wants l to be large, while the potential term wants l to be small. The competition $dM/dl = 0$ gives $v^2/l \sim l\lambda v^4$, thus fixing $l \sim (\lambda v^2)^{-\frac{1}{2}} \sim 1/\mu$. The mass comes out to be $\sim \mu v^2 \sim \mu(\mu^2/\lambda)$.

We have a lump of energy spread over a region of length l of the order of the Compton wavelength of the χ meson. By translation invariance, the center of the lump x_0 can be anywhere. Furthermore, since the theory is Lorentz invariant, we can always boost to send the lump moving at any velocity we like. Recalling a famous retort in the annals of American politics ("It walks like a duck, quacks like a duck, so Mr. Senator, why don't you want to call it a duck?") we have here a particle, known as a kink or a soliton, with mass $\sim \mu(\mu^2/\lambda)$ and size $\sim l$. Perhaps because of the way the soliton was discovered, many physicists think of it as a big lumbering object, but as we have seen, the size of a soliton $l \sim 1/\mu$ can be made as small as we like by increasing μ. So a soliton could look like a point particle. We will come back to this point in chapter VI.3 when we discuss duality.

Topological stability

While the kink and the meson are the same size, for small λ the kink is much more massive than the meson. Nevertheless, the kink cannot decay into mesons because it costs an infinite amount of energy to undo the kink [by "lifting" $\varphi(x)$ over the potential energy barrier to change it from $+v$ to $-v$ for x from some point $\gtrsim x_0$ to $+\infty$, for example]. The kink is said to be topologically stable.

The stability is formally guaranteed by the conserved current

$$J^\mu = \frac{1}{2v}\varepsilon^{\mu\nu}\partial_\nu\varphi \tag{2}$$

with the charge

$$Q = \int_{-\infty}^{+\infty} dx\, J^0(x) = \frac{1}{2v}[\varphi(+\infty) - \varphi(-\infty)]$$

Mesons, which are small localized packets of oscillations in the field clearly have $Q = 0$, while the kink has $Q = 1$. Thus, the kink cannot decay into a bunch of mesons. Incidentally, the charge density $J^0 = (1/2v)(d\varphi/dx)$ is concentrated at x_0 where φ changes most rapidly, as you would expect.

Note that $\partial_\mu J^\mu = 0$ follows immediately from the antisymmetric symbol $\varepsilon^{\mu\nu}$ and does not depend on the equation of motion. The current J^μ is known as a "topological current." Its existence does not follow from Noether's theorem (chapter I.10) but from topology.

Our discussion also makes clear the existence of an antikink with $Q = -1$ and described by a configuration with $\varphi(-\infty) = +v$ and $\varphi(+\infty) = -v$. The name is justified by considering the configuration pictured in figure V.6.2 containing a kink and an antikink far apart. As the kink and the antikink move closer to each other, they clearly can annihilate into mesons, since the configuration shown in figure V.6.2 and the vacuum configuration with $\varphi(x) = +v$ everywhere are separated by a finite amount of energy.

A nonperturbative phenomenon

That the mass of the kink comes out inversely proportional to the coupling λ is a clear sign that field theorists could have done perturbation theory in λ till they were blue in the face without ever discovering the kink. Feynman diagrams could not have told us about it.

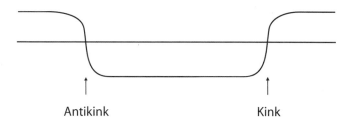

Antikink   Kink

Figure V.6.2

You can calculate the mass of a kink by minimizing

$$M = \int dx \left[\frac{1}{2}\left(\frac{d\varphi}{dx}\right)^2 + \frac{\lambda}{4}\left(\varphi^2 - v^2\right)^2 \right]$$

$$= \left(\frac{\mu^2}{\lambda}\right)\mu \int dy \left[\frac{1}{2}\left(\frac{df}{dy}\right)^2 + \frac{1}{4}\left(f^2 - 1\right)^2 \right]$$

where in the last step we performed the obvious scaling $\varphi(x) \to vf(y)$ and $y = \mu x$. This scaling argument immediately showed that the mass of the kink $M = a(\mu^2/\lambda)\mu$ with a a pure number: The heuristic estimate of the mass proved to be highly trustworthy. The actual function $\varphi(x)$ and hence a can be computed straightforwardly with standard variational methods.

Bogomol'nyi inequality

More cleverly, observe that the energy density (1) is the sum of two squares. Using $a^2 + b^2 \geq 2|ab|$ we obtain

$$M \geq \int dx \left(\frac{\lambda}{2}\right)^{\frac{1}{2}} \left|\left(\frac{d\varphi}{dx}\right)\left(\varphi^2 - v^2\right)\right| \geq \left(\frac{\lambda}{2}\right)^{\frac{1}{2}} \left|\left[\frac{1}{3}\varphi^3 - v^2\varphi^2\right]_{-\infty}^{+\infty}\right| = \left|\frac{4}{3\sqrt{2}}\mu\left(\frac{\mu^2}{\lambda}\right)Q\right|$$

We have the elegant result

$$M \geq |Q| \tag{3}$$

with mass M measured in units of $(4/3\sqrt{2})\mu(\mu^2/\lambda)$. This is an example of a Bogomol'nyi inequality, which plays an important role in string theory.

Exercises

V.6.1 Show that if $\varphi(x)$ is a solution of the equation of motion, then so is $\varphi[(x - vt)/\sqrt{1 - v^2}]$.

V.6.2 Discuss the solitons in the so-called sine-Gordon theory $\mathcal{L} = \frac{1}{2}(\partial\varphi)^2 - g\cos(\beta\varphi)$. Find the topological current. Is the $Q = 2$ soliton stable or not?

V.6.3 Compute the mass of the kink by the brute force method and check the result from the Bogomol'nyi inequality.

V.7 | Vortices, Monopoles, and Instantons

Vortices

The kink is merely the simplest example of a large class of topological objects in quantum field theory.

Consider the theory of a complex scalar field in $(2 + 1)$-dimensional spacetime $\mathcal{L} = \partial \varphi^\dagger \partial \varphi - \lambda(\varphi^\dagger \varphi - v^2)^2$ with the now familiar Mexican hat potential. With some minor changes in notation, this is the theory we used to describe interacting bosons and superfluids. (We choose to study the relativistic rather than the nonrelativistic version but as you will see the issue does not enter for the questions I want to discuss here.)

Are there solitons, that is, objects like the kink, in this theory?

Given some time-independent configuration $\varphi(x)$ let us look at its mass or energy

$$M = \int d^2x [\partial_i \varphi^\dagger \partial_i \varphi + \lambda(\varphi^\dagger \varphi - v^2)^2]. \tag{1}$$

The integrand is a sum of two squares, each of which must give a finite contribution. In particular, for the contribution of the second term to be finite the magnitude of φ must approach v at spatial infinity.

This finite energy requirement does not fix the phase of φ however. Using polar coordinates (r, θ) we will consider the Ansatz $\varphi \xrightarrow[r \to \infty]{} ve^{i\theta}$. Writing $\varphi = \varphi_1 + i\varphi_2$, we see that the vector $(\varphi_1, \varphi_2) = v(\cos \theta, \sin \theta)$ points radially outward at infinity. Recall the definition of the current $J_i = i(\partial_i \varphi^\dagger \varphi - \varphi^\dagger \partial_i \varphi)$ in a bosonic fluid given in chapter III.5. The flow whirls about at spatial infinity, and thus this configuration is known as the vortex.

By explicit differentiation or dimensional analysis, we have $\partial_i \varphi \sim v(1/r)$ as $r \to \infty$. Now look at the first term in M. Oops, the energy diverges logarithmically as $v^2 \int d^2x(1/r^2)$.

Is there a way out? Not unless we change the theory.

Vortex into flux tubes

Suppose we gauge the theory by replacing $\partial_i \varphi$ by $D_i \varphi = \partial_i \varphi - ie A_i \varphi$. Now we can achieve finite energy by requiring that the two terms in $D_i \varphi$ knock each other out so that $D_i \varphi \xrightarrow[r \to \infty]{} 0$ faster than $1/r$. In other words, $A_i \xrightarrow[r \to \infty]{} -(i/e)(1/|\varphi|^2)\varphi^\dagger \partial_i \varphi = (1/e)\partial_i \theta$. Immediately, we have

$$\text{Flux} \equiv \int d^2x \, F_{12} = \oint_C dx_i A_i = \frac{2\pi}{e} \tag{2}$$

where C is an infinitely large circle at spatial infinity and we have used Stokes' theorem. Thus, in a gauged $U(1)$ theory the vortex carries a magnetic flux inversely proportional to the charge. When I say magnetic, I am presuming that A represents the electromagnetic gauge potential. The vortex discussed here appears as a flux tube in so-called type II superconductors. It is worth remarking that this fundamental unit of flux (2) is normally written in the condensed matter physics literature in unnatural units as

$$\Phi_0 = \frac{hc}{e} \tag{3}$$

very pleasingly uniting three fundamental constants of Nature.

Homotopy groups

Since spatial infinity in 2 dimensional space is topologically a unit circle S^1 and since the field configuration with $|\varphi| = v$ also forms a circle S^1, this boundary condition can be characterized as a map $S^1 \to S^1$. Since this map cannot be smoothly deformed into the trivial map in which S^1 is mapped onto a point in S^1, the corresponding field configuration is indeed topologically stable. (Think of wrapping a loop of string around a ring.)

Mathematically, maps of S^n into a manifold M are classified by the homotopy group $\Pi_n(M)$, which counts the number of topologically inequivalent maps. You can look up the homotopy groups for various manifolds in tables.[1] In particular, for $n \geq 1$, $\Pi_n(S^n) = Z$, where Z is the mathematical notation for the set of all integers. The simplest example $\Pi_1(S^1) = Z$ is proved almost immediately by exhibiting the maps $\varphi \xrightarrow[r \to \infty]{} v e^{im\theta}$, with m any integer (positive or negative), using the context and notation of our discussion for convenience. Clearly, this map wraps one circle around the other m times.

The language of homotopy groups is not just to impress people, but gives us a unifying language to discuss topological solitons. Indeed, looking back you can now see that the kink is a physical manifestation of $\Pi_0(S^0) = Z_2$, where Z_2 denotes the multiplicative group consisting of $\{+1, -1\}$ (since the 0-dimensional sphere $S^0 = \{+1, -1\}$ consists of just two points and is topologically equivalent to the spatial infinity in 1-dimensional space).

---

[1] See tables 6.V and 6.VI, S. Iyanaga and Y. Kawada, eds., *Encyclopedic Dictionary of Mathematics*, p. 1415.

Hedgehogs and monopoles

If you absorbed all this, you are ready to move up to $(3 + 1)$-dimensional spacetime. Spatial infinity is now topologically S^2. By now you realize that if the scalar field lives on the manifold M, then we have at infinity the map of $S^2 \to M$. The simplest choice is thus to take S^2 for M. Hence, we are led to scalar fields φ^a $(a = 1, 2, 3)$ transforming as a vector $\vec{\varphi}$ under an internal symmetry group $O(3)$ and governed by $\mathcal{L} = \frac{1}{2}\partial\vec{\varphi} \cdot \partial\vec{\varphi} - V(\vec{\varphi} \cdot \vec{\varphi})$. (There should be no confusion in using the arrow to indicate a vector in the internal symmetry group.)

Let us choose $V = \lambda(\vec{\varphi}^2 - v^2)^2$. The story unfolds much as the story of the vortex. The requirement that the mass of a time independent configuration

$$M = \int d^3x[\tfrac{1}{2}(\partial\vec{\varphi})^2 + \lambda(\vec{\varphi}^2 - v^2)^2] \tag{4}$$

be finite forces $|\vec{\varphi}| = v$ at spatial infinity so that $\vec{\varphi}(r = \infty)$ indeed lives on S^2.

The identity map $S^2 \to S^2$ indicates that we should consider a configuration such that

$$\varphi^a \xrightarrow[r \to \infty]{} v\frac{x^a}{r} \tag{5}$$

This equation looks a bit strange at first sight since it mixes the index of the internal symmetry group with the index of the spatial coordinates (but in fact we have already encountered this phenomenon in the vortex). At spatial infinity, the field $\vec{\varphi}$ is pointing radially outward, so this configuration is known picturesquely as a hedgehog. Draw a picture if you don't get it!

As in the vortex story, the requirement that the first term in (4) be finite forces us to introduce an $O(3)$ gauge potential A_μ^b so that we can replace the ordinary derivative $\partial_i\varphi^a$ by the covariant derivative $D_i\varphi^a = \partial_i\varphi^a + e\varepsilon^{abc}A_i^b\varphi^c$. We can then arrange $D_i\varphi^a$ to vanish at infinity. Simple arithmetic shows that with (5) the gauge potential has to go as

$$A_i^b \xrightarrow[r \to \infty]{} \frac{1}{e}\varepsilon^{bij}\frac{x^j}{r^2} \tag{6}$$

Imagine yourself in a lab at spatial infinity. Inside a small enough lab, the $\vec{\varphi}$ field at different points are all pointing in approximately the same direction. The gauge group $O(3)$ is broken down to $O(2) \simeq U(1)$. The experimentalists in this lab observe a massless gauge field associated with the $U(1)$, which they might as well call the electromagnetic field ("quacks like a duck"). Indeed, the gauge invariant tensor field

$$\mathcal{F}_{\mu\nu} \equiv \frac{F_{\mu\nu}^a\varphi^a}{|\varphi|} - \frac{\varepsilon^{abc}\varphi^a(D_\mu\varphi)^b(D_\nu\varphi)^c}{e|\varphi|^3} \tag{7}$$

can be identified as the electromagnetic field (see exercise V.7.5).

There is no electric field since the configuration is time independent and $A_0^b = 0$. We can only have a magnetic field $\vec{B}$ which you can immediately calculate since you know A_i^b, but by symmetry we already see that $\vec{B}$ can point only in the radial direction.

This is the fabled magnetic monopole first postulated by Dirac!

The presence of magnetic monopoles in spontaneously broken gauge theory was discovered by 't Hooft and Polyakov. If you calculate the total magnetic flux coming out of the monopole $\int d\vec{S} \cdot \vec{B}$, where as usual $d\vec{S}$ denotes a small surface element at infinity pointing radially outward, you will find that it is quantized in suitable units, exactly as Dirac had stated, as it must (recall chapter IV.4).

We can once again write a Bogomol'nyi inequality for the mass of the monopole

$$M = \int d^3x \left[\tfrac{1}{4}(\vec{F}_{ij})^2 + \tfrac{1}{2}(D_i\vec{\varphi})^2 + V(\vec{\varphi}) \right] \tag{8}$$

[$\vec{F}_{ij}$ transforms as a vector under $O(3)$; recall (IV.5.17).] Observe that

$$\tfrac{1}{4}(\vec{F}_{ij})^2 + \tfrac{1}{2}(D_i\vec{\varphi})^2 = \tfrac{1}{4}(\vec{F}_{ij} \pm \varepsilon_{ijk}D_k\vec{\varphi})^2 \mp \tfrac{1}{2}\varepsilon_{ijk}\vec{F}_{ij} \cdot D_k\vec{\varphi}$$

Thus

$$M \geq \int d^3x \left[\mp\tfrac{1}{2}\varepsilon_{ijk}\vec{F}_{ij} \cdot D_k\vec{\varphi} + V(\vec{\varphi}) \right] \tag{9}$$

We next note that

$$\int d^3x \, \tfrac{1}{2}\varepsilon_{ijk}\vec{F}_{ij} \cdot D_k\vec{\varphi} = \int d^3x \, \tfrac{1}{2}\varepsilon_{ijk}\partial_k(\vec{F}_{ij} \cdot \vec{\varphi}) = v \int d\vec{S} \cdot \vec{B} = 4\pi v g$$

has an elegant interpretation in terms of the magnetic charge g of the monopole. Furthermore, if we can throw away $V(\vec{\varphi})$ while keeping $|\vec{\varphi}| \xrightarrow[r\to\infty]{} v$, then the inequality $M \geq 4\pi v|g|$ is saturated by $\vec{F}_{ij} = \pm\varepsilon_{ijk}D_k\vec{\varphi}$. The solutions of this equation are known as Bogomol'nyi-Prasad-Sommerfeld or BPS states.

It is not difficult to construct an electrically charged magnetic monopole, known as a dyon. We simply take $A_0^b = (x^b/r)f(r)$ with some suitable function $f(r)$.

One nice feature of the topological monopole is that its mass comes out to be $\sim M_W/\alpha \sim 137 M_W$ (exercise V.7.11), where M_W denotes the mass of the intermediate vector boson of the weak interaction. We are anticipating chapter VII.2 a bit in that the gauge boson that becomes massive by the Anderson-Higgs mechanism of chapter IV.6 may be identified with the intermediate vector boson. This explains naturally why the monopole has not yet been discovered.

Instanton

Consider a nonabelian gauge theory, and rotate the path integral to 4-dimensional Euclidean space. We might wish to evaluate $Z = \int DAe^{-S(A)}$ in the steepest descent approximation, in which case we would have to find the extrema of

$$S(A) = \int d^4x \frac{1}{2g^2} \operatorname{tr} F_{\mu\nu}F_{\mu\nu}$$

with finite action. This implies that at infinity $|x| = \infty$, $F_{\mu\nu}$ must vanish faster than $1/|x|^2$, and so the gauge potential A_μ must be a pure gauge: $A = gdg^\dagger$ for g an element of the gauge group [see (IV.5.6)]. Configurations for which this is true are known as instantons.

We see that the instanton is yet one more link in the "great chain of being": kink-vortex-monopole-instanton.

Choose the gauge group $SU(2)$ to be definite. In the parametrization $g = x_4 + i\vec{x} \cdot \vec{\sigma}$ we have by definition $g^\dagger g = 1$ and $\det g = 1$, thus implying $x_4^2 + \vec{x}^2 = 1$. We learn that the group manifold of $SU(2)$ is S^3. Thus, in an instanton, the gauge potential at infinity $A \underset{|x| \to \infty}{\longrightarrow} g d g^\dagger + O(1/|x|^2)$ defines a map $S^3 \to S^3$. Sound familiar? Indeed, you've already seen $S^0 \to S^0$, $S^1 \to S^1$, $S^2 \to S^2$ playing a role in field theory.

Recall from chapter IV.5 that tr $F^2 = d \operatorname{tr}(AdA + \frac{2}{3}A^3)$. Thus

$$\int \operatorname{tr} F^2 = \int_{S^3} \operatorname{tr}(AdA + \tfrac{2}{3}A^3) = \int_{S^3} \operatorname{tr}(AF - \tfrac{1}{3}A^3) = -\frac{1}{3}\int_{S^3} \operatorname{tr}(gdg^\dagger)^3 \tag{10}$$

where we used the fact that F vanishes at infinity. This shows explicitly that $\int \operatorname{tr} F^2$ depends only on the homotopy of the map $S^3 \to S^3$ defined by g and is thus a topological quantity. Incidentally, $\int_{S^3} \operatorname{tr}(gdg^\dagger)^3$ is known to mathematicians as the Pontryagin index (see exercise V.7.12).

I mentioned in chapter IV.7 that the chiral anomaly is not affected by higher-order quantum fluctuations. You are now in position to give an elegant topological proof of this fact (exercise V.7.13).

Kosterlitz-Thouless transition

We were a bit hasty in dismissing the vortex in the nongauged theory $\mathcal{L} = \partial\varphi^\dagger \partial\varphi - \lambda(\varphi^\dagger\varphi - v^2)^2$ in $(2+1)$-dimensional spacetime. Around a vortex $\varphi \sim ve^{i\theta}$ and so it is true, as we have noted, that the energy of a single vortex diverges logarithmically. But what about a vortex paired with an antivortex?

Picture a vortex and an antivortex separated by a distance R large compared to the distance scales in the theory. Around an antivortex $\varphi \sim ve^{-i\theta}$. The field φ winds around the vortex one way and around the antivortex the other way. Convince yourself by drawing a picture that at spatial infinity φ does not wind at all: It just goes to a fixed value. The winding one way cancels the winding the other way.

Thus, a configuration consisting of a vortex-antivortex pair does not cost infinite energy. But it does cost a finite amount of energy: In the region between the vortex and the antivortex φ is winding around, in fact roughly twice as fast (as you can see by drawing a picture). A rough estimate of the energy is thus $v^2 \int d^2x (1/r^2) \sim v^2 \log(R/a)$, where we integrated over a region of size R, the relevant physical scale in the problem. (To make sense of the problem we divide R by the size a of the vortex.) The vortex and the antivortex attract each other with a logarithmic potential. In other words, the configuration cannot be static: The vortex and the antivortex want to get together and annihilate each other in a fiery embrace and release that finite amount of energy $v^2 \log(R/a)$. (Hence the term antivortex.)

All of this is at zero temperature, but in condensed matter physics we are interested in the free energy $F = E - TS$ (with S the entropy) at some temperature T rather than the energy E. Appreciating this elementary point, Kosterlitz and Thouless discovered a phase transition as the temperature is raised. Consider a gas of vortices and antivortices at some nonzero temperature. Because of thermal agitation, the vortices and antivortices moving around may or may not find each other to annihilate. How high do we have to crank up the temperature for this to happen?

Let us do a heuristic estimate. Consider a single vortex. Herr Boltzmann tells us that the entropy is the logarithm of the "number" of ways in which we can put the vortex inside a box of size L (which we will let tend to infinity). Thus, $S \sim \log(L/a)$. The entropy S is to battle the energy $E \sim v^2 \log(L/a)$. We see that the free energy $F \sim (v^2 - T) \log(L/a)$ goes to infinity if $T \lesssim v^2$, which we identify as essentially the critical temperature T_c.

A single vortex cannot exist below T_c. Vortices and antivortices are tightly bound below T_c but are liberated above T_c.

Black hole

The discovery in the 1970s of these topological objects that cannot be seen in perturbation theory came as a shock to the generation of physicists raised on Feynman diagrams and canonical quantization. People (including yours truly) were taught that the field operator $\varphi(x)$ is a highly singular quantum operator and has no physical meaning as such, and that quantum field theory is defined perturbatively by Feynman diagrams. Even quite eminent physicists asked in puzzlement what a statement such as $\varphi \xrightarrow[r \to \infty]{} v e^{i\theta}$ would mean. Learned discussions that in hindsight are totally irrelevant ensued. As I said in introducing chapter V.6, I like to refer to this historical process as "field theorists breaking the shackles of Feynman diagrams."

It is worth mentioning one argument physicists at that time used to convince themselves that solitons do exist. After all, the Schwarzschild black hole, defined by the metric $g_{\mu\nu}(x)$ (see chapter I.11), had been known since 1916. Just what are the components of the metric $g_{\mu\nu}(x)$? They are fields in exactly the same way that our scalar field $\varphi(x)$ and our gauge potential $A_\mu(x)$ are fields, and in a quantum theory of gravity $g_{\mu\nu}$ would have to be replaced by a quantum operator just like φ and A_μ. So the objects discovered in the 1970s are conceptually no different from the black hole known in the 1910s. But in the early 1970s most particle theorists were not particularly aware of quantum gravity.

Exercises

V.7.1 Explain the relation between the mathematical statement $\Pi_0(S^0) = Z_2$ and the physical result that there are no kinks with $|Q| \geq 2$.

V.7.2 In the vortex, study the length scales characterizing the variation of the fields φ and A. Estimate the mass of the vortex.

V.7.3 Consider the vortex configuration in which $\varphi \xrightarrow[r \to \infty]{} v e^{i v \theta}$, with v an integer. Calculate the magnetic flux. Show that the magnetic flux coming out of an antivortex (for which $v = -1$) is opposite to the magnetic flux coming out of a vortex.

V.7.4 Mathematically, since $g(\theta) \equiv e^{i v \theta}$ may be regarded as an element of the group $U(1)$, we can speak of a map of S^1, the circle at spatial infinity, onto the group $U(1)$. Calculate $(i/2\pi) \int_{S^1} g \, dg^\dagger$, thus showing that the winding number is given by this integral of a 1-form.

V.7.5 Show that within a region in which φ^a is constant, $\mathcal{F}_{\mu\nu}$ as defined in the text is the electromagnetic field strength. Compute $\vec{B}$ far from the center of a magnetic monopole and show that Dirac quantization holds.

V.7.6 Display explicitly the map $S^2 \to S^2$, which wraps one sphere around the other twice. Verify that this map corresponds to a magnetic monopole with magnetic charge 2.

V.7.7 Write down the variational equations that minimize (8).

V.7.8 Find the BPS solution explicitly.

V.7.9 Discuss the dyon solution. Work it out in the BPS limit.

V.7.10 Verify explicitly that the magnetic monopole is rotation invariant in spite of appearances. By this is meant that all physical gauge invariant quantities such as $\vec{B}$ are covariant under rotation. Gauge variant quantities such as A_i^b can and do vary under rotation. Write down the generators of rotation.

V.7.11 Show that the mass of the magnetic monopole is about $137 M_W$.

V.7.12 Evaluate $n \equiv -(1/24\pi^2) \int_{S^3} \mathrm{tr}(g \, dg^\dagger)^3$ for the map $g = e^{i \vec{\theta} \cdot \vec{\sigma}}$. [Hint: By symmetry, you need calculate the integrand only in a small neighborhood of the identity element of the group or equivalently the north pole of S^3. Next, consider $g = e^{i(\theta_1 \sigma_1 + \theta_2 \sigma_2 + m \theta_3 \sigma_3)}$ for m an integer and convince yourself that m measures the number of times S^3 wraps around S^3.] Compare with exercise V.7.4 and admire the elegance of mathematics.

V.7.13 Prove that higher order corrections do not change the chiral anomaly $\partial_\mu J_5^\mu = [1/(4\pi)^2] \varepsilon^{\mu\nu\lambda\sigma} \, \mathrm{tr} \, F_{\mu\nu} F_{\lambda\sigma}$ (I have rescaled $A \to (1/g)A$). [Hint: Integrate over spacetime and show that the left hand side is given by the number of right-handed fermion quanta minus the number of left-handed fermion quanta, so that both sides are given by integers.]

Part VI | Field Theory and Condensed Matter

VI.1 Fractional Statistics, Chern-Simons Term, and Topological Field Theory

Fractional statistics

The existence of bosons and fermions represents one of the most profound features of quantum physics. When we interchange two identical quantum particles, the wave function acquires a factor of either $+1$ or -1. Leinaas and Myrheim, and later Wilczek independently, had the insight to recognize that in $(2 + 1)$-dimensional spacetime particles can also obey statistics other than Bose or Fermi statistics, a statistics now known as fractional or anyon statistics. These particles are now known as anyons.

To interchange two particles, we can move one of them half-way around the other and then translate both of them appropriately. When you take one anyon half-way around another anyon going anticlockwise, the wave function acquires a factor of $e^{i\theta}$ where θ is a real number characteristic of the particle. For $\theta = 0$, we have bosons and for $\theta = \pi$, fermions. Particles half-way between bosons and fermions, with $\theta = \pi/2$, are known as semions.

After Wilczek's paper came out, a number of distinguished senior physicists were thoroughly confused. Thinking in terms of Schrödinger's wave function, they got into endless arguments about whether the wave function must be single valued. Indeed, anyon statistics provides a striking example of the fact that the path integral formalism is sometimes significantly more transparent. The concept of anyon statistics can be formulated in terms of wave functions but it requires thinking clearly about the configuration space over which the wave function is defined.

Consider two indistinguishable particles at positions x_1^i and x_2^i at some initial time that end up at positions x_1^f and x_2^f a time T later. In the path integral representation for $\langle x_1^f, x_2^f | e^{-iHT} | x_1^i, x_2^i \rangle$ we have to sum over all paths. In spacetime, the worldlines of the two particles braid around each other (see fig. VI.1.1). (We are implicitly assuming that the particles cannot go through each other, which is the case if there is a hard core repulsion between them.) Clearly, the paths can be divided into topologically distinct classes, characterized by an integer n equal to the number of times the worldlines of the

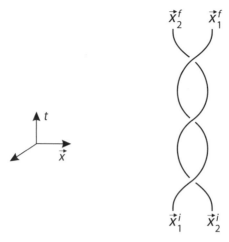

Figure VI.1.1

two particles braid around each other. Since the classes cannot be deformed into each other, the corresponding amplitudes cannot interfere quantum mechanically, and with the amplitudes in each class we are allowed to associate an additional phase factor $e^{i\alpha_n}$ beyond the usual factor coming from the action.

The dependence of α_n on n is determined by how quantum amplitudes are to be combined. Suppose one particle goes around the other through an angle $\Delta\varphi_1$, a history to which we assign an additional phase factor $e^{if(\Delta\varphi_1)}$ with f some as yet unknown function. Suppose this history is followed by another history in which our particle goes around the other by an additional angle $\Delta\varphi_2$. The phase factor $e^{if(\Delta\varphi_1+\Delta\varphi_2)}$ we assign to the combined history clearly has to satisfy the composition law $e^{if(\Delta\varphi_1+\Delta\varphi_2)} = e^{if(\Delta\varphi_1)}e^{if(\Delta\varphi_2)}$. In other words, $f(\Delta\varphi)$ has to be a linear function of its argument.

We conclude that in $(2+1)$-dimensional spacetime we can associate with the quantum amplitude corresponding to paths in which one particle goes around the other anticlockwise through an angle $\Delta\varphi$ a phase factor $e^{i(\theta/\pi)\Delta\varphi}$, with θ an arbitrary real parameter. Note that when one particle goes around the other clockwise through an angle $\Delta\varphi$ the quantum amplitude acquires a phase factor $e^{-i\frac{\theta}{\pi}\Delta\varphi}$.

When we interchange two anyons, we have to be careful to specify whether we do it "anticlockwise" or "clockwise," producing factors $e^{i\theta}$ and $e^{-i\theta}$, respectively. This indicates immediately that parity P and time reversal invariance T are violated.

Chern-Simons theory

The next important question is whether all this can be incorporated in a local quantum field theory. The answer was given by Wilczek and Zee, who showed that the notion of fractional statistics can result from the effect of coupling to a gauge potential. The significance of a field theoretic formulation is that it demonstrates conclusively that the idea of anyon statistics is fully compatible with the cherished principles that we hold dear and that go into the construction of quantum field theory.

Given a Lagrangian $\mathcal{L}_0$ with a conserved current j^μ, construct the Lagrangian

$$\mathcal{L} = \mathcal{L}_0 + \gamma \varepsilon^{\mu\nu\lambda} a_\mu \partial_\nu a_\lambda + a_\mu j^\mu \tag{1}$$

Here $\varepsilon^{\mu\nu\lambda}$ denotes the totally antisymmetric symbol in $(2+1)$-dimensional spacetime and γ is an arbitrary real parameter. Under a gauge transformation $a_\mu \to a_\mu + \partial_\mu \Lambda$, the term $\varepsilon^{\mu\nu\lambda} a_\mu \partial_\nu a_\lambda$, known as the Chern-Simons term, changes by $\varepsilon^{\mu\nu\lambda} a_\mu \partial_\nu a_\lambda \to \varepsilon^{\mu\nu\lambda} a_\mu \partial_\nu a_\lambda + \varepsilon^{\mu\nu\lambda} \partial_\mu \Lambda \partial_\nu a_\lambda$. The action changes by $\delta S = \gamma \int d^3 x \varepsilon^{\mu\nu\lambda} \partial_\mu (\Lambda \partial_\nu a_\lambda)$ and thus, if we are allowed to drop boundary terms, as we assume to be the case here, the Chern-Simons action is gauge invariant. Note incidentally that in the language of differential form you learned in chapters IV.5 and IV.6 the Chern-Simons term can be written compactly as ada.

Let us solve the equation of motion derived from (1):

$$2\gamma \varepsilon^{\mu\nu\lambda} \partial_\nu a_\lambda = -j^\mu \tag{2}$$

for a particle sitting at rest (so that $j_i = 0$). Integrating the $\mu = 0$ component of (2), we obtain

$$\int d^2 x (\partial_1 a_2 - \partial_2 a_1) = -\frac{1}{2\gamma} \int d^2 x j^0 \tag{3}$$

Thus, the Chern-Simons term has the effect of endowing the charged particles in the theory with flux. (Here the term charged particles simply means particles that couple to the gauge potential a_μ. In this context, when we refer to charge and flux, we are obviously not referring to the charge and flux associated with the ordinary electromagnetic field. We are simply borrowing a useful terminology.)

By the Aharonov-Bohm effect (chapter IV.4), when one of our particles moves around another, the wave function acquires a phase, thus endowing the particles with anyon statistics with angle $\theta = 1/4\gamma$ (see exercise VI.1.5).

Strictly speaking, the term "fractional statistics" is somewhat misleading. First, a trivial remark: The statistics parameter θ does not have to be a fraction. Second, statistics is not directly related to counting how many particles we can put into a state. The statistics between anyons is perhaps better thought of as a long ranged phase interaction between them, mediated by the gauge potential a.

The appearance of $\varepsilon^{\mu\nu\lambda}$ in (1) signals the violation of parity P and time reversal invariance T, something we already know.

Hopf term

An alternative treatment is to integrate out a in (1). As explained in chapter III.4, and as in any gauge theory, the inverse of the differential operator $\varepsilon \partial$ is not defined: It has a zero mode since $(\varepsilon^{\mu\nu\lambda} \partial_\nu)(\partial_\lambda F(x)) = 0$ for any smooth function $F(x)$. Let us choose the Lorenz gauge $\partial_\mu a^\mu = 0$. Then, using the fundamental identity of field theory (see appendix A) we obtain the nonlocal Lagrangian

$$\mathcal{L}_{\text{Hopf}} = \frac{1}{4\gamma} \left(j_\mu \frac{\varepsilon^{\mu\nu\lambda}\partial_\nu}{\partial^2} j_\lambda \right) \tag{4}$$

known as the Hopf term.

To determine the statistics parameter θ consider a history in which one particle moves half-way around another sitting at rest. The current j is then equal to the sum of two terms describing the two particles. Plugging into (4) we evaluate the quantum phase $e^{iS} = e^{i \int d^3x \mathcal{L}_{\text{Hopf}}}$ and obtain $\theta = 1/(4\gamma)$.

Topological field theory

There is something conceptually new about the pure Chern-Simons theory

$$S = \gamma \int_M d^3x \varepsilon^{\mu\nu\lambda} a_\mu \partial_\nu a_\lambda \tag{5}$$

It is topological.

Recall from chapter I.11 that a field theory written in flat spacetime can be immediately promoted to a field theory in curved spacetime by replacing the Minkowski metric $\eta^{\mu\nu}$ by the Einstein metric $g^{\mu\nu}$ and including a factor $\sqrt{-g}$ in the spacetime integration measure.

But in the Chern-Simons theory $\eta^{\mu\nu}$ does not appear! Lorentz indices are contracted with the totally antisymmetric symbol $\varepsilon^{\mu\nu\lambda}$. Furthermore, we don't need the factor $\sqrt{-g}$, as I will now show. Recall also from chapter I.11 that a vector field transforms as $a_\mu(x) = (\partial x'^\lambda/\partial x^\mu)a'_\lambda(x')$ and so for three vector fields

$$\varepsilon^{\mu\nu\lambda} a_\mu(x)b_\nu(x)c_\lambda(x) = \varepsilon^{\mu\nu\lambda}\frac{\partial x'^\sigma}{\partial x^\mu}\frac{\partial x'^\tau}{\partial x^\nu}\frac{\partial x'^\rho}{\partial x^\lambda} a'_\sigma(x')b'_\tau(x')c'_\rho(x')$$

$$= \det\left(\frac{\partial x'}{\partial x}\right) \varepsilon^{\sigma\tau\rho} a'_\sigma(x')b'_\tau(x')c'_\rho(x')$$

On the other hand, $d^3x' = d^3x \det(\partial x'/\partial x)$. Observe, then,

$$d^3x \varepsilon^{\mu\nu\lambda} a_\mu(x)b_\nu(x)c_\lambda(x) = d^3x' \varepsilon^{\sigma\tau\rho} a'_\sigma(x')b'_\tau(x')c'_\rho(x')$$

which is invariant without the benefit of $\sqrt{-g}$.

So, the Chern-Simons action in (5) is invariant under general coordinate transformation—it is already written for curved spacetime. The metric $g_{\mu\nu}$ does not enter anywhere. The Chern-Simons theory does not know about clocks and rulers! It only knows about the topology of spacetime and is rightly known as a topological field theory. In other words, when the integral in (5) is evaluated over a closed manifold M the property of the field theory $\int Da e^{iS(a)}$ depends only on the topology of the manifold, and not on whatever metric we might put on the manifold.

Ground state degeneracy

Recall from chapter I.11 the fundamental definition of energy and momentum. The energy-momentum tensor is defined by the variation of the action with respect to $g_{\mu\nu}$, but hey, the action here does not depend on $g_{\mu\nu}$. The energy-momentum tensor and hence the Hamiltonian is identically zero! One way of saying this is that to define the Hamiltonian we need clocks and rulers.

What does it mean for a quantum system to have a Hamiltonian $H = 0$? Well, when we took a course on quantum mechanics, if the professor assigned an exam problem to find the spectrum of the Hamiltonian 0, we could do it easily! All states have energy $E = 0$. We are ready to hand it in.

But the nontrivial problem is to find how many states there are. This number is known as the ground state degeneracy and depends only on the topology of the manifold M.

Massive Dirac fermions and the Chern-Simons term

Consider a gauge potential a_μ coupled to a massive Dirac fermion in $(2+1)$-dimensional spacetime: $\mathcal{L} = \bar{\psi}(i\slashed{\partial} + \slashed{a} - m)\psi$. You did an exercise way back in chapter II.1 discovering the rather surprising phenomenon that in $(2+1)$-dimensional spacetime the Dirac mass term violates P and T. (What? You didn't do it? You have to go back.) Thus, we would expect to generate the P and T violating the Chern-Simons term $\varepsilon^{\mu\nu\lambda}a_\mu\partial_\lambda a_\nu$ if we integrate out the fermion to get the term $\operatorname{tr}\log(i\slashed{\partial} + \slashed{a} - m)$ in the effective action, along the lines discussed in chapter IV.3.

In one-loop order we have the vacuum polarization diagram (diagrammatically exactly the same as in chapter III.7 but in a spacetime with one less dimension) with a Feynman integral proportional to

$$\int \frac{d^3p}{(2\pi)^3} \operatorname{tr}\left(\gamma^\nu \frac{1}{\slashed{p}+\slashed{q}-m}\gamma^\mu \frac{1}{\slashed{p}-m}\right) \tag{6}$$

As we will see, the change $4 \to 3$ makes all the difference in the world. I leave it to you to evaluate (6) in detail (exercise VI.1.7) but let me point out the salient features here. Since the ∂_λ in the Chern-Simons term corresponds to q_λ in momentum space, in order to identify the coefficient of the Chern-Simons term we need only differentiate (6) with respect to q_λ and set $q \to 0$:

$$\int \frac{d^3p}{(2\pi)^3} \operatorname{tr}(\gamma^\nu \frac{1}{\slashed{p}-m}\gamma^\lambda \frac{1}{\slashed{p}-m}\gamma^\mu \frac{1}{\slashed{p}-m})$$
$$= \int \frac{d^3p}{(2\pi)^3} \frac{\operatorname{tr}\left[\gamma^\nu(\slashed{p}+m)\gamma^\lambda(\slashed{p}+m)\gamma^\mu(\slashed{p}+m)\right]}{(p^2-m^2)^3} \tag{7}$$

I will simply focus on one piece of the integral, the piece coming from the term in the trace proportional to m^3:

$$\varepsilon^{\mu\nu\lambda} m^3 \int \frac{d^3 p}{(p^2 - m^2)^3} \tag{8}$$

As I remarked in exercise II.1.12, in $(2+1)$-dimensional spacetime the γ^μ's are just the three Pauli matrices and thus $\mathrm{tr}(\gamma^\nu \gamma^\lambda \gamma^\mu)$ is proportional to $\varepsilon^{\mu\nu\lambda}$: The antisymmetric symbol appears as we expect from P and T violation.

By dimensional analysis, we see that the integral in (8) is up to a numerical constant equal to $1/m^3$ and so m cancels.

But be careful! The integral depends only on m^2 and doesn't know about the sign of m. The correct answer is proportional to $1/|m|^3$, not $1/m^3$. Thus, the coefficient of the Chern-Simons term is equal to $m^3/|m|^3 = m/|m| = \text{sign of } m$, up to a numerical constant. An instructive example of an important sign! This makes sense since under P (or T) a Dirac field with mass m is transformed into a Dirac field with mass $-m$. In a parity-invariant theory, with a doublet of Dirac fields with masses m and $-m$ a Chern-Simons term should not be generated.

Exercises

VI.1.1   In a nonrelativistic theory you might think that there are two separate Chern-Simons terms, $\varepsilon_{ij} a_i \partial_0 a_j$ and $\varepsilon_{ij} a_0 \partial_i a_j$. Show that gauge invariance forces the two terms to combine into a single Chern-Simons term $\varepsilon^{\mu\nu\lambda} a_\mu \partial_\nu a_\lambda$. For the Chern-Simons term, gauge invariance implies Lorentz invariance. In contrast, the Maxwell term would in general be nonrelativistic, consisting of two terms, f_{0i}^2 and f_{ij}^2, with an arbitrary relative coefficient between them (with $f_{\mu\nu} = \partial_\mu a_\nu - \partial_\nu a_\mu$ as usual).

VI.1.2   By thinking about mass dimensions, convince yourself that the Chern-Simons term dominates the Maxwell term at long distances. This is one reason that relativistic field theorists find anyon fluids so appealing. As long as they are interested only in long distance physics they can ignore the Maxwell term and play with a relativistic theory (see exercise VI.1.1). Note that this picks out $(2+1)$-dimensional spacetime as special. In $(3+1)$-dimensional spacetime the generalization of the Chern-Simons term $\varepsilon^{\mu\nu\lambda\sigma} f_{\mu\nu} f_{\lambda\sigma}$ has the same mass dimension as the Maxwell term f^2. In $(4+1)$-dimensional space the term $\varepsilon^{\rho\mu\nu\lambda\sigma} a_\rho f_{\mu\nu} f_{\lambda\sigma}$ is less important at long distances than the Maxwell term f^2.

VI.1.3   There is a generalization of the Chern-Simons term to higher dimensional spacetime different from that given in exercise IV.1.2. We can introduce a p-form gauge potential (see chapter IV.4). Write the generalized Chern-Simons term in $(2p+1)$-dimensional spacetime and discuss the resulting theory.

VI.1.4   Consider $\mathcal{L} = \gamma a \varepsilon \, \partial a - (1/4g^2) f^2$. Calculate the propagator and show that the gauge boson is massive. Some physicists puzzled by fractional statistics have reasoned that since in the presence of the Maxwell term the gauge boson is massive and hence short ranged, it can't possibly generate fractional statistics, which is manifestly an infinite ranged interaction. (No matter how far apart the two particles we are interchanging are, the wave function still acquires a phase.) The resolution is that the information is in fact propagated over an infinite range by a $q = 0$ pole associated with a gauge degree of freedom. This apparent paradox is intimately connected with the puzzlement many physicists felt when they first heard of the Aharonov-Bohm effect. How can a particle in a region with no magnetic field whatsoever and arbitrarily far from the magnetic flux know about the existence of the magnetic flux?

VI.1.5 Show that $\theta = 1/4\gamma$. There is a somewhat tricky factor[1] of 2. So if you are off by a factor of 2, don't despair. Try again.

VI.1.6 Find the nonabelian version of the Chern-Simons term ada. [Hint: As in chapter IV.6 it might be easier to use differential forms.]

VI.1.7 Using the canonical formalism of chapter I.8 show that the Chern-Simons Lagrangian leads to the Hamiltonian $H = 0$.

VI.1.8 Evaluate (6).

---

[1] X.G. Wen and A. Zee, *J. de Physique*, 50: 1623, 1989.

VI.2 | Quantum Hall Fluids

Interplay between two pieces of physics

Over the last decade or so, the study of topological quantum fluids (of which the Hall fluid is an example) has emerged as an interesting subject. The quantum Hall system consists of a bunch of electrons moving in a plane in the presence of an external magnetic field B perpendicular to the plane. The magnetic field is assumed to be sufficiently strong so that the electrons all have spin up, say, so they may be treated as spinless fermions. As is well known, this seemingly innocuous and simple physical situation contains a wealth of physics, the elucidation of which has led to two Nobel prizes. This remarkable richness follows from the interplay between two basic pieces of physics.

1. Even though the electron is pointlike, it takes up a finite amount of room.

Classically, a charged particle in a magnetic field moves in a Larmor circle of radius r determined by $evB = mv^2/r$. Classically, the radius is not fixed, with more energetic particles moving in larger circles, but if we quantize the angular momentum mvr to be $h = 2\pi$ (in units in which $\hbar$ is equal to unity) we obtain $eBr^2 \sim 2\pi$. A quantum electron takes up an area of order $\pi r^2 \sim 2\pi^2/eB$.

2. Electrons are fermions and want to stay out of each other's way.

Not only does each electron insist on taking up a finite amount of room, each has to have its own room. Thus, the quantum Hall problem may be described as a sort of housing crisis, or as the problem of assigning office space at an Institute for Theoretical Physics to visitors who do not want to share offices.

Already at this stage, we would expect that when the number of electrons N_e is just right to fill out space completely, namely when $N_e \pi r^2 \sim N_e(2\pi^2/eB) \sim A$, the area of the system, something special happens.

Landau levels and the integer Hall effect

These heuristic considerations could be made precise, of course. The textbook problem of a single spinless electron in a magnetic field

$$-[(\partial_x - ieA_x)^2 + (\partial_y - ieA_y)^2]\psi = 2mE\psi$$

was solved by Landau decades ago. The states occur in degenerate sets with energy $E_n = \left(n + \frac{1}{2}\right)\frac{eB}{m}$, $n = 0, 1, 2, \cdots$, known as the nth Landau level. Each Landau level has degeneracy $BA/2\pi$, where A is the area of the system, reflecting the fact that the Larmor circles may be placed anywhere. Note that one Landau level is separated from the next by a finite amount of energy (eB/m).

Imagine putting in noninteracting electrons one by one. By the Pauli exclusion principle, each succeeding electron we put in has to go into a different state in the Landau level. Since each Landau level can hold $BA/2\pi$ electrons it is natural (see exercise VI.2.1) to define a filling factor $\nu \equiv N_e/(BA/2\pi)$. When ν is equal to an integer, the first ν Landau levels are filled. If we want to put in one more electron, it would have to go into the $(\nu + 1)$st Landau level, costing us more energy than what we spent for the preceding electron.

Thus, for ν equal to an integer the Hall fluid is incompressible. Any attempt to compress it lessens the degeneracy of the Landau levels (the effective area A decreases and so the degeneracy $BA/2\pi$ decreases) and forces some of the electrons to the next level, costing us lots of energy.

An electric field E_y imposed on the Hall fluid in the y direction produces a current $J_x = \sigma_{xy}E_y$ in the x direction with $\sigma_{xy} = \nu$ (in units of e^2/h). This is easily understood in terms of the Lorentz force law obeyed by electrons in the presence of a magnetic field. The surprising experimental discovery was that the Hall conductance σ_{xy} when plotted against B goes through a series of plateaus, which you might have heard about. To understand these plateaus we would have to discuss the effect of impurities. I will touch upon the fascinating subject of impurities and disorder in chapter VI.8.

So, the integer quantum Hall effect is relatively easy to understand.

Fractional Hall effect

After the integer Hall effect, the experimental discovery of the fractional Hall effect, namely that the Hall fluid is also incompressible for filling factor ν equal to simple odd-denominator fractions such as $\frac{1}{3}$ and $\frac{1}{5}$, took theorists completely by surprise. For $\nu = \frac{1}{3}$, only one-third of the states in the first Landau level are filled. It would seem that throwing in a few more electrons would not have that much effect on the system. Why should the $\nu = \frac{1}{3}$ Hall fluid be incompressible?

Interaction between electrons turns out to be crucial. The point is that saying the first Landau level is one-third filled with noninteracting spinless electrons does not define a unique many-body state: there is an enormous degeneracy since each of the electrons can go into any of the $BA/2\pi$ states available subject only to Pauli exclusion. But as soon as we turn on a repulsive interaction between the electrons, a presumably unique ground state is picked out within the vast space of degenerate states. Wen has described the fractional Hall state as an intricate dance of electrons: Not only does each electron occupy a finite amount of room on the dance floor, but due to the mutual repulsion, it has to be careful not to bump into another electron. The dance has to be carefully choreographed, possible only for certain special values of ν.

Impurities also play an essential role, but we will postpone the discussion of impurities to chapter VI.6.

In trying to understand the fractional Hall effect, we have an important clue. You will remember from chapter V.7 that the fundamental unit of flux is given by 2π, and thus the number of flux quanta penetrating the plane is equal to $N_\phi = BA/2\pi$. Thus, the puzzle is that something special happens when the number of flux quanta per electron $N_\phi/N_e = \nu^{-1}$ is an odd integer.

I arranged the chapters so that what you learned in the previous chapter is relevant to solving the puzzle. Suppose that ν^{-1} flux quanta are somehow bound to each electron. When we interchange two such bound systems there is an additional Aharonov-Bohm phase in addition to the (-1) from the Fermi statistics of the electrons. For ν^{-1} odd these bound systems effectively obey Bose statistics and can be described by a complex scalar field φ. The condensation of φ turns out to be responsible for the physics of the quantum Hall fluid.

Effective field theory of the Hall fluid

We would like to derive an effective field theory of the quantum Hall fluid, first obtained by Kivelson, Hansson, and Zhang. There are two alternative derivations, a long way and a short way.

In the long way, we start with the Lagrangian describing spinless electrons in a magnetic field in the second quantized formalism (we will absorb the electric charge e into A_μ),

$$\mathcal{L} = \psi^+ i(\partial_0 - i A_0)\, \psi + \frac{1}{2m}\, \psi^\dagger (\partial_i - i A_i)^2 \psi + V(\psi^\dagger \psi) \tag{1}$$

and massage it into the form we want. In the previous chapter, we learned that by introducing a Chern-Simons gauge field we can transform ψ into a scalar field. We then invoke duality, which we will learn about in the next chapter, to represent the phase degree of freedom of the scalar field as a gauge field. After a number of steps, we will discover that the effective theory of the Hall fluid turns out to be a Chern-Simons theory.

Instead, I will follow the short way. We will argue by the "what else can it be" method or, to put it more elegantly, by invoking general principles.

Let us start by listing what we know about the Hall system.

1. We live in $(2+1)$-dimensional spacetime (because the electrons are restricted to a plane.)
2. The electromagnetic current J_μ is conserved: $\partial_\mu J^\mu = 0$.

These two statements are certainly indisputable; when combined they tell us that the current can be written as the curl of a vector potential

$$J^\mu = \frac{1}{2\pi}\epsilon^{\mu\nu\lambda}\partial_\nu a_\lambda \tag{2}$$

The factor of $1/(2\pi)$ defines the normalization of a_μ. We learned in school that in 3-dimensional spacetime, if the divergence of something is zero, then that something is the curl of something else. That is precisely what (2) says. The only sophistication here is that what we learned in school works in Minkowskian space as well as Euclidean space—it is just a matter of a few signs here and there.

The gauge potential comes looking for us

Observe that when we transform a_μ by $a_\mu \to a_\mu - \partial_\mu \Lambda$, the current is unchanged. In other words, a_μ is a gauge potential.

We did not go looking for a gauge potential; the gauge potential came looking for us! There is no place to hide. The existence of a gauge potential follows from completely general considerations.

3. We want to describe the system field theoretically by an effective local Lagrangian.

4. We are only interested in the physics at long distance and large time, that is, at small wave number and low frequency.

Indeed, a field theoretic description of a physical system may be regarded as a means of organizing various aspects of the relevant physics in a systematic way according to their relative importance at long distances and according to symmetries. We classify terms in a field theoretic Lagrangian according to powers of derivatives, powers of the fields, and so forth. A general scheme for classifying terms is according to their mass dimensions, as explained in chapter III.2. The gauge potential a_μ has dimension 1, as is always the case for any gauge potential coupled to matter fields according to the gauge principle, and thus (2) is consistent with the fact that the current has mass dimension 2 in $(2+1)$-dimensional spacetime.

5. Parity and time reversal are broken by the external magnetic field.

This last statement is just as indisputable as statements 1 and 2. The experimentalist produces the magnetic field by driving a current through a coil with the current flowing either clockwise or anticlockwise.

Given these five general statements we can deduce the form of the effective Lagrangian.

Since gauge invariance forbids the dimension-2 term $a_\mu a^\mu$ in the Lagrangian, the simplest possible term is in fact the dimension-3 Chern-Simons term $\epsilon^{\mu\nu\lambda}a_\mu\partial_\nu a_\lambda$. Thus, the Lagrangian is simply

$$\mathcal{L} = \frac{k}{4\pi}a\epsilon\partial a + \cdots \tag{3}$$

where k is a dimensionless parameter to be determined.

We have introduced and will use henceforth the compact notation $\epsilon a \partial b \equiv \epsilon^{\mu\nu\lambda} a_\mu \partial_\nu b_\lambda = \epsilon b \partial a$ for two vector fields a_μ and b_μ.

The terms indicated by ($\cdots$) in (3) include the dimension-4 Maxwell term $(1/g^2)(f_{0i}^2 - \beta f_{ij}^2)$ and other terms with higher dimensions. (Here β is some constant; see exercise VI.1.1.) The important observation is that these higher dimensional terms are less important at long distances. The long distance physics is determined purely by the Chern-Simons term. In general the coefficient k may well be zero, in which case the physics is determined by the short distance terms represented by the ($\cdots$) in (3). Put differently, a Hall fluid may be defined as a 2-dimensional electron system for which the coefficient of the Chern-Simons term does not vanish, and consequently is such that its long distance physics is largely independent of the microscopic details that define the system. Indeed, we can classify 2-dimensional electron systems according to whether k is zero or not.

Coupling the system to an "external" or "additional" electromagnetic gauge potential A_μ and using (2) we obtain (after integrating by parts and dropping a surface term)

$$\mathcal{L} = \frac{k}{4\pi} \epsilon^{\mu\nu\lambda} a_\mu \partial_\nu a_\lambda - \frac{1}{2\pi} \epsilon^{\mu\nu\lambda} A_\mu \partial_\nu a_\lambda = \frac{k}{4\pi} \epsilon^{\mu\nu\lambda} a_\mu \partial_\nu a_\lambda - \frac{1}{2\pi} \epsilon^{\mu\nu\lambda} a_\mu \partial_\nu A_\lambda \tag{4}$$

Note that the gauge potential of the magnetic field responsible for the Hall effect should not be included in A_μ; it is implicitly contained already in the coefficient k.

The notion of quasiparticles or "elementary" excitations is basic to condensed matter physics. The effects of a many-body interaction may be such that the quasiparticles in the system are no longer electrons. Here we define the quasiparticles as the entities that couple to the gauge potential and thus write

$$\mathcal{L} = \frac{k}{4\pi} a \epsilon \partial a + a_\mu j^\mu - \frac{1}{2\pi} \epsilon^{\mu\nu\lambda} a_\mu \partial_\nu A_\lambda \cdots \tag{5}$$

Defining $\tilde{j}_\mu \equiv j_\mu - (1/2\pi) \epsilon_{\mu\nu\lambda} \partial^\nu A^\lambda$ and integrating out the gauge field we obtain (see VI.1.4)

$$\mathcal{L} = \frac{\pi}{k} \tilde{j}_\mu \left(\frac{\epsilon^{\mu\nu\lambda} \partial_\nu}{\partial^2} \right) \tilde{j}_\lambda \tag{6}$$

Fractional charge and statistics

We can now simply read off the physics from (6). The Lagrangian contains three types of terms: AA, Aj, and jj. The AA term has the schematic form $A(\epsilon \partial \epsilon \partial \epsilon \partial / \partial^2)A$. Using $\epsilon \partial \epsilon \partial \sim \partial^2$ and canceling between numerator and denominator, we obtain

$$\mathcal{L} = \frac{1}{4\pi k} A \epsilon \partial A \tag{7}$$

Varying with respect to A we determine the electromagnetic current

$$J_{\text{em}}^\mu = \frac{1}{4\pi k} \epsilon^{\mu\nu\lambda} \partial_\nu A_\lambda \tag{8}$$

We learn from the $\mu = 0$ component of this equation that an excess density δn of electrons is related to a local fluctuation of the magnetic field by $\delta n = (1/2\pi k)\delta B$; thus we can identify the filling factor ν as $1/k$, and from the $\mu = i$ components that an electric field produces a current in the orthogonal direction with $\sigma_{xy} = (1/k) = \nu$.

The Aj term has the schematic form $A(\epsilon\partial\epsilon\partial/\partial^2)j$. Canceling the differential operators, we find

$$\mathcal{L} = \frac{1}{k} A_\mu j^\mu \tag{9}$$

Thus, the quasiparticle carries electric charge $1/k$.

Finally, the quasiparticles interact with each other via

$$\mathcal{L} = \frac{\pi}{k} j^\mu \frac{\epsilon_{\mu\nu\lambda}\partial}{\partial^2} j^\lambda \tag{10}$$

We simply remove the twiddle sign in (6). Recalling chapter VI.1 we see that quasiparticles obey fractional statistics with

$$\frac{\theta}{\pi} = \frac{1}{k} \tag{11}$$

By now, you may well be wondering that while all this is fine and good, what would actually tell us that ν^{-1} has to be an odd integer?

We now argue that the electron or hole should appear somewhere in the excitation spectrum. After all, the theory is supposed to describe a system of electrons and thus far our rather general Lagrangian does not contain any reference to the electron!

Let us look for the hole (or electron). We note from (9) that a bound object made up of k quasiparticles would have charge equal to 1. This is perhaps the hole! For this to work, we see that k has to be an integer. So far so good, but k doesn't have to be odd yet.

What is the statistics of this bound object? Let us move one of these bound objects halfway around another such bound object, thus effectively interchanging them. When one quasiparticle moves around another we pick up a phase given by $\theta/\pi = 1/k$ according to (11). But here we have k quasiparticles going around k quasiparticles and so we pick up a phase

$$\frac{\theta}{\pi} = \frac{1}{k}k^2 = k \tag{12}$$

For the hole to be a fermion we must require θ/π to be an odd integer. This fixes k to be an odd integer.

Since $\nu = 1/k$, we have here the classic Laughlin odd-denominator Hall fluids with filling factor $\nu = \frac{1}{3}, \frac{1}{5}, \frac{1}{7}, \cdots$. The famous result that the quasiparticles carry fractional charge and statistics just pops out [see (9) and (11)].

This is truly dramatic: a bunch of electrons moving around in a plane with a magnetic field corresponding to $\nu = \frac{1}{3}$, and lo and behold, each electron has fragmented into three pieces, each piece with charge $\frac{1}{3}$ and fractional statistics $\frac{1}{3}$!

A new kind of order

The goal of condensed matter physics is to understand the various states of matter. States of matter are characterized by the presence (or absence) of order: a ferromagnet becomes ordered below the transition temperature. In the Landau-Ginzburg theory, as we saw in chapter V.3, order is associated with spontaneous symmetry breaking, described naturally with group theory. Girvin and MacDonald first noted that the order in Hall fluids does not really fit into the Landau-Ginzburg scheme: We have not broken any obvious symmetry. The topological property of the Hall fluids provides a clue to what is going on. As explained in the preceding chapter, the ground state degeneracy of a Hall fluid depends on the topology of the manifold it lives on, a dependence group theory is incapable of accounting for. Wen has forcefully emphasized that the study of topological order, or more generally quantum order, may open up a vast new vista on the possible states of matter.[1]

Comments and generalization

Let me conclude with several comments that might spur you to explore the wealth of literature on the Hall fluid.

1. The appearance of integers implies that our result is robust. A slick argument can be made based on the remark in the previous chapter that the Chern-Simons term does not know about clocks and rulers and hence can't possibly depend on microphysics such as the scattering of electrons off impurities which cannot be defined without clocks and rulers. In contrast, the physics that is not part of the topological field theory and described by $(\cdots)$ in (3) would certainly depend on detailed microphysics.

2. If we had followed the long way to derive the effective field theory of the Hall fluid, we would have seen that the quasiparticle is actually a vortex constructed (as in chapter V.7) out of the scalar field representing the electron. Given that the Hall fluid is incompressible, just about the only excitation you can think of is a vortex with electrons coherently whirling around.

3. In the previous chapter we remarked that the Chern-Simons term is gauge invariant only upon dropping a boundary term. But real Hall fluids in the laboratory live in samples with boundaries. So how can (3) be correct? Remarkably, this apparent "defect" of the theory actually represents one of its virtues! Suppose the theory (3) is defined on a bounded 2-dimensional manifold, a disk for example. Then as first argued by Wen there must be physical degrees of freedom living on the boundary and represented by an action whose change under a gauge transformation cancels the change of $\int d^3x (k/4\pi) a\epsilon \, \partial a$. Physically, it is clear that an incompressible fluid would have edge excitations[2] corresponding to waves on its boundary.

---

[1] X. G. Wen, *Quantum Field Theory of Many-Body Systems*.
[2] The existence of edge currents in the integer Hall fluid was first pointed out by Halperin.

4. What if we refuse to introduce gauge potentials? Since the current J_μ has dimension 2, the simplest term constructed out of the currents, $J_\mu J^\mu$, is already of dimension 4; indeed, this is just the Maxwell term. There is no way of constructing a dimension 3 local interaction out of the currents directly. To lower the dimension we are forced to introduce the inverse of the derivative and write schematically $J(1/\epsilon\partial)J$, which is of course just the non-local Hopf term. Thus, the question "why gauge field?" that people often ask can be answered in part by saying that the introduction of gauge fields allows us to avoid dealing with nonlocal interactions.

5. Experimentalists have constructed double-layered quantum Hall systems with an infinitesimally small tunneling amplitude for electrons to go from one layer to the other. Assuming that the current J_I^μ ($I = 1, 2$) in each layer is separately conserved, we introduce two gauge potentials by writing $J_I^\mu = \frac{1}{2\pi}\epsilon^{\mu\nu\lambda}\partial_\nu a_{I\lambda}$ as in (2). We can repeat our general argument and arrive at the effective Lagrangian

$$\mathcal{L} = \sum_{I,J} \frac{K_{IJ}}{4\pi} a_I \epsilon \partial a_J + \cdots \tag{13}$$

The integer k has been promoted to a matrix K. As an exercise, you can derive the Hall conductance, the fractional charge, and the statistics of the quasiparticles. You would not be surprised that everywhere $1/k$ appears we now have the matrix inverse K^{-1} instead. An interesting question is what happens when K has a zero eigenvalue. For example, we could have $K = \begin{pmatrix} 1 & 1 \\ 1 & 1 \end{pmatrix}$. Then the low energy dynamics of the gauge potential $a_- \equiv a_1 - a_2$ is not governed by the Chern-Simons term, but by the Maxwell term in the $(\cdots)$ in (13). We have a linearly dispersing mode and thus a superfluid! This striking prediction[3] was verified experimentally.

6. Finally, an amusing remark: In this formalism electron tunneling corresponds to the nonconservation of the current $J_-^\mu \equiv J_1^\mu - J_2^\mu = (1/2\pi)\epsilon^{\mu\nu\lambda}\partial_\nu a_{-\lambda}$. The difference $N_1 - N_2$ of the number of electrons in the two layers is not conserved. But how can $\partial_\mu J_-^\mu \neq 0$ even though J_-^μ is the curl of $a_{-\lambda}$ (as I have indicated explicitly)? Recalling chapter IV.4, you the astute reader say, aha, magnetic monopoles! Tunneling in a double-layered Hall system in Euclidean spacetime can be described as a gas of monopoles and antimonopoles.[4] (Think, why monopoles and antimonopoles?) Note of course that these are not monopoles in the usual electromagnetic gauge potential but in the gauge potential $a_{-\lambda}$.

What we have given in this section is certainly a very slick derivation of the effective long distance theory of the Hall fluid. Some would say too slick. Let us go back to our five general statements or principles. Of these five, four are absolutely indisputable. In fact, the most questionable is the statement that looks the most innocuous to the casual reader, namely statement 3. In general, the effective Lagrangian for a condensed matter system would be nonlocal. We are implicitly assuming that the system does not contain a massless field, the exchange of which would lead to a nonlocal interaction.[5] Also implicit

---

[3] X. G. Wen and A. Zee, *Phys. Rev. Lett.* 69: 1811, 1992.

[4] X. G. Wen and A. Zee, *Phys. Rev.* B47: 2265, 1993.

[5] A technical remark: Vortices (i.e., quasiparticles) pinned to impurities in the Hall fluid can generate an interaction nonlocal in time.

in (3) is the assumption that the Lagrangian can be expressed completely in terms of the gauge potential a. A priori, we certainly do not know that there might not be other relevant degrees of freedom. The point is that as long as these degrees of freedom are not gapless they can be safely integrated out.

Exercises

VI.2.1 To define filling factor precisely, we have to discuss the quantum Hall system on a sphere rather than on a plane. Put a magnetic monopole of strength G (which according to Dirac can be only a half-integer or an integer) at the center of a unit sphere. The flux through the sphere is equal to $N_\phi = 2G$. Show that the single electron energy is given by $E_l = (\frac{1}{2}\hbar\omega_c)\left[l(l+1) - G^2\right]/G$ with the Landau levels corresponding to $l = G, G+1, G+2, \ldots$, and that the degeneracy of the lth level is $2l+1$. With L Landau levels filled with noninteracting electrons ($\nu = L$) show that $N_\phi = \nu^{-1}N_e - \mathcal{S}$, where the topological quantity $\mathcal{S}$ is known as the shift.

VI.2.2 For a challenge, derive the effective field theory for Hall fluids with filling factor $\nu = m/k$ with k an odd integer, such as $\nu = \frac{2}{5}$. [Hint: You have to introduce m gauge potentials $a_{I\lambda}$ and generalize (2) to $J^\mu = (1/2\pi)\epsilon^{\mu\nu\lambda}\partial_\nu \sum_{I=1}^{m} a_{I\lambda}$. The effective theory turns out to be

$$
\mathcal{L} = \frac{1}{4\pi} \sum_{I,J=1}^{m} a_I K_{IJ} \epsilon \partial a_J + \sum_{I=1}^{m} a_{I\mu} \tilde{j}^{I\mu} + \cdots
$$

with the integer k replaced by a matrix K. Compare with (13).]

VI.2.3 For the Lagrangian in (13), derive the analogs of (8), (9), and (11).

VI.3 | Duality

A far reaching concept

Duality is a profound and far reaching concept[1] in theoretical physics, with origins in electromagnetism and statistical mechanics. The emergence of duality in recent years in several areas of modern physics, ranging from the quantum Hall fluids to string theory, represents a major development in our understanding of quantum field theory. Here I touch upon one particular example just to give you a flavor of this vast subject.

My plan is to treat a relativistic theory first, and after you get the hang of the subject, I will go on to discuss the nonrelativistic theory. It makes sense that some of the interesting physics of the nonrelativistic theory is absent in the relativistic formulation: A larger symmetry is more constraining. By the same token, the relativistic theory is actually much easier to understand if only because of notational simplicity.

Vortices

Couple a scalar field in $(2 + 1)$-dimensions to an external electromagnetic gauge potential, with the electric charge q indicated explicitly for later convenience:

$$\mathcal{L} = \tfrac{1}{2}|(\partial_\mu - iqA_\mu)\varphi|^2 - V(\varphi^\dagger \varphi) \tag{1}$$

We have already studied this theory many times, most recently in chapter V.7 in connection with vortices. As usual, write $\varphi = |\varphi|e^{i\theta}$. Minimizing the potential V at $|\varphi| = v$ gives the ground state field configuration. Setting $\varphi = ve^{i\theta}$ in (1) we obtain

$$\mathcal{L} = \tfrac{1}{2}v^2(\partial_\mu \theta - qA_\mu)^2 \tag{2}$$

---

[1] For a first introduction to duality, I highly recommend J. M. Figueroa-O'Farrill, *Electromagnetic Duality for Children*, http://www.maths.ed.ac.uk/~jmf/Teaching/Lectures/EDC.html.

which upon absorbing θ into A by a gauge transformation we recognize as the Meissner Lagrangian. For later convenience we also introduce the alternative form

$$\mathcal{L} = -\frac{1}{2v^2}\xi_\mu^2 + \xi^\mu(\partial_\mu\theta - qA_\mu) \tag{3}$$

We recover (2) upon eliminating the auxiliary field ξ^μ (see appendix A and chapter III.5).

In chapter V.7 we learned that the excitation spectrum includes vortices and anti-vortices, located where $|\varphi|$ vanishes. If, around the zero of $|\varphi|$, θ changes by 2π, we have a vortex. Around an antivortex, $\Delta\theta = -2\pi$. Recall that around a vortex sitting at rest, the electromagnetic gauge potential has to go as

$$qA_i \to \partial_i\theta \tag{4}$$

at spatial infinity in order for the energy of the vortex to be finite, as we can see from (2). The magnetic flux

$$\int d^2x\,\varepsilon_{ij}\partial_i A_j = \oint d\vec{x}\cdot\vec{A} = \frac{\Delta\theta_{\text{vortex}}}{q} = \frac{2\pi}{q} \tag{5}$$

is quantized in units of $2\pi/q$.

Let us pause to think physically for a minute. On a distance scale large compared to the size of the vortex, vortices and antivortices appear as points. As discussed in chapter V.7, the interaction energy of a vortex and an antivortex separated by a distance R is given by simply plugging into (2). Ignoring the probe field A_μ, which we can take to be as weak as possible, we obtain $\sim \int_a^R dr\, r(\nabla\theta)^2 \sim \log(R/a)$ where a is some short distance cutoff. But recall that the Coulomb interaction in 2-dimensional space is logarithmic since by dimensional analysis $\int d^2k(e^{i\vec{k}\cdot\vec{x}}/k^2) \sim \log(|\vec{x}|/a)$ (with a^{-1} some ultraviolet cutoff). Thus, a gas of vortices and antivortices appears as a gas of point "charges" with a Coulomb interaction between them.

Vortex as charge in a dual theory

Duality is often made out by some theorists to be a branch of higher mathematics but in fact it derives from an entirely physical idea. In view of the last paragraph, can we not rewrite the theory so that vortices appear as point "charges" of some as yet unknown gauge field? In other words, we want a dual theory in which the fundamental field creates and annihilates vortices rather than φ quanta. We will explain the word "dual" in due time.

Remarkably, the rewriting can be accomplished in just a few simple steps. Proceeding physically and heuristically, we picture the phase field θ as smoothly fluctuating, except that here and there it winds around 2π. Write $\partial_\mu\theta = \partial_\mu\theta_{\text{smooth}} + \partial_\mu\theta_{\text{vortex}}$. Plugging into (3) we write

$$\mathcal{L} = -\frac{1}{2v^2}\xi_\mu^2 + \xi^\mu(\partial_\mu\theta_{\text{smooth}} + \partial_\mu\theta_{\text{vortex}} - qA_\mu) \tag{6}$$

Integrate over θ_{smooth} and obtain the constraint $\partial_\mu\xi^\mu = 0$, which can be solved by writing

$$\xi^\mu = \varepsilon^{\mu\nu\lambda}\partial_\nu a_\lambda \tag{7}$$

a trick we used earlier in chapter VI.2. As in that chapter, a gauge potential comes looking for us, since the change $a_\lambda \to a_\lambda + \partial_\lambda \Lambda$ does not change ξ^μ. Plugging into (6), we find

$$\mathcal{L} = -\frac{1}{4v^2} f_{\mu\nu}^2 + \varepsilon^{\mu\nu\lambda} \partial_\nu a_\lambda (\partial_\mu \theta_{\text{vortex}} - q A_\mu) \tag{8}$$

where $f_{\mu\nu} = \partial_\mu a_\nu - \partial_\nu a_\mu$.

Our treatment is heuristic because we ignore the fact that $|\varphi|$ vanishes at the vortices. Physically, we think of the vortices as almost pointlike so that $|\varphi| = v$ "essentially" everywhere. As mentioned in chapter V.6, by appropriate choice of parameters we can make solitons, vortices, and so on as small as we like. In other words, we neglect the coupling between θ_{vortex} and $|\varphi|$. A rigorous treatment would require a proper short distance cutoff by putting the system on a lattice.[2] But as long as we capture the essential physics, as we assuredly will, we will ignore such niceties.

Note for later use that the electromagnetic current J^μ, defined as the coefficient of $-A_\mu$ in (8), is determined in terms of the gauge potential a_λ to be

$$J^\mu = q \varepsilon^{\mu\nu\lambda} \partial_\nu a_\lambda \tag{9}$$

Let us integrate the term $\varepsilon^{\mu\nu\lambda} \partial_\nu a_\lambda \partial_\mu \theta_{\text{vortex}}$ in (8) by parts to obtain $a_\lambda \varepsilon^{\lambda\mu\nu} \partial_\mu \partial_\nu \theta_{\text{vortex}}$. According to Newton and Leibniz, ∂_μ commutes with ∂_ν, and so apparently we get zero. But ∂_μ and ∂_ν commute only when acting on a globally defined function, and, heavens to Betsy, θ_{vortex} is not globally defined since it changes by 2π when we go around a vortex. In particular, consider a vortex at rest and look at the quantity a_0 couples to in (8) namely $\varepsilon^{ij} \partial_i \partial_j \theta_{\text{vortex}} = \vec{\nabla} \times (\vec{\nabla} \theta_{\text{vortex}})$ in the notation of elementary physics. Integrating this over a region containing the vortex gives $\int d^2 x \vec{\nabla} \times (\vec{\nabla} \theta_{\text{vortex}}) = \oint d\vec{x} \cdot \vec{\nabla} \theta_{\text{vortex}} = 2\pi$. Thus, we recognize $(1/2\pi) \varepsilon^{ij} \partial_i \partial_j \theta_{\text{vortex}}$ as the density of vortices, the time component of some vortex current $j_{\text{vortex}}^\lambda$. By Lorentz invariance, $j_{\text{vortex}}^\lambda = (1/2\pi) \varepsilon^{\lambda\mu\nu} \partial_\mu \partial_\nu \theta_{\text{vortex}}$.

Thus, we can now write (8) as

$$\mathcal{L} = -\frac{1}{4v^2} f_{\mu\nu}^2 + (2\pi) a_\mu j_{\text{vortex}}^\mu - A_\mu (q \varepsilon^{\mu\nu\lambda} \partial_\nu a_\lambda) \tag{10}$$

Lo and behold, we have accomplished what we set out to do. We have rewritten the theory so that the vortex appears as an "electric charge" for the gauge potential a_μ. Sometimes this is called a dual theory, but strictly speaking, it is more accurate to refer to it as the dual representation of the original theory (1).

Let us introduce a complex scalar field Φ, which we will refer to as the vortex field, to create and annihilate the vortices and antivortices. In other words, we "elaborate" the description in (10) to

$$\mathcal{L} = -\frac{1}{4v^2} f_{\mu\nu}^2 + \frac{1}{2} |(\partial_\mu - i(2\pi) a_\mu) \Phi|^2 - W(\Phi) - A_\mu (q \varepsilon^{\mu\nu\lambda} \partial_\nu a_\lambda) \tag{11}$$

---

[2] For example, M. P. A. Fisher, "Mott Insulators, Spin Liquids, and Quantum Disordered Superconductivity," cond-mat/9806164, appendix A.

The potential $W(\Phi)$ contains terms such as $\lambda(\Phi^\dagger\Phi)^2$ describing the short distance interaction of two vortices (or a vortex and an antivortex.) In principle, if we master all the short distance physics contained in the original theory (1) then these terms are all determined by the original theory.

Vortex of a vortex

Now we come to the most fascinating aspect of the duality representation and the reason why the word "dual" is used in the first place. The vortex field Φ is a complex scalar field, just like the field φ we started with. Thus we can perfectly well form a vortex out of Φ, namely a place where Φ vanishes and around which the phase of Φ goes through 2π. Amusingly, we are forming a vortex of a vortex, so to speak.

So, what is a vortex of a vortex?

The duality theorem states that the vortex of a vortex is nothing but the original charge, described by the field φ we started out with! Hence the word duality.

The proof is remarkably simple. The vortex in the theory (11) carries "magnetic flux." Referring to (11) we see that $2\pi a_i \to \partial_i\theta$ at spatial infinity. By exactly the same manipulation as in (5), we have

$$2\pi \int d^2x\, \varepsilon_{ij}\partial_i a_j = 2\pi \oint d\vec{x}\cdot\vec{a} = 2\pi \tag{12}$$

Note that I put quotation marks around the term "magnetic flux" since as is evident I am talking about the flux associated with the gauge potential a_μ and not the flux associated with the electromagnetic potential A_μ. But remember that from (9) the electromagnetic current $J^\mu = q\varepsilon^{\mu\nu\lambda}\partial_\nu a_\lambda$ and in particular $J^0 = q\varepsilon^{ij}\partial_i a_j$. Hence, the electric charge (note no quotation marks) of this vortex of a vortex is equal to $\int d^2x\, J^0 = q$, precisely the charge of the original complex scalar field φ. This proves the assertion.

Here we have studied vortices, but the same sort of duality also applies to monopoles. As I remarked in chapter IV.4, duality allows us a glimpse into field theories in the strongly coupled regime. We learned in chapter V.7 that certain spontaneously broken nonabelian gauge theories in $(3 + 1)$-dimensional spacetime contains magnetic monopoles. We can write a dual theory in terms of the monopole field out of which we can construct monopoles. The monopole of a monopole turns out be none other than the charged fields of the original gauge theory. This duality was first conjectured many years ago by Olive and Montonen and later shown to be realized in certain supersymmetric gauge theories by Seiberg and Witten. The understanding of this duality was a "hot" topic a few years ago as it led to deep new insight about how certain string theories are dual to each other.[3] In contrast, according to one of my distinguished condensed matter colleagues, the important notion of duality is still underappreciated in the condensed matter physics community.

---

[3] For example, D. I. Olive and P. C. West, eds., *Duality and Supersymmetric Theories*.

Meissner begets Maxwell and so on

We will close by elaborating slightly on duality in $(2 + 1)$-dimensional spacetime and how it might be relevant to the physics of 2-dimensional materials. Consider a Lagrangian $\mathcal{L}(a)$ quadratic in a vector field a_μ. Couple an external electromagnetic gauge potential A_μ to the conserved current $\varepsilon^{\mu\nu\lambda}\partial_\nu a_\lambda$:

$$\mathcal{L} = \mathcal{L}(a) + A_\mu(\varepsilon^{\mu\nu\lambda}\partial_\nu a_\lambda) \tag{13}$$

Let us ask: For various choices of $\mathcal{L}(a)$, if we integrate out a what is the effective Lagrangian $\mathcal{L}(A)$ describing the dynamics of A?

If you have gotten this far in the book, you can easily do the integration. The central identity of quantum field theory again! Given

$$\mathcal{L}(a) \sim aKa \tag{14}$$

we have

$$\mathcal{L}(A) \sim (\varepsilon\partial A)\frac{1}{K}(\varepsilon\partial A) \sim A(\varepsilon\partial\frac{1}{K}\varepsilon\partial)A \tag{15}$$

We have three choices for $\mathcal{L}(a)$ to which I attach various illustrious names:

$\mathcal{L}(a) \sim a^2$	Meissner
$\mathcal{L}(a) \sim a\varepsilon\partial a$	Chern-Simons
$\mathcal{L}(a) \sim f^2 \sim a\partial^2 a$	Maxwell

$$(16)$$

Since we are after conceptual understanding, I won't bother to keep track of indices and irrelevant overall constants. (You can fill them in as an exercise.) For example, given $\mathcal{L}(a) = f_{\mu\nu}f^{\mu\nu}$ with $f_{\mu\nu} = \partial_\mu a_\nu - \partial_\nu a_\mu$ we can write $\mathcal{L}(a) \sim a\partial^2 a$ and so $K = \partial^2$. Thus, the effective dynamics of the external electromagnetic gauge potential is given by (15) as $\mathcal{L}(A) \sim A[\varepsilon\partial(1/\partial^2)\varepsilon\partial]A \sim A^2$, the Meissner Lagrangian! In this "quick and dirty" way of making a living, we simply set $\varepsilon\varepsilon \sim 1$ and cancel factors of ∂ in the numerator against those in the denominator. Proceeding in this way, we construct the following table:

Dynamics of a	K	Effective Lagrangian $\mathcal{L}(A) \sim A[\varepsilon\partial(1/K)\varepsilon\partial]A$	Dynamics of the external probe A
Meissner a^2	1	$A(\varepsilon\partial\varepsilon\partial)A \sim A\partial^2 A$	Maxwell F^2
Chern-Simons $a\varepsilon\partial a$	$\varepsilon\partial$	$A(\varepsilon\partial\frac{1}{\varepsilon\partial}\varepsilon\partial)A \sim A\varepsilon\partial A$	Chern-Simons $A\varepsilon\partial A$
Maxwell $f^2 \sim a\partial^2 a$	∂^2	$A(\varepsilon\partial\frac{1}{\partial^2}\varepsilon\partial)A \sim AA$	Meissner A^2

$$(17)$$

Meissner begets Maxwell, Chern-Simons begets Chern-Simons, and Maxwell begets Meissner. I find this beautiful and fundamental result, which represents a form of duality, very striking. Chern-Simons is self-dual: It begets itself.

Going nonrelativistic

It is instructive to compare the nonrelativistic treatment of duality.[4] Go back to the superfluid Lagrangian of chapter V.1:

$$\mathcal{L} = i\varphi^\dagger \partial_0 \varphi - \frac{1}{2m} \partial_i \varphi^\dagger \partial_i \varphi - g^2 (\varphi^\dagger \varphi - \bar{\rho})^2 \tag{18}$$

As before, substitute $\varphi \equiv \sqrt{\rho} e^{i\theta}$ to obtain

$$\mathcal{L} = -\rho \partial_0 \theta - \frac{\rho}{2m} (\partial_i \theta)^2 - g^2 (\rho - \bar{\rho})^2 + \cdots \tag{19}$$

which we rewrite as

$$\mathcal{L} = -\xi_\mu \partial^\mu \theta + \frac{m}{2\rho} \xi_i^2 - g^2 (\rho - \bar{\rho})^2 + \cdots \tag{20}$$

In (19) we have dropped a term $\sim (\partial_i \rho^{1/2})^2$. In (20) we have defined $\xi_0 \equiv \rho$. Integrating out ξ_i in (20) we recover (19).

All proceeds as before. Writing $\theta = \theta_{\text{smooth}} + \theta_{\text{vortex}}$ and integrating out θ_{smooth}, we obtain the constraint $\partial^\mu \xi_\mu = 0$, solved by writing $\xi^\mu = \epsilon^{\mu\nu\lambda} \partial_\nu \hat{a}_\lambda$. The hat on $\hat{a}_\lambda$ is for later convenience. Note that the density

$$\xi_0 \equiv \rho = \epsilon_{ij} \partial_i \hat{a}_j \equiv \hat{f} \tag{21}$$

is the "magnetic" field strength while

$$\xi_i = \epsilon_{ij} (\partial_0 \hat{a}_j - \partial_j \hat{a}_0) \equiv \epsilon_{ij} \hat{f}_{0j} \tag{22}$$

is the "electric" field strength.

Putting all of this into (20) we have

$$\mathcal{L} = \frac{m}{2\rho} \hat{f}_{0i}^2 - g^2 (\hat{f} - \bar{\rho})^2 - 2\pi \hat{a}_\mu j_{\text{vortex}}^\mu + \cdots \tag{23}$$

To "subtract out" the background "magnetic" field $\bar{\rho}$, an obviously sensible move is to write

$$\hat{a}_\mu = \bar{a}_\mu + a_\mu \tag{24}$$

where we define the background gauge potential by $\bar{a}_0 = 0$, $\partial_0 \bar{a}_j = 0$ (no background "electric" field) and

$$\epsilon_{ij} \partial_i \bar{a}_j = \bar{\rho} \tag{25}$$

---

[4] The treatment given here follows essentially that given by M. P. A. Fisher and D. H. Lee.

The Lagrangian (23) then takes on the cleaner form

$$\mathcal{L} = \left(\frac{m}{2\bar{\rho}} f_{0i}^2 - g^2 f^2 \right) - 2\pi a_\mu j_{\text{vortex}}^\mu - 2\pi \bar{a}_i j_{\text{vortex}}^i + \cdots \tag{26}$$

We have expanded $\rho \sim \bar{\rho}$ in the first term. As in (10) the first two terms form the Maxwell Lagrangian, and the ratio of their coefficients determines the speed of propagation

$$c = \left(\frac{2g^2 \bar{\rho}}{m} \right)^{1/2} \tag{27}$$

In suitable units in which $c = 1$, we have

$$\mathcal{L} = -\frac{m}{4\bar{\rho}} f_{\mu\nu} f^{\mu\nu} - 2\pi a_\mu j_{\text{vortex}}^\mu - 2\pi \bar{a}_i j_{\text{vortex}}^i + \cdots \tag{28}$$

Compare this with (10).

The one thing we missed with our relativistic treatment is the last term in (28), for the simple reason that we didn't put in a background. Recall that the term like $A_i J_i$ in ordinary electromagnetism means that a moving particle associated with the current J_i sees a magnetic field $\vec{\nabla} \times \vec{A}$. Thus, a moving vortex will see a "magnetic field"

$$\epsilon_{ij} \partial_i (\bar{a} + a)_j = \bar{\rho} + \epsilon_{ij} \partial_i a_j \tag{29}$$

equal to the sum of $\bar{\rho}$, the density of the original bosons, and a fluctuating field.

In the Coulomb gauge $\partial_i a_i = 0$ we have $(f_{0i})^2 = (\partial_0 a_i)^2 + (\partial_i a_0)^2$, where the cross term $(\partial_0 a_i)(\partial_i a_0)$ effectively vanishes upon integration by parts. Integrating out the Coulomb field a_0, we obtain

$$\mathcal{L} = -\frac{\bar{\rho}}{2m} (2\pi)^2 \iint d^2x d^2y \left[j_0(\vec{x}) \log \frac{|\vec{x} - \vec{y}|}{a} j_0(\vec{y}) \right]$$
$$+ \frac{m}{2\bar{\rho}} (\partial_0 a_i)^2 - g^2 f^2 + 2\pi (a_i + \bar{a}_i) j_i^{\text{vortex}} \tag{30}$$

The vortices repel each other by a logarithmic interaction $\int d^2k (e^{i\vec{k}\cdot\vec{x}}/k^2) \sim \log(|\vec{x}|/a)$ as we have known all along.

A self-dual theory

Interestingly, the spatial part f^2 of the Maxwell Lagrangian comes from the short ranged repulsion between the original bosons.

If we had taken the bosons to interact by an arbitrary potential $V(x)$ we would have, instead of the last term in (20),

$$\iint d^2x d^2y [\rho(x) - \bar{\rho}] V(x - y) [\rho(y) - \bar{\rho}] \tag{31}$$

It is easy to see that all the steps go through essentially as before, but now the second term in (26) becomes

$$\iint d^2x d^2y f(x) V(x - y) f(y) \tag{32}$$

Thus, the gauge field propagates according to the dispersion relation

$$\omega^2 = (2\bar{\rho}/m)V(k)\vec{k}^2 \tag{33}$$

where $V(k)$ is the Fourier transformation of $V(x)$. In the special case $V(x) = g^2\delta^{(2)}(x)$ we recover the linear dispersion given in (27). Indeed, we have a linear dispersion $\omega \propto |\vec{k}|$ as long as $V(x)$ is sufficiently short ranged for $V(\vec{k} = 0)$ to be finite.

An interesting case is when $V(x)$ is logarithmic. Then $V(k)$ goes as $1/k^2$ and so $\omega \sim$ constant: The gauge field a_i becomes massive and drops out. The low energy effective theory consists of a bunch of vortices with a logarithmic interaction between them. Thus, a theory of bosons with a logarithmic repulsion between them is self dual in the low energy limit.

The dance of vortices and antivortices

Having gone through this nonrelativistic discussion of duality, let us reward ourselves by deriving the motion of vortices in a fluid. Let the bulk of the fluid be at rest. According to (28) the vortex behaves like a charged particle in a background magnetic field $\bar{b}$ proportional to the mean density of the fluid $\bar{\rho}$. Thus, the force acting on a vortex is the usual Lorentz force $\vec{v} \times \vec{B}$, and the equation of motion of the vortex in the presence of a force F is then just

$$\bar{\rho}\epsilon_{ij}\dot{x}_j = F_i \tag{34}$$

This is the well-known result that a vortex, when pushed, moves in a direction perpendicular to the force.

Consider two vortices. According to (30) they repel each other by a logarithmic interaction. They move perpendicular to the force. Thus, they end up circling each other. In contrast, consider a vortex and an antivortex, which attract each other. As a result of this attraction, they both move in the same direction, perpendicular to the straight line joining them (see fig. VI.3.1). The vortex and antivortex move along in step, maintaining the distance between them. This in fact accounts for the famous motion of a smoke ring. If we cut a smoke ring through its center and perpendicular to the plane it lies in, we have just

(a)

(b)

Figure VI.3.1

a vortex with an antivortex for each section. Thus, the entire smoke ring moves along in a direction perpendicular to the plane it lies in.

All of this can be understood by elementary physics, as it should be. The key observation is simply that vortices and antivortices produce circular flows in the fluid around them, say clockwise for vortices and anticlockwise for antivortices. Another basic observation is that if there is a local flow in the fluid, then any object, be it a vortex or an antivortex, caught in it would just flow along in the same direction as the local flow. This is a consequence of Galilean invariance. By drawing a simple picture you can see that this produces the same pattern of motion as discussed above.

VI.4 | The σ Models as Effective Field Theories

The Lagrangian as a mnemonic

Our beloved quantum field theory has had two near death experiences. The first started around the mid-1930s when physical quantities came out infinite. But it roared back to life in the late 1940s and early 1950s, thanks to the work of the generation that included Feynman, Schwinger, Dyson, and others. The second occurred toward the late 1950s. As we have already discussed, quantum field theory seemed totally incapable of addressing the strong interaction: The coupling was far too strong for perturbation theory to be of any use. Many physicists—known collectively as the S-matrix school—felt that field theory was irrelevant for studying the strong interaction and advocated a program of trying to derive results from general principles without using field theory. For example, in deriving the Goldberger-Treiman relation, we could have foregone any mention of field theory and Feynman diagrams.

Eventually, in a reaction against this trend, people realized that if some results could be obtained from general considerations such as notions of spontaneous symmetry breaking and so forth, any Lagrangian incorporating these general properties had to produce the same results. At the very least, the Lagrangian provides a mnemonic for any physical result derived without using quantum field theory. Thus was born the notion of long distance or low energy effective field theory, which would prove enormously useful in both particle and condensed matter physics (as we have already seen and as we will discuss further in chapter VIII.3).

The strong interaction at low energies

One of the earliest examples is the σ model of Gell-Mann and Lévy, which describes the interaction of nucleons and pions. We now know that the strong interaction has to be described in terms of quarks and gluons. Nevertheless, at long distances, the degrees of freedom are the two nucleons and the three pions. The proton and the neutron transform as a spinor $\psi \equiv \binom{p}{n}$ under the $SU(2)$ of isospin. Consider the kinetic energy term $\bar{\psi} i \gamma \partial \psi = \bar{\psi}_L i \gamma \partial \psi_L + \bar{\psi}_R i \gamma \partial \psi_R$. We note that this term has the larger symmetry $SU(2)_L \times SU(2)_R$, with the left handed field ψ_L and the right handed field ψ_R transforming as a doublet under $SU(2)_L$ and $SU(2)_R$, respectively. [The $SU(2)$ of isospin is the diagonal subgroup of $SU(2)_L \times SU(2)_R$.] We can write $\psi_L \sim (\frac{1}{2}, 0)$ and $\psi_R \sim (0, \frac{1}{2})$.

Now we see a problem immediately: The mass term $m\bar{\psi}\psi = m(\bar{\psi}_L \psi_R + \text{h.c.})$ is not allowed since $\bar{\psi}_L \psi_R \sim (\frac{1}{2}, \frac{1}{2})$, a 4-dimensional representation of $SU(2)_L \times SU(2)_R$, a group locally isomorphic to $SO(4)$.

At this point, lesser physicists would have said, what is the problem, we knew all along that the strong interaction is invariant only under the $SU(2)$ of isospin, which we will write as $SU(2)_I$. Under $SU(2)_I$ the bilinears constructed out of $\bar{\psi}_L$ and ψ_R transform as $\frac{1}{2} \times \frac{1}{2} = 0 + 1$, the singlet being $\bar{\psi}\psi$ and the triplet $\bar{\psi} i \gamma^5 \tau^a \psi$. With only $SU(2)_I$ symmetry, we can certainly include the mass term $\bar{\psi}\psi$.

To say it somewhat differently, to fully couple to the four bilinears we can construct out of $\bar{\psi}_L$ and ψ_R, namely $\bar{\psi}\psi$ and $\bar{\psi} i \gamma^5 \tau^a \psi$, we need four meson fields transforming as the vector representation under $SO(4)$. But only the three pion fields are known. It seems clear that we only have $SU(2)_I$ symmetry.

Nevertheless, Gell-Mann and Lévy boldly insisted on the larger symmetry $SU(2)_L \times SU(2)_R \simeq SO(4)$ and simply postulated an additional meson field, which they called σ, so that $(\sigma, \vec{\pi})$ form the 4-dimensional representation. I leave it to you to verify that $\bar{\psi}_L(\sigma + i\vec{\tau} \cdot \vec{\pi})\psi_R + \text{h.c.} = \bar{\psi}(\sigma + i\vec{\tau} \cdot \vec{\pi}\gamma_5)\psi$ is invariant. Hence, we can write down the invariant Lagrangian

$$\mathcal{L} = \bar{\psi}[i\gamma \partial + g(\sigma + i\vec{\tau} \cdot \vec{\pi}\gamma_5)]\psi + \mathcal{L}(\sigma, \vec{\pi}) \tag{1}$$

where the part not involving the nucleons reads

$$\mathcal{L}(\sigma, \vec{\pi}) = \frac{1}{2}\left((\partial\sigma)^2 + (\partial\vec{\pi})^2\right) + \frac{\mu^2}{2}(\sigma^2 + \vec{\pi}^2) - \frac{\lambda}{4}(\sigma^2 + \vec{\pi}^2)^2 \tag{2}$$

This is known as the linear σ model.

The σ model would have struck most physicists as rather strange at the time it was introduced: The nucleon does not have a mass and there is an extra meson field. Aha, but you would recognize (2) as precisely the Lagrangian (IV.1.2) (for $N = 4$) that we studied, which exhibits spontaneous symmetry breaking. The four scalar fields $(\varphi_4, \varphi_1, \varphi_2, \varphi_3)$ in (IV.1.2) correspond to $(\sigma, \vec{\pi})$. With no loss of generality, we can choose the vacuum expectation value of φ to point in the 4th direction, namely the vacuum in which $\langle 0| \sigma |0 \rangle = \sqrt{\mu^2/\lambda} \equiv v$ and $\langle 0| \vec{\pi} |0 \rangle = 0$. Expanding $\sigma = v + \sigma'$ we see immediately that the nucleon has a mass $M = gv$. You should not be surprised that the pion comes out massless. The meson associated with the field σ', which we will call the σ meson, has no reason to be massless and indeed is not.

Can the all-important parameter v be related to a measurable quantity? Indeed. From chapter I.10 you will recall that the axial current is given by Noether's theorem as $J_{\mu 5}^a = \bar{\psi}\gamma_\mu\gamma_5(\tau^a/2)\psi + \pi^a\partial_\mu\sigma - \sigma\partial_\mu\pi^a$. After σ acquires a vacuum expectation value,

$J^a_{\mu 5}$ contains a term $-v\partial_\mu \pi^a$. This term implies that the matrix element $\langle 0| J^a_{\mu 5} |\pi^b\rangle = i v k_\mu$, where k denotes the momentum of the pion, and thus v is proportional to the f defined in chapter IV.2. Indeed, we recognize the mass relation $M = gv$ as precisely the Goldberger-Treiman relation (IV.2.7) with $F(0) = 1$ (see exercise VI.4.4).

The nonlinear σ model

It was eventually realized that the main purpose in life of the potential in $\mathcal{L}(\sigma, \vec{\pi})$ is to force the vacuum expectation values of the fields to be what they are, so the potential can be replaced by a constraint $\sigma^2 + \vec{\pi}^2 = v^2$. A more physical way of thinking about this point is by realizing that the σ meson, if it exists at all, must be very broad since it can decay via the strong interaction into two pions. We might as well force it out of the low energy spectrum by making its mass large. By now, you have learned from chapters IV.1 and V.1 that the mass of the σ meson, namely $\sqrt{2}\mu$, can be taken to infinity while keeping v fixed by letting μ^2 and λ tend to infinity, keeping their ratio fixed.

We will now focus on $\mathcal{L}(\sigma, \vec{\pi})$. Instead of thinking abut $\mathcal{L}(\sigma, \vec{\pi}) = \frac{1}{2}[(\partial\sigma)^2 + (\partial\vec{\pi})^2]$ with the constraint $\sigma^2 + \vec{\pi}^2 = v^2$, we can simply solve the constraint and plug the solution $\sigma = \sqrt{v^2 - \vec{\pi}^2}$ into the Lagrangian, thus obtaining what is known as the nonlinear σ model:

$$\mathcal{L} = \frac{1}{2}\left[(\partial\vec{\pi})^2 + \frac{(\vec{\pi} \cdot \partial\vec{\pi})^2}{f^2 - \vec{\pi}^2}\right] = \frac{1}{2}(\partial\vec{\pi})^2 + \frac{1}{2f^2}(\vec{\pi} \cdot \partial\vec{\pi})^2 + \cdots \tag{3}$$

Note that $\mathcal{L}$ can be written in the form $\mathcal{L} = (\partial\pi^a)G^{ab}(\vec{\pi})(\partial\pi^b)$; some people like to think of G^{ab} as a "metric" in field space. [Incidentally, recall that way back in chapter I.3 we restricted ourselves to the simplest possible kinetic energy term $\frac{1}{2}(\partial\varphi)^2$, rejecting possibilities such as $U(\varphi)(\partial\varphi)^2$. But recall also that in chapter IV.3 we noted that such a term would arise by quantum fluctuations.]

In accordance with the philosophy that introduced this chapter, any Lagrangian that captures the correct symmetry properties should describe the same low energy physics.[1] This means that anybody, including you, can introduce his or her own parametrization of the fields.

The nonlinear σ model is actually an example of a broad class of field theories whose Lagrangian has a simple form but with the fields appearing in it subject to some nontrivial constraint. An example is the theory defined by

$$\mathcal{L}(U) = \frac{f^2}{4} \operatorname{tr}(\partial_\mu U^\dagger \cdot \partial^\mu U) \tag{4}$$

with $U(x)$ a matrix-valued field and an element of $SU(2)$. Indeed, if we write $U = e^{(i/f)\vec{\pi}\cdot\vec{\tau}}$ we see that $\mathcal{L}(U) = \frac{1}{2}(\partial\vec{\pi})^2 + (1/2f^2)(\vec{\pi} \cdot \partial\vec{\pi})^2 + \cdots$, identical to (3) up to the terms indicated. The $\vec{\pi}$ field here is related to the one in (3) by a field redefinition.

There is considerably more we can say about the nonlinear σ models and their applications in particle and condensed matter physics, but a thorough discussion would take us

[1] S. Weinberg, *Physica* 96A: 327, 1979.

far beyond the scope of this book. Instead, I will develop some of their properties in the exercises and in the next chapter will sketch how they can arise in one class of condensed matter systems.

Exercises

VI.4.1   Show that the vacuum expectation value of $(\sigma, \vec{\pi})$ can indeed point in any direction without changing the physics. At first sight, this statement seems strange since, by virtue of its γ_5 coupling to the nucleon, the pion is a pseudoscalar field and cannot have a vacuum expectation without breaking parity. But $(\sigma, \vec{\pi})$ are just Greek letters. Show that by a suitable transformation of the nucleon field parity is conserved, as it should be in the strong interaction.

VI.4.2   Calculate the pion-pion scattering amplitude up to quadratic order in the external momenta, using the nonlinear σ model (3). [Hint: For help, see S. Weinberg, *Phys. Rev. Lett.* 17: 616, 1966.]

VI.4.3   Calculate the pion-pion scattering amplitude up to quadratic order in the external momenta, using the linear σ model (2). Don't forget the Feynman diagram involving σ meson exchange. You should get the same result as in exercise VI.4.2.

VI.4.4   Show that the mass relation $M = gv$ amounts to the Goldberger-Treiman relation.

Magnetic moments

In chapters IV.1 and V.3 I discussed how the concept of the Nambu-Goldstone boson originated as the spin wave in a ferromagnetic or an antiferromagnetic material. A cartoon description of such materials consists of a regular lattice on each site of which sits a local magnetic moment, which we denote by a unit vector $\vec{n}_j$ with j labeling the site. In a ferromagnetic material the magnetic moments on neighboring sites want to point in the same direction, while in an antiferromagnetic material the magnetic moments on neighboring sites want to point in opposite directions. In other words, the energy is $\mathcal{H} = J \sum_{<ij>} \vec{n}_i \cdot \vec{n}_j$, where i and j label neighboring sites. For antiferromagnets $J > 0$, and for ferromagnets $J < 0$. I will merely allude to the fully quantum description formulated in terms of a spin $\vec{S}_j$ operator on each site j; the subject lies far beyond the scope of this text.

In a more microscopic treatment, we would start with a Hamiltonian (such as the Hubbard Hamiltonian) describing the hopping of electrons and the interaction between them. Within some approximate mean field treatment the classical variable $\vec{n}_j$ would then emerge as the unit vector pointing in the direction of $\langle c_j^\dagger \vec{\sigma} c_j \rangle$ with $c_j^\dagger$ and c_j the electron creation and annihilation operators, respectively. But this is not a text on solid state physics.

First versus second order in time

Here we would like to derive an effective low energy description of the ferromagnet and antiferromagnet in the spirit of the σ model description of the preceding chapter. Our treatment will be significantly longer than the standard discussion given in some field theory texts, but has the slight advantage of being correct.

The somewhat subtle issue is what kinetic energy term we have to add to $-\mathcal{H}$ to form the Lagrangian L. Since for a unit vector $\vec{n}$ we have $\vec{n} \cdot (d\vec{n}/dt) = (d(\vec{n} \cdot \vec{n})/dt) = 0$, we cannot make do with one time derivative. With two derivatives we can form $(d\vec{n}/dt) \cdot (d\vec{n}/dt)$ and so

$$L_{\text{wrong}} = \frac{1}{2g^2} \sum_j \frac{\partial \vec{n}_j}{\partial t} \cdot \frac{\partial \vec{n}_j}{\partial t} - J \sum_{<ij>} \vec{n}_i \cdot \vec{n}_j \tag{1}$$

A typical field theory text would then pass to the continuum limit and arrive at the Lagrangian density

$$\mathcal{L} = \frac{1}{2g^2} \left(\frac{\partial \vec{n}}{\partial t} \cdot \frac{\partial \vec{n}}{\partial t} - c_s^2 \sum_l \frac{\partial \vec{n}}{\partial x^l} \cdot \frac{\partial \vec{n}}{\partial x^l} \right) \tag{2}$$

with the constraint $[\vec{n}(x, t)]^2 = 1$. This is another example of a nonlinear σ model. Just as in the nonlinear σ model discussed in chapter VI.4, the Lagrangian looks free, but the nontrivial dynamics comes from the constraint. The constant c_s (which is determined in terms of the microscopic variable J) is the spin wave velocity, as you can see by writing down the equation of motion $(\partial^2/\partial t^2)\vec{n} - c_s^2 \nabla^2 \vec{n} = 0$.

But you can feel that something is wrong. You learned in a quantum mechanics course that the dynamics of a spin variable $\vec{S}$ is first order in time. Consider the most basic example of a spin in a constant magnetic field described by $H = \mu \vec{S} \cdot \vec{B}$. Then $d\vec{S}/dt = i[H, \vec{S}] = \mu \vec{B} \times \vec{S}$. Besides, you might remember from a solid state physics course that in a ferromagnet the dispersion relation of the spin wave has the nonrelativistic form $\omega \propto k^2$ and not the relativistic form $\omega^2 \propto k^2$ implied by (2).

The resolution of this apparent paradox is based on the Pauli-Hopf identity: Given a unit vector $\vec{n}$ we can always write $\vec{n} = z^\dagger \vec{\sigma} z$, where $z = \begin{pmatrix} z_1 \\ z_2 \end{pmatrix}$ consists of two complex numbers such that $z^\dagger z \equiv z_1^\dagger z_1 + z_2^\dagger z_2 = 1$. Verify this! (A mathematical aside: Writing z_1 and z_2 out in terms of real numbers we see that this defines the so-called Hopf map $S^3 \rightarrow S^2$.) While we cannot form a term quadratic in $\vec{n}$ and linear in time derivative, we can write a term quadratic in the complex doublet z and linear in time derivative. Can you figure it out before looking at the next line?

The correct version of (1) is

$$L_{\text{correct}} = i \sum_j z_j^\dagger \frac{\partial z_j}{\partial t} + \frac{1}{2g^2} \sum_j \frac{\partial \vec{n}_j}{\partial t} \cdot \frac{\partial \vec{n}_j}{\partial t} - J \sum_{<ij>} \vec{n}_i \cdot \vec{n}_j \tag{3}$$

The added term is known as the Berry's phase term and has deep topological meaning. You should derive the equation of motion using the identity

$$\int dt \, \delta \left(z_j^\dagger \frac{\partial z_j}{\partial t} \right) = \frac{1}{2} i \int dt \, \delta \vec{n}_j \cdot \left(\vec{n}_j \times \frac{\partial \vec{n}_j}{\partial t} \right) \tag{4}$$

Remarkably, although $z_j^\dagger(\partial z_j/\partial t)$ cannot be written simply in terms of $\vec{n}_j$, its variation can be.

Low energy modes in the ferromagnet and the antiferromagnet

In the ground state of a ferromagnet, the magnetic moments all point in the same direction, which we can choose to be the z-direction. Expanding the equation of motion in small fluctuations around this ground state $\vec{n}_j = \hat{e}_z + \delta\vec{n}_j$ (where evidently $\hat{e}_z$ denotes the appropriate unit vector) and Fourier transforming, we obtain

$$\begin{pmatrix} -\frac{\omega^2}{g^2} + h(k) & -\frac{1}{2}i\omega \\ \frac{1}{2}i\omega & -\frac{\omega^2}{g^2} + h(k) \end{pmatrix} \begin{pmatrix} \delta n_x(k) \\ \delta n_y(k) \end{pmatrix} = 0 \tag{5}$$

linking the two components $\delta n_x(k)$ and $\delta n_y(k)$ of $\delta\vec{n}(k)$. The condition $\vec{n}_j \cdot \vec{n}_j = 1$ says that $\delta n_z(k) = 0$. Here a is the lattice spacing and $h(k) \equiv 4J[2 - \cos(k_x a) - \cos(k_y a)] \simeq 2Ja^2 k^2$ for small k. (I am implicitly working in two spatial dimensions as evidenced by k_x and k_y.)

At low frequency the Berry term $i\omega$ dominates the naive term ω^2/g^2, which we can therefore throw away. Setting the determinant of the matrix equal to zero, we see that we get the correct quadratic dispersion relation $\omega \propto k^2$.

The treatment of the antiferromagnet is interestingly different. The so-called Néel state[1] for an antiferromagnet is defined by $\vec{n}_j = (-1)^j \hat{e}_z$. Writing $\vec{n}_j = (-1)^j \hat{e}_z + \delta\vec{n}_j$, we obtain

$$\begin{pmatrix} -\frac{\omega^2}{g^2} + f(k) & -\frac{1}{2}i\omega \\ \frac{1}{2}i\omega & -\frac{\omega^2}{g^2} + f(k) \end{pmatrix} \begin{pmatrix} \delta n_x(k) \\ \delta n_y(k+Q) \end{pmatrix} = 0 \tag{6}$$

linking δn_x and δn_y evaluated at different momenta. Here $f(k) = 4J [2 + \cos(k_x a) + \cos(k_y a)]$ and $Q = [\pi/a, \pi/a]$. The appearance of Q is due to $(-1)^j = e^{iQaj}$. (I will let you figure out the somewhat overly compact notation.) The antiferromagnetic factor $(-1)^j$ explicitly breaks translation invariance and kicks in the momentum Q whenever it occurs. A similar equation links $\delta n_y(k)$ and $\delta n_x(k+Q)$. Solving these equations, you will find that there is a high frequency branch that we are not interested in and a low frequency branch with the linear dispersion $\omega \propto k$.

Thus, the low frequency dynamics of the antiferromagnet can be described by the nonlinear σ model (2), which when the spin wave velocity is normalized to 1 can be written in the relativistic form:

$$\mathcal{L} = \frac{1}{2g^2} \partial_\mu \vec{n} \cdot \partial^\mu \vec{n} \tag{7}$$

Exercises

VI.5.1 Work out the two branches of the spin wave spectrum in the ferromagnetic case, paying particular attention to the polarization.

VI.5.2 Verify that in the antiferromagnetic case the Berry's phase term merely changes the spin wave velocity and does not affect the spectrum qualitatively as in the ferromagnetic case.

---

[1] Note that while the Néel state describes the lowest energy configuration for a classical antiferromagnet, it does not describe the ground state of a quantum antiferromagnet. The terms $S_i^+ S_j^- + S_i^- S_j^+$ in the Hamiltonian $J \sum_{<ij>} \vec{S}_i \cdot \vec{S}_j$ flip the spins up and down.

VI.6 | Surface Growth and Field Theory

In this chapter I will discuss a topic, rather unusual for a field theory text, taken from non-equilibrium statistical mechanics, one of the hottest growth fields in theoretical physics over the last few years. I want to introduce you to yet another area in which field theoretic concepts are of use.

Imagine atoms being deposited randomly on some surface. This is literally how some novel materials are grown. The height $h(x, t)$ of the surface as it grows is governed by the Kardar-Parisi-Zhang equation

$$\frac{\partial h}{\partial t} = \nu \nabla^2 h + \frac{\lambda}{2}(\nabla h)^2 + \eta(\vec{x}, t) \tag{1}$$

This equation describes a deceptively simple prototype of nonequilibrium dynamics and has a remarkably wide range of applicability.

To understand (1), consider the various terms on the right-hand side. The term $\nu \nabla^2 h$ (with $\nu > 0$) is easy to understand: Positive in the valleys of h and negative on the peaks, it tends to smooth out the surface. With only this term the problem would be linear and hence trivial. The nonlinear term $(\lambda/2)(\nabla h)^2$ renders the problem highly nontrivial and interesting; I leave it to you as an exercise to convince yourself of the geometric origin of this term. The third term describes the random arrival of atoms, with the random variable $\eta(\vec{x}, t)$ usually assumed to be Gaussian distributed, with zero mean,[1] and correlations

$$\langle \eta(\vec{x}, t)\eta(\vec{x}', t') \rangle = 2\sigma^2 \delta^D(\vec{x} - \vec{x}')\delta(t - t') \tag{2}$$

In other words, the probability distribution for a particular $\eta(\vec{x}, t)$ is given by

$$P(\eta) \propto e^{-\frac{1}{2\sigma^2} \int d^D x dt \, \eta(\vec{x}, t)^2}$$

Here $\vec{x}$ represents coordinates in D-dimensional space. Experimentally, $D = 2$ for the situation I described, but theoretically we are free to investigate the problem for any D.

---

[1] There is no loss of generality here since an additive constant in η can be absorbed by the shift $h \to h + ct$.

Typically, condensed matter physicists are interested in calculating the correlation between the height of the surface at two different positions in space and time:

$$\langle[h(\vec{x}, t) - h(\vec{x}', t')]^2\rangle = |\vec{x} - \vec{x}'|^{2\chi} f\left(\frac{|\vec{x} - \vec{x}'|^z}{|t - t'|}\right), \tag{3}$$

The bracket $\langle\cdots\rangle$ here and in (2) denotes averaging over different realizations of the random variable $\eta(\vec{x}, t)$. On the right-hand side of (3) I have written the dynamic scaling form typically postulated in condensed matter physics, where χ and z are the so-called roughness and dynamic exponents. The challenge is then to show that the scaling form is correct and to calculate χ and z. Note that the dynamic exponent z (which in general is not an integer) tells us, roughly speaking, how many powers of space is worth one power of time. (For $\lambda = 0$ we have simple diffusion for which $z = 2$.) Here f denotes an unknown function.

I will not go into more technical details. Our interest here is to see how this problem, which does not even involve quantum mechanics, can be converted into a quantum field theory. Start with

$$Z \equiv \int \mathcal{D}h \int \mathcal{D}\eta \, e^{-\frac{1}{2\sigma^2}\int d^D x dt \, \eta(\vec{x},t)^2} \delta\left[\frac{\partial h}{\partial t} - \nu\nabla^2 h - \frac{\lambda}{2}(\nabla h)^2 - \eta(\vec{x}, t)\right]$$

$$\tag{4}$$

Integrating over η, we obtain $Z = \int \mathcal{D}h \, e^{-S(h)}$ with the action

$$S(h) = \frac{1}{2\sigma^2} \int d^D \vec{x} \, dt \left[\frac{\partial h}{\partial t} - \nu\nabla^2 h - \frac{\lambda}{2}(\nabla h)^2\right]^2 \tag{5}$$

You will recognize that this describes a nonrelativistic field theory of a scalar field $h(\vec{x}, t)$. The physical quantity we are interested in is then given by

$$\langle[h(\vec{x}, t) - h(\vec{x}', t')]^2\rangle = \frac{1}{Z} \int \mathcal{D}h \, e^{-S(h)}[h(\vec{x}, t) - h(\vec{x}', t')]^2 \tag{6}$$

Thus, the challenge of determing the roughness and dynamic exponents in statistical physics is equivalent to the problem of determining the propagator

$$D(\vec{x}, t) \equiv \frac{1}{Z} \int \mathcal{D}h \, e^{-S(h)} h(\vec{x}, t) h(\vec{0}, 0)$$

of the scalar field h.

Incidentally, by scaling $t \to t/\nu$ and $h \to \sqrt{\sigma^2/\nu}\, h$, we can write the action as

$$S(h) = \frac{1}{2} \int d^D \vec{x} \, dt \left[\left(\frac{\partial}{\partial t} - \nabla^2\right) h - \frac{g}{2}(\nabla h)^2\right]^2 \tag{7}$$

with $g^2 \equiv \lambda^2\sigma^2/\nu^3$. Expanding the action in powers of h as usual

$$S(h) = \frac{1}{2} \int d^D \vec{x} \, dt \left\{\left[\left(\frac{\partial}{\partial t} - \nabla^2\right) h\right]^2 - g(\nabla h)^2 \left(\frac{\partial}{\partial t} - \nabla^2\right) h + \frac{g^2}{4}(\nabla h)^4\right\}, \tag{8}$$

we recognize the quadratic term as giving us the rather unusual propagator $1/(\omega^2 + k^4)$ for the scalar field h, and the cubic and quartic term as describing the interaction. As always,

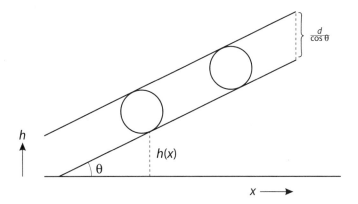

Figure VI.6.1

to calculate the desired physical quantity we evaluate the functional or "path" integral

$$Z = \int \mathcal{D}h \; e^{-S(h) + \int d^D x \, dt \, J(x,t) h(x,t)}$$

and then functionally differentiate repeatedly with respect to J.

My intent here is not so much to teach you nonequilibrium statistical mechanics as to show you that quantum field theory can emerge in a variety of physical situations, including those involving only purely classical physics. Note that the "quantum fluctuations" here arise from the random driving term. Evidently, there is a close methodological connection between random dynamics and quantum physics.

Exercises

VI.6.1   An exercise in elementary geometry: Draw a straight line tilted at an angle θ with respect to the horizontal. The line represents a small segment of the surface at time t. Now draw a number of circles of diameter d tangent to and on top of this line. Next draw another line tilted at angle θ with respect to the horizontal and lying on top of the circles, namely tangent to them. This new line represents the segment of the surface some time later (see fig. VI.6.1). Note that $\Delta h = d/\cos\theta \simeq d(1 + \frac{1}{2}\theta^2)$. Show that this generates the nonlinear term $(\lambda/2)(\nabla h)^2$ in the KPZ equation (1). For applications of the KPZ equation, see for example, T. Halpin–Healy and Y.-C. Zhang, *Phys. Rep.* 254: 215, 1995; A. L. Barabasi and H. E. Stanley, *Fractal Concepts in Surface Growth.*

VI.6.2   Show that the scalar field h has the propagator $1/(\omega^2 + k^4)$.

VI.6.3   Field theory can often be cast into apparently rather different forms by a change of variable. Show that by writing $U = e^{\frac{1}{2}gh}$ we can change the action (7) to

$$S = \frac{2}{g^2} \int d^D \vec{x} \, dt \left(U^{-1} \frac{\partial}{\partial t} U - U^{-1} \nabla^2 U \right)^2 \tag{9}$$

a kind of nonlinear σ model.

Impurities and random potential

An important area in condensed matter physics involves the study of disordered systems, a subject that has been the focus of a tremendous amount of theoretical work over the last few decades. Electrons in real materials scatter off the impurities inevitably present and effectively move in a random potential. In the spirit of this book I will give you a brief introduction to this fascinating subject, showing how the problem can be mapped into a quantum field theory.

The prototype problem is that of a quantum particle obeying the Schrödinger equation $H\psi = [-\nabla^2 + V(x)]\psi = E\psi$, where $V(x)$ is a random potential (representing the impurities) generated with the Gaussian white noise probability distribution $P(V) = \mathcal{N}e^{-\int d^D x (1/2g^2) V(x)^2}$ with the normalization factor $\mathcal{N}$ determined by $\int DV P(V) = 1$. The parameter g measures the strength of the impurities: the larger g, the more disordered the system. This of course represents an idealization in which interaction between electrons and a number of other physical effects are neglected.

As in statistical mechanics we think of an ensemble of systems each of which is characterized by a particular function $V(x)$ taken from the distribution $P(V)$. We study the average or typical properties of the system. In particular, we might want to know the averaged density of states defined by $\rho(E) = \langle \text{tr } \delta(E - H) \rangle = \langle \sum_i \delta(E - E_i) \rangle$, where the sum runs over the ith eigenstate of H with corresponding eigenvalue E_i. We denote by $\langle O(V) \rangle \equiv \int DV P(V) O(V)$ the average of any functional $O(V)$ of $V(x)$. Clearly, $\int_{E^*}^{E^*+\delta E} dE\, \rho(E)$ counts the number of states in the interval from E^* to $E^* + \delta E$, an important quantity in, for example, tunneling experiments.

Anderson localization

Another important physical question is whether the wave functions at a particular energy E extend over the entire system or are localized within a characteristic length scale $\xi(E)$. Clearly, this issue determines whether the material is a conductor or an insulator. At first sight, you might think that we should study

$$S(x, y; E) \equiv \left\langle \sum_i \delta(E - E_i)\psi_i^*(x)\psi_i(y) \right\rangle$$

which might tell us how the wave function at x is correlated with the wave function at some other point y, but S is unsuitable because $\psi_i^*(x)\psi_i(y)$ has a phase that depends on V. Thus, S would vanish when averaged over disorder. Instead, the correct quantity to study is

$$K(x - y; E) \equiv \left\langle \sum_i \delta(E - E_i)\psi_i^*(x)\psi_i(y)\psi_i^*(y)\psi_i(x) \right\rangle$$

since $\psi_i^*(x)\psi_i(y)\psi_i^*(y)\psi_i(x)$ is manifestly positive. Note that upon averaging over all possible $V(x)$ we recover translation invariance so that K does not depend on x and y separately, but only on the separation $|x - y|$. As $|x - y| \to \infty$, if $K(x - y; E) \sim e^{-|x-y|/\xi(E)}$ decreases exponentially the wave functions around the energy E are localized over the so-called localization length $\xi(E)$. On the other hand, if $K(x - y; E)$ decreases as a power law of $|x - y|$, the wave functions are said to be extended.

Anderson and his collaborators made the surprising discovery that localization properties depend on D, the dimension of space, but not on the detailed form of $P(V)$ (an example of the notion of universality). For $D = 1$ and 2 all wave functions are localized, regardless of how weak the impurity potential might be. This is a highly nontrivial statement since a priori you might think, as eminent physicists did at the time, that whether the wave functions are localized or not depends on the strength of the potential. In contrast, for $D = 3$, the wave functions are extended for E in the range $(-E_c, E_c)$. As E approaches the energy E_c (known as the mobility edge) from above, the localization length $\xi(E)$ diverges as $\xi(E) \sim 1/(E - E_c)^\mu$ with some critical exponent[1] μ. Anderson received the Nobel Prize for this work and for other contributions to condensed matter theory.

Physically, localization is due to destructive interference between the quantum waves scattering off the random potential.

When a magnetic field is turned on perpendicular to the plane of a $D = 2$ electron gas the situation changes dramatically: An extended wave function appears at $E = 0$. For nonzero E, all wave functions are still localized, but with the localization length diverging as $\xi(E) \sim 1/|E|^\nu$. This accounts for one of the most striking features of the quantum Hall effect (see chapter VI.2): The Hall conductivity stays constant as the Fermi energy increases but then suddenly jumps by a discrete amount due to the contribution of the extended state

---

[1] This is an example of a quantum phase transition. The entire discussion is at zero temperature. In contrast to the phase transition discussed in chapter V.3, here we vary E instead of the temperature.

as the Fermi energy passes through $E = 0$. Understanding this behavior quantitatively poses a major challenge for condensed matter theorists. Indeed, many consider an analytic calculation of the critical exponent ν as one of the "Holy Grails" of condensed matter theory.

Green's function formalism

So much for a lightning glimpse of localization theory. Fascinating though the localization transition might be, what does quantum field theory have to do with it? This is after all a field theory text. Before proceeding we need a bit of formalism. Consider the so-called Green's function $G(z) \equiv \langle \mathrm{tr}[1/(z - H)] \rangle$ in the complex z-plane. Since $\mathrm{tr}[1/(z - H)] = \sum_i 1/(z - E_i)$, this function consists of a sum of poles at the eigenvalues E_i. Upon averaging, the poles merge into a cut. Using the identity (I.2.13) $\lim_{\varepsilon \to 0} \mathrm{Im}[1/(x + i\varepsilon)] = -\pi \delta(x)$, we see that

$$\rho(E) = -\frac{1}{\pi} \lim_{\varepsilon \to 0} \mathrm{Im}\, G(E + i\varepsilon) \tag{1}$$

So if we know $G(z)$ we know the density of states.

The infamous denominator

I can now explain how quantum field theory enters into the problem. We start by taking the logarithm of the identity (A.15)

$$J^\dagger \cdot K^{-1} \cdot J = \log(\int D\varphi^\dagger D\varphi e^{-\varphi^\dagger \cdot K \cdot \varphi + J^\dagger \cdot \varphi + \varphi^\dagger \cdot J})$$

(where as usual we have dropped an irrelevant term). Differentiating with respect to $J^\dagger$ and J and then setting $J^\dagger$ and J equal to 0 we obtain an integral representation for the inverse of a hermitean matrix:

$$(K^{-1})_{ij} = \frac{\int D\varphi^\dagger D\varphi e^{-\varphi^\dagger \cdot K \cdot \varphi} \varphi_i \varphi_j^\dagger}{\int D\varphi^\dagger D\varphi e^{-\varphi^\dagger \cdot K \cdot \varphi}} \tag{2}$$

(Incidentally, you may recognize this as essentially related to the formula (I.7.14) for the propagator of a scalar field.) Now that we know how to represent $1/(z - H)$ we have to take its trace, which means setting $i = j$ in (2) and summing. In our problem, $H = -\nabla^2 + V(x)$ and the index i corresponds to the continuous variable x and the summation to an integration over space. Replacing K by $i(z - H)$ (and taking care of the appropriate delta function) we obtain

$$\mathrm{tr}\, \frac{-i}{z - H} = \int d^D y \left\{ \frac{\int D\varphi^\dagger D\varphi e^{i \int d^D x \{\partial \varphi^\dagger \partial \varphi + [V(x) - z]\varphi^\dagger \varphi\}} \varphi(y)\varphi^\dagger(y)}{\int D\varphi^\dagger D\varphi e^{i \int d^D x \{\partial \varphi^\dagger \partial \varphi + [V(x) - z]\varphi^\dagger \varphi\}}} \right\} \tag{3}$$

This is starting to look like a scalar field theory in D-dimensional Euclidean space with the action $S = \int d^D x \{\partial \varphi^\dagger \partial \varphi + [V(x) - z] \varphi^\dagger \varphi\}$. [Note that for (3) to be well defined z has to be in the lower half-plane.]

But now we have to average over $V(x)$, that is, integrate over V with the probability distribution $P(V)$. We immediately run into the difficulty that confounded theorists for a long time. The denominator in (3) stops us cold: If that denominator were not there, then the functional integration over $V(x)$ would just be the Gaussian integral you have learned to do over and over again. Can we somehow lift this infamous denominator into the numerator, so to speak? Clever minds have come up with two tricks, known as the replica method and the supersymmetric method, respectively. If you can come up with another trick, fame and fortune might be yours.

Replicas

The replica trick is based on the well-known identity $(1/x) = \lim\limits_{n \to 0} x^{n-1}$, which allows us to write that much disliked denominator as

$$\lim_{n \to 0} \left(\int D\varphi^\dagger D\varphi \, e^{i \int d^D x \{\partial \varphi^\dagger \partial \varphi + [V(x) - z]\varphi^\dagger \varphi\}} \right)^{n-1}$$

$$= \lim_{n \to 0} \int \prod_{a=2}^{n} D\varphi_a^\dagger D\varphi_a \, e^{i \int d^D x \sum_{a=2}^{n} \{\partial \varphi_a^\dagger \partial \varphi_a + [V(x) - z]\varphi_a^\dagger \varphi_a\}}$$

Thus (3) becomes

$$\text{tr} \, \frac{1}{z - H} = \lim_{n \to 0} i \int d^D y \int \left(\prod_{a=1}^{n} D\varphi_a^\dagger D\varphi_a \right) e^{i \int d^D x \sum_{a=1}^{n} \{\partial \varphi_a^\dagger \partial \varphi_a + [V(x) - z]\varphi_a^\dagger \varphi_a\}} \varphi_1(y) \varphi_1^\dagger(y) \tag{4}$$

Note that the functional integral is now over n complex scalar fields φ_a. The field φ has been replicated. For positive integers, the integrals in (4) are well defined. We hope that the limit $n \to 0$ will not blow up in our face.

Averaging over the random potential, we recover translation invariance; thus the integrand for $\int d^D y$ does not depend on y and $\int d^D y$ just produces the volume $\mathcal{V}$ of the system. Using (A.13) we obtain

$$\left\langle \text{tr} \, \frac{1}{z - H} \right\rangle = i\mathcal{V} \lim_{n \to 0} \int \left(\prod_{a=1}^{n} D\varphi_a^\dagger D\varphi_a \right) e^{i \int d^D x \mathcal{L}} \varphi_1(0) \varphi_1^\dagger(0) \tag{5}$$

where

$$\mathcal{L}(\varphi) \equiv \sum_{a=1}^{n} (\partial \varphi_a^\dagger \partial \varphi_a - z\varphi_a^\dagger \varphi_a) + \frac{ig^2}{2} \left(\sum_{a=1}^{n} \varphi_a^\dagger \varphi_a \right)^2 \tag{6}$$

We obtain a field theory (with a peculiar factor of i) of n scalar fields with a good old φ^4 interaction invariant under $O(n)$ (known as the replica symmetry.) Note the wisdom of replacing K by $i(z - H)$; if we didn't include the i the functional integral would diverge at large φ, as you can easily check. For z in the upper half-plane we would replace K by $-i(z - H)$. The quantity from which we can extract the desired averaged density of states is given by the propagator of the scalar field. Incidentally, we can replace $\varphi_1(0)\varphi_1^\dagger(0)$ in (5) by the more symmetric expression $(1/n) \sum_{b=1}^n \varphi_b^\dagger \varphi_b$.

Absorbing $\mathcal{V}$ so that we are calculating the density of states per unit volume, we find

$$G(z) = i \lim_{n \to 0} \int \left(\prod_{a=1}^n D\varphi_a^\dagger D\varphi_a \right) e^{iS(\varphi)} \left(\frac{1}{n} \sum_{b=1}^n \varphi_b^\dagger(0)\varphi_b(0) \right) \tag{7}$$

For positive integer n the field theory is perfectly well defined, so the delicate step in the replica approach is in taking the $n \to 0$ limit. There is a fascinating literature on this limit. (Consult a book devoted to spin glasses.)

Some particle theorists used to speak disparagingly of condensed matter physics as dirt physics, and indeed the influence of impurities and disorder on matter is one of the central concerns of modern condensed matter physics. But as we see from this example, in many respects there is no mathematical difference between averaging over randomness and summing over quantum fluctuations. We end up with a φ^4 field theory of the type that many particle theorists have devoted considerable effort to studying in the past. Furthermore, Anderson's surprising result that for $D = 2$ any amount of disorder, no matter how small, localizes all states means that we have to understand the field theory defined by (6) in a highly nontrivial way. The strength of the disorder shows up as the coupling g^2, so no amount of perturbation theory in g^2 can help us understand localization. Anderson localization is an intrinsically nonperturbative effect.

Grassmannian approach

As I mentioned earlier, people have dreamed up not one, but two, tricks in dealing with the nasty denominator. The second trick is based on what we learned in chapter II.5 on integration over Grassmann variables: Let $\eta(x)$ and $\bar{\eta}(x)$ be Grassmann fields, then

$$\int D\eta D\bar{\eta} e^{-\int d^4x \bar{\eta} K \eta} = C \det K = \tilde{C} \left(\int D\varphi D\varphi^\dagger e^{-\int d^4x \varphi^\dagger K \varphi} \right)^{-1}$$

where C and $\tilde{C}$ are two uninteresting constants that we can absorb into the definition of $D\eta D\bar{\eta}$. With this identity we can write (3) as

$$\text{tr} \frac{1}{z - H} = i \int d^D y \int D\varphi^\dagger D\varphi D\eta D\bar{\eta} e^{i \int d^D x \{ \{ \partial \varphi^\dagger \partial \varphi + [V(x) - z]\varphi^\dagger \varphi \} + \{ \partial \bar{\eta} \partial \eta + [V(x) - z]\bar{\eta}\eta \} \}} \varphi(y)\varphi^\dagger(y) \tag{8}$$

and then easily average over the disorder to obtain (per unit volume)

$$\left\langle \text{tr} \frac{1}{z - H} \right\rangle = i \int D\varphi^\dagger D\varphi D\eta D\bar{\eta} e^{i \int d^D x \mathcal{L}(\bar{\eta}, \eta, \varphi^\dagger, \varphi)} \varphi(0)\varphi^\dagger(0) \tag{9}$$

with

$$\mathcal{L}(\bar{\eta}, \eta, \varphi^\dagger, \varphi) = \partial \varphi^\dagger \partial \varphi + \partial \bar{\eta} \partial \eta - z(\varphi^\dagger \varphi + \bar{\eta}\eta) + \frac{ig^2}{2}(\varphi^\dagger \varphi + \bar{\eta}\eta)^2 \tag{10}$$

We end up with a field theory with bosonic (commuting) fields $\varphi^\dagger$ and φ and fermionic (anticommuting) fields $\bar{\eta}$ and η interacting with a strength determined by the disorder. The action S exhibits an obvious symmetry rotating bosonic fields into fermionic fields and vice versa, and hence this approach is known in the condensed matter physics community as the supersymmetric method. (It is perhaps worth emphasizing that $\bar{\eta}$ and η are not spinor fields, which we underline by not writing them as $\bar{\psi}$ and ψ. The supersymmetry here, perhaps better referred to as Grassmannian symmetry, is quite different from the supersymmetry in particle physics to be discussed in chapter VIII.4.)

Both the replica and the supersymmetry approaches have their difficulties, and I was not kidding when I said that if you manage to invent a new approach without some of these difficulties it will be met with considerable excitement by condensed matter physicists.

Probing localization

I have shown you how to calculate the averaged density of states $\rho(E)$. How do we study localization? I will let you develop the answer in an exercise. From our earlier discussion it should be clear that we have to study an object obtained from (3) by replacing $\varphi(y)\varphi^\dagger(y)$ by $\varphi(x)\varphi^\dagger(y)\varphi(y)\varphi^\dagger(x)$. If we choose to think of the replica field theory in the language of particle physics as describing the interaction of some scalar meson, then rather pleasingly, we see that the density of states is determined by the meson propagator and localization is determined by meson-meson scattering.

Exercises

VI.7.1 Work out the field theory that will allow you to study Anderson localization. [Hint: Consider the object

$$\left\langle \left(\frac{1}{z - H}\right)(x, y) \left(\frac{1}{w - H}\right)(y, x) \right\rangle$$

for two complex numbers z and w. You will have to introduce two sets of replica fields, commonly denoted by φ_a^+ and φ_a^-.] {Notation: $[1/(z - H)](x, y)$ denotes the xy element of the matrix or operator $[1/(z - H)]$.}

VI.7.2 As another example from the literature on disorder, consider the following problem. Place N points randomly in a D-dimensional Euclidean space of volume V. Denote the locations of the points by $\vec{x}_i$ $(i = 1, \ldots, N)$. Let

$$f(\vec{x}) = (-) \int \frac{d^D k}{(2\pi)^D} \frac{e^{i\vec{k}\vec{x}}}{k^2 + m^2}$$

Consider the N by N matrix $H_{ij} = f\left(\vec{x}_i - \vec{x}_j\right)$. Calculate $\rho(E)$, the density of eigenvalues of H as we average over the ensemble of matrices, in the limit $N \to \infty$, $V \to \infty$, with the density of points $\rho \equiv N/V$ (not to be confused with $\rho(E)$ of course) held fixed. [Hint: Use the replica method and arrive at the field theory action

$$S(\varphi) = \int d^D x \left[\sum_{a=1}^{n}(|\nabla\varphi_a|^2 + m^2|\varphi_a|^2) - \rho e^{-(1/z)\sum_{a=1}^{n}|\varphi_a|^2}\right]$$

This problem is not entirely trivial; if you need help consult M. Mézard et al., *Nucl. Phy.* B559: 689, 2000, cond-mat/9906135.

VI.8 Renormalization Group Flow as a Natural Concept in High Energy and Condensed Matter Physics

Therefore, conclusions based on the renormalization group arguments . . . are dangerous and must be viewed with due caution. So is it with all conclusions from local relativistic field theories.
— J. Bjorken and S. Drell, 1965

It is not dangerous

The renormalization group represents the most important conceptual advance in quantum field theory over the last three or four decades. The basic ideas were developed simultaneously in both the high energy and condensed matter physics communities, and in some areas of research renormalization group flow has become part of the working language.

As you can easily imagine, this is an immensely rich and multifaceted subject, which we can discuss from many different points of view, and a full exposition would require a book in itself. Unfortunately, there has never been a completely satisfactory and comprehensive treatment of the subject. The discussions in some of the older books are downright misleading and confused, such as the well-known text from which I learned quantum field theory and from which the quote above was taken. In the limited space available here, I will attempt to give you a flavor of the subject rather than all the possible technical details. I will first approach it from the point of view of high energy physics and then from that of condensed matter physics. As ever, the emphasis will be on the conceptual rather than the computational. As you will see, in spite of the order of my presentation, it is easier to grasp the role of the renormalization group in condensed matter physics than in high energy physics.

I laid the foundation for the renormalization group in chapter III.1—I do plan ahead! Let us go back to our experimentalist friend with whom we were discussing $\lambda \varphi^4$ theory. We will continue to pretend that our world is described by a simple $\lambda \varphi^4$ theory and that an approximation to order λ^2 suffices.

What experimentalists insist on

Our experimentalist friend was not interested in the coupling constant λ we wrote down on a piece of paper, a mere Greek letter to her. She insisted that she would accept only quantities she and her experimental colleagues can actually measure, even if only in principle. As a result of our discussion with her we sharpened our understanding of what a coupling constant is and learned that we should define a physical coupling constant by [see (III.1.4)]

$$\lambda_P(\mu) = \lambda - 3C\lambda^2 \log\left(\frac{\Lambda^2}{\mu^2}\right) + O(\lambda^3) \tag{1}$$

At her insistence, we learned to express our result for physical amplitudes in terms of $\lambda_P(\mu)$, and not in terms of the theoretical construct λ. In particular, we should write the meson-meson scattering amplitude as

$$\mathcal{M} = -i\lambda_P(\mu) + iC\lambda_P(\mu)^2 \left[\log\left(\frac{\mu^2}{s}\right) + \log\left(\frac{\mu^2}{t}\right) + \log\left(\frac{\mu^2}{u}\right)\right]$$
$$+ O[\lambda_P(\mu)^3] \tag{2}$$

What is the physical significance of $\lambda_P(\mu)$? To be sure, it measures the strength of the interaction between mesons as reflected in (2). But why one particular choice of μ? Clearly, from (2) we see that $\lambda_P(\mu)$ is particularly convenient for studying physics in the regime in which the kinematic variables s, t, and u are all of order μ^2. The scattering amplitude is given by $-i\lambda_P(\mu)$ plus small logarithmic corrections. (Recall from a footnote in chapter III.3 that the renormalization point $s_0 = t_0 = u_0 = \mu^2$ is adopted purely for theoretical convenience and cannot be reached in actual experiments. For our conceptual understanding here this is not a relevant issue.) In short, $\lambda_P(\mu)$ is known as the coupling constant appropriate for physics at the energy scale μ.

In contrast, if we were so idiotic as to use the coupling constant $\lambda_P(\mu')$ while exploring physics in the regime with s, t, and u of order μ^2, with μ' vastly different from μ, then we would have a scattering amplitude

$$\mathcal{M} = -i\lambda_P(\mu') + iC\lambda_P(\mu')^2 \left[\log\left(\frac{\mu'^2}{s}\right) + \log\left(\frac{\mu'^2}{t}\right) + \log\left(\frac{\mu'^2}{u}\right)\right]$$
$$+ O[\lambda_P(\mu')^3] \tag{3}$$

in which the second term [with $\log(\mu'^2/\mu^2)$ large] can be comparable to or larger than the first term. The coupling constant $\lambda_P(\mu')$ is not a convenient choice. Thus, for each energy scale μ there is an "appropriate" coupling constant $\lambda_P(\mu)$.

Subtracting (2) from (3) we can easily relate $\lambda_P(\mu)$ and $\lambda_P(\mu')$ for $\mu \sim \mu'$:

$$\lambda_P(\mu') = \lambda_P(\mu) + 3C\lambda_P(\mu)^2 \log\left(\frac{\mu'^2}{\mu^2}\right) + O[\lambda_P(\mu)^3] \tag{4}$$

We can express this as a differential "flow equation"

$$\mu\frac{d}{d\mu}\lambda_P(\mu) = 6C\lambda_P(\mu)^2 + O(\lambda_P^3) \tag{5}$$

As you have already seen repeatedly, quantum field theory is full of historical misnomers. The description of how $\lambda_P(\mu)$ changes with μ is known as the renormalization group. The only appearance of a group concept here is the additive group of transformation $\mu \to \mu + \delta\mu$.

For the conceptual discussion in chapter III.1 and here, we don't need to know what the constant C happens to be. If C happens to be negative, then the coupling $\lambda_P(\mu)$ will decrease as the energy scale μ increases, and the opposite will occur if C happens to be positive. (In fact, the sign is positive, so that as we increase the energy scale, λ_P flows away from the origin.)

Flow of the electromagnetic coupling

The behavior of λ is typical of coupling constants in 4-dimensional quantum field theories. For example, in quantum electrodynamics, the coupling e or equivalently $\alpha = e^2/4\pi$, measures the strength of the electromagnetic interaction. The story is exactly as that told for the $\lambda\varphi^4$ theory: Our experimentalist friend is not interested in the Latin letter e, but wants to know the actual interaction when the relevant momenta squared are of the order μ^2. Happily, we have already done the computation: We can read off the effective coupling at momentum transferred squared $q^2 = \mu^2$ from (III.7.14):

$$e_P(\mu)^2 = e^2 \frac{1}{1 + e^2\Pi(\mu^2)} \simeq e^2[1 - e^2\Pi(\mu^2) + O(e^4)]$$

Take μ much larger than the electron mass m but much smaller than the cutoff mass M. Then from (III.7.13)

$$\mu\frac{d}{d\mu}e_P(\mu) = -\frac{1}{2}e^3\mu\frac{d}{d\mu}\Pi(\mu^2) + O(e^5) = +\frac{1}{12\pi^2}e_P^3 + O(e_P^5) \tag{6}$$

We learn that the electromagnetic coupling increases as the energy scale increases. Electromagnetism becomes stronger as we go to higher energies, or equivalently shorter distances.

Physically, the origin of this phenomenon is closely related to the physics of dielectrics. Consider a photon interacting with an electron, which we will call the test electron to avoid confusion in what follows. Due to quantum fluctuations, as described way back in chapter I.1, spacetime is full of electron-positron pairs, popping in and out of existence. Near the test electron, the electrons in these virtual pairs are repelled by the test electron and thus tend to move away from the test electron while the positrons tend to move toward the test electron. Thus, at long distances, the charge of the test electron is shielded to some extent by the cloud of positrons, causing a weaker coupling to the photon, while at short distances the coupling to the photon becomes stronger. The quantum vacuum is just as much a dielectric as a lump of actual material.

You may have noticed by now that the very name "coupling constant" is a terrible misnomer due to the fact that historically much of physics was done at essentially one energy scale, namely "almost zero"! In particular, people speak of the fine structure "constant" $\alpha = 1/137$ and crackpots continue to try to "derive" the number 137 from numerology or some fancier method. In fact, α is merely the coupling "constant" of the electromagnetic interaction at very low energies. It is an experimental fact that α, more properly written as $\alpha_P(\mu) \equiv e_P^2(\mu)/4\pi$, varies with the energy scale μ we are exploring. But alas, we are probably stuck with the name "coupling constant."

Renormalization group flow

In general, in a quantum field theory with a coupling constant g, we have the renormalization group flow equation

$$\mu \frac{dg}{d\mu} = \beta(g) \tag{7}$$

which is sometimes written as $dg/dt = \beta(g)$ upon defining $t \equiv \log(\mu/\mu_0)$. I will now suppress the subscript P on physical coupling constants. If the theory happens to have several coupling constants g_i, $i = 1, \cdots, N$, then we have

$$\frac{dg_i}{dt} = \beta_i(g_1, \cdots, g_N) \tag{8}$$

We can think of $(g_1, \cdots, g_N)$ as the coordinate of a particle in N-dimensional space, t as time, and $\beta_i(g_1, \cdots, g_N)$ a position dependent velocity field. As we increase μ or t we would like to study how the particle moves or flows. For notational simplicity, we will now denote $(g_1, \cdots, g_N)$ collectively as g. Clearly, those couplings at which $\beta_i(g^*)$ (for all i) happen to vanish are of particular interest: g^* is known as a fixed point. If the velocity field around a fixed point g^* is such that the particle moves toward that point (and once reaching it stays there since its velocity is now zero) the fixed point is known as attractive or stable. Thus, to study the asymptotic behavior of a quantum field theory at high energies we "merely" have to find all its attractive fixed points under the renormalization group flow. In a given theory, we can typically see that some couplings are flowing toward larger values while others are flowing toward zero.

Unfortunately, this wonderful theoretical picture is difficult to implement in practice because we essentially have no way of calculating the functions $\beta_i(g)$. In particular, g^* could well be quite large, associated with what is known as a strong coupling fixed point, and perturbation theory and Feynman diagrams are of no use in determining the properties of the theory there. Indeed, we know the fixed point structure of very few theories.

Happily, we know of one particularly simple fixed point, namely $g^* = 0$, at which perturbation theory is certainly applicable. We can always evaluate (8) perturbatively: $dg_i/dt = c_i^{jk} g_j g_k + d_i^{jkl} g_j g_k g_l + \cdots$. (In some theories, the series starts with quadratic terms and in others, with cubic terms. Sometimes there is also a linear term.) Thus, as we have already seen in a couple of examples, the asymptotic or high energy behavior of the theory depends on the sign of β_i in (8).

Let us now join the film "Physics History" already in progress. In the late 1960s, experimentalists studying the so-called deep inelastic scattering of electrons on protons discovered that their data seemed to indicate that after being hit by a highly energetic electron, one of the quarks inside the proton would propagate freely without interacting strongly with the other quarks. Normally, of course, the three quarks inside the proton are strongly bound to each other to form the proton. Eventually, a few theorists realized that this puzzling state of affairs could be explained if the theory of strong interaction is such that the coupling flows toward the fixed point $g^* = 0$. If so, then the strong interaction between quarks would actually weaken at higher and higher energy scales.

All of this is of course now "obvious" with the benefit of hindsight, but dear students, remember that at that time field theory was pronounced as possibly unsuitable for young minds and the renormalizable group was considered "dangerous" even in a field theory text!

The theory of strong interaction was unknown. But if we were so bold as to accept the dangerous renormalization group ideas then we might even find the theory of the strong interaction by searching for asymptotically free theories, which is what theories with an attractive fixed point at $g^* = 0$ became known as.

Asymptotically free theories are clearly wonderful. Their behavior at high energies can be studied using perturbative methods. And so in this way the fundamental theory of the strong interaction, now known as quantum chromodynamics, about which more later, was found.

Looking at physics on different length scales

The need for renormalization groups is really transparent in condensed matter physics. Instead of generalities, let me focus on a particularly clear example, namely surface growth. Indeed, that was why I chose to introduce the Kardar-Parisi-Zhang equation in chapter VI.6. We learned that to study surface growth we have to evaluate the functional or path integral

$$Z(\Lambda) = \int_{\Lambda} \mathcal{D}h \, e^{-S(h)}. \tag{9}$$

with, you will recall,

$$S(h) = \frac{1}{2} \int d^D \vec{x} \, dt \left(\frac{\partial h}{\partial t} - \nabla^2 h - \frac{g}{2} (\nabla h)^2 \right)^2. \tag{10}$$

This defines a field theory. As with any field theory, and as I indicate, a cutoff Λ has to be introduced. We integrate over only those field configurations $h(\vec{x}, t)$ that do not contain Fourier components with $\vec{k}$ and ω larger than Λ. (In principle, since this is a nonrelativistic theory we should have different cutoffs for $\vec{k}$ and for ω, but for simplicity of exposition let us just refer to them together generically as Λ.) The appearance of the cutoff is completely physical and necessary. At the very least, on length scales comparable to the size of the relevant molecules, the continuum description in terms of the field $h(\vec{x}, t)$ has long since broken down.

Physically, since the random driving term $\eta(\vec{x}, t)$ is a white noise, that is, η at $\vec{x}$ and at $\vec{x}'$ (and also at different times) are not correlated at all, we expect the surface to look

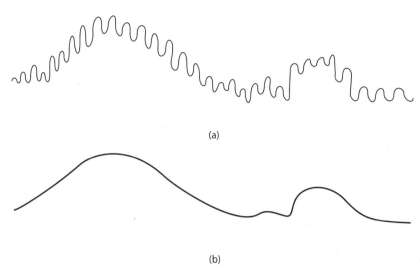

(a)

(b)

Figure VI.8.1

very uneven on a microscopic scale, as depicted in figure VI.8.1a. But suppose we are not interested in the detailed microscopic structure, but more in how the surface behaves on a larger scale. In other words, we are content to put on blurry glasses so that the surface appears as in figure VI.8.1b. This is a completely natural way to study a physical system, one that we are totally familiar with from day one in studying physics. We may be interested in physics over some length scale L and do not care about what happens on length scales much less than L.

The renormalization group is the formalism that allows us to relate the physics on different length scales or, equivalently, physics on different energy scales. In condensed matter physics, one tends to think of length scales, and in particle physics, of energy scales. The modern approach to renormalization groups came out of the study of critical phenomena by Kadanoff, Fisher, Wilson, and others, as mentioned in chapter V.3. Consider, for example, the Ising model, with the spin at each site either up or down and with a ferromagnetic interaction between neighboring spins. At high temperatures, the spins point randomly up and down. As the temperature drops toward the ferromagnetic transition point, islands of up spins (we say up spins to be definite, we could just as easily talk of down spins) start to appear. They grow ever larger in size until the critical temperature T_c at which all the spins in the entire system point up. The characteristic length scale of the physics at any particular temperature is given by the typical size of the islands. The physically motivated block spin method of Kadanoff et al. treats blocks of up spin as one single effective up spin, and similarly blocks of down spins. The notion of a renormalization group is then the natural one for describing these effective spins by an effective Hamiltonian appropriate to that length scale.

It is more or less clear how to implement this physical idea of changing length scales in the functional integral (9). We are supposed to integrate over those $h(\vec{k}, \omega)$, with $\vec{k}$ and ω less than Λ. Suppose we do only a fraction of what we are supposed to do. Let us integrate over those $h(\vec{k}, \omega)$ with $\vec{k}$ and ω larger than $\Lambda - \delta\Lambda$ but smaller than Λ. This is precisely what we mean when we say that we don't care about the fluctuations of $h(\vec{x}, t)$ on length and time scales less than $(\Lambda - \delta\Lambda)^{-1}$.

Putting on blurry glasses

For the sake of simplicity, let us go back to our favorite, the $\lambda\varphi^4$ theory, instead of the surface growth problem. Recall from the preceding chapters the importance of the Euclidean $\lambda\varphi^4$ theory in modern condensed matter theory. So, continue the $\lambda\varphi^4$ theory to Euclidean space and stare at the integral

$$Z(\Lambda) = \int_{\Lambda} \mathcal{D}\varphi e^{-\int d^dx \mathcal{L}(\varphi)} \tag{11}$$

The notation $\int_{\Lambda}$ instructs us to include only those field configurations $\varphi(x) = \int [d^dk/(2\pi)^d]e^{ikx}\varphi(k)$ such that $\varphi(k) = 0$ for $|k| \equiv (\sum_{i=1}^d k_i^2)^{\frac{1}{2}}$ larger than Λ. As explained in the text this amounts to putting on blurry glasses with resolution $L = 1/\Lambda$: We do not admit or see fluctuations with length scales less than L.

Evidently, the $O(d)$ invariance, namely the Euclidean equivalent of Lorentz invariance, will make our lives considerably easier. In contrast, for the surface growth problem we will need special glasses that blur space and time differently.[1]

We are now ready to make our glasses blurrier by letting $\Lambda \to \Lambda - \delta\Lambda$ (with $\delta\Lambda > 0$). Write $\varphi = \varphi_s + \varphi_w$ (s for "smooth" and w for "wriggly"), defined such that the Fourier components $\varphi_s(k)$ and $\varphi_w(k)$ are nonzero only for $|k| \leq (\Lambda - \delta\Lambda)$ and $(\Lambda - \delta\Lambda) \leq |k| \leq \Lambda$, respectively. (Obviously, the designation "smooth" and "wriggly" is for convenience.) Plugging into (11) we can write

$$Z(\Lambda) = \int_{\Lambda-\delta\Lambda} \mathcal{D}\varphi_s e^{-\int d^dx \mathcal{L}(\varphi_s)} \int \mathcal{D}\varphi_w e^{-\int d^dx \mathcal{L}_1(\varphi_s, \varphi_w)} \tag{12}$$

where all the terms in $\mathcal{L}_1(\varphi_s, \varphi_w)$ depend on φ_w. (What we are doing here is somewhat reminiscent of what we did in chapter IV.3.) Imagine doing the integral over φ_w. Call the result

$$e^{-\int d^dx \delta\mathcal{L}(\varphi_s)} \equiv \int \mathcal{D}\varphi_w e^{-\int d^dx \mathcal{L}_1(\varphi_s, \varphi_w)}$$

and thus we have

$$Z(\Lambda) = \int_{\Lambda-\delta\Lambda} \mathcal{D}\varphi_s e^{-\int d^dx [\mathcal{L}(\varphi_s)+\delta\mathcal{L}(\varphi_s)]} \tag{13}$$

There, we have done it! We have rewritten the theory in terms of the "smooth" field φ_s.

Of course, this is all formal, since in practice the integral over φ_w can only be done perturbatively assuming that the relevant couplings are small. If we could do the integral

---

[1] In condensed matter physics, the so-called dynamical exponent z measures this difference. More precisely, in the context of the surface growth problem, the correlator (introduced in chapter VI.6) satisfies the dynamic scaling form given in (VI.6.3). Naively, the dynamical exponent z should be 2. (For a brief review of all this, see M. Kardar and A. Zee, *Nucl. Phys.* B464[FS]: 449, 1996, cond-mat/9507112.)

over φ_w exactly, we might as well just do the integral over φ and then we would have no need for all this renormalization group stuff.

For pedagogical purposes, consider more generally $\mathcal{L} = \frac{1}{2}(\partial\varphi)^2 + \sum_n \lambda_n \varphi^n + \cdots$ (so that λ_2 is the usual $\frac{1}{2}m^2$ and λ_4 the usual λ.) Since terms such as $\partial\varphi_s \partial\varphi_w$ integrate to zero, we have

$$\int d^d x \mathcal{L}_1(\varphi_s, \varphi_w) = \int d^d x \left(\frac{1}{2}(\partial\varphi_w)^2 + \frac{1}{2}m^2\varphi_w^2 + \cdots \right)$$

with φ_s hiding in the $(\cdots)$. This describes a field φ_w interacting with both itself and a background field $\varphi_s(x)$. By symmetry considerations $\delta\mathcal{L}(\varphi_s)$ has the same form as $\mathcal{L}(\varphi_s)$ but with different coefficients. Adding $\delta\mathcal{L}(\varphi_s)$ to $\mathcal{L}(\varphi_s)$ thus shifts[2] the couplings λ_n [and the coefficient of $\frac{1}{2}(\partial\varphi_s)^2$.] These shifts generate the flow in the space of couplings I described earlier.

We could have perfectly well left (13) as our end result. But suppose we want to compare (13) with (11). Then we would like to change the $\int_{\Lambda - \delta\Lambda}$ in (13) to $\int_\Lambda$. For convenience, introduce the real number $b < 1$ by $\Lambda - \delta\Lambda = b\Lambda$. In $\int_{\Lambda - \delta\Lambda}$ we are told to integrate over fields with $|k| \leq b\Lambda$. So all we have to do is make a trivial change of variable: Let $k = bk'$ so that $|k'| \leq \Lambda$. But then correspondingly we have to change $x = x'/b$ so that $e^{ikx} = e^{ik'x'}$. Plugging in, we obtain

$$\int d^d x \mathcal{L}(\varphi_s) = \int d^d x' b^{-d} \left[\frac{1}{2}b^2(\partial'\varphi_s)^2 + \sum_n \lambda_n \varphi_s^n + \cdots \right] \tag{14}$$

where $\partial' = \partial/\partial x' = (1/b)\partial/\partial x$. Define φ' by $b^{2-d}(\partial'\varphi_s)^2 = (\partial'\varphi')^2$ or in other words $\varphi' = b^{\frac{1}{2}(2-d)}\varphi_s$. Then (14) becomes

$$\int d^d x' \left[\frac{1}{2}(\partial'\varphi')^2 + \sum_n \lambda_n b^{-d+(n/2)(d-2)}\varphi'^n + \cdots \right]$$

Thus, if we define the coefficient of φ'^n as λ'_n we have

$$\lambda'_n = b^{(n/2)(d-2)-d}\lambda_n \tag{15}$$

an important result in renormalization group theory.

Relevant, irrelevant, and marginal

Let us absorb what this means. (For the time being, let us ignore $\delta\mathcal{L}(\varphi_s)$ to keep the discussion simple.) As we put on blurrier glasses, in other words, as we become interested in physics over longer distance scales, we can once again write $Z(\Lambda)$ as in (11) except that the couplings λ_n have to be replaced by λ'_n. Since $b < 1$ we see from (15) that the λ_n's with $(n/2)(d-2) - d > 0$ get smaller and smaller and can eventually be neglected. A dose of jargon here: The corresponding operators φ^n (for historical reasons we revert for an instant

---

[2] Terms such as $(\partial\varphi)^4$ can also be generated and that is why I wrote $\mathcal{L}(\varphi)$ with the $(\cdots)$ under which terms such as these can be swept. You can check later that for most applications these terms are irrelevant in the technical sense to be defined below.

from the functional integral language to the operator language) are called irrelevant. They are the losers. Conversely, the winners, namely the φ^n's for which $(n/2)(d-2) - d < 0$, are called relevant. Operators for which $(n/2)(d-2) - d = 0$ are called marginal.

For example, take $n = 2$: $m'^2 = b^{-2}m^2$ and the mass term is always relevant in any dimension. On the other hand, take $n = 4$, and we see that $\lambda' = b^{d-4}\lambda$ and φ^4 is relevant for $d < 4$, irrelevant for $d > 4$, and marginal at $d = 4$. Similarly, $\lambda'_6 = b^{2d-6}\lambda$ and φ^6 is marginal at $d = 3$ and becomes irrelevant for $d > 3$.

We also see that $d = 2$ is special: All the φ^n's are relevant.

Now all this may ring a bell if you did the exercises religiously. In exercise III.2.1 you showed that the coupling λ_n has mass dimension $[\lambda_n] = (n/2)(2-d) + d$. Thus, the quantity $(n/2)(d-2) - d$ is just the length dimension of λ_n. For example, for $d = 4$, λ_6 has mass dimension -2 and thus as explained in chapter III.2 the φ^6 interaction is nonrenormalizable, namely that it has nasty behavior at high energy. But condensed matter physicists are interested in the long distance limit, the opposite limit from the one that interests particle physicsts. Thus, it is the nasty guys like φ^6 that become irrelevant in the long distance limit.

One more piece of jargon: Given a scalar field theory, the dimension d at which the most relevant interaction becomes marginal is known as the critical dimension in condensed matter physics. For example, the critical dimension for a φ^6 theory is 3. It is now just a matter of "high school arithmetic" to translate (15) into differential form. Write $\lambda'_n = \lambda_n + \delta\lambda_n$; then from $b = 1 - (\delta\Lambda/\Lambda)$ we have $\delta\lambda_n = -[\frac{n}{2}(d-2) - d]\lambda_n(\delta\Lambda/\Lambda)$.

Let us now be extra careful about signs. As I have already remarked, for $(n/2)(d-2) - d > 0$ the coupling λ_n (which, for definiteness, we will think of as positive) get smaller, as is evident from (15). But since we are decreasing Λ to $\Lambda - \delta\Lambda$, a positive $\delta\Lambda$ actually corresponds to the resolution of our blurry glasses $L = \Lambda^{-1}$ changing to $L + L(\delta\Lambda/\Lambda)$. Thus we obtain

$$L\frac{d\lambda_n}{dL} = -\left[\frac{n}{2}(d-2) - d\right]\lambda_n, \tag{16}$$

so that for $(n/2)(d-2) - d > 0$ a positive λ_n would decrease as L increases.[3]

In particular, for $n = 4$, $L(d\lambda/dL) = (4 - d)\lambda$. In most condensed matter physics applications, $d \leq 3$ and so λ increases as the length scale of the physics under study increases. The φ^4 coupling is relevant as noted above.

The $\delta\mathcal{L}(\varphi_s)$, which we provisionally neglected, contributes an additional term, which we call dynamical in contrast to the geometrical or "trivial" term displayed, to the right-hand side of (16). Thus, in general $L(d\lambda_n/dL) = -[(n/2)(d-2) - d]\lambda_n + K(d, n, \cdots, \lambda_j, \cdots)$, with the dynamical term K depending not only on d and n, but also on all the other couplings. [For example, in (5) the "trivial" term vanishes since we are in 4-dimensional spacetime; there is only a dynamical contribution.]

---

[3] Note that what appears on the right-hand side is minus the length dimension of λ_n, not the length dimension $(n/2)(d-2) - d$ as one might have guessed naively.

As you can see from this discussion, a more descriptive name for the renormalization group might be "the trick of doing an integral a little bit at a time."

Exploiting symmetry

To determine the renormalization group flow of the coupling g in the surface growth problem we can repeat the same type of computation we did to determine the flow of the coupling λ and of e in our two previous examples, namely we would calculate, to use the language of particle physics, the amplitude for h-h scattering to one loop order. But instead, let us follow the physical picture of Kadanoff et al. In $Z(\Lambda) = \int \mathcal{D}h \, e^{-S(h)}$ we integrate over only those $h(\vec{k}, \omega)$ with $\vec{k}$ and ω larger than $\Lambda - \delta\Lambda$ but less than Λ.

I will now show you how to exploit the symmetry of the problem to minimize our labor. The important thing is not necessarily to learn about the dynamics of surface growth, but to learn the methodology that will serve you well in other situations. I have picked a particularly "difficult" nonrelativistic problem whose symmetries are not manifest, so that if you master the renormalization group for this problem you will be ready for almost anything.

Imagine having done this partial integration and call the result $\int \mathcal{D}h \, e^{-\tilde{S}(h)}$. At this point you should work out the symmetries of the problem as indicated in the exercises. Then you can argue that $\tilde{S}(h)$ must have the form

$$\tilde{S}(h) = \frac{1}{2} \int d^D x \, dt \left[\left(\alpha \frac{\partial}{\partial t} - \beta \nabla^2 \right) h - \alpha \frac{g}{2} (\nabla h)^2 \right]^2 + \cdots, \tag{17}$$

depending on two parameters α and β. The $(\cdots)$ indicates terms involving higher powers of h and its derivatives. The simplifying observation is that the same coefficient α multiplies both $\partial h/\partial t$ and $(g/2)(\nabla h)^2$. Once we know α and β then by suitable rescaling we can bring the action $\tilde{S}(h)$ back into the same form as $S(h)$ and thus find out how g changes. Therefore, it suffices to look at the $(\partial h/\partial t)^2$ and $(\nabla^2 h)^2$ terms in the action, or equivalently at the propagator, which is considerably simpler to calculate. As Rudolf Peierls once said[4] to the young Hans Bethe, "Erst kommt das Denken, dann das Integral." (Roughly, "First think, then do the integral.") We will not do the computation here. Suffice it to note that g has the high school dimension of (length)$^{\frac{1}{2}(D-2)}$ (see exercise VI.8.5). Thus, according to the preceding discussion we should have

$$L \frac{dg}{dL} = \frac{1}{2}(2 - D)g + c_D g^3 + \cdots \tag{18}$$

A detailed calculation is needed to determine the coefficient c_D, which obviously depends on the dimension of space D since the Feynman integrals depend on D. The equation tells us how g, an effective measure of nonlinearity in the physics of

---

[4] John Wheeler gave me similar advice when I was a student: "Never calculate without first knowing the answer."

surface growth, changes when we change the length scale L. For the record, $c_D = [S(D)/4(2\pi)^D](2D-3)/D$, with $S(D)$ the D-dimensional solid angle. The interesting factor is of course $(2D-3)$, changing sign between[5] $D=1$ and 2.

Localization

As I said earlier, renormalization group flow has literally become part of the language of condensed matter and high energy physics. Let me give you another example of the power of the renormalization group. Go back to Anderson localization (chapter VI.7), which so astonished the community at the time. People were surprised that the localization behavior depends so drastically on the dimension of space D, and perhaps even more so, that for $D=2$ all states are localized no matter how weak the strength of the disorder. Our usual physical intuition would say that there is a critical strength. As we will now see, both features are quite naturally accounted for in the renormalization group language. Already, you see in (18) that D enters in an essential way.

I now offer you a heuristic but beautiful (at least to me) argument given by Abrahams, Anderson, Licciardello, and Ramakrishnan, who as a result became known to the condensed matter community as the "Gang of Four." First, you have to understand the difference between conductivity σ and conductance G in solid state physics lingo. Conductivity[6] is defined by $\vec{J} = \sigma \vec{E}$, where $\vec{J}$ measures the number of electrons passing through a unit area per unit time. Conductance G is the inverse of resistance (the mnemonic: the two words rhyme). Resistance R is the property of a lump of material and defined in high school physics by $V = IR$, where the current I measures the number of electrons passing by per unit time. To relate σ and G, consider a lump of material, taken to be a cube of size L, with a voltage drop V across it. Then $I = JL^2 = \sigma E L^2 = \sigma (V/L)L^2 = \sigma LV$ and thus[7] $G(L) = 1/R = I/V = \sigma L$. Next, let us go to two dimensions. Consider a thin sheet of material of length and width L and thickness $a \ll L$. (We are doing real high school physics, not talking about some sophisticated field theorist's idea of two dimensional space!) Again, apply a voltage drop V over the length $L : I = J(aL) = \sigma EaL = \sigma(V/L)aL = \sigma Va$ and so $G(L) = 1/R = I/V = \sigma a$. I will let you go on to one dimension: Consider a wire of length L and width and thickness a. In this way, we obtain $G(L) \propto L^{D-2}$. Incidentally, condensed matter physicists customarily define a dimensionless conductance $g(L) \equiv \hbar G(L)/e^2$.

---

[5] Incidentally, the theory is exactly solvable for $D=1$ (with methods not discussed in this book).

[6] Over the years I have asked a number of high energy theorists how is it possible to obtain $\vec{J} = \sigma \vec{E}$, which manifestly violates time reversal invariance, if the microscopic physics of an electron scattering on an impurity atom perfectly well respects time reversal invariance. Very few knew the answer. The resolution of this apparent paradox is in the order of limits! Condensed matter theorists calculate a frequency and wave vector dependent conductivity $\sigma(\omega, \vec{k})$ and then take the limit $\omega, \vec{k} \to 0$ and $\vec{k}^2/\omega \to 0$. Before the limit is taken, time reversal invariance holds. The time it takes the particle to find out that it is in a box of size of order $1/k$ is of order $1/(Dk^2)$ (with D the diffusion constant). The physics is that this time has to be much longer than the observation time $\sim 1/\omega$.

[7] Sam Treiman told me that when he joined the U.S. Army as a radio operator he was taught that there were three forms of Ohm's law: $V = IR$, $I = V/R$, and $R = V/I$. In the second equality here we use the fourth form.

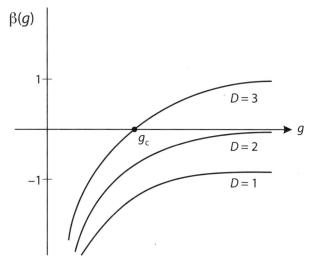

Figure VI.8.2

We also know the behavior of $g(L)$ when $g(L)$ is small or, in other words, when the material is an insulator for which we expect $g(L) \sim ce^{-L/\xi}$, with ξ some length characteristic of the material and determined by the microscopic physics. Thus, for $g(L)$ small, $L(dg/dL) = -(L/\xi)g(L) = g(L)[\log g(L) - \log c]$, where the constant $\log c$ is negligible in the regime under consideration.

Putting things together, we obtain

$$\beta(g) \equiv \frac{L}{g}\frac{dg}{dL} = \begin{cases} (D-2) + \cdots & \text{for large } g \\ \log g + \cdots & \text{for small } g \end{cases} \tag{19}$$

First, a trivial note: in different subjects, people define $\beta(g)$ differently (without affecting the physics of course). In localization theory, $\beta(g)$ is traditionally defined as $d \log g / d \log L$ as indicated here. Given (19) we can now make a "most plausible" plot of $\beta(g)$ as shown in figure VI.8.2. You see that for $D = 2$ (and $D = 1$) the conductance $g(L)$ always flows toward 0 as we go to long distances (macroscopic measurements on macroscopic materials) regardless of where we start. In contrast, for $D = 3$, if g_0 the initial value of g is greater than a critical g_c then $g(L)$ flows to infinity (presumably cut off by physics we haven't included) and the material is a metal, while if $g_0 < g_c$, the material is an insulator. Incidentally, condensed matter theorists often speak of a critical dimension D_c at which the long distance behavior of a system changes drastically; in this case $D_c = 2$.

Effective description

In a sense, the renormalization group goes back to a basic notion of physics, that the effective description can and should change as we move from one length scale to another. For example, in hydrodynamics we do not have to keep track of the detailed interaction among water molecules. Similarly, when we apply the renormalization group flow to the strong interaction, starting at high energies and moving toward low energies, the effective description goes from a theory of quarks and gluons to a theory of nucleons and mesons. In this more general picture then, we no longer think of flowing in a space of coupling constants, but in "the space of Hamiltonians" that some condensed matter physicists like to talk about.

Exercises

VI.8.1   Show that the solution of $dg/dt = -bg^3 + \cdots$ is given by

$$\frac{1}{\alpha(t)} = \frac{1}{\alpha(0)} + 8\pi bt + \cdots \tag{20}$$

where we defined $\alpha(t) = g(t)^2/4\pi$.

VI.8.2   In our discussion of the renormalization group, in $\lambda\varphi^4$ theory or in QED, for the sake of simplicity we assumed that the mass m of the particle is much smaller than μ and thus set m equal to zero. But nothing in the renormalization group idea tells us that we can't flow to a mass scale below m. Indeed, in particle physics many orders of magnitude separate the top quark mass m_t from the up quark mass m_u. We might want to study how the strong interaction coupling flows from some mass scale far above m_t down to some mass scale μ below m_t but still large compared to m_u. As a crude approximation, people often set any mass m below μ equal to zero and any mass m above μ to infinity (i.e., not contributing to the renormalization group flow). In reality, as μ approaches m from above the particle starts to contribute less and drops out as μ becomes much less than m. Taking either the $\lambda\varphi^4$ theory or QED study this so-called threshold effect.

VI.8.3   Show that (10) is invariant under the so-called Galilean transformation

$$h(\vec{x}, t) \rightarrow h'(\vec{x}, t) = h\left(\vec{x} + g\vec{u}t, t\right) + \vec{u} \cdot \vec{x} + \frac{g}{2}u^2 t \tag{21}$$

Show that because of this symmetry only two parameters α and β appear in (17).

VI.8.4   In $\bar{S}(h)$ only derivatives of the field h can appear and not the field itself. (Since the transformation $h(\vec{x}, t) \rightarrow h(\vec{x}, t) + c$ with c a constant corresponds to a trivial shift of where we measure the surface height from, the physics must be invariant under this transformation.) Terms involving only one power of h cannot appear since they are all total divergences. Thus, $\bar{S}(h)$ must start with terms quadratic in h. Verify that the $\bar{S}(h)$ given in (17) is indeed the most general. A term proportional to $(\nabla h)^2$ is also allowed by symmetries and is in fact generated. However, such a term can be eliminated by transforming to a moving coordinate frame $h \rightarrow h + ct$.

VI.8.5   Show that g has the high school dimension of $(\text{length})^{\frac{1}{2}(D-2)}$. [Hint: The form of $S(h)$ implies that t has the dimension of length squared and so h has the dimension $(\text{length})^{\frac{1}{2}(2-D)}$.] Comparing the terms $\nabla^2 h$ and $g(\nabla h)^2$ we determine the dimension of g.]

VI.8.6   Calculate the h propagator to one loop order. Extract the coefficients of the ω^2 and k^4 terms in a low frequency and wave number expansion of the inverse propagator and determine α and β.

VI.8.7   Study the renormalization group flow of g for $D = 1, 2, 3$.

Part VII | Grand Unification

VII.1 | Quantizing Yang-Mills Theory and Lattice Gauge Theory

One reason that Yang-Mills theory was not immediately taken up by physicists is that people did not know how to calculate with it. At the very least, we should be able to write down the Feynman rules and calculate perturbatively. Feynman himself took up the challenge and concluded, after looking at various diagrams, that extra fields with ghostlike properties had to be introduced for the theory to be consistent. Nowadays we know how to derive this result more systematically.

The story goes that Feynman wanted to quantize gravity but Gell-Mann suggested to him to first quantize Yang-Mills theory as a warm-up exercise.

Consider pure Yang-Mills theory—it will be easy to add matter fields later. Follow what we have learned. Split the Lagrangian $\mathcal{L} = \mathcal{L}_0 + \mathcal{L}_1$ as usual into two pieces (we also choose to scale $A \to gA$):

$$\mathcal{L}_0 = -\tfrac{1}{4}(\partial_\mu A_\nu^a - \partial_\nu A_\mu^a)^2 \tag{1}$$

and

$$\mathcal{L}_1 = -\tfrac{1}{2}g(\partial_\mu A_\nu^a - \partial_\nu A_\mu^a)f^{abc}A^{b\mu}A^{c\nu} - \tfrac{1}{4}g^2 f^{abc}f^{ade}A_\mu^b A_\nu^c A^{d\mu}A^{e\nu} \tag{2}$$

Then invert the differential operator in the quadratic piece (1) to obtain the propagator. This part looks the same as the corresponding procedure for quantum electrodynamics, except for the occurrence of the index a. Just as in electrodynamics, the inverse does not exist and we have to fix a gauge.

I built up the elaborate Faddeev-Popov method to quantize quantum electrodynamics and as I noted, it was a bit of overkill in that context. But here comes the payoff: We can now turn the crank. Recall from chapter III.4 that the Faddeev-Popov method would give us

$$\mathcal{Z} = \int DA e^{iS(A)} \Delta(A)\delta[f(A)] \tag{3}$$

with $\Delta(A) \equiv \{\int Dg\delta[f(A_g)]\}^{-1}$ and $S(A) = \int d^4x \mathcal{L}$ the Yang-Mills action. (As in chapter III.4, $A_g \equiv gAg^{-1} - i(\partial g)g^{-1}$ denotes the gauge transform of A. Here $g \equiv g(x)$ denotes the group element that defines the gauge transformation at x and is obviously not to be confused with the coupling constant.)

Since $\Delta(A)$ appears in (3) multiplied by $\delta[f(A)]$, in the integral over g we expect, for a reasonable choice of $f(A)$, only infinitesimal g to be relevant. Let us choose $f(A) = \partial A - \sigma$. Under an infinitesimal transformation, $A_\mu^a \to A_\mu^a - f^{abc}\theta^b A_\mu^c + \partial_\mu \theta^a$ and thus

$$\Delta(A) = \{\int D\theta \delta[\partial A^a - \sigma^a - \partial^\mu(f^{abc}\theta^b A_\mu^c - \partial_\mu \theta^a)]\}^{-1} \tag{4}$$

$$\text{“} = \text{”} \{\int D\theta \delta[\partial^\mu(f^{abc}\theta^b A_\mu^c - \partial_\mu \theta^a)]\}^{-1}.$$

where the "effectively equal sign" follows since $\Delta(A)$ is to be multiplied later by $\delta[f(A)]$.

Let us write formally

$$\partial^\mu(f^{abc}\theta^b A_\mu^c - \partial_\mu \theta^a) = \int d^4y K^{ab}(x, y)\theta^b(y) \tag{5}$$

thus defining the operator $K^{ab}(x, y) = \partial^\mu(f^{abc}A_\mu^c - \partial_\mu \delta^{ab})\delta^{(4)}(x - y)$. Note that in contrast to electromagnetism here K depends on the gauge potential. The elementary result $\int d\theta \delta(K\theta) = 1/K$ for θ and K real numbers can be generalized to $\int d\theta \delta(K\theta) = 1/\det K$ for θ a real vector and K a nonsingular matrix. Regarding $K^{ab}(x, y)$ as a matrix, we obtain $\Delta(A) = \det K$, but we know from chapter II.5 how to represent the determinant as a functional integral over Grassmann variables: Write $\Delta(A) = \int Dc Dc^\dagger e^{iS_{\text{ghost}}(c^\dagger, c)}$, with

$$S_{\text{ghost}}(c^\dagger, c) = \int d^4x d^4y c_a^\dagger(x) K^{ab}(x, y) c_b(y)$$

$$= \int d^4x[\partial c_a^\dagger(x)\partial c_a(x) - \partial^\mu c_a^\dagger(x) f^{abc} A_\mu^c(x) c_b(x)]$$

$$= \int d^4x \partial c_a^\dagger(x) D c_a(x) \tag{6}$$

and with D the covariant derivative for the adjoint representation, to which the fields c_a and $c_a^\dagger$ belong just like A_μ^a. The fields c_a and $c_a^\dagger$ are known as ghost fields because they violate the spin-statistics connection: Though scalar, they are treated as anticommuting. This "violation" is acceptable because they are not associated with physical particles and are introduced merely to represent $\Delta(A)$ in a convenient form.

This takes care of the $\Delta(A)$ factor in (3). As for the $\delta[f(A)]$ factor, we use the same trick as in chapter III.4 and integrate $\mathcal{Z}$ over $\sigma^a(x)$ with a Gaussian weight $e^{-(i/2\xi)\int d^4x \sigma^a(x)^2}$ so that $\delta[f(A)]$ gets replaced by $e^{-(i/2\xi)\int d^4x (\partial A^a)^2}$.

Putting it all together, we obtain

$$\mathcal{Z} = \int DA Dc Dc^\dagger e^{iS(A)-(i/2\xi)\int d^4x(\partial A)^2 + iS_{\text{ghost}}(c^\dagger, c)} \tag{7}$$

with ξ a gauge parameter. Comparing with the corresponding expression for an abelian gauge theory in chapter III.4, we see that in nonabelian gauge theories we have a ghost action S_{ghost} in addition to the Yang-Mills action. Thus, $\mathcal{L}_0$ and $\mathcal{L}_1$ are changed to

$$\mathcal{L}_0 = -\frac{1}{4}(\partial_\mu A_\nu^a - \partial_\nu A_\mu^a)^2 - \frac{1}{2\xi}(\partial^\mu A_\mu^a)^2 + \partial c_a^\dagger \partial c_a \tag{8}$$

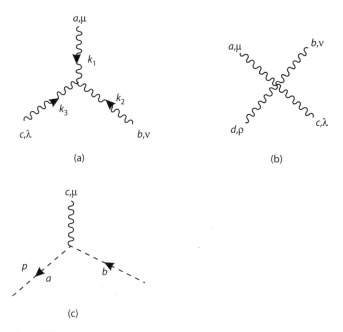

Figure VII.1.1

and

$$\mathcal{L}_1 = -\tfrac{1}{2}g(\partial_\mu A_\nu^a - \partial_\nu A_\mu^a)f^{abc}A^{b\mu}A^{c\nu} + \tfrac{1}{4}g^2 f^{abc}f^{ade}A_\mu^b A_\nu^c A^{d\mu}A^{e\nu} - \partial^\mu c_a^\dagger g f^{abc}A_\mu^c c_b(x) \qquad (9)$$

We can now read off the propagators for the gauge boson and for the ghost field immediately from (8). In particular, we see that except for the group index a the terms quadratic in the gauge potential are exactly the same as the terms quadratic in the electromagnetic gauge potential in (III.4.8). Thus, the gauge boson propagator is

$$\frac{(-i)}{k^2}\left[g_{\nu\lambda} - (1-\xi)\frac{k_\nu k_\lambda}{k^2}\right]\delta_{ab} \qquad (10)$$

Compare with (III.4.9). From the term $\partial c_a^\dagger \partial c_a$ in (8) we find the ghost propagator to be $(i/k^2)\delta_{ab}$.

From $\mathcal{L}_1$ we see that there is a cubic and a quartic interaction between the gauge bosons, and an interaction between the gauge boson and the ghost field, as illustrated in figure VII.1.1. The cubic and the quartic couplings can be easily read off as

$$g f^{abc}[g_{\mu\nu}(k_1 - k_2)_\lambda + g_{\nu\lambda}(k_2 - k_3)_\mu + g_{\lambda\mu}(k_3 - k_1)_\nu] \qquad (11)$$

and

$$-ig^2[f^{abe}f^{cde}(g_{\mu\lambda}g_{\nu\rho} - g_{\mu\rho}g_{\nu\lambda}) + f^{ade}f^{cbe}(g_{\mu\lambda}g_{\nu\rho} - g_{\mu\nu}g_{\rho\lambda})$$
$$+ f^{ace}f^{bde}(g_{\mu\nu}g_{\lambda\rho} - g_{\mu\rho}g_{\nu\lambda})] \qquad (12)$$

respectively. The coupling to the ghost field is

$$g f^{abc}p^\mu \qquad (13)$$

Obviously, we can exploit various permutation symmetries in writing these down. For instance, in (12) the second term is obtained from the first by the interchange $\{c, \lambda\} \leftrightarrow \{d, \rho\}$, and the third and fourth terms are obtained from the first and second by the interchange $\{a, \mu\} \leftrightarrow \{c, \lambda\}$.

Unnatural act

In a highly symmetric theory such as Yang-Mills, perturbating is clearly an unnatural act as it involves brutally splitting $\mathcal{L}$ into two parts: a part quadratic in the fields and the rest. Consider, for example, an exactly soluble single particle quantum mechanics problem, such as the Schrödinger equation with $V(x) = 1 - (1/\cosh x)^2$. Imagine writing $V(x) = \frac{1}{2}x^2 + W(x)$ and treating $W(x)$ as a perturbation on the harmonic oscillator. You would have a hard time reproducing the exact spectrum, but this is exactly how we brutalize Yang-Mills theory in the perturbative approach: We took the "holistic entity" tr $F_{\mu\nu}F^{\mu\nu}$ and split it up into the "harmonic oscillator" piece $\mathrm{tr}(\partial_\mu A_\nu - \partial_\nu A_\mu)^2$ and a "perturbation."

If Yang-Mills theory ever proves to be exactly soluble, the perturbative approach with its mangling of gauge invariance is clearly not the way to do it.

Lattice gauge theory

Wilson proposed a way out: Do violence to Lorentz invariance rather than to gauge invariance. Let us formulate Yang-Mills theory on a hypercubic lattice in 4-dimensional Euclidean spacetime. As the lattice spacing $a \to 0$ we expect to recover 4-dimensional rotational invariance and (by a Wick rotation) Lorentz invariance. Wilson's formulation, known as lattice gauge theory, is easy to understand, but the notation is a bit awkward, due to the lack of rotational invariance. Denote the location of the lattice sites by the vector x_i. On each link, say the one going from x_i to one of its nearest neighbors x_j, we associate an N by N simple unitary matrix U_{ij}. Consider the square, known as a plaquette, bounded by the four corners x_i, x_j, x_k, and x_l (with these nearest neighbors to each other.) See figure VII.1.2. For each plaquette P we associate the quantity $S(P) = \mathrm{Re\,tr}\, U_{ij}U_{jk}U_{kl}U_{li}$, constructed to be invariant under the local transformation

$$U_{ij} \to V_i^\dagger U_{ij} V_j \tag{14}$$

The symmetry is local because for each site x_i we can associate an independent V_i.

Wilson defined Yang-Mills theory by

$$\mathcal{Z} = \int \Pi dU \, e^{(1/2f^2) \sum_P S(P)} \tag{15}$$

where the sum is taken over all the plaquettes in the lattice. The coupling strength f controls how wildly the unitary matrices U_{ij}'s fluctuate. For small f, large values of $S(P)$ are favored, and so the U_{ij}'s are all approximately equal to the unit matrix (up to an irrelevant global transformation.)

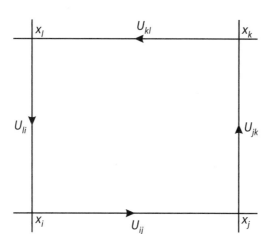

Figure VII.1.2

Without doing any arithmetic, we can argue by symmetry that in the continuum limit $a \to 0$, Yang-Mills theory as we know it must emerge: The action is manifestly invariant under local $SU(N)$ transformation. To actually see this, define a field $A_\mu(x)$ with $\mu = 1, 2, 3, 4$, permeating the 4-dimensional Euclidean space the lattice lives in, by

$$U_{ij} = V_i^\dagger e^{iaA_\mu(x)} V_j \tag{16}$$

where $x = \frac{1}{2}(x_i + x_j)$ (namely the midpoint of the link U_{ij} lives on) and μ is the direction connecting x_i to x_j (namely $\hat{\mu} \equiv (x_j - x_i)/a$ is the unit vector in the μ direction.) The V's just reflect the gauge freedom in (14) and obviously do not enter into the plaquette action $S(P)$ by construction. I will let you show in an exercise that

$$\text{tr}\, U_{ij} U_{jk} U_{kl} U_{li} = \text{tr}\, e^{ia^2 F_{\mu\nu} + O(a^3)} \tag{17}$$

with $F_{\mu\nu}$ the Yang-Mills field strength evaluated at the center of the plaquette. Indeed, we could have discovered the Yang-Mills field strength in this way. I hope that you start to see the deep geometric significance of $F_{\mu\nu}$. Continuing the exercise you will find that the action on each plaquette comes out to be

$$\begin{aligned} S(P) &= \text{Re}\, \text{tr}\, e^{ia^2 F_{\mu\nu} + O(a^3)} \\ &= \text{Re}\, \text{tr}[1 + ia^2 F_{\mu\nu} - \tfrac{1}{2}a^4 F_{\mu\nu} F_{\mu\nu} + O(a^5)] = \text{tr}\, 1 - \tfrac{1}{2}a^4 \, \text{tr}\, F_{\mu\nu} F_{\mu\nu} + \cdots \end{aligned} \tag{18}$$

and so up to an irrelevant additive constant we recover in (15) the Yang-Mills action in the continuum limit. Again, it is worth emphasizing that without going through any arithmetic we could have fixed the a^4 term in (18) (up to an overall constant) by dimensional analysis and gauge invariance.[1]

---

[1] The sign can be easily checked against the abelian case.

The Wilson formulation is beautiful in that none of the hand-wringing over gauge fixing, Faddeev-Popov determinant, ghost fields, and so forth is necessary for (15) to make sense. Recalling chapter V.3 you see that (15) defines a statistical mechanics problem like any other. Instead of integrating over some spin variables say, we integrate over the group $SU(N)$ for each link. Most importantly, the lattice gauge formulation opens up the possibility of computing the properties of a highly nontrivial quantum field theory numerically. Lattice gauge theory is a thriving area of research. For a challenge, try to incorporate fermions into lattice gauge theory: This is a difficult and ongoing problem because fermions and spinor fields are naturally associated with $SO(4)$, which does not sit well on a lattice.

Wilson loop

Field theorists usually deal with local observables, that is, observables defined at a space-time point x, such as $J^\mu(x)$ or tr $F_{\mu\nu}(x)F^{\mu\nu}(x)$, but of course we can also deal with nonlocal observables, such as $e^{i \oint_C dx^\mu A_\mu}$ in electromagnetism, where the line integral is evaluated over a closed curve C. The gauge invariant quantity in the exponential is equal to the electromagnetic flux going through the surface bounded by C. (Indeed, recall chapter IV.4.)

Wilson pointed out that lattice gauge theory contains a natural gauge invariant but nonlocal observable $W(C) \equiv$ tr $U_{ij}U_{jk} \ldots U_{nm}U_{mi}$, where the set of links connecting x_i to x_j to x_k et cetera and eventually to x_m and back to x_i traces out a loop called C. Referring to (16) we see that $W(C)$, known as the Wilson loop, is the trace of a product of many factors of e^{iaA_μ}. Thus, in the continuum limit $a \to 0$, we have evidently

$$W(C) \equiv \text{tr } P e^{i \oint_C dx^\mu A_\mu} \tag{19}$$

with C now an arbitrary curve in Euclidean spacetime. Here P denotes path ordering, clearly necessary since the A_μ's associated with different segments of C, being matrices, do not commute with each other. [Indeed, P is defined by the lattice definition of $W(C)$.]

To understand the physical meaning of the Wilson loop, Recall chapters I.4 and I.5. To obtain the potential energy E between two oppositely charged lumps we have to compute

$$\lim_{T \to \infty} \frac{1}{\mathcal{Z}} \int DA e^{iS_{\text{Maxwell}}(A) + i \int d^4x A_\mu J^\mu} = e^{-iET}$$

For two lumps held at a distance R apart we plug in

$$J^\mu(x) = \eta^{\mu 0}\{\delta^{(3)}(\vec{x}) - \delta^{(3)}[\vec{x} - (R, 0, 0)]\}$$

and see that we are actually computing the expectation value $\langle e^{i(\int_{C_1} dx^\mu A_\mu - \int_{C_2} dx^\mu A_\mu)} \rangle$ in a fluctuating electromagnetic field, where C_1 and C_2 denote two straight line segments at $\vec{x} = (0, 0, 0)$ and $\vec{x} = (R, 0, 0)$, respectively. It is convenient to imagine bringing the two lumps together in the far future (and similarly in the far past). Then we deal instead with the manifestly gauge invariant quantity $\langle e^{i \oint_C dx^\mu A_\mu} \rangle$, where C is the rectangle shown

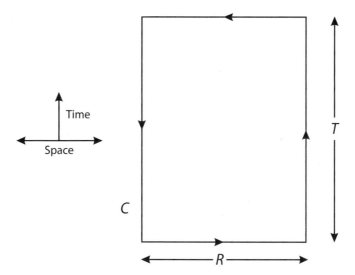

Figure VII.1.3

in figure VII.1.3. Note that for T large $\log\langle e^{i\oint_C dx^\mu A_\mu}\rangle \sim -iE(R)T$, which is essentially proportional to the perimeter length of the rectangle C.

As we will discuss in chapter VII.3 and as you have undoubtedly heard, the currently accepted theory of the strong interaction involves quarks coupled to a nonabelian Yang-Mills gauge potential A_μ. Thus, to determine the potential energy $E(R)$ between a quark and an antiquark held fixed at a distance R from each other we "merely" have to compute the expectation value of the Wilson loop

$$\langle W(C)\rangle = \frac{1}{\mathcal{Z}}\int \Pi dU e^{-(1/2f^2)\sum_P S(P)} W(C) \tag{20}$$

In lattice gauge theory we could compute $\log\langle W(C)\rangle$ for C the large rectangle in figure VII.1.3 numerically, and extract $E(R)$. (We lost the i because we are living in Euclidean spacetime for the purpose of this discussion.)

Quark confinement

You have also undoubtedly heard that since free quarks have not been observed, quarks are generally believed to be permanently confined. In particular, it is believed that the potential energy between a quark and an antiquark grows linearly with separation $E(R) \sim \sigma R$. One imagines a string tying the quark to the antiquark with a string tension σ. If this conjecture is correct, then $\log\langle W(C)\rangle \sim \sigma RT$ should go as the area RT enclosed by C. Wilson calls this behavior the area law, in contrast to the perimeter law characteristic of familiar theories such as electromagnetism. To prove the area law in Yang-Mills theory is one of the outstanding challenges of theoretical physics.

Exercises

VII.1.1 The gauge choice in the text preserves Lorentz invariance. It is often useful to choose a gauge that breaks Lorentz invariance, for example, $f(A) = n^\mu A_\mu(x)$ with n some fixed 4-vector. This class of gauge choices, known as the axial gauge, contains various popular gauges, each of which corresponds to a particular choice of n. For instance, in light-cone gauge, $n = (1, 0, 0, 1)$, in space-cone gauge, $n = (0, 1, i, 0)$. Show that for any given $A(x)$ we can find a gauge transformation so that $n \cdot A'(x) = 0$.

VII.1.2 Derive (17) and relate f to the coupling g in the continuum formulation of Yang-Mills theory. [Hint: Use the Baker-Campbell-Hausdorff formula

$$e^A e^B = e^{A+B+\frac{1}{2}[A,B]+\frac{1}{12}([A,[A,B]]+[B,[B,A]])+\cdots}$$

VII.1.3 Consider a lattice gauge theory in $(D + 1)$-dimensional space with the lattice spacing a in D-dimensional space and b in the extra dimension. Obtain the continuum D-dimensional field theory in the limit $a \to 0$ with b kept fixed.

VII.1.4 Study in (2) the alternative limit $b \to 0$ with a kept fixed so that you obtain a theory on a spatial lattice but with continuous time.

VII.1.5 Show that for lattice gauge theory the Wilson area law holds in the limit of strong coupling. [Hint: Expand (20) in powers of f^{-2}.]

VII.2 | Electroweak Unification

The scourge of massless spin 1 particles

With the benefit of hindsight, we now know that Nature likes Yang-Mills theory. In the late 1960s and early 1970s, the electromagnetic and weak interactions were unified into an electroweak interaction, described by a nonabelian gauge theory based on the group $SU(2) \otimes U(1)$. Somewhat later, in the early 1970s, it was realized that the strong interaction can be described by a nonabelian gauge theory based on the group $SU(3)$. Nature literally consists of a web of interacting Yang-Mills fields.

But when the theory was first proposed in 1954, it seemed to be totally inconsistent with observations as they were interpreted at that time. As Yang and Mills themselves pointed out in their paper, the theory contains massless spin 1 particles, which were certainly not known experimentally. Thus, except for interest on the part of a few theorists (Schwinger, Glashow, Bludman, and others) who found the mathematical structure elegantly attractive and felt that nonabelian gauge theory must somehow be relevant for the weak interaction, the theory gradually sank into oblivion and was not part of the standard graduate curriculum in particle physics in the 1960s.

Again with the benefit of hindsight, it would seem that there are only two logical solutions to the difficulty that experimentalists do not see any massless spin 1 particles except for the photon: (1) the Yang-Mills particles somehow acquire mass, or (2) the Yang-Mills particles are in fact massless but are somehow not observed. We now know that the first possibility was realized in the electroweak interaction and the second in the strong interaction.

Constructing the electroweak theory

We now discuss electroweak unification. It is perhaps pedagogically clearest to motivate how we would go about constructing such a theory. As I have said before, this is not a textbook on particle physics and I necessarily will have to keep the discussion of particle physics to the bare minimum. I gave you a brief introduction to the structure of the weak interaction in chapter IV.2. The other salient fact is that weak interaction violates parity, as mentioned in chapter II.1. In particular, the left handed electron field e_L and the right handed electron field e_R, which transform into each other under parity, enter into the weak interaction quite differently.

Let us start with the weak decay of the muon, $\mu^- \to e^- + \bar{\nu} + \nu'$, with ν and ν' the electron neutrino and muon neutrino, respectively. The relevant term in the Lagrangian is $\bar{\nu}'_L \gamma^\mu \mu_L \bar{e}_L \gamma_\mu \nu_L$, with the left hand electron field e_L, the electron neutrino field (which is left handed) ν_L, and so forth. The field μ_L annihilates a muon, the field $\bar{e}_L$ creates an electron, and so on. (Henceforth, we will suppress the word field.) As you probably know, the elementary constituents of matter form three families, with the first family consisting of ν, e, and the up u and down d quarks, the second of ν', μ, and the charm c and strange s quarks, and so on. For our purposes here, we will restrict our attention to the first family. Thus, we start with $\bar{\nu}_L \gamma^\mu e_L \bar{e}_L \gamma_\mu \nu_L$.

As I remarked in chapter III.2, a Fermi interaction of this type can be generated by the exchange of an intermediate vector boson W^+ with the coupling $W^+_\mu \bar{\nu}_L \gamma^\mu e_L + W^-_\mu \bar{e}_L \gamma_\mu \nu_L$.

The idea is then to consider an $SU(2)$ gauge theory with a triplet of gauge bosons denoted by W^a_μ, with $a = 1, 2, 3$. Put ν_L and e_L into the doublet representation and the right handed electron field e_R into a singlet representation, thus

$$\psi_L \equiv \begin{pmatrix} \nu \\ e \end{pmatrix}_L , \quad e_R \tag{1}$$

(The notation is such that the upper component of ψ_L is ν_L and the lower component is e_L.)

The fields ν_L and e_L, but not e_R, listen to the gauge bosons W^a_μ. Indeed, according to (IV.5.21) the Lagrangian contains

$$W^a_\mu \bar{\psi}_L \tau^a \gamma^\mu \psi_L = (W^{1-i2}_\mu \bar{\psi}_L \tfrac{1}{2} \tau^{1+i2} \gamma^\mu \psi_L + \text{h.c.}) + W^3_\mu \bar{\psi}_L \tau^3 \gamma^\mu \psi_L$$

where $W^{1-i2}_\mu \equiv W^1_\mu - i W^2_\mu$ and so forth. We recognize $\tau^{1+i2} \equiv \tau^1 + i\tau^2$ as the raising operator and the first two terms as $(W^{1-i2}_\mu \bar{\nu}_L \gamma^\mu e_L + \text{h.c.})$, precisely what we want. By design, the exchange of $W^\pm_\mu$ generates the desired term $\bar{\nu}_L \gamma^\mu e_L \bar{e}_L \gamma_\mu \nu_L$.

We need more room

We would hope that the boson W^3 we were forced to introduce would turn out to be the photon so that electromagnetism is included. But alas, W^3 couples to the current $\bar{\psi}_L \tau^3 \gamma^\mu \psi_L = (\bar{\nu}_L \gamma^\mu \nu_L - \bar{e}_L \gamma_\mu e_L)$, not the electromagnetic current $-(\bar{e}_L \gamma_\mu e_L + \bar{e}_R \gamma_\mu e_R)$. Oops!

Another problem lurks. To generate a mass term for the electron, we need a doublet Higgs field $\varphi \equiv \begin{pmatrix} \varphi^+ \\ \varphi^0 \end{pmatrix}$ in order to construct the $SU(2)$ invariant term $f\bar{\psi}_L \varphi e_R$ in the Lagrangian so that when φ acquires the vacuum expectation value $\begin{pmatrix} 0 \\ v \end{pmatrix}$ we will have

$$f\bar{\psi}_L\varphi e_R \to f(\bar{v},\bar{e})_L \begin{pmatrix} 0 \\ v \end{pmatrix} e_R = fv\bar{e}_L e_R \tag{2}$$

But none of the $SU(2)$ transformations leaves $\begin{pmatrix} 0 \\ v \end{pmatrix}$ invariant: The vacuum expectation value of φ spontaneously breaks the entire $SU(2)$ symmetry, leaving all three W bosons massive. There is no room for the photon in this failed theory. Aagh!

We need more room. Remarkably, we can avoid both the oops and the aagh by extending the gauge symmetry to $SU(2) \otimes U(1)$. Denoting the generator of $U(1)$ by $\frac{1}{2}Y$ (called the hypercharge) and the associated gauge potential by B_μ [and their counterparts T^a and W^a_μ for $SU(2)$] we have the covariant derivative $D_\mu = \partial_\mu - igW^a_\mu T^a - ig'B_\mu \frac{Y}{2}$. With four gauge bosons, we dare to hope that one of them might turn out to be the photon.

The gauge potentials are normalized by the corresponding kinetic energy terms, $\mathcal{L} = -\frac{1}{4}(B_{\mu\nu})^2 - \frac{1}{4}(W^a_{\mu\nu})^2 + \cdots$ with the abelian $B_{\mu\nu} = \partial_\mu B_\nu - \partial_\nu B_\mu$ and nonabelian field strength $W^a_{\mu\nu} = \partial_\mu W^a_\nu - \partial_\nu W^a_\mu + \varepsilon^{abc}W^b_\mu W^c_\nu$. The generators T^a are of course normalized by the commutation relations that define $SU(2)$. In contrast, there is no commutation relation in the abelian algebra $U(1)$ to fix the normalization of the generator $\frac{1}{2}Y$. Until this is fixed, the normalization of the $U(1)$ gauge coupling g' is not fixed.

How do we fix the normalization of the generator $\frac{1}{2}Y$? By construction, we want spontaneous symmetry breaking to leave a linear combination of T_3 and $\frac{1}{2}Y$ invariant, to be identified as the generator the massless photon couples to, namely the charge operator Q. Thus, we write

$$Q = T_3 + \frac{1}{2}Y \tag{3}$$

Once we know T_3 and $\frac{1}{2}Y$ of any field, this equation tells us its charge. For example, $Q(v_L) = \frac{1}{2} + \frac{1}{2}Y(v_L)$ and $Q(e_L) = -\frac{1}{2} + \frac{1}{2}Y(e_L)$. In particular, we see that the coefficient of T_3 in (3) must be 1 since the charges of v_L and e_L differ by 1. The relation (3) fixes the normalization of $\frac{1}{2}Y$.

Determining the hypercharge

The next step is to determine the hypercharge of various multiplets in the theory, which in turn determines how B_μ couples to these multiplets. Consider ψ_L. For e_L to have charge -1, the doublet ψ_L must have $\frac{1}{2}Y = -\frac{1}{2}$. In contrast, the field e_R has $\frac{1}{2}Y = -1$ since $T_3 = 0$ on e_R.

Given the hypercharge of ψ_L and e_R we see that the invariance of the term $f\bar{\psi}_L\varphi e_R$ under $SU(2) \otimes U(1)$ forces the Higgs field φ to have $\frac{1}{2}Y = +\frac{1}{2}$. Thus, according to (3) the upper component of φ has electric charge $Q = +\frac{1}{2} + \frac{1}{2} = +1$ and the lower component $Q = -\frac{1}{2} + \frac{1}{2} = 0$. Thus, we write $\varphi = \begin{pmatrix} \varphi^+ \\ \varphi^0 \end{pmatrix}$. Recall that φ has the vacuum expectation value $\begin{pmatrix} 0 \\ v \end{pmatrix}$. The fact that the electrically neutral field φ^0 acquires a vacuum expectation value but the charged field φ^+ does not provide a consistency check.

The theory works itself out

Now that the couplings of the gauge bosons to the various fields, in particular, the Higgs field, are determined, we can easily work out the mass spectrum of the gauge bosons, as indeed, let me remind you, you have already done in exercise IV.6.3!

Upon spontaneous symmetry breaking $\varphi \to (1/\sqrt{2})\binom{0}{v}$ (the normalization is conventional): We simply plug in

$$\mathcal{L} = (D_\mu \varphi)^\dagger (D^\mu \varphi) \to \frac{g^2 v^2}{4} W_\mu^+ W^{-\mu} + \frac{v^2}{8}(g W_\mu^3 - g' B_\mu)^2 \tag{4}$$

I trust that this is what you got! Thus, the linear combination $g W_\mu^3 - g' B_\mu$ becomes massive while the orthogonal combination remains massless and is identified with the photon. It is clearly convenient to define the angle θ by $\tan \theta = g'/g$. Then,

$$Z_\mu = \cos \theta W_\mu^3 - \sin \theta B_\mu \tag{5}$$

describes a massive gauge boson known as the Z boson, while the electromagnetic potential is given by $A_\mu = \sin \theta W_\mu^3 + \cos \theta B_\mu$. Combine (4) and (5) and verify that the mass squared of the Z boson is $M_Z^2 = v^2(g^2 + g'^2)/4$, and thus by elementary trigonometry obtain the relation

$$M_W = M_Z \cos \theta \tag{6}$$

The exchange of the W boson generates the Fermi weak interaction

$$\mathcal{L} = -\frac{g^2}{2M_W^2} \bar{\nu}_L \gamma^\mu e_L \bar{e}_L \gamma_\mu \nu_L = -\frac{4G}{\sqrt{2}} \bar{\nu}_L \gamma^\mu e_L \bar{e}_L \gamma_\mu \nu_L$$

where the second equality merely gives the historical definition of the Fermi coupling G. Thus,

$$\frac{G}{\sqrt{2}} = \frac{g^2}{8M_W^2} \tag{7}$$

Next, we write the relevant piece of the covariant derivative

$$g W_\mu^3 T^3 + g' B_\mu \frac{Y}{2} = g(\cos \theta Z_\mu + \sin \theta A_\mu)T^3 + g'(-\sin \theta Z_\mu + \cos \theta A_\mu)\frac{Y}{2}$$

in terms of the physically observed Z and A. The coefficient of A_μ works out to be $g \sin \theta T^3 + g' \cos \theta (Y/2) = g \sin \theta (T^3 + Y/2)$; the fact that the combination $Q = T^3 + Y/2$ emerges provides a nice check on the formalism. Furthermore, we obtain

$$e = g \sin \theta \tag{8}$$

Meanwhile, it is convenient to write $g \cos \theta T^3 - g' \sin \theta (Y/2)$, the coefficient of Z_μ in the covariant derivative, in terms of the physically familiar electric charge Q rather than the theoretical hypercharge Y: Thus,

$$g \cos \theta T^3 - g' \sin \theta (Q - T^3) = \frac{g}{\cos \theta}(T^3 - \sin^2 \theta Q)$$

In other words, we have determined the coupling of the Z boson to an arbitrary fermion field Ψ in the theory:

$$\mathcal{L} = \frac{g}{\cos\theta} Z_\mu \bar{\Psi} \gamma^\mu (T^3 - \sin^2\theta Q)\Psi \tag{9}$$

For example, using (9) we can immediately write the coupling of Z to leptons:

$$\mathcal{L} = \frac{g}{\cos\theta} Z_\mu[\tfrac{1}{2}(\bar{\nu}_L\gamma^\mu\nu_L - \bar{e}_L\gamma^\mu e_L) + \sin^2\theta\,\bar{e}\gamma^\mu e] \tag{10}$$

Including quarks

How to include the hadrons is now almost self evident. Given that only left handed fields participate in the weak interaction, we put the quarks of the first generation into $SU(2) \otimes U(1)$ multiplets as follows:

$$q_L^\alpha \equiv \begin{pmatrix} u^\alpha \\ d^\alpha \end{pmatrix}_L, \quad u_R^\alpha, d_R^\alpha \tag{11}$$

where $\alpha = 1, 2, 3$ denotes the color index, which I will discuss in the next chapter. The right handed quarks u_R^α and d_R^α are put into singlets so that they do not hear the weak bosons W^a. Recall that the up quark u and the down quark d have electric charges $\frac{2}{3}$ and $-\frac{1}{3}$ respectively. Referring to (3) we see $\frac{1}{2}Y = \frac{1}{6}, \frac{2}{3}$, and $-\frac{1}{3}$ for q_L^α, u_R^α, and d_R^α, respectively. From (9) we can immediately read off the coupling of the Z boson to the quarks:

$$\mathcal{L} = \frac{g}{\cos\theta} Z_\mu[\tfrac{1}{2}(\bar{u}_L\gamma^\mu u_L - \bar{d}_L\gamma^\mu d_L) - \sin^2\theta J_{em}^\mu] \tag{12}$$

Finally, I leave it to you to verify that of the four degrees of freedom contained in φ (since φ^+ and φ^0 are complex) three are eaten by the W and Z bosons, leaving one physical degree of freedom H corresponding to the elusive Higgs particle that experimenters are still searching for as of this writing.

The neutral current

By virtue of its elegantly economical gauge group structure, this $SU(2) \otimes U(1)$ electroweak theory of Glashow, Salam, and Weinberg ushered in the last great predictive era of theoretical particle physics. Writing (10) and (12) as

$$\mathcal{L} = \frac{g}{\cos\theta} Z_\mu (J_{\text{leptons}}^\mu + J_{\text{quarks}}^\mu)$$

and using (6) we see that Z boson exchange generates a hitherto unknown neutral current interaction

$$\mathcal{L}_{\text{neutral current}} = -\frac{g^2}{2M_W^2} (J_{\text{leptons}} + J_{\text{quarks}})^\mu (J_{\text{leptons}} + J_{\text{quarks}})_\mu$$

between leptons and quarks. By studying various processes described by $\mathcal{L}_{\text{neutral current}}$ we can determine the weak angle θ. Once θ is determined, we can predict g from (8). Once g

is determined, we can predict M_W from (7). Once M_W is determined, we can predict M_Z from (6).

Concluding remarks

As I mentioned, there are three families of leptons and quarks in Nature, consisting of (ν_e, e, u, d), (ν_μ, μ, c, s), and (ν_τ, τ, t, b). The appearance of this repetitive family structure, about which the $SU(2) \otimes U(1)$ theory has nothing to say, represents one of the great unsolved puzzles of particle physics. The three families, with the appropriate rotation angles between them, are simply incorporated into the theory by repeating what we wrote above.

A more logical approach than the one given here would be to start with an $SU(2) \otimes U(1)$ theory with a doublet Higgs field with some hypercharge, and to say, "Behold, upon spontaneous symmetry breaking, one linear combination of generators remains unbroken with a corresponding massless gauge field." I think that our quasi-historical approach is clearer.

As I have mentioned on several occasions, Fermi's theory of the weak interaction is nonrenormalizable. In 1999, 't Hooft and Veltman were awarded the Nobel Prize for showing that the $SU(2) \otimes U(1)$ electroweak theory is renormalizable, thus paving the way for the triumph of nonabelian gauge theories in describing the strong, electromagnetic, and weak interactions. I cannot go into the details of their proof here, but I would like to mention that the key is to start with the nonabelian analog of the unitary gauge (recall chapter IV.6) and proceed to the R_ξ gauge. At large momenta, the massive gauge boson propagators go as $\sim (k_\mu k_\nu / k^2)$ in the unitary gauge, but as $\sim (1/k^2)$ in the R_ξ gauge. The theory is then renormalizable by power counting.

Exercises

VII.2.1 Unfortunately, the mass of the elusive Higgs particle H depends on the parameters in the double well potential $V = -\mu^2 \varphi^\dagger \varphi + \lambda (\varphi^\dagger \varphi)^2$ responsible for the spontaneous symmetry breaking. Assuming that H is massive enough to decay into $W^+ + W^-$ and $Z + Z$, determine the rates for H to decay into various modes.

VII.2.2 Show that it is possible to stay with the $SU(2)$ gauge group and to identify W^3 as the photon A, but at the cost of inventing some experimentally unobserved lepton fields. This theory does not describe our world: For one thing, it is essentially impossible to incorporate the quarks. Show this! [Hint: We have to put the leptons into a triplet of $SU(2)$ instead of a doublet.]

VII.3 | Quantum Chromodynamics

Quarks

Quarks come in six flavors, known as up, down, strange, charm, bottom, and top, denoted by u, d, s, c, b, and t. The proton, for example, is made of two up quarks and a down quark $\sim (uud)$, while the neutral pion corresponds to $\sim (u\bar{u} - d\bar{d})/\sqrt{2}$. Please consult any text on particle physics for details.

By the late 1960s the notion of quarks was gaining wide acceptance, but two separate lines of evidence indicated that a crucial element was missing. In studying how hadrons are made of quarks, people realized that the wave function of the quarks in a nucleon does not come out to be antisymmetric under the interchange of any pair of quarks, as required by the Pauli exclusion principle. At around the same time, it was realized that in the ideal world we used to derive the Goldberger-Treiman relation and in which the pion is massless we can calculate the decay rate for the process $\pi^0 \to \gamma + \gamma$, as mentioned in chapter IV.7. Puzzlingly enough, the calculated rate came out smaller than the observed rate by a factor of $9 = 3^2$.

Both puzzles could be resolved in one stroke by having quarks carry a hitherto unknown internal degree of freedom that Gell-Mann called color. For any specified flavor, a quark comes in one of three colors. Thus, the up quark can be red, blue, or yellow. In a nucleon, the wave function of the three quarks will then contain a factor referring to color, besides the factors referring to orbital motion, spin, and so on. We merely have to make the color part of the wave function antisymmetric; in fact, we simply take it to be $\varepsilon^{\alpha\beta\gamma}$, where α, β, and γ denote the colors carried by the three quarks. With quarks in three colors, we have to multiply the amplitude for π^0 decay by a factor of 3, thus neatly resolving the discrepancy between theory and experiment.

Asymptotic freedom

As I mentioned in chapter VI.6, the essential clue came from studying deep inelastic scattering of electrons off nucleons. Experimentalists made the intriguing discovery that when hit hard the quarks in the nucleons act as if they hardly interact with each other, in other words, as if they are free. On the other hand, since quarks are never seen as isolated entities, they appear to be tightly bound to each other within the nucleon. As I have explained, this puzzling and apparently contradictory behavior of the quarks can be understood if the strong interaction coupling flows to zero in the large momentum (ultraviolet) limit and to infinity or at least to some large value in the small momentum (infrared) limit. A number of theorists proposed searching for theories whose couplings would flow to zero in the ultraviolet limit, now known as asymptotic free theories. Eventually, Gross, Wilczek, and Politzer discovered that Yang-Mills theory is asymptotically free.

This result dovetails perfectly with the realization that quarks carry color. The nonabelian gauge transformation would take a quark of one color into a quark of another color. Thus, to write down the theory of the strong interaction we simply take the result of exercise IV.5.6,

$$\mathcal{L} = -\frac{1}{4g^2} F_{\mu\nu}^a F^{a\mu\nu} + \bar{q}(i\gamma^\mu D_\mu - m)q \tag{1}$$

with the covariant derivative $D_\mu = \partial_\mu - iA_\mu$. The gauge group is $SU(3)$ with the quark field q in the fundamental representation. In other words, the gauge fields $A_\mu = A_\mu^a T^a$, where T^a $(a = 1, \ldots, 8)$ are traceless hermitean 3 by 3 matrices. Explicitly, $(A_\mu q)^\alpha = A_\mu^a (T^a)^\alpha_\beta q^\beta$, where $\alpha, \beta = 1, 2, 3$. The theory is known as quantum chromodynamics, or QCD for short, and the nonabelian gauge bosons are known as gluons. To incorporate flavor, we simply write $\sum_{j=1}^f \bar{q}_j (i\gamma^\mu D_\mu - m_j)q_j$ for the second term in (1), where the index j goes over the f flavors. Note that quarks of different flavors have different masses.

Infrared slavery

The flip side of asymptotic freedom is infrared slavery. We cannot follow the renormalization group flow all the way down to the low momentum scale characteristic of the quarks bound inside hadrons since the coupling g becomes ever stronger and our perturbative calculation of $\beta(g)$ is no longer adequate. Nevertheless, it is plausible although never proven that g goes to infinity and that the gluons keep the quarks and themselves in permanent confinement. The Wilson loop introduced in chapter VII.1 provides the order parameter for confinement.

In elementary physics forces decrease with the separation between interacting objects, so permanent confinement is a rather bizarre concept. Are there any other instances of permanent confinement?

Consider a magnetic monopole in a superconductor. We get to combine what we learned in chapters IV.4 and V.4 (and even VI.2)! A quantized amount of magnetic flux comes out of the monopole, but according to the Meissner effect a superconductor expels magnetic flux. Thus, a single magnetic monopole cannot live inside a superconductor.

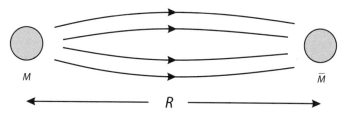

Figure VII.3.1

Now consider an antimonopole a distance R away (figure VII.3.1). The magnetic flux coming out of the monopole can go into the antimonopole, forming a tube connecting the monopole and the antimonopole and obliging the superconductor to give up being a superconductor in the region of the flux tube. In the language of chapter V.4, it is no longer energetically favorable for the field or order parameter φ to be constant everywhere; instead it vanishes in the region of the flux tube. The energy cost of this arrangement evidently grows as R (consistent with Wilson's area law).

In other words, an experimentalist living inside a superconductor would find that it costs more and more energy to pull a monopole and an antimonopole apart. This confinement of monopoles inside a superconductor is often taken to be a model of the yet-to-be-proven confinement of quarks. Invoking electromagnetic duality we can imagine a magnetic superconductor in contrast to the usual electric superconductor. Inside a magnetic superconductor, electric charges would be permanently confined. Our universe may be likened to a color magnetic superconductor in which quarks (the analog of electric charges) are confined.

On distance scales large compared to the radius of the color flux tube connecting a quark to an antiquark, the tube can be thought of as a string. Historically, that was how string theory originated. The challenge, boys and girls, is to prove that the ground state or vacuum of (1) is a color magnetic superconductor.

Symmetries of the strong interaction

Now that we have a theory of the strong interaction, we can understand the origin of the symmetries of the strong interaction, namely the isospin symmetry of Heisenberg and the chiral symmetry that when spontaneously broken leads to the appearance of the pion as a Nambu-Goldstone boson (as discussed in chapters IV.2 and VI.4).

Consider a world with two flavors, which is all that is relevant for a discussion of the pion. Introduce the notation $u \equiv q_1, d \equiv q_2$, and $q = \binom{u}{d}$ so that we can write the Lagrangian as

$$\mathcal{L} = -\frac{1}{4g^2} F^a_{\mu\nu} F^{a\mu\nu} + \bar{q}(i\gamma^\mu D_\mu - m)q$$

with

$$m = \begin{pmatrix} m_u & 0 \\ 0 & m_d \end{pmatrix}$$

where m_u and m_d are the masses of the up and down quarks, respectively. If $m_u = m_d$, the Lagrangian is invariant under $q \to e^{i\theta \cdot \tau} q$, corresponding to Heisenberg's· isospin symmetry.

In the limit in which m_u and m_d vanish, the Lagrangian is invariant under $q \to e^{i\varphi \cdot \tau \gamma_5} q$, known as the chiral $SU(2)$ symmetry, chiral because the right handed quarks q_R and the left handed quarks q_L transform differently. To the extent that m_u and m_d are both much smaller than the energy scale of the strong interaction, chiral $SU(2)$ is an approximate symmetry.

The pion is the Nambu-Goldstone boson associated with the spontaneous breaking of the chiral $SU(2)$. Indeed, this is an example of dynamical symmetry breaking since there is no elementary scalar field around to acquire a vacuum expectation. Instead, the strong interaction dynamics is supposed to drive the composite scalar fields $\bar{u}u$ and $\bar{d}d$ to "condense into the vacuum" so that $\langle 0| \bar{u}u |0 \rangle = \langle 0| \bar{d}d |0 \rangle$ become nonvanishing, where the equality between the two vacuum expectation values ensures that Heisenberg's isospin is not spontaneously broken, an experimental fact since there are no corresponding Nambu-Goldstone bosons. In terms of the doublet field q, the QCD vacuum is supposed to be such that $\langle 0| \bar{q}q |0 \rangle \neq 0$ while $\langle 0| \bar{q}\vec{\tau}q |0 \rangle = 0$.

Renormalization group flow

The renormalization group flow of the QCD coupling is governed by

$$\frac{dg}{dt} = \beta(g) = -\frac{11}{3} T_2(G) \frac{g^3}{16\pi^2} \tag{2}$$

with the all-crucial minus sign. Here

$$T_2(G)\delta^{ab} = f^{acd} f^{bcd} \tag{3}$$

I will not go through the calculation of $\beta(g)$ here, but having mastered chapters VI.8 and VII.1 you should feel that you can do it if you want to.[1] At the very least, you should understand the factor g^3 and $T_2(G)$ by drawing the relevant Feynman diagrams.

When fermions are included,

$$\frac{dg}{dt} = \beta(g) = \left[-\frac{11}{3} T_2(G) + \frac{4}{3} T_2(F) \right] \frac{g^3}{16\pi^2} \tag{4}$$

where

$$T_2(F)\delta^{ab} = \text{tr}[T^a(F)T^b(F)] \tag{5}$$

I do expect you to derive (4) given (2). For $SU(N)$ $T_2(F) = \frac{1}{2}$ for each fermion in the fundamental representation. Note that asymptotic freedom is lost when there are too many fermions.

---

[1] For a detailed calculation, see, e.g., S. Weinberg, *The Quantum Theory of Fields*, Vol. 2, sec. 18.7.

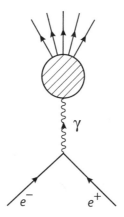

Figure VII.3.2

You already solved an equation like (4) in exercise VI.8.1. Let us define, in analogy to quantum electrodynamics, $\alpha_S(\mu) \equiv g(\mu)^2/4\pi$, the strong coupling at the momentum scale μ. From (4) we obtain[2]

$$\alpha_S(Q) = \frac{\alpha_S(\mu)}{1 + (1/4\pi)(11 - \frac{2}{3}n_f)\alpha_S(\mu) \log(Q^2/\mu^2)} \tag{6}$$

showing explicitly that $\alpha_S(Q) \to 0$ logarithmically as $Q \to \infty$.

Electron-positron annihilation

I have space to show you only one physical application. Experimentalists have measured the cross section σ of e^+e^- annihilation into hadrons as a function of the total center-of-mass energy E. The amplitude is shown in figure VII.3.2. To calculate the cross section in terms of the amplitude, we have to go through what some people call "boring kinematics," such as normalizing everything correctly, dividing by the flux of the two beams, and so forth (see the appendix to chapter II.6). For the good of your soul, you should certainly go through this type of calculation at least once. Believe me, I did it more times than I care to remember. But happily, as I will now show you, we can avoid most of this grunge labor. First, consider the ratio

$$R(E) \equiv \frac{\sigma(e^+e^- \to \text{hadrons})}{\sigma(e^+e^- \to \mu^+\mu^-)}$$

The kinematic stuff cancels out. In figure VII.3.2 the half of the diagram involving the electron positron lines and the photon propagator also appears in the Feynman diagram $e^+e^- \to \mu^+\mu^-$ (figure VII.3.3) and so cancels out in $R(E)$. The blob in figure VII.3.2, which hides all the complexity of the strong interaction, is given by $\langle 0| J^\mu(0) |h\rangle$, where J^μ is the

---

[2] For the accumulated experimental evidence on $\alpha_S(Q)$, see figure 14.3 in F. Wilczek, in: V. Fitch et al., eds., *Critical Problems in Physics*, p. 281.

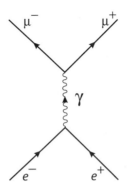

Figure VII.3.3

electromagnetic current and the state $|h\rangle$ can contain any number of hadrons. To obtain the cross section we have to square the amplitude, include a δ-function for momentum conservation, and sum over all $|h\rangle$, thus arriving at

$$\sum_h (2\pi)^4 \delta^4(p_h - p_{e^+} - p_{e^-}) \langle 0| J^\mu(0) |h\rangle \langle h| J^\nu(0) |0\rangle \tag{7}$$

[with $q \equiv p_{e^+} + p_{e^-} = (E, \vec{0})$]. This quantity can be written as

$$\int d^4x e^{iqx} \langle 0| J^\mu(x) J^\nu(0) |0\rangle = \int d^4x e^{iqx} \langle 0| [J^\mu(x), J^\nu(0)] |0\rangle$$

$$= 2 \operatorname{Im}(i \int d^4x e^{iqx} \langle 0| T J^\mu(x) J^\nu(0) |0\rangle)$$

(The first equality follows from $E > 0$ and the second was explained in chapter III.8.) To determine this quantity, we would have to calculate an infinite number of Feynman diagrams involving lots of quarks and gluons. A typical diagram is shown in figure VII.3.4. Completely hopeless!

This is where asymptotic freedom rides to the rescue! From chapter VI.7 you learned that for a process at energy E the appropriate coupling strength to use is $g(E)$. But as we crank up E, $g(E)$ gets smaller and smaller. Thus diagrams such as figure VII.3.4 involving many powers of $g(E)$ all fall away, leaving us with the diagrams with no power of $g(E)$ (fig. VII.3.5a) and two powers of $g(E)$ (figs. VII.3.5b,c,d). No calculation is necessary to obtain the leading term in $R(E)$, since the diagram in figure VII.3.5a is the same one that enters into $e^+e^- \to \mu^+\mu^-$: We merely replace the quark propagator by the muon propagator (quark and muon masses are negligible compared to E). At high energy, the quarks are free and $R(E)$ merely counts the square of the charge Q_a of the various quarks contributing at that energy. We predict

$$R(E) \underset{E \to \infty}{\longrightarrow} 3 \sum_a Q_a^2 \tag{8}$$

The factor of 3 accounts for color.

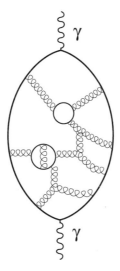

Figure VII.3.4

Not only does QCD turn itself off at high energies, it tells us how fast it is turning itself off. Thus, we can determine how the limit in (8) is approached:

$$R(E) = \left(3 \sum_a Q_a^2\right) \left(1 + C \frac{2}{(11 - \frac{2}{3} n_f) \log(E/\mu)} + \cdots\right) \tag{9}$$

I will leave it to you to calculate C.

Dreams of exact solubility

An analytic solution of quantum chromodynamics is something of a "Holy Grail" for field theorists (a grail that now carries a prize of one million dollars: see www.ams.org/claymath/). Many field theorists have dreamed that at least "pure" QCD, that is QCD without quarks, might be exactly soluble. After all, if any 4-dimensional quantum field

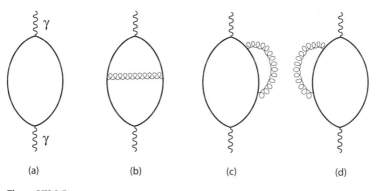

(a)   (b)   (c)   (d)

Figure VII.3.5

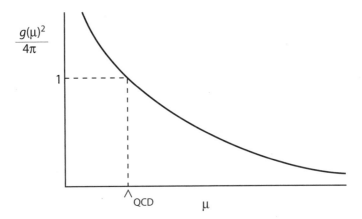

Figure VII.3.6

theory turns out to be exactly soluble, pure Yang-Mills, with all its fabulous symmetries, is the most likely possibility. (Perhaps an even more likely candidate for solubility is supersymmetric Yang-Mills theory. We will touch on supersymmetry in chapter VIII.4.)

Let me be specific about what it means to solve QCD. Consider a world with only up and down quarks with m_u and m_d both set equal to zero, namely a world described by

$$\mathcal{L} = -\frac{1}{4g^2} F^a_{\mu\nu} F^{a\mu\nu} + \bar{q} i \gamma^\mu D_\mu q \tag{10}$$

The goal would be to calculate something like the ratio of the mass of the ρ meson m_ρ to the mass of the proton m_P.

To make progress, theoretical physicists typically need to have a small parameter to expand in, but in trying to solve (10) we are confronted with the immediate difficulty that there is no such parameter. You might think that g is a parameter, but you would be mistaken. The renormalization group analysis taught us that $g(\mu)$ is a function of the energy scale μ at which it is measured. Thus, there is no particular dimensionless number we can point to and say that it measures the strength of QCD. Instead, the best we can do is to point to the value of μ at which $(g(\mu)^2/4\pi)$ becomes of order 1. This is the energy, known as Λ_{QCD}, at which the strong interaction becomes strong as we come down from high energy (fig. VII.3.6). But Λ_{QCD} merely sets the scale against which other quantities are to be measured. In other words, if you manage to calculate m_P it better come out proportional to Λ_{QCD} since Λ_{QCD} is the only quantity with dimension of mass around. Similarly for m_ρ. Put in precise terms, if you publish a paper with a formula giving m_ρ/m_P in terms of pure numbers such as 2 and π, the field theory community will hail you as a conquering hero who has solved QCD exactly.

The apparent trade of a dimensionless coupling g for a dimensional mass scale Λ_{QCD} is known as dimensional transmutation, of which we will see another example in the next chapter.

Exercises

VII.3.1 Calculate C in (9). [Hint: If you need help, consult T. Appelquist and H. Georgi, *Phys. Rev.* D8: 4000, 1973; and A. Zee, *Phys. Rev.* D8: 4038, 1973.]

VII.3.2 Calculate (2).

Inventing an expansion parameter

Quantum chromodynamics is a zero-parameter theory, so it is difficult to give even a first approximation. In desperation, field theorists invented a parameter in which to expand QCD. Suppose instead of three colors we have N colors. 't Hooft[1] noticed that as $N \to \infty$ remarkable simplifications occur. The idea is that if we can calculate m_ρ/m_P, for example, in the large N limit the result may be close to the actual value. People sometimes joke that particle physicists regard 3 as a large number, but actually the correction to the large N limit is typically of order $1/N^2$, about 10% in the real world. Particle physicists would be more than happy to be able to calculate hadron masses to this degree of accuracy.

As with spontaneous symmetry breaking and a number of other important concepts, the large N expansion came out of condensed matter physics but nowadays is used routinely in all sorts of contexts. For example, people have tried a large N approach to solve high-temperature superconductivity and to fold RNA.[2]

Scaling the QCD coupling

So, let the color group be $U(N)$ and write

$$\mathcal{L} = -\frac{N^a}{2g^2} \operatorname{tr} F_{\mu\nu} F^{\mu\nu} + \bar{\psi}[i(\not{\partial} - i \not{A}) - m]\psi \tag{1}$$

Note that we have replaced g^2 by g^2/N^a. For finite N this change has no essential significance. The point is to choose the power a so that interesting simplifications occur in the limit $N \to \infty$ with g^2 held fixed. The cubic and quartic interaction vertices of the gluons

---

[1] G. 't Hooft, *Under the Spell of the Gauge Principle*, p. 378.

[2] M. Bon, G. Vernizzi, H. Orland, and A. Zee, "Topological classification of RNA structures," *J. Mol. Biol.* 379:900, 2008.

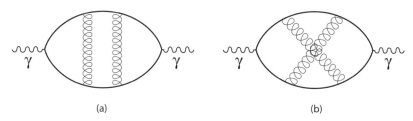

Figure VII.4.1

are proportional to N^a. On the other hand, since the gluon propagator goes as the inverse of the quadratic terms in $\mathcal{L}$, it is proportional to $1/N^a$. The coupling of the gluon to the quark does not depend on N.

To fix a, let us focus on a specific application, the calculation of $\sigma(e^+e^- \to \text{hadrons})$ discussed in the last chapter. Suppose we want to calculate this cross section at low energies. Consider the two-gluon exchange diagrams shown in figures VII.4.1a and b. The two diagrams are of order g^4 and we would have to calculate both. Note that 1b is nonplanar: Since one gluon crosses over the other, the diagram cannot be drawn on the plane if we insist that lines cannot go through each other.

Now the double-line formalism introduced in chapter IV.5 shines. In this formalism the diagrams figure VII.4.1a and b are redrawn as in figure VII.4.2a and b. The two gluon propagators common to both diagrams give a factor $1/N^{2a}$. Now comes the punchline. We sum over three independent color indices in 2a, thus getting a factor N^3. Grab some crayons and try to color each line in 2a with a different color: you will need three crayons. In contrast, we sum over only one independent index in 2b, getting only a factor of N. In other words, 2a dominates 2b by a factor N^2. In the large N limit we can throw 2b away.

Clearly, the rule is to associate one factor of N with each loop. Thus, the lowest order diagram, shown in 2c, with N different colors circulating in it, scales as N; 2a scales as N^3/N^{2a}. We want 2a and 2c to scale in the same way and thus we choose $a = 1$.

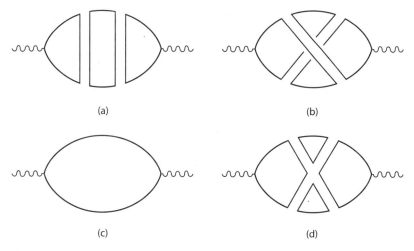

Figure VII.4.2

By drawing more diagrams [e.g., 2d scales as $N(1/N^4)N^4$, with the three factors coming from the quartic coupling, the propagators, and the sum over colors, respectively], you can convince yourself that planar diagrams dominate in the large N limit, all scaling as N. For a challenge, try to prove it. Evidently, there is a topological flavor to all this.

The reduction to planar diagrams is a vast simplification but there are still an infinite number of diagrams. At this stage in our mastery of field theory, we still can't solve large N QCD. (As I started writing this book, there were tantalizing clues, based on insight and techniques developed in string theory, that a solution of large N QCD might be within sight. As I now go through the final revision, that hope has faded.)

The double-line formalism has a natural interpretation. Group theoretically, the matrix gauge potential A^i_j transforms just like $\bar{q}^i q_j$ (but assuredly we are not saying that the gluon is a quark-antiquark bound state) and the two lines may be thought of as describing a quark and an antiquark propagating along, with the arrows showing the direction in which color is flowing.

Random matrix theory

There is a much simpler theory, structurally similar to large N QCD, that actually can be solved. I am referring to random matrix theory.

Exaggerating a bit, we can say that quantum mechanics consists of writing down a matrix known as the Hamiltonian and then finding its eigenvalues and eigenvectors. In the early 1950s, when confronted with the problem of studying the properties of complicated atomic nuclei, Eugene Wigner proposed that instead of solving the true Hamiltonian in some dubious approximation we might generate large matrices randomly and study the distribution of the eigenvalues—a sort of statistical quantum mechanics. Random matrix theory has since become a rich and flourishing subject, with an enormous and growing literature and applications to numerous areas of theoretical physics and even to pure mathematics (such as operator algebra and number theory.)[3] It has obvious applications to disordered condensed matter systems and less obvious applications to random surfaces and hence even to string theory. Here I will content myself with showing how 't Hooft's observation about planar diagrams works in the context of random matrix theory.

Let us generate N by N hermitean matrices φ randomly according to the probability

$$P(\varphi) = \frac{1}{Z} e^{-N \, \mathrm{tr} \, V(\varphi)} \tag{2}$$

with $V(\varphi)$ a polynomial in φ. For example, let $V(\varphi) = \frac{1}{2}m^2\varphi^2 + g\varphi^4$. The normalization $\int d\varphi \, P(\varphi) = 1$ fixes

$$Z = \int d\varphi \, e^{-N \, \mathrm{tr} \, V(\varphi)} \tag{3}$$

The limit $N \to \infty$ is always understood.

---

[3] For a glimpse of the mathematical literature, see D. Voiculescu, ed., *Free Probability Theory*.

As in chapter VI.7 we are interested in $\rho(E)$, the density of eigenvalues of φ. To make sure that you understand what is actually meant, let me describe what we would do were we to evaluate $\rho(E)$ numerically. For some large integer N, we would ask the computer to generate a hermitean matrix φ with the probability $P(\varphi)$ and then to solve the eigenvalue equation $\varphi v = E v$. After this procedure had been repeated many times, the computer could plot the distribution of eigenvalues in a histogram that eventually approaches a smooth curve, called the density of eigenvalues $\rho(E)$.

We already developed the formalism to compute $\rho(E)$ in (VI.7.1): Compute the real analytic function $G(z) \equiv \langle (1/N) \, \text{tr}[1/(z - \varphi)] \rangle$ and $\rho(E) = -(1/\pi) \lim_{\varepsilon \to 0} \text{Im} \, G(E + i\varepsilon)$. The average $\langle \cdots \rangle$ is taken with the probability $P(\varphi)$:

$$\langle O(\varphi) \rangle = \frac{1}{Z} \int D\varphi \, e^{-N \text{tr} V(\varphi)} O(\varphi)$$

You see that my choice of notation, φ for the matrix and $V(\varphi) = \frac{1}{2} m^2 \varphi^2 + g \varphi^4$ as an example, is meant to be provocative. The evaluation of Z is just like the evaluation of a path integral, but for an action $S(\varphi) = N \, \text{tr} \, V(\varphi)$ that does not involve $\int d^d x$. Random matrix theory can be thought of as a quantum field theory in $(0 + 0)$-dimensional spacetime!

Various field theoretic methods, such as Feynman diagrams, can all be applied to random matrix theory. But life is sweet in $(0 + 0)$-dimensional spacetime: There is no space, no time, no energy, and no momentum and hence no integral to do in evaluating Feynman diagrams.

The Wigner semicircle law

Let us see how this works for the simple case $V(\varphi) = \frac{1}{2} m^2 \varphi^2$ (we can always absorb m into φ but we won't). Instead of $G(z)$, it is slightly easier to calculate

$$G^i_j(z) \equiv \left\langle \left(\frac{1}{z - \varphi} \right)^i_j \right\rangle = \delta^i_j G(z)$$

The last equality follows from invariance under unitary transformations:

$$P(\varphi) = P(U^\dagger \varphi U) \tag{4}$$

Expand

$$G^i_j(z) = \sum_{n=0}^{\infty} \frac{1}{z^{2n+1}} \langle (\varphi^{2n})^i_j \rangle \tag{5}$$

Do the Gaussian integral

$$\frac{1}{Z} \int d\varphi \, e^{-N \, \text{tr} \, \frac{1}{2} m^2 \varphi^2} \varphi^i_k \varphi^l_j = \frac{1}{Z} \int d\varphi \, e^{-N \frac{1}{2} m^2 \sum_{p,q} \varphi^p_q \varphi^q_p} \varphi^i_k \varphi^l_j = \delta^i_j \delta^l_k \frac{1}{N m^2} \tag{6}$$

Setting $k = l$ and summing, we find the $n = 1$ term in (5) is equal to $(1/z^3) \delta^i_j (1/m^2)$.

Just as in any field theory we can associate a Feynman diagram with each of the terms in (5). For the $n = 1$ term, we have figure VII.4.3. The matrix character of φ lends itself naturally to 't Hooft's double-line formalism and thus we can speak of quark and gluon

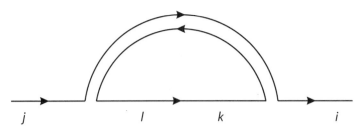

Figure VII.4.3

propagators with a good deal of ease. The Feynman rules are given in figure VII.4.4. We recognize φ as the gluon field and (5) as the gluon propagator. Indeed, we can formulate our problem as follows: Given the bare quark propagator $1/z$, compute the true quark propagator $G(z)$ with all interaction effects taken into account.

Let us now look at the $n = 2$ term in (5) $1/z^5 < \varphi_h^i \varphi_k^h \varphi_l^k \varphi_j^l >$, which we represent in figure VII.4.5a. With a bit of thought you can see that the index i can be contracted with k, l, or j, thus giving rise to figures VII.4.5b, c, d. Summing over color indices, just as in QCD, we see that the planar diagrams in 5b and 5d dominate the diagram in 5c by a factor N^2. We can take over 't Hooft's observation that planar diagrams dominate.

Incidentally, in this example, you see how large N is essential, allowing us to get rid of nonplanar diagrams. After all, if I ask you to calculate the density of eigenvalues for say $N = 7$ you would of course protest saying that the general formula for solving a degree-7 polynomial equation is not even known.

The simple example in figure VII.4.5 already indicates how all possible diagrams could be constructed. In 5b the same "unit" is repeated, while in 5d the same "unit" is nested inside a more basic diagram. A more complicated example is shown in 5e. You can convince yourself that for $N = \infty$ all diagrams contributing to $G(z)$ can be generated by either "nesting" existing diagrams inside an overarching gluon propagator or "repeating" an

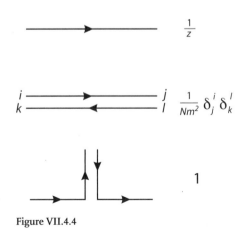

Figure VII.4.4

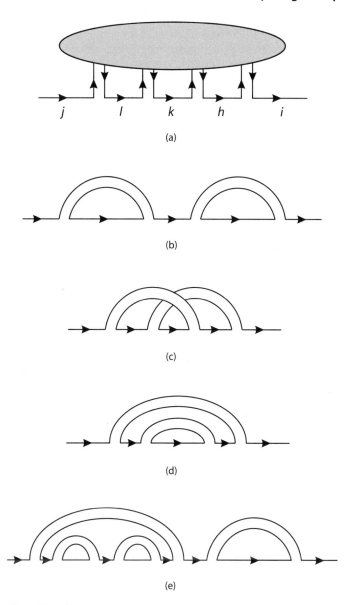

j l k h i

(a)

(b)

(c)

(d)

(e)

Figure VII.4.5

existing structure over and over again. Translate the preceding sentence into two equations: "Repeat" (see figure VII.4.6a),

$$G(z) = \frac{1}{z} + \frac{1}{z}\Sigma(z)\frac{1}{z} + \frac{1}{z}\Sigma(z)\frac{1}{z}\Sigma(z)\frac{1}{z} + \cdots$$
$$= \frac{1}{z - \Sigma(z)} \tag{7}$$

and "nest" (see figure VII.4.6b),

$$\Sigma(z) = \frac{1}{m^2}G(z) \tag{8}$$

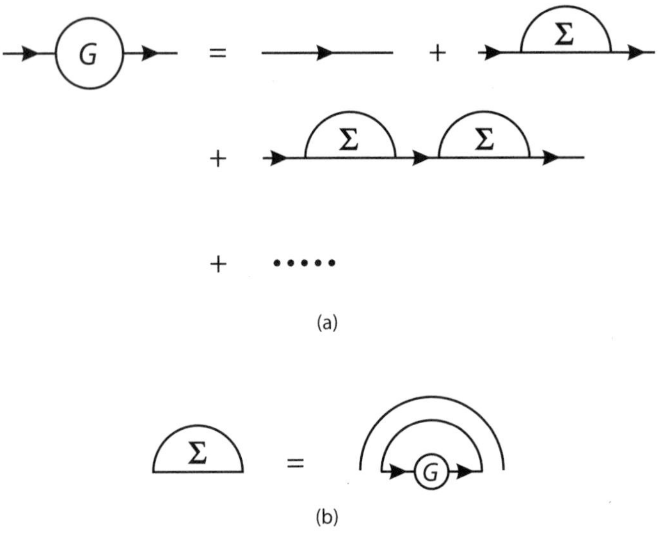

(a)

(b)

Figure VII.4.6

Combining these two equations we obtain a simple quadratic equation for $G(z)$ that we can immediately solve to obtain

$$G(z) = \frac{m^2}{2}\left(z - \sqrt{z^2 - \frac{4}{m^2}}\right) \tag{9}$$

(From the definition of $G(z)$ we see that $G(z) \to 1/z$ for large z and thus we choose the negative root.) We immediately deduce that

$$\rho(E) = \frac{2}{\pi a^2}\sqrt{a^2 - E^2} \tag{10}$$

where $a^2 = 4/m^2$. This is a famous result known as Wigner's semicircle law.

The Dyson gas

I hope that you are struck by the elegance of the large N planar diagram approach. But you might have also noticed that the gluons do not interact. It is as if we have solved quantum electrodynamics while we have to solve quantum chromodynamics. What if we have to deal with $V(\varphi) = \frac{1}{2}m^2\varphi^2 + g\varphi^4$? The $g\varphi^4$ term causes the gluons to interact with each other, generating horrible diagrams such as the one in figure VII.4.7. Clearly, diagrams proliferate and as far as I know nobody has ever been able to calculate $G(z)$ using the Feynman diagram approach.

Happily, $G(z)$ can be evaluated using another method known as the Dyson gas approach. The key is to write

$$\varphi = U^\dagger \Lambda U \tag{11}$$

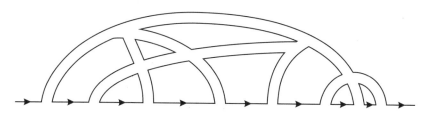

Figure VII.4.7

where Λ denotes the N by N diagonal matrix with diagonal elements equal to λ_i, $i = 1, \ldots, N$. Change the integration variable in (3) from φ to U and Λ:

$$Z = \int dU \int (\Pi_i d\lambda_i) \, J e^{-N \sum_k V(\lambda_k)} \tag{12}$$

with J the Jacobian. Since the integrand does not depend on U we can throw away the integral over U. It just gives the volume of the group $SU(N)$. Does this remind you of chapter VII.1? Indeed, in (11) U corresponds to the unphysical gauge degrees of freedom—the relevant degrees of freedom are the eigenvalues $\{\lambda_i\}$. As an exercise you can use the Faddeev-Popov method to calculate J.

Instead, we will follow the more elegant tack of determining J by arguing from general principles. The change of integration variables in (11) is ill defined when any two of the λ_i's are equal, at which point J must vanish. (Recall that the change from Cartesian coordinates to spherical coordinates is ill defined at the north and south poles and indeed the Jacobian in $\sin\theta d\theta d\varphi$ vanishes at $\theta = 0$ and π.) Since the λ_i's are created equal, interchange symmetry dictates that $J = [\Pi_{m>n}(\lambda_m - \lambda_n)]^\beta$. The power β can be fixed by dimensional analysis. With N^2 matrix elements $d\varphi$ obviously has dimension λ^{N^2} while $(\Pi_i d\lambda_i) J$ has dimension $\lambda^N \lambda^{\beta N(N-1)/2}$; thus $\beta = 2$.

Having determined J, let us rewrite (12) as

$$Z = \int (\Pi_i d\lambda_i)[\Pi_{m>n}(\lambda_m - \lambda_n)]^2 e^{-N \sum_k V(\lambda_k)}$$
$$= \int (\Pi_i d\lambda_i) e^{-N \sum_k V(\lambda_k) + \frac{1}{2} \sum_{m \neq n} \log(\lambda_m - \lambda_n)^2} \tag{13}$$

Dyson pointed out that in this form $Z = \int (\Pi_i d\lambda_i) e^{-NE(\lambda_1, \ldots, \lambda_N)}$ is just the partition function of a classical 1-dimensional gas (recall chapter V.2). Think of λ_i, a real number, as the position of the ith molecule. The energy of a configuration

$$E(\lambda_1, \ldots, \lambda_N) = \sum_k V(\lambda_k) - \frac{1}{2N} \sum_{m \neq n} \log(\lambda_m - \lambda_n)^2 \tag{14}$$

consists of two terms with obvious physical interpretations. The gas is confined in a potential well $V(x)$ and the molecules repel[4] each other with the two-body potential $-(1/N) \log(x - y)^2$. Note that the two terms in E are of the same order in N since each

---

[4] Note that this corresponds to the repulsion between energy levels in quantum mechanics.

sum counts for a power of N. In the large N limit (we can think of N as the inverse temperature), we evaluate Z by steepest descent and minimize E, obtaining

$$V'(\lambda_k) = \frac{2}{N} \sum_{n \neq k} \frac{1}{\lambda_k - \lambda_n} \tag{15}$$

which in the continuum limit, as the poles in (15) merge into a cut, becomes $V'(\lambda) = 2\mathcal{P} \int d\mu[\rho(\mu)/(\lambda - \mu)]$, where $\rho(\mu)$ is the unknown function we want to solve for and $\mathcal{P}$ denotes principal value.

Defining as before $G(z) = \int d\mu[\rho(\mu)/(z - \mu)]$ we see that our equation for $\rho(\mu)$ can be written as $\text{Re } G(\lambda + i\varepsilon) = \frac{1}{2}V'(\lambda)$. In other words, $G(z)$ is a real analytic function with cuts along the real axis. We are given the real part of $G(z)$ on the cut and are to solve for the imaginary part. Brézin, Itzykson, Parisi, and Zuber have given an elegant solution of this problem. Assume for simplicity that $V(z)$ is an even polynomial and that there is only one cut (see exercise VII.4.7). Invoke symmetry and, incorporating what we know, postulate the form

$$G(z) = \frac{1}{2}\left[V'(z) - P(z)\sqrt{z^2 - a^2}\right]$$

with $P(z)$ an unknown even polynomial. Remarkably, the requirement $G(z) \to 1/z$ for large z completely determines $P(z)$. Pedagogically, it is clearest to go to a specific example, say $V(z) = \frac{1}{2}m^2 z^2 + g z^4$. Since $V'(z)$ is a cubic polynomial in z, $P(z)$ has to be a quadratic (even) polynomial in z. Taking the limit $z \to \infty$ and requiring the coefficients of z^3 and of z in $G(z)$ to vanish and the coefficient of $1/z$ to be 1 gives us three equations for three unknowns [namely a and the two unknowns in $P(z)$]. The density of eigenvalues is then determined to be $\rho(E) = (1/\pi)P(E)\sqrt{a^2 - E^2}$.

I think the lesson to take away here is that Feynman diagrams, in spite of their historical importance in quantum electrodynamics and their usefulness in helping us visualize what is going on, are vastly overrated. Surely, nobody imagines that QCD, even large N QCD, will one day be solved by summing Feynman diagrams. What is needed is the analog of the Dyson gas approach for large N QCD. Conversely, if a reader of this book manages to calculate $G(z)$ by summing planar diagrams (after all, the answer is known!), the insight he or she gains might conceivably be useful in seeing how to deal with planar diagrams in large N QCD.

Field theories in the large N limit

A number of field theories have also been solved in the large N expansion. I will tell you about one example, the Gross-Neveu model, partly because it has some of the flavor of QCD. The model is defined by

$$S(\psi) = \int d^2x \left[\sum_{a=1}^{N} \bar{\psi}_a i \slashed{\partial} \psi_a + \frac{g^2}{2N}\left(\sum_{a=1}^{N} \bar{\psi}_a \psi_a\right)^2\right] \tag{16}$$

Recall from chapter III.3 that this theory should be renormalizable in $(1+1)$-dimensional spacetime. For some finite N, say $N = 3$, this theory certainly appears no easier to solve than any other fully interacting field theory. But as we will see, as $N \to \infty$ we can extract a lot of interesting physics.

Using the identity (A.14) we can rewrite the theory as

$$S(\psi, \sigma) = \int d^2x \left[\sum_{a=1}^{N} \bar{\psi}_a (i \slashed{\partial} - \sigma) \psi_a - \frac{N}{2g^2} \sigma^2 \right] \tag{17}$$

By introducing the scalar field $\sigma(x)$ we have "undone" the four-fermion interaction. (Recall that we used the same trick in chapter III.5.) You will note that the physics involved is similar to that behind the introduction of the weak boson to generate the Fermi interaction. Using what we learned in chapters II.5 and IV.3 we can immediately integrate out the fermion fields to obtain an action written purely in terms of the σ field

$$S(\sigma) = - \int d^2x \frac{N}{2g^2} \sigma^2 - iN \, \mathrm{tr} \log(i\slashed{\partial} - \sigma) \tag{18}$$

Note the factor of N in front of the tr log term coming from the integration over N fermion fields. With the malice of forethought we, or rather Gross and Neveu, have introduced an explicit factor of $1/N$ in the coupling strength in (16), so that the two terms in (18) both scale as N. Thus, the path integral $Z = \int D\sigma \, e^{iS(\sigma)}$ may be evaluated by the steepest descent or stationary phase method in the large N limit. We simply extremize $S(\sigma)$.

Incidentally, we can see the judiciousness of the choice $a = 1$ in large N QCD in the same way. Integrating out the quarks in (1) we get

$$S = - \int d^4x \frac{N}{2g^2} \, \mathrm{tr} \, F_{\mu\nu} F^{\mu\nu} + N \, \mathrm{tr} \log(i(\slashed{\partial} - i\slashed{A}) - m)$$

and thus the two terms both scale as N and can balance each other. The increase in the number of degrees of freedom has to be offset by a weakening of the coupling.

To study the ground state behavior of the theory, we restrict our attention to field configurations $\sigma(x)$ that do not depend on x. (In other words, we are not expecting translation symmetry to be spontaneously broken.) We can immediately take over the result you got in exercise IV.3.3 and write the effective potential

$$\frac{1}{N} V(\sigma) = \frac{1}{2g(\mu)^2} \sigma^2 + \frac{1}{4\pi} \sigma^2 \left(\log \frac{\sigma^2}{\mu^2} - 3 \right) \tag{19}$$

We have imposed the condition $(1/N)[d^2 V(\sigma)/d\sigma^2]|_{\sigma=\mu} = 1/g(\mu)^2$ as the definition of the mass scale dependent coupling $g(\mu)$ (compare IV.3.18). The statement that $V(\sigma)$ is independent of μ immediately gives

$$\frac{1}{g(\mu)^2} - \frac{1}{g(\mu')^2} = \frac{1}{\pi} \log \frac{\mu}{\mu'} \tag{20}$$

As $\mu \to \infty$, $g(\mu) \to 0$. Remarkably, this theory is asymptotically free, just like QCD. If we want to, we can work backward to find the flow equation

$$\mu \frac{d}{d\mu} g(\mu) = -\frac{1}{2\pi} g(\mu)^3 + \cdots \tag{21}$$

The theory in its different incarnations, (16), (17), and (18), enjoys a discrete Z_2 symmetry under which $\psi_a \to \gamma^5 \psi_a$ and $\sigma \to -\sigma$. As in chapter IV.3, this symmetry is dynamically broken by quantum fluctuations. The minimum of $V(\sigma)$ occurs at $\sigma_{\min} = \mu e^{1-\pi/g(\mu)^2}$ and so according to (17) the fermions acquire a mass

$$m_F = \sigma_{\min} = \mu e^{1-\pi/g(\mu)^2} \tag{22}$$

Note that this highly nontrivial result can hardly be seen by staring at (16) and we have no way of proving it for finite N. In the spirit of the large N approach, however, we expect that the fermion mass might be given by $m_F = \mu e^{1-\pi/g(\mu)^2} + O(1/N^2)$ so that (22) would be a decent approximation even for say, $N = 3$. Since m_F is physically measurable, it better not depend on μ. You can check that.

This theory also exhibits dimensional transmutation as described in the previous chapter. We start out with a theory with a dimensionless coupling g and end up with a dimensional fermion mass m_F. Indeed, any other quantity with dimension of mass would have to be equal to m_F times a pure number.

Dynamically generated kinks

I discuss the existence of kinks and solitons in chapter V.6. You clearly understood that the existence of such objects follows from general considerations of symmetry and topology, rather than from detailed dynamics. Here we have a $(1 + 1)$-dimensional theory with a discrete Z_2 symmetry, so we certainly expect a kink, namely a time independent configuration $\sigma(x)$ (henceforth x will denote only the spatial coordinate and will no longer label a generic point in spacetime) such that $\sigma(-\infty) = -\sigma_{\min}$ and $\sigma(+\infty) = \sigma_{\min}$. [Obviously, there is also the antikink with $\sigma(-\infty) = \sigma_{\min}$ and $\sigma(+\infty) = -\sigma_{\min}$.]

At first sight, it would seem almost impossible to determine the precise shape of the kink. In principle, we have to evaluate tr $\log[i\slashed{\partial} - \sigma(x)]$ for an arbitrary function $\sigma(x)$ such that $\sigma(+\infty) = -\sigma(-\infty)$ (and as I explained in chapter IV.3, this involves finding all the eigenvalues of the operator $i\slashed{\partial} - \sigma(x)$, summing over the logarithm of the eigenvalues), and then varying this functional of $\sigma(x)$ to find the optimal shape of the kink.

Remarkably, the shape can actually be determined thanks to a clever observation.[5] In analogy with the steps leading to (IV.3.24) we note that

$$\text{tr } \log[i\slashed{\partial} - \sigma(x)] = \text{tr } \log \gamma^5[i\slashed{\partial} - \sigma(x)]\gamma^5 = \text{tr } \log(-1)[i\slashed{\partial} + \sigma(x)]$$

and thus up to an irrelevant additive constant

$$\text{tr } \log(i\slashed{\partial} - \sigma(x)) = \tfrac{1}{2} \text{ tr } \log[i\slashed{\partial} - \sigma(x)][i\slashed{\partial} + \sigma(x)]$$
$$= \tfrac{1}{2} \text{ tr } \log\left\{-\partial^2 + i\gamma^1\sigma'(x) - [\sigma(x)]^2\right\} \tag{23}$$

---

[5] C. Callan, S. Coleman, D. Gross, and A. Zee, (unpublished). See D. J. Gross, "Applications of the Renormalization Group to High-Energy Physics," in: R. Balian and J. Zinn-Justin, eds., *Methods in Field Theory*, p. 247. By the way, I recommend this book to students of field theory.

Since γ^1 has eigenvalues $\pm i$, this is equal to

$$\tfrac{1}{2} \left\{ \text{tr} \log\{-\partial^2 + \sigma'(x) - [\sigma(x)]^2\} + \text{tr} \log\{-\partial^2 - \sigma'(x) - [\sigma(x)]^2\} \right\}$$

but these two terms are equal by parity (space reflection) and hence

$$\text{tr} \log[i\slashed{\partial} - \sigma(x)] = \text{tr} \log\{-\partial^2 - \sigma'(x) - [\sigma(x)]^2\}$$

Referring to (18) we see that $S(\sigma)$ is the sum of two terms, a term quadratic in $\sigma(x)$ and a term that depends only on the combination $\sigma'(x) + [\sigma(x)]^2$. But we know that $\sigma_{\min}$ minimizes $S(\sigma)$. Thus, the soliton is given by the solution of the ordinary differential equation

$$\sigma'(x) + [\sigma(x)]^2 = \sigma_{\min}^2 \tag{24}$$

namely $\sigma(x) = \sigma_{\min} \tanh \sigma_{\min} x$. The soliton would be observed as an object of size $1/\sigma_{\min} = 1/m_F$. I leave it to you to show that its mass is given by

$$m_S = \frac{N}{\pi} m_F \tag{25}$$

Precisely as theorized in the last chapter, the ratio m_S/m_F comes out to be a pure number, N/π, as it must.

By an even more clever method that I do not have space to describe, Dashen, Hasslacher, and Neveu were able to study time dependent configurations of σ and determine the mass spectrum of this model.

Exercises

VII.4.1 Since the number of gluons only differs by one, it is generally argued that it does not make any difference whether we choose to study the $U(N)$ theory or the $SU(N)$ theory. Discuss how the gluon propagator in a $U(N)$ theory differs from the gluon propagator in an $SU(N)$ theory and decide which one is easier.

VII.4.2 As a challenge, solve large N QCD in $(1+1)$-dimensional spacetime. [Hint: The key is that in $(1+1)$-dimensional spacetime with a suitable gauge choice we can integrate out the gauge potential A_μ.] For help, see 't Hooft, *Under the Spell of the Gauge Principle*, p. 443.

VII.4.3 Show that if we had chosen to calculate $G(z) \equiv \langle (1/N) \text{tr}(1/z - \varphi) \rangle$, we would have to connect the two open ends of the quark propagator. We see that figures VII.4.5b and d lead to the same diagram. Complete the calculation of $G(z)$ in this way.

VII.4.4 Suppose the random matrix φ is real symmetric rather than hermitean. Show that the Feynman rules are more complicated. Calculate the density of eigenvalues. [Hint: The double-line propagator can twist.]

VII.4.5 For hermitean random matrices φ, calculate

$$G_c(z, w) \equiv \left\langle \frac{1}{N} \text{tr} \frac{1}{z - \varphi} \frac{1}{N} \text{tr} \frac{1}{w - \varphi} \right\rangle - \left\langle \frac{1}{N} \text{tr} \frac{1}{z - \varphi} \right\rangle \left\langle \frac{1}{N} \text{tr} \frac{1}{w - \varphi} \right\rangle$$

for $V(\varphi) = \frac{1}{2} m^2 \varphi^2$ using Feynman diagrams. [Note that this is a much simpler object to study than the object we need to study in order to learn about localization (see exercise VI.6.1).] Show that by taking suitable imaginary parts we can extract the correlation of the density of eigenvalues with itself. For help, see E. Brézin and A. Zee, *Phys. Rev.* E51: p. 5442, 1995.

VII.4.6 Use the Faddeev-Popov method to calculate J in the Dyson gas approach.

VII.4.7 For $V(\varphi) = \frac{1}{2}m^2\varphi^2 + g\varphi^4$, determine $\rho(E)$. For m^2 sufficiently negative (the double well potential again) we expect the density of eigenvalues to split into two pieces. This is evident from the Dyson gas picture. Find the critical value m_c^2. For $m^2 < m_c^2$ the assumption of $G(z)$ having only one cut used in the text fails. Show how to calculate $\rho(E)$ in this regime.

VII.4.8 Calculate the mass of the soliton (25).

VII.5 | Grand Unification

Crying out for unification

A gauge theory is specified by a group and the representations the matter fields belong to. Let us go back to chapter VII.2 and make a catalogue for the $SU(3) \otimes SU(2) \otimes U(1)$ theory. For example, the left handed up and down quarks are in a doublet $\left(\begin{smallmatrix} u^\alpha \\ d^\alpha \end{smallmatrix} \right)_L$ with hypercharge $\frac{1}{2}Y = \frac{1}{6}$. Let us denote this by $(3, 2, \frac{1}{6})_L$, with the three numbers indicating how these fields transform under $SU(3) \otimes SU(2) \otimes U(1)$. Similarly, the right handed up quark is $(3, 1, \frac{2}{3})_R$. The leptons are $(1, 2, -\frac{1}{2})_L$ and $(1, 1, -1)_R$, where the "1" in the first entry indicates that these fields do not participate in the strong interaction. Writing it all down, we see that the quarks and leptons of each family are placed in

$$(3, 2, \tfrac{1}{6})_L, (3, 1, \tfrac{2}{3})_R, (3, 1, -\tfrac{1}{3})_R, (1, 2, -\tfrac{1}{2})_L, \text{ and } (1, 1, -1)_R \tag{1}$$

This motley collection of representations practically cries out for further unification. Who would have constructed the universe by throwing this strange looking list down?

What we would like to have is a larger gauge group G containing $SU(3) \otimes SU(2) \otimes U(1)$, such that this laundry list of representations is unified into (ideally) one great big representation. The gauge bosons in G [but not in $SU(3) \otimes SU(2) \otimes U(1)$ of course] would couple the representations in (1) to each other.

Before we start searching for G, note that since gauge transformations commute with the Lorentz group, these desired gauge transformations cannot change left handed fields to right handed fields. So let us change all the fields in (1) to left handed fields. Recall from exercise II.1.9 that charge conjugation changes left handed fields to right handed fields and vice versa. Thus, instead of (1) we can write

$$(3, 2, \tfrac{1}{6}), (3^*, 1, -\tfrac{2}{3}), (3^*, 1, \tfrac{1}{3}), (1, 2, -\tfrac{1}{2}), \text{ and } (1, 1, 1) \tag{2}$$

We now omit the subscripts L and R: everybody is left handed.

A perfect fit

The smallest group that contains $SU(3) \otimes SU(2) \otimes U(1)$ is $SU(5)$. (If you are shaky about group theory, study appendix B now.) Recall that $SU(5)$ has $5^2 - 1 = 24$ generators. Explicitly, the generators are represented by 5 by 5 hermitean traceless matrices acting on five objects we denote by ψ^μ with $\mu = 1, 2, \ldots, 5$. [These five objects form the fundamental or defining representation of $SU(5)$.]

It is now obvious how we can fit $SU(3)$ and $SU(2)$ into $SU(5)$. Of the 24 matrices that generate $SU(5)$, eight have the form $\left(\begin{smallmatrix} A & 0 \\ 0 & 0 \end{smallmatrix} \right)$ and three the form $\left(\begin{smallmatrix} 0 & 0 \\ 0 & B \end{smallmatrix} \right)$, where A represents 3 by 3 hermitean traceless matrices (of which there are $3^2 - 1 = 8$, the so-called Gell-Mann matrices) and B represents 2 by 2 hermitean traceless matrices (of which there are $2^2 - 1 = 3$, namely the Pauli matrices). Clearly, the former generate an $SU(3)$ and the latter an $SU(2)$. Furthermore, the 5 by 5 hermitean traceless matrix

$$\tfrac{1}{2}Y = \begin{pmatrix} -\tfrac{1}{3} & 0 & 0 & 0 & 0 \\ 0 & -\tfrac{1}{3} & 0 & 0 & 0 \\ 0 & 0 & -\tfrac{1}{3} & 0 & 0 \\ 0 & 0 & 0 & \tfrac{1}{2} & 0 \\ 0 & 0 & 0 & 0 & \tfrac{1}{2} \end{pmatrix} \tag{3}$$

generates a $U(1)$. Without being coy about it, we have already called this matrix the hypercharge $\tfrac{1}{2}Y$.

In other words, if we separate the index $\mu = \{\alpha, i\}$ with $\alpha = 1, 2, 3$ and $i = 4, 5$, then the $SU(3)$ acts on the index α and the $SU(2)$ acts on the index i. Thus, the three objects ψ^α transform as a 3-dimensional representation under $SU(3)$ and hence could be a 3 or a 3*. Let us choose ψ^α as transforming as 3; we will see shortly that this is the right choice with $Y/2$ given as in (3). The three objects ψ^α do not transform under $SU(2)$ and hence each of them belongs to the singlet 1 representation. Furthermore, they carry hypercharge $-\tfrac{1}{3}$ as we can read off from (3). To sum up, ψ^α transform as $(3, 1, -\tfrac{1}{3})$ under $SU(3) \otimes SU(2) \otimes U(1)$. On the other hand, the two objects ψ^i transform as 1 under $SU(3)$ and 2 under $SU(2)$, and carry hypercharge $\tfrac{1}{2}$; thus they transform as $(1, 2, \tfrac{1}{2})$. In other words, we embed $SU(3) \otimes SU(2) \otimes U(1)$ into $SU(5)$ by specifying how the defining representation of $SU(5)$ decomposes into representations of $SU(3) \otimes SU(2) \otimes U(1)$

$$5 \to (3, 1, -\tfrac{1}{3}) \oplus (1, 2, \tfrac{1}{2}) \tag{4}$$

Taking the conjugate we see that

$$5^* \to (3^*, 1, \tfrac{1}{3}) \oplus (1, 2, -\tfrac{1}{2}) \tag{5}$$

Inspecting (2), we see that $(3^*, 1, \tfrac{1}{3})$ and $(1, 2, -\tfrac{1}{2})$ appear on the list. We are on the right track! The fields in these two representations fit snugly into 5^*.

This accounts for five of the fields contained in (2); we still have the ten fields

$$(3, 2, \tfrac{1}{6}), (3^*, 1, -\tfrac{2}{3}), \text{ and } (1, 1, 1) \tag{6}$$

Consider the next representation of $SU(5)$ in order of size, namely the antisymmetric tensor representation $\psi^{\mu\nu}$. Its dimension is $(5 \times 4)/2 = 10$, precisely the number we want, if only the quantum numbers under $SU(3) \otimes SU(2) \otimes U(1)$ work out!

Since we know that $5 \to (3, 1, -\frac{1}{3}) \oplus (1, 2, \frac{1}{2})$, we simply (again, see appendix B!) have to work out the antisymmetric product of $(3, 1, -\frac{1}{3}) \oplus (1, 2, \frac{1}{2})$ with itself, namely the direct sum of (where $\otimes_A$ denotes the antisymmetric product)

$$(3, 1, -\tfrac{1}{3}) \otimes_A (3, 1, -\tfrac{1}{3}) = (3^*, 1, -\tfrac{2}{3}) \tag{7}$$

$$(3, 1, -\tfrac{1}{3}) \otimes_A (1, 2, \tfrac{1}{2}) = (3, 2, -\tfrac{1}{3} + \tfrac{1}{2}) = (3, 2, \tfrac{1}{6}) \tag{8}$$

and

$$(1, 2, \tfrac{1}{2}) \otimes_A (1, 2, \tfrac{1}{2}) = (1, 1, 1) \tag{9}$$

[I will walk you through (7): In $SU(3)$ $3 \otimes_A 3 = 3^*$ (remember ε_{ijk} from appendix B?), in $SU(2)$ $1 \otimes_A 1 = 1$, and in $U(1)$ the hypercharges simply add $-\frac{1}{3} - \frac{1}{3} = -\frac{2}{3}$.]

Lo and behold, these $SU(3) \otimes SU(2) \otimes U(1)$ representations form exactly the collection of representations in (6). In other words,

$$10 \to (3, 2, \tfrac{1}{6}) \oplus (3^*, 1, -\tfrac{2}{3}) \oplus (1, 1, 1) \tag{10}$$

The known quark and lepton fields in a given family fit perfectly into the 5* and 10 representations of $SU(5)$!

I have just described the $SU(5)$ grand unified theory of Georgi and Glashow. In spite of the fact that the theory has not been directly verified by experiment, it is extremely difficult for me and for many other physicists not to believe that $SU(5)$ is at least structurally correct, in view of the perfect group theoretic fit.

It is often convenient to display the contents of the representation 5* and 10, using the names given to the various fields historically. We write 5* as a column vector

$$\psi_\mu = \begin{pmatrix} \psi_\alpha \\ \psi_i \end{pmatrix} = \begin{pmatrix} \bar{d}_\alpha \\ \nu \\ e \end{pmatrix} \tag{11}$$

and the 10 as an antisymmetric matrix

$$\psi^{\mu\nu} = \{\psi^{\alpha\beta}, \psi^{\alpha i}, \psi^{ij}\}$$

$$= \begin{pmatrix} 0 & \bar{u} & -\bar{u} & d & u \\ -\bar{u} & 0 & \bar{u} & d & u \\ \bar{u} & -\bar{u} & 0 & d & u \\ -d & -d & -d & 0 & \bar{e} \\ -u & -u & -u & -\bar{e} & 0 \end{pmatrix} \tag{12}$$

(I suppressed the color indices.)

Deepening our understanding of physics

Aside from its esthetic appeal, grand unification deepens our understanding of physics enormously.

1. Ever wondered why electric charge is quantized? Why don't we see particles with charge equal to $\sqrt{\pi}$ times the electron's charge? In quantum electrodynamics, you could perfectly well write down

$$\mathcal{L} = \bar{\psi}[i(\slashed{\partial} - i\slashed{A}) - m]\psi + \bar{\psi}'[i(\slashed{\partial} - i\sqrt{\pi}\,\slashed{A}) - m']\psi' + \cdots \tag{13}$$

In contrast, in grand unified theory A_μ couples to a generator of the grand unifying gauge group, and you know that the generators of any group such as $SU(N)$ (that is not given by the direct product of $U(1)$ with other groups) are forced by the nontrivial commutation relations $[T_a, T_b] = if_{abc}T_c$ to assume quantized values. For example, the eigenvalues of T_3 in $SU(2)$, which depend on the representation of course, must be multiples of $\frac{1}{2}$. Within $SU(3) \times SU(2) \times U(1)$, we cannot understand charge quantization: The generator of $U(1)$ is not quantized. But upon grand unification into $SU(5)$ [or more generally any group without $U(1)$ factors] electric charge is quantized.

The result here is deeply connected to Dirac's remark (chapter IV.4) that electric charge is quantized if the magnetic monopole exists. We know from chapter V.7 that spontaneously broken nonabelian gauge theories such as the $SU(5)$ theory contain the monopole.

2. Ever wondered why the proton charge is exactly equal and opposite to the electron charge? This important fact allows us to construct the universe as we know it. Atoms must be electrically neutral to some fantastic degree of accuracy for standard cosmology to work; otherwise, electrostatic forces between macroscopic matter would tear the universe apart.

This remarkable fact is nicely incorporated into $SU(5)$. It is fun to see how it goes. Evaluating $\text{tr}\,Q = 0$ over the 5* implies that $3Q_{\bar{d}} = -Q_{e^-}$. I have used the fact that the strong interaction commutes with electromagnetism and hence quarks with different color have the same charge. Now let us calculate the proton charge Q_P:

$$Q_P = 2Q_u + Q_d = 2(Q_d + 1) + Q_d = 3Q_d + 2 = Q_{e^-} + 2 \tag{14}$$

If $Q_{e^-} = -1$, then $Q_P = -Q_{e^-}$, as is indeed the case!

3. Recall that in electroweak theory we defined $\tan\theta = g_1/g_2$, with the coupling of the gauge bosons $g_2 A_\mu^a T_a + g_1 B_\mu(Y/2)$. Since the normalization of A_μ^a and B_μ is fixed by their respective kinetic energy term, the relative strength of g_2 and g_1 is determined by the normalization of $Y/2$ relative to T_3. Let us evaluate $\text{tr}\,T_3^2$ and $\text{tr}(Y/2)^2$ on the defining representation $5 : \text{tr}\,T_3^2 = (\frac{1}{2})^2 + (\frac{1}{2})^2 = \frac{1}{2}$ and $\text{tr}(Y/2)^2 = (\frac{1}{3})^2 3 + (\frac{1}{2})^2 2 = \frac{5}{6}$.

Thus, T_3 and $\sqrt{3/5}(Y/2)$ are normalized equally. So the correct grand unified combination is $A_\mu^a T_a + B_\mu\sqrt{3/5}(Y/2)$, and therefore $\tan\theta = g_1/g_2 = \sqrt{3/5}$ or

$$\sin^2\theta = \frac{3}{8} \tag{15}$$

at the grand unification scale. To compare with the experimental value of $\sin^2 \theta$ we would have to study how the couplings g_2 and g_1 flow under the renormalization group down to low energies. We will postpone this discussion until the next chapter.

Freedom from anomaly

Recall from chapter VII.2 that the key to proving renormalizability of nonabelian gauge theory is the ability to pass freely between the unitary gauge and the R_ξ gauge. The crucial ingredient is gauge invariance and the resulting Ward-Takahashi identities (see chapter II.7).

Suddenly you start to worry. What about the chiral anomaly? The existence of the anomaly means that some Ward-Takahashi identities fail to hold. For our theories to make sense, they had better be free from anomalies. I remarked in chapter IV.7 that the historical name "anomaly" makes it sound like some kind of sickness. Well, in a way, it is.

We should have already checked the $SU(3) \otimes SU(2) \otimes U(1)$ theory for anomalies, but we didn't. I will let you do it as an exercise. Here I will show that the $SU(5)$ theory is healthy. If the $SU(5)$ theory is anomaly-free, then a fortiori so is the $SU(3) \otimes SU(2) \otimes U(1)$ theory.

In chapter IV.7 I computed the anomaly in an abelian theory but as I remarked there clearly all we have to do to generalize to a nonabelian theory is to insert a generator T_a of the gauge group at each vertex of the triangle diagram in figure IV.7.1. Summing over the various fermions running around the loop, we see that the anomaly is proportional to $A_{abc}(R) \equiv \text{tr}(T_a\{T_b, T_c\})$, where R denotes the representation to which the fermions belong. We have to sum $A_{abc}(R)$ over all the representations in the theory, remembering to associate opposite signs to left handed and right handed fermion fields. (It may be helpful to remind yourself of remark 3 in chapter IV.7 and exercise IV.7.6.)

We are now ready to give the $SU(5)$ theory a health check. First, all fermion fields in (2) are left handed. Second, convince yourself (simply imagine calculating A_{abc} for all possible abc) that it suffices to set T_a, T_b, and T_c all equal to

$$T \equiv \begin{pmatrix} 2 & 0 & 0 & 0 & 0 \\ 0 & 2 & 0 & 0 & 0 \\ 0 & 0 & 2 & 0 & 0 \\ 0 & 0 & 0 & -3 & 0 \\ 0 & 0 & 0 & 0 & -3 \end{pmatrix}$$

a multiple of the hypercharge. Let us now evaluate $\text{tr}\, T^3$ on the 5^* representation,

$$\text{tr}\, T^3|_{5^*} = 3(-2)^3 + 2(+3)^3 = 30 \tag{16}$$

and on the 10,

$$\text{tr}\, T^3|_{10} = 3(+4)^3 + 6(-1)^3 + (-6)^3 = -30 \tag{17}$$

An apparent miracle! The anomaly cancels.

This remarkable cancellation between sums of cubes of a strange list of numbers suggests strongly, to say the least, that $SU(5)$ is not the end of the story. Besides, it would be nice if the 5* and 10 could be unified into a single representation.

Exercises

VII.5.1 Write down the charge operator Q acting on 5, the defining representation ψ^μ. Work out the charge content of the $10 = \psi^{\mu\nu}$ and identify the various fields contained therein.

VII.5.2 Show that for any grand unified theory, as long as it is based on a simple group, we have at the unification scale

$$\sin^2 \theta = \frac{\sum T_3^2}{\sum Q^2} \tag{18}$$

where the sum is taken over all fermions.

VII.5.3 Check that the $SU(3) \otimes SU(2) \otimes U(1)$ theory is anomaly-free. [Hint: The calculation is more involved than in $SU(5)$ since there are more independent generators. First show that you only have to evaluate tr $Y\{T_a, T_b\}$ and tr Y^3, with T_a and Y the generators of $SU(2)$ and $U(1)$, respectively.]

VII.5.4 Construct grand unified theories based on $SU(6)$, $SU(7)$, $SU(8)$, . . . , until you get tired of the game. People used to get tenure doing this. [Hint: You would have to invent fermions yet to be experimentally discovered.]

VII.6 | Protons Are Not Forever

Proton decay

Charge conservation guarantees the stability of the electron, but what about the stability of the proton? Charge conservation allows $p \to \pi^0 + e^+$. No fundamental principle says that the proton lives forever, but yet the proton is known for its longevity: It has been around essentially since the universe began.

The stability of the proton had to be decreed by an authority figure: Eugene Wigner was the first to proclaim the law of baryon number conservation. The story goes that when Wigner was asked how he knew that the proton lives forever he quipped, "I can feel it in my bones." I take the remark to mean that just from the fact that we do not glow in the dark we can set a fairly good lower bound on the proton's life span.

As soon as we start grand unifying, we better start worrying. Generically, when we grand unify we put quarks and leptons into the same representation of some gauge group [see (VII.5.11 and VII.5.12)]. This miscegenation immediately implies that there are gauge bosons transforming quarks into leptons and vice versa. The bag of three quarks known as the proton could very well get turned into leptons upon the exchange of these gauge bosons. In other words, the proton, the rock on which our world is built upon, may not be forever! Thus, grand unification runs the risk of being immediately falsified.

Let M_X denote generically the masses of those gauge bosons transforming quarks into leptons and vice versa. Then the amplitude for proton decay is of order g^2/M_X^2, with g the coupling strength of the grand unifying gauge group, and the proton decay rate Γ is given by $(g^2/M_X^2)^2$ times a phase space factor controlled essentially by the proton mass m_P since the pion and positron masses are negligible compared to the proton mass. By dimensional analysis, we determine that $\Gamma \sim (g^2/M_X^2)^2 m_P^5$. Since the proton is known to live for something like at least 10^{31} years, M_X had better be huge compared to the kind of energy scales we can reach experimentally.

The mass M_X is of the same order as the mass scale M_{GUT} at which the grand unified theory is spontaneously broken down to $SU(3) \otimes SU(2) \otimes U(1)$. Specifically, in the $SU(5)$

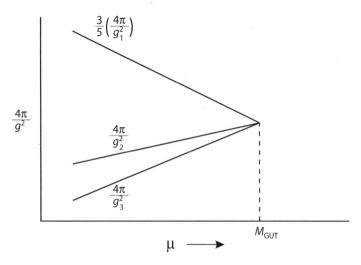

Figure VII.6.1

theory, a Higgs field H_ν^μ transforming as the adjoint 24, with its vacuum expectation value $\langle H_\nu^\mu \rangle$ equal to the diagonal matrix with elements $(-\frac{1}{3}, -\frac{1}{3}, -\frac{1}{3}, \frac{1}{2}, \frac{1}{2})$ times some v, can do the job, as was discussed in chapter IV.6. The gauge bosons in $SU(3) \otimes SU(2) \otimes U(1)$ remain massless while the other gauge bosons acquire mass M_X of order gv.

To determine M_{GUT}, we apply renormalization group flow to g_3, g_2, and g_1, the couplings of $SU(3)$, $SU(2)$, and $U(1)$, respectively. The idea is that as we move up in the mass or energy scale μ the two asymptotically free couplings $g_3(\mu)$ and $g_2(\mu)$ decrease while $g_1(\mu)$ increases. Thus, at some mass scale M_{GUT} they will meet and that is where $SU(3) \otimes SU(2) \otimes U(1)$ is unified into $SU(5)$ (see figure VII.6.1). Because of the extremely slow logarithmic running (it should be called walking or even crawling but again for historical reasons we are stuck with running) of the coupling constant, we anticipate that the unification mass scale M_{GUT} will come out to be much larger than any scale we were used to in particle physics prior to grand unification. In fact, M_{GUT} will turn out to have an enormous value of the order 10^{14-15}Gev and the idea of grand unification passes its first hurdle.

Stability of the world implies the weakness of electromagnetism

Using the result of exercise VI.8.1 we obtain (here $\alpha_S \equiv g_3^2/4\pi$ and $\alpha_{\text{GUT}} \equiv g^2/4\pi$ denote the strong interaction and grand unification analog of the fine structure constant α, respectively, with F the number of families)

$$\frac{4\pi}{[g_3(\mu)]^2} \equiv \frac{1}{\alpha_S(\mu)} = \frac{1}{\alpha_{\text{GUT}}} + \frac{1}{6\pi}(4F - 33)\log\frac{M_{\text{GUT}}}{\mu} \tag{1}$$

$$\frac{4\pi}{[g_2(\mu)]^2} \equiv \frac{\sin^2\theta(\mu)}{\alpha(\mu)} = \frac{1}{\alpha_{\text{GUT}}} + \frac{1}{6\pi}(4F - 22)\log\frac{M_{\text{GUT}}}{\mu} \tag{2}$$

$$\frac{3}{5}\frac{4\pi}{[g_1(\mu)]^2} \equiv \frac{3}{5}\frac{\cos^2\theta(\mu)}{\alpha(\mu)} = \frac{1}{\alpha_{\text{GUT}}} + \frac{1}{6\pi}4F\log\frac{M_{\text{GUT}}}{\mu} \tag{3}$$

By $\theta(\mu)$ we mean the value of θ at the scale μ. At $\mu = M_{GUT}$, the three couplings are related through $SU(5)$.

We evaluate these equations for some experimentally accessible value of μ, plugging in measured values of α_S and α. With three equations, we not only manage to determine the unification scale M_{GUT} and coupling α_{GUT}, but we can predict θ. In other words, unless the ratio g_1 to g_2 is precisely right, the three lines in figure VII.6.1 will not meet at one point.

Note that the number of fermion families F contributes equally to (1), (2), and (3). This is as it should be since the fermions are effectively massless for the purpose of this calculation and do not "know" that the unifying group has been broken into $SU(3) \otimes SU(2) \otimes U(1)$. These equations are derived assuming that all fermion masses are small compared to μ.

Rearranging these equations somewhat, we find

$$\sin^2 \theta = \frac{1}{6} + \frac{5\alpha(\mu)}{9\alpha_S(\mu)} \tag{4}$$

$$\frac{\sin^2 \theta}{\alpha(\mu)} = \frac{1}{\alpha_S(\mu)} + \frac{1}{6\pi} 11 \log \frac{M_{GUT}}{\mu} \tag{5}$$

$$\frac{1}{\alpha(\mu)} = \frac{8}{3} \frac{1}{\alpha_{GUT}} + \frac{1}{6\pi} \left(\frac{32}{3} F - 22 \right) \log \frac{M_{GUT}}{\mu} \tag{6}$$

We obtain in (4) a prediction for $\sin^2 \theta(\mu)$ independent of M_{GUT} and of the number of families.

Note that (5) gives the bound

$$\frac{1}{\alpha(\mu)} \geq \frac{1}{6\pi} 11 \log \frac{M_{GUT}}{\mu} \tag{7}$$

A lower bound on the proton lifetime (and hence on M_{GUT}) translates into an upper bound on the fine structure constant. Amusingly, the stability of the world implies the weakness of electromagnetism.

As I noted earlier, plugging in the measured value of α_S, we obtain a huge value for M_{GUT}. I regard this as a triumph of grand unification: M_{GUT} could have come out to have a much lower scale, leading to an immediate contradiction with the observed stability of the proton, but it didn't. Another way of looking at it is that if we are somehow given M_{GUT} and α_{GUT}, grand unification fixes the couplings of all three nongravitational interactions! The point is not that this simplest try at grand unification doesn't quite agree with experiment: The miracle is that it works at all.

It is beyond the scope of this book to discuss in detail the comparison of (4), (5), and (6) with experiment. To do serious phenomenology, one has to include threshold effects (see exercise VII.6.1), higher order corrections, and so on. To make a long story short, after grand unified theory came out there was enormous excitement over the possibility of proton decay. Alas, the experimental lower bound on the proton lifetime was eventually pushed above the prediction. This certainly does not mean the demise of the notion of grand unification. Indeed, as I mentioned earlier, the perfect fit is enough to convince most particle theorists of the essential correctness of the idea. Over the years people have proposed adding various hypothetical particles to the theory to promote proton longevity.

The idea is that these particles would affect the renormalization group flow and hence M_{GUT}. The proton lifetime is actually not the most critical issue. With more accurate measurements of α_S and of θ, it was found that the three couplings do not quite meet at a point. Indeed, for believers in low energy supersymmetry, part of their faith is founded on the fact that with supersymmetric particles included, the three coupling constants do meet.[1] But skeptics of course can point to the extra freedom to maneuver.

Branching ratios

You may have realized that (1), (2), and (3) are not specific for $SU(5)$: they hold as long as $SU(3) \otimes SU(2) \otimes U(1)$ is unified into some simple group (simple so that there is only one gauge coupling g).

Let us now focus on $SU(5)$. Recall that we decompose the $SU(5)$ index μ, which can take on five values, into two types. In other words, the index μ is labeled by $\{\alpha, i\}$, where α takes on three values and i takes on two values. The gauge bosons in $SU(5)$ correspond to the 24 independent components of the traceless hermitean field A^μ_ν ($\mu, \nu = 1, 2, \ldots, 5$) transforming as the adjoint representation. Focusing on the group theory of $SU(5)$, we will suppress Lorentz indices, spinor indices, etc. Clearly, the eight gauge bosons in $SU(3)$ transform an index of type α into an index of type α, while the three gauge bosons in $SU(2)$ transform an index of type i into an index of type i. Then there is the $U(1)$ gauge boson that couples to the hypercharge $\frac{1}{2}Y$. (Of course, you know what I mean by my somewhat loose language: The $SU(3)$ gauge bosons transform fields carrying a color index into a field carrying a color index.)

The fun comes with the gauge bosons A^α_i and A^i_α, which transform the index α into the index i and vice versa. Since α takes on three values, and i takes on two values, there are $6 + 6 = 12$ such gauge bosons, thus accounting for all the gauge bosons in $SU(5)$. In other words, $24 \to (8, 1) + (1, 3) + (1, 1) + (3, 2) + (3^*, 2)$. We will now see explicitly that the exchange of these bosons between quarks and leptons leads to proton decay.

We merely have to write down the terms in the Lagrangian involving the coupling of the bosons A^α_i and A^i_α to fermions and draw the appropriate Feynman diagrams. I will go through part of the group theoretic analysis, leaving you to work out the rest. Simply by contracting indices we see that the boson A^μ_ν acting on ψ_μ takes it to ψ_ν and acting on $\psi^{\nu\rho}$ takes it to $\psi^{\mu\rho}$. Let us look at what A^5_α does, using your result from exercise VII.5.1. It takes

$$\psi_5 = e^- \to \psi_\alpha = \bar{d} \tag{8}$$

$$\psi^{\alpha\beta} = \bar{u} \to \psi^{5\beta} = u \tag{9}$$

and

$$\psi^{\alpha 4} = d \to \psi^{54} = e^+ \tag{10}$$

---

[1] See, e.g., F. Wilczek, in: V. Fitch et al., eds., *Critical Problems in Physics*, p. 297.

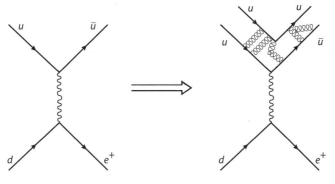

Figure VII.6.2

Thus, the exchange of A_α^5 generates the process (figure VII.6.2) $u + d \to \bar{u} + e^+$, leading to proton decay $p(uud) \to \pi^0(u\bar{u}) + e^+$. Observe that while the decay $p \to \pi^0 + e^+$ violates both baryon number B and lepton number L, it conserves the combination $B - L$.

In exercise VII.6.2 you will work out the branching ratios for various decay modes. Too bad experimentalists have not yet measured them.

Fermion masses

We might hope that with grand unification we would gain new understanding of quark and lepton masses. Unfortunately, the situation on fermion masses in $SU(5)$ is muddled, and to this day nobody understands the origin of quark and lepton masses.

Introducing a Higgs field φ^μ transforming as the 5 (as indicated by the notation) we can write the coupling

$$\psi_\mu C \psi^{\mu\nu} \varphi_\nu \tag{11}$$

and

$$\psi^{\mu\nu} C \psi^{\lambda\rho} \varphi^\sigma \varepsilon_{\mu\nu\lambda\rho\sigma} \tag{12}$$

(with φ_ν the conjugate 5*), reflecting the group theoretic fact (see appendix C) that 5* $\otimes$ 10 contains the 5 and 10 $\otimes$ 10 contains the 5*.

Since $5 \to (3, 1, -\frac{1}{3}) \oplus (1, 2, \frac{1}{2})$ we see that this Higgs field is just the natural extension of the $SU(2) \otimes U(1)$ Higgs doublet $(1, 2, \frac{1}{2})$. Not wanting to break electromagnetism, we allow only the electrically neutral fourth component of φ to acquire a vacuum expectation value. Setting $\langle \varphi^4 \rangle = v$, we obtain (up to uninteresting overall constants)

$$\psi_\alpha C \psi^{\alpha 4} + \psi_5 C \psi^{54} \implies m_d = m_e \tag{13}$$

and

$$\psi^{\alpha\beta} C \psi^{\gamma 5} \varepsilon_{\alpha\beta\gamma} \implies m_u \neq 0 \tag{14}$$

The larger symmetry yields a mass relation $m_d = m_e$ at the unification scale; we again have to apply the renormalization group flow. It is worth noting that the mass relation $m_d = m_e$ comes about because as far as the fermions are concerned, $SU(5)$ has been only broken down to $SU(4)$ by φ. The trouble is that we obtain more or less the same relation for each of the three families, since most of the running occurs between the unification scale M_{GUT} and the top quark mass so that threshold effects give only a small correction. Putting in numbers one gets something like

$$\frac{m_b}{m_\tau} \sim \frac{m_s}{m_\mu} \sim \frac{m_d}{m_e} \sim 3 \tag{15}$$

Let us use this to predict the down sector quark masses in terms of the lepton masses. The formula $m_b \sim 3m_\tau$ works rather well and provides indirect evidence that there can only be three families since the renormalization group flow depends on F. The formula $m_s \sim 3m_\mu$ is more or less in the ballpark, depending on what "experimental" value one takes for m_s. The formula for m_d, on the other hand, is downright embarrassing. People mumble something about the first family being so light and hence other effects, such as one-loop corrections might be important. At the cost of making the theory uglier, people also concoct various schemes by introducing more Higgs fields, such as the 45, to give mass to fermions.

Note that in one respect $SU(5)$ is not as "economical" as $SU(2) \otimes U(1)$, in which the same Higgs field that gives mass to the gauge bosons also gives mass to the fermions.

The universe is not empty, but almost

I mention in passing another triumph of grand unification: its ability to explain the origin of matter in our universe. It has long behooved physicists to understand two fundamental facts about the universe: (1) the universe is not empty, and (2) the universe is almost empty. To physicists, (1) means that the universe is not symmetric between matter and antimatter, that is, the net baryon number N_B is nonzero; and (2) is quantified by the strikingly small observed value $N_B/N_\gamma \sim 10^{-10}$ of the ratio of the number of baryons to the number of photons.

Suppose we start with a universe with equal quantities of matter and antimatter. For the universe to evolve into the observed matter dominated universe, three conditions must be satisfied: (1) The laws of the universe must be asymmetric between matter and antimatter. (2) The relevant physical processes had to be out of equilibrium so that there was an arrow of time. (3) Baryon number must be violated.

We know for a fact that conditions (1) and (2) indeed hold in the world: There is CP violation in the weak interaction and the early universe expanded rapidly. As for (3), grand unification naturally violates baryon number. Furthermore, while proton decay (suppressed by a factor of $1/M_{\text{GUT}}^2$ in amplitude) proceeds at an agonizingly slow rate (for those involved in the proton decay experiment!), in the early universe, when the X and Y bosons are produced in abundance, their fast decays could easily drive baryon number violation. The suppression factor $1/M_{\text{GUT}}^2$ does not come in. I have no doubt that eventually the number 10^{-10} measuring "the amount of dirt in the universe" will be calculated in some grand unified theory.

Hierarchy

I promised you that the Weisskopf phenomenon would come back to haunt us. That the grand unification mass scale M_{GUT} naturally comes out so large counts as a triumph, but it also leads to a problem known as the hierarchy problem. The hierarchy refers to the enormous ratio $M_{\text{GUT}}/M_{\text{EW}}$, where M_{EW} denotes the electroweak unification scale, of order 10^2 Gev. I will sketch this rather murky subject. Look at the Higgs field φ responsible for breaking electroweak theory. We don't know its renormalized or physical mass precisely, but we do know that it is of order M_{EW}. Imagine calculating the bare perturbation series in some grand unified theory—the precise theory does not enter into the discussion—starting with some bare mass μ_0 for φ. The Weisskopf phenomenon tells us that quantum correction shifts μ_0^2 by a huge quadratically cutoff dependent amount $\delta\mu_0^2 \sim f^2\Lambda^2 \sim f^2 M_{\text{GUT}}^2$, where we have substituted for Λ the only natural mass scale around, namely M_{GUT}, and where f denotes some dimensionless coupling. To have the physical mass squared $\mu^2 = \mu_0^2 + \delta\mu_0^2$ come out to be of order M_{EW}^2, something like 28 orders of magnitude smaller than M_{GUT}^2, would require an extremely fine-tuned and highly unnatural cancellation between μ_0^2 and $\delta\mu_0^2$. How this could happen "naturally" poses a severe challenge to theoretical physicists.

Naturalness

The hierarchy problem is closely connected with the notion of naturalness dear to the theoretical physics community. We naturally expect that dimensionless ratios of parameters in our theories should be of order unity, where the phrase "order unity" is interpreted liberally between friends, say anywhere from 10^{-2} or 10^{-3} to 10^2 or 10^3. Following 't Hooft, we can formulate a technical definition of naturalness: The smallness of a dimensionless parameter η would be considered natural only if a symmetry emerges in the limit $\eta \rightarrow 0$. Thus, fermion masses could be naturally small, since, as you will recall from chapter II.1, a chiral symmetry emerges when a fermion mass is set equal to zero. On the other hand, no particular symmetry emerges when we set either the bare or renormalized mass of a scalar field equal to zero. This represents the essence of the hierarchy problem.

Exercises

VII.6.1 Suppose there are F' new families of quarks and leptons with masses of order M'. Adopting the crude approximation described in exercise VI.8.2 of ignoring these families for μ below M' and of treating M'

as negligible for μ above M', run the renormalization group flow and discuss how various predictions, such as proton lifetime, are changed.

VII.6.2 Work out proton decay in detail. Derive relations between the following decay rates: $\Gamma(p \to \pi^0 e^+)$, $\Gamma(p \to \pi^+ \bar{\nu})$, $\Gamma(n \to \pi^- e^+)$, and $\Gamma(n \to \pi^0 \bar{\nu})$.

VII.6.3 Show that $SU(5)$ conserves the combination $B - L$. For a challenge, invent a grand unified theory that violates $B - L$.

SO(10) Unification

Each family into a single representation

At the end of chapter VII.5 we felt we had good reason to think that $SU(5)$ unification is not the end of the story. Let us ask if we might be able to fit the 5 and 10* into a single representation of a bigger group G containing $SU(5)$.

It turns out that there is a natural embedding of $SU(5)$ into the orthogonal $SO(10)$ that works,[1] but to explain that I have to teach you some group theory. The starting point is perhaps somewhat surprising: We go back to chapter II.3, where we learned that the Lorentz group $SO(3, 1)$, or its Euclidean cousin $SO(4)$, has spinor representations. We will now generalize the concept of spinors to d-dimensional Euclidean space. I will work out the details for d even and leave the odd dimensions as an exercise for you. You might also want to review appendix B now.

Clifford algebra and spinor representations

Start with an assertion. For any integer n we claim that we can find $2n$ hermitean matrices γ_i ($i = 1, 2, \cdots, 2n$) that satisfy the Clifford algebra

$$\{\gamma_i, \gamma_j\} = 2\delta_{ij} \tag{1}$$

In other words, to prove our claim we have to produce $2n$ hermitean matrices γ_i that anticommute with each other and square to the identity matrix. We will refer to the γ_i's as the γ matrices for $SO(2n)$.

For $n = 1$, it is a breeze: $\gamma_1 = \tau_1$ and $\gamma_2 = \tau_2$. There you are.

---

[1] Howard Georgi told me that he actually found $SO(10)$ before $SU(5)$.

Now iterate. Given the $2n$ γ matrices for $SO(2n)$ we construct the $(2n + 2)$ γ matrices for $SO(2n + 2)$ as follows

$$\gamma_j^{(n+1)} = \gamma_j^{(n)} \otimes \tau_3 = \begin{pmatrix} \gamma_j^{(n)} & 0 \\ 0 & -\gamma_j^{(n)} \end{pmatrix}, \quad j = 1, 2, \cdots, 2n \tag{2}$$

$$\gamma_{2n+1}^{(n+1)} = 1 \otimes \tau_1 = \begin{pmatrix} 0 & 1 \\ 1 & 0 \end{pmatrix} \tag{3}$$

$$\gamma_{2n+2}^{(n+1)} = 1 \otimes \tau_2 = \begin{pmatrix} 0 & -i \\ i & 0 \end{pmatrix} \tag{4}$$

(Throughout this book 1 denotes a unit matrix of the appropriate size.) The superscript in parentheses is obviously for us to keep track of which set of γ matrices we are talking about. Verify that if the $\gamma^{(n)}$'s satisfy the Clifford algebra, the $\gamma^{(n+1)}$'s do as well. For example,

$$\{\gamma_j^{(n+1)}, \gamma_{2n+1}^{(n+1)}\} = (\gamma_j^{(n)} \otimes \tau_3) \cdot (1 \otimes \tau_1) + (1 \otimes \tau_1) \cdot (\gamma_j^{(n)} \otimes \tau_3)$$

$$= \gamma_j^{(n)} \otimes \{\tau_3, \tau_1\} = 0$$

This iterative construction yields for $SO(2n)$ the γ matrices

$$\gamma_{2k-1} = 1 \otimes 1 \otimes \cdots \otimes 1 \otimes \tau_1 \otimes \tau_3 \otimes \tau_3 \otimes \cdots \otimes \tau_3 \tag{5}$$

and

$$\gamma_{2k} = 1 \otimes 1 \otimes \cdots \otimes 1 \otimes \tau_2 \otimes \tau_3 \otimes \tau_3 \otimes \cdots \otimes \tau_3 \tag{6}$$

with 1 appearing $k - 1$ times and τ_3 appearing $n - k$ times. The γ's are evidently 2^n by 2^n matrices. When and if you feel confused at any point in this discussion you should work things out explicitly for $SO(4)$, $SO(6)$, and so on.

In analogy with the Lorentz group, we define $2n(2n - 1)/2 = n(2n - 1)$ hermitean matrices

$$\sigma_{ij} \equiv \frac{i}{2}[\gamma_i, \gamma_j] \tag{7}$$

Note that σ_{ij} is equal to $i\gamma_i\gamma_j$ for $i \neq j$ and vanishes for $i = j$. The commutation of the σ's with each other is thus easy to work out. For example,

$$[\sigma_{12}, \sigma_{23}] = -[\gamma_1\gamma_2, \gamma_2\gamma_3] = -\gamma_1\gamma_2\gamma_2\gamma_3 + \gamma_2\gamma_3\gamma_1\gamma_2 = -[\gamma_1, \gamma_3] = 2i\sigma_{13}$$

Roughly speaking, the γ_2's in σ_{12} and σ_{23} knock each other out. Thus, you see that the $\frac{1}{2}\sigma_{ij}$'s satisfy the same commutation relations as the generators J^{ij}'s of $SO(2n)$ (as given in appendix B). The $\frac{1}{2}\sigma_{ij}$'s represent the J^{ij}'s.

As 2^n by 2^n matrices, the σ's act on an object ψ with 2^n components that we will call the spinor ψ. Consider the unitary transformation $\psi \to e^{i\omega_{ij}\sigma_{ij}}\psi$ with $\omega_{ij} = -\omega_{ji}$ a set of real numbers. Then

$$\psi^\dagger\gamma_k\psi \to \psi^\dagger e^{-i\omega_{ij}\sigma_{ij}}\gamma_k e^{i\omega_{ij}\sigma_{ij}}\psi = \psi^\dagger\gamma_k\psi - i\omega_{ij}\psi^\dagger[\sigma_{ij}, \gamma_k]\psi + \cdots$$

for ω_{ij} infinitesimal. Using the Clifford algebra we easily evaluate the commutator as $[\sigma_{ij}, \gamma_k] = -2i(\delta_{ik}\gamma_j - \delta_{jk}\gamma_i)$. (If k is not equal to either i or j then γ_k clearly commutes with σ_{ij}, and if k is equal to either i or j, then we use $\gamma_k^2 = 1$.) We see that the set of objects $v_k \equiv \psi^\dagger \gamma_k \psi$, $k = 1, \cdots, 2n$ transforms as a vector in $2n$-dimensional space, with $4\omega_{ij}$ the infinitesimal rotation angle in the ij plane:

$$v_k \to v_k - 2(\omega_{kj}v_j - \omega_{ik}v_i) = v_k - 4\omega_{kj}v_j \qquad (8)$$

(in complete analogy to $\bar{\psi}\gamma^\mu\psi$ transforming as a vector under the Lorentz group.) This gives an alternative proof that $\frac{1}{2}\sigma_{ij}$ represents the generators of $SO(2n)$.

We define the matrix $\gamma^{\mathrm{FIVE}} = (-i)^n \gamma_1 \gamma_2 \cdots \gamma_{2n}$, which in the basis we are using has the explicit form

$$\gamma^{\mathrm{FIVE}} = \tau_3 \otimes \tau_3 \otimes \cdots \otimes \tau_3 \qquad (9)$$

with τ_3 appearing n times. By analogy with the Lorentz group we define the "left handed" spinor $\psi_L \equiv \frac{1}{2}(1 - \gamma^{\mathrm{FIVE}})\psi$ and the "right handed" spinor $\psi_R \equiv \frac{1}{2}(1 + \gamma^{\mathrm{FIVE}})\psi$, such that $\gamma^{\mathrm{FIVE}}\psi_L = -\psi_L$ and $\gamma^{\mathrm{FIVE}}\psi_R = \psi_R$. Under $\psi \to e^{i\omega_{ij}\sigma_{ij}}\psi$, we have $\psi_L \to e^{i\omega_{ij}\sigma_{ij}}\psi_L$ and $\psi_R \to e^{i\omega_{ij}\sigma_{ij}}\psi_R$ since γ^{FIVE} commutes with σ_{ij}. The projection into left and right handed spinors cut the number of components into halves and thus we arrive at the important conclusion that the two irreducible spinor representations of $SO(2n)$ have dimension 2^{n-1}. (Convince yourself that the representation cannot be reduced further.) In particular, the spinor representation of $SO(10)$ is $2^{10/2-1} = 2^4 = 16$-dimensional. We will see that the 5* and 10 of $SU(5)$ can be fit into the 16 of $SO(10)$.

Embedding unitary groups into orthogonal groups

The unitary group $SU(5)$ can be naturally embedded into the orthogonal group $SO(10)$. In fact, I will now show you that embedding $SU(n)$ into $SO(2n)$ is as easy as $z = x + iy$.

Consider the $2n$-dimensional real vectors $x = (x_1, \cdots, x_n, y_1, \cdots, y_n)$ and $x' = (x'_1, \cdots, x'_n, y'_1, \cdots, y'_n)$. By definition, $SO(2n)$ consists of linear transformations on these two real vectors leaving their scalar product $x'x = \sum_{j=1}^{n}(x'_j x_j + y'_j y_j)$ invariant.

Now out of these two real vectors we can construct two n-dimensional complex vectors $z = (x_1 + iy_1, \cdots, x_n + iy_n)$ and $z' = (x'_1 + iy'_1, \cdots, x'_n + iy'_n)$. The group $U(n)$ consists of transformations on the two n-dimensional complex vectors z and z' leaving invariant their scalar product

$$(z')^* z = \sum_{j=1}^{n}(x'_j + iy'_j)^*(x_j + iy_j)$$

$$= \sum_{j=1}^{n}(x'_j x_j + y'_j y_j) + i\sum_{j=1}^{n}(x'_j y_j - y'_j x_j)$$

In other words, $SO(2n)$ leaves $\sum_{j=1}^{n}(x'_j x_j + y'_j y_j)$ invariant, but $U(n)$ consists of the subset of those transformations in $SO(2n)$ that leave invariant not only $\sum_{j=1}^{n}(x'_j x_j + y'_j y_j)$ but also $\sum_{j=1}^{n}(x'_j y_j - y'_j x_j)$.

Now that we understand this natural embedding of $U(n)$ into $SO(2n)$, we see that the defining or vector representation of $SO(2n)$, which we will call simply $2n$, decomposes upon restriction to $U(n)$ into the two defining representations of $U(n)$, n and n^*; thus

$$2n \to n \oplus n^* \tag{10}$$

In other words, $(x_1, \cdots, x_n, y_1, \cdots, y_n)$ can be written as $(x_1 + iy_1, \cdots, x_n + iy_n)$ and $(x_1 - iy_1, \cdots, x_n - iy_n)$ Note that this is the analog of (VII.5.4) indicating that the defining representation of $SU(5)$ decomposes into representations of $SU(3) \otimes SU(2) \otimes U(1)$:

$$5 \to (3^*, 1, \tfrac{1}{3}) \oplus (1, 2, -\tfrac{1}{2}). \tag{11}$$

Given the decomposition law (10), we can now figure out how other representations of $SO(2n)$ decompose when restricted to the natural subgroup $U(n)$. The tensor representations of $SO(2n)$ are easy, since they are constructed out of the vector representation. [This is precisely what we did in going from (VII.5.4) to (VII.5.7, 8, and 9).] For example, the adjoint representation of $SO(2n)$, which has dimension $2n(2n-1)/2 = n(2n-1)$, transforms as an antisymmetric 2-index tensor $2n \otimes_A 2n$ and so decomposes into

$$2n \otimes_A 2n \to (n \oplus n^*) \otimes_A (n \oplus n^*) \tag{12}$$

according to (10). The antisymmetric product $\otimes_A$ on the right hand side is, of course, to be evaluated within $U(n)$. For instance, $n \otimes_A n$ is the $n(n-1)/2$ representation of $U(n)$. In this way, we see that

$$
\begin{aligned}
n(2n-1) \to\ & n^2 - 1 \text{ (the adjoint)} \\
& \oplus 1 \text{ (the singlet)} \\
& \oplus n(n-1)/2 \\
& \oplus (n(n-1)/2)^*
\end{aligned} \tag{13}
$$

As a check, the total dimension of the representations of $U(n)$ on the right hand side adds up to $(n^2 - 1) + 1 + 2n(n-1)/2 = n(2n-1)$. In particular, for $SO(10) \supset SU(5)$, we have $45 \to 24 \oplus 1 \oplus 10 \oplus 10^*$ and of course $24 + 1 + 10 + 10 = 45$.

Decomposing the spinor

It is more difficult to figure out how the spinor representation of $SO(2n)$ decompose upon restriction to $U(n)$. I give here a heuristic argument that satisfies most physicists, but certainly not mathematicians. I will just do $SO(10) \supset SU(5)$ and let you work out the general case. The question is how the 16 falls apart. Just from numerology and from knowing the dimensions of the smaller representations of $SU(5)$ (1, 5, 10, 15) we see there are only so many possibilities, some of them rather unlikely, for example, the 16 falling apart into 16 1's.

Picture the spinor 16 of $SO(10)$ breaking up into a bunch of representations of $SU(5)$. By definition, the 45 generators of $SO(10)$ scramble all these representations together. Let us ask what the various pieces of 45, namely $24 \oplus 1 \oplus 10 \oplus 10^*$, do to these representations.

The 24 transform each of the representations of $SU(5)$ into itself, of course, because they are the 24 generators of $SU(5)$ and that is what generators were born to do. The generator

1 can only multiply each of these representations by a real number. (In other words, the corresponding group element multiplies each of the representations by a phase factor.)

What does the 10, which as you recall from chapter VII.5 is represented as an antisymmetric tensor with two upper indices and hence also known as [2], do to these representations? Suppose the bunch of representations that S breaks up into contains the singlet $[0] = 1$ of $SU(5)$. The $10 = [2]$ acting on [0] gives the $[2] = 10$. (Almost too obvious for words! An antisymmetric tensor of two indices combined with a tensor with no indices is an antisymmetric tensor of two indices.) What about $10 = [2]$ acting on [2]? The result is a tensor with four upper indices. It certainly contains the [4], which is equivalent to $[1]^* = 5^*$. But look, $1 \oplus 10 \oplus 5^*$ already add up to 16. Thus, we have accounted for everybody. There can't be more. So we conclude

$$S^+ \to [0] \oplus [2] \oplus [4] = 1 \oplus 10 \oplus 5^* \tag{14}$$

The 5^* and the 10 of $SU(5)$ fit inside the 16^+ of $SO(10)$!

We will learn later that the two spinor representations of $SO(10)$ are conjugate to each other. Indeed, you may have noticed that I snuck a superscript plus on the letter S. The conjugate spinor S^- breaks up into the conjugate of the representations in (14):

$$S^- \to [1] \oplus [3] \oplus [5] = 5 \oplus 10^* \oplus 1^* \tag{15}$$

The long lost antineutrino

The fit would be perfect if we introduce one more field transforming as a 1, that is, a singlet under $SU(5)$ and hence a fortiori a singlet under $SU(3) \otimes SU(2) \otimes U(1)$. In other words, this field does not participate in the strong, weak, and electromagnetic interactions, or in plain English, it describes a lepton with no electric charge and is not involved in the known weak interaction. Thus, this field can be identified as the "long lost" antineutrino field ν_L^c. This guy does not listen to any of the known gauge bosons.

Recall that we are using a convention in which all fermion fields are left handed, and hence we have written ν_L^c. By a conjugate transformation, as explained earlier, this is equivalent to the right handed neutrino field ν_R.

Since ν_R is an $SU(5)$ singlet, we can give it a Majorana mass M without breaking $SU(5)$. Hence we expect M to be larger than or of the same order of magnitude as the mass scale at which $SU(5)$ is broken, which as we saw in chapter VII.5 is much higher than the mass scales that have been explored experimentally. This explains why ν_R has not been seen.

On the other hand, with the presence of ν_R we can have a Dirac mass term $m(\bar{\nu}_L \nu_R +$ h.c.). Since this term breaks $SU(2) \otimes U(1)$ just like the mass terms for the quarks and leptons we know, we expect m to be of the same order of magnitude[2] as the known quark and lepton masses (which for reasons unknown span an enormous range).

---

[2] Explicitly, with ν_R now available we can add to the $SU(2) \otimes U(1)$ theory of chapter VII.2 the term $f' \tilde{\varphi} \bar{\nu}_R \psi_L$, where $\tilde{\varphi} \equiv \tau^2 \varphi^\dagger$. In the absence of any indication to the contrary, we might suppose that f' is of the same order of magntiude as the coupling f that leads to the electron mass.

Thus, in the space spanned by (ν, ν^c) we have the (Majorana) mass matrix

$$\mathcal{M} = \begin{pmatrix} 0 & m \\ m & M \end{pmatrix} \tag{16}$$

with $M \gg m$. Since the trace and determinant of $\mathcal{M}$ are M and $-m^2$, respectively, $\mathcal{M}$ has a large eigenvalue $\sim M$ and a small eigenvalue $\sim m^2/M$. A tiny mass $\sim m^2/M$, suppressed relative to the usual quark and lepton masses by the factor m/M, is naturally generated for the (observed) left handed neutrino. This rather attractive scenario, known as the seesaw mechanism for obvious reason, was discovered independently by Minkowski and by Glashow, and somewhat later by Yanagida and by Gell-Mann, Ramond, and Slansky.

Again, the tight fit of the 5^* and the 10 of $SU(5)$ inside the 16^+ of $SO(10)$ has convinced many physicists that it is surely right.

A binary code for the world

Given the product form of the γ matrices in (5) and (6), and hence of σ_{ij}, we can write the states of the spinor representations as

$$|\varepsilon_1 \varepsilon_2 \cdots \varepsilon_n\rangle \tag{17}$$

where each of the ε's takes on the values ± 1. For example, for $n = 1$, $\tau_1 |+\rangle = |-\rangle$ and $\tau_1 |-\rangle = |+\rangle$, while $\tau_2 |+\rangle = i |-\rangle$ and $\tau_2 |-\rangle = -i |+\rangle$. From (9) we see that

$$\gamma^{\text{FIVE}} |\varepsilon_1 \varepsilon_2 \cdots \varepsilon_n\rangle = (\Pi_{j=1}^n \varepsilon_j) |\varepsilon_1 \varepsilon_2 \cdots \varepsilon_n\rangle \tag{18}$$

The right handed spinor S^+ consists of those states $|\varepsilon_1 \varepsilon_2 \cdots \varepsilon_n\rangle$ with $(\Pi_{j=1}^n \varepsilon_j) = +1$, and the left handed spinor S^- those states with $(\Pi_{j=1}^n \varepsilon_j) = -1$. Indeed, the spinor representations have dimension 2^{n-1}.

Thus, in $SO(10)$ unification the fundamental quarks and leptons are described by a five-bit binary code, with states such as $|++--+\rangle$ and $|-+---\rangle$. Personally, I find this a rather pleasing picture of the world.

Let us work out the states explicitly. This also gives me a chance to make sure that you understand the group theory presented in this chapter. Start with the much simpler case of $SO(4)$. The spinor S^+ consists of $|++\rangle$ and $|--\rangle$ while the spinor S^- consists of $|+-\rangle$ and $|-+\rangle$. As discussed in chapter II.3, $SO(4)$ contains two distinct $SU(2)$ subgroups. Removing a few factors of i from the discussion in chapter II.3 we see that the third generator of $SU(2)$, call it σ_3, can be taken to be either $\sigma_{12} - \sigma_{34}$ or $\sigma_{12} + \sigma_{34}$. The two choices correspond to the two distinct $SU(2)$ subgroups. We choose (arbitrarily) $\sigma_3 = \frac{1}{2}(\sigma_{12} - \sigma_{34})$. From (5) and (6) we have $\sigma_{12} = i\gamma_1\gamma_2 = i(\tau_1 \otimes \tau_3)(\tau_2 \otimes \tau_3) = -\tau_3 \otimes 1$ and $\sigma_{34} = -1 \otimes \tau_3$, and so $\sigma_3 = \frac{1}{2}(-\tau_3 \otimes 1 + 1 \otimes \tau_3)$. To figure out how the four states $|++\rangle$, $|--\rangle$, $|+-\rangle$, and $|-+\rangle$ transform under our chosen $SU(2)$, let us act on them with σ_3. For example,

$$\sigma_3 |++\rangle = \frac{1}{2}(-\tau_3 \otimes 1 + 1 \otimes \tau_3) |++\rangle = \frac{1}{2}(-1 + 1) |++\rangle = 0$$

and

$$\sigma_3|-+\rangle = \tfrac{1}{2}(-\tau_3 \otimes 1 + 1 \otimes \tau_3)|-+\rangle = \tfrac{1}{2}(1+1)|-+\rangle = |-+\rangle$$

Aha, under $SU(2)$ $|++\rangle$ and $|--\rangle$ are two singlets while $|+-\rangle$ and $|-+\rangle$ make up a doublet.

Note that this is consistent with the generalization of (14) and (15), namely that upon the restriction of $SO(2n)$ to $U(n)$ the spinors decompose as

$$S^+ \to [0] \oplus [2] \oplus \cdots \qquad (19)$$

and

$$S^- \to [1] \oplus [3] \oplus \cdots \qquad (20)$$

I have not indicated the end of the two sequences: A moment's reflection indicates that it depends on whether n is even or odd. In our example, $n = 2$, and thus $2^+ \to [0] \oplus [2] = 1 \oplus 1$ and $2^- \to [1] = 2$. Similarly, for $n = 3$, upon the restriction of $SO(6)$ to $U(3)$, $4^+ \to [0] \oplus [2] = 1 \oplus 3^*$ and $4^- \to [1] \oplus [3] = 3 \oplus 1$. (Our choice of which triplet representation of $U(3)$ to call 3 or 3* is made to conform to common usage, as we will see presently.)

We are now ready to figure out the identity of each of the 16 states such as $|++--+\rangle$ in $SO(10)$ unification. First of all, (18) tells us that under the subgroup $SO(4) \otimes SO(6)$ of $SO(10)$ the spinor 16^+ decomposes as (since $\Pi_{j=1}^5 \varepsilon_j = +1$ implies $\varepsilon_1\varepsilon_2 = \varepsilon_3\varepsilon_4\varepsilon_5$)

$$16^+ \to (2^+, 4^+) \oplus (2^-, 4^-) \qquad (21)$$

We identify the natural $SU(2)$ subgroup of $SO(4)$ as the $SU(2)$ of the electroweak interaction and the natural $SU(3)$ subgroup of $SO(6)$ as the color $SU(3)$ of the strong interaction. Thus, according to the preceding discussion, $(2^+, 4^+)$ are the $SU(2)$ singlets of the standard $U(1) \otimes SU(2) \otimes SU(3)$ model, while $(2^-, 4^-)$ are the $SU(2)$ doublets. Here is the lineup (all fields being left handed as usual):

$SU(2)$ doublets:
$$\nu = |-+---\rangle$$
$$e^- = |+----\rangle$$
$$u = |-+++-\rangle, \ |-++-+\rangle, \ \text{and} \ |-+-++\rangle$$
$$d = |+-++-\rangle, \ |+-+-+\rangle, \ \text{and} \ |+--++\rangle$$

$SU(2)$ singlets:
$$\nu^c = |+++++\rangle$$
$$e^+ = |--+++\rangle$$
$$u^c = |+++--\rangle, \ |++-+-\rangle, \ \text{and} \ |++--+\rangle$$
$$d^c = |--+--\rangle, \ |---+-\rangle, \ \text{and} \ |-----+\rangle.$$

I assure you that this is a lot of fun to work out and I urge you to reconstruct this table without looking at it. Here are a few hints if you need help. From our discussion of $SU(2)$ I know that $\nu = |-+\varepsilon_3\varepsilon_4\varepsilon_5\rangle$ and $e^- = |+-\varepsilon_3\varepsilon_4\varepsilon_5\rangle$, but how do I know that $\varepsilon_3 = \varepsilon_4 = \varepsilon_5 = -1$? First, I know that $\varepsilon_3\varepsilon_4\varepsilon_5 = -1$. I also know that $4^- \to 3 \oplus 1$ upon restricting $SO(6)$ to color $SU(3)$. Well, of the four states $|---\rangle$, $|++-\rangle$, $|+-+\rangle$, and $|-++\rangle$ the "odd man out" is clearly $|---\rangle$. By the same heuristic argument, among the 16 possible states $|+++++\rangle$ is the "odd man out" and so must be ν^c.

There are lots of consistency checks. For example, once I identify $\nu = |-+---\rangle$, $e^- = |+----\rangle$, and $\nu^c = |+++++\rangle$, I can figure out the electric charge Q, which, since it transforms as a singlet under color $SU(3)$, must have the value $Q = a\varepsilon_1 + b\varepsilon_2 + c(\varepsilon_3 + \varepsilon_4 + \varepsilon_5)$ when acting on the state $|\varepsilon_1\varepsilon_2\varepsilon_3\varepsilon_4\varepsilon_5\rangle$. The constants a, b, and c can be determined from the three equations $Q(\nu) = -a + b - 3c = 0$, $Q(e^-) = -1$, and $Q(\nu^c) = 0$. Thus, $Q = -\frac{1}{2}\varepsilon_1 + \frac{1}{6}(\varepsilon_3 + \varepsilon_4 + \varepsilon_5)$.

Living in the computer age, I find it intriguing that the fundamental constituents of matter are coded by five bits. You can tell your condensed matter colleagues that their beloved electron is composed of the binary strings $+----$ and $--+++$. An intriguing possibility[3] suggests itself, that quarks and leptons may be composed of five different species of fundamental fermionic objects. We construct composites, writing a $+$ if that species is present, and a $-$ if it is absent. For example, from the expression for Q given above, we see that species 1 carries electric charge $-\frac{1}{2}$, species 2 is neutral, and species 3, 4, and 5 carry charge $\frac{1}{6}$. A more or less concrete model can even be imagined by binding these fundamental fermionic objects to a magnetic monopole.

I emphasize that particles transforming in 16^-, such as $|+-+++\rangle$, have not been observed experimentally.

A speculation on the origin of families

One of the great unsolved puzzles in particle physics is the family problem. Why do quarks and leptons come in three generations $\{\nu_e, e, u, d\}$, $\{\nu_\mu, \mu, c, s\}$, and $\{\nu_\tau, \tau, t, b\}$? The way we incorporate this experimental fact into our present day theory can only be described as pathetic: We repeat the fermionic sector of the Lagrangian three times without any understanding whatsoever. Three generations living together gives rise to a nagging family problem.

Our binary code view of the world suggests a wildly speculative (perhaps too speculative to mention in a textbook?) approach to the family problem: We add more bits. To me, a reasonable possibility is to "hyperunify" into an $SO(18)$ theory, putting all fermions into a single spinorial representation $S^+ = 256^+$, which upon the breaking of $SO(18)$ to $SO(10) \otimes SO(8)$ decomposes as

$$256^+ \to (16^+, 8^+) \oplus (16^-, 8^-) \tag{22}$$

We have a lot of 16^+'s. Unhappily, we see that group theory [see also (21)] dictates that we also get a bunch of unwanted 16^-'s. One suggestion is that Nature might repeat the trick

---

[3] For further details, see F. Wilczek and A. Zee, *Phys. Rev.* D25: 553, 1982, Section IV.

She uses with color $SU(3)$, whose strong force confines fields that are not color singlets (chapter VII.3). Interestingly, we can exploit a striking feature of $SO(8)$, which some people regard as the most beautiful of all groups. In particular, the two spinorial representations $8^{\pm}$ have the same dimension as the vectorial representation 8^v (the equation $2^{n-1} = 2n$ has the unique solution $n = 4$). There is a transformation that cyclically rotates these three representations 8^+, 8^-, and 8^v into each other (in the jargon, the group $SO(8)$ admits an outer automorphism). Thus, there exists a subgroup $SO(5)$ of $SO(8)$ such that when we break $SO(8)$ into that $SO(5)$ 8^+ behaves like 8^v while 8^- behaves like a spinor, namely

$$8^+ \rightarrow 5 \oplus 1 \oplus 1 \oplus 1 \qquad \text{and} \qquad 8^- \rightarrow 4 \oplus 4^* \tag{23}$$

If we call this[4] $SO(5)$ hypercolor and assumes that the strong force associated with it confines all fields that are not hypercolor singlets, then only three 16^+'s remain! Unfortunately, as the relevant physics occurs in the energy regime above grand unification, our knowledge of the dynamics of symmetry breaking is far too paltry for us to make any further statements.

Charge conjugation

The product $\otimes$ notation we use here allows us to construct the conjugation matrix C explicitly. By definition $C^{-1}\sigma_{ij}^* C = -\sigma_{ij}$ (so that C changes $e^{i\theta_{ij}\sigma_{ij}}$ into its complex conjugate.) From (2), (3), and (4) we see that we can construct

$$C^{(n+1)} = \{ \begin{array}{ll} C^{(n)} \otimes \tau_1 & \text{if} \quad n \text{ odd} \\ C^{(n)} \otimes \tau_2 & \text{if} \quad n \text{ even} \end{array} \tag{24}$$

You can check that this gives $C^{-1}\gamma_j^* C = (-1)^n \gamma_j$ and hence the desired result.

Explicitly, C is a direct product of an alternating sequence of τ_1 and τ_2 and so we deduce an important property. Acting on $|\varepsilon_1 \varepsilon_2 \cdots \varepsilon_n\rangle$, C flips the sign of all the ε's. Thus C changes the sign of $(\Pi_{j=1}^n \varepsilon_j)$ for n odd, and does not for n even. For n odd, the two spinor representations S^+ and S^- are conjugates of each other, while for n even, they are conjugates of themselves, or in other words, they are real. This can also be seen directly from $C^{-1}\gamma^{\text{FIVE}}C = (-1)^n\gamma^{\text{FIVE}}$. You can check this with all the explicit examples we have encountered: $SO(2)$, $SO(4)$, $SO(6)$, $SO(8)$, $SO(10)$, and $SO(18)$. See also exercise VII.7.3.

Anomalies

What about anomalies in $SO(2n)$ grand unification? According to the discussion in chapter VII.5 we have to evaluate $A^{ijklmn} \equiv \text{tr}(J^{ij} \{J^{kl}, J^{mn}\})$ over the fermion representation.

---

[4] The reader savvy with group theory would recognize that $SO(5)$ is isomorphic with the symplectic group $Sp(4)$ and that the Dynkin diagram of $SO(8)$ is the most symmetric of all.

Applying an $SO(2n)$ transformation $J^{ij} \to O^T J^{ij} O$ we see easily that A^{ijklmn} is an invariant tensor. Can we construct an invariant 6-index tensor with the appropriate symmetry properties (e.g., $A^{ijklmn} = -A^{jiklmn}$) in $SO(2n)$? We can't, except in $SO(6)$, for which we have ε^{ijklmn}. Thus, A^{ijklmn} vanishes except in $SO(6)$, where it is proportional to ε^{ijklmn}. An elegant one line proof that any grand unified theory based on $SO(2n)$ for $n \neq 3$ is free from anomaly!

The cancellation of the anomaly between 5* and 10 at the end of chapter VII.5 doesn't seem so miraculous any more. Miracles tend to fade away as we gain deeper understanding.

Amusingly, by discussing a physics question, namely whether a gauge theory is renormalizable or not, we have discovered a mathematical fact. What is so special about $SO(6)$? See exercise VII.7.5.

Exercises

VII.7.1 Work out the Clifford algebra in d-dimensional space for d odd.

VII.7.2 Work out the Clifford algebra in d-dimensional Minkowski space.

VII.7.3 Show that the Clifford algebra for $d = 4k$ and for $d = 4k + 2$ have somewhat different properties. (If you need help with this and the two preceding exercises, look up F. Wilczek and A. Zee, *Phys. Rev.* D25: 553, 1982.)

VII.7.4 Discuss the Higgs sector of the $SO(10)$. What do you need to give mass to the quarks and leptons?

VII.7.5 The group $SO(6)$ has $6(6-1)/2 = 15$ generators. Notice that the group $SU(4)$ also has $4^2 - 1 = 15$ generators. Substantiate your suspicion that $SO(6)$ and $SU(4)$ are isomorphic. Identify some low dimensional representations.

VII.7.6 Show that (unfortunately) the number of families we get in $SO(18)$ depends on which subgroup of $SO(8)$ we take to be hypercolor.

VII.7.7 If you want to grow up to be a string theorist, you need to be familiar with the Dirac equation in various dimensions but especially in 10. As a warm up, study the Dirac equation in 2-dimensional spacetime. Then proceed to study the Dirac equation in 10-dimensional spacetime.

Part VIII | Gravity and Beyond

VIII.1 | Gravity as a Field Theory and the Kaluza-Klein Picture

Including gravity

Field theory texts written a generation ago typically do not even mention gravity. The gravitational interaction, being so much weaker than the other three interactions, was simply not included in the education of particle physicists. The situation has changed with a vengeance: The main drive of theoretical high energy physics today is the unification of gravity with the other three interactions, with string theory the main candidate for a unified theory.

From a course on general relativity you would have learned about the Einstein-Hilbert action for gravity

$$S = \frac{1}{16\pi G} \int d^4x \sqrt{-g}\, R \equiv \int d^4x \sqrt{-g}\, M_P^2 R \tag{1}$$

where $g = \det g_{\mu\nu}$ denotes the determinant of the curved metric $g_{\mu\nu}$ of spacetime, R is the scalar curvature, and G is Newton's constant. Let me remind you that the Riemann curvature tensor

$$R^\lambda_{\ \mu\nu\kappa} = \partial_\nu \Gamma^\lambda_{\mu\kappa} - \partial_\kappa \Gamma^\lambda_{\mu\nu} + \Gamma^\sigma_{\mu\kappa}\Gamma^\lambda_{\nu\sigma} - \Gamma^\sigma_{\mu\nu}\Gamma^\lambda_{\kappa\sigma} \tag{2}$$

is constructed out of the Riemann-Christoffel symbol (recall chapter I.11):

$$\Gamma^\lambda_{\mu\nu} = \tfrac{1}{2} g^{\lambda\rho}(\partial_\nu g_{\rho\mu} + \partial_\mu g_{\rho\nu} - \partial_\rho g_{\mu\nu}) \tag{3}$$

The Ricci tensor is defined by $R_{\mu\kappa} = R^\nu_{\ \mu\nu\kappa}$ and the scalar curvature by $R = g^{\mu\nu} R_{\mu\nu}$. Varying S gives us[1] the Einstein field equation

$$R_{\mu\nu} - \tfrac{1}{2} g_{\mu\nu} R = -8\pi G T_{\mu\nu} \tag{4}$$

The Einstein-Hilbert action is uniquely determined if we require the action to be coordinate invariant and to involve two powers of spacetime derivative. As you can see from

---

[1] See, e.g., S. Weinberg, *Gravitation and Cosmology*, p. 364.

(2) and (3) the scalar curvature R involves two powers of derivative and the dimensionless field $g_{\mu\nu}$ and thus has mass dimension 2. Hence G^{-1} must have mass dimension 2. The second form in (1) emphasizes this point and is often preferred in modern work on gravity. (The modified Planck mass $M_P \equiv 1/\sqrt{16\pi G}$ differs from the usual Planck mass by a trivial factor, much like the relation between h and $\hbar$.)

The theory sprang from Einstein's profound intuition regarding the curvature of spacetime and is manifestly formulated in terms of geometric concepts. In many textbooks, Einstein's theory is developed, and rightly so, in purely geometric terms.

On the other hand, as I hinted back in chapter I.6, gravity can be treated on the same footing as the other interactions. After all, the graviton may be regarded as just another elementary particle like the photon. The action (1), however, does not look anything like the field theories we have studied thus far. I will now show you that in fact it does have the same kind of structure.

Gravity as a field theory

Let us write $g_{\mu\nu} = \eta_{\mu\nu} + h_{\mu\nu}$, where $\eta_{\mu\nu}$ denotes the flat Minkowski metric and $h_{\mu\nu}$ the deviation from the flat metric. Expand the action in powers of $h_{\mu\nu}$. In order not to drown in a sea of Lorentz indices, let us suppress them for a first go-around. Merely from the fact that the scalar curvature R involves two derivatives ∂ in its definition, we see that the expansion must have the schematic form

$$S = \int d^4x \frac{1}{16\pi G} (\partial h \partial h + h \partial h \partial h + h^2 \partial h \partial h + \cdots) \tag{5}$$

after dropping total divergences. As I remarked in chapter I.11, the field $h^{\mu\nu}(x)$ describes a graviton in flat space and is to be treated like any other field. The first term $\partial h \partial h$, which governs how the graviton propagates, is conceptually no different than the first term in the action for a scalar field $\partial \varphi \partial \varphi$ or for the photon field $\partial A \partial A$. The terms cubic and higher in h determine the interaction of the graviton with itself.

The Einstein-Hilbert action in the weak field expansion is structurally reminiscent of the Yang-Mills action, which may be written in schematic form as $S = \int d^4x (1/g^2)(\partial A \partial A + A^2 \partial A + A^4)$. As I explained in chapter IV.5, we understand the self interaction of the Yang-Mills bosons physically: The bosons themselves carry the charge to which they couple. We can understand the self interaction of the graviton similarly: The graviton couples to anything carrying energy and momentum, and it certainly carries energy and momentum. In contrast, the photon does not couple to itself.

We say that Yang-Mills and Einstein theories are nonlinear, while Maxwell theory is linear. The former are hard, the latter easy.

But while the Yang-Mills action terminates, the Einstein-Hilbert action, because of the presence of $\sqrt{-g}$ and of the inverse of $g_{\mu\nu}$, is an infinite series in the graviton field $h_{\mu\nu}$.

The other major difference is that while Yang-Mills theory is renormalizable, gravity is notoriously nonrenormalizable, as we argued by dimensional analysis in chapter III.2. We are now in a position to see this explicitly. Consider the self energy correction to the graviton

(a)   (b)

Figure VIII.1.1

propagator shown in figure VIII.1.1a. We see from the second term in (5) that the three-graviton coupling involves two powers of momentum. Thus the Feynman integral goes as $\int d^4k(kkkk/k^2k^2)$, with four powers of k in the numerator from the two vertices and four powers in the denominator from the propagators. Taking out two powers of momentum to extract the coefficient of $\partial h \partial h$, we see that the correction to $1/G$ is quadratically divergent. Because of the explicit powers of momentum in the coupling, the divergence gets worse and worse as we go to higher and higher order. Compare figure VIII.1.1b to 1a: We have three more propagators, worth $\sim 1/k^6$, and one more loop integration $\int d^4k$, but two more vertices $\sim k^4$. The degree of divergence goes up by 2. Of course we already knew all this by dimensional analysis.

As mentioned in chapter I.11 the fundamental definition

$$T^{\mu\nu}(x) = -\frac{2}{\sqrt{-g}}\frac{\delta S_M}{\delta g_{\mu\nu}(x)}$$

tells us that coupling of the graviton to matter (in the weak field limit) can be included by adding the term

$$-\int d^4x \, \tfrac{1}{2}h_{\mu\nu}T^{\mu\nu} \tag{6}$$

to the action, where $T^{\mu\nu}$ stands for the (flat spacetime) stress-energy tensor of all the matter fields of the world, a matter field being any field that is not the graviton field. Thus, with the inclusion of matter (5) is modified schematically[2] to

$$S = \int d^4x[\frac{1}{16\pi G}(\partial h \partial h + h \partial h \partial h + h^2 \partial h \partial h + \cdots) + (hT + \cdots)] \tag{7}$$

In chapter IV.5 I noted that we can bring Yang-Mills theory into the same convention commonly used in Maxwell theory by a trivial rescaling $A \to gA$. Similarly, we can also bring Einstein theory into the same convention by rescaling the graviton field $h^{\mu\nu} \to \sqrt{G}h^{\mu\nu}$ so that the action becomes (to ease writing we absorb 16π into G whenever we feel like it)

$$S = \int d^4x \, (\partial h \partial h + \sqrt{G}h\partial h\partial h + Gh^2\partial h\partial h + \cdots + \sqrt{G}hT)$$

---

[2] If this is to represent an expansion of S in powers of h, then strictly speaking, if we display the terms cubic and quartic in the Einstein-Hilbert action, we should also display the contribution coming from the terms of higher order in h contained in $T^{\mu\nu}(x) = -(2/\sqrt{-g})\delta S_M/\delta g_{\mu\nu}(x)$.

We see explicitly that $\sqrt{16\pi G} = 1/M_P$ measures the strength of the graviton coupling to itself and to all other fields. Once again, the enormity of M_P (compared to the scale of the strong interaction, say) indicates the feebleness of gravity.

Here we expanded $g_{\mu\nu}$ around a flat metric but we could just as well expand $g_{\mu\nu} = \bar{g}_{\mu\nu} + h_{\mu\nu}$, with $\bar{g}_{\mu\nu}$ a curved metric, that of a black hole (see chapter V.7) for instance.

Determining the weak field action

After this index free survey we are ready to tackle the indices. We would like to determine the first term $\partial h \partial h$ in (7) so that we can obtain the graviton propagator. Thus, we have to expand the action $S \equiv M_P^2 \int d^4x \sqrt{-g} g^{\mu\nu} R_{\mu\nu}$ up to and including order h^2. From (2) and (3) we see that the Ricci tensor $R_{\mu\nu}$ starts in $O(h)$ so that it suffices to evaluate $\sqrt{-g} g^{\mu\nu}$ to $O(h)$. That's easy: As we have already seen in chapter I.11, $g = -[1 + \eta^{\mu\nu} h_{\mu\nu} + O(h^2)]$ and $g^{\mu\nu} = \eta^{\mu\nu} - h^{\mu\nu} + O(h^2)$ so that $\sqrt{-g} g^{\mu\nu} = \eta^{\mu\nu} - h^{\mu\nu} + \frac{1}{2}\eta^{\mu\nu}h + O(h^2)$, where we have defined $h \equiv \eta^{\mu\nu} h_{\mu\nu}$. We now must calculate $R_{\mu\nu}$ to $O(h^2)$, a straightforward but tedious task starting from (2) and (3).

In line with the spirit of this book, which is to avoid tedious calculation whenever possible, I will now show you how to get around this. We invoke symmetry considerations! Under a general coordinate transformation $x^\mu \to x'^\mu = x^\mu - \varepsilon^\mu(x)$ the metric changes to $g'^{\mu\nu} = (\partial x'^\mu/\partial x^\sigma)(\partial x'^\nu/\partial x^\tau)g^{\sigma\tau}$. Plugging in $g^{\mu\nu} = \eta^{\mu\nu} - h^{\mu\nu} + \cdots$, lowering the indices (with $\eta_{\mu\nu}$ to this order), and using $(\partial x'^\mu/\partial x^\sigma) = \delta^\mu_\sigma - \partial_\sigma \varepsilon^\mu$, we find, treating $\partial_\mu \varepsilon_\nu$ as of the same order as $h^{\mu\nu}$:

$$h'_{\mu\nu} = h_{\mu\nu} + \partial_\mu \varepsilon_\nu + \partial_\nu \varepsilon_\mu \tag{8}$$

Note the structural similarity to the electromagnetic gauge transformation $A'_\mu = A_\mu - \partial_\mu \Lambda$. Very nice! We will explore the sense in which gravity can be regarded as a gauge theory in more detail later.

We are looking for the terms in the action quadratic in h and quadratic in ∂. Lorentz invariance tells us that there are four possible terms (To see this, first write down terms with the indices on the two ∂ matching, then the terms with the index on a ∂ matching an index on an h, and so on):

$$S = \int d^4x (a \partial_\lambda h^{\mu\nu} \partial^\lambda h_{\mu\nu} + b \partial_\lambda h^\mu_\mu \partial^\lambda h^\nu_\nu + c \partial_\lambda h^{\lambda\nu} \partial^\mu h_{\mu\nu} + d h^\lambda_\lambda \partial^\mu \partial^\nu h_{\mu\nu})$$

with four unknown constants a, b, c, and d. Now vary S with $\delta h_{\mu\nu} = \partial_\mu \varepsilon_\nu + \partial_\nu \varepsilon_\mu$, integrating by parts freely. For example,

$$\delta(\partial_\lambda h^{\mu\nu} \partial^\lambda h_{\mu\nu}) = 2[\partial_\lambda(2\partial^\mu \varepsilon^\nu)](\partial^\lambda h_{\mu\nu}) \text{``} = \text{''} 4\varepsilon^\nu \partial^2 \partial^\mu h_{\mu\nu}$$

Since there are three objects linear in h, linear in ε, and cubic in ∂ (namely $\varepsilon^\nu \partial^2 \partial_\nu h$ and $\varepsilon^\nu \partial_\nu \partial^\lambda \partial^\mu h_{\lambda\mu}$ in addition to the one already shown) the condition $\delta S = 0$ gives three equations, just enough to fix the action up to an overall constant, corresponding to Newton's constant. The invariant combination turns out to be

$$\mathcal{I} \equiv \tfrac{1}{2}\partial_\lambda h^{\mu\nu} \partial^\lambda h_{\mu\nu} - \tfrac{1}{2}\partial_\lambda h^\mu_\mu \partial^\lambda h^\nu_\nu - \partial_\lambda h^{\lambda\nu} \partial^\mu h_{\mu\nu} + \partial^\nu h^\lambda_\lambda \partial^\mu h_{\mu\nu} \tag{9}$$

Thus, even if we had never heard of the Einstein-Hilbert action we could still determine the action for gravity in the weak field limit by requiring that the action be invariant under the transformation (8). This is hardly surprising since coordinate invariance determines the Einstein-Hilbert action. Still, it is nice to construct gravity "from scratch."

Referring to (6), we can now write the weak field expansion of S as

$$S_{\text{wfg}} = \int d^4x \left(\frac{1}{32\pi G} \mathcal{I} - \frac{1}{2} h_{\mu\nu} T^{\mu\nu} \right)$$

without having to expand R to $O(h^2)$. The coefficient of $\mathcal{I}$ is fixed by the requirement that we reproduce the usual Newtonian gravity (see later).

The graviton propagator

As we anticipated in (5) the action S_{wfg} indeed has the same quadratic structure of all the field theories we have studied, and so as usual the graviton propagator is just the inverse of a differential operator. But just as in Maxwell and Yang-Mills theories the relevant differential operator in Einstein-Hilbert theory does not have an inverse because of the "gauge invariance" in (8).

No problem. We have already developed the Faddeev-Popov method to deal with this difficulty. In fact, for my limited purposes here, to derive the graviton propagator in flat spacetime, I don't even need the full-blown Faddeev-Popov formalism with ghosts and all.[3] Indeed, recall from chapter III.4 that for the Feynman gauge ($\xi = 1$) we simply add $(\partial A)^2$ to the invariant

$$\tfrac{1}{2} F^{\mu\nu} F_{\mu\nu} = \partial^\mu A^\nu (\partial_\mu A_\nu - \partial_\nu A_\mu) \text{``} = \text{''} - A^\mu \eta_{\mu\nu} \partial^2 A^\nu - (\partial A)^2$$

thus canceling the last term. Inverting the differential operator $-\eta_{\mu\nu}\partial^2$ we obtain the photon propagator in the Feynman gauge $-i\eta_{\mu\nu}/k^2$. We play the same "trick" for gravity. After staring at

$$\mathcal{I} = \tfrac{1}{2}\partial_\lambda h^{\mu\nu}\partial^\lambda h_{\mu\nu} - \tfrac{1}{2}\partial_\lambda h^\mu_\mu \partial^\lambda h^\nu_\nu - \partial_\lambda h^{\lambda\nu}\partial^\mu h_{\mu\nu} + \partial^\nu h^\lambda_\lambda \partial^\mu h_{\mu\nu}$$

for a while, we see that by adding $(\partial^\mu h_{\mu\nu} - \tfrac{1}{2}\partial_\nu h^\lambda_\lambda)^2$ we can knock off the last two terms in $\mathcal{I}$ so that S_{wfg} effectively becomes

$$S_{\text{wfg}} = \int d^4x \frac{1}{2} \left[\frac{1}{32\pi G} \left(\partial_\lambda h^{\mu\nu}\partial^\lambda h_{\mu\nu} - \tfrac{1}{2}\partial_\lambda h \partial^\lambda h \right) - h_{\mu\nu} T^{\mu\nu} \right] \qquad (10)$$

In other words, the freedom in choosing $h_{\mu\nu}$ in (8) allows us to impose the so-called harmonic gauge condition

$$\partial_\mu h^\mu_\nu = \tfrac{1}{2}\partial_\nu h^\lambda_\lambda \qquad (11)$$

(the linearized version of $\partial_\mu(\sqrt{-g}\,g^{\mu\nu}) = 0$.)

---

[3] This is because (8) does not involve the field $h_{\mu\nu}$, just as in the Maxwell case but unlike the Yang-Mills case. Since we do not intend to calculate loop diagrams in quantum gravity, we do not need the full power of the Faddeev-Popov method.

Writing (10) in the form

$$S = \frac{1}{32\pi G} \int d^4x \left[h^{\mu\nu} K_{\mu\nu;\lambda\sigma}(-\partial^2) h^{\lambda\sigma} + O(h^3) \right]$$

we see that we have to invert the matrix

$$K_{\mu\nu;\lambda\sigma} \equiv \tfrac{1}{2}(\eta_{\mu\lambda}\eta_{\nu\sigma} + \eta_{\mu\sigma}\eta_{\nu\lambda} - \eta_{\mu\nu}\eta_{\lambda\sigma})$$

regarding $\mu\nu$ and $\lambda\sigma$ as the two indices. Note that we have to maintain the symmetry of $h^{\mu\nu}$. In other words, we are dealing with matrices acting in a linear space spanned by symmetric two-index tensors. Thus, the identity matrix is actually

$$I_{\mu\nu;\lambda\sigma} \equiv \tfrac{1}{2}(\eta_{\mu\lambda}\eta_{\nu\sigma} + \eta_{\mu\sigma}\eta_{\nu\lambda})$$

You can check that $K_{\mu\nu;\lambda\sigma} K^{\lambda\sigma}{}_{;\rho\omega} = I_{\mu\nu;\rho\omega}$ so that $K^{-1} = K$. Thus, in the harmonic gauge the graviton propagator in flat spacetime is given by (scaling out Newton's constant)

$$D_{\mu\nu,\lambda\sigma}(k) = \frac{1}{2}\frac{\eta_{\mu\lambda}\eta_{\nu\sigma} + \eta_{\mu\sigma}\eta_{\nu\lambda} - \eta_{\mu\nu}\eta_{\lambda\sigma}}{k^2 + i\varepsilon} \tag{12}$$

Newton from Einstein

Varying (10) with respect to $h_{\mu\nu}$ we obtain the Euler-Lagrange equation of motion[4] $\frac{1}{32\pi G}(-2\partial^2 h_{\mu\nu} + \eta_{\mu\nu}\partial^2 h) - T_{\mu\nu} = 0$. Taking the trace, we find $\partial^2 h = 16\pi GT$ (with $T \equiv \eta_{\mu\nu}T^{\mu\nu}$) and so we obtain[5]

$$\partial^2 h_{\mu\nu} = -16\pi G(T_{\mu\nu} - \tfrac{1}{2}\eta_{\mu\nu}T) \tag{13}$$

In the static limit, T_{00} is the dominant component[6] of the stress-energy tensor and (13) reduces to $\vec{\nabla}^2\phi = 4\pi GT_{00}$ upon recalling from chapter I.5 that the Newtonian gravitational potential $\phi \equiv \tfrac{1}{2}h_{00}$. We have just derived Poisson's equation for ϕ.

Incidentally, this suggests another way of avoiding the tedious task of expanding the Einstein-Hilbert action (and hence R) to $O(h^2)$ if you are willing to accept the Einstein field equation (4) as given. You need expand $R_{\mu\nu}$ only to $O(h)$ to obtain (13) from (4), and from (13) you can reconstruct the action to $O(h^2)$. Indeed, from (2) and (3) you easily get

$$R_{\mu\nu} = \tfrac{1}{2}(-\partial^2 h_{\mu\nu} + \partial_\mu\partial_\lambda h^\lambda_\nu + \partial_\nu\partial_\lambda h^\lambda_\mu - \partial_\mu\partial_\nu h^\lambda_\lambda) + O(h^2) \rightarrow -\tfrac{1}{2}\partial^2 h_{\mu\nu} + O(h^2)$$

with the further simplification in harmonic gauge. But this is not quite fair since considerable technology[7] (Palatini identity and all the rest) is needed to derive (4) from (1).

---

[4] Note that the flat spacetime energy momentum conservation $\partial^\mu T_{\mu\nu} = 0$ together with the equation of motion implies $\partial^2(\partial^\mu h_{\mu\nu} - \tfrac{1}{2}\partial_\nu h) = 0$.

[5] Thus, the Einstein equation in vacuum $R_{\mu\nu} = 0$ reduces to $\partial^2 h_{\mu\nu} = 0$; hence the name "harmonic."

[6] Note that, in contrast to T_{00}, h_{00} does not dominate the other components of $h_{\mu\nu}$.

[7] See S. Weinberg, *Gravitation and Cosmology*, pp. 290 and 364.

Einstein's theory and the deflection of light

Consider two particles with stress-energy tensors $T_{(1)}^{\mu\nu}$ and $T_{(2)}^{\mu\nu}$ respectively interacting via the exchange of a graviton. The scattering amplitude is then (up to some overall constant not essential for our purposes here) given by

$$GT_{(1)}^{\mu\nu} D_{\mu\nu,\lambda\sigma}(k) T_{(2)}^{\lambda\sigma} = \frac{G}{2k^2}(2T_{(1)}^{\mu\nu} T_{(2)\mu\nu} - T_{(1)}T_{(2)})$$

For nonrelativistic matter T^{00} is much larger than the other components T^{0j} and T^{ij} (as I have just remarked), so the scattering amplitude between two lumps of nonrelativistic matter (say, the earth and you) is proportional to

$$\frac{G}{2k^2}(2T_{(1)}^{00}T_{(2)}^{00} - T_{(1)}^{00}T_{(2)}^{00}) = \frac{G}{2k^2}T_{(1)}^{00}T_{(2)}^{00}$$

As explained way back in chapters I.4 and I.5, the interaction potential is given by the Fourier transform of the scattering amplitude, namely

$$G \iint d^3x d^3x' T^{(1)00}(x) T^{(2)00}(x') \int d^3k e^{i\vec{k}\cdot(\vec{x}-\vec{x}')} \frac{1}{\vec{k}^2}$$

and thus for two well-separated objects we recover the Newtonian potential $GM_{(1)}M_{(2)}/r$.

We are now able to address the issue raised at the end of chapter I.5. Suppose a particle theorist, Dr. Gravity, wants to propose a theory of gravity to rival Einstein's theory. Dr. G claims that gravity is due to the exchange of a spin 2 particle with a teeny mass m_G coupled to the stress-energy tensor $T^{\mu\nu}$. In chapter I.5 we worked out the propagator of a massive spin 2 particle, namely

$$D_{\mu\nu,\lambda\sigma}^{\text{spin 2}}(k) = \tfrac{1}{2}(G_{\mu\lambda}G_{\nu\sigma} + G_{\mu\sigma}G_{\nu\lambda} - \tfrac{2}{3}G_{\mu\nu}G_{\lambda\sigma})/(k^2 - m_G^2 + i\varepsilon)$$

with $G_{\mu\nu} = \eta_{\mu\nu} - k_\mu k_\nu/m_G^2$ (after a trivial notational adjustment). Since the particle is coupled to a conserved source $k_\mu T^{\mu\nu} = 0$ we can replace $G_{\mu\nu}$ by $\eta_{\mu\nu}$. Thus, in the limit $m_G \to 0$ we have the propagator

$$D_{\mu\nu,\lambda\sigma}^{\text{spin 2}}(k) = \frac{1}{2}\frac{\eta_{\mu\lambda}\eta_{\nu\sigma} + \eta_{\mu\sigma}\eta_{\nu\lambda} - \tfrac{2}{3}\eta_{\mu\nu}\eta_{\lambda\sigma}}{k^2 + i\varepsilon} \tag{14}$$

Compare this with (12). Dr. G's propagator differs from Einstein's: $\tfrac{2}{3}$ versus 1. Remarkably, gravity is not generated by an almost massless spin 2 particle. The "$\tfrac{2}{3}$ discontinuity" between (12) and (14) was discovered in 1970 independently by Iwasaki, by van Dam and Veltman, and by Zakharov.

In Dr. G's theory (with his own gravitational coupling G_G), the interaction between two particles is given by

$$G_G T_{(1)}^{\mu\nu} D_{\mu\nu,\lambda\sigma}(k) T_{(2)}^{\lambda\sigma} = \frac{G_G}{2k^2}(2T_{(1)}^{\mu\nu} T_{(2)\mu\nu} - \frac{2}{3}T_{(1)}T_{(2)})$$

For two lumps of nonrelativistic matter this becomes

$$\frac{G_G}{2k^2}(2T_{(1)}^{00}T_{(2)}^{00} - \frac{2}{3}T_{(1)}^{00}T_{(2)}^{00}) = \frac{4}{3}\frac{G_G}{2k^2}T_{(1)}^{00}T_{(2)}^{00}$$

Dr. G simply takes his $G_G = \tfrac{3}{4}G$ and his theory passes all experimental tests.

But wait! There is also the famous 1919 observation of the deflection of starlight by the sun, and the photon is definitely not a lump of nonrelativistic matter. Indeed, recall from chapter I.11 (or from your course on electromagnetism) that $T \equiv T^\mu_\mu$ vanishes for the photon. Thus, taking $T^{\mu\nu}_{(1)}$ and $T^{\mu\nu}_{(2)}$ to be the stress-energy tensor of the sun and of the photon respectively, Einstein would have for the scattering amplitude $(G/2k^2)2T^{\mu\nu}_{(1)}T_{(2)\mu\nu}$ while Dr. G would have $(G_G/2k^2)2T^{\mu\nu}_{(1)}T_{(2)\mu\nu} = \frac{3}{4}(G/2k^2)2T^{\mu\nu}_{(1)}T_{(2)\mu\nu}$. Dr. G would have predicted a deflection angle of $3GM/R$ instead of $4GM/R$ (with M and R the mass and radius of the sun). On the Brazilian island of Sobral in 1919 Einstein triumphed over Dr. G.

As explained in chapter I.5, while a massive spin 2 particle has 5 degrees of freedom the massless graviton has only 2. (I give an analysis of the helicity ± 2 structure of one graviton exchange in appendix 2.) The 5 degrees of freedom may be thought of as consisting of the helicity ± 2 degrees of freedom we want plus 2 helicity ± 1 and a helicity 0 degrees of freedom. The coupling of the helicity ± 1 degrees of freedom vanishes because $k_\mu T^{\mu\nu} = 0$. Thus, effectively, we are left with an extra scalar coupling to the trace $T \equiv \eta_{\mu\nu}T^{\mu\nu}$ of the stress-energy tensor; as we can see plainly the discrepancy indeed resides in the last term of (12) and (14).

You should be disturbed that a measurement of the deflection of starlight can show that a physical quantity, the graviton mass m_G, is mathematically zero rather than less than some extremely small value. This apparent paradox was resolved by A. Vainshtein in 1972.[8] He found that Dr. G's theory contains a distance scale

$$r_V = \left(\frac{GM}{m_G^4}\right)^{\frac{1}{5}}$$

in the gravitational field around a body of mass M. The helicity 0 degree of freedom becomes effective only on the distance scales $r \gg r_V$. Inside the Vainshtein radius r_V, the gravitational field is the same as in Einstein's theory and experiments cannot distinguish between Einstein's and Dr. G's theories. With the current astrophysical bound $m_G \ll (10^{24} \text{ cm})^{-1}$ and M the mass of the sun, r_V comes out to be much larger than the size of the solar system. In other words, the apparent paradox arose because of an interchange of limits: We can take either the characteristic distance of the measurement r_{obs} (the radius of the sun in the deflection of starlight) or the Vainshtein radius r_V to infinity first.

So all is well: Dr. G's theory is consistent with current measurements provided that he takes m_G small enough. What he is not allowed to do is use the one graviton exchange approximation. Instead, he should solve the massive analog of Einstein's field equation (4) around a massive body such as the sun, as Vainshtein did. This is equivalent to expanding to all orders in the graviton field h and resumming: In Feynman diagram language we

---

[8] A. I. Vainshtein, *Phys. Lett.* 39B:393, 1972; see also C. Deffayet, G. Dvali, G. Gabadadze, and A. I. Vainshtein, *Phys. Rev.* D65:044026, 2002.

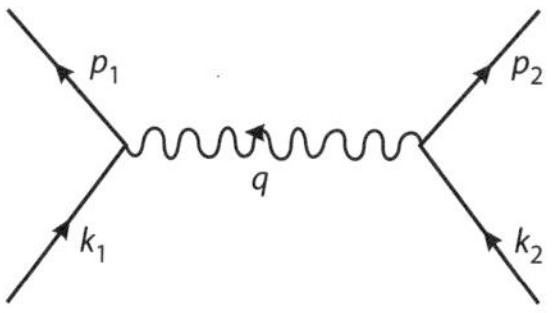

Figure VIII.1.2

have an infinite number of diagrams corresponding to the sun emitting 1, 2, 3, $\cdots$, ∞ gravitons respectively. The paradox is formally resolved by noting that the higher orders are increasingly singular as $m_G \to 0$.

The gravity of light

At this point, you are ready to do perturbative quantum gravity: You have the graviton propagator (12), and you can read off the interaction between gravitons from the detailed version of (7) and the interaction between the graviton and any other field from the term $-\frac{1}{2}h_{\mu\nu}T^{\mu\nu}$. The only trouble is that you might "drown in a sea of indices" if you don't watch out, as I have already warned you.

I know of one calculation (in fact one of my favorites in theoretical physics) in which we can beat the indices down easily. An interesting question: Einstein said that light is deflected by a massive object, but is light deflected gravitationally by light? Tolman, Ehrenfest, and Podolsky discovered that in the weak field limit two light beams moving in the same direction do not interact gravitationally, but two light beams moving in the opposite directions do. Surprising, eh?

The scattering of two photons $k_1 + k_2 \to p_1 + p_2$ via the exchange of a graviton is given by the Feynman diagram in figure VIII.1.2, with the momentum transfer $q \equiv p_1 - k_1$, plus another diagram with p_1 and p_2 interchanged. The Feynman rule for coupling a graviton to two photons can be read off from

$$h^{\mu\nu}T_{\mu\nu} = -h^{\mu\nu}(F_{\mu\lambda}F_\nu{}^\lambda - \tfrac{1}{4}\eta_{\mu\nu}F_{\rho\lambda}F^{\rho\lambda})$$

but all we need is that the interaction involve two powers of spacetime derivatives ∂ acting on the electromagnetic potential A_μ so that the graviton-photon-photon vertex involves 2 powers of momenta, one from each photon. Hence the scattering amplitude (with all Lorentz indices suppressed) has the schematic form $\sim (k_1 p_1)D(k_2 p_2)$. The η's in the graviton propagator D tie the indices on $(k_1 p_1)$ and $(k_2 p_2)$ together. (We have suppressed the polarization vectors of the photons, imagining that they are to be averaged over in the amplitude squared.) Referring to (12), we see that the amplitude is the sum of three terms such as $\sim (k_1 \cdot p_1)(k_2 \cdot p_2)/q^2$, $\sim (k_1 \cdot k_2)(p_1 \cdot p_2)/q^2$, and $\sim (k_1 \cdot p_2)(k_2 \cdot p_1)/q^2$. Since according to Fourier the long distance part of the interaction potential is given by the small q behavior of the scattering amplitude, we need only evaluate these terms in the limit $q \to 0$. We can throw almost everything away! For example,

$$k_1 \cdot p_1 \to k_1 \cdot k_1 = 0, \quad k_1 \cdot p_2 = k_1 \cdot (k_1 + k_2 - p_1) \to k_1 \cdot k_2$$

Just imagining contracting all those indices in our heads is good enough: We obtain the amplitude $\sim (k_1 \cdot k_2)(p_1 \cdot p_2)/q^2$.

If k_1 and k_2 point in the same direction, $k_1 \cdot k_2 \propto k_1 \cdot k_1 = 0$. Two photons moving in the same direction do not interact gravitationally.

Of course, this result is not of any practical importance since electromagnetic effects are far more important, but this is not an engineering text. In appendix 1 I give an alternative derivation of this amusing result.

Kaluza-Klein compactification

You have probably read about how excited Einstein was when he heard of the proposal of Kaluza and of Klein to extend the dimension of spacetime to 5 and thus unify electromagnetism and gravity. The 5th dimension is supposed to be compactified into a tiny circle of radius a far smaller than what experimentalists can see; in other words, x^5 is an angular variable with $x^5 = x^5 + 2\pi a$. You have surely heard that string theory, at least in some version, is based on the Kaluza-Klein idea. Strings live in 10-dimensional spacetime, with 6 of the dimensions compactified.

I can now show you how the Kaluza-Klein mechanism works. Start with the action

$$S = \frac{1}{16\pi G_5} \int d^5x \sqrt{-g_5} R_5 \tag{15}$$

in 5-dimensional spacetime. The subscript 5 serves to indicate the 5-dimensional quantities. We denote the 5-dimensional metric by g_{AB} with the indices A and B running over 0, 1, 2, 3, 5.

Assume that g_{AB} does not depend on x^5. Plug into S, integrate over x^5, and compute the effective 4-dimensional action. Since R_5 and the 4-dimensional scalar curvature R both involve two powers of ∂ and g_{AB} contains $g_{\mu\nu}$, we must have (exercise VIII.1.5) $R_5 = R + \cdots$. Thus, (15) contains the Einstein-Hilbert action with Newton's gravitational constant $G \sim G_5/a$.

What else do we get? We don't even have to work through the arithmetic. We can argue by symmetry. Under the 5-dimensional coordinate transformation $x^A \to x'^A = x^A + \varepsilon^A(x)$, we have [see (8)] $h'_{AB} = h_{AB} - \partial_A \varepsilon_B - \partial_B \varepsilon_A$. Let us choose $\varepsilon_\mu = 0$ and $\varepsilon_5(x)$ to be independent of x_5: We go around and rotate each of the tiny circles attached to every point in our spacetime a tiny bit. Well, we have $h'_{\mu\nu} = h_{\mu\nu}$ and $h'_{55} = h_{55}$, but $h'_{\mu 5} = h_{\mu 5} - \partial_\mu \varepsilon_5$. But if we give the Lorentz 4-vector $h_{\mu 5}$ and 4-scalar ε_5 new names, call them A_μ and Λ, this just says $A'_\mu = A_\mu - \partial_\mu \Lambda$, the usual electromagnetic gauge transformation!

Since we know that the 5-dimensional action (15) is invariant under $x^A \to x'^A = x^A + \varepsilon^A(x)$, the resulting 4-dimensional action must be invariant under $A_\mu \to A'_\mu = A_\mu - \partial_\mu \Lambda$

and hence must contain the Maxwell action. Note once again the power of symmetry considerations. No need to do tedious calculations.

Electromagnetism comes out of gravity!

Differential geometry of Riemannian manifolds

I hinted earlier at a deep connection between general coordinate transformation and gauge transformation. Let us flesh this out by looking at differential geometry and gravity. For this sketch we will consider locally Euclidean (rather than Minkowskian) spaces.

The differential geometry of Riemannian manifolds can be elegantly summarized in the language of differential forms. Consider a Riemannian manifold (such as a sphere) with the metric $g_{\mu\nu}(x)$. Locally, the manifold is Euclidean by definition, which means

$$g_{\mu\nu}(x) = e_\mu^a(x)\delta_{ab}e_\nu^b(x) \tag{16}$$

where the matrix $e(x)$ may be thought of as a similarity transformation that diagonalizes $g_{\mu\nu}$ and scales it to the unit matrix. Thus, for a D-dimensional manifold there exist D "world vectors" $e_\mu^a(x)$ obviously dependent on x and labeled by the index $a = 1, 2, \cdots, D$. The functions $e_\mu^a(x)$ are known as "vielbeins" (meaning "many legs" in German, vierbeins = four legs for $D = 4$, dreibeins = three legs for $D = 3$, and so on.) In some sense, the vielbeins can be thought as the "square root" of the metric.

Let us clarify by a simple example. The familiar two-sphere (of unit radius) has the line element[9] $ds^2 = d\theta^2 + \sin^2\theta d\varphi^2$. From the metric ($g_{\theta\theta} = 1$, $g_{\varphi\varphi} = \sin^2\theta$) we can read off $e_\theta^1 = 1$ and $e_\varphi^2 = \sin\theta$ (all other components are zero). We are invited to define D 1-forms $e^a = e_\mu^a dx^\mu$. (In our example, $e^1 = d\theta$, $e^2 = \sin\theta d\varphi$.)

On a curved manifold, when we parallel transport a vector, the vector changes when expressed in terms of the locally Euclidean coordinate frame. (This is just the familiar statement that on a curved manifold such as the surface of the earth the notion of a vector pointing straight north is a local concept: When we move infinitesimally away keeping our "north vector" pointing in the same direction, it will end up being infinitesimally rotated away from the "north vector" defined at the point we have just moved to.) This infinitesimal rotation of the vielbeins is described by

$$de^a = -\omega^{ab}e^b \tag{17}$$

Note that since ω generates an infinitesimal rotation it is an antisymmetric matrix: $\omega^{ab} = -\omega^{ba}$. Since de^a is a 2-form, ω is a 1-form, known as the connection: It "connects" the locally Euclidean frames at nearby points. (Since the indices a, b, etc. are associated with the Euclidean metric δ_{ab} we do not have to distinguish between upper and lower indices. When we do write upper or lower indices a, b, etc. it is for typographical convenience.) In

---

[9] Note that this represents the square of an infinitesimal distance element and not an area element, and so a quantity such as $d\theta^2$ is literally the square of $d\theta$ and not the wedge product $d\theta d\theta$ (of chapter IV.4), which would have been identically zero.

the simple example of the sphere, $de^1 = 0$ and $de^2 = \cos\theta\, d\theta\, d\varphi$ and so the connection has only one nonvanishing component $\omega^{12} = -\omega^{21} = -\cos\theta\, d\varphi$.

At any point, we are free to rotate the vielbeins: If you use the vielbeins e^a_μ I am free to use some other vielbeins e'^a_μ instead, as long as mine are related to yours by a rotation $e^a_\mu(x) = O^a_b(x)e'^b_\mu(x)$. [You can check that $g_{\mu\nu}(x) = e^a_\mu(x)\delta_{ab}e^b_\nu(x) = e'^a_\mu(x)\delta_{ab}e'^b_\nu(x)$ if $O^T O = 1$.] The connection ω' is defined by $de'^a = -\omega'^{ab}e'^b$. You can readily work out that (suppressing indices)

$$\omega = O\omega' O^T - (dO)O^T \tag{18}$$

The local curvature of the manifold is a measure of how the connection varies from point to point. We would like the curvature to be invariant under the local rotation O (or at least to transform as a tensor so that by contracting it with vectors we can form a scalar). The desired object is the 2-form $R^{ab} = d\omega^{ab} + \omega^{ac}\omega^{cb}$. You can check that $R = OR'O^T$. (For the sphere, $R^{12} = d\omega^{12} + \omega^{1c}\omega^{c2} = \sin\theta\, d\theta\, d\varphi$.) Written out in components, $R^{ab} = R^{ab}_{\mu\nu}dx^\mu dx^\nu$. I leave it to you to verify that $R^{ab}_{\mu\nu}e^\lambda_a e^\sigma_b$ is the usual Riemann curvature tensor $R^{\lambda\sigma}_{\mu\nu}$, where e^λ_a is the inverse of the matrix e^a_λ. In particular, $R^{ab}_{\mu\nu}e^\mu_a e^\nu_b$ is the scalar curvature, which in our convention works out to be $+1$ for the sphere.

Thus, Riemannian geometry can be elegantly summarized by the two statements (again suppressing indices)

$$de + \omega e = 0 \tag{19}$$

and

$$R = d\omega + \omega^2 \tag{20}$$

Look familiar? You should be struck by the similarity between (20) and the expression for the field strength in nonabelian gauge theories $F = dA + A^2$. Note ω transforms [see (18)] exactly the same way as the gauge potential A. But one nagging difference, namely the lack of an analog of e in gauge theory, has long bothered some theoretical physicists (but is shrugged off by most as inconsequential). Also, note that Einstein theory is linear in R while Yang-Mills theory is quadratic in F.

Gravity and Yang-Mills

We can make the connection between gravity and Yang-Mills theory more explicit by looking at the derivative of a vector field. Yang-Mills theory was born of the requirement that a field φ and its derivative $\partial_\mu\varphi$ transform in the same way under a spacetime-dependent internal symmetry transformation (IV.5.1). In Einstein gravity a vector field $W^\mu(x)$ transforms as $W'^\mu(x') = S^\mu_\nu(x)W^\nu(x)$ with $S^\mu_\nu(x) = \partial x'^\mu/\partial x^\nu$. Since the matrix S depends on the spacetime coordinate x, we see that $\partial_\lambda W^\mu$ could not possibly transform like a tensor with one upper and one lower index, as we would like naively just by looking at indices. We would have to introduce a covariant derivative. Not surprisingly, this closely parallels the discussion in chapter IV.5. Historically, Yang and Mills were inspired by Einstein gravity.

Using the chain rule and the product rule, we have

$$\partial'_\lambda W'^\mu(x') = \frac{\partial W'^\mu(x')}{\partial x'^\lambda} = \frac{\partial x^\rho}{\partial x'^\lambda}\frac{\partial}{\partial x^\rho}[S^\mu_\nu(x)W^\nu(x)] = (S^{-1})^\rho_\lambda S^\mu_\nu \partial_\rho W^\nu + [(S^{-1})^\rho_\lambda \partial_\rho S^\mu_\nu]W^\nu \tag{21}$$

Were the second term in (21), which comes from differentiating S, not there, the naive guess, that $\partial_\lambda W^\mu$ transforms like a tensor, would be valid. The fact that the transformation S varies from place to place has negated the naive guess.

What is happening is quite clear: as the vector W varies from a given point to a neighboring point, the coordinate axes that define the components of W also change. This suggests that we could define a more suitable derivative, called the covariant derivative and written as $D_\lambda W^\mu$, to take this effect into account, so that $D_\lambda W^\mu$ would indeed transform like a tensor. Exactly as in Yang-Mills theory (IV.5.1), we have to add an extra term to knock out the second term in (21).

Just the way the indices hang together immediately suggests the correct construction. The factor $(S^{-1})^\rho_\lambda \partial_\rho S^\mu_\nu$ in the unwanted second term in (21) has one upper index and two lower indices, so we need an object with this set of indices. Lo, the Riemann-Christoffel symbol $\Gamma^\mu_{\lambda\nu}$ in (3) (and introduced in chapter I.11) fits the bill perfectly. I will let you have the fun of verifying that the covariant derivative defined by

$$D_\lambda W^\mu \equiv \partial_\lambda W^\mu + \Gamma^\mu_{\lambda\nu} W^\nu \tag{22}$$

indeed transforms like a tensor (note that Γ was normalized correctly for this purpose).

I end with a technical remark about the coupling of gravity to spin $\frac{1}{2}$ fields. First, we of course have to Wick rotate so that the vierbein e^a_μ erects a locally Minkowskian rather than a Euclidean coordinate frame. The indices a, b, etc. are now contracted with the Minkowskian metric η_{ab}. The slight subtlety is that the Dirac gamma matrices γ^a are associated with the Lorentz rotation of the vierbein $e^a_\mu(x) = O^a_b(x)e'^b_\mu(x')$ and thus carry the Lorentz index a rather than the "world" index μ. Similarly, the Dirac spinor $\psi(x)$ is defined relative to the local Lorentz frame specified by the vierbein, and thus its covariant derivative has to be defined in terms of the connection ω rather than the symbol Γ. Hence the flat space Dirac action $\int d^4x\, \bar\psi(i\gamma^\mu\partial_\mu - m)\psi$ must be generalized to $\int d^4x\, \sqrt{-g}\,\bar\psi(i\gamma^a\eta_{ab}e^{b\mu}\mathcal{D}_\mu - m)\psi$, where the covariant derivative $\mathcal{D}_\mu\psi = \partial_\mu\psi - \frac{i}{4}\omega_{\mu ab}\sigma^{ab}\psi$ expresses the rotation of the local Lorentz frame as we move from a point x to a neighboring point. In contrast to the action for integer spin fields in curved spacetime (see chapter I.11), the Dirac action in curved spacetime involves the vierbein explicitly.

Appendix 1: Light on light again

The stress-energy tensor $T^{\mu\nu}$ of a light beam moving in the x-direction has four nonzero components: the energy density T^{00} of course, then $T^{0x} = T^{00}$ since photons carry the same energy and momentum, next $T^{x0} = T^{0x}$ by symmetry, and finally $T^{xx} = T^{00}$ since the stress-energy tensor of the electromagnetic field is traceless (chapter I.11). Without having to solve Einstein's equations in the weak field limit (13) explicitly we know immediately that $h^{00} = h^{0x} = h^{x0} = h^{xx} \equiv h$. The metric around the light beam is given by $g_{00} = 1 + h$,

$g_{0x} = g_{x0} = -h$, and $g_{xx} = -1 + h$ (and of course $g_{yy} = g_{zz} = -1$ plus a bunch of vanishing components). Consider a photon moving parallel to the light beam. Its worldline is determined by (recall chapter I.11)

$$\frac{d^2 x^\rho}{d\zeta^2} = -\Gamma^\rho_{\mu\nu} \frac{dx^\mu}{d\zeta} \frac{dx^\nu}{d\zeta}$$

Let's calculate $d^2 y/d\zeta^2$ and $d^2 z/d\zeta^2$ with $(dy/d\zeta)$, $(dz/d\zeta) \ll (dt/d\zeta)$, $(dx/d\zeta)$. Using (3) we find (with μ, ν restricted to $0, x$)

$$\frac{d^2 y}{d\zeta^2} = \frac{1}{2}(\partial_\nu g_{y\mu} + \partial_\mu g_{y\nu} - \partial_y g_{\mu\nu}) \frac{dx^\mu}{d\zeta} \frac{dx^\nu}{d\zeta}$$

$$= -\frac{1}{2}(\partial_y h)\left[(\frac{dt}{d\zeta})^2 + (\frac{dx}{d\zeta})^2 - 2\frac{dt}{d\zeta}\frac{dx}{d\zeta}\right] = -\frac{1}{2}(\partial_y h)(\frac{dt}{d\zeta} - \frac{dx}{d\zeta})^2$$

For a photon moving in the same direction as the light beam $dt = dx$ and $d^2 y/d\zeta^2 = d^2 z/d\zeta^2 = 0$. We have once again derived the Tolman-Ehrenfest-Podolsky effect. Note we never had to solve for h.

Incidentally, if you are a bit unsure of $dt = dx$, the condition $ds = 0$ for a light beam moving in the x-direction amounts to $(1+h)dt^2 - 2h dt dx - (1-h)dx^2 = 0$. Upon division by dt^2 we obtain $-(1+h) + 2hv + (1-h)v^2 = 0$, with $v \equiv dx/dt$. The quadratic equation has two roots $v = \mp(1 \pm h)/(1-h)$. The negative root gives $v = 1$, and thus for a photon moving in the same direction as the light beam $dx/dt = 1$. In contrast, the positive root $v = -(1+h)/(1-h)$ describes a photon moving in the opposite direction.

Appendix 2: The helicity structure of gravity

To gain a deeper understanding of the difference between Einstein's and Dr. G's theories let us look at the helicity structure of the interaction in the two cases. To warm up, consider the interaction between two conserved currents due to the exchange of a spin 1 particle of momentum k and mass m : $J^\mu_{(1)} J_{(2)\mu} = J^0_{(1)} J^0_{(2)} - J^i_{(1)} J^i_{(2)}$. Use current conservation $k_\mu J^\mu = 0$ to eliminate $J^0 = k^i J^i/\omega$ (with $\omega \equiv k^0$). We obtain $(k^i k^j/\omega^2 - \delta^{ij}) J^i_{(1)} J^j_{(2)}$. Let $\vec{k}$ point in the 3rd direction and use $\vec{k}^2 = \omega^2 - m^2$ to write this as $-[(m^2/\omega^2) J^3_{(1)} J^3_{(2)} + J^1_{(1)} J^1_{(2)} + J^2_{(1)} J^2_{(2)}]$. We see that as $m \to 0$ the longitudinal component of the current J^3 indeed decouples as explained in chapter II.7 and we obtain $-\frac{1}{2}(J^{1+i2}_{(1)} J^{1-i2}_{(2)} + J^{1-i2}_{(1)} J^{1+i2}_{(2)})$, showing explicitly that the photon has helicity ± 1. (Obvious notation: $J^{1+i2} \equiv J^1 + i J^2$ etc.)

Onward to gravity. Consider the interaction $T^{\mu\nu}_{(1)} T_{(2)\mu\nu} - \xi T_{(1)} T_{(2)}$, where $\xi = \frac{1}{2}$ for Einstein and $\frac{1}{3}$ for Dr. G. For ease of writing I will now omit the subscripts (1) and (2). Conservation $k_\mu T^{\mu\nu} = 0$ allows us to eliminate $T^{0i} = k^j T^{ji}/\omega$ and $T^{00} = k^j k^l T^{jl}/\omega^2$. Again taking $\vec{k}$ to point in the 3rd direction we obtain the mess

$$\left(\frac{m}{\omega}\right)^4 T^{33} T^{33} + 2\left(\frac{m}{\omega}\right)^2 (T^{13} T^{13} + T^{23} T^{23}) + T^{11} T^{11} + T^{22} T^{22} + 2 T^{12} T^{12}$$

$$- \xi \left[\left(\frac{m}{\omega}\right)^2 T^{33} + T^{11} + T^{22}\right]\left[\left(\frac{m}{\omega}\right)^2 T^{33} + T^{11} + T^{22}\right]$$

which simplifies in the limit $m \to 0$ to

$$T^{11} T^{11} + T^{22} T^{22} + 2 T^{12} T^{12} - \xi(T^{11} + T^{22})(T^{11} + T^{22})$$

In Einstein's theory, $\xi = \frac{1}{2}$ and this becomes

$$\frac{1}{2}(T^{11} - T^{22})(T^{11} - T^{22}) + 2 T^{12} T^{12}$$

which lo and behold is equal to $\frac{1}{2}(T^{1+i2,1+i2} T^{1-i2,1-i2} + T^{1-i2,1-i2} T^{1+i2,1+i2})$, showing that indeed the graviton carries helicity ± 2. In Dr. G's theory, this would not be the case.

Exercises

VIII.1.1 Work out $T^{\mu\nu}$ for a scalar field. Draw the Feynman diagram for the contribution of one-graviton exchange to the scattering of two scalar mesons. Calculate the amplitude and extract the interaction energy between two mesons sitting at rest, thus deriving Newton's law of gravity.

VIII.1.2 Work out $T^{\mu\nu}$ for the Yang-Mills field.

VIII.1.3 Show that if $h_{\mu\nu}$ does not satisfy the harmonic gauge, we can always make a gauge transformation with ε_ν determined by $\partial^2 \varepsilon_\nu = \partial_\mu h^\mu_\nu - \frac{1}{2}\partial_\nu h^\lambda_\lambda$ so that it does. All of this should be conceptually familiar from your study of electromagnetism.

VIII.1.4 Count the number of degrees of polarization of a graviton. [Hint: Consider a plane wave $h_{\mu\nu}(x) = h_{\mu\nu}(k)e^{ikx}$ just because it is a bit easier to work in momentum space. A symmetric tensor has 10 components and the harmonic gauge $k_\mu h^\mu_\nu = \frac{1}{2}k_\nu h^\lambda_\lambda$ imposes 4 conditions. Oops, we are left with 6 degrees of freedom. What is going on?] [Hint: You can make a further gauge transformation and still stay in the harmonic gauge. The graviton should have only 2 degrees of polarization.]

VIII.1.5 The Kaluza-Klein result that we argued by symmetry considerations can of course be derived explicitly. Let me sketch the calculation for you. Consider the metric

$$ds^2 = g_{\mu\nu}dx^\mu dx^\nu - a^2[d\theta + A_\mu(x)dx^\mu]^2$$

where θ denotes an angular variable $0 \leq \theta < 2\pi$. With $A_\mu = 0$, this is just the metric of a curved spacetime, which has a circle of radius a attached at every point. The transformation $\theta \to \theta + \Lambda(x)$ leaves ds invariant provided that we also transform $A_\mu(x) \to A_\mu(x) - \partial_\mu \Lambda(x)$. Calculate the 5-dimensional scalar curvature R_5 and show that $R_5 = R_4 - \frac{1}{4}a^2 F_{\mu\nu}F^{\mu\nu}$. Except for the precise coefficient $\frac{1}{4}$ this result follows entirely from symmetry considerations and from the fact that R_5 involves two derivatives on the 5-dimensional metric, as explained in the text. After some suitable rescaling this is the usual action for gravity plus electromagnetism. Note that the 5-dimensional metric has the explicit form

$$g^5_{AB} = \begin{pmatrix} g_{\mu\nu} - a^2 A_\mu A_\nu & -a^2 A_\mu \\ -a^2 A_\nu & -a^2 \end{pmatrix} \tag{23}$$

VIII.1.6 Generalize the Kaluza-Klein construction by replacing the circles by higher dimensional spheres. Show that Yang-Mills fields emerge.

VIII.1.7 Starting with the connection 1-form $\omega^{12} = -\cos\theta d\varphi$ for the sphere, show that the scalar curvature is a constant independent of θ and φ.

VIII.1.8 The vielbeins for a spacetime with Minkowski metric is defined by $g_{\mu\nu}(x) = e^a_\mu(x)\eta_{ab}e^b_\nu(x)$, where the Minkowski metric η_{ab} replaces the Euclidean metric δ_{ab}. The indices a and b are to be contracted with η_{ab}. For example, $R^{ab} = d\omega^{ab} + \omega^{ac}\eta_{cd}\omega^{db}$. Show that everything goes through as expected.

VIII.2 | The Cosmological Constant Problem and the Cosmic Coincidence Problems

The force that knows too much

The word paradox has been debased by loose usage in the physics literature. A real paradox should involve a major and clear-cut discrepancy between theoretical expectation and experimental measurement. The ultraviolet catastrophe, for example, is a paradox, the resolution of which around the dawn of the twentieth century ushered in quantum physics. I now come to the most egregious paradox of present day physics.

The electromagnetic force knows about the particles carrying charge, and the strong force knows about the particles carrying color. And the gravitational force? It knows everybody! More precisely, anybody carrying energy and momentum.

Within a particle physics frame of mind, which is the only frame of mind we have in exploring the fundamental structure of physics, the graviton can be regarded as just another particle. Indeed, given that a massless spin 2 particle couples to the stress-energy tensor, one can reconstruct Einstein's theory.

Nevertheless, there is an uncomfortable feel to this whole picture. Gravity has to do with the curvature of spacetime, the arena in which all fields and particles live. The graviton is not just another particle.

This in essence is the root origin[1] of the paradox of the cosmological constant. The graviton is not just another particle—it knows too much!

The cosmological constant

In the absence of gravity, the addition of a constant Λ to the Lagrangian $\mathcal{L} \to \mathcal{L} - \Lambda$ has no effect whatsoever. In classical physics the Euler-Lagrange equations of motion depend

---

[1] For more along this line, see A. Zee, hep-th/0805.2183 in *Proceedings of the Conference in Honor of C. N. Yang's 85th Birthday*, World Scientific, Singapore 2008, p. 131.

only on the variation of the Lagrangian. In quantum field theory we have to evaluate the functional integral $Z = \int D\varphi e^{i \int d^4x \mathcal{L}(x)}$, which upon the inclusion of Λ merely acquires a multiplicative factor. As we have seen repeatedly, a multiplicative factor in Z does not enter into the calculation of Green's function and scattering amplitudes.

Gravity, however, knows about Λ. Physically, the inclusion of Λ corresponds to a shift in the Hamiltonian $H \to H + \int d^3x \Lambda$. Thus, the "cosmological constant" Λ describes a constant energy or mass per unit volume permeating the universe, and of course gravity knows about it.

More technically, the term in the action $- \int d^4x \Lambda$ is not invariant under a coordinate transformation $x \to x'(x)$. In the presence of gravity, general coordinate invariance requires that the term $- \int d^4x \Lambda$ in the action S be modified to $- \int d^4x \sqrt{g}\Lambda$, as I explained way back in chapter I.11. Thus, the gravitational field $g_{\mu\nu}$ knows about Λ, the infamous cosmological constant introduced by Einstein and lamented by him as his biggest mistake. This often quoted lament is itself a mistake. The introduction of the cosmological constant is not a mistake: It should be there.

Symmetry breaking generates vacuum energy

In our discussion on spontaneous symmetry breaking, we repeatedly ignored an additive term $\mu^4/4\lambda$ that appears in $\mathcal{L}$.

Particle physics is built on a series of spontaneous symmetry breaking. As the universe cools, grand unified symmetry is spontaneously broken, followed by electroweak symmetry breaking, then chiral symmetry breaking, just to mention a few that we have discussed. At every stage a term like $\mu^4/4\lambda$ appears in the Lagrangian, and gravity duly takes note.

How large do we expect the cosmological constant Λ to be? As we will see, for our purposes the roughest order of magnitude estimate suffices. Let us take λ to be of order 1. As for μ, for the three kinds of symmetry breaking I just mentioned, μ is of order 10^{17}, 10^2, and 1 Gev, respectively. We thus expect the cosmological constant Λ to be roughly $\mu^4 = \mu/(\mu^{-1})^3$, where the last form of writing μ^4 reminds us that Λ is a mass or energy density: An energy of order μ packed into a cube of size μ^{-1}. But this is outrageous even if we take the smallest value for μ: We know that the universe is not permeated with a mass density of the order of 1 Gev in every cube of size 1 (Gev)$^{-1}$.

We don't have to put in actual numbers to see that there is a humongous discrepancy between theoretical expectation and observational reality. If you want numbers, the current observational bound on the cosmological constant is $\lesssim (10^{-3}\text{ev})^4$. With the grand unification energy scale, we are off by $(17+9+3) \times 4 = 116$ orders of magnitude. This is the mother of all discrepancies!

With the Planck mass $M_{\text{Pl}} \sim 10^{19}\text{Gev}$ the natural scale of gravity, we would expect $\Lambda \sim M_{\text{Pl}}^4$ if it is of gravitational origin. We are then off by 124 orders of magnitude. We are not talking about the crummy calculation of some pitiful theorist not fitting some experimental curve by a factor of 2.

We can imagine the universe starting out with a negative cosmological constant, fined tuned to cancel the cosmological constant generated by the various episodes of spontaneous symmetry breaking. Or there must be a dynamical mechanism that adjusts the cosmological constant to zero.

Notice I say zero, because the cosmological constant problem is basically an enormous mismatch between the units natural to particle physics and natural to cosmology. Measured in units of Gev^4 the cosmological constant is so incredibly tiny that particle physicists have traditionally assumed that it must be zero and have looked in vain for a plausible mechanism to drive it to zero. One of the disappointments of string theory is its inability to resolve the cosmological constant problem. As of the writing of this chapter around the turn of the millennium, the brane world scenarios (chapter I.6) have generated a great deal of excitement by offering a glimmer of a hope. Roughly, the idea is that the gravitational dynamics of the larger space that our universe is embedded in may cancel the effect of the cosmological constant.

Cosmic coincidence

But Nature has a big surprise for us. While theorists racked their brains trying to come up with a convincing argument that $\Lambda = 0$, observational cosmologists steadily refined their measurements and discovered dark energy. The "cleanest" explanation of dark energy by far is that it represents the cosmological constant. Assuming that this is the case (and who knows?), the upper bound on the cosmological constant would be changed to an approximate equality

$$\Lambda \sim (10^{-3}\text{ev})^4 !!! \tag{1}$$

The cosmological constant paradox deepens. Theoretically, it is easier to explain why some quantity is mathematically 0 than why it happens to be $\sim 10^{-124}$ in the units natural (?) to the problem.

To make things worse, $(10^{-3}\text{ev})^4$ happens to be the same order of magnitude as the present matter density of the universe ρ_M. More precisely, dark energy accounts for $\sim 74\%$ of the mass content of the universe, dark matter for $\sim 22\%$, and ordinary matter for $\sim 4\%$. First, the ordinary matter we know and love is reduced to an almost negligibly small component of the universe. Second, why should ρ_M be comparable to Λ to within a factor of 3? This is sometimes referred to as the cosmic coincidence problem.

Now the cosmological constant Λ is, within our present understanding, a parameter in the Lagrangian. On the other hand, since most of the mass density of the universe resides in rest mass, as the universe expands $\rho_M(t)$ decreases as $[1/R(t)]^3$, where $R(t)$ denotes the scale size of the universe.[2] In the far past, ρ_M was much larger than Λ, and in the

---

[2] For an easy introduction to cosmology, see A. Zee, *Unity of Forces in the Universe*, vol. II, chap. 10.

far future, it will be much smaller. It just so happens that, in this particular epoch of the universe, when you and I are around, $\rho_M \sim \Lambda$. Or to be less anthropocentric, the epoch when $\rho_M \sim \Lambda$ happens to be when galaxy formation has been largely completed.

Very bizarre!

In their desperation, some theorists have even been driven to invoke anthropic selection.[3]

[3] For a recent review, see A. Vilenkin, hep-th/0106083.

VIII.3 Effective Field Theory Approach to Understanding Nature

Low energy manifestation

The pioneers of quantum field theory, Dirac for example, tended to regard field theory as a fundamental description of Nature, complete in itself. As I have mentioned several times, in the 1950s, after the success of quantum electrodynamics many leading particle physicists rejected quantum field theory as incapable of dealing with the strong and weak interactions, not to mention gravity. Then came the great triumph of field theory in the early 1970s. But after particle physicists retrieved field theory from the dust bin of theoretical physics, they realized that the field theories they were studying might be "merely" the low energy manifestation of a deeper structure, a structure first identified as a grand unified theory and later as a string theory. Thus was developed an outlook known as the effective field theory approach, pace Dirac.

The general idea is that we can use field theory to say something about physics at low energies or equivalently long distances even if we don't know anything about the ultimate theory, be it a theory built on strings or some as yet undreamed of structure. An important consequence of this paradigm shift was that nonrenormalizable field theories became acceptable. I will illuminate these remarks with specific examples.

The emergence of this effective field theory philosophy, championed especially by Wilson, marks another example of cross fertilization between condensed matter and particle physics. Toward the late 1960s, Wilson and others developed a powerful effective field theory approach to understanding critical phenomena, culminating in his Nobel Prize. The situation in condensed matter physics is in many ways the opposite of that in particle physics at least as particle physics was understood in the 1960s. Condensed matter physicists know the short distance physics, namely the quantum mechanics of electrons and ions. But it certainly doesn't help in most cases to write down the Schrödinger equation for the electrons and ions. Rather, what one would like to have is an effective description of how a system would respond when probed at low frequency and small wave vector. A striking example is the effective theory of the quantum Hall fluid as described in chapter VI.2: The relevant degree of freedom is a gauge field, certainly a far cry from the underlying electron. As in the σ model description (chapter VI.4) of quantum chromodynamics, it is fair to say that without experimental guidance theorists would have a terribly hard time deciding what the relevant low energy long distance degrees of freedom might be. You have seen numerous other examples in condensed matter physics, from the Landau-Ginzburg theory of superconductivity to Peierls instability.

The threshold of ignorance

In our discussion of renormalization, I espouse the philosophy that a quantum field theory provides an effective description of physics up to a certain energy scale Λ, a threshold of ignorance beyond which physics not included in the theory comes into play. In a nonrenormalizable theory, various physical quantities that we might wish to calculate will come out dependent on Λ, thus indicating that the physics at or beyond the scale Λ is essential for understanding the low energy physics we are interested in. Nonrenormalizable theories suffer from not being totally predictive, but nevertheless they may be useful. After all, the Fermi theory of the weak interaction described experiments and even foretold its own demise.

In a renormalizable theory, various physical quantities come out independent of Λ, provided that the calculated results are expressed in terms of physical coupling constants and masses, rather than in terms of some not particularly meaningful bare coupling constants and masses. Low energy physics is not sensitive to what happens at high energies, and we are able to parametrize our ignorance of high energy physics in terms of a few physical constants.

From the late 1960s to the 1970s, one main thrust of fundamental physics was to classify and study renormalizable theories. As we know, this program was "more than spectacularly successful." It allowed us to pin down the theory of the strong, the weak, and the electromagnetic interactions.

Renormalization group flow and dimensional analysis

The effective field theory philosophy is intrinsically tied to renormalization group flow. In a given field theory, as we flow toward low energies, some couplings may tend to zero while others do not (and if they tend to infinity as in QCD, then we are unable to figure out the effective theory without experimental input). Thus, the first step is to calculate the renormalization group flow. A simple example is given in exercise VIII.3.1.

In many cases, we can simply use dimensional analysis. As I explained in our earlier discussion on renormalization theory, couplings with negative dimensions of mass are not important at low energies. To be specific, suppose we add a $g\varphi^6$ term to a $\lambda\varphi^4$ theory. The coupling g has the dimension of inverse mass squared. Let us define $M^2 \equiv 1/g$. At low energies, the effect of the $g\varphi^6$ term is suppressed by $(E/M)^2$.

How do we understand Schwinger's spectacular calculation of the anomalous magnetic moment of the electron in the effective field theory philosophy?

Let me first tell the traditional (i.e., pre-Wilsonian) version of the story. A student could have asked, "Professor Schwinger, why didn't you include the term $(1/M)\bar{\psi}\sigma^{\mu\nu}\psi F_{\mu\nu}$ in the Lagrangian?"

The answer is that we better not. Otherwise, we would lose our prediction for the anomalous magnetic moment; it would depend on M. Recall that $[\psi] = \frac{3}{2}$ and $[A_\mu] = 1$, and hence $\bar{\psi}\sigma^{\mu\nu}\psi F_{\mu\nu}$ has mass dimension $\frac{3}{2} + \frac{3}{2} + 1 + 1 = 5 > 4$. The requirement of renormalizability, that the Lagrangian be restricted to contain operators of dimension 4 or less, provides the rationale for excluding this term.

Actually, the "real" punchline of my story is that Schwinger probably would not have answered the question. When I took Schwinger's field theory class, it was well known among the students that it was forbidden to ask questions. Schwinger would simply ignore any raised hands. There was no opportunity to ask questions after class either: As he uttered his last sentence of an invariably beautifully prepared lecture, he would sail majestically out of the room. Dirac dealt with questions differently. I was too young to have witnessed it, but the story goes that when a student asked, "Professor Dirac, I did not understand . . . ," Dirac replied, "That is an assertion, not a question."

The modern retelling of the magnetic moment story turns it around. We now regard the Lagrangian of quantum electrodynamics as an effective Lagrangian which should include an infinite sequence of terms of ever higher dimensions, with coefficients parametrizing our threshold of ignorance. The physics of electrons and photons is now described by

$$\mathcal{L} = \bar{\psi}(i\gamma^\mu(\partial_\mu - ieA_\mu) - m)\psi - \frac{1}{4}F_{\mu\nu}F^{\mu\nu} + \frac{1}{M}\bar{\psi}\sigma^{\mu\nu}\psi F_{\mu\nu} + \cdots$$

Yes, the term $(1/M)\bar{\psi}\sigma^{\mu\nu}\psi F_{\mu\nu}$ is there, with some unknown M having the dimension of a mass. Schwinger's result, that quantum fluctuations generate a term $(\alpha/2\pi)(1/2m_e)\bar{\psi}\sigma^{\mu\nu}\psi F_{\mu\nu}$, should then be interpreted as saying that the anomalous magnetic moment of the electron is predicted to be $[(\alpha/2\pi)(1/2m_e) + 1/M]$. The close agreement of $(\alpha/2\pi)(1/2m_e)$ with the experimental value of the anomalous magnetic moment can then be turned around to set a lower bound on $M \gg (4\pi/\alpha)m_e$.

Equivalently, Schwinger's result predicts the anomalous magnetic moment of the electron if we have independent evidence that M is much larger than $[(\alpha/2\pi)(1/2m_e)]^{-1}$. I want to emphasize that all of this makes total physical sense. For example, if you speculate that the electron has some finite size a, then you would expect $M \sim 1/a$. The anomalous magnetic moment calculation gives an upper bound for a, telling us that the electron must be pointlike down to some small scale. Alternatively, we could have had independent evidence, from electron scattering for example, that a has to be smaller than a certain length, thus giving us a lower bound on M.

To underscore this point, imagine that in 1948 we followed Schwinger and quickly calculated the anomalous magnetic moment of the proton. We could literally have done it in 3 seconds, since all we have to do is replace m_e by m_p in the Lagrangian, thus obtaining $(\alpha/2\pi)(1/2m_p)\bar{\psi}\sigma^{\mu\nu}\psi F_{\mu\nu}$, which would of course disagree resoundingly with experiment. The disagreement tells us that we had not included all the relevant physics, namely that the proton interacts strongly and is not pointlike. Indeed, we now know that the anomalous magnetic moment of the proton gets contributions from the anomalous magnetic moments and the orbital motion of the quarks inside the proton.

Effective theory of proton decay

It may seem that with the effective field theory approach we lose some predictive power. But effective field theories can also be surprisingly predictive. Let me give a specific example. Suppose we had never heard of grand unified theory. All we know is the $SU(3) \otimes SU(2) \otimes U(1)$ theory. An experimentalist tells us that he is planning to see if the proton would decay.

Without the foggiest notion about what would cause the proton to decay we can still write down a field theory to describe proton decay. The Lagrangian $\mathcal{L}$ is to be constructed out of quark q and lepton l fields and must satisfy the symmetries that we know. Three quarks disappear, so we write down schematically qqq, but three spinors do not a Lorentz scalar make. We have to include a lepton field and write $qqql$.

Since four fermion fields are involved, the terms $qqql$ have mass dimension 6 and so in $\mathcal{L}$ they have to appear as $(1/M^2)qqql$ with some mass M, corresponding to the mass scale of the physics responsible for proton decay. The experimental lower bound on the lifetime of the proton sets a lower bound on M.

It is instructive to contrast this analysis with an (imagined) effective field theory analysis of proton decay long before the concept of quarks was invented, say around 1950. We would construct an effective Lagrangian out of the available fields, namely the proton field p, the electron field e, and the pion field π, and thus write down the dimension 4 operator $f \bar{p} e^{+} \pi^0$ with some dimensionless constant f. To estimate f, we would naively compare this operator with the one describing pion-nucleon coupling (chapter IV.2) $g \bar{p} n \pi^+$ in the effective Lagrangian. Since $f \bar{p} e^{+} \pi^0$ violates isospin invariance, we might expect $f \sim \alpha g$, namely the same order as g multiplied by some measure of isospin breaking, say the fine structure constant. But this would give an unacceptably short lifetime to the proton. We are forced to set f to a ridiculously small number, which seems highly unnatural. Thus, at least in hindsight, we can say that the extremely long lifetime of the proton almost points to the existence of quarks. The key, as we saw above, is to promote of the mass dimension of the term in the effective Lagrangian responsible for proton decay from 4 to 6. (Can the cosmological constant puzzle be solved in the same way?)

Another way of saying this is that $SU(3) \otimes SU(2) \otimes U(1)$ plus renormalizability predicts one of the most striking facts of the universe, the stability of the proton. In contrast, the old pion-nucleon theory glaringly failed to explain this experimental fact.

In accordance with our philosophy, $\mathcal{L}$ must be invariant under $SU(3) \otimes SU(2) \otimes U(1)$, under which quark and lepton fields transform rather idiosyncratically, as we saw in chapter VII.5. To construct $\mathcal{L}$ we have to sit down and list all Lorentz invariant $SU(3) \otimes SU(2) \otimes U(1)$ terms of the form $qqql$.

Sitting down, we would find that, assuming only one family of quarks and leptons for simplicity, there are only four terms we can write down for proton decay, which I list here for the sake of completeness: $(\widetilde{l}_L C q_L)(u_R C d_R)$, $(e_R C u_R)(\widetilde{q}_L C q_L)$, $(\widetilde{l}_L C q_L)(\widetilde{q}_L C q_L)$, and $(e_R C u_R)(u_R C d_R)$. Here $l_L = \binom{\nu}{e}_L$ and $q_L = \binom{u}{d}_L$ denote the lepton and quark doublet of $SU(2) \otimes U(1)$, the twiddle is defined by $\widetilde{l}^j = l_i \varepsilon^{ij}$ with $SU(2)$ indices $i, j = 1, 2$ (see appendix B), and C denotes the charge conjugation matrix. Color indices on the quark fields are contracted in the only possible way. The effective Lagrangian is then given by the sum of these four terms, with four unknown coefficients.

The effective field theory tells us that all possible baryon number violating decay processes can be determined in terms of four unknowns. We expect that these predictions will hold to an accuracy of order $(M_W/M)^2$. (If M_W were zero, $SU(3) \otimes SU(2) \otimes U(1)$ would be exact.)

Of course, we can increase our predictive power by making further assumptions. For example, if we think that proton decay is mediated by a vector particle, as in a generic grand unified theory, then only the first two terms in the above list are allowed. In a specific grand unified theory, such as the $SU(5)$ theory, the two unknown coefficients are determined in terms of the grand unified coupling and the mass of the X boson.

To appreciate the predictive power of the effective field theory approach, inspect the list of the four possible operators. We can immediately predict that while proton decay violates both baryon number B and lepton number L, it conserves the combination $B - L$. I emphasize that this is not at all obvious before doing the analysis. Could you have told the experimentalist which of the two possible modes $n \to e^+ \pi^-$ or $n \to e^- \pi^+$ he should expect? A priori, it could well be that $B + L$ is conserved.

Note that Fermi's theory of the weak interaction would be called an effective field theory these days. Of course, in contrast to proton decay, beta decay was actually seen, and the prediction from this sort of symmetry analysis, namely the existence of the neutrino, was triumphantly confirmed.

Along the same line, we could construct an effective field theory of neutrino masses. Surely one of the most exciting experimental discoveries in particle physics of recent years was that neutrinos are not massless. Let us construct an $SU(2) \otimes U(1)$ invariant effective theory. Since ν_L resides inside l_L, without doing any detailed analysis we can see that a dimension-5 operator is required: schematically $l_L l_L$ contains the desired neutrino bilinear but it carries hypercharge $Y/2 = -1$; on the other hand, the Higgs doublet φ carries hypercharge $+\frac{1}{2}$, and so the lowest dimensional operator we can form is of the form $l l \varphi \varphi$ with dimension $\frac{3}{2} + \frac{3}{2} + 1 + 1 = 5$. Thus, the effective $\mathcal{L}$ must contain a term $(1/M) l l \varphi \varphi$, with M the mass scale of the new physics responsible for the neutrino mass. Thus, by dimensional analysis we can estimate $m_\nu \sim m_l^2/M$, with m_l some typical charged lepton mass. If we take m_l to be the muon mass $\sim 10^2 \mathrm{Mev}$ and $m_\nu \sim 10^{-1} \mathrm{ev}$, we find $M \sim (10^2 \mathrm{Mev})^2/10^{-1}(10^{-6}\mathrm{Mev}) = 10^8 \mathrm{Gev}$.

The philosophy of effective field theories valid up to a certain energy scale Λ seems so obvious by now that it is almost difficult to imagine that at one time many eminent physicists demanded much more of quantum field theory: that it be fundamental up to arbitrarily high energy scales.

Indeed, we now regard all quantum field theories as effective field theory. For all we know, spacetime on some short distance does consist of a lattice, and so the Yang-Mills action is but the leading term in an expansion of the Wilson lattice action. The Einstein-Hilbert Lagrangian, being nonrenormalizable, is a fortiori "merely" the leading term in an effective field theory

$$\mathcal{L} = \sqrt{-g}(M_\Lambda^4 + M_P^2 R + c_1 R^2 + c_2 R_{\mu\nu} R^{\mu\nu} + c_3 R_{\mu\nu\sigma\rho} R^{\mu\nu\sigma\rho} + \frac{1}{M^2}(d_1 R^3 + \cdots) + \cdots)$$

Here $c_{1,2,3}$ and d_1 are dimensionless numbers presumably of order 1. The three terms quadratic in the curvature involve four powers of derivatives versus the two powers in the Einstein-Hilbert term, and hence their effects relative to the leading terms are suppressed by $(E/M_P)^2$ with E an energy scale characteristic of the process we are studying. Thus, these so-called Weyl-Eddington terms could be safely ignored in any conceivable experiment. [A technical aside: The Gauss-Bonnet theorem implies that the combination $(R^2 - 4R_{\mu\nu} R^{\mu\nu} + R_{\mu\nu\sigma\rho} R^{\mu\nu\sigma\rho})$ is a total derivative, so c_3 can be effectively set to 0, but that is besides the point here.] We have indicated only one representative dimension 6 term R^3 (out of many). Its coefficient, in accordance with high school dimensional analysis, is suppressed by two powers of some mass M.

What do we expect the mass scale M to be? Suppose we live in a universe with only gravity (and of course we don't, actually) then once again, we could risk being presumptuous and take M to be the intrinsic mass scale of gravity, namely the Planck mass M_P, but we have not yet recovered from our third-degree burn from supposing that $M_\Lambda \sim M_P$. If we could ignore the cosmological constant problem for a moment, then the standard (but quite possibly wrong!) consensus is that in a universe of pure gravity our theory of gravity is an effective expansion in powers of $(E/M_P)^2$.

Alternatively, we could treat $\mathcal{L}$ as the effective theory of gravity after we integrate out all the matter degree of freedom. In that case, M would be of order m_e (imagine gravitons coupled to an electron loop; see exercise VIII.3.5), or perhaps even m_ν (generated by a neutrino loop).

Effective field theory of the blue sky

As another application of the effective field theory philosophy, consider the scattering of electromagnetic waves on an electrically neutral spinless particle described by a scalar field Φ. Since Φ is neutral, the lowest dimension gauge invariant term that can be added to $\mathcal{L} = \partial \Phi^\dagger \partial \Phi + m^2 \Phi^\dagger \Phi + \cdots$ is $(1/M^2)\Phi^\dagger \Phi F_{\mu\nu} F^{\mu\nu}$. A factor of $1/M^2$, with M some mass scale, has to be included with the dimension $1 + 1 + 2 + 2 = 6$ operator to bring the high school dimension down to 4. The two powers of derivative in $F_{\mu\nu} F^{\mu\nu}$ tell us immediately that the amplitude for photon scattering on this neutral particle goes like $\mathcal{M} \propto \omega^2$, with ω the frequency of the electromagnetic wave. Thus we conclude that the scattering cross section varies like $\sigma(\omega) \propto \omega^4$.

We have arrived at Rayleigh's celebrated explanation of the color of the sky. In passing through the atmosphere red light scatters less than blue light on air molecules and hence the sky is blue.

For application to spinless atoms or molecules, we can pass to the nonrelativistic limit as described in chapter III.5, setting $\Phi = (1/\sqrt{2m})e^{-imt}\varphi$, so that the effective Lagrangian now reads

$$\mathcal{L} = \varphi^\dagger i\partial_0\varphi - \frac{1}{2m}\partial_i\varphi^\dagger\partial_i\varphi + \frac{1}{mM^2}\varphi^\dagger\varphi(c_1\vec{E}^2 - c_2\vec{B}^2) + \cdots$$

In this case, since we understand the microscopic physics governing atoms and molecules, we know perfectly well what the mass scale M represents. The coupling of a photon to an electrically neutral system such as an atom or a molecule must vanish like the characteristic size d of the system, since as $d \to 0$ the positive and negative charges are on top of each other, giving a vanishing net coupling to the photon. Rotational invariance implies that the coupling $\sim \vec{k} \cdot \vec{d}$. The scattering amplitude then goes like $\mathcal{M} \propto (\omega d)^2$, since the coupling has to act twice, once for the incoming photon and once for the out-going photon. (Note that by rotational invariance the expectation value of the operator $\vec{d}$ vanishes, but we are doing second order perturbation theory so that we have to evaluate the expectation value of a quantity quadratic in $\vec{d}$.)[1] Squaring $\mathcal{M}$ and invoking some elementary quantum mechanics and dimensional analysis, we obtain the cross section $\sigma(\omega) \sim d^6\omega^4$.

Appendix: Reshuffling terms in effective field theory

The Lagrangian of an effective field theory consists of an infinite sequence of terms arranged in an orderly progression of higher and higher mass dimension, constrained only by the assumed symmetries of the theory. In fact, some terms could be effectively eliminated. To explain this, we focus on a toy example:

$$\mathcal{L} = \frac{1}{2}(\partial\varphi)^2 - \lambda\varphi^4 + \frac{1}{M^2}(a\varphi^6 + b\varphi^3\partial^2\varphi + c(\partial^2\varphi)^2) + O\left(\frac{1}{M^4}\right) \tag{1}$$

We are secretly dealing with the action and thus we freely integrate by parts. For arithmetical simplicity, we did not include a mass term, so that to leading order in $1/M$ the equation of motion reads simply $\partial^2\varphi = 0$. The three possible dimension 6 terms are shown explicitly [we integrate by parts to get rid of the term $\varphi^2(\partial\varphi)^2$].

Are we allowed to use the equation of motion to eliminate the two dimension 6 terms that are proportional to $\partial^2\varphi$?

We know that we could make a field redefinition without changing the on shell amplitudes, so let us redefine $\varphi \to \varphi + (1/M^2)F$. Then $\frac{1}{2}(\partial\varphi)^2 \to \frac{1}{2}(\partial\varphi)^2 - (1/M^2)F\partial^2\varphi + O(1/M^4)$ and $\lambda\varphi^4 \to \lambda(\varphi^4 + (1/M^2)\varphi^3 F + O(1/M^4))$. Set $F = p\varphi^3 + q\partial^2\varphi$. We see that with an appropriate choice of p and q we can cancel off b and c. Notice that in the process we also change a to some other value.

The answer to the question is yes, but the naive statement that the equation of motion $\partial^2\varphi = 0$ empowers us to simply set $\partial^2\varphi$ to zero in the nonleading terms in the effective field theory is, legalistically speaking, incorrect, or at least misleading. We see that we actually generated $O(1/M^4)$ terms and changed the φ^6 term. Thus, more correctly, a field redefinition allows us to shuffle terms around and to higher order. The net effect, however, is the same as if we trusted the naive statement and set $\partial^2\varphi$ to zero in the nonleading terms.

---

[1] For details, see, for example, J. J. Sakurai, *Advanced Quantum Mechanics*, Addison-Wesley, New York, 1967, p. 47.

This procedure works for fermions also. As an example, consider the effective Lagrangian $\mathcal{L} = \bar{\psi}(i\gamma^\mu\partial_\mu - m)\psi + (1/M^3)\bar{\psi}(i\gamma^\mu\partial_\mu - m)\psi(\bar{\psi}\psi) + \cdots$. Then the field redefinition $\psi \to \psi - (1/2M^3)\psi(\bar{\psi}\psi)$ gets rid of the dimension 7 term shown.

We could also apply what we just learned to the effective theory of gravity if without any understanding we set the cosmological constant to zero. Also, use the Gauss-Bonnet theorem to get rid of the $R_{\mu\nu\sigma\rho}R^{\mu\nu\sigma\rho}$ term, so that we have

$$\mathcal{L} = \sqrt{-g}\left(M_P^2 R + c_1 R^2 + c_2 R_{\mu\nu}R^{\mu\nu} + \frac{1}{M^2}(d_1 R^3 + \cdots) + \cdots\right) \tag{2}$$

Make a field redefinition $g_{\mu\nu} \to g_{\mu\nu} + \delta g_{\mu\nu}$ and use

$$\delta \int d^4x \sqrt{-g}\, R = -\int d^4x \sqrt{-g}\,(R^{\mu\nu} - \frac{1}{2}g^{\mu\nu}R)\delta g_{\mu\nu}$$

Set $\delta g_{\mu\nu} = pR^{\mu\nu} + qg^{\mu\nu}R$. Then we can cancel off c_1 and c_2 with a judicious choice of p and q. I emphasize that this works only if we set the cosmological constant to zero without any ado.

Exercises

VIII.3.1 Consider

$$\mathcal{L} = \frac{1}{2}\left[(\partial\varphi_1)^2 + (\partial\varphi_2)^2\right] - \lambda(\varphi_1^4 + \varphi_2^4) - g\varphi_1^2\varphi_2^2 \tag{3}$$

We have taken the $O(2)$ theory from chapter I.10 and broken the symmetry explicitly. Work out the renormalization group flow in the $(\lambda - g)$ plane and draw your own conclusions.

VIII.3.2 Assuming the nonexistence of the right handed neutrino field ν_R (i.e., assuming the minimal particle content of the standard model) write down all $SU(2) \otimes U(1)$ invariant terms that violate lepton number L by 2 and hence construct an effective field theory of the neutrino mass. Of course, by constructing a specific theory one can be much more predictive. Out of the product $l_L l_L$ we can form a Lorentz scalar transforming as either a singlet or triplet under $SU(2)$. Take the singlet case and construct a theory. [Hint: For help, see A. Zee, *Phys. Lett.* 93B: p. 389, 1980.]

VIII.3.3 Let A, B, C, D denote four spin $\frac{1}{2}$ fields and label their handedness by a subscript: $\gamma^5 A_h = hA_h$ with $h = \pm 1$. Thus, A_+ is right handed, A_- left handed, and so on. Show that

$$(A_h B_h)(C_{-h}D_{-h}) = -\frac{1}{2}(A_h\gamma^\mu D_{-h})(C_{-h}\gamma_\mu B_h) \tag{4}$$

This is an example of a broad class of identities known as Fierz identities (some of which we will need in discussing supersymmetry.) Argue that if proton decay proceeds in lowest order from the exchange of a vector particle then only the terms $(\widetilde{l}_L Cq_L)(u_R Cd_R)$ and $(e_R Cu_R)(\widetilde{q}_L Cq_L)$ are allowed in the Lagrangian.

VIII.3.4 Given the conclusion of the previous exercise show that the decay rate for the processes $p \to \pi^+ + \bar{\nu}$, $p \to \pi^0 + e^+$, $n \to \pi^0 + \bar{\nu}$, and $n \to \pi^- + e^+$ are proportional to each other, with the proportionality factors determined by a single unknown constant [the ratio of the coefficients of $(\widetilde{l}_L Cq_L)(u_R Cd_R)$ and $(e_R Cu_R)(\widetilde{q}_L Cq_L)$].

For help on these last three exercises see S. Weinberg, *Phys. Rev. Lett.* 43: 1566, 1979; F. Wilczek and A. Zee, ibid. p. 1571; H. A. Weldon and A. Zee, *Nucl. Phys.* B173: 269, 1980.

VIII.3.5 Imagine a mythical (and presumably impossible) race of physicists who only understand physics at energies less than the electron mass m_e. They manage to write down the effective field theory for the one particle they know, the photon,

$$\mathcal{L} = -\frac{1}{4}F_{\mu\nu}F^{\mu\nu} + \frac{1}{m_e^4}\{a(F_{\mu\nu}F^{\mu\nu})^2 + b(F_{\mu\nu}\tilde{F}^{\mu\nu})^2\} + \cdots \tag{5}$$

with $\tilde{F}^{\mu\nu} = \frac{1}{2}\varepsilon^{\mu\nu\rho\sigma}F_{\rho\sigma}$ the dual field strength as usual and a and b two dimensionless constants presumably of order unity.

(a) Show that $\mathcal{L}$ respects charge conjugation ($A \to -A$ in this context), parity, and time reversal, (and of course gauge invariance.)

(b) Draw the Feynman diagrams that give rise to the two dimension 8 terms shown. The coefficients a and b were calculated by Euler and Kockel in 1935 and by Heisenberg and Euler in 1936, quite a feat since they did not know about Feynman diagrams and any of the modern quantum field theory set up.

(c) Explain why dimensional 6 terms are absent in $\mathcal{L}$. [Hint: One possible term is $\partial_\lambda F_{\mu\nu}\partial^\lambda F^{\mu\nu}$.]

(d) Our mythical physicists do not know about the electron, but they are getting excited. They are going to start doing photon-photon scattering experiments with a machine called LPC that could produce photons with energy greater than m_e. Discuss what they will see. Apply unitarity and the Cutkosky rules.

VIII.3.6 Use the effective field theory approach to show that the scattering cross section of light on an electrically neutral spin $\frac{1}{2}$ particle (such as the neutron) goes like $\sigma \propto \omega^2$ to leading order, not ω^4. Argue further that the constant of proportionality can be fixed in terms of the magnetic moment μ of the particle. [Historical note: This result was first obtained in 1954 by F. Low (*Phys. Rev.* 96: 1428) and by M. Gell-Mann and Murph L. Goldberger (*Phys. Rev.* 96: 1433) using much more elaborate arguments.]

VIII.4 | Supersymmetry: A Very Brief Introduction

Unifying bosons and fermions

Let me start with a few of the motivations for supersymmetry. (1) All experimentally known symmetries relate bosons to bosons and fermions to fermions. We would like to have a symmetry, supersymmetry, relating bosons and fermions. (2) It is natural for fermions to be massless (recall chapter VII.6), but not for bosons. Perhaps by pairing the Higgs field with a fermion field we can resolve the hierarchy problem mentioned in chapter VII.6. (3) Recalling from chapter II.5 that fermions contribute negatively to the vacuum energy, you might be tempted to speculate that the cosmological constant problem could be solved if we could get the fermion contribution to cancel the boson contribution.

Disappointingly, it has been more than 30 years[1] since the conception of supersymmetry (Golfand and Likhtman constructed the first supersymmetric field theory in 1971) and direct experimental evidence is still lacking. All existing supersymmetric theories pair known bosons with unknown fermions and known fermions with unknown bosons. Supersymmetry has to be broken at some mass scale M beyond the regime already explored experimentally, but then (as explained in chapter VIII.2) we might expect a cosmological constant of order M^4.

Be that as it may, supersymmetric field theories have many nice properties (hardly surprising since the relevant symmetry is much larger). Supersymmetry has thus attracted a multitude of devotees. I give you here as brief an introduction to supersymmetry as I can write. In the spirit of a first exposure, I will avoid mentioning any subtleties and caveats, hoping that this brief introduction will be helpful to students before they tackle the tomes out there.

---

[1] For a fascinating account of the early history of supersymmetry, see G. Kane and M. Shifman, eds., *The Supersymmetric World: The Beginning of the Theory.*

Inventing supersymmetry

Suppose one day you woke up wanting to invent a field theory with a symmetry relating bosons to fermions. The first thing you would need is the same number of fermionic and bosonic degrees of freedom. The simplest fermion field is the two-component Weyl spinor ψ. You would now have one complex degree of freedom,[2] so you would have to throw in a complex scalar field φ. You could proceed by trial and error: Write down a Lagrangian including all terms with dimension up to four and then adjust the various parameters in the Lagrangian until the desired symmetry appears. For instance, you might adjust μ in the mass terms $\mu^2 \varphi^\dagger \varphi + m(\psi\psi + \bar\psi\bar\psi)$ until the theory becomes more symmetrical so that the boson and the fermion have the same mass.

If you were to try to play the game by using a Dirac spinor Ψ and a complex scalar φ you would be doomed to failure from the very start since there would be twice as many fermionic degrees of freedom as bosonic degrees of freedom. I believe that the development of supersymmetry was very much retarded by the fact that until the early 1970s most field theorists, having grown up with Dirac spinors, had little knowledge of Weyl spinors. That was a hint that now is the time for you to get thoroughly familiar with the dotted and undotted notation of appendix E. To read this chapter, you need to be fluent with that notation.

Supersymmetric algebra

It is perfectly feasible to construct this supersymmetric field theory, known as the Wess-Zumino model, by trial and error, but instead I will show you an elegant but more abstract approach known as the superspace and superfield formalism, invented by Salam and Strathdee. We will have to develop a considerable amount of formal machinery. Everything is very super here.

Write the supersymmetry generator taking us from φ to ψ_α as Q_α (known as the supercharge). The statement that Q_α transforms as a Weyl spinor means $[J^{\mu\nu}, Q_\alpha] = -i(\sigma^{\mu\nu})_\alpha{}^\beta Q_\beta$, where $J^{\mu\nu}$ denotes the generators of the Lorentz group. Of course, since Q_α is independent of the spacetime coordinates $[P^\mu, Q_\alpha] = 0$. From appendix E we denote the conjugate of Q_α by $\bar Q_{\dot\alpha}$ and $[J^{\mu\nu}, \bar Q^{\dot\alpha}] = -i(\bar\sigma^{\mu\nu})^{\dot\alpha}{}_{\dot\beta} \bar Q^{\dot\beta}$.

We have to write down the anticommutation relation between the Grassman objects Q_α and $\bar Q_{\dot\beta}$ and now the work we did in appendix E really pays off. The supersymmetry algebra is given by

$$\{Q_\alpha, \bar Q_{\dot\beta}\} = 2(\sigma^\mu)_{\alpha\dot\beta} P_\mu \tag{1}$$

---

[2] One complex degree of freedom on mass shell and two complex degrees of freedom off mass shell. See the superfield formalism below.

We argue by the "what else can it be?" method. The right-hand side must carry the indices α and $\dot{\beta}$ and we know that the only object that carries these indices is σ^μ. The Lorentz index μ has to be contracted and the only vector around is P_μ. The factor of 2 fixes the normalization of Q.

By the same kind of argument we must have $\{Q_\alpha, Q^\beta\} = c_1(\sigma^{\mu\nu})_\alpha{}^\beta J_{\mu\nu} + c_2 \delta_\alpha^\beta$. Commuting with P^λ we see that the constant c_1 must vanish. Recalling that $Q_\gamma = \varepsilon_{\gamma\beta} Q^\beta$, we have $\{Q_\alpha, Q_\gamma\} = c_2 \varepsilon_{\gamma\alpha}$; but since the left-hand side is symmetric in α and γ we have $c_2 = 0$. Thus, $\{Q_\alpha, Q_\beta\} = 0$ and $\{\bar{Q}_{\dot{\alpha}}, \bar{Q}_{\dot{\beta}}\} = 0$ (see exercise VIII.4.2).

A basic theorem

An important physical fact follows immediately from (1). Contracting with $(\bar{\sigma}^\nu)^{\dot{\beta}\alpha}$ we obtain

$$4P^\nu = (\bar{\sigma}^\nu)^{\dot{\beta}\alpha}\{Q_\alpha, \bar{Q}_{\dot{\beta}}\} \tag{2}$$

In particular the time component tells us about the Hamiltonian

$$4H = \sum_\alpha \{Q_\alpha, \bar{Q}_{\dot{\alpha}}\} = \sum_\alpha \{Q_\alpha, Q_\alpha^\dagger\} = \sum_\alpha (Q_\alpha Q_\alpha^\dagger + Q_\alpha^\dagger Q_\alpha) \tag{3}$$

We obtain the important theorem that in a supersymmetric field theory any physical state $|S\rangle$ must have nonnegative energy:

$$\langle S|H|S\rangle = \frac{1}{2}\sum_\alpha \sum_{S'} |\langle S'|Q_\alpha|S\rangle|^2 \geq 0 \tag{4}$$

Superspace

Now that we have constructed the supersymmetric algebra let us keep in mind our goal of constructing supersymmetric field theories. To do that, we need to figure out and classify how fields transform under this supersymmetric algebra. We have to go through a lot of formalism, the necessity for which will become clear in due course.

Imagine that you are trying to invent the superspace formalism. Let us motivate it by staring at the basic relation (1) $\{Q_\alpha, \bar{Q}_{\dot{\beta}}\} = 2(\sigma^\mu)_{\alpha\dot{\beta}} P_\mu$. A supersymmetric transformation Q followed by its conjugate $\bar{Q}_{\dot{\beta}}$ generates a translation P_μ. Hmm, let's see, $P_\mu \equiv i(\partial/\partial x^\mu)$ generates translation in x^μ, so perhaps Q_α, being Grassmannian, would generate translation in some abstract Grassmannian coordinate θ^α? (Similarly, $\bar{Q}_{\dot{\beta}}$ would generate translation in $\bar{\theta}^{\dot{\beta}}$.)

Salam and Strathdee invented the notion of a superspace with bosonic and fermionic coordinates $\{x^\mu, \theta^\alpha, \bar{\theta}^{\dot{\beta}}\}$ with the supersymmetry algebra represented by translations in this space.

So let us try Q_α and $\bar{Q}_{\dot{\beta}}$ being something like $\partial/\partial\theta^\alpha$ and $\partial/\partial\bar{\theta}^{\dot{\beta}}$, respectively. But then $\{Q_\alpha, \bar{Q}_{\dot{\beta}}\} = 0$ and we don't get (1). We have to keep playing around modifying Q_α and $\bar{Q}_{\dot{\beta}}$. You may already see what we need. If we add a term such as $\theta\sigma^\mu\partial_\mu$ to $\bar{Q}_{\dot{\beta}}$, then the $\partial/\partial\theta^\alpha$

in Q_α acting on $\theta\sigma^\mu\partial_\mu$ will produce something like the right-hand side of (1). Similarly, we will want to add a term such as $\bar{\theta}\sigma^\mu\partial_\mu$ to Q_α. (Once again, the dotted and undotted notation we worked hard to develop fixes what we must write, namely $(\sigma^\mu)_{\alpha\dot\alpha}\bar{\theta}^{\dot\alpha}\partial_\mu$ so that the indices match and obey the "southwest to northeast" rule.) Thus, we represent the supercharges as

$$Q_\alpha = \frac{\partial}{\partial\theta^\alpha} - i(\sigma^\mu)_{\alpha\dot\alpha}\bar{\theta}^{\dot\alpha}\partial_\mu \tag{5}$$

and

$$\bar{Q}_{\dot\beta} = -\frac{\partial}{\partial\bar{\theta}^{\dot\beta}} + i\theta^\beta (\sigma^\mu)_{\beta\dot\beta}\partial_\mu \tag{6}$$

You see that (1) is now satisfied. Interestingly, when we translate in the fermionic direction we have to translate a bit in the bosonic direction as well.

Superfield

A superfield $\Phi(x^\mu, \theta^\alpha, \bar{\theta}^{\dot\beta})$, as the name suggests, is just a field living in superspace. An infinitesimal supersymmetry transformation takes

$$\Phi \to \Phi' = (1 + i\xi^\alpha Q_\alpha + i\bar{\xi}_{\dot\alpha}\bar{Q}^{\dot\alpha})\Phi \tag{7}$$

with ξ and $\bar{\xi}$ two Grassmannian parameters.

It turns out that we can impose some condition on Φ and restrict this rather broad definition a bit. After staring at (5) and (6) for a while, you may realize that there are two other objects,

$$D_\alpha = \frac{\partial}{\partial\theta^\alpha} + i(\sigma^\mu)_{\alpha\dot\alpha}\bar{\theta}^{\dot\alpha}\partial_\mu$$

and

$$\bar{D}_{\dot\beta} = -\left[\frac{\partial}{\partial\bar{\theta}^{\dot\beta}} + i\theta^\beta (\sigma^\mu)_{\beta\dot\beta}\partial_\mu\right]$$

that we can define, sort of the combinations orthogonal to Q_α and $\bar{Q}_{\dot\beta}$. Clearly, D_α and $\bar{D}_{\dot\beta}$ anticommute with Q_α and $\bar{Q}_{\dot\beta}$. The significance of this fact is that if we impose the condition $\bar{D}_{\dot\beta}\Phi = 0$ on the superfield Φ, then according to (7) its transform Φ' also satisfies the condition.

A superfield Φ satisfying the condition $\bar{D}_{\dot\beta}\Phi = 0$ is known as a chiral superfield. The condition is actually easy to implement:[3] Observe that if we define $y^\mu \equiv (x^\mu + i\theta^\alpha (\sigma^\mu)_{\alpha\dot\alpha}\bar{\theta}^{\dot\alpha})$ (note we are adding two bosonic quantities here), then

$$\bar{D}_{\dot\beta}y^\mu = -\left[\frac{\partial}{\partial\bar{\theta}^{\dot\beta}} + i\theta^\beta(\sigma^\nu)_{\beta\dot\beta}\partial_\nu\right]y^\mu = -\left[-i\theta^\alpha(\sigma^\mu)_{\alpha\dot\beta} + i\theta^\beta(\sigma^\mu)_{\beta\dot\beta}\right] = 0$$

Thus, a superfield $\Phi(y, \theta)$ that depends on y and θ only is a chiral superfield.

---

[3] This is analogous to the problem of constructing a function $f(x, y)$ satisfying the condition $Lf = 0$ with $L \equiv [x(\partial/\partial y) - y(\partial/\partial x)]$. We define $r \equiv (x^2 + y^2)^{\frac{1}{2}}$ and observe that $Lr = 0$. Then any f that only depends on r satisfies the desired condition.

Let us expand Φ in powers of θ holding y fixed. Remember that θ contains two components (θ^1, θ^2). Thus, we can form an object with at most two powers of θ, namely $\theta\theta$, which you worked out in exercise E.3. Thus, as usual, power series in Grassmannian variables terminate, and we have

$$\Phi(y, \theta) = \varphi(y) + \sqrt{2}\theta\psi(y) + \theta\theta F(y)$$

with $\varphi(y)$, $\psi(y)$, and $F(y)$ merely coefficients in the series at this stage. We can Taylor expand once more around x:

$$\begin{aligned}
\Phi(y, \theta) = {}& \varphi(x) + \sqrt{2}\theta\psi(x) + \theta\theta F(x) \\
& + i\theta\sigma^\mu\bar{\theta}\partial_\mu\varphi(x) - \tfrac{1}{2}\theta\sigma^\mu\bar{\theta}\theta\sigma^\nu\bar{\theta}\partial_\mu\partial_\nu\varphi(x) + \sqrt{2}i\theta\sigma^\mu\bar{\theta}\partial_\mu\psi(x)
\end{aligned} \tag{8}$$

We see that a chiral superfield Φ contains a Weyl fermion field ψ, and two complex scalar fields φ and F.

Finding a total divergence

Let's do a bit of dimensional analysis for fun and profit. Given that P_μ has the dimension of mass, which we write as $[P_\mu] = 1$ using the same notation as in chapter III.2, then (1), (5), and (6) tell us that $[Q] = [\bar{Q}] = \tfrac{1}{2}$ and $[\theta] = [\bar{\theta}] = -\tfrac{1}{2}$. Given $[\varphi] = 1$, then (8) tells us $[\psi] = \tfrac{3}{2}$, which we know already, and $[F] = 2$, which we didn't know. In fact, we have never met a Lorentz scalar field with mass dimension 2. How can we have a kinetic energy term for F in $\mathcal{L}$ with dimension 4? We can't. The term $F^\dagger F$ already has dimension 4, and any derivative is going to make the dimension even higher. Also, didn't we say that with φ and ψ we balance the same number of bosonic and fermionic degrees of freedom?

The field $F(x)$ definitely has something strange about him. What is he doing in our theory?

Under an infinitesimal supersymmetric transformation the superfield changes by $\delta\Phi = i(\xi Q + \bar{\xi}\bar{Q})\Phi$. Referring to (8), (5), and (6), you can work out how the component fields φ, ψ, and F transform (see exercise VIII.4.5). But we can go a long way invoking symmetry and dimensional analysis. For example, δF is linear in ξ or $\bar{\xi}$, which by dimensional analysis must multiply something with dimension $[\tfrac{5}{2}]$ since $[F] = 2$ and $[\xi] = [\bar{\xi}] = -\tfrac{1}{2}$. The only thing around with dimension $[\tfrac{5}{2}]$ is $\partial_\mu\psi$, which carries an undotted index. Note it can't be $\partial_\mu\bar{\psi}$ since Φ does not contain $\bar{\psi}$. By Lorentz invariance we have to find something carrying the index μ, and that can only be $(\sigma^\mu)_{\alpha\dot{\alpha}}$. The dotted index on $(\sigma^\mu)_{\alpha\dot{\alpha}}$ can only be contracted with $\bar{\xi}$. So everything is fixed except for an overall constant:

$$\delta F \sim \partial_\mu\psi^\alpha(\sigma^\mu)_{\alpha\dot{\alpha}}\bar{\xi}^{\dot{\alpha}} \tag{9}$$

Arguing along the same lines you can easily show that $\delta\varphi \sim \xi\psi$ and $\delta\psi \sim \xi F + \partial_\mu\varphi\sigma^\mu\bar{\xi}$.

The important point here is not the overall constant in (9) but that δF is a total divergence.

Given any superfield Φ let us denote by $[\Phi]_F$ the coefficient of $\theta\theta$ in an expansion of Φ [as in (8)]. What we have learned is that under a supersymmetric transformation $\delta([\Phi]_F)$ is a total divergence and thus $\int d^4x[\Phi]_F$ is invariant under supersymmetry.

Our next observation is that if $\bar{D}_{\dot{\beta}}\Phi = 0$, then $\bar{D}_{\dot{\beta}}\Phi^2 = 0$ also. In other words, if Φ is a chiral superfield, then so is Φ^2 (and by extension, Φ^3, Φ^4, and so forth).

Supersymmetric action

What do we want to achieve anyway? We want to construct an action invariant under supersymmetry.

Finally, after all this formalism we are ready. In fact, it is almost staring us in the face: $\int d^4x [\frac{1}{2}m\Phi^2 + \frac{1}{3}g\Phi^3 + \cdots]_F$ is invariant under supersymmetry by virtue of the last two paragraphs. Squaring (8) and extracting the coefficient of $\theta\theta$ we see by inspection that $[\Phi^2]_F = (2F\varphi - \psi\psi)$. Similarly, $[\Phi^3]_F = 3(F\varphi^2 - \varphi\psi\psi)$. Now do exercise VIII.4.6.

Looks like we have generated a mass term for the Weyl fermion ψ and its coupling to the scalar field φ, but where are the kinetic energy terms, such as $\bar{\psi}_{\dot{\alpha}}(\bar{\sigma}^\mu)^{\dot{\alpha}\alpha}\partial_\mu\psi_\alpha$?

Vector superfield

The kinetic energy terms contain $\bar{\psi}_{\dot{\alpha}}$, which does not appear in Φ. To get the conjugate field $\bar{\psi}_{\dot{\alpha}}$, we obviously have to use $\Phi^\dagger$, and so we are led to consider $\Phi^\dagger\Phi$. More formalism here! We call a superfield $V(x, \theta, \bar{\theta})$ a vector superfield if $V = V^\dagger$. For example, $\Phi^\dagger\Phi$ is a vector superfield.

Imagine expanding $\Phi^\dagger\Phi = \varphi^\dagger\varphi + \cdots$ or any vector superfield V in powers of θ and $\bar{\theta}$. The highest power is uniquely $\bar{\theta}\bar{\theta}\theta\theta$ since by the properties of Grassmannian variables the only object we can form is $\bar{\theta}^1\bar{\theta}^2\theta_1\theta_2$. Any object quadratic in θ and quadratic in $\bar{\theta}$, such as $(\theta\sigma^\mu\bar{\theta})(\theta\sigma_\mu\bar{\theta})$, can be beaten down to $\bar{\theta}\bar{\theta}\theta\theta$ by using the kind of identities you discovered in the exercises in appendix E. Let $[V]_D$ denote the coefficient of $\bar{\theta}\bar{\theta}\theta\theta$ in the expansion of V.

Again, dimensional analysis can carry us a long way. If V has mass dimension n, then $[V]_D$ has mass dimension $n + 2$ since θ and $\bar{\theta}$ each has mass dimension $-\frac{1}{2}$. Let us study how $[V]_D$ changes under an infinitesimal supersymmetry transformation $\delta V = i(\xi Q + \bar{\xi}\bar{Q})V$. We use the same kind of argument as before: $\delta([V]_D)$ is linear in ξ or $\bar{\xi}$, which by dimensional analysis must multiply something with dimension $n + \frac{5}{2}$ since $[\xi] = [\bar{\xi}] = -\frac{1}{2}$. This can only be the derivative ∂ of something with dimension $n + \frac{3}{2}$, namely the coefficients of $\bar{\theta}\bar{\theta}\theta$ and $\bar{\theta}\theta\theta$ in the expansion of V. We conclude that $\delta([V]_D)$ has to have the form $\partial_\mu(\ldots)$, namely that $\delta([V]_D)$ is a total divergence. This is the same type of argument that allows us to conclude that $\delta([\Phi]_F)$ is a total divergence.

Thus, the action $\int d^4x [\Phi^\dagger\Phi]_D$ is invariant under supersymmetry.

Staring at (8), which I repeat for your convenience,

$$\Phi(y, \theta) = \varphi(x) + \sqrt{2}\theta\psi(x) + \theta\theta F(x) \tag{10}$$

$$+ i\theta\sigma^\mu\bar{\theta}\partial_\mu\varphi(x) - \frac{1}{2}\theta\sigma^\mu\bar{\theta}\theta\sigma^\nu\bar{\theta}\partial_\mu\partial_\nu\varphi(x) + \sqrt{2}i\theta\theta\sigma^\mu\bar{\theta}\partial_\mu\psi(x)$$

we see that $\int d^4x [\Phi^\dagger \Phi]_D$ contains $\int d^4x \varphi^\dagger \partial^2 \varphi$ (from multiplying the first term in $\Phi^\dagger$ with the fifth term in Φ), $\int d^4x \partial \varphi^\dagger \partial \varphi$ (from multiplying the fourth term in $\Phi^\dagger$ with the fourth term in Φ), $\int d^4x \bar{\psi} \bar{\sigma}^\mu \partial_\mu \psi$ (from multiplying the second term in $\Phi^\dagger$ with the sixth term in Φ), and finally $\int d^4x F^\dagger F$ (from multiplying the third term in $\Phi^\dagger$ with the third term in Φ). It is quite amusing how derivatives of fields arise in supersymmetric field theories: Note that the action $\int d^4x [\Phi^\dagger \Phi]_D$ does not contain derivatives explicitly.

To summarize, given a chiral superfield Φ we have constructed the supersymmetric action

$$S = \int d^4x \{ [\Phi^\dagger \Phi]_D - ([W(\Phi)]_F + \text{h.c.}) \} \tag{11}$$

Explicitly, with the choice $W(\Phi) = \frac{1}{2}m\Phi^2 + \frac{1}{3}g\Phi^3$, we have

$$S = \int d^4x \{ \partial \varphi^\dagger \partial \varphi + i \bar{\psi} \bar{\sigma}^\mu \partial_\mu \psi + F^\dagger F - (mF\varphi - \tfrac{1}{2}m\psi\psi + gF\varphi^2 - g\varphi\psi\psi + \text{h.c.}) \} \tag{12}$$

An auxiliary field

From the very beginning the field F seemed strange. Since $[F] = 2$ we anticipated that it cannot have a kinetic energy term with mass dimension 4 and indeed it doesn't. We see that it is not a dynamical field that propagates—it is an auxiliary field (just like σ in chapter III.5 and ξ_μ in chapter VI.3) and can be integrated out in the path integral $\int DF^\dagger DF e^{iS}$. Indeed, collect the terms that depend on F in S, namely

$$F^\dagger F - F(m\varphi + g\varphi^2) - F^\dagger(m\varphi^\dagger + g\varphi^{\dagger 2}) = |F - (m\varphi + g\varphi^2)^\dagger|^2 - |m\varphi + g\varphi^2|^2$$

So, integrate over F and $F^\dagger$ and get

$$S = \int d^4x \{ \partial \varphi^\dagger \partial \varphi + i \bar{\psi} \bar{\sigma}^\mu \partial_\mu \psi - |m\varphi + g\varphi^2|^2 + (\tfrac{1}{2}m\psi\psi - g\varphi\psi\psi + \text{h.c.}) \} \tag{13}$$

Note that the scalar potential $V(\varphi^\dagger, \varphi) = |m\varphi + g\varphi^2|^2 \geq 0$ in accordance with (4) and vanishes at its minimum, giving a zero cosmological constant. Note that we are no longer free to add an arbitrary constant to $V(\varphi^\dagger, \varphi)$ as we could in a nonsupersymmetric field theory.

As expected, supersymmetric field theories are much more restrictive than ordinary field theories, and, duh, also much more symmetric. The formalism described here can be extended to construct supersymmetric Yang-Mills theory.

Another important generalization is to introduce, instead of one supercharge Q_α, $\mathcal{N}$ supercharges Q_α^I, with $I = 1, \cdots, \mathcal{N}$ (exercise VIII.4.2). Since each charge Q_α^I transforms like the $S_z = \frac{1}{2}$ component of a spin $\frac{1}{2}$ operator, it takes a state with $S_z = m$ in a super-multiplet to a state with $S_z = m + \frac{1}{2}$. Thus the integer $\mathcal{N}$ is bounded from above. For supersymmetric Yang-Mills theory, the maximum number of supersymmetry generators is $\mathcal{N} = 4$ if we do not want to introduce fields with spin ≥ 1. Similarly, the most supersymmetric supergravity theory we could construct (exercise VIII.4.3) has $\mathcal{N} = 8$.

As mentioned in chapter VII.3, if any nontrivial 4-dimensional quantum field theory turned out to be exactly soluble, the supersymmetric $\mathcal{N} = 4$ Yang-Mills theory is probably our best bet. In all likelihood, the first relativistic quantum field theory to be solved exactly would be $\mathcal{N} = 4$ Yang-Mills in the planar large N limit of chapter VII.4.

I hope that this brief introduction gave you a flavor of supersymmetry and will enable you to go on to specialized treatises.

Exercises

VIII.4.1 Construct the Wess-Zumino Lagrangian by the trial and error approach.

VIII.4.2 In general there may be $\mathcal{N}$ supercharges Q_α^I, with $I = 1, \ldots, \mathcal{N}$. Show that we can have $\{Q_\alpha^I, Q_\beta^J\} = \varepsilon_{\alpha\beta} Z^{IJ}$, where Z^{IJ} denotes c-numbers known as central charges.

VIII.4.3 From the fact that we do not know how to write consistent quantum field theories with fields having spin greater than 2 show that the $\mathcal{N}$ in the previous exercise cannot exceed 8. Theories with $\mathcal{N} = 8$ supersymmetry are said to be maximally supersymmetric. Show that if we do not want to include gravity, $\mathcal{N}$ cannot be greater than 4. Supersymmetric $\mathcal{N} = 4$ Yang-Mills theory has many remarkable properties.

VIII.4.4 Show that $\partial\theta_\alpha / \partial\theta^\beta = \varepsilon_{\alpha\beta}$.

VIII.4.5 Work out $\delta\varphi$, $\delta\psi$, and δF precisely by computing $\delta\Phi = i(\xi^\alpha Q_\alpha + \bar{\xi}_{\dot\alpha} \bar{Q}^{\dot\alpha})\Phi$.

VIII.4.6 For any polynomial $W(\Phi)$ show that $[W(\Phi)]_F = F[dW(\varphi)/d\varphi] +$ terms not involving F. Show that for the theory (11) the potential energy is given by $V(\varphi^\dagger, \varphi) = |\partial W(\varphi)/\partial\varphi|^2$.

VIII.4.7 Construct a field theory in which supersymmetry is spontaneously broken. [Hint: You need at least three chiral superfields.]

VIII.4.8 If we can construct supersymmetric quantum field theory, surely we can construct supersymmetric quantum mechanics. Indeed, consider $Q_1 \equiv \frac{1}{2}[\sigma_1 P + \sigma_2 W(x)]$ and $Q_2 \equiv \frac{1}{2}[\sigma_2 P - \sigma_1 W(x)]$, where the momentum operator $P = -i(d/dx)$ as usual. Define $Q \equiv Q_1 + i Q_2$. Study the properties of the Hamiltonian H defined by $\{Q, Q^\dagger\} = 2H$.

Geometrical action for the bosonic string

In this chapter, I will try to give you a tiny glimpse into string theory. Needless to say, you can get only the merest whiff of the subject here, but fortunately excellent texts do exist and I believe that this book has prepared you for them. My main purpose is to show you that perhaps surprisingly the basic formulation of string theory is naturally phrased in terms of a 2-dimensional field theory.

In chapter I.11 I described a point particle tracing out a world line given by $X^\mu(\tau)$ in $D-$ dimensional spacetime. Recall that the action is given geometrically by the length of the world line

$$S = -m \int d\tau \sqrt{\frac{dX^\mu}{d\tau}\frac{dX_\mu}{d\tau}} \tag{1}$$

and remains unchanged under reparametrization $\tau \to \tau'(\tau)$. Recall also that classically, S is equivalent to

$$S_{\text{imp}} = -\frac{1}{2} \int d\tau \left(\frac{1}{\gamma}\frac{dX^\mu}{d\tau}\frac{dX_\mu}{d\tau} + \gamma m^2 \right) \tag{2}$$

Now consider a string sweeping out a world sheet given by $X^\mu(\tau, \sigma)$ in D-dimensional spacetime, which we have already encountered in chapter IV.4 in connection with differential forms. In analogy with (1), Nambu and Goto proposed an action given geometrically by the area of the world sheet

$$S_{NG} = T \int d\tau d\sigma \sqrt{\det(\partial_a X^\mu \partial_b X_\mu)} \tag{3}$$

where $\partial_1 X^\mu \equiv \partial X^\mu/\partial\tau$, $\partial_2 X^\mu \equiv \partial X^\mu/\partial\sigma$, and $(\partial_a X^\mu \partial_b X_\mu)$ denotes the ab element of a 2 by 2 matrix. Here, as in (1), μ ranges over D values: $0, 1, \ldots, D-1$. The constant $T (\equiv 1/2\pi\alpha'$ with α' the slope of the Regge trajectory in particle phenomenology) corresponds to the string tension since stretching the string to enlarge the world sheet costs an extra amount of action proportional to T.

In a precise parallel with the discussion for the point particle, it is preferable to avoid the square root and instead use the action

$$S = \tfrac{1}{2} T \int d\tau d\sigma \, \gamma^{\frac{1}{2}} \gamma^{ab} (\partial_a X^\mu \partial_b X_\mu) \tag{4}$$

with $\gamma = \det \gamma_{ab}$ in the path integral to quantize the string. We will now show that S is equivalent classically to S_{NG}.

As in (2), we vary S with respect to the auxiliary variable γ_{ab}, which we then eliminate. For a matrix M, $\delta M^{-1} = -M^{-1}(\delta M)M^{-1}$ and $\delta \det M = \delta e^{\mathrm{tr} \log M} = e^{\mathrm{tr} \log M} \mathrm{tr} M^{-1} \delta M = (\det M) \mathrm{tr} M^{-1} \delta M$. Thus, $\delta \gamma^{ab} = -\gamma^{ac} \delta \gamma_{cd} \gamma^{db}$ and $\delta \gamma = \gamma \gamma^{ba} \delta \gamma_{ab}$. For ease of writing, define $h_{ab} \equiv \partial_a X^\mu \partial_b X_\mu$. The variation of the integrand in (4) thus gives

$$\delta[\gamma^{\frac{1}{2}} \gamma^{ab} h_{ab}] = \gamma^{\frac{1}{2}} [\tfrac{1}{2} \gamma^{dc} \delta \gamma_{cd} (\gamma^{ab} h_{ab}) - \gamma^{ac} \delta \gamma_{cd} \gamma^{db} h_{ab}]$$

Setting the coefficient of $\delta \gamma_{cd}$ equal to 0 we obtain

$$h_{cd} = \tfrac{1}{2} \gamma_{cd} (\gamma^{ab} h_{ab}) \tag{5}$$

where the indices on h are raised and lowered by the metric γ. Multiplying (5) by h^{dc} (and summing over repeated indices) we find $\gamma^{ab} h_{ab} = 2$ and thus $\gamma_{cd} = h_{cd}$. Plugging this into (4) we find that $S = T \int d\tau d\sigma (\det h)^{\frac{1}{2}}$. Thus, S and S_{NG} are indeed equivalent classically. The action (4), first discovered by Brink, Di Vecchia, and Howe and by Deser and Zumino, is known as the Polyakov action.

Note that (5) determines γ_{ab} only up to an arbitrary local rescaling known as a Weyl transformation:

$$\gamma_{ab}(\tau, \sigma) \to e^{2\omega(\tau, \sigma)} \gamma_{ab}(\tau, \sigma) \tag{6}$$

Thus, the action (4) must be invariant under the Weyl transformation.

Staring at the string action (4), you will recognize that it is just the action for a quantum field theory of D massless scalar fields $X^\mu(\tau, \sigma)$ in 2-dimensional spacetime with coordinates (τ, σ), albeit with some unusual signs. The index μ plays the role of an internal index, and Poincaré invariance in our original D-dimensional spacetime now appears as an internal symmetry. Indeed, a good deal of string theory is devoted to the study of quantum field theories in 2-dimensional spacetime! It is amusing how quantum field theory manages to stay on the stage.

To this bosonic string theory we can add fermionic variables in such a way as to make the action supersymmetric. The result, as you surely have heard, is superstring theory, thought by some to be the theory of everything.[1]

---

[1] "To understand macroscopic properties of matter based on understanding these microscopic laws is just unrealistic. Even though the microscopic laws are, in a strict sense, controlling what happens at the larger scale, they are not the right way to understand that. And that is why this phrase, "theory of everything," sounds sleazy."— J. Schwarz, one of the founders of string theory.

This infinitesimal introduction to string theory is all I can give you here, but I hope that this book has prepared you adequately to begin studying various specialized texts on string theory.[2]

---

[2] For a brief but authoritative introduction, see E. Witten, "Reflections on the Fate of Spacetime," *Physics Today*, April 1996, p. 24.

Closing Words

As I confessed in the preface, I started out intending to write a concise introduction to quantum field theory, but the book grew and grew. The subject is simply too rich. As I mentioned, after a period of almost being abandoned, quantum field theory came roaring back. To quote my thesis advisor Sidney Coleman, the triumph of quantum field theory was veritably "a victory parade" that made "the spectator gasp with awe and laugh with joy."

String theory is beautiful and marvellous, but until it is verified, quantum field theory remains the true theory of everything. All of physics can now be said to be derivable from field theory. To start with, quantum field theory contains quantum mechanics as a $(0 + 1)$-dimensional field theory, and to end (perhaps) with, string theory may be formulated as a $(1 + 1)$-dimensional field theory.

Quantum field theory can arguably be regarded as the pinnacle of human thought. (Hush, you hear the distant howls of the mathematicians, English professors, philosophers, and perhaps even a few stuck-up musicologists?) It is a distillation of basic notions from the very beginning of the physics: Newton's realization that energy is the square of momentum appears in field theory as the two powers of spatial derivative. But yet—you knew that was coming, didn't you, with field theory set up as the pinnacle et cetera?— but yet, field theory in its present form is in my opinion still incomplete and surely some bright young minds will see how to develop it further.

For one thing, field theory has not progressed much beyond the harmonic paradigm, as I presaged in the first chapter. The discovery of the soliton and instanton opened up a new vista, showing in no uncertain terms that Feynman diagrams ain't everything, contrary to what some field theorists thought. Duality offers one way of linking perturbative weak coupling theory to strong coupling, but as yet practically nothing is known of the strong coupling regime. When speaking of renormalization groups, we bravely speak of flowing to a strong coupling fixed point, but we merely have the boat ticket: We have little idea of what the destination looks like. Perhaps in the not too distant future, lattice field theorists can extract the field configurations that dominate.

Another restriction is to two powers of the derivative, a restriction going back to Newton as I remarked above. In modern applications of field theory to problems far beyond particle physics, there is no reason at all to impose this restriction. For example, in studying visual perception, one encounters field theories much more involved than those we have studied in this book. (See the appendix for a brief description.) These field theories are Euclidean in any case and the corresponding functional integral with higher derivatives certainly makes sense: It is only in Minkowskian theories that we do not know how to handle higher derivatives. Newton again—certainly economists consider the rate of change of the acceleration as well as acceleration. Another innovative application is the formulation[1] of a class of problems in nonequilibrium statistical mechanics as field theories. Typically, various objects wander around and react when they meet. This class of problems appears in areas ranging from chemical reactions to population biology.

We can go far beyond the restriction on the number of derivatives in the Lagrangian. Who said that we can only have integrands of the form "exponential of a spacetime integral"? Most modifications you can think of might immediately run afoul of some basic principles (for example, $\int D\varphi e^{-\int d^4x \mathcal{L}(\varphi) - [\int d^4x \tilde{\mathcal{L}}(\varphi)]^2}$ would violate locality), but surely others might not. Another speculative thought I like to entertain goes along the following line: Classical and quantum physics are formulated in terms of differential equations and functional integrals, respectively. But how are differential equations contained in integrals? The answer is that the integrals $\int D\varphi e^{-(1/\hbar) \int d^4x \mathcal{L}(\varphi)}$ contain a parameter $\hbar$ so that in the limit $\hbar$ going to zero the evaluation of the integrals amounts to solving partial differential equations. Can we go beyond quantum field theory by finding a mathematical operation that in the limit of some parameter k going to zero reduces to doing the integral $\int D\varphi e^{-(1/\hbar) \int d^4x \mathcal{L}(\varphi)}$?

The arena of local field theory has always been restricted to the set of d real numbers x^μ. The recent excitement over noncommutative field theory promises to take us beyond. (I was tempted to discuss noncommutative field theory too, but then the nutshell would truly burst.)

But perhaps the most unsatisfying feature of field theory is the present formulation of gauge theories. Gauge "symmetry" does not relate two different physical states, but two descriptions of the same physical state. We have this strange language full of redundancy we can't live without. We start with unneeded baggage that we then gauge-fix away. We even know how to avoid this redundancy from the start but at the price of discretizing spacetime. This redundancy of description is particularly glaring in the manufactured gauge theories now fashionable in condensed matter physics, in which the gauge symmetry is not there to begin with. Also, surely the way we calculate in nonabelian gauge theories by cutting the Yang-Mills action up into pieces and doing violence to gauge invariance will be held

---

[1] By M. Doi, L. Peliti, J. C. Cardy, and others. See for example J. C. Cardy, cond-mat/9607163, "Renormalisation Group Approach to Reaction-Diffusion Problems," in: J.-B. Zuber, ed., *Mathematical Beauty of Physics*, p. 113.

up to ridicule a hundred years from now. I would not be surprised if a brilliant reader of this book finds a more elegant formulation of what we now call gauge theories.

Look at the development of the very first field theory, namely Maxwell's theory of electromagnetism. By the end of the nineteenth century it had been thoroughly studied and the overwhelming consensus was that at least the mathematical structure was completely understood. Yet the big news of the early twentieth century was that the theory, surprise surprise, contains two hidden symmetries, Lorentz invariance and gauge invariance: two symmetries that, as we now know, literally hold the key to the secrets of the universe. Might not our present day theory also contain some unknown hidden symmetries, symmetries even more lovely than Lorentz and gauge invariance? I think that most physicists would say that the nineteenth-century greats missed these two crucial symmetries because of their lousy notation[2] and tendency to use equations of motion instead of the action. Some of these same people would doubt that we could significantly improve our notation and formalism, but the dotted-undotted notation looks clunky to me and I have a nagging feeling[3] that a more powerful formalism will one day replace the path integral formalism.

Since the point of good pedagogy is to make things look easy, students sometimes do not fully appreciate that symmetries do not literally leap out at you. If someone had written a supersymmetric Yang-Mills theory in the mid-1950s, it would certainly have been a long time before people realized that it contained a hidden symmetry. So it is entirely possible that an insightful reader could find a hitherto unknown symmetry hidden in our well-studied field theories.

It is not just a matter of clearer notation and formalism that caused the nineteenth-century greats to miss two important symmetries; it is also that they did not possess the mind set for symmetry. The old paradigm "experiments → action → symmetry" had to be replaced[4] in fundamental physics by the new paradigm "symmetry → action → experiments," the new paradigm being typified by grand unified theory and later by string theory. Surely, some future physicists will remark archly that we of the early twenty-first century did not possess the right mind set.

In physics textbooks, many subjects have a finished completed feel to them, but not quantum field theory. Some people say to me, what else is there to say about field theory? I would like to remind those people that a large portion of the material in this book was unknown 30 years ago. Of course, while I feel that further developments are possible, I have no idea what—otherwise I would have published it—so I can't tell you what. But let me mention two recent developments that I find extremely intriguing. (1) Some field theories may be dual to string theories. (2) In dimensional deconstruction a d-dimensional field theory may look $(d + 1)$-dimensional in some range of the energy scale: the field theory can literally generate a spatial dimension. These developments suggest that quantum field

---

[2] It is said, and I agree, that one of Einstein's great contributions is the repeated indices summation convention. Try to read Maxwell's treatises and you will appreciate the importance of good notation.

[3] I once asked Feynman how he would solve the finite square well using the path integral.

[4] A. Zee, *Fearful Symmetry*, chap. 6.

theories contain considerable hidden structures waiting to be uncovered. Perhaps another golden age is in store for quantum field theory.

So boys and girls, the parade is over, and now it's up to you to get another parade going.[5]

Appendix

An image presented to the visual system can be described as a 2-dimensional Euclidean field $\varphi_0(x)$, with φ_0 representing the gray scale from black ($\varphi_0 = -\infty$) to white ($\varphi_0 = +\infty$). [You can see that color might be included by going to a field $\vec{\varphi}$ transforming under some internal $SO(2)$ group for example.] The image actually perceived, $\varphi(x)$, is the actual image $\varphi_0(x)$ distorted to $\varphi_0[y(x)]$ plus some noise $\eta(x)$. Distortion is described by a map $x \to y(x)$ of the 2-dimensional Euclidean plane. Your brain's task is to decide whether the actual image is $\varphi_0(x)$ or some other $\varphi_1(x)$. Your ability to discriminate between images depends on the functional integral

$$Z = \int Dy(x) \int D\eta(x) e^{-W[y(x)]-(1/2C)\int d^2x \eta(x)^2} \delta\{\varphi_0[y(x)] + \eta(x) - \varphi(x)\} \tag{1}$$

$$= \int Dy(x) e^{-W[y(x)]-(1/2C)\int d^2x \{\varphi(x)-\varphi_0[y(x)]\}^2}$$

where for simplicity I have taken the noise, measured by the parameter C, to be Gaussian and white. The weighting function $W[y(x)]$ is presumably hard wired by evolution into our visual system, telling us that certain distortions (translations, rotations, and dilations) are much more likely than others. Writing $y(x) = x + A(x)$ we note that Z defines a field theory of the 2-component field $A_i(x)$, which can always be written as $A_i = \partial_i \eta + \varepsilon_{ij}\partial_j \chi$. Note that the field $A_i(x)$ appears "inside" an "external" field φ_0. From symmetry considerations we might argue that

$$W = -\int d^2x \left(\frac{1}{g^2} \eta \partial^6 \eta + \frac{1}{f^2} \chi \partial^6 \chi \right)$$

with two coupling constants f and g. I can give here only the briefest of sketches and refer the interested reader to the literature.[6] Clearly, one can think of other examples. This particular example serves only to show that there are many more field theories than those described in standard texts.

---

[5] As the Beatles said, quantum fields forever!

[6] W. Bialek and A. Zee, Statistical mechanics and invariant perception, *Phys. Rev. Lett.* 58: 741, 1987; Understanding the efficiency of human perception, *Phys. Rev. Lett.* 61: 1512, 1988.

Part N

While quantum field theory was discovered and developed in the twentieth century, I will introduce in this part, added in the second edition of this text, some topics that have been worked out in the twenty-first century. At the rate these topics are rapidly evolving, I may be quite foolish to include them here. But I am taking the plunge as I think that I would serve my readers better by letting them have a taste of the twenty-first century rather than expanding on the twentieth. More likely than not, by the time this second edition is published, there will be better ways of treating the material contained here. You should read part N in this spirit and regard what is given here as an entry key to a fast growing[1] research literature.

---

[1] Indeed, by the time the manuscript was copyedited (April 2009) it had been discovered that the amplitudes discussed in chapters N.2–4 could be written even more simply using a twistor and dual twistor formalism. See p. 494 and N. Arkani-Hamed, P. Cachazo, C. Cheung, and J. Kaplan, arXiv:09032110.

N.1 | Gravitational Waves and Effective Field Theory

An unfinished symphony

One astounding prediction of Einstein gravity is the existence of ripples crisscrossing the fabric of spacetime, what one writer refers to as Einstein's unfinished symphony.[1] Massive detectors have been built, with more to come, in a human "curious George" effort to tune in to the song of the cosmos.

Consider a black hole of size $r_S = 2Gm$ (its Schwarzschild radius—see chapter I.10, with m its mass) a distance r_O from another black hole, moving with velocity v. As the black holes spiral into each other they emit gravitational waves, with a characteristic wavelength determined by the orbital period $\lambda = 2\pi r_O / v$. Thus the physics contains three distance scales: r_S, r_O, and λ. We will stay within the simple post Newtonian regime $r_S \ll r_O \ll \lambda$. Toward the end, as $r_O \simeq r_S$ and $v \simeq 1$, relativistic effects rear their nasty heads, from which we will prudently stay away.

In the Closing Words to the first edition of this text, I mention that one intriguing development over the last few decades has been the use of effective field theory to describe situations involving more than one energy scale (or equivalently, length and time scales). The physics at the high energy scale M is then represented in the low energy effective Lagrangian by higher dimensional terms, suppressed by powers of M but constrained by the symmetries we know. Examples abound in this book, from the quantum Hall effect to surface growth to proton decay. The latter provides a classic example: while we profess ignorance of the physics responsible for proton decay, we can nevertheless make useful predictions by adding 4-fermion interactions invariant under the low energy gauge group $SU(3) \otimes SU(2) \otimes U(1)$, as shown in chapter VIII.3.

An interesting recent development is an elegant description of the emission of gravitational waves by inspiraling black holes using effective field theory. Here, in contrast to proton decay, we actually know the short distance physics involved. Effective field theory

---

[1] M. Bartusiak, *Einstein's Unfinished Symphony.*

nevertheless offers an efficient and sensible way to organize and compartmentalize physics on the various distance scales. We will merely touch upon one aspect of this approach.

Finite size objects in general relativity

Since $r_S \ll r_O$, the leading approximation would be to treat the black hole as a point particle using the action (I.11.12) $S_{pp} = -m \int d\tau = -m \int \sqrt{g_{\mu\nu}dX^\mu dX^\nu} = -m \int d\tau \sqrt{g_{\mu\nu}\dot{X}^\mu \dot{X}^\nu}$, where $\dot{X}^\mu = dX^\mu/d\tau$. Let us now include the corrections due to the finite size of the black hole. As you will see, the following discussion actually applies not only to black holes but to any finite sized object, including you.

In the spirit of effective field theory, we add to S_{pp} higher dimensional terms to be formed out of the point particle degree of freedom $\dot{X}^\mu$ and the ambient $g_{\mu\nu}$ the particle moves in, subject to local coordinate invariance, of course. The invariant tensors we can form out of $g_{\mu\nu}$ are, to leading order, the scalar curvature R, the Ricci curvature $R_{\mu\nu}$, and the Riemann curvature tensor $R_{\mu\lambda\nu\rho}$. You might start with the scalar curvature and the Ricci curvature, and add to S_{pp} the terms $S_{drop} = \int d\tau (c_S R(X) + c_R R_{\mu\nu}(X)\dot{X}^\mu \dot{X}^\nu)$. The curvatures $R(X)$ and $R_{\mu\nu}(X)$ are evaluated on the worldline $X^\mu(\tau)$ of the particle of course.

Einstein's equation of motion $R^{\mu\nu} - \frac{1}{2}g^{\mu\nu}R = 0$ implies $R_{\mu\nu}(X) = 0$ and thus also $R(X) = 0$. Following the discussion in chapter VIII.3, we now show that, as we might intuitively feel, we are allowed to drop S_{drop}. For our problem we have the total action $S = S_{EH} + S_{pp} + S_{drop}$ with the Einstein-Hilbert action (VIII.1.1) $S_{EH} = \int d^4x \sqrt{-g}M_P^2 R$. Under a field redefinition $g_{\mu\nu} \to g_{\mu\nu} + \delta g_{\mu\nu}$,

$$\delta S_{EH} = \int d^4x \sqrt{-g}M_P^2 (R^{\mu\nu} - \frac{1}{2}g^{\mu\nu}R)\delta g_{\mu\nu} \tag{1}$$

Note that was how we would have derived the equation of motion for gravity, by varying $g_{\mu\nu}$. Here we are not making an arbitrary variation, but rather our goal is to choose a specific $\delta g_{\mu\nu}$ so that the resulting δS_{EH} negates S_{drop}. Since S_{drop} consists of an integral over the worldline of the particle while δS_{EH} is given by an integral over spacetime, we need a delta function in $\delta g_{\mu\nu}$ to switch from one kind of integral to another. The choice

$$\delta g_{\mu\nu}(x) = \frac{1}{\sqrt{-g(x)}M_P^2} \int d\tau \delta^4(x - X(\tau))[ag_{\mu\nu}(X) + b\dot{X}_\mu \dot{X}_\nu] \tag{2}$$

gives

$$\delta S = \int d\tau (-aR(X) + b\left[R_{\mu\nu}(X) - \frac{1}{2}g_{\mu\nu}R(X)\right]\dot{X}^\mu \dot{X}^\nu) \tag{3}$$

So, with some appropriate values of a and b, we can indeed cancel off[2] S_{drop}, thus vindicating our intuition that the particle does not feel the Ricci and scalar curvatures for the obvious reason that they vanish.

---

[2] A technicality: field redefinition also induces a contact interaction of the form $\int d\tau(1/\sqrt{-g})\delta^4(X_1(\tau) - X_2(\tau))$ between the two massive objects. Going from field theory to the point particle description represents a conceptual step backward, so that we should expect delta function effects at the location of the point particles.

How about terms we can construct out of the Riemann curvature tensor $R_{\mu\lambda\nu\rho}$? For your convenience, I list its symmetry properties[3] here: $R_{\tau\rho\mu\nu} = -R_{\tau\rho\nu\mu} = -R_{\rho\tau\mu\nu}$, $R_{\tau\rho\mu\nu} = R_{\mu\nu\tau\rho}$, and $R_{\tau\rho\mu\nu} + R_{\tau\mu\nu\rho} + R_{\tau\nu\rho\mu} = 0$. Thus, due to the antisymmetry, we are not able to contract all four indices of $R_{\mu\lambda\nu\rho}$ with $\dot{X}^\mu$. We can contract at most two indices, to form the two objects $E_{\mu\nu}(X) \equiv R_{\mu\lambda\nu\rho}(X)\dot{X}^\lambda\dot{X}^\rho$ and $B_{\mu\nu}(X) \equiv \tilde{R}_{\mu\lambda\nu\rho}(X)\dot{X}^\lambda\dot{X}^\rho$, where

$$\tilde{R}_{\mu\lambda\nu\rho}(x) \equiv \frac{1}{2\sqrt{-g}}\varepsilon_{\mu\lambda\sigma\eta}R^{\sigma\eta}{}_{\nu\rho}(x)$$

denotes the dual of the curvature tensor. These are 2-indexed tensors and we need to square them to form scalars to put into the action. Hence, to next order the particle action becomes

$$S_p = \int d\tau(-m + c_E E_{\mu\nu}E^{\mu\nu} + c_B B_{\mu\nu}B^{\mu\nu} + \cdots) \tag{4}$$

Note that the unknown constants c_E and c_B have dimension of inverse mass cubed.

Before we explore the physical content of this effective action, let us understand the meaning of E and B by retreating to the more familiar case of a point particle moving in an electromagnetic field $F_{\mu\nu}$ (in flat space). Form $E_\mu \equiv F_{\mu\nu}\dot{X}^\nu$ and $B_\mu \equiv \tilde{F}_{\mu\nu}\dot{X}^\nu$, where $\tilde{F}_{\mu\nu}(x) \equiv \frac{1}{2}\varepsilon_{\mu\nu\sigma\eta}F^{\sigma\eta}$. Going to the rest frame of the particle, where $\dot{X}^0 = 1$ and $\dot{X}^i = 0$, we see that, as the notation suggests, this is just the familiar decomposition of electromagnetism into electricity and magnetism. Similarly, $E_{\mu\nu}$ and $B_{\mu\nu}$ represent the decomposition of curvature into its "electric" and "magnetic" components.

In chapter I.11 we varied the first term in (4) to obtain the standard geodesic equation that is at the heart of Einstein's theory. Here we obtain

$$\frac{d^2X^\rho}{d\tau^2} + \Gamma^\rho_{\mu\nu}(X(\tau))\frac{dX^\mu}{d\tau}\frac{dX^\nu}{d\tau} = f^\rho(X(\tau)) \tag{5}$$

where $f^\mu(X(\tau))$ comes from varying the E and B terms in (4). A finite sized body experiences a tidal force f^μ due to the varying gravitational force acting on it. It no longer follows a geodesic.

The fact that we had to square the electric and magnetic components of the curvature to form the effective action (4) means that the effects of these correction terms are highly suppressed. Since Riemann curvature contains two derivatives, the correction terms involve four derivatives. To estimate the magnitude of c_E and c_B we exploit a rather cute argument as follows.

Consider the scattering of a graviton off this point particle (which, remember, is a black hole in the problem we are studying) generated by the couplings in (4): $i\mathcal{M} \sim \cdots + ic_{E,B}\omega^4/M_P^2 + \cdots$ where ω denotes the energy of the graviton. The powers of ω follows from the four derivatives just mentioned. (If you don't understand the powers of M_P you need to read chapter VIII.1 again.) Here $c_{E,B}$ denotes the two unknown couplings $c_E \sim c_B$ generically. Imagine calculating the total scattering cross section for a graviton on a black hole. Squaring the amplitude $\mathcal{M}$ etc., we would end up with $\sigma(\omega) \sim \cdots + c_{E,B}^2\omega^8/M_P^4 + \cdots$.

---

[3] S. Weinberg, *Gravitation and Cosmology*, p. 141.

This treatment of the black hole as a point particle is only valid for $\omega r_S \ll 1$ of course. The $(\cdots)$ in $i\mathcal{M}$ represents diagrams we have not included, for example, the one originating from the first term in (4) (namely the term responsible for keeping us down to earth!). A nice feature of the argument I am about to give is that we don't even need to know what the terms in $(\cdots)$ are.

On the other hand, we argue that by dimensional analysis the cross section must have the form $\sigma(\omega) = r_S^2 f(\omega r_S)$ since the only length scale in the Schwarzschild metric is r_S. Expanding the unknown function $f(\omega r_S)$ in powers of its argument we have $\sigma(\omega) = \cdots + \alpha \omega^8 r_S^{10} + \cdots$ with α some constant. (A technical aside: the massless graviton could produce infrared factors like $\log \omega r_S$, which we ignore for our purposes.)

Requiring that the two expressions agree, we obtain $c_{E,B} \sim M_P^2 r_S^5$. Indeed, as expected, the couplings $c_{E,B}$ are highly suppressed as $r_S \to 0$.

Exercise

N.1.1 Using considerations similar to those in the text, show that the scattering cross section for a photon of frequency ω on an atom or a molecule vanishes like ω^4 as $\omega \to 0$, a result which, as mentioned in chapter VIII.3, underlies the well-known explanation of why the sky is blue.

N.2 | Gluon Scattering in Pure Yang-Mills Theory

Boil and toil with Feynman diagrams

You might think that after some 50 years, there could not possibly be any novelty in calculating Feynman amplitudes. But you would be wrong. Over the last dozen years or so, and largely since the first edition of this text, a group of intrepid searchers have found some amazingly powerful methods of tackling Feynman diagrams. As I said at the beginning of chapters VIII.4 and 5, I can only give you an introduction to this subject, telling you just enough for you to explore this fast-growing literature.

To best appreciate this new development, you should do a little calculation before reading further. Consider pure Yang-Mills theory, by consensus the nicest field theory we have, simple to write down and perfumed with symmetries. Not even any fermions around to mess things up. Call the gauge bosons gluons for convenience. Now calculate 5-gluon scattering at tree level as shown in figure N.2.1. No loops, just trees. The Feynman rules are given in chapter VII.1 and also in appendix C.

You really must calculate before reading on. I will wait for you. You think to yourself, this is easy, just a bunch of tree diagrams. In fact, to make it easier, put all the external gluons on-shell, that is, set $p_i^2 = 0$, $i = 1, 2, \cdots, 5$.

This calculation is not merely an idle exercise, but is in fact phenomenologically important. At an accelerator such as the soon-to-be-operational Large Hadron Collider, two

Figure N.2.1

Result of a brute force calculation (actually only a small part of it):

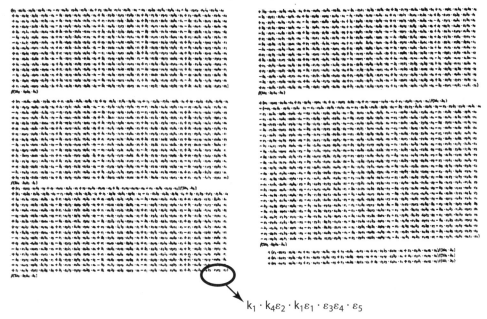

$$k_1 \cdot k_4 \varepsilon_2 \cdot k_1 \varepsilon_1 \cdot \varepsilon_3 \varepsilon_4 \cdot \varepsilon_5$$

Figure N.2.2

protons are smashed together at high energies. Two gluons, one from each proton, collide and produce three gluons, which then materialize into three jets of hadrons. Because of asymptotic freedom, at high energies the effective coupling g becomes small enough for perturbative field theory to be relevant, and the tree amplitude you are busily calculating provides a key ingredient for the phenomenological models used to study the experimental measurements.

Time's up! A small part of the answer is shown in figure N.2.2, taken from a lecture by Zvi Bern.[1] You really should take a look in order to appreciate, to be grateful for even, the formalism to be explained in this chapter. You know that the amplitude is linear in each of the five polarization vectors ϵ_i. The 3-gluon vertex (VII.1.11) is linear in momentum and there are three of them in a typical diagram. Thus a typical term in the numerator of the Feynman amplitude would be, as shown in the figure, $p_1 \cdot p_4 \epsilon_2 \cdot p_1 \epsilon_1 \cdot \epsilon_3 \epsilon_4 \cdot \epsilon_5$. A rough estimate shows that there are almost 10,000 such terms. That's why, in spite of my admonition, you didn't finish the calculation before reading ahead. Incidentally, you could see that even 4-gluon scattering at tree level, though doable by hand, is rather involved.

[1] Z. Bern, "Magic Tricks for Scattering Amplitudes," http://online.itp.ucsb.edu/online/colloq/bern1/pdf/Bern1.pdf.

New technology for Feynman diagrams

In practice, phenomenologists studying jet production have developed elaborate computer codes based on numerical recursion and these prove to be quite efficient. In this introductory text, however, we are not after numerical efficiency but a deeper understanding of the structure of multi-gluon amplitudes. I have set you up so that, surely, after your abortive attempt to calculate the 5-gluon amplitude you now fully appreciate the need for new ways of approaching Feynman diagrams. I will now explain some of the novel methods people have invented over the last 15 years or so.

A relatively simple first step is to strip the color off the amplitude. Evidently, it is much better to use the matrix notation of (IV.5.16) than the index notation of (IV.5.17). Instead of the structure constants f^{abc} and their products in the "indexed" Feynman rules in (VII.1.11–13) we have colored Feynman rules (read appendix 1 now) with objects like $\text{tr}(T^a[T^b, T^c])$ and $\text{tr}([T^a, T^b][T^c, T^d])$ for the cubic and quartic coupling vertex, respectively, where T^a denotes the matrix representing the suitably normalized generators of the gauge group. Denote the color matrices carried by the external gluons by $T^{a_1}, T^{a_2}, \cdots T^{a_n}$ (with $n = 5$ in the example you failed to do).

In calculating a multi-gluon scattering amplitude in tree approximation, you would find each term multiplied by the product of a bunch of color traces, such as $\text{tr}(T^e A)\text{tr}(T^e B)$ where A and B denote products of T's. Here the index e is carried by a virtual gluon and hence is summed over. We now use the group theoretic identity (with e summed over) for the gauge group $SU(N)$ [recall (IV.5.19)]

$$(T^e)^i_j (T^e)^k_l = \frac{1}{2}\left(\delta^i_l \delta^k_j - \frac{1}{N}\delta^i_j \delta^k_l\right) \tag{1}$$

(Here $e = 1, \cdots, N^2 - 1$ and the indices $i, j, k, l = 1, \cdots, N$, of course.) The second term takes care of the traceless condition $\text{tr } T^e = 0$. However, we can drop it, since if we extend the gauge group to $U(N)$, that extra gluon does not couple to the other gluons anyway. Thus $\text{tr}(T^e A)\text{tr}(T^e B) = \frac{1}{2}\text{tr}(AB)$. Repeating this procedure, we reduce the product of traces to a single trace of n T^a's multiplied together in some specific order.

Indeed, the astute reader will have noted that had we used the double-line formalism of figure IV.5.2, this entire discussion would not even have been necessary. As we also saw in chapter VII.4, the double-line formalism does offer many advantages.

The other simplifying step is to specify the helicity of the gluon instead of writing the amplitude in terms of polarization vectors. You recall, from way back when, that a massless spin 1 particle moving along the third-direction $k = \omega(1, 0, 0, 1)$ can have helicity $h = +$, corresponding to the polarization vector $\epsilon = 1/(\sqrt{2})(0, 1, i, 0)$, or helicity $h = -$, corresponding to the polarization vector $\epsilon = (1/\sqrt{2})(0, 1, -i, 0)$. We specify the external gluons by momentum, helicity, and color: $(p_1, h_1, a_1, p_2, h_2, a_2, \cdots, p_n, h_n, a_n)$.

Thus we can write the n-gluon amplitude as

$$\mathcal{M} = i \sum_{\text{permutations}} \text{tr}(T^{a_1} T^{a_2} \cdots T^{a_n}) A(1, 2, \cdots, n) \tag{2}$$

Following the literature, we have compressed the notation further and denote $\{p_i, h_i\}$ by i. The sum is over all possible permutations of the n gluons. We can now focus on the "color stripped" amplitude $A(1, 2, \cdots, n)$.

First, a triviality. It is convenient to treat the gluons as all outgoing (or if you prefer, as all incoming), so that $\sum_i p_i = 0$ and the time component of some of the momenta can be negative. We can then obtain the physically desired amplitude by crossing. Keep in mind that under crossing $p \to -p$ and $\epsilon \to \epsilon^*$, that is, the helicity flips.

The spinor helicity formalism

Now we are ready to return to the expression in figure N.2.2. The technical term for this expression is an "unholy mess." It turns out that the key to unraveling this hopeless morass can be found in exercise II.3.1: that the Lorentz vector sits in the representation $(\frac{1}{2}, \frac{1}{2})$ and thus can be constructed as a product of two spinors, one from the representation $(\frac{1}{2}, 0)$, the other from $(0, \frac{1}{2})$. You did the exercise, didn't you? So you know how to write, for example, the momentum vector p^μ as a product of two spinors. To go on, you should also read appendices B and E.

I am now ready to explain the spinor helicity formalism designed to exploit this peculiar property of the Lorentz vector. Or, to say it a bit more mysteriously, I am going to show you how to take the square root of the momentum.

Now you appreciate the power of the undotted-dotted notation introduced in appendix E. The undotted index goes with $(\frac{1}{2}, 0)$, and the dotted with $(0, \frac{1}{2})$. We are looking for an object transforming like $(\frac{1}{2}, 0) \otimes (0, \frac{1}{2})$ to represent a vector. The problem can then be stated as follows: instead of writing momentum as p^μ, we want to write it as $p_{\alpha\dot{\alpha}}$, an object carrying an undotted and a dotted index, namely a 2 by 2 matrix in cruder language.

We merely have to flip through appendix E and look for an object carrying the desired indices. There it is, $(\sigma^\mu)_{\alpha\dot{\alpha}}$, and indeed, its μ index is begging to be contracted with p^μ. Thus, with no further work, we can write [since $\sigma^\mu = (I, \vec{\sigma})$]

$$p_{\alpha\dot{\alpha}} \equiv p_\mu (\sigma^\mu)_{\alpha\dot{\alpha}} = (p^0 I - p^i \sigma^i)_{\alpha\dot{\alpha}} = \begin{pmatrix} (p^0 - p^3) & -(p^1 - ip^2) \\ -(p^1 + ip^2) & (p^0 + p^3) \end{pmatrix}_{\alpha\dot{\alpha}} \tag{3}$$

We have succeeded in writing the momentum as a 2 by 2 matrix. You may recognize this as nothing but the matrix X_M (with some trivial change in notation) used in appendix B to construct the covering of $SO(3, 1)$ by $SL(2, C)$.

Given two vectors p and q, their scalar product is given by

$$p \cdot q = \varepsilon^{\alpha\beta} \varepsilon^{\dot{\alpha}\dot{\beta}} p_{\alpha\dot{\alpha}} q_{\beta\dot{\beta}} \tag{4}$$

which you can check explicitly, writing the right-hand side as a trace and once again using $\sigma_2 \sigma_i^T = -\sigma_i \sigma_2$, as we did in appendix E. For $q = p$, this reduces to $p \cdot p = \varepsilon^{\alpha\beta} \varepsilon^{\dot{\alpha}\dot{\beta}} p_{\alpha\dot{\alpha}} p_{\beta\dot{\beta}} = \det p$; here we recognized a definition of the determinant. [Of course, you could also evaluate the determinant of (3) by inspection, or recall that this was also used in appendix B.]

Clearly, there is an unavoidable notational overload: the single letter p denotes both the vector and the matrix, but you should be able to tell from the context which object is being referred to.

Here we are going to apply this formalism to massless gluons with lightlike momenta. Things simplify considerably: for p lightlike, $\det p = 0$ and thus the matrix p generically has one 0 eigenvalue. (In fancy talk, the matrix has rank 1 rather than 2.) From elementary linear algebra we recall that a 2 by 2 matrix m of rank 1 can always be written as $m_{ij} = v_i w_j$, with v and w two 2-component vectors, (obviously since the vector orthogonal to w provides the 0 eigenvector.) Thus, for a lightlike vector, we can write

$$p_{\alpha\dot\alpha} = \lambda_\alpha \tilde{\lambda}_{\dot\alpha} \qquad (5)$$

in terms of two 2-component spinors λ and $\tilde{\lambda}$.

For physical momentum, the components p^μ are real, of course. I invite you to verify, however, that everything we just did from (3) to (5) goes through even if p^μ are complex. It turns out that in the next chapters we will find it convenient to consider complex momentum.

Upon first exposure, the formalism appears quite opaque, but actually, like a lot of formalisms, it is fairly simple or perhaps even trivial. If you are confused at any point in the following exposition, just work things out explicitly. For example, consider a physical momentum with $p^0 = E > 0$. With no loss of generality, you can choose $\vec{p}$ to point along the third direction, so that (with a trivial abuse of notation $p = |\vec{p}|$)

$$p = \begin{pmatrix} E - p & 0 \\ 0 & E + p \end{pmatrix}$$

which for p lightlike collapses to the rank 1 matrix

$$p = 2E \begin{pmatrix} 0 & 0 \\ 0 & 1 \end{pmatrix} = 2E \begin{pmatrix} 0 \\ 1 \end{pmatrix} (0\ 1)$$

Thus, in this case, λ and $\tilde{\lambda}$ are both equal to

$$\sqrt{2E} \begin{pmatrix} 0 \\ 1 \end{pmatrix}$$

numerically. (To make sure you get it, work this out for $\vec{p}$ pointing in some other direction.)

You can think of the Pauli spinors λ and $\tilde{\lambda}$ as the "square root" of the Lorentz vector p^μ. Note how the group theory discussion in chapter II.3 foreordained this rather nontrivial possibility. After all, there we saw how a Lorentz vector can be constructed out of two Dirac spinors u and u'.

Interestingly, in discussing ferromagnets and antiferromagnets in chapter VI.5, we used a poor man's version of (3), namely $\vec{n} = z^\dagger \vec{\sigma} z$.

You learned in school that the ordinary square root has a sign ambiguity. Analogously, in (5) p does not determine λ and $\tilde{\lambda}$ uniquely. We can always rescale $\lambda \to u\lambda$ and $\tilde{\lambda} \to \frac{1}{u}\tilde{\lambda}$ for any complex number u. (You might have wondered what fixed the overall constant in λ and $\tilde{\lambda}$ in the simple example above: I made an arbitrary choice.)

For real momentum, the matrix $p_{\alpha\dot\alpha} = p_\mu(\sigma^\mu)_{\alpha\dot\alpha}$ is hermitean, which implies that $\tilde\lambda = \lambda^*$ is the complex conjugate of λ. The spinor $\tilde\lambda$ is not independent of λ, and so the rescaling parameter u is restricted to be a phase factor $e^{i\gamma}$. [Also, recall from appendix B how X_M transforms under $SL(2, C)$ and you will see that it is all consistent.] In this case, the condition that p has rank 1 allows for two solutions: $p_{\alpha\dot\alpha} = \pm\lambda_\alpha\tilde\lambda_{\dot\alpha}$, with the two possible signs corresponding to whether $p^0 > 0$ or not.

A side remark at this point: We will see that it is useful to consider the group $SO(2, 2)$ instead of the Lorentz group $SO(3, 1)$. Thus, as the discussion in appendix B indicates, you can also take the square root of an $SO(2, 2)$ vector and write $p_{\alpha\dot\alpha} = \lambda_\alpha\tilde\lambda_{\dot\alpha}$, but with λ and $\tilde\lambda$ two independent real spinors, as is consistent with the local isomorphism between $SO(2, 2)$ and $SL(2, R) \otimes SL(2, R)$. The rescaling mentioned above is now restricted to u being a real number.

It is instructive to count the number of real degrees of freedom for these different cases. A complex lightlike momentum depends on $4 \times 2 - 2 = 6$ real numbers, since the condition p^2 now amounts to two real conditions, while λ and $\tilde\lambda$ each contains 2 complex numbers, but with rescaling we are left with $2 \times 2 - 1 = 3$ complex numbers, that is, 6 real numbers. A real lightlike momentum depends on $4 - 1 = 3$ real numbers, but now $\tilde\lambda$ is tied to λ containing 2 complex numbers, which get reduced to 3 real numbers after rescaling by a phase factor. For a (real) lightlike vector transforming under $SO(2, 2)$, we have 2 real spinors, which after rescaling contains 3 real numbers. So it all works out, of course.

I mention all this here for future use. It should be evident to you, for the rest of this chapter, which statements hold for complex momenta and which hold only for real momenta. At the end of the day, when we arrive at a physical quantity, such as the amplitude, we will of course set the momenta contained therein to be real.

For two lightlike vectors p and q, write $p_{\alpha\dot\alpha} = \lambda_\alpha\tilde\lambda_{\dot\alpha}$ and $q_{\alpha\dot\alpha} = \mu_\alpha\tilde\mu_{\dot\alpha}$, then we have

$$p \cdot q = (\varepsilon^{\alpha\beta}\lambda_\alpha\mu_\beta)(\varepsilon^{\dot\alpha\dot\beta}\tilde\lambda_{\dot\alpha}\tilde\mu_{\dot\beta}) \equiv \langle\lambda, \mu\rangle[\tilde\lambda, \tilde\mu] \tag{6}$$

Here we have defined the two Lorentz invariants

$$\langle\lambda, \mu\rangle \equiv \varepsilon^{\alpha\beta}\lambda_\alpha\mu_\beta = -\langle\mu, \lambda\rangle \tag{7}$$

and

$$[\tilde\lambda, \tilde\mu] \equiv \varepsilon^{\dot\alpha\dot\beta}\tilde\lambda_{\dot\alpha}\tilde\mu_{\dot\beta} = -[\tilde\mu, \tilde\lambda] \tag{8}$$

(treating the spinors as c-number objects.) Note in passing that with our convention, $\lambda_1 = \lambda^2$ and $\lambda_2 = -\lambda^1$, and so $\langle\lambda, \mu\rangle = -\lambda_1\mu_2 + \lambda_2\mu_1 = -\varepsilon_{\alpha\beta}\lambda^\alpha\mu^\beta$.

We have already verified in (E.13) that $\langle\lambda, \mu\rangle$ is invariant, but for the sake of total pedagogical clarity let us check it once more, this time using infinitesimal transformations. Write (E.4) more compactly as $\delta\lambda_\alpha = \sigma_\alpha^\beta\lambda_\beta$, where σ denotes some linear combination of Pauli matrices. Noting that $\langle\lambda, \mu\rangle$ is nothing but $\lambda\sigma_2\mu$ up to some irrelevant overall constant, we have indeed $\delta(\lambda\sigma_2\mu) = (\lambda\sigma^T\sigma_2\mu + \lambda\sigma_2\sigma\mu) = 0$.

A notational remark: the twiddles in $[\tilde\lambda, \tilde\mu]$ are redundant. The square bracket is defined only for spinors transforming like $(0, \frac{1}{2})$. Henceforth, we will write $[\lambda, \mu] \equiv \varepsilon^{\dot\alpha\dot\beta}\tilde\lambda_{\dot\alpha}\tilde\mu_{\dot\beta}$.

For real physical momenta, $\tilde{\lambda} = \lambda^*$ so that $\langle \lambda, \mu \rangle = [\lambda, \mu]^*$. Then $p \cdot q = \langle \lambda, \mu \rangle [\lambda, \mu]$ implies that $\langle \lambda, \mu \rangle = \sqrt{p \cdot q} e^{i\phi}$ and $[\lambda, \mu] = \sqrt{p \cdot q} e^{-i\phi}$, with some phase factor $e^{i\phi}$. We thus conclude that the two spinorial products may be regarded as the (two) square roots of the Lorentz dot product $p \cdot q$ up to a phase factor.

You could now raise an interesting question: how do we write the polarization vectors $\epsilon(p)$ of a massless gluon?

The requirement that $\epsilon(p) \cdot p = 0$ can be satisfied, according to (4), by setting $\epsilon_{\alpha\dot{\alpha}} = d^{-1} \lambda_\alpha \tilde{\mu}_{\dot{\alpha}}$, for an arbitrary $\tilde{\mu}_{\dot{\alpha}}$ and with the factor d determined as follows. We require that, for an arbitrary complex number w, scaling $\tilde{\mu} \to w\tilde{\mu}$ does not change ϵ (since $\tilde{\mu}$ is arbitrary after all). Thus d has to be linear in $\tilde{\mu}$. The further requirement that d be Lorentz invariant implies, as we just learned, that $d = [x, \mu]$, where $\tilde{x}$ is some $(0, \frac{1}{2})$ spinor. The only spinor available is $\tilde{\lambda}$ and hence we obtain

$$\epsilon_{\alpha\dot{\alpha}}^- = \frac{\lambda_\alpha \tilde{\mu}_{\dot{\alpha}}}{[\lambda, \mu]} \tag{9}$$

By convention, we will call this polarization negative helicity.

The arbitrary choice of $\tilde{\mu}_{\dot{\alpha}}$ represents the freedom inherent in a gauge theory. Indeed, we see that gauge transformation corresponds to the spinorial shift $\tilde{\mu} \to \tilde{\mu} + y\tilde{\lambda}$ (for some arbitrary number y) under which $\epsilon_{\alpha\dot{\alpha}} \to \epsilon_{\alpha\dot{\alpha}} + y\lambda_\alpha \tilde{\lambda}_{\dot{\alpha}}$, which translates into the usual shift of ϵ by some multiple of p.

The positive helicity polarization is given by the other possible choice

$$\epsilon_{\alpha\dot{\alpha}}^+ = \frac{\mu_\alpha \tilde{\lambda}_{\dot{\alpha}}}{\langle \mu, \lambda \rangle} \tag{10}$$

Check that it works. Gauge transformation now corresponds to the shift $\mu \to \mu + y\lambda$. Note that the polarization vectors are normalized as $\epsilon^+ \cdot \epsilon^- = \langle \mu\lambda \rangle [\mu\lambda] / (\langle \mu\lambda \rangle [\mu\lambda]) = 1$.

Taming the unholy mess

Consider the tree-level scattering amplitude with $n \geq 4$ outgoing massless gluons. (In this and the next sections, we can take all momenta to be real.) The color-stripped amplitude is then characterized by a string of helicities $(h_1, \ldots, h_n)$. Take for example the amplitude with $(+ + + \cdots + +)$. Upon crossing, it describes two gluons, each with helicity $-$, going into $n - 2$ gluons all with helicity $+$. Both incoming gluons flip their helicity and thus this amplitude is said to be maximal helicity violating. Your intuition may tell you that this amplitude ought to be suppressed, since highly energetic massless particles tend to maintain their helicities. If you try to verify this using traditional Feynman diagrams, you would once again encounter a big mess.

The spinor helicity formalism rides to the rescue. Consider the amplitude $A(h_1, \cdots, h_n)$. For each of the n gluons, we have $p_{i\alpha\dot{\alpha}} = \lambda_{i\alpha} \tilde{\lambda}_{i\dot{\alpha}}$, and an arbitrary spinor that we are free to choose (subject to some conditions), namely either $\mu_{i\alpha}$ or $\tilde{\mu}_{i\dot{\alpha}}$, depending on whether the corresponding helicity is $+$ or $-$, respectively. There are quite a few indices, but fortunately, in computing amplitudes, we encounter only Lorentz invariants, such as

$\epsilon_i \cdot \epsilon_j = \varepsilon^{\alpha\beta}\varepsilon^{\dot{\alpha}\dot{\beta}}\epsilon_{i\alpha\dot{\alpha}}\epsilon_{j\beta\dot{\beta}}$ (be sure to distinguish between the two varieties of epsilon here!), and thus the spinor indices will be contracted over and disappear. In particular, we have (omitting the comma in the angled and square brackets)

$$\epsilon_i^+ \cdot \epsilon_j^+ = \frac{\langle \mu_i \mu_j \rangle [\lambda_i \lambda_j]}{\langle \mu_i \lambda_i \rangle \langle \mu_j \lambda_j \rangle} \tag{11}$$

$$\epsilon_i^- \cdot \epsilon_j^- = \frac{\langle \lambda_i \lambda_j \rangle [\mu_i \mu_j]}{[\lambda_i \mu_i][\lambda_j \mu_j]} \tag{12}$$

$$\epsilon_i^- \cdot \epsilon_j^+ = \frac{\langle \lambda_i \mu_j \rangle [\mu_i \lambda_j]}{[\lambda_i \mu_i]\langle \mu_j \lambda_j \rangle} \tag{13}$$

We also list for convenience

$$\epsilon_i^+ \cdot p_j = \frac{\langle \mu_i \lambda_j \rangle [\lambda_i \lambda_j]}{\langle \mu_i \lambda_i \rangle} \tag{14}$$

and

$$\epsilon_i^- \cdot p_j = \frac{\langle \lambda_i \lambda_j \rangle [\mu_i \lambda_j]}{[\lambda_i \mu_i]} \tag{15}$$

Evidently, in this formalism, flipping helicity corresponds to interchanging the brackets $\langle \cdots \rangle$ and $[\cdots]$.

We need one more important observation. Obviously, in a tree-level diagram for n-gluon scattering, you cannot have as many 3-gluon vertices as you like. Draw the tree diagrams for $n = 4$ for example (see figure N.2.3). The number of 3-gluon vertices could be either 0 or 2. In general, the number of 3-gluon vertices can be at most $n - 2$. You are asked to verify this in exercise N.2.2. As remarked earlier, while the 4-gluon vertex does not involve momentum, the 3-gluon vertex is linear in momentum. Thus, in the numerator of the Feynman amplitude, we have n polarization vectors ϵ_i but at most $n - 2$ momenta. We are to form a scalar out of these Lorentz vectors by taking dot products. Clearly, there are at least two polarization vectors who have to dance with each other. Therefore we conclude that the tree amplitude must contain at least one power of $\epsilon_i \cdot \epsilon_j$. (In the $n = 5$ case that power was actually 2, as we saw.)

Now we are ready to rock. For the amplitude $A(+ + \cdots +)$ (suppressing the momentum labels), we simply choose the spinors μ_i representing the gauge degrees of freedom to all be equal. Then all dot products $\epsilon_i^+ \cdot \epsilon_j^+$ between polarization vectors vanish according to (11). But we just argued that the tree amplitude must contain at least one power of $\epsilon_i \cdot \epsilon_j$. Remarkably, we have shown that the maximal helicity-violating amplitude vanishes for any n! Our intuition suggested that these amplitudes are suppressed, but in fact they vanish.

What about the next-to-maximal helicity-violating amplitudes with one negative helicity, namely $A(- + + \cdots + +)$? Label the gluon with negative helicity as 1. Once again, for $i = 2, \cdots n$, choose μ_i all equal to λ_1. Then $\epsilon_i^+ \cdot \epsilon_j^+ \propto \langle \mu_i \mu_j \rangle = 0$, for $i, j \neq 1$. Furthermore, $\epsilon_1^- \cdot \epsilon_i^+ \propto \langle \lambda_1 \mu_i \rangle = \langle \lambda_1 \lambda_1 \rangle = 0$ for $i \neq 1$. The amplitude $A(- + + \cdots + +)$ also vanishes!

Clearly, this "cheap" trick of exploiting gauge freedom no longer works for the next amplitude with two negative helicities. To see why the trick does not work any more, look at $A(- - + \cdots + +)$ for instance. Once again we could, for $i = 3, \cdots n$, choose μ_i all equal

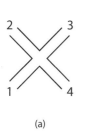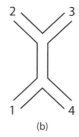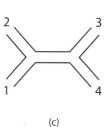

(a)　　　　　　　　　(b)　　　　　　　　　(c)

Figure N.2.3

so that $\epsilon_i^+ \cdot \epsilon_j^+ = 0$ for i, $j \geq 3$, but then we don't have enough freedom to make all the other polarization dot products vanish. In fact, at some point, we better have some nonvanishing amplitudes. In the literature, these amplitudes with two negative helicities are called maximal helicity-violating amplitudes. Upon crossing two of the gluons, they describe two gluons producing $n - 2$ gluons, with helicities $++ \rightarrow + + \cdots +$, $-+ \rightarrow - + \cdots +$, and $-- \rightarrow - - + \cdots +$.

Explicit calculation of A(1,2,3,4)

The $n = 4$ case is the simplest. Take a deep breath and try to calculate $A(1^-, 2^-, 3^+, 4^+)$ and $A(1^-, 2^+, 3^-, 4^+)$. For 4-gluon scattering these two are the only nonvanishing tree amplitudes, since by parity the amplitudes with three minuses are related to the amplitudes with three pluses (which we know vanish), and so on.

The bad news is that the calculation is fairly involved. The good news is that we can still exploit gauge freedom mercilessly and that the final answer is surprisingly simple.

Tackle $A(1^-, 2^-, 3^+, 4^+)$ first. The relevant diagrams are shown in figure N.2.3. Let us simplify the notation as much as possible: write $\langle 12 \rangle = \langle \lambda_1 \lambda_2 \rangle$, $[12] = [\lambda_1 \lambda_2]$, and so forth.

Now we need the colored Feynman rules in the form given in appendix 1. In line with the preceding discussion let us choose $\tilde{\mu}_1 = \tilde{\mu}_2 = \tilde{\lambda}_3$ and $\mu_3 = \mu_4 = \lambda_2$. Then all but one of the polarization dot products vanish. For instance, $\epsilon_2^- \cdot \epsilon_3^+ \propto \langle \lambda_2 \mu_3 \rangle [\mu_2 \lambda_3] \propto \langle \lambda_2 \lambda_2 \rangle = 0$. The only nonzero product is $\epsilon_1^- \cdot \epsilon_4^+ = \langle \lambda_1 \mu_4 \rangle [\mu_1 \lambda_4]/([\lambda_1 \mu_1]\langle \mu_4 \lambda_4 \rangle) = \langle 12 \rangle [34]/([13]\langle 24 \rangle)$, where the second equality follows from our gauge choice. This implies that the quartic diagram N.2.3a vanishes, since it involves the product of two polarization dot products.

We notice that there are only two more diagrams (fig. N.2.3b,c) rather than three. With the traditional Feynman rules there is a diagram with 1 and 3 on the same cubic vertex. Here we see another advantage of color stripping. We are looking at the coefficient of $\mathrm{tr}(T^{a_1} T^{a_2} T^{a_3} T^{a_4})$. The diagram we just described has T^{a_1} next to T^{a_3} and so does not contribute to this particular color ordering.

Next, the diagram in figure N.2.3b vanishes. Look at the cubic vertex involving 2, 3, and v (for the virtual gluon): $(\epsilon_2 \cdot \epsilon_3 \epsilon_v \cdot p_2 + \epsilon_3 \cdot \epsilon_v \epsilon_2 \cdot p_3 + \epsilon_v \cdot \epsilon_2 \epsilon_3 \cdot p_v)$, with ϵ_v understood as a "placeholder" to be contracted with the ϵ_v^* from the other cubic vertex. The first term vanishes because $\epsilon_2 \cdot \epsilon_3 = 0$, the second term because $\epsilon_2 \cdot p_3 \propto [\mu_2 \lambda_3] = [\lambda_3 \lambda_3] = 0$, and the third term because $\epsilon_3 \cdot p_\nu = -\epsilon_3 \cdot (p_2 + p_3) = -\epsilon_3 \cdot p_2 \propto \langle \mu_3 \lambda_2 \rangle = \langle \lambda_2 \lambda_2 \rangle = 0$. Our gauge choice was wise indeed!

Only one diagram (fig. N.2.3c) left to calculate. The cubic vertex ($\epsilon_1 \cdot \epsilon_2 \epsilon_\nu \cdot p_1 + \epsilon_2 \cdot \epsilon_\nu \epsilon_1 \cdot p_2 + \epsilon_\nu \cdot \epsilon_1 \epsilon_2 \cdot p_\nu$) is to be contracted with the other cubic vertex ($\epsilon_3 \cdot \epsilon_4 \epsilon_\nu^* \cdot p_3 + \epsilon_4 \cdot \epsilon_\nu^* \epsilon_3 \cdot p_4 + \epsilon_\nu^* \cdot \epsilon_3 \epsilon_4 \cdot (-p_\nu)$). In each of these vertices, the first term vanishes, since the only nonzero polarization product is $\epsilon_1 \cdot \epsilon_4$. To obtain the amplitude we replace the polarization product $\epsilon_\nu^\rho \epsilon_\nu^{\omega*}$ for the placeholder by the propagator $-ig^{\rho\omega}/(p_1 + p_2)^2 = -ig^{\rho\omega}/(2p_1 \cdot p_2)$.

Again, since all but one of the polarization dot products vanish, only one term survives the contraction with $g^{\rho\omega}$. We obtain $A(1^-, 2^-, 3^+, 4^+) = \epsilon_1 \cdot \epsilon_4 \epsilon_2 \cdot p_1 \epsilon_3 \cdot p_4/p_1 \cdot p_2$.

Since we are after conceptual understanding more than anything else, now and henceforth, in this and the next two chapters, we will suppress overall factors to keep various expressions as uncluttered as possible.

We have already calculated $\epsilon_1 \cdot \epsilon_4$, so it remains to evaluate $\epsilon_2 \cdot p_1 = \langle 21 \rangle [31]/[23]$, $\epsilon_3 \cdot p_4 = \langle 24 \rangle [34]/\langle 23 \rangle$, and $p_1 \cdot p_2 = \langle 12 \rangle [12]$. Thus $A = \langle 12 \rangle [34]^2/([12][23]\langle 23 \rangle)$.

We can now use various identities to write this in a more symmetric form. First, momentum conservation gives $\sum_i p_{\alpha\dot\alpha}^{(i)} = \sum_i \lambda_\alpha^{(i)} \tilde\lambda_{\dot\alpha}^{(i)} = 0$. Multiplying this by $\varepsilon^{\beta\alpha} \varepsilon^{\dot\alpha\dot\gamma} \lambda_\beta^{(j)} \tilde\lambda_{\dot\gamma}^{(k)}$ we obtain $\sum_i \langle ji \rangle [ik] = 0$ for any j and k. Second, we have $\langle 34 \rangle [34] = p_3 \cdot p_4 = p_1 \cdot p_2 = \langle 12 \rangle [12]$. Finally, the spinors can be regarded as 2-dimensional vectors and so any two spinors μ and ν span the space. Thus a third spinor λ can always be expanded as a linear combination of the other two, viz, $\lambda = (\langle \lambda\nu \rangle \mu - \langle \lambda\mu \rangle \nu)/\langle \mu\nu \rangle$, with the coefficients determined easily by contracting with μ and ν. Contracting with a fourth spinor η then yields

$$\langle \lambda\eta \rangle \langle \mu\nu \rangle = \langle \lambda\nu \rangle \langle \mu\eta \rangle - \langle \lambda\mu \rangle \langle \nu\eta \rangle \tag{16}$$

known as the Schouten identity.

Using these identities we now massage A into shape. Multiply the numerator and denominator of A by $\langle 34 \rangle$ to obtain $\langle 12 \rangle^2 [34]/(\langle 23 \rangle \langle 34 \rangle [23])$. Next, multiply the numerator and denominator by $\langle 12 \rangle^2$. In the denominator write $\langle 12 \rangle [23] = -\langle 14 \rangle [43]$. Finally, we obtain (suppressing overall phase factors, as promised)

$$A(1^-, 2^-, 3^+, 4^+) = \frac{\langle 12 \rangle^4}{\langle 12 \rangle \langle 23 \rangle \langle 34 \rangle \langle 41 \rangle} = \frac{p_1 \cdot p_2}{p_2 \cdot p_3} \tag{17}$$

Compare this with figure N.2.2. You should be impressed, even though here we are doing the $n = 4$ rather than the $n = 5$ case.

Recall that we have another amplitude $A(1^-, 2^+, 3^-, 4^+)$ yet to calculate, in which the two negative-helicity gluons are not adjacent in color. You should work this out as an exercise, but it turns out that we can use a trick. Write the analog of (2) for 4-gluon scattering

$$\mathcal{M} = i \sum_{\text{permutations}} \text{tr}(T^{a_1} T^{a_2} T^{a_3} T^{a_4}) A(1^-, 2^+, 3^-, 4^+) \tag{18}$$

We have already remarked that if we extend the gauge group from $SU(N)$ to $U(N)$, the extra gluon (known in the literature, perhaps confusingly, as the "photon") does not couple to the other gluons (because the couplings in Yang-Mills theory all involve commutators; see appendix 1.) Thus if we replace, say, T^{a_2}, by the identity matrix, the entire sum should vanish. The six terms in the sum then break up into two groups, multiplied by either

$\text{tr}(T^{a_1}T^{a_3}T^{a_4})$ or $\text{tr}(T^{a_1}T^{a_4}T^{a_3})$. Since the two traces are independent, the two groups vanish separately. The traces in the three terms $\text{tr}(T^{a_1}T^{a_2}T^{a_3}T^{a_4})A(1^-, 2^+, 3^-, 4^+) + \text{tr}(T^{a_1}T^{a_3}T^{a_2}T^{a_4})A(1^-, 3^-, 2^+, 4^+) + \text{tr}(T^{a_1}T^{a_3}T^{a_4}T^{a_2})A(1^-, 3^-, 4^+, 2^+)$ all become $\text{tr}(T^{a_1}T^{a_3}T^{a_4})$. We thus obtain the so-called photon decoupling identity $A(1^-, 2^+, 3^-, 4^+) + A(1^-, 3^-, 2^+, 4^+) + A(1^-, 3^-, 4^+, 2^+) = 0$, relating the desired amplitude to two amplitudes already known from (17). Thus

$$\begin{aligned} A(1^-, 2^+, 3^-, 4^+) &= -(A(1^-, 3^-, 2^+, 4^+) + A(1^-, 3^-, 4^+, 2^+)) \\ &= -\langle 13 \rangle^4 \left(\frac{1}{\langle 13 \rangle \langle 32 \rangle \langle 24 \rangle \langle 41 \rangle} + \frac{1}{\langle 13 \rangle \langle 34 \rangle \langle 42 \rangle \langle 21 \rangle} \right) \\ &= \frac{\langle 13 \rangle^4}{\langle 12 \rangle \langle 23 \rangle \langle 34 \rangle \langle 41 \rangle} \end{aligned} \tag{19}$$

where we used the Schouten identity.

Remarkably, the two amplitudes $A(1^-, 2^-, 3^+, 4^+)$ and $A(1^-, 2^+, 3^-, 4^+)$ have the same form. It is tempting to conjecture that for n-gluon scattering, the maximal helicity-violating amplitudes in which two of the gluons carry negative helicity and the rest positive helicity is given by the elegant expression (for $n \geq 4$)

$$A(1^+, 2^+, \cdots j^-, \cdots, k^- \cdots n^+) = \frac{\langle jk \rangle^4}{\langle 12 \rangle \langle 23 \rangle \langle 34 \rangle \cdots \langle (n-1)n \rangle \langle n1 \rangle} \tag{20}$$

This conjecture was first put forward by Parke and Taylor and proved by Berends and Giele (using an off-shell recursion method and a precursor to the on-shell recursion method to be explained in the next chapter.) We will prove it in the next chapter.

Meanwhile, we note that one way of arguing for the conjecture's validity is to verify that the proposed amplitude satisfies all the symmetry requirements. Besides Lorentz invariance (obviously satisfied), amplitudes at tree level in a massless theory like pure Yang-Mills should also satisfy scale and conformal invariance.

One interesting check is to count, for each i, the powers of λ_i minus the powers of $\tilde{\lambda}_i$. Call this quantity Λ_i. Then since momentum has the form $\sim \lambda \tilde{\lambda}$, it contributes 0 to Λ_i. In contrast, for negative helicity $\epsilon_{\alpha\dot{\alpha}}^- = \lambda_\alpha \tilde{\mu}_{\dot{\alpha}} / [\lambda, \mu] \sim \lambda / \tilde{\lambda}$. For positive helicity we have the opposite: $\epsilon_{\alpha\dot{\alpha}}^+ = \mu_\alpha \tilde{\lambda}_{\dot{\alpha}} / \langle \mu, \lambda \rangle \sim \tilde{\lambda} / \lambda$. Thus we have $\Lambda_i = -2h_i$.

We checked that indeed, in (20), we have $\Lambda_i = 2$ for $i = j, k$ and $\Lambda_i = -2$ for $i \neq j, k$. Keeping track of Λ_i during the calculation also provides us with a useful check.

Note that the $n = 5$ scattering amplitude, which we started this chapter with, is completely determined, since there are only two independent nonzero amplitudes: $A(1^-, 2^-, 3^+, 4^+, 5^+)$ and $A(1^-, 2^+, 3^-, 4^+, 5^+)$.

Further developments

The astonishing simplicity of (20) has sparked a surge of interest and further developments. Here I will be content to mention some of them.

Once the tree amplitudes are done, one can calculate loop amplitudes by using a more sophisticated version of the unitarity methods and of the Cutkosky cutting rules of chapter II.8. Proceeding in this way, various authors have bootstrapped their way up to multiloop amplitudes. While the actual computational labor can quickly get out of hand, it is still enormously less than the labor needed with traditional Feynman methods.

What about the basic cubic vertex of the theory? We will work it out in appendix 2 and show that it fits nicely into the form in (20) with one important caveat.

Surely you, the astute reader, feel that there must be some deep reason for the astonishing simplification from the mess in figure (1) to the elegant expression in (20). Indeed, tree amplitudes in gauge theories (and in gravity) turn out to be even simpler when written in terms of the twistors studied by Penrose decades ago. As this exciting development[2] occurred while this book was going to press, I have to balance my desire to make the book as up-to-date as possible against pagination constraints. Thus I can provide here only an ultra-concise (and hence perhaps somewhat cryptic) key to the literature, giving you no more than a flavor of what is involved.

Include the momentum conservation delta function with the amplitudes of the type studied here and define $M(\cdots, \lambda_i, \tilde{\lambda}_i, \cdots) \equiv A(\lambda, \tilde{\lambda})\delta^{(4)}(\sum_{j=1}^{n} \lambda_j \tilde{\lambda}_j)$. Due to space constraints, I will suppress the kinematic dependence of M on all but the particle i and write simply $M(\lambda_i, \tilde{\lambda}_i)$. Let us Fourier transform M in two possible ways (and overuse the letter M somewhat):

$$M(W_i) = \int d^2\lambda_i \, \exp(i\tilde{\mu}_i^{\alpha}\lambda_{i\alpha})M(\lambda_i, \tilde{\lambda}_i). \tag{21a}$$

and

$$M(Z_i) = \int d^2\tilde{\lambda}_i \, \exp(i\mu_i^{\dot{\alpha}}\tilde{\lambda}_{i\dot{\alpha}})M(\lambda_i, \tilde{\lambda}_i). \tag{21b}$$

where $W \equiv (\tilde{\mu}, \tilde{\lambda})$ and $Z \equiv (\lambda, \mu)$ denote two 4−component objects which may be regarded for the time being as column "vectors." The intent here is to transform M sequentially for $i = 1, 2, \cdots n$ using either (21a) or (21b). Consider $SO(2, 2)$ here instead of $SO(3, 1)$, so that the spinors λ and $\tilde{\lambda}$ are real, and hence we can take μ and $\tilde{\mu}$ to be real as well. Thus, these integral transforms are no more and no less than the Fourier transforms you have long been familiar with, and the variable μ is conjugate to the variable $\tilde{\lambda}$ in the same sense that p is conjugate to q in quantum mechanics. The objects W and Z, known as a twistor and a dual twistor and conjugate to each other, each consisting of 4 real components, naturally transform under the group $SL(4, R)$ (namely the set of all 4 by 4 matrices with real entries and unit determinant), with the invariant $W \cdot Z = \tilde{\mu}\lambda + \tilde{\lambda}\mu$. Given more than one W's and Z's we also have the Lorentz invariants $Z_1 I Z_2 \equiv \langle \lambda_1, \lambda_2 \rangle$ and $W_1 I W_2 \equiv [\lambda_1, \lambda_2]$. (Here I, in a slightly abused notation used in the literature, evidently denotes the 4 by 4 matrix containing the 2 by 2 identity matrix either in its upper left corner or in its lower right corner depending on whether it acts on W or Z, with all other entries equal to zero.)

We have (displaying the helicity h of particle i while suppressing the index i) $M(tW, h) = \int d^2\lambda \, \exp(it\tilde{\mu}\lambda)M(\lambda, t\tilde{\lambda}, h) = t^{-2} \int d^2\lambda' \, \exp(i\tilde{\mu}\lambda')M(t^{-1}\lambda', t\tilde{\lambda}, h) = t^{2(h-1)}M(W, h)$ where we used the observation earlier that $\Lambda = -2h$, namely that $M(t^{-1}\lambda, t\tilde{\lambda}) = t^{2h}M(\lambda, \tilde{\lambda})$. Similarly, $M(tZ, h) = t^{-2(h+1)}M(Z, h)$. This scaling result, which you realize comes from the little group (see p. 186), indicates that we should favor a mixed or am-

---

[2] The literature on twistors could be traced starting with the paper mentioned on p. 477.

bitwistor representation for the scattering amplitude, using W when the particle carries $+$ helicity and Z when the particle carries $-$ helicity.

For example, for the basic Yang-Mills cubic vertex with helicities $(+ + -)$ (see appendix 2) we write $M(W_1^+, W_2^+, Z_3^-)$. The scaling relation just derived imposes powerful constraints on this amplitude, namely $M(W_1, W_2, Z_3) = M(t W_1, W_2, Z_3) = M(W_1, t W_2, Z_3) = M(W_1, W_2, t Z_3)$, which implies that in the ambitwistor representation the defining vertex for Yang-Mills theory is apparently, up to an irrelevant overall constant, just 1! More precisely, $M(W_1, W_2, Z_3)$ depends on the three possible invariants $W_1 \cdot Z_3$, $W_2 \cdot Z_3$, and $W_1 I W_2$. The scaling relations (note that t could be either positive or negative) then force M to have the amazingly simple form

$$M(W_1^+, W_2^+, Z_3^-) = \text{sign}(W_1 \cdot Z_3)\text{sign}(W_2 \cdot Z_3)\text{sign}(W_1 I W_2)$$

In different kinematic regions, the basic Yang-Mills vertex is numerically equal to ± 1.

Tree amplitudes live naturally in ambitwistor space. As another example, the 4-gluon scattering amplitude (19) we worked hard to get becomes simply

$$M(W_1^+, Z_2^-, W_3^+, Z_4^-) = \text{sign}(W_1 \cdot Z_2)\text{sign}(Z_2 \cdot W_3)\text{sign}(W_3 \cdot Z_4)\text{sign}(Z_4 \cdot W_1)$$

Let's anticipate a bit and write the basic cubic vertex for gravity to be given in (N.3.20) in this ambitwistor representation. Indeed, the scaling relations derived above could be immediately applied to the graviton, for which $h = \pm 2$. We obtain $M(t W, ++) = t^2 M(W, ++)$ and $M(t Z, --) = t^2 M(Z, --)$, thus immediately fixing the cubic vertex for gravity to be

$$M(W_1^{++}, W_2^{++}, Z_3^{--}) = |(W_1 \cdot Z_3)(W_2 \cdot Z_3)(W_1 I W_2)|$$

Going from Yang-Mills to Einstein-Hilbert, we merely have to replace the sign function by the absolute value!

Clearly, the take-home message is that quantum field theory possesses hidden structures that the traditional Feynman diagram approach would likely have no hope of uncovering.

Appendix 1: Colored Feynman rules for Yang-Mills theory

Using the double line formalism of chapter IV.5, we can draw the cubic and quartic vertices in Yang-Mills theory as in figure IV.5.2. Our conventions for the generators of $SU(N)$ are $[T^a, T^b] = i f^{abc} T^c$ and $\text{tr}(T^a T^b) = \frac{1}{2}\delta^{ab}$. Thus $f^{abc} = -2i\text{tr}([T^a, T^b]T^c)$. Start with the Feynman rule for the quartic vertex given in chapter VII.1 and appendix C. First, $f^{abe} f^{cde} = -4\text{tr}([T^a, T^b][T^c, T^d])$. Next we multiply by polarization vectors and obtain the colored rule for the quartic vertex:

$$4 i g^2 \text{tr}(T^a T^b T^c T^d)(\epsilon_1 \cdot \epsilon_2 \epsilon_3 \cdot \epsilon_4 - \epsilon_4 \cdot \epsilon_1 \epsilon_2 \cdot \epsilon_3) \tag{22}$$

The two other terms are obtained by permutation. Similarly, the cubic vertex in (C.18) becomes (with a trivial change $k \to p$)

$$-4 i g \text{tr}(T^a T^b T^c)(\epsilon_1 \cdot \epsilon_2 \epsilon_3 \cdot p_1 + \epsilon_2 \cdot \epsilon_3 \epsilon_1 \cdot p_2 + \epsilon_3 \cdot \epsilon_1 \epsilon_2 \cdot p_3) \tag{23}$$

As described in the text, we can now strip off the color factors $\text{tr}(T^a T^b T^c)$ and $\text{tr}(T^a T^b T^c T^d)$.

Color stripped amplitudes satisfy a number of useful identities. For example, the color stripped amplitude for n-gluon scattering satisfies the reflection identity $A(1, 2, \cdots, n) = (-1)^n A(n, \cdots, 2, 1)$. To show this, note that the stripped quartic vertex $(\epsilon_1 \cdot \epsilon_2 \epsilon_3 \cdot \epsilon_4 - \epsilon_4 \cdot \epsilon_1 \epsilon_2 \cdot \epsilon_3)$ does not change sign under the reflection $1234 \to 4321$, while the stripped cubic vertex changes sign under $123 \to 321$. From exercise N.2.2, $V_3 + 2V_4 = n - 2$, and thus V_3 is odd or even according to whether n is odd or even.

Appendix 2: The cubic vertex in the spinor helicity formalism

A rather natural question to ask is what the cubic vertex (23) looks like in the spinor helicity formalism.

The first observation is that if we put all momenta on shell, $p_1^2 = p_2^2 = p_3^2 = 0$, then the cubic vertex actually vanishes. By momentum conservation, we have $p_1^2 = (p_2 + p_3)^2 = p_2 \cdot p_3 = 0$. The conditions $p_i \cdot p_j = 0$ then imply all three lightlike momenta point in the same direction, so that $p_i = E_i(1, 0, 0, 1)$, $i = 1, 2, 3$. But this means that, for example, $\epsilon_3 \cdot p_1 \propto \epsilon_3 \cdot p_3 = 0$, and thus the cubic vertex (23) vanishes.

Now you see the motivation for allowing the momenta to be complex. Then the conditions $p_i \cdot p_j = 0$ no longer force all three lightlike momenta to point in the same direction, and we can have a nonvanishing cubic vertex on shell. As explained in the text, to complexify momentum, we simply remove the constraint $\tilde{\lambda} = \lambda^*$. By the way, by complexifying the momenta here, we are anticipating the discussion in the next chapter a bit.

As always, we are free to choose the μ spinors to our advantage. A good choice here is $\mu_1 = \mu_2$ and $\mu_3 = \lambda_1$. Referring to (12) and (13), we then have $\epsilon_1^- \cdot \epsilon_2^- \propto [\mu_1 \mu_2] = 0$ and $\epsilon_1^- \cdot \epsilon_3^+ \propto \langle \lambda_1 \mu_3 \rangle = 0$. The cubic vertex collapses to

$$A(1^-, 2^-, 3^+) = \epsilon_2^- \cdot \epsilon_3^+ \epsilon_1^- \cdot p_2 = \left(\frac{\langle \lambda_2 \mu_3 \rangle [\mu_2 \lambda_3]}{[\lambda_2 \mu_2] \langle \mu_3 \lambda_3 \rangle} \right) \left(\frac{\langle \lambda_1 \lambda_2 \rangle [\mu_1 \lambda_2]}{[\lambda_1 \mu_1]} \right) = \frac{\langle 12 \rangle^2}{\langle 13 \rangle} \frac{[\mu_1 \lambda_3]}{[\mu_1 \lambda_1]} \tag{24}$$

As in the text, we are ignoring all overall factors.

To get rid of the unphysical μ_1, we need a variant of the momentum conservation identity given in the text. Multiplying $\sum_i p_{\alpha\dot\alpha}^{(i)} = \sum_i \lambda_\alpha^{(i)} \tilde{\lambda}_{\dot\alpha}^{(i)} = 0$ by $\varepsilon^{\beta\alpha} \varepsilon^{\dot\alpha\dot\gamma} \lambda_\beta^{(j)} \tilde{\mu}_{\dot\gamma}$, we obtain $\sum_i [\mu \lambda_i] \langle \lambda_i \lambda_j \rangle = 0$ for any j, which for $j = 2$ implies $[\mu_1 \lambda_3] \langle \lambda_3 \lambda_2 \rangle = -[\mu_1 \lambda_1] \langle \lambda_1 \lambda_2 \rangle$.

Multiplying (24) by $\langle \lambda_3 \lambda_2 \rangle / \langle \lambda_3 \lambda_2 \rangle$ and applying the identity just derived we finally obtain the "mostly minus" cubic vertex

$$A(1^-, 2^-, 3^+) = \frac{\langle 12 \rangle^4}{\langle 12 \rangle \langle 23 \rangle \langle 31 \rangle} \tag{25}$$

Satisfyingly, we have obtained an expression consistent with (20) (which we have not yet proven) but keep in mind that (25) holds only for complex momenta. I leave it to you to obtain the "mostly plus" cubic vertex

$$A(1^+, 2^+, 3^-) = \frac{[12]^4}{[12][23][31]} \tag{26}$$

which also follows from the rule about flipping helicities stated in the text.

What about the "all plus" and "all minus" vertices? By now you should be able to determine them as a simple exercise.

Exercises

N.2.1 Work out the two polarization vectors for general μ and $\tilde{\mu}$ for a gluon moving along the third direction.

N.2.2 Show that the number of cubic vertices in tree-level n-gluon scattering can be at most $n - 2$.

N.2.3 Show that the result in (17) satisfies the reflection identity $A(1^-, 2^-, 3^+, 4^+) = A(4^+, 3^+, 2^-, 1^-)$.

N.2.4 Show that the "all plus" and "all minus" cubic Yang-Mills vertices (see appendix 2) vanish. [Hint: Choose the μ spinors wisely.]

N.2.5 Why doesn't the argument in the text that $A(- + + \cdots + +)$ vanish apply to $A(- + +)$?

N.2.6 Insert the expression for the cubic vertex into (21) and derive $M(W_1^+, W_2^+, Z_3^-)$.

N.2.7 Show that $M(W_1^+, Z_2^-, W_3^+, Z_4^-)$ reproduces (19).

N.2.8 Show that $SL(4, R)$ is locally isomorphic to the conformal group. [Hint: Identify the $15 = 4^2 - 1$ generators of the conformal group (3 rotations J^i, 3 boosts K^i, 1 dilation D, 4 translations P^μ, and 4 conformal transformations K^μ) with the 15 traceless real 4 by 4 matrices.]

N.3 | Subterranean Connections in Gauge Theories

Excess baggage

This text, like all texts on field theory, sings the praise of gauge theories—hey, Nature loves them regardless of what physicists like—but, unlike many texts, emphasizes repeatedly that gauge symmetry is strictly speaking not a symmetry, but a redundancy in description. Extra degrees of freedom are introduced only to be gauge fixed away. In the first edition of this book, I expressed in the Closing Words the hope that in the future physics will find a more elegant way of formulating this peculiar concept of local invariance. Perhaps that hope is being realized sooner rather than later!

In our current formulation of gauge theories, for a process involving n massless gauge bosons (photons or gluons) we are instructed to laboriously calculate an off-shell amplitude $\mathcal{M}^{\mu_1\mu_2\cdots\mu_n}$.

But experimentalists don't know about amplitudes carrying Lorentz indices! SE from chapter III.1 speaks up again. "My gauge bosons are specified by their helicities h_i, $i = 1, \cdots n$, not a Lorentz index."

Come to think of it, we theorists do go through a strange two-step procedure involving a lot of excess baggage. After toiling to obtain $\mathcal{M}^{\mu_1\mu_2\cdots\mu_n}$ with external momenta off shell, we then set external momenta on shell and contract with polarization vectors to determine the scattering amplitude for gluons in specified polarization states $\mathcal{M}^{\lambda_1\lambda_2\cdots\lambda_n} \equiv \epsilon^{\lambda_1}_{\mu_1}\epsilon^{\lambda_2}_{\mu_2}\cdots\epsilon^{\lambda_n}_{\mu_n}\mathcal{M}^{\mu_1\mu_2\cdots\mu_n}|_{\text{onshell}}$. In effect, in step 2 we wash away much of the unnecessary information in $\mathcal{M}^{\mu_1\mu_2\cdots\mu_n}$ we worked hard to get in step 1.

The cancellation in the 5-gluon scattering in the preceding chapter, with $\sim$10,000 terms boiling down to a single term, should have convinced you that the traditional Feynman way may not be so good. In your study of physics, you surely have had the pleasure of watching terms canceling against each other toward the end of a calculation, but 10,000 terms down to 1, that was the mother of all cancellations.

The key is that in gauge theories there is a kind of secret subterranean connection between different Feynman diagrams, and cancellations are routine. Gauge invariance tells us that $p^1_{\mu_1}(\epsilon^2_{\mu_2} \cdots \epsilon^n_{\mu_n} \mathcal{M}^{\mu_1\mu_2\cdots\mu_n}|_{\text{onshell}})$, for example, vanishes. Thus, the many diagrams that go into $\mathcal{M}^{\mu_1\mu_2\cdots\mu_n}$ must know about each other in some intricate way. (We saw a glimpse of that way back in chapter II.7 when we proved gauge invariance.)

The S-matrix reloaded

In spite of the tremendous difficulties lying ahead, I feel that S-matrix theory is far from dead and that . . . much new interesting mathematics will be created by attempting to formalize it.

—T. Regge[1]

The garbage of the past often becomes the treasure of the present (and vice versa).

—A. Polyakov[2]

As discussed in chapter III.8, back in the 1950s and 1960s, dispersion theorists[3] tried to forge ahead by studying the analytic properties of various amplitudes as functions of their external Lorentz invariants, namely the Mandelstam variables s and t for 2-to-2 scattering and q^2 in our simple vacuum polarization example. But once one gets past 2-to-2 scattering, the analytic structure becomes unwieldy. The program failed and was swept into the dustbin of physics history. (However, you might know that, through a rather convoluted process, this massive effort eventually gave birth to string theory.)

Remarkably, some features of this program are being revived. In particular, in this chapter we will discuss the notion of complexifying physical variables. In an interesting twist, it turns out to be better to complexify the external momenta (to be explained below) rather than invariants like s and t. A historical aside: Landau apparently suggested on one occasion that it might be useful to consider complex momenta.

Consider the amplitude $\mathcal{M}(p_i, h_i)$ for tree-level scattering of n massless particles with momentum and helicity (p_i, h_i), $i = 1, \ldots, n$, with $p_i^2 = 0$. (For a gauge theory, we will define the amplitude with the color factors already stripped away. Also, suppress the trivial multiplicative coupling constant dependence and drop all such overall factors as we move along.)

The novel idea is to pick two external momenta p_r, p_s, complexifying them while keeping them on shell and maintaining momentum conservation. We take all momenta as incoming. At this stage we can keep the discussion general and not even specify the theory except to stipulate that it contains only massless particles. But to fix ideas, you can imagine a gauge theory. For some complex number z, replace p_r and p_s by

$$p_r(z) = p_r + zq \quad \text{and} \quad p_s(z) = p_s - zq \tag{1}$$

[1] T. Regge, *Publ. RIMS*, Kyoto University, 12 suppl.: pp. 367–375, 1977.
[2] A. Polyakov, *Gauge Fields and Strings*, CRC, 1987, p. 1.
[3] See, for example, G. Barton, *Dispersion Techniques in Field Theory*, W. A. Benjamin, 1965.

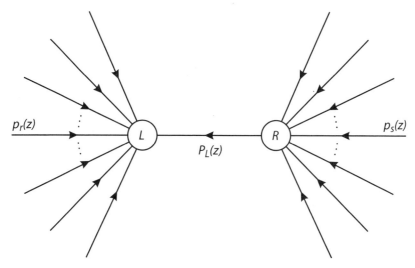

Figure N.3.1

To keep $p_r(z)^2 = 0$ and $p_s(z)^2 = 0$, we need $q \cdot p_{r,s} = 0$ and $q^2 = 0$, which is possible only if we make q complex. To be explicit, go to a frame in which $p_r + p_s$ has only a time component and use units so that the time component is equal to 2. Then

$$p_r = (1, 0, 0, 1), \qquad p_s = (1, 0, 0, -1), \qquad q = (0, 1, i, 0) \tag{2}$$

A technical aside. This is why I mentioned $SO(2, 2)$ in appendix B and in the preceding chapter: with a $(+ + - -)$ signature one could satisfy the on-shell constraint without having a complex q. Here I will stick with the more physical $SO(3, 1)$ and consider complex momenta as explained in the preceding chapter. Another side remark: As you will see, the discussion goes through for any spacetime dimension $d \geq 4$.

With this set up the scattering amplitude $\mathcal{M}(z)$ becomes an analytic function of z. Think of the complex momentum zq flowing into the diagram with $p_r(z)$, cruising through some of the internal lines, and then flowing out with $-p_s(z)$. Let us turn on our pole detector. At tree level, a pole can arise only from a propagator carrying momentum $zq + \cdots$. Thus the tree amplitude $\mathcal{M}(z)$ has only simple poles, coming from diagrams of the type shown in figure N.3.1. Divide p_i into two sets L and R, with those in L flowing into a blob on the left-hand side and those in R flowing into a blob on the right-hand side. The two blobs are connected by a single propagator carrying momentum $P_L(z)$, which by arbitrary convention we choose to flow into blob L. The two blobs are themselves tree amplitudes in the theory. Let n_L and n_R be the number of external momenta in sets L and R, respectively (with $n_L + n_R = n$, of course, and $n_L \geq 2$, $n_R \geq 2$). Then the left-hand blob represents tree scattering of $n_L + 1$ particles, with n_L particles on shell and one particle with momentum $P_L(z)$ off shell, with an amplitude $\mathcal{M}_L(z)$. Similarly, the right-hand blob represents tree scattering of $n_R + 1$ particles, with n_R particles on shell and one particle with momentum $-P_L(z)$ off shell, with an amplitude $\mathcal{M}_R(z)$.

Clearly, the momentum $P_L(z)$ depends on z only if $p_r(z)$ and $p_s(z)$ do not appear in the same set. With no loss of generality let $p_r(z)$ belong to the set L and $p_s(z)$ to the set R. Then $P_L(z) = -((\sum_{i \in L} p_i) + zq) = P_L(0) - zq$ and $P_L(z)^2 = P_L(0)^2 - 2zq \cdot P_L(0) = -2q \cdot P_L(0)(z - z_L)$, where $z_L = P_L(0)^2/(2q \cdot P_L(0))$. Thus $\mathcal{M}$ has a pole at $z = z_L$, which, since q is complex, is in general complex.

The amplitude $\mathcal{M}$ has poles all over the complex z-plane, at $z = z_L$, one for each valid partition of the external momenta into $L + R$. The residue has the factorized form $\mathcal{R}_L = \mathcal{M}_L(z_L)\mathcal{M}_R(z_L)/(2q \cdot P_L(0))$, where $\mathcal{M}_L(z_L)$ and $\mathcal{M}_R(z_L)$ are now both on-shell amplitudes, since the particle carrying momentum $P_L(z_L)$ is now on shell. As always, we suppress all inessential overall factors.

If, and that is a crucial if, $\mathcal{M}(z) \to 0$ as $z \to \infty$, then $\oint_C (dz/z)\mathcal{M}(z) = 0$, where the contour C is a circle of infinite radius running along infinity. We then shrink the contour, picking up the pole at $z = 0$, which contributes $\mathcal{M}(0)$ to the contour integral, and a bunch of poles at $z = z_L$, contributing a sum of terms consisting of the residue at each pole, multiplied by $1/z_L$. We thus determine the scattering amplitude to be

$$\mathcal{M}(0) = -\sum_{L,h} \frac{\mathcal{R}_L}{z_L} = -\sum_{L,h} \frac{\mathcal{M}_L(z_L)\mathcal{M}_R(z_L)}{P_L(0)^2} \tag{3}$$

Note that the sum also runs over the helicity h carried by the intermediate particle P_L.

The notation has been a bit compact, but suffices to get the essential point across without cluttering the page with bloated expressions. But let us now make the notation a bit more precise. To start with, z_L of course depends on the specific partition L through the momentum P_L. To make sure you follow, let us describe $\mathcal{M}_L(z_L)$ more explicitly. It is an on-shell amplitude with $(n_L + 1)$ particles coming in, respectively carrying momentum and helicity $(p_r(z_L), h_r)$, (p_i, h_i) for $i \in L$, $i \neq r$, and $(P_L(z_L), h)$. Two of the momenta are complex, namely $p_r(z_L)$ and $P_L(z_L)$. Let us emphasize that by construction $P_L(z_L)^2 = 0$ and so all particles are on shell. Similarly, $\mathcal{M}_R(z_L)$ is an on-shell amplitude with $(n_R + 1)$ particles coming in, respectively carrying momentum and helicity $(p_s(z_L), h_s)$, (p_i, h_i) for $i \in R$, $i \neq s$, and $(-P_L(z_L), -h)$.

The crucial point is that, amazingly, as was discovered by Britto, Cachazo, Feng, and Witten, we can determine the n-point tree amplitude $\mathcal{M}(z)$ in terms of lower point on-shell tree amplitudes, specifically (3) as a sum over products of the $(n_L + 1)$-point amplitude $\mathcal{M}_L(z_L)$ and $(n_R + 1)$-point amplitude $\mathcal{M}_R(z_L)$. Note that $n - 1 \geq n_L + 1 \geq 3$ (similarly for $n_R + 1$), and thus by applying these so-called BCFW recursion relations repeatedly, we can calculate any on-shell tree amplitude in Yang-Mills theory and in gravity in terms of an irreducible 3-point amplitude. Furthermore, in the primitive 3-point on-shell amplitude, all Lorentz invariants constructed out of the momenta vanish, since $p_i \cdot p_j = \frac{1}{2}(p_i + p_j)^2 = 0$.

To determine the loop amplitudes, Bern, Dixon, and Kosower have generalized the unitarity methods sketched earlier in this text and alluded to in the preceding chapter. With these methods, one can calculate all amplitudes, trees and loops, and thus determine the theory completely in terms of the helicity dependence of the 3-point on-shell amplitude.

Amazingly, the old dream of the S-matrix school comes true! Everything within perturbation theory is determined without our ever having to refer to a Lagrangian.

Note that to obtain physical amplitudes we need only $\mathcal{M}(z=0)$ but to recurse to higher point amplitudes we need to know $\mathcal{M}(z \neq 0)$. As we will see, once we have $\mathcal{M}(z=0)$ we can obtain $\mathcal{M}(z \neq 0)$ by analytic continuation.

As emphasized in the preceding chapter, the decomposition of a lightlike vector in terms of spinors

$$p_{\alpha\dot\alpha} \equiv p_\mu (\sigma^\mu)_{\alpha\dot\alpha} = \lambda_\alpha \tilde\lambda_{\dot\alpha} \tag{4}$$

works equally well for complex lightlike vectors. In that case, as already explained in chapter N.2, the two spinors λ and $\tilde\lambda$ are independent of each other.

Another side remark: The deformation (1) considered here has a nice form in the helicity spinor formalism of the preceding chapter. Let $p_r = \lambda_r \tilde\lambda_r$ and $p_s = \lambda_s \tilde\lambda_s$ (with spinor indices suppressed). Then the spinor deformation $\tilde\lambda_r \to \tilde\lambda_r + z\tilde\lambda_s$ and $\lambda_s \to \lambda_s - z\lambda_r$ (leaving λ_r and $\tilde\lambda_s$ unchanged) gives the desired momentum deformation with $q = \lambda_r \tilde\lambda_s$, which we see is not hermitean and hence corresponds to a complex momentum. This is consistent with the discussion in the preceding chapter, since the deformation obviously does not respect the equality between $\tilde\lambda$ and λ^* necessary for real momenta.

The naive person about to recurse

Imagine that you woke up one morning and had the wonderful idea of complexifying momentum. Then suppose you had enough wits, after a bow to Cauchy, to discover these marvellous recursion relations. But after you calmed down, you wanted to try the recursion out on some theory. Naturally, you first chose a scalar field theory, say a φ^3 or a φ^4 theory.

Your enthusiasm dies immediately. In these theories, the basic vertex is just a number, the coupling. For an n-point amplitude, there are always some Feynman diagrams in which $p_r(z)$ and $p_s(z)$ meet at one of the basic vertices, and the entire diagram does not even depend on z. The crucial assumption that $\mathcal{M}(z) \to 0$ as $z \to \infty$ is simply not true.

Most physicists might give up at this point, but suppose you were possessed of strength of character and decided to take a look at Yang-Mills theory, thinking that, after all, it seemed much more fundamental than some dumb scalar field theory. But a quick look convinces you that things are even worse. Consider the diagram in figure N.3.2a contributing to the n-gluon amplitude. Put $p_r(z)$ and $p_s(z)$ as "far apart" as possible to maximize the number of propagators between them. There are $(n-3)$ propagators, contributing a factor of $1/z^{n-3}$ to the amplitude as $z \to \infty$. But alas, this is overwhelmed by $(n-2)$ cubic vertices with each vertex linear in momentum, thus contributing a factor of z^{n-2}.

You have not yet included the polarization vectors, which for $z=0$ are given by $\epsilon_r^- = q$, $\epsilon_r^+ = q^*$ and $\epsilon_s^- = q^*$, $\epsilon_s^+ = q$ (note that $q \leftrightarrow q^*$ under $r \leftrightarrow s$, since the two momenta

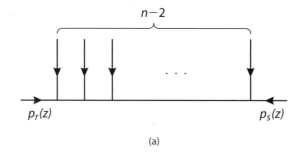

(a)

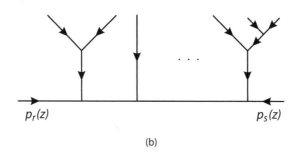

(b)

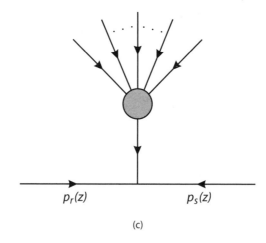

(c)

Figure N.3.2

point in opposite directions). We also have to deform them to maintain their orthogonality with the corresponding momentum vectors:

$$\epsilon_r^-(z) = q, \; \epsilon_r^+(z) = q^* + z p_s \tag{5}$$

and

$$\epsilon_s^-(z) = q^* - z p_r, \; \epsilon_s^+(z) = q \tag{6}$$

You should check that all conditions are satisfied, for example, $\epsilon_r^+(z) \cdot p_r(z) = (q^* + z p_s)(p_r + z q) = z(q^* q + p_s p_r) = 0$. [In the notation of (N.2.9, 10), the polarization vectors here correspond to the choice $\mu_r(z) = \mu_r(0) = \lambda_s$, $\tilde{\mu}_r(z) = \tilde{\mu}_r(0) = \tilde{\lambda}_s$, $\mu_s(z) = \mu_s(0) = \lambda_r$, $\tilde{\mu}_s(z) = \tilde{\mu}_s(0) = \tilde{\lambda}_r$. The first equal sign in each relation simply emphasizes that we choose not to deform the μ's and $\tilde{\mu}$'s.]

Note the peculiar asymmetry between r and s after deformation: in particular, two of the polarization vectors, ϵ_r^+ and ϵ_s^-, grow with z and thus worsen the large z behavior. Putting it all together and referring to (5) and (6), you would conclude [with the notation $\mathcal{M}^{h_r h_s}(z)$] that

$$\mathcal{M}_{\text{naive}}^{-+}(z) \to \frac{z^{n-2}}{z^{n-3}} = z, \; \mathcal{M}_{\text{naive}}^{--\text{or}++}(z) \to z^2, \; \mathcal{M}_{\text{naive}}^{+-}(z) \to z^3 \tag{7}$$

which most certainly do not $\to 0$.

A seemingly unimportant comment that will become important later: Of course, some of the gluons other than r and s could first interact among themselves as shown in figure N.3.2b. This merely reduces the effective n in the discussion above for those particular diagrams, and we reach the same naive estimates.

Reality more benign than expectation

Reality turns out to be much more benign than our naive expectation! Actually, amplitudes in Yang-Mills theory behave better than amplitudes in scalar field theory, the opposite of what we thought.

This fact is either astonishing or not so astonishing, depending on how jaded you are. I have to admit that it sounds a bit less amazing after learning in the preceding chapter that $\sim 10,000$ terms can cancel down to a single term.

We can even concoct a heuristic physical argument. Go back to Yang-Mills theory and call the particles gluons, as before. Apply crossing to gluon s, so that we have an incoming gluon r with a huge momentum $p_r(z) \sim z q$ in the large z limit, emerging as a gluon with the huge momentum $-p_s(z) \sim z q$. The other $(n-2)$ gluons have fixed momenta and are thus soft. We have a hard gluon blasting through a soft gluon background, something like a high energy gamma ray blasting through a magnetic field, and thus we do not expect much scattering as $z \to \infty$, and even less scattering that would flip the helicity of the hard gluon. (The situation is conceptually similar to electron scattering in an external Coulomb potential, discussed in chapter II.6, except that here the field excitation being scattered is of the same type as the background field.)

But not so fast! Even though you and I have studied physics for years, we haven't built up much intuition about complex momenta. At least I speak for myself. Alternatively, we could go to $SO(2, 2)$ and deal with real momenta, but we haven't much experience with signature $(+ + --)$ spacetime, either.

Background field method

Nevertheless, the picture of a hard gluon blasting through a soft background turns out to be helpful in guiding us toward an elegant formulation of the problem. We split the Yang-Mills gauge potential (which we write as $\mathcal{A}$ on this occasion) into two pieces, $\mathcal{A}(x) = A(x) + a(x)$, a background potential A without high momentum components in its Fourier transform and a fluctuating potential a with high momentum components. (You would do exactly the same split when studying a laser beam passing through a laboratory magnetic field.) To develop this so-called background field method (which is useful for other problems besides this one), it pays to use the differential form notation used in chapter IV.5.

We split the transformation law $\mathcal{A} \to U\mathcal{A}U^\dagger + UdU^\dagger$ into $A \to UAU^\dagger + UdU^\dagger$ and $a \to UaU^\dagger$. In other words, the background A transforms like a Yang-Mills potential, while the fluctuating a transforms like a matter field in the adjoint representation. Plugging into the field strength $\mathcal{F} = d\mathcal{A} + \mathcal{A}^2 = d(A + a) + (A + a)^2$, we find $\mathcal{F}$ equal to the sum of the background field strength $F = dA + A^2$ and the 2-form $da + Aa + aA + a^2 = (\partial_\mu a_\nu + [A_\mu, a_\nu] + a_\mu a_\nu)dx^\mu dx^\nu \equiv \frac{1}{2}(D_\mu a_\nu - D_\nu a_\mu + [a_\mu, a_\nu])dx^\mu dx^\nu$. Switching back from math to physics notation and defining the shorthand notation $D_{[\mu}a_{\nu]} \equiv D_\mu a_\nu - D_\nu a_\mu$, we have $\mathcal{F}_{\mu\nu} = F_{\mu\nu} + D_{[\mu}a_{\nu]} - i[a_\mu, a_\nu]$. Here $D_\mu a_\nu = \partial_\mu a_\nu - i[A_\mu, a_\nu]$ is the covariant derivative (with respect to the background potential A) of the adjoint field a.

Since we have only two hard gluons interacting with the soft background, it suffices to expand the Yang-Mills Lagrangian to quadratic order in a:

$$\mathcal{L} = -\frac{1}{2g^2}\mathrm{tr}\mathcal{F}_{\mu\nu}\mathcal{F}^{\mu\nu}$$

$$= -\frac{1}{2g^2}\mathrm{tr}\left(F_{\mu\nu}F^{\mu\nu} + D_{[\mu}a_{\nu]}D^{[\mu}a^{\nu]} + 2F^{\mu\nu}D_{[\mu}a_{\nu]} - 2i F^{\mu\nu}[a_\mu, a_\nu]\right) + O(a^3) \tag{8}$$

Since in the action we integrate $\mathcal{L}$ over spacetime, we are effectively allowed to integrate by parts. Thus the third term in the parenthesis, $\mathrm{tr}(F^{\mu\nu}D_\mu a_\nu) = \mathrm{tr}(F^{\mu\nu}(\partial_\mu a_\nu - i[A_\mu, a_\nu]))$ "=" $\mathrm{tr}((D_\mu F^{\mu\nu})a_\nu)$. Since the background field satisfies the field equation $D_\mu F^{\mu\nu} = 0$, this term vanishes. (You should not be surprised that the term linear in a in the action is linear in the field equation.)

Thus, to study the propagation of a through the background A, we can focus on the Lagrangian quadratic in a: $\mathcal{L}_{\mathrm{quad}} = -(1/g^2)\mathrm{tr}\left((D_\mu a_\nu - D_\nu a_\mu)D^\mu a^\nu - i F^{\mu\nu}[a_\mu, a_\nu]\right)$ As always, we need to fix the gauge. Upon integration by parts we have

$$\mathrm{tr}\, D_\nu a_\mu D^\mu a^\nu = \mathrm{tr}(D_\mu a_\mu D^\nu a^\nu + i F^{\mu\nu}[a_\mu, a_\nu]) \tag{9}$$

Note that, unlike ordinary derivatives, when the gauge derivatives D_ν and D^μ pass each other, they produce the field strength F^μ. [Verify this! You might recall the more mathematical form (IV.5.13).] Thus a convenient way of fixing the gauge is to add $\text{tr}(D_\mu a_\mu D^\nu a^\nu)$ so that the gauge-fixed Lagrangian becomes

$$\mathcal{L}_{\text{quad}} = -\frac{1}{g^2}\text{tr}(D_\mu a_\nu D^\mu a^\nu - 2i\,F^{\mu\nu}[a_\mu, a_\nu]) \tag{10}$$

[Incidentally, this parallels precisely what was done in (III.4.8) to obtain the Feynman gauge with $\xi = 1$.]

Return now to our problem of studying the large-z behavior of the scattering amplitude $\mathcal{M}^{\lambda\rho}$. Recall from (7) that we obtain $\mathcal{M}^{\lambda\rho} \to z^{n-2}/z^{n-3} = z$. (Also recall that polarization vectors have not yet been included, and they could multiply this behavior by z^0, z^1, or z^2.)

The culprit is the derivative in the cubic vertex $\sim Aa\partial a$ sitting inside the first term in (10). In contrast, the $\sim AaAa$ piece in the first term and the second term $\text{tr}\,F^{\mu\nu}[a_\mu, a_\nu]$ in (10) insert quartic vertices that do not grow with z.

The situation confronting us is now best discussed by the clever trick of renaming indices. First, understand that Lorentz invariance is broken by the presence of the background field A_μ, to be regarded as given and fixed. (This is the same as in chapter VI.2: the presence of a background magnetic field means that parity and time reversal are broken.) But now suppose we simply relabel indices and write

$$\mathcal{L}_{\text{quad}} = -\frac{1}{_\epsilon g^2}\text{tr}(\eta^{ab} D_\mu a_a D^\mu a_b - 2i\,F^{ab}[a_a, a_b]) \tag{11}$$

where η^{ab} is nothing but the humble Minkowski metric.

The first term by itself enjoys a hidden "enhanced Lorentz" symmetry: an $SO(3, 1)$ transformation on the indices a, b leaves the Lagrangian invariant. We now exploit this hidden symmetry. Since the leading behavior of $\mathcal{M}^{ab}$ for large z comes from repeated insertion of the cubic vertex $\eta^{ab}a_a A^\mu \partial_\mu a_b$ contained in the first term in (11), we conclude that the leading behavior must be proportional to η^{ab}.

In contrast, with one insertion of the quartic vertex from the second term in (11), we decrease the power of z by one, since it does not contain a derivative on the field a. But we also break the hidden "enhanced Lorentz" symmetry, since F^{ab} is fixed. On the other hand, there is an extra bit of information: we know that it is antisymmetric in (ab). [Note that an insertion of the quartic vertex $\eta^{ab}a_a A^\mu A_\mu a_b$ contained in the first term in (11) also decreases the power of z by one, but its contribution is proportional to η^{ab}.]

Thus the hidden "enhanced Lorentz" symmetry tells us that the amplitude expanded in powers of z must have the form

$$\mathcal{M}^{ab} = (cz + \cdots)\eta^{ab} + A^{ab} + \frac{1}{z}B^{ab} + \cdots \tag{12}$$

with c some unknown constant. The only thing we know about the matrix A^{ab} is that it is antisymmetric in (ab). (I am following the notation in the literature. If you are confused between this matrix A and the background gauge potential $A(x)$ you need to go back to square 1.)

We still have gauge invariance in the form $p_{ra}(z)\mathcal{M}^{ab}(z)\varepsilon_{sb}(z) = 0$ and $\varepsilon_{ra}(z)\mathcal{M}^{ab}(z)p_{sb}(z) = 0$, giving us valuable information. For example, looking up the form $p_r(z) = p_r + zq$, we obtain $q_a\mathcal{M}^{ab}(z)\varepsilon_{sb}(z) = -(1/z)p_{ra}\mathcal{M}^{ab}(z)\varepsilon_{sb}(z)$, but since from (5) $\epsilon_r^-(z) = q$, this means that $\epsilon_{ra}^-(z)\mathcal{M}^{ab}(z)\varepsilon_{sb}(z) = -(1/z)p_{ra}\mathcal{M}^{ab}(z)\varepsilon_{sb}(z)$.

Let us now look at the specific helicity combinations for which we had naive expectations in (7). Recall that we expected $\mathcal{M}^{-+}(z) \to z$. In fact, since $\epsilon_s^+(z) = q$ and $p_r \cdot q = 0$, we have

$$\mathcal{M}^{-+}(z) = \epsilon_{ra}^-(z)\mathcal{M}^{ab}(z)\varepsilon_{sb}^+(z) = -\frac{1}{z}p_{ra}\{(cz + \cdots)\eta^{ab} + A^{ab} + \frac{1}{z}B^{ab} + \cdots\}q_b$$

$$= -\frac{1}{z}p_{ra}A^{ab}q_b + O(\frac{1}{z^2}) \to \frac{1}{z} \tag{13}$$

This amplitude behaves better than naive expectation by two powers of $(1/z)$!

Next, $\mathcal{M}^{--}(z) \to z^2$ naively, but in fact

$$\mathcal{M}^{--}(z) = \epsilon_{ra}^-(z)\mathcal{M}^{ab}(z)\varepsilon_{sb}^-(z) = -\frac{1}{z}p_{ra}\{(cz + \cdots)\eta^{ab} + A^{ab} + \frac{1}{z}B^{ab} + \cdots\}(q_b^* - zp_{rb})$$

$$= -\frac{1}{z}(p_{ra}A^{ab}q_b^* + p_{ra}B^{ab}p_{rb}) + O(\frac{1}{z^2}) \to \frac{1}{z}, \tag{14}$$

three powers better than naive expectation. Similarly, $\mathcal{M}^{++}(z) \to 1/z$. Note that these conclusions hold for any n. If you have the strength, you might want to witness the cancellations by explicitly calculating the various $\mathcal{M}$'s for low values of n.

But not all helicity amplitudes behave better than naive expectation. We finally come to $\mathcal{M}^{+-}(z)$, which $\to z^3$ naively. Looking at (5) and (6) we already see trouble, since both $\epsilon_{ra}^+(z)$ and $\varepsilon_{sb}^-(z)$ grow like z. Now we have

$$\mathcal{M}^{+-}(z) = \epsilon_{ra}^+(z)\mathcal{M}^{ab}(z)\varepsilon_{sb}^-(z) = (q_a^* + zp_{sa})\{(cz + \cdots)\eta^{ab} + A^{ab} + \frac{1}{z}B^{ab} + \cdots\}(q_b^* - zp_{rb})$$

$$= -cp_s \cdot p_r z^3 + O(z^2) \to z^3 \tag{15}$$

Incidentally, note that our intuition about complex momenta is a bit shaky. The helicity-conserving amplitude $(+ \to +)$ [namely $\mathcal{M}^{+-}(z)$ by crossing; recall that $\mathcal{M}$ was defined with all momenta going in] behaves worse than the $(+ \to -)$ amplitude $\mathcal{M}^{++}(z)$, the $(- \to +)$ amplitude $\mathcal{M}^{--}(z)$, and the $(- \to -)$ amplitude $\mathcal{M}^{-+}(z)$. The polarization vectors are continued for complex momentum in a nonsymmetric fashion.

Confusio suddenly speaks up! "You haven't yet exploited the gauge invariance of the background field," he says.

We forgot that he often appears in the company of SE. Indeed, he is right. Very good—Confusio did not become an assistant professor for nothing.

Indeed, let us look at the cubic vertex in figure N.3.2a more carefully: we have a hard gluon carrying momentum $zq + \cdots$ scattering off a background gluon carrying some small momentum p into a hard gluon with momentum $zq + \cdots$. The coupling comes from the term $\text{tr}\partial^\mu a^\nu[A_\mu, a_\nu]$ in the Lagrangian, and thus to leading order in z the vertex is proportional to $zq^\mu \cdot A_\mu(p)$. According to exercise VII.1.1, we can choose a gauge in which $q^\mu \cdot A_\mu(p) = A_{2+i3}(p) = 0$, known as the Chalmers-Siegel space cone gauge.

We should check to see if this is possible, but to streamline the exposition let us merely do the abelian case. With $A_\mu(x) \to A_\mu(x) - \partial_\mu\Lambda(x)$, the desired gauge choice requires $q \cdot A(p) = iq \cdot p\Lambda(p)$, and thus we can solve for $\Lambda(p)$ as long as $q \cdot p \neq 0$. While $q \cdot p_{r,s} = 0$ by construction, generically there is no reason for $q \cdot p_i$ to vanish for $i \neq r, s$. So we conclude that indeed we can get rid of the offending cubic vertex.

But not so fast! What about figure N.3.2c, in which all the soft gluons interact with each other to form one single soft gluon carrying momentum $\sum_{i \neq r,s} p_i = -(p_r + p_s)$? Since $q \cdot (p_r + p_s) = 0$, we cannot set $q \cdot A(p_r + p_s) = 0$ and the diagram in figure N.3.2c remains. Thus, even though we managed to get rid of the cubic vertices in figure N.3.2a, b, our previous conclusion about the large-z behavior of $\mathcal{M}$ still stands.

"Wait! What about the color factor?" Confusio yells. Let us look at the color structure we stripped off. From figure IV.5.2b we see that the cubic vertex in figure N.3.2c requires that the two hard gluons be adjacent in color. It is easiest to explain the terminology by an example: a red-green gluon and a blue-yellow gluon are not adjacent in color, but they are both adjacent to a red-yellow gluon (and to a blue-green gluon). Note that the coupling $\mathrm{tr} \, F^{\mu\nu}[a_\mu, a_\nu]$ also requires that the two hard gluons be color adjacent.

Thus, if the two hard gluons are not color adjacent, the large-z behavior of $\mathcal{M}$ is somewhat better, since now $c = 0$ and $A^{ab} = O(1/z)$. Then $\mathcal{M}^{-+} \to 1/z^2$ instead of $1/z$, $\mathcal{M}^{+-} \to z^2$ instead of z^3, while $\mathcal{M}^{--}$ and $\mathcal{M}^{++}$ are not improved. Confusio deserves credit for his partial triumph, and perhaps eventually should be given tenure.

The bottom line is that, contrary to naive expectation, amplitudes in gauge theory behave well enough for the BCFW recursion program to work. We don't even mind that $\mathcal{M}^{+-}$ behaves badly; it suffices for the program that $\mathcal{M}^{-, \text{any helicity}}$ vanishes for large z. In particular, in appendix 1 we will show how to complete the calculation started in the preceding chapter.

As indicated earlier, once we determine the tree amplitudes, we can in principle obtain all loop amplitudes by using unitarity. In this modern revival of the S-matrix spirit, we deal with only on-shell amplitudes. The message here is that traditional Feynman diagrams carry around an enormous amount of unnecessary off-shell baggage. A dramatic example is furnished by this innocuous looking Feynman integral

$$\int \frac{d^4 l}{(2\pi)^4} \frac{l^\mu l^\nu l^\rho l^\lambda}{l^2 (l-k)^2 (l-p)^2 (l-q)^2} \tag{16}$$

which you can evaluate most conveniently using dimensional regularization. Try it. The integral looks similar to the integrals we did back in chapters III.6,7, but looks are deceptive. The answer, if printed on a page, is a total black smudge (see http://online.kitp.ucsb.edu/online/colloq/bern2/oh/05.html). After all, this integral is just one piece of a physical amplitude and by itself does not possess any nice qualities, such as gauge invariance.

All possible Lorentz invariant theories

Remarkably, not only does BCFW recursion allow us to determine all n-point on-shell amplitudes in terms of a primitive 3-point on-shell amplitude, it also restricts all possible theories for which the recursion works. Let us sketch how this is possible. We anticipate here, as we will explain in the next chapter, that the recursion program works for massless spin 2 as well as for spin 1 particles. Consider a 4-point on-shell amplitude $\mathcal{M}$. The point is that we are free to deform different pairs (r, s) to determine $\mathcal{M}$. Suppose we pick $(r, s) = (1, 4)$. Then $\mathcal{M}$ is the sum of two pieces, one with a pole in $s = (p_1 + p_2)^2 = (p_3 + p_4)^2$ and another with a pole in $t = (p_1 + p_3)^2 = (p_2 + p_4)^2$. But we could have also picked $(1, 2)$ for example. That the physical 4-point on-shell amplitude $\mathcal{M}(z = 0)$ constructed in different ways must agree imposes powerful self-consistency conditions on the primitive 3-point on-shell amplitude.

Perhaps not surprisingly, for spin 2 massless particles, Einstein gravity is the only possible theory, while for spin 1 massless particles, Yang-Mills gauge theory. Indeed, this result was proven long ago by Weinberg using rather general arguments. But it is still instructive to see how the same result emerges from a strikingly different formalism. These self-consistency conditions also allow one to explore and search for other possible theories.

It is crucial that the primitive 3-point on-shell amplitude $\mathcal{M}_3$ is evaluated for complex momenta, which allow more freedom than garden-variety everyday real lightlike momenta. (As already noted in appendix 1 to chapter N.2, the Yang-Mills cubic vertex vanishes for real lightlike momenta.) I remind you again that for complex momenta $p_i = \lambda_i \tilde{\lambda}_i$, the two spinors λ_i and $\tilde{\lambda}_i$ are independent of each other. Recall from chapter N.2 that $\langle ij \rangle = \langle \lambda_i \lambda_j \rangle \equiv \varepsilon^{\alpha\beta} \lambda_{i\alpha} \lambda_{j\beta}$ and $[ij] = [\lambda_i \lambda_j] = [\tilde{\lambda}_i \tilde{\lambda}_j] \equiv \varepsilon^{\dot{\alpha}\dot{\beta}} \tilde{\lambda}_{i\dot{\alpha}} \tilde{\lambda}_{j\dot{\beta}}$. Also, $p_i \cdot p_j = \langle ij \rangle [ij]$.

The on-mass shell conditions $p_i \cdot p_j = 0$ then become $\langle 12 \rangle [12] = 0$, $\langle 23 \rangle [23] = 0$, and $\langle 31 \rangle [31] = 0$. Apparently there are several possible solutions. For example, we could have all three square brackets vanish with all three angled brackets nonzero, or we could have two square brackets vanish, say $[12] = [23] = 0$, with $\langle 31 \rangle = 0$. But there are only two independent 2-component spinors, so three spinors cannot be linearly independent [take $\tilde{\lambda}_1 \propto (0, 1)$ and $\tilde{\lambda}_2 \propto (1, w)$, then the third spinor $\tilde{\lambda}_3$ is necessarily a linear combination of the other two]. Thus, $[12] = 0$ and $[23] = 0$ mean that $\tilde{\lambda}_1 \propto \tilde{\lambda}_2$ and $\tilde{\lambda}_2 \propto \tilde{\lambda}_3$, respectively, which implies that $\tilde{\lambda}_3 \propto \tilde{\lambda}_1$ and $[31] = 0$. Of course, the discussion can be repeated with square and angled brackets interchanged. Thus we conclude that

$$\text{either } \langle 12 \rangle = \langle 23 \rangle = \langle 31 \rangle = 0 \text{ or } [12] = [23] = [31] = 0 \tag{17}$$

(For example, if $[12] = [23] = [31] = 0$, then $\tilde{\lambda}_2 = \alpha_2 \tilde{\lambda}_1$ and $\tilde{\lambda}_3 = \alpha_3 \tilde{\lambda}_1$, and momentum conservation $\sum_i p_i = \sum_i \lambda_i \tilde{\lambda}_i = 0$ implies $\lambda_1 + \alpha_2 \lambda_2 + \alpha_3 \lambda_3 = 0$. The information here is in the coefficients, since three 2-component spinors are always linearly dependent.)

Thus, depending on the helicities, either $\mathcal{M}_3 = \mathcal{M}_H(\langle 12 \rangle, \langle 23 \rangle, \langle 31 \rangle)$ or $\mathcal{M}_3 = \mathcal{M}_A([12], [23], [31])$.

Recall from the preceding chapter that $\Lambda_i = -2h_i$, where Λ_i counts the powers of λ_i minus the powers of $\tilde{\lambda}_i$. But acting on $\mathcal{M}_H$, this just counts the powers of λ_i. Write $\mathcal{M}_H = \langle 12 \rangle^{d_3} \langle 23 \rangle^{d_1} \langle 31 \rangle^{d_2}$ and solve for the unknown d's using $\Lambda_1 = d_2 + d_3 = -2h_i$, etc. Then $d_1 = h_1 - h_2 - h_3$, $d_2 = h_2 - h_3 - h_1$, and $d_3 = h_3 - h_1 - h_2$. For example, suppose the theory contains spin 1 massless particles, with different varieties labeled by an index a whose range we need not specify. Then we have, for example,

$$\mathcal{M}_3(1_a^-, 2_b^-, 3_c^+) = f_{abc} \left(\frac{\langle 12 \rangle^3}{\langle 23 \rangle \langle 31 \rangle} \right) \tag{18}$$

since the helicities $h_1 = h_2 = -1$ and $h_3 = +1$ imply that $d_1 = d_2 = -1$ and $d_3 = 3$. At this stage f_{abc} is some unknown coefficient that depends on the particle variety. As required, we have two positive powers of λ_1 and λ_2 and two negative powers of λ_3. This confirms what we obtained in appendix 1 to the preceding chapter.

Several remarks follow.

1. From $p_i = \lambda_i \tilde{\lambda}_i$ the spinors λ and $\tilde{\lambda}$ have mass dimension $\frac{1}{2}$. Thus $\mathcal{M}_3$ has mass dimension 1, as expected (recall that the cubic coupling in gauge theory has the form $\sim \epsilon \cdot \epsilon \epsilon \cdot p$).

2. We obtain the 3-point amplitude

$$\mathcal{M}_3(1_a^+, 2_b^+, 3_c^-) = f_{abc} \left(\frac{[12]^3}{[23][31]} \right) \tag{19}$$

by flipping helicities, which, as we have learned in the preceding chapter, amounts to interchanging the roles played by λ and $\tilde{\lambda}$, so that it given by square instead of angled brackets.

3. Note the power of the spinor helicity formalism. We can immediately generalize to higher integer spin s by scaling the Λ_i's and hence the d's up by a factor of s. Thus we simply raise the round parenthesis in $\mathcal{M}_3(1_a^-, 2_b^-, 3_c^+)$ to power s. The cubic vertex for spin 2 is thus given by $\mathcal{M}_3(1_a^{--}, 2_b^{--}, 3_c^{++}) = f_{abc}(\langle 12 \rangle^3/(\langle 23 \rangle \langle 31 \rangle))^2$.

4. Interchanging 1 and 2, we see that $f_{abc} = -f_{bac}$ for s odd. Thus for s odd ($s = 1$, for example), we cannot have a theory with only one variety of particles. We are compelled to introduce the index a (and call it color!).

5. For s even ($s = 2$, for example) we can get away with only one variety. Call it the graviton. The coefficient f_{abc} can be omitted and one of the two basic cubic vertices for Einstein gravity is simply given by

$$\mathcal{M}_3(1^{--}, 2^{--}, 3^{++}) = \left(\frac{\langle 12 \rangle^3}{\langle 23 \rangle \langle 31 \rangle} \right)^2 \tag{20}$$

(The other vertex is of course obtained by replacing angled brackets by square brackets.) More on the 3-graviton vertex in appendix 2.

6. Check out the power of the self-consistency argument sketched above. Consider the 4-point amplitude $\mathcal{M}(1_a, 2_b, 3_c, 4_d)$ in a theory with a variety of spin 1 massless particles. Apply the recursion to construct $\mathcal{M}$ as the sum of an amplitude with an s channel pole, evidently proportional to $f_{abe} f_{cde}$ with an implicit sum over the label e of the intermediate particle, and an amplitude with a t channel pole proportional to $f_{ace} f_{bde}$. Requiring $\mathcal{M}$ constructed with different choices of (r, s) in the recursion to be the same then gives the constraint

$$f_{abe} f_{cde} + f_{ace} f_{bde} + f_{ade} f_{bce} = 0 \tag{21}$$

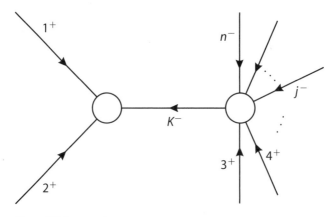

Figure N.3.3

But we recognize this as just the defining relation (B.19) for the generators of a Lie algebra $[T^a, T^b] = i f_{abc} T^c$ written out in the adjoint representation! The coefficients f_{abc} that appear in the primitive 3-point on-shell amplitude are the structure constants of the algebra. If this is too abstract for you, verify it for $SU(2)$.

The recursion program produces Einstein gravity and Yang-Mills theory as the unique low energy theory for massless spin 2 and spin 1 particles, respectively, with sufficiently good large-z behavior for the recursion relations to be valid. Of course, we also know that in the Lagrangian formalism, the powerful constraints of local coordinate invariance and local gauge invariance fix the actions for Einstein gravity and Yang-Mills completely.

Appendix 1

Here, as promised, we use the recursion approach to prove the result conjectured in the preceding chapter, that for n-gluon scattering, the maximal helicity-violating amplitude is given by

$$A(1^+, 2^+, \cdots j^-, \cdots, n^-) = \frac{\langle jn \rangle^4}{\langle 12 \rangle \langle 23 \rangle \langle 34 \rangle \cdots \langle (n-1)n \rangle \langle n1 \rangle} \tag{22}$$

(Using the cyclicity of the amplitude, we have with no loss of generality let gluon n carry negative helicity.)
  We take $r = n$ and $s = 1$, and deform $\tilde{\lambda}_n \to \tilde{\lambda}_n + z\tilde{\lambda}_1$ and $\lambda_1 \to \lambda_1 - z\lambda_n$ (leaving λ_n and $\tilde{\lambda}_1$ unchanged), in other words, $p_n \to p_n + zq$ and $p_1 \to p_1 - zq$ with $q = \lambda_n \tilde{\lambda}_1$. Write (3) as (see fig. N.3.3)

$$A(1^+, 2^+, \cdots j^-, \cdots, n^-) = A_3(\hat{1}^+, 2^+, \hat{K}^-) A_{n-1}(-\hat{K}^+, 3^+, \cdots, j^-, \cdots, \hat{n}^-)/P_L(0)^2 \tag{23}$$

Here we define $K(z) = P_L(z)$ to simplify writing. We use a hat to indicate that the corresponding momentum has been complexified. Thus $\hat{1}, \hat{n}, \hat{K}$ remind us that $p_1(z)$, $p_n(z)$, and $K(z) = -(p_1(z) + p_2)$ (evaluated at $z = z_L$) are the three complex momenta in the problem.
  In the spirit of recursion, we are supposing that A_3 and A_{n-1} are given by (22) (and the corresponding expression with all helicities flipped and angled brackets replaced by square brackets). Note that (22) does not refer to any possible relation between the untwiddled λ and twiddled $\tilde{\lambda}$ spinors and thus makes sense for both complex and real momenta. Notice that the sum over partition L and over h in (3) collapses to one term in (23). We have used the result $A(+ + \cdots +) = 0$ and $A(- + \cdots +) = 0$, and the second half of (17) to eliminate a diagram similar to that in fig. N.3.3 but with particles $(n - 1)$ and n participating in the cubic vertex instead of 1 and 2.

Recursing and using $P_L(0)^2 = 2p_1 \cdot p_2 = 2\langle 12\rangle[12]$, we obtain [see (19)]

$$A(1^+, 2^+, \cdots j^-, \cdots, n^-) = \frac{[\hat{1}2]^3}{[2, \hat{K}][\hat{K}, \hat{1}]} \frac{\langle j\hat{n}\rangle^4}{\langle \hat{K}3\rangle\langle 34\rangle \cdots \langle (n-1)\hat{n}\rangle\langle \hat{n}\hat{K}\rangle} \frac{1}{\langle 12\rangle[12]} \tag{24}$$

As before, we suppress overall constants.

The trick consists of taking various hats off or leaving them on. Since $\tilde{\lambda}_1$ is unchanged, we can remove the hat on $\hat{1}$ when it appears in a square bracket. Similarly, since λ_n is unchanged, we can remove the hat on $\hat{n}$ when it appears in an angled bracket. On the other hand, we should leave the hat on $\hat{K}$. Instead, we use momentum conservation $-\lambda_K\tilde{\lambda}_K = \lambda_1\tilde{\lambda}_1 + \lambda_2\tilde{\lambda}_2$ so that $\langle \hat{K}, 3\rangle[\hat{K}, 1] = -\langle 3, \hat{K}\rangle[\hat{K}, 1] = \langle 32\rangle[21]$ since $[11] = 0$. (You might note that this is the same sort of manipulation used to derive the first identity we needed to massage A_4 into shape in the preceding chapter.) Similarly, $\langle n\hat{K}\rangle[2, \hat{K}] = -\langle n\hat{K}\rangle[\hat{K}, 2] = \langle n1\rangle[12]$.

Doing all this to (24) we obtain

$$A(1^+, 2^+, \cdots j^-, \cdots, n^-) = \frac{[12]^3\langle jn\rangle^4}{\langle 12\rangle[12]\langle n1\rangle[12]\langle 32\rangle[21]\langle 34\rangle \cdots \langle (n-1)n\rangle}$$

$$= \frac{\langle jn\rangle^4}{\langle 12\rangle\langle 23\rangle\langle 34\rangle \cdots \langle (n-1)n\rangle\langle n1\rangle}, \tag{25}$$

precisely the conjectured result.

Note how much more powerful the recursion approach is compared to the explicit spinor helicity calculation we did to obtain $A(1, 2, 3, 4)$ in the preceding chapter, which in turn is so much more powerful than the traditional Feynman diagram calculation. Thus theoretical physics marches on.

You might be puzzled that the quartic vertex (figure IV.5.1c) of Yang-Mills theory is not needed in the recursion program. Does this nonparticipation in the program mean that we can multiply the quartic term in the Lagrangian by an arbitrary coefficient (including 0)? The resolution of this apparent paradox can be traced to the fact that the (perturbative) physical states of the gluon are built into the recursion relations. The quartic term is needed to guarantee gauge invariance and hence the two helicity states of the gluon.

Appendix 2

By showing you the mess in figure N.2.2 I have already plenty impressed upon you that the traditional Feynman diagram approach is almost hopeless when it comes to gluons. The situation with gravity is far worse. Consider the 3-graviton vertex. Conceptually it is easy to understand: we write $g_{\mu\nu} = \eta_{\mu\nu} + h_{\mu\nu}$ and expand the Einstein-Hilbert action (VIII.1.1) to $O(h^3)$. There it is, with indices suppressed, the cubic term $h\partial h\partial h$ in (VIII.1.5). Of course, this actually represents many terms with the eight indices contracted every which way, but which you can readily work out. Next, pick the harmonic gauge for example, and derive the Feynman rule for the 3-graviton vertex $G_{\mu\alpha, \nu\beta, \sigma\gamma}(p_1, p_2, p_3)$, namely the analog of the 3-gluon vertex in (C.18). Each of the three gravitons, say the one carrying momentum p_1, can be created by any one of the three h's in $h\partial h\partial h$, and thus many terms are generated simply by permuting. The two derivatives give two powers of momentum. Thus, a typical term has the form $p_{1\beta}p_{2\mu}\eta_{\alpha\nu}\eta_{\sigma\gamma}$.

Keep working! In all, $G_{\mu\alpha, \nu\beta, \sigma\gamma}(p_1, p_2, p_3)$ contains about 100 terms. Now imagine calculating the one-loop contribution to graviton-graviton scattering. You get the point.

By now, you fully appreciate that the traditional Feynman approach carries an enormous amount of unnecessary off-shell information. Already, if we put p_1, p_2, and p_3 on shell and contract $G_{\mu\alpha, \nu\beta, \sigma\gamma}(p_1, p_2, p_3)$ with the polarization vectors $\epsilon_1^{\mu\alpha}$, $\epsilon_2^{\nu\beta}$, and $\epsilon_3^{\sigma\gamma}$, the 3-graviton vertex simplifies enormously to

$$G(p_1, p_2, p_3) = \epsilon_1^{\mu\alpha}\epsilon_2^{\nu\beta}\epsilon_3^{\sigma\gamma}(p_{1\sigma}\eta_{\mu\nu} + \text{cyclic})(p_{1\gamma}\eta_{\alpha\beta} + \text{cyclic}) \tag{26}$$

Quite naturally, we can write the polarization vector for a spin 2 massless particle in terms of the polarization vector for a spin 1 massless particle: $\epsilon^{\mu\alpha}(p) = \epsilon^\mu(p)\epsilon^\alpha(p)$. This form satisfies all that is required of a polarization vector for spin 2: $\epsilon^{\mu\alpha}(p)p_\mu = 0$, $\epsilon^{\mu\alpha}(p) = \epsilon^{\alpha\mu}(p)$, and $\eta_{\mu\alpha}\epsilon^{\mu\alpha}(p) = 0$. Thus, indeed, the 3-graviton vertex $G(p_1, p_2, p_3) = [\epsilon_1^\mu\epsilon_2^\nu\epsilon_3^\sigma(p_{1\sigma}\eta_{\mu\nu} + \text{cyclic})]^2$ is the square of the 3-gluon vertex (N.2.23), in confirmation of (20), which of course is just the same statement couched in another notation.

Exercises

N.3.1 Show that the structure of Lie algebra (21) emerges naturally.

N.3.2 In appendix 1 we recursed by complexifying the momenta of two external lines with helicity $+$ and $-$. In the derivation of the recursion relation (3) we could have picked any two external lines to complexify. Determine the amplitude calculated directly in chapter N.2, namely $A(1^-, 2^-, 3^+, 4^+)$, by complexifying lines 1 and 2. This is an example of the self-consistency argument sketched in the text.

Particle physics experimentalists are fond of saying that yesterday's spectacular discovery is today's calibration and tomorrow's annoying background. The canonical example is the Nobel-winning discovery of the CP-violating decay of the K_L meson into two pions. In theoretical physics, yesterday's discovery is today's homework exercise and tomorrow's trivium.

N.3.3 Using the explicit forms given for $A(1^-, 2^-, 3^+, 4^+)$ and $A(1^-, 2^+, 3^-, 4^+)$ in the preceding chapter, check the estimated large z behavior in (13–15).

N.3.4 Worry about the sloppy handling of factors of 2 in appendix 1. [Hint: The final result is correct because the polarization vectors in (5–6) are normalized to $|\varepsilon|^2 = 2$ for convenience.]

N.4 | Is Einstein Gravity Secretly the Square of Yang-Mills Theory?

Gravity and gauge theory

Quantum gravity has baffled generations of theoretical physicists, as you have no doubt heard. One aspect of this puzzle is the relationship between gravity and gauge theory, which describes the other fundamental interactions. While gravity and gauge theory are both born of local invariance, the Einstein-Hilbert action $\int d^4x \sqrt{-g}R$ and the Yang-Mills action $\int d^4x \operatorname{tr}(F_{\mu\nu}F^{\mu\nu})$ look completely different.

Perturbatively, gravity is afflicted with an infinite number of interaction terms, as was explained in chapter VIII.1, and hence gravity is not renormalizable, in stark contrast to gauge theory. On the other hand, the two field theories enjoy many conceptual similarities between them. Yang-Mills theory is the unique low energy effective theory of a spin 1 massless field, just as Einstein gravity is the unique low energy effective theory of a spin 2 massless field.

String theory unifies gravity and gauge theory. This remarkable fact alone points to a deep connection between gravity and gauge theory, even though within field theory the connection is totally obscure. One important clue is that the oscillator spectrum of the open string contains only the gauge field but not the graviton, which appears in the spectrum of the closed string. However, the closed string spectrum could be described as two copies of an open string spectrum, thus leading Kawai, Lewellen, and Tye to discover relations between graviton scattering and gauge boson scattering.[1] In the limit of the string energy scale going to infinity, we know that string theory reduces to field theory and thus a shadow of these KLT relations should survive in field theory. (As you might know, not all theorists are convinced that string theory corresponds to reality. If string theory eventually fails, its ultimate value might well turn out to be the light it sheds on the hidden structure of quantum field theory.)

---

[1] It is definitely beyond the scope of this book to explain these statements. See, for example, J. Polchinski, *String Theory*, p. 27.

In any case, the bottom line is that string theory strongly hints that graviton amplitudes can be expressed as products of Yang-Mills amplitudes, schematically $\mathcal{M}_{gravitons} \sim \mathcal{M}_{gauge} \times \mathcal{M}_{gauge}$. The first reaction of many theoretical physicists when first told this is puzzled skepticism. How is this possible, they ask quite reasonably, since Yang-Mills contains an internal symmetry group while gravity doesn't?

Now that we have learned to strip color, a connection between amplitudes no longer strikes us as so implausible, particularly if we stick to on-shell scattering amplitudes for gluons in specified polarization states $\mathcal{M}^{\lambda_1 \lambda_2 \cdots \lambda_n}$, namely amplitudes that experimentalists can measure, rather than amplitudes $\mathcal{M}^{\mu_1 \mu_2 \cdots \mu_n}$ carrying Lorentz indices that theorists using traditional methods play with. As we saw in the preceding chapter, the color stripped tree-level on-shell helicity amplitude for gauge boson scattering boils down to the $\langle \cdots \rangle$ and $[\cdots]$ products of two component spinors. We did not do the analogous calculation of the tree-level on-shell helicity amplitude for graviton scattering, but we could anticipate that the result would again be expressed in terms of the $\langle \cdots \rangle$ and $[\cdots]$ products. The spinor helicity formalism is intrinsic to the Lorentz group $SO(3, 1)$, not tied to a specific theory. In particular, the interaction vertices in Einstein gravity are again given in terms of scalar products of momenta and polarization vectors. Quite suggestively, the graviton polarization vectors can be written, as mentioned in appendix 1 to the preceding chapter, as $\epsilon^{\mu\nu} = \epsilon^{\mu} \epsilon^{\nu}$, a product of the gauge theory polarization vectors.

Indeed, I have already given part of the mystery away in the preceding chapter. We saw that the basic cubic interaction vertex of three gravitons (with complex momenta) is given by the square of the corresponding quantity for three gluons.

In summary, thanks to our string theory friends, we now know that there exists a secret structural connection between gravity and gauge theory that is totally opaque at the Lagrangian level.

Deformed graviton polarizations

In this closing chapter, I give a brief introduction to the exciting quest for this secret connection. I will be content to look at one specific calculation.

Go back to the BCFW recursion (chapter N.3). It would work for gravity if the complexified scattering amplitude $\mathcal{M}(z)$ vanishes as $z \to \infty$. But naively, it would seem that the situation for gravity is even worse than the situation for gauge theory, since the cubic graviton vertex is quadratic in momentum and thus goes like z^2. (Recall the two powers of derivative in the scalar curvature; see chapter VIII.1.) Repeat the calculation in the preceding chapter for n-graviton on shell scattering. Go back to figure N.3.2 and interpret the lines as gravitons. The $(n - 2)$ cubic vertices give a factor of $z^{2(n-2)}$ for large z, easily overwhelming the factor of $1/z^{n-3}$ from the $(n - 3)$ propagators. This nasty behavior occurs even before we include the polarization of the two hard gravitons.

The graviton carries helicity ± 2 (appendix 2 of chapter VIII.1) and hence a polarization "vector" $\epsilon^{\mu\nu}$, given by a symmetric and traceless tensor. We can naturally construct $\epsilon^{\mu\nu} =$

$\epsilon^\mu \epsilon^\nu$ out of the polarization vectors for a massless spin 1 particle (as already explained in the preceding chapter). Thus, after deformation,

$$\epsilon_r^{++\mu\nu}(z) = \epsilon_r^{+\mu}\epsilon_r^{+\nu} = (q^* + zp_s)^\mu(q^* + zp_s)^\nu, \quad \epsilon_r^{--\mu\nu}(z) = \epsilon_r^{-\mu}\epsilon_r^{-\nu} = q^\mu q^\nu \tag{1}$$

and

$$\epsilon_s^{++\mu\nu}(z) = \epsilon_s^{+\mu}\epsilon_s^{+\nu} = q^\mu q^\nu, \quad \epsilon_s^{--\mu\nu}(z) = \epsilon_s^{-\mu}\epsilon_s^{-\nu} = (q^* - zp_r)^\mu(q^* - zp_r)^\nu \tag{2}$$

Note that the $\epsilon^{\mu\nu}(z)$'s are in fact traceless and could go as either z^0 or z^2 for large z.

Putting it together, we obtain the naive estimate

$$\mathcal{M}_{\text{naive}}^{--,++}(z) \to \frac{z^{2(n-2)}}{z^{n-3}} = z^{n-1}, \quad \mathcal{M}_{\text{naive}}^{--,-- \text{ or } ++,++}(z) \to z^{n+1}, \quad \mathcal{M}_{\text{naive}}^{++,--}(z) \to z^{n+3} \tag{3}$$

The escalating behavior as n increases is the hallmark of a nonrenormalizable theory, as explained in chapter III.2.

Hard graviton in a soft spacetime

Once again, we hope that real life is cushier than naive expectation. By the same reasoning used for gauge theory, we study a hard graviton blasting through a gravitational field, that is, a background of soft gravitons. So write the metric of spacetime as $\mathcal{G}_{\mu\nu} = g_{\mu\nu} + h_{\mu\nu}$. Plug this into the Einstein-Hilbert action (VIII.1.1) and extract the terms quadratic in h. While the calculation is straightforward, it does involve some heavy lifting. To avoid the labor, we note that using the harmonic gauge, we did this calculation in (VIII.1.10) but only for the special case $g_{\mu\nu} = \eta_{\mu\nu}$ (in other words, we expanded around flat Minkowski spacetime rather than a general curved spacetime). We had

$$\mathcal{L} = \frac{1}{64\pi G}(\eta^{\mu\nu}\eta^{\lambda\rho}\eta^{\sigma\tau}\partial_\mu h_{\lambda\sigma}\partial_\nu h_{\rho\tau} - \frac{1}{2}\eta^{\mu\nu}\partial_\mu h \partial_\nu h) \tag{4}$$

with the trace degree of freedom $h \equiv \eta^{\mu\nu}h_{\mu\nu}$. Henceforth, we set $64\pi G = 1$.

Armed with symmetry considerations and our knowledge of gravity (chapter VIII.1), we can almost immediately guess that when we go from a flat $\eta_{\mu\nu}$ to a curved $g_{\mu\nu}$ background, this quadratic Lagrangian generalizes to

$$\mathcal{L} = \sqrt{-g}(g^{\mu\nu}g^{\lambda\rho}g^{\sigma\tau}D_\mu h_{\lambda\sigma}D_\nu h_{\rho\tau} - \frac{1}{2}g^{\mu\nu}D_\mu h D_\nu h - 2R^{\lambda\rho\sigma\tau}h_{\lambda\sigma}h_{\rho\tau}) \tag{5}$$

with h now defined as $h \equiv g^{\mu\nu}h_{\mu\nu}$. Here D denotes the covariant derivative with respect to the curved metric $g_{\mu\nu}$ introduced in chapter VIII.1 and $R^{\lambda\rho\sigma\tau}$ the Riemann curvature tensor constructed out of $g_{\mu\nu}$. I trust you not to confuse this D associated with the curved background with the covariant derivative in Yang-Mills theory used in the preceding chapter and mentioned below in passing.

Let us go through the various features of (5). The $\sqrt{-g}$ goes with the spacetime volume and is common to any Lagrangian in curved spacetime, as we learned way back in (I.11.2). We also learned there to promote any Lagrangian from flat to curved spacetime by replacing $\eta_{\mu\nu}$ with $g_{\mu\nu}$ and the ordinary derivative by a covariant derivative (see chapter IV.5). As you can see, everything pretty much works out in parallel with how things work out for gauge theory. The new feature is the term involving the Riemann curvature tensor $R^{\lambda\rho\sigma\tau}$, which vanishes upon restriction to flat spacetime. But you are not surprised that such a term could pop up, given $\operatorname{tr} F^{\mu\nu}[a_\mu, a_\nu]$ in (N.3.8). Indeed, the only thing we can't determine without doing the actual calculation is the numerical coefficient (-2) of this term. That particular number will play no role in the following discussion.

Also, in the preceding chapter we dropped a term linear in a_μ because of the equation of motion $D_\mu F^{\mu\nu} = 0$. Here, analogously, we dropped a term linear in $h_{\mu\nu}$ because of Einstein's equation of motion $R_{\mu\nu} = 0$. This also explains why terms involving the Ricci tensor $R_{\mu\nu}$ and the scalar curvature R do not appear in (5).

We want to calculate the large-z behavior of various scattering amplitudes $\mathcal{M}^{--,++}$, etc., and compare with the naive expectation (3). We hope that the same trick we used for the gauge theory case would also work for gravity. Now the string theory hint, that $\mathcal{M}_{\text{gravitons}} \sim \mathcal{M}_{\text{gauge}} \times \mathcal{M}_{\text{gauge}}$, suggests a factorized structure in the graviton amplitude, and so more or less naturally leads to the guess that the first index λ and the second index σ of $h_{\lambda\sigma}$ are somehow associated respectively with the two copies of $\mathcal{M}_{\text{gauge}}$.

The key to breaking the problem apart is the Bern transformation unlinking these two indices. Examining (5), we see that the only term that links the first index with the second index of $h_{\lambda\sigma}$ appears in the term $g^{\mu\nu} D_\mu h D_\nu h$, since $h = g^{\mu\nu} h_{\mu\nu}$ does precisely that. How to get rid of this term? The trick, following Bern and Grant, is to introduce a scalar field ϕ and add the term $2g^{\mu\nu}\partial_\mu\phi\partial_\nu\phi$. We are allowed to do this, since ϕ does not appear in the tree level graviton scattering amplitude we are studying. (Of course, the theory is changed from pure gravity, and ϕ does circulate in loop diagrams for graviton scattering. Some readers may also know that in string theory the graviton appears with a scalar ϕ, the experimentally unobserved dilaton.)

For pedagogical clarity in explaining what we are going to do next, it is best to retreat to the case of the flat background. Focus on the parenthesis in (4), now modified to $(\partial_\mu h_{\lambda\sigma}\partial^\mu h^{\lambda\sigma} - \frac{1}{2}\partial_\mu h\partial^\mu h + 2\partial_\mu\phi\partial^\mu\phi)$. (The normalization of ϕ is, in this context, just chosen for convenience.) Since we can always make a field redefinition (see the appendix to chapter VIII.3) without affecting on-shell scattering amplitudes, we let $h_{\lambda\sigma} \to h_{\lambda\sigma} + \eta_{\lambda\sigma}\phi$ (and hence $h \to h + 4\phi$) and $\phi \to \phi + \frac{1}{2}h$. You can verify that our parenthesis changes to $(\partial_\mu h_{\lambda\sigma}\partial^\mu h^{\lambda\sigma} - 2\partial_\mu\phi\partial^\mu\phi)$. Since in this manipulation the role of $\eta_{\lambda\sigma}$ is merely to convert $h_{\lambda\sigma}$ into h, the same transformation works when $\eta_{\lambda\sigma}$ is promoted to $g_{\lambda\sigma}$.

The upshot is that we can effectively rewrite (5) as

$$\mathcal{L} = \sqrt{-g}\,(g^{\mu\nu}g^{\lambda\rho}g^{\sigma\tau}D_\mu h_{\lambda\sigma}D_\nu h_{\rho\tau} - 2R^{\lambda\rho\sigma\tau}h_{\lambda\sigma}h_{\rho\tau}) \tag{6}$$

Now that ϕ has done his job, we have unceremoniously thrown him out since he doesn't contribute to the on-shell tree amplitudes we are interested in. We have thus dropped the term $g^{\mu\nu}\partial_\mu\phi\partial_\mu\phi$.

There has been quite a bit of formal development and perhaps the reader has lost sight of what we are trying to do. Recall that we want to study the amplitude of a hard graviton blasting through spacetime. Although a multitude of indices have appeared, as is always the case with gravity, you should recognize that this Lagrangian is conceptually simple: it is quadratic in the quantum field h describing the hard graviton and contains some given c-number tensors $g^{\lambda\rho}(x)$ and $R^{\lambda\rho\sigma\tau}(x)$ pertaining to the background.

Unlinked melody

The important point is that the two indices carried by $h_{\lambda\sigma}$ are now unlinked from each other in the first term in (6). In chapter VIII.1, we learned to trade a "world index" like λ for a locally flat Lorentz index a by using the vierbein $e^a_\lambda(x)$. Here we are invited to introduce two sets of vierbein, e and $\tilde{e}$, with their associated connections ω and $\tilde{\omega}$, and write $h_{\lambda\sigma} \equiv e^a_\lambda \tilde{e}^{\tilde{a}}_\sigma h_{a\tilde{a}}$. In reality, of course $e = \tilde{e}$ and $\omega = \tilde{\omega}$, but this notation keeps track of the fact that the two sets of indices carried by $h_{\lambda\sigma}$ are unlinked. Note that $h_{\lambda\sigma}$ is treated in our quadratic Lagrangian as just some tensor field living in a curved spacetime specified by $g_{\lambda\sigma} \equiv e^a_\lambda \tilde{e}^{\tilde{a}}_\sigma \eta_{a\tilde{a}}$.

Also in chapter VIII.1 we emphasized that the covariant derivative acting on vectors carrying a world index and on vectors carrying a locally flat Lorentz index assumes different forms, $D_\mu V_\nu = \partial_\mu V_\nu - \Gamma^\lambda_{\mu\nu} V_\lambda$ and $\mathcal{D}_\mu V_a = \partial_\mu V_a - \omega^b_{\mu a} V_b$, respectively. For pedagogical clarity, I will use two different symbols D and $\mathcal{D}$ to denote what is conceptually the same operation. For our problem we have $D_\lambda h_{\mu\nu} = e^a_\mu \tilde{e}^{\tilde{a}}_\nu \mathcal{D}_\lambda h_{a\tilde{a}}$, with $\mathcal{D}_\lambda h_{a\tilde{a}} = \partial_\lambda h_{a\tilde{a}} - \omega^b_{\lambda a} h_{b\tilde{a}} - \tilde{\omega}^{\tilde{b}}_{\lambda\tilde{a}} h_{a\tilde{b}}$.

With this notation, the relevant Lagrangian becomes

$$\mathcal{L} = \sqrt{-g}(g^{\mu\nu}\eta^{ab}\tilde{\eta}^{\tilde{a}\tilde{b}}\mathcal{D}_\mu h_{a\tilde{a}}\mathcal{D}_\nu h_{b\tilde{b}} - 2R^{ab\tilde{a}\tilde{b}}h_{a\tilde{a}}h_{b\tilde{b}}) \tag{7}$$

We are now ready to study the large z behavior of the scattering amplitude of a hard graviton carrying momentum $zq + \cdots$ blasting through a curved background spacetime $g_{\mu\nu}$.

The analysis proceeds much as in the Yang-Mills case discussed in the preceding chapter. Focus on the first term: $\sqrt{-g}g^{\mu\nu}\eta^{ab}\tilde{\eta}^{\tilde{a}\tilde{b}}(\partial_\mu h_{a\tilde{a}} - \omega^c_{\mu a}h_{c\tilde{a}} - \tilde{\omega}^{\tilde{c}}_{\mu\tilde{a}}h_{a\tilde{c}})(\partial_\nu h_{b\tilde{b}} - \omega^d_{\nu b}h_{d\tilde{b}} - \tilde{\omega}^{\tilde{d}}_{\nu\tilde{b}}h_{b\tilde{d}})$. The leading $O(z^2)$ behavior comes from the piece containing two derivatives in the first term, namely $\mathcal{L}_{\text{lead}} \equiv \sqrt{-g}g^{\mu\nu}\eta^{ab}\tilde{\eta}^{\tilde{a}\tilde{b}}\partial_\mu h_{a\tilde{a}}\partial_\nu h_{b\tilde{b}}$, and thus contributes to the amplitude a term proportional to $\eta^{ab}\tilde{\eta}^{\tilde{a}\tilde{b}}$. In the Yang-Mills case, the Lagrangian contains a hidden "enhanced Lorentz" symmetry. Here the situation is even better: we have not one, but two hidden "enhanced Lorentz" symmetries. The term $\mathcal{L}_{\text{lead}}$ is evidently left invariant by two separate $SO(3,1)$ Lorentz transformations, one operating on the a, b indices, the other on the $\tilde{a}, \tilde{b}$ indices.

The subleading $O(z)$ behavior comes from the pieces in the first term containing one derivative and one factor of either ω or $\tilde{\omega}$, for example $\sqrt{-g}g^{\mu\nu}\eta^{ab}\tilde{\eta}^{\tilde{a}\tilde{b}}\partial_\mu h_{a\tilde{a}}(\omega^d_{\nu b}h_{d\tilde{b}} + \tilde{\omega}^{\tilde{d}}_{\nu\tilde{b}}h_{b\tilde{d}})$. In this way we find that $\mathcal{M}^{ab,\tilde{a}\tilde{b}} \to cz^2\eta^{ab}\tilde{\eta}^{\tilde{a}\tilde{b}} + z(\eta^{ab}\tilde{A}^{\tilde{a}\tilde{b}} + A^{ab}\tilde{\eta}^{\tilde{a}\tilde{b}}) + \cdots$, with A^{ab} and $\tilde{A}^{\tilde{a}\tilde{b}}$ two matrices antisymmetric in their indices. To see this, consider for example the piece involving ω (after some relabeling of indices): $\sqrt{-g}g^{\mu\nu}\eta^{ac}\tilde{\eta}^{\tilde{a}\tilde{b}}(\partial_\mu h_{a\tilde{a}})\omega^b_{\nu c}h_{b\tilde{b}}$. This gives rise to the term $A^{ab}\tilde{\eta}^{\tilde{a}\tilde{b}}$ in $\mathcal{M}^{ab,\tilde{a}\tilde{b}}$. Note that since the matrix A^{ab} depends on the spin connection ω^{ab}_ν of the background, all we can say is that it is antisymmetric in its two indices $a\,b$. Recall that this is quite analogous to what we did in the Yang-Mills case.

Before reading further, you could now flex your mental muscle and push ahead to obtain the sub-subleading $O(z^0)$ behavior. This comes from the pieces in the first term containing two factors of ω and $\tilde\omega$, for example (again, after some relabeling of indices) $\sqrt{-g}\, g^{\mu\nu} \eta^{cd} \tilde\eta^{\tilde a \tilde b} \omega^a_{\mu c} h_{a\tilde a} \omega^b_{vd} h_{b\tilde b}$. All we can now conclude is that this contributes to the scattering amplitude a term of the form $B^{ab} \tilde\eta^{\tilde a \tilde b}$, with B^{ab} an arbitrary matrix. To this order, the second term in (7) also contributes, breaking the "enhanced Lorentz" symmetries completely. Nevertheless, we can still exploit the known symmetry properties of the Riemann curvature tensor under interchange of its indices to say something about its contribution to the scattering amplitude. Putting it all together we conclude that

$$\mathcal{M}^{ab,\tilde a \tilde b} \to cz^2 \eta^{ab}\tilde\eta^{\tilde a \tilde b} + z(\eta^{ab}\tilde A^{\tilde a \tilde b} + A^{ab}\tilde\eta^{\tilde a \tilde b}) + A^{ab\tilde a\tilde b} + (\eta^{ab}\tilde B^{\tilde a \tilde b} + B^{ab}\tilde\eta^{\tilde a \tilde b}) + O\left(\frac{1}{z}\right) \tag{8}$$

Compare this with (N.3.12), which states that the amplitude for the scattering of a hard gluon off a background of soft gluons goes like $\mathcal{M}^{ab} = (cz + \cdots)\eta^{ab} + A^{ab} + \frac{1}{z}B^{ab} + \cdots$. Amazingly, you can see that the large-z behavior for the scattering of a hard graviton off a background of soft gravitons can be obtained by "squaring" the large-z behavior for the scattering of a hard gluon off a background of soft gluons! In other words, $\mathcal{M}^{ab,\tilde a \tilde b} \sim \mathcal{M}^{ab}\mathcal{M}^{\tilde a \tilde b}$, as far as the large-$z$ behavior is concerned.

Just as in the Yang-Mills case, by exploiting gauge identities like $p_{ra}(z)\mathcal{M}^{a\tilde a, b\tilde b}\epsilon_{sb\tilde b}(z) = 0$, we can determine the large-z behavior of various helicity amplitudes. For your convenience I remind you that (from the preceding chapter) $p_r(z) = p_r + zq$ and $p_s(z) = p_s - zq$. Thus the gauge identity just displayed says that $q_a\mathcal{M}^{a\tilde a, b\tilde b}\epsilon_{sb\tilde b}(z) = -(1/z)p_{ra}\mathcal{M}^{a\tilde a, b\tilde b}\epsilon_{sb\tilde b}(z)$. Recalling [see (1)] that $\epsilon_r^{--\mu\nu}(z) = q^\mu q^\nu$ we see that in calculating the amplitude $\mathcal{M}^{--,h}(z)$ we can effectively replace $\epsilon_r^{--\mu\nu}(z)$ by $(1/z^2)p_r^\mu p_r^\nu$. Thus we can immediately conclude, since $\epsilon_s^{++\mu\nu}(z) \sim z^0$ for large z [see (2)], that for example

$$\mathcal{M}^{--,++}(z) \to \frac{1}{z^2} \tag{9}$$

which is far better than the naive expectation $\mathcal{M}_{\text{naive}}^{--,++}(z) \to z^{n-1}$. Indeed, the horrible ever-escalating behavior with increasing n has disappeared. Even more remarkably, the large-z behavior of graviton scattering amplitudes is consistent with the string-inspired notion that gravity is "the square of Yang-Mills." Recall that in gauge theory $\mathcal{M}^{-+}(z) \to 1/z$. Thus, for large z, indeed $\mathcal{M}^{--,++}(z) \sim (\mathcal{M}^{-+}(z))^2$.

The bottom line here is that the large-z behavior of gravity is surprisingly benign and vanishes fast enough for the recursion program to work.

Gravity is a square?

So, is Einstein gravity secretly the square of Yang-Mills theory?

Already we have seen in the preceding chapter, anticipating that the recursion program works, that the primitive 3-point amplitude for gravity (for one helicity configuration) $(\langle 12\rangle^3/\langle 23\rangle\langle 31\rangle)^2$ is the square of the primitive 3-point amplitude for gauge theory

($\langle 12 \rangle^3 / \langle 23 \rangle \langle 31 \rangle$), something that you could have never suspected by staring at $\sqrt{-g} R$ and $\text{tr} F_{\mu\nu} F^{\mu\nu}$ untill you are blue in the face.

The calculations in the previous section show that the large-z behavior for the scattering of a hard graviton off a background of soft gravitons could be obtained by "squaring" the large z behavior for the scattering of a hard gluon off a background of soft gluons, certainly something that nobody could have anticipated by looking at Lagrangians.

Further evidence that the answer to the title of this chapter is "yes" comes from a recent calculation by Bern, Carrasco, and Johansson. Interestingly, they do not strip the color from a Yang-Mills theory, but instead show that they can write the "color-dressed" tree amplitudes in the form

$$\mathcal{A}^{\text{tree}}(1, 2, \cdots, n) = \sum_a \frac{n_a c_a}{(\Pi_j p_j^2)_a} \tag{10}$$

It is beyond the scope of this book to explain in detail how this expression is obtained. I merely state that the index a labels an individual diagram. For each diagram, the amplitude may be written as the product of a kinematic function n_a of the momenta and a color factor c_a, divided by the product of the momenta p_j carried by the internal lines. (I do not explain here how n_a and c_a are defined.) The tree amplitude is then given by a sum over all tree diagrams.

Bern et al. then conjecture that the n-graviton scattering amplitude at the tree level is given, amazingly, by

$$\mathcal{M}^{\text{tree}}_{\text{gravity}}(1, 2, \cdots, n) = \sum_a \frac{n_a n_a}{(\Pi_j p_j^2)_a} \tag{11}$$

They have checked by explicit computation that their conjecture in fact holds up to $n = 8$. Furthermore, they have also verified that the conjecture, suitably generalized, also holds for the various supercousins of Einstein gravity and Yang-Mills theory.

Thus the evidence is extremely strong that, yes indeed, Einstein gravity is secretly the square of Yang-Mills theory, at least at the level of tree amplitudes. However, as of this writing (February 2009), there is no definitive understanding within field theory. The final word on the subject has yet to be said, and it is not even clear what the final path to the final word might be. I would be foolish indeed to discuss this further in a textbook when the entire subject is being rapidly developed. By the time this book is published, the conjecture that Einstein gravity is the square of Yang-Mills theory may well have been proved. If not, then nothing would please me more than if a reader of this textbook could go on and prove it, hopefully not just at the tree level, but to all orders.

What is the simplest field theory?

The uninitiated would likely answer φ^4 theory. Indeed, field theory texts almost all start with some kind of scalar field theory. Even I am not able to do any better. But the sophisticated, namely you, now that you have reached the end of this text, realize that the more symmetry the theory has the better. To theoretical physicists, simplicity actually secretly means symmetry. Incidentally, I have always hated scalar field theories, and have ventured to say so publicly. It is hard to like the action $\mathcal{L} = \frac{1}{2}(\partial\varphi)^2 - \lambda\varphi^4$, so barren of color and flavor. Some of the major problems facing particle physics, such as the hierarchy problem, may eventually turn out to stem from our not having mastered scalar field theory.

Of course, scalar field theory is the simplest in the superficial sense that you need to know the least to approach it. As I said in chapter I.12, once one is familiar with scalar field theory the rest consists of "merely" decorating the field with various indices describing spacetime or internal symmetries. But the symmetry and the resulting structure provide us with handles to grab on to. Both Yang-Mills theory and Einstein gravity have an internal logic sorely lacking in scalar field theory. As I mentioned in chapters VII.3 and VIII.4, the consensus view is that the first exactly soluble field theory would almost certainly be $\mathcal{N} = 4$ supersymmetric Yang-Mills theory, the supercousin of pure Yang-Mills theory. The remarkable recent developments described in the last three chapters have only reinforced this view. Almost beyond belief, even gravity may be simpler than we had long thought. For large complexified momentum, graviton scattering for some helicity arrangements actually behaves better than gluon scattering. The evidence is mounting that Einstein gravity may in some sense be the square of Yang-Mills theory. So now we are left with the amusing thought that the simplest field theory may well end up being gravity or $\mathcal{N} = 8$ supergravity with its maximal supersymmetry. (At this point, a friend of mine who works with $\mathcal{N} = 8$ supergravity pipes up, "It sure doesn't look simpler if you are the guy doing the calculation!" It is clear from the simple dimensional argument of chapter III.2 that as one goes to higher order, the numerator of the Feynman integrand quickly becomes extremely involved.)

Only time will tell who will win the simplest field theory contest, but we do have two convincing candidates.

More Closing Words

In the closing words to the first edition of this book, I wrote that Yang-Mills theory was almost begging for a better notation that would lay bare the deeper structure of the theory. Oy, the excess baggage we have to carry! Ten thousand terms instead of one. In some respects, the spinor helicity formalism and the recursion program explained in chapters N.2–N.4 provide a partial answer to that pious wish.

Imagine some theorist idly wondering, after 1865, if there were a better notation to describe the six fields E_x, E_y, E_z, B_x, B_y, B_z for which Maxwell had written 20 equations (since he did not use vector notation). We can even fantasize that by fooling around with numerology ("Look, $4 \cdot 3/2 = 6$!"), this "crackpot" came up with an antisymmetric 4 by 4 matrix he called F. Shoehorning Maxwell's equations in vacuum (some of them stating that the time variation of E and B is related to the space variation of E and B) into this strange notation, this guy could even stumble on a secret connection between space and time.

The spinor helicity formalism and the recursion program, though elegant, are still rooted in the perturbative expansion of the 1940s. Can they be pushed into the nonperturbative regime? There have been attempts in that direction.

In all previous revolutions in physics, a formerly cherished concept has to be jettisoned. If we are poised before another conceptual shift, something else might have to go. Lorentz invariance, perhaps? More likely, we may have to abandon strict locality. Again, in closing words I mumble something (from steepest descent to integral to what?) about modifying the form of the path integral. The recursion program and the resuscitated S-matrix approach might be a step in this direction, formulating field theory while avoiding mention of a local Lagrangian. But we need analyticity, and of course analyticity follows from locality and causality, as far as we understand. We know also that even local field theory could spawn non-local constructs, most notoriously the horizon of a black hole. But there the dynamics bends the causal structure of spacetime out of whack. The lack of strict locality is not built into the laws of physics.

Of course, we also know how to imbue physics with non-locality right from the start. We have Wilson's lattice formulation of gauge theory, and more recently, Wen's intriguing lattice formulation of gravity.

When I showed the last three chapters to our friend SE, she mused, after some reflection, "Now I see what theorists could always do when in doubt: enhance the symmetry and make it local, complexify and bow to Cauchy, and take a square root when possible!"

I nodded, "These are the three ways of the warrior theorist: I call them the Einstein way, the Heisenberg way, and the Dirac way. They were wildly successful in the past, and perhaps they will work in the future as well."

With this edition of my textbook, I can no doubt count on a new group of readers to come up with fresh insights into field theory. As these new chapters suggest, there may still be plenty of secret structures to uncover. And thus field theory marches on.

Finally, I reveal the origin of the quote at the start of the preface to the second edition. As a kid, Feynman came across a calculus book[1] that proclaimed "What one fool can do, another can." He was thus inspired to master calculus. Now that *you* have mastered quantum field theory, you can switch from the "understand" in the preface to the "do" in these closing words.

---

[1] Silvanus P. Thompson (1851–1916), *Calculus Made Easy*, 1910, updated by Martin Gardner, St. Martin's Press, (1998). I am kind of trying to do for quantum field theory what Thompson did for calculus.

The basic Gaussian:

$$\int_{-\infty}^{+\infty} dx\, e^{-\frac{1}{2}x^2} = \sqrt{2\pi} \tag{1}$$

The scaled Gaussian:

$$\int_{-\infty}^{+\infty} dx\, e^{-\frac{1}{2}ax^2} = \left(\frac{2\pi}{a}\right)^{\frac{1}{2}} \tag{2}$$

Moments:

$$\int_{-\infty}^{+\infty} dx\, e^{-\frac{1}{2}ax^2} x^{2n} = \left(\frac{2\pi}{a}\right)^{\frac{1}{2}} \frac{1}{a^n}(2n-1)(2n-3)\cdots 5\cdot 3\cdot 1, n \geq 1 \tag{3}$$

Gaussian with source:

$$\int_{-\infty}^{+\infty} dx\, e^{-\frac{1}{2}ax^2+Jx} = \left(\frac{2\pi}{a}\right)^{\frac{1}{2}} e^{J^2/2a} \tag{4}$$

$$\int_{-\infty}^{+\infty} dx\, e^{-\frac{1}{2}ax^2+iJx} = \left(\frac{2\pi}{a}\right)^{\frac{1}{2}} e^{-J^2/2a} \tag{5}$$

$$\int_{-\infty}^{+\infty} dx\, e^{\frac{1}{2}iax^2+iJx} = \left(\frac{2\pi i}{a}\right)^{\frac{1}{2}} e^{-iJ^2/2a} \tag{6}$$

$$\int_{-\infty}^{+\infty} \int_{-\infty}^{+\infty} \cdots \int_{-\infty}^{+\infty} dx_1 dx_2 \cdots dx_N\, e^{\frac{i}{2}x\cdot A\cdot x+iJ\cdot x} = \left(\frac{(2\pi i)^N}{\det[A]}\right)^{\frac{1}{2}} e^{-(i/2)J\cdot A^{-1}\cdot J} \tag{7}$$

$$\int_{-\infty}^{+\infty} \int_{-\infty}^{+\infty} \cdots \int_{-\infty}^{+\infty} dx_1 dx_2 \cdots dx_N\, e^{-\frac{1}{2}x\cdot A\cdot x+J\cdot x} = \left(\frac{(2\pi)^N}{\det[A]}\right)^{\frac{1}{2}} e^{\frac{1}{2}J\cdot A^{-1}\cdot J} \tag{8}$$

In what follows, we omit an overall factor.
Central identity of quantum field theory:

$$\int D\varphi\, e^{-\frac{1}{2}\varphi\cdot K\cdot\varphi-V(\varphi)+J\cdot\varphi} = e^{-V(\delta/\delta J)} e^{\frac{1}{2}J\cdot K^{-1}\cdot J} \tag{9}$$

A trivial variation:

$$\int D\varphi e^{-\frac{1}{2}\varphi \cdot K \cdot \varphi + J \cdot \varphi} = e^{\frac{1}{2}J \cdot K^{-1} \cdot J} \tag{10}$$

Variations:

$$\int D\varphi e^{(i/2)\varphi \cdot K \cdot \varphi + i J \cdot \varphi} = e^{-(i/2)J \cdot K^{-1} \cdot J} \tag{11}$$

$$\int D\varphi e^{i \int d^d x [\frac{1}{2}\varphi(x) K \varphi(x) + J(x)\varphi(x)]} = e^{i \int d^d x [-\frac{1}{2}J(x) K^{-1} J(x)]} \tag{12}$$

$$\int D\varphi e^{-\int d^d x [\frac{1}{2}\varphi(x) K \varphi(x) + J(x)\varphi(x)]} = e^{\int d^d x [\frac{1}{2}J(x) K^{-1} J(x)]} \tag{13}$$

(where K or K^{-1} or both may be nonlocal)
   A specific example:

$$\int D\varphi e^{i \int d^d x [(\lambda/2)\varphi^2 + \varphi \bar{\psi}\psi]} = e^{i \int d^d x [-(1/2\lambda)(\bar{\psi}\psi)^2]} \tag{14}$$

For K hermitean with φ complex:

$$\int D\varphi^\dagger D\varphi e^{-\varphi^\dagger \cdot K \cdot \varphi + J^\dagger \cdot \varphi + \varphi^\dagger \cdot J} = e^{J^\dagger \cdot K^{-1} \cdot J} \tag{15}$$

As noted earlier, various numerical factors have been swept under the integration measure. In applying these formulas, be sure that these factors are not relevant for your purposes.

Appendix B | A Brief Review of Group Theory

I give here a brief review of the group theory I will need in the text. I assume that you have been exposed to some group theory, otherwise this instant review might not be intelligible. Most of the concepts are illustrated with examples, and it goes without saying that you should work out all the examples and verify the assertions made without proof.

SO(N)

The special orthogonal group $SO(N)$ consists of all N by N real matrices O that are orthogonal

$$O^T O = 1 \tag{1}$$

and have unit determinant

$$\det O = 1 \tag{2}$$

We denote the element in the ith row and jth column by O^{ij}. The group $SO(N)$ consists of rotations in N-dimensional Euclidean space and its defining or fundamental representation is given by the N component vector $\vec{v} = \{v^j, j = 1, \ldots, N\}$, which transforms under the action of the group element O according to (as always, all repeated indices are summed over)

$$v^i \rightarrow v'^i = O^{ij} v^j \tag{3}$$

We define tensors as objects that transform as if they are equal to the product of vectors. For example, the tensor T^{ijk} transforms according to

$$T^{ijk} \rightarrow T'^{ijk} = O^{il} O^{jm} O^{kn} T^{lmn} \tag{4}$$

as if it is equal to the product $v^i v^j v^k$. The emphasis is on the phrase "as if": T^{ijk} is not to be thought of as being equal to $v^i v^j v^k$.

It is important to develop some "feel" or intuition for groups and their representations. Some people find it helpful to picture a certain number of objects being acted upon by the group and transformed into linear combinations of each other. Thus, picture T^{ijk} as N^3 objects being scrambled together.

Tensors furnish representations of the group. In our particular example, each group element is represented by an N^3 by N^3 matrix acting on the N^3 objects T^{ijk}. The number of objects in a tensor is called the dimension of the representation.

It may well be that any given object in a representation does not transform, under all the elements of the group, into a linear combination of all the other objects, but only into a subset of them. Let me illustrate with an example.

Consider $T^{ij} \to T'^{ij} = O^{il} O^{jm} T^{lm}$. Form the symmetric $S^{ij} \equiv \frac{1}{2}(T^{ij} + T^{ji})$ and antisymmetric combinations $A^{ij} \equiv \frac{1}{2}(T^{ij} - T^{ji})$. The symmetric combination S^{ij} transforms into $O^{il} O^{jm} S^{lm}$, which is obviously symmetric. Similarly, A^{ij} transforms into $O^{il} O^{jm} A^{lm}$, which is obviously antisymmetric. In other words, the set of N^2 objects contained in T^{ij} split into two sets: $\frac{1}{2}N(N+1)$ objects contained in S^{ij} and $\frac{1}{2}N(N-1)$ objects contained in A^{ij}. The S^{ij}'s transform among themselves and the A^{ij}'s transform among themselves.

The representation furnished by T^{ij} is said to be reducible: It breaks apart into two representations. Obviously, representations that do not break apart are called irreducible.

We just exploited the obvious fact that the symmetry properties of a tensor under permutation of its indices is not changed by the group transformation, namely that the indices on a tensor transform independently, as in (4). The various possible symmetry properties may be classified with Young tableaux, which is useful in a general treatment of group theory. Fortunately, in the field theory literature one rarely encounters a tensor with such complex symmetry properties that one has to learn about Young tableaux.

Another way of saying this is that we can restrict our attention to tensors with definite symmetry properties under permutation of their indices. In our specific example, we can always take T^{ij} to be either symmetric or antisymmetric under the exchange of i and j.

We have yet to use the properties (1) and (2). Given a symmetric tensor T^{ij} consider the combination $T \equiv \delta^{ij} T^{ij}$, known as the trace. Then $T \to \delta^{ij} T'^{ij} = \delta^{ij} O^{il} O^{jm} T^{lm} = (O^T)^{li} \delta^{ij} O^{jm} T^{lm} = \delta^{lm} T^{lm} = T$, where we used (1). In other words, T transforms into itself. We can subtract the trace from T^{ij} forming the traceless tensor $Q^{ij} \equiv T^{ij} - (1/N)\delta^{ij} T$. The $\frac{1}{2}N(N+1) - 1$ objects contained in Q^{ij} transform among themselves.

To summarize, given two vectors v and w, we can form a tensor, and decompose the tensor into a symmetric traceless combination, a trace, and an antisymmetric tensor. This process is written as

$$N \otimes N = [\tfrac{1}{2}N(N+1) - 1] \oplus 1 \oplus \tfrac{1}{2}N(N-1) \tag{5}$$

In particular, for $SO(3)$, $3 \otimes 3 = 5 \oplus 1 \oplus 3$, a relation you should be familiar with from courses on mechanics and electromagnetism.

There are two conventions for naming representations. We can simply give the dimension of the representation. (This can occasionally be ambiguous: Two distinct representations may happen to have the same dimension.) Alternatively, we can specify the symmetry properties of the tensor furnishing the representation. For instance, the representation furnished by a totally antisymmetric tensor of n indices is often denoted by $[n]$ and the representation furnished by a totally symmetric traceless tensor of n indices by $\{n\}$. Obviously, $[1] = \{1\}$. In this notation, the decomposition in (5) can be written as $\{1\} \otimes \{1\} = \{2\} \oplus \{0\} \oplus [2]$. For the group $SO(3)$, with its long standing in physics, the confusion over names is almost worse than in reading Russian novels: For instance, $\{1\}$ is also known as p and $\{2\}$ as d.

We have yet to use (2). Using the antisymmetric symbol $\varepsilon^{123\ldots N}$, we write (2) as

$$\varepsilon^{i_1 i_2 \ldots i_N} O^{i_1 1} O^{i_2 2} \ldots O^{i_N N} = 1 \tag{6}$$

or equivalently

$$\varepsilon^{i_1 i_2 \ldots i_N} O^{i_1 j_1} O^{i_2 j_2} \ldots O^{i_N j_N} = \varepsilon^{j_1 j_2 \ldots j_N} \tag{7}$$

By multiplying (7) by O^T repeatedly, we can obviously generate more identities. Instead of drowning in a sea of indices, let me explain this point by specializing to say $N = 3$. Thus, multiplying (7) by $(O^T)^{j_N k_N}$, we obtain

$$\varepsilon^{i_1 i_2 i_3} O^{i_1 j_1} O^{i_2 j_2} = \varepsilon^{j_1 j_2 j_3} (O^T)^{j_3 i_3}$$

Speaking loosely, we can think of moving some of the O's on the left hand side of (7) to the right hand side, where they become O^T's.

Using these identities, you can easily show that $[n]$ is equivalent to $[N - n]$, that is, these two representations transform in the same way. For example, as is well known, in $SO(3)$ the antisymmetric 2-index tensor is equivalent to the vector. (The cross product of two vectors is a vector.)

Any orthogonal matrix can be written as $O = e^A$. The conditions (1) and (2) imply that A is real and antisymmetric, so that A may be expressed as a linear combination of $N(N-1)/2$ antisymmetric matrices denoted by iJ^{ij}: $O = e^{i\theta^{ij} J^{ij}}$ (with repeated indices summed over). We have defined J^{ij} as imaginary and antisymmetric and hence hermitean. Since the commutator $[J^{ij}, J^{kl}]$ is antihermitean, it can be written as a linear combination of the iJ's.

Ironically, some students are confused at this point because of their familiarity with $SO(3)$, which has special properties that do not generalize to $SO(N)$.

In speaking about rotations in 3-dimensional space we can specify a rotation as either around say the third axis, with the corresponding generator J^3, or as in the (1-2)-plane, with the corresponding generator $J^{12} = -J^{21}$.

In higher dimensions, for example 10-dimensional space, we can speak of a rotation in the (6-7)-plane, with the corresponding generator $J^{67} = -J^{76}$, but it is nonsense to speak of a rotation around the fifth axis. Thus, to generalize to higher dimensions we should write the standard commutation relation $[J^1, J^2] = i J^3$ for $SO(3)$ as $[J^{23}, J^{31}] = i J^{12}$, which can be generalized immediately to

$$[J^{ij}, J^{kl}] = i(\delta^{ik}J^{jl} - \delta^{jk}J^{il} + \delta^{jl}J^{ik} - \delta^{il}J^{jk}) \tag{8}$$

The right hand side reflects the antisymmetric character of $J^{ij} = -J^{ji}$. A potential confusion some students may have about the notation: J^{ij} denotes a matrix generating rotation in the $(i\text{-}j)$-plane, a matrix with element $(J^{ij})^{kl}$ in the k-th row and l-th column. The indices i, j, k, and l all run from 1 to N, but in $(J^{ij})^{kl}$ the set $\{ij\}$ and the set $\{kl\}$ should be distinguished conceptually: The former labels the generator and the latter are matricial indices when the generator is regarded as a matrix. As an exercise, write down $(J^{ij})^{kl}$ explicitly and obtain (8) by direct computation.

In studying group theory, as I have already remarked, one source of confusion comes from the fact that some of the smaller groups, which we tend to encounter first in our studies, have special properties that do not generalize. The special property of $SO(3)$ we just noted is due to the fact that the antisymmetric symbol ε^{ijk} carries three indices and thus J^{ij} may be written as $J^k \equiv \frac{1}{2}\varepsilon^{ijk}J^{ij}$. For $SO(4)$ the antisymmetric symbol ε^{ijkl} carries four indices and we can form the combinations $\frac{1}{2}(J^{ij} \pm \frac{1}{2}\varepsilon^{ijkl}J^{kl})$. Define $J^1_\pm \equiv \frac{1}{2}(J^{23} \pm J^{14})$, $J^2_\pm \equiv \frac{1}{2}(J^{31} \pm J^{24})$, and $J^3_\pm \equiv \frac{1}{2}(J^{12} \pm J^{34})$. By explicit computation, show that $[J^i_+, J^j_+] = i\varepsilon^{ijk}J^k_+$, $[J^i_-, J^j_-] = i\varepsilon^{ijk}J^k_-$, and $[J^i_+, J^j_-] = 0$. This proves the well-known theorem that $SO(4)$ is locally isomorphic to $SO(3) \otimes SO(3)$.

I assume that you know that $SO(3)$ is locally isomorphic to $SU(2)$. If you don't, I give a brief review below.

With a few i's included here and there, these two results prove the statement that the Lorentz group $SO(3, 1)$ is locally isomorphic to $SU(2) \otimes SU(2)$, which we proved explicitly in chapter II.3. The Lorentz group can be thought of as an "analytic continuation" of the rotation group $SO(4)$. See below for a more precise statement.

One highly non-obvious result of group theory is that $SO(N)$ contains representations other than vector and tensor. I develop the relevant group theory for the spinor representations in chapter VII.7.

SU(N)

We next turn to the special unitary group $SU(N)$ consisting of all N by N matrices U that are unitary

$$U^\dagger U = 1 \tag{9}$$

and have unit determinant

$$\det U = 1 \tag{10}$$

The story of $SU(N)$ has more or less the same plot as the story of $SO(N)$ with the crucial difference that the tensors of the unitary groups can carry both upper and lower indices. We denote the element in the ith row and jth column by U^i_j; the wisdom of this notation will soon become apparent.

The defining or fundamental representation of $SU(N)$ consists of N objects φ^j, $j = 1, \ldots, N$, that transform under the action of the group element U according to

$$\varphi^i \to \varphi'^i = U^i_j \varphi^j \tag{11}$$

Taking the complex conjugate of (11) we have

$$\varphi^{*i} \to (U^i_j)^* \varphi^{*j} = (U^\dagger)^j_i \varphi^{*j} \tag{12}$$

We invite ourselves to define an object we write as φ_i that transforms in the same way as φ^{*i}; thus

$$\varphi_i \to \varphi'_i = (U^\dagger)^j_i \varphi_j \tag{13}$$

Note that we did not say that φ_i is equal to φ^{*i}; we merely said that φ_i and φ^{*i} transform in the same way.

As before, we can have tensors. The tensor φ^{ij}_k, for example, transforms as if it is equal to the product $\varphi^i \varphi^j \varphi_k$:

$$\varphi^{ij}_k \to \varphi'^{ij}_k = U^i_l U^j_m (U^\dagger)^n_k \varphi^{lm}_n \tag{14}$$

Again, we emphasize that we did not say that φ_k^{ij} is equal to $\varphi^i \varphi^j \varphi_k$. (In some books φ^i is called a covariant vector and φ_i a contravariant vector. A tensor $\varphi^{\cdots}_{\cdots}$ with m upper indices and n lower indices is defined to transform as if it is equal to the product of m covariant vectors and n contravariant vectors.)

The possibility of complex conjugation in $SU(N)$ leads naturally to having indices "upstairs" and "downstairs." Note that (9) can be written out explicitly as $(U^\dagger)_i^k U_k^j = \delta_i^j$ and thus the Kronecker delta in $SU(N)$ carries one upper and one lower index. It is important when taking traces that we set an upper index equal to a lower index and sum over them: for example, we can consider $\delta_j^k \varphi_k^{ij} \equiv \varphi_j^{ij}$, which transforms as

$$\varphi_j^{ij} \to U_l^i U_m^j (U^\dagger)_j^n \varphi_n^{lm} = U_l^i \varphi_m^{lm} \tag{15}$$

where we have used (9). In other words, φ_j^{ij}, the trace of φ_k^{ij}, denote N objects that transform into linear combinations of each other in the same way as φ^i. Thus, given a tensor, we can always subtract out its trace.

As in the discussion for $SO(N)$, tensors furnish representations of the group. The discussion proceeds as before. The symmetry properties of a tensor under permutation of its indices are not changed by the group transformation.

Another way of saying this is that given a tensor we can always take it to have definite symmetry properties under permutation of its upper indices and under permutation of its lower indices. In our specific example, we can always take φ_k^{ij} to be either symmetric or antisymmetric under the exchange of i and j and to be traceless. Thus, the symmetric traceless tensor φ_k^{ij} furnishes a representation with dimension $\frac{1}{2} N^2(N+1) - N$ and the antisymmetric traceless tensor φ_k^{ij} a representation with dimension $\frac{1}{2} N^2(N-1) - N$.

Thus, in summary, the irreducible representations of $SU(N)$ are realized by traceless tensors with definite symmetry properties under permutation of indices. For example, in $SU(5)$, some commonly encountered representations are φ^i, φ^{ij} (antisymmetric), φ^{ij} (symmetric), φ_j^i, φ_k^{ij} (antisymmetric in the upper indices and traceless) with dimensions 5, 10, 15, 24, and 45, respectively. Convince yourself that for $SU(N)$ the dimensions of the representations defined by these tensors are N, $N(N-1)/2$, $N(N+1)/2$, $N^2 - 1$, and $\frac{1}{2} N^2(N-1) - N$, respectively.

The representation defined by the traceless tensor φ_j^i is known as the adjoint representation. By definition, it transforms according to $\varphi_j^i \to \varphi_j'^i = U_l^i (U^\dagger)_j^n \varphi_n^l = U_l^i \varphi_n^l (U^\dagger)_j^n$. We are thus invited to regard φ_j^i as a matrix transforming according to

$$\varphi \to \varphi' = U \varphi U^\dagger \tag{16}$$

Note that if φ is hermitean it stays hermitean, and thus we can take φ to be a hermitean traceless matrix. (If φ is antihermitean we can always multiply it by i.) Another way of saying this is that given a hermitean traceless matrix X, $U X U^\dagger$ is also hermitean and traceless if U is an element of $SU(N)$.

As in the $SO(N)$ story, representations of $SU(N)$ have many names. For example, we can refer to the representation furnished by a tensor with m upper and n lower indices as (m, n). Alternatively, we can refer to them by their dimensions, with an asterisk to distinguish representations with mostly lower indices from the representations with mostly upper indices. For example, an alias for $(1, 0)$ is N and for $(0, 1)$ is N^*. A square bracket is used to indicate that the indices are antisymmetric and a curly bracket indicate that the indices are symmetric. Thus, the 10 of $SU(5)$ is also known as $[2, 0] = [2]$, where as indicated the 0 (no lower index) is suppressed. Similarly, 10^* is also known as $[0, 2] = [2]^*$.

The condition (10) can be written as either

$$\varepsilon_{i_1 i_2 \ldots i_N} U_1^{i_1} U_2^{i_2} \ldots U_N^{i_N} = 1 \tag{17}$$

or

$$\varepsilon^{i_1 i_2 \ldots i_N} U_{i_1}^1 U_{i_2}^2 \ldots U_{i_N}^N = 1 \tag{18}$$

Thus, we have two antisymmetric symbols $\varepsilon_{i_1 i_2 \ldots i_N}$ and $\varepsilon^{i_1 i_2 \ldots i_N}$ that we can use to raise and lower indices. Again, we can immediately generalize (17) to

$$\varepsilon_{i_1 i_2 \ldots i_N} U_{j_1}^{i_1} U_{j_2}^{i_2} \ldots U_{j_N}^{i_N} = \varepsilon_{j_1 j_2 \ldots j_N}$$

and multiplying this identity by $(U^\dagger)_{p_N}^{j_N}$ and summing over j_N we obtain

$$\varepsilon_{i_1 i_2 \ldots p_N} U_{j_1}^{i_1} U_{j_2}^{i_2} \ldots U_{j_{N-1}}^{i_{N-1}} = \varepsilon_{j_1 j_2 \ldots j_N} (U^\dagger)_{p_N}^{j_N}$$

Clearly, by repeating this process, we can peel off the U's on the left hand side and put them back as $U^\dagger$'s on the right hand side. We can play a similar game with (18).

To avoid drowning in a sea of indices, let me show you how to raise and lower indices in a specific example rather than in general. Consider the tensor φ_k^{ij} in $SU(4)$. We expect that the tensor $\varphi_{kpq} \equiv \varphi_k^{ij} \varepsilon_{ijpq}$ will transform as a tensor with three lower indices. Indeed,

$$\varphi_{kpq} \equiv \varphi_k^{ij}\varepsilon_{ijpq} \to \varepsilon_{ijpq}U_l^i U_m^j (U^\dagger)_k^n \varphi_n^{lm} = \varepsilon_{lmst}(U^\dagger)_p^s (U^\dagger)_q^t (U^\dagger)_k^n \varphi_n^{lm} = (U^\dagger)_k^n (U^\dagger)_p^s (U^\dagger)_q^t \varphi_{nst}$$

As in $SO(N)$ we can look at the generators of $SU(N)$ by noting that any unitary matrix can be written as $U = e^{iH}$, with H hermitean and traceless as required by (9) and (10). There are $(N^2 - 1)$ linearly independent N by N hermitean traceless matrices T^a ($a = 1, 2, \ldots, N^2 - 1$). Any N by N hermitean traceless matrix can be written as a linear combination of the T^a's and thus we can write $U = e^{i\theta^a T^a}$, where θ^a are real numbers and the index a is summed over.

Since the commutator $[T^a, T^b]$ is antihermitean and traceless, it can also be written as a linear combination of the T^a's:

$$[T^a, T^b] = if^{abc}T^c \tag{19}$$

(with the index c summed over.) The commutation relations (19) define the Lie algebra of $SU(N)$, and f^{abc} are known as the structure constants. For $SU(2)$ the structure constants f^{abc} are simply given by the antisymmetric symbol ε^{abc}.

Sometimes students are confused by how the generators act. Consider an infinitesimal transformation $U \simeq 1 + i\theta^a T^a$. On the defining representation, $\varphi^i \to U_j^i \varphi^j \simeq \varphi^i + i\theta^a (T^a)_j^i \varphi^j$. Thus, the ath generator acting on the defining representation gives $T^a \varphi$. Now consider the adjoint representation (16)

$$\varphi \to \varphi' \simeq (1 + i\theta^a T^a)\varphi(1 + i\theta^a T^a)^\dagger \simeq \varphi + i\theta^a T^a \varphi - \varphi i\theta^a T^a = \varphi + i\theta^a [T^a, \varphi] \tag{20}$$

In other words, the ath generator acting on the adjoint representation gives $[T^a, \varphi]$. Perhaps some students are confused by the fact that φ is used as a generic symbol to denote different objects.

Since the adjoint representation φ is hermitean and traceless it can also be written as a linear combination of the generators, thus $\varphi = \varphi^b T^b$. Using (19) we can thus also write (20) as $\varphi^c \to \varphi'^c \simeq \varphi^c - f^{abc}\theta^a \varphi^b$. In particular for $SU(2)$, the three objects φ^a transform as a 3-vector. (Note the notation: φ^a is not to be confused with φ^i: in $SU(2)$ the index $a = 1, 2, 3$ while $i = 1, 2$.)

This last remark essentially amounts to a proof that $SU(2)$ is locally isomorphic to $SO(3)$. I will now give a somewhat more formal proof. Any 2 by 2 hermitean traceless matrix X can be written as a linear combination of the three Pauli matrices $X = \vec{x} \cdot \vec{\sigma}$ with three real coefficients (x^1, x^2, x^3), which we regard as the components of a 3-vector $\vec{x}$. For any element U of $SU(2)$, $X' \equiv U^\dagger X U$ is hermitean and traceless, so that we can write $X' = \vec{x}' \cdot \vec{\sigma}$. Note that we have implicitly used the first defining property of an $SU(2)$ matrix (9). By explicit computation, we find $\det X = -\vec{x}^2$. Invoking the second defining property of an $SU(2)$ matrix (10), we obtain $\det X' = \det X$ and thus $\vec{x}'^2 = \vec{x}^2$. The 3-vector $\vec{x}$ is rotated into the 3-vector $\vec{x}'$. Thus we can associate a rotation with any given U. Since U and $-U$ are associated with the same rotation, this gives a double covering of $SO(3)$ by $SU(2)$. A physicist would just say that when a spin $\frac{1}{2}$ particle is rotated through 2π, its wave function changes sign. The map clearly preserves group multiplication: if two elements U_1 and U_2 of $SU(2)$ are mapped to the rotations R_1 and R_2 respectively, then the element $U_1 U_2$ is mapped to the rotation $R_1 R_2$. Alternatively, noting that $\text{tr} X^2 = \vec{x}^2$ and $\text{tr} X'^2 = \text{tr} X^2$, we obtain the same conclusion.

Once again, the two special unitary groups that most students learn first, namely $SU(2)$ and $SU(3)$, have special properties that do not generalize to $SU(N)$, just as $SO(3)$ has special properties that do not generalize to $SO(N)$, possibly leading to confusion.

For $SU(2)$, because the antisymmetric symbol ε^{ij} and ε_{ij} carry two indices, it suffices to consider only tensors with upper indices, all symmetrized: We can raise all lower indices of any tensor by contracting with ε^{ij} repeatedly. After this is done, we can remove any pair of indices in which the tensor is antisymmetric by contracting with ε_{ij}.

In particular, $\varphi^i = \varepsilon^{ij}\varphi_j$, which can be stated equivalently in terms of a special property of the Pauli matrices

$$\sigma_2 \sigma_a^* \sigma_2 = -\sigma_a \tag{21}$$

so that

$$\sigma_2(e^{i\vec{\theta}\vec{\sigma}})^*\sigma_2 = e^{i\vec{\theta}\vec{\sigma}} \qquad (22)$$

For $SU(2)$ (11) becomes

$$\varphi^i \rightarrow \varphi'^i = (e^{i\vec{\theta}\vec{\sigma}})^i_j \varphi^j$$

Complex conjugating, we obtain

$$\varphi^{*i} \rightarrow [(e^{i\vec{\theta}\vec{\sigma}})^i_j]^* \varphi^{*j} = [(-i\sigma_2)e^{i\vec{\theta}\vec{\sigma}}(i\sigma_2)]^i_j \varphi^{*j}$$

and so

$$i\sigma_2\varphi^* \rightarrow e^{i\vec{\theta}\vec{\sigma}}(i\sigma_2\varphi^*)$$

We learn that $i\sigma_2\varphi^*$ transform in the same way as φ. Recall that we define φ_i to transform in the same way as φ^{*i}. Thus, $\varepsilon^{ij}\varphi_j$ transforms in the same way as φ^i. In the jargon, $SU(2)$ is said to have only real and pseudoreal representations, but not complex representations. A pseudoreal representation is equivalent to its complex conjugate upon a similarity transformation. Recall that (21) figures into our discussion of charge conjugation in chapter II.1 and of the Higgs doublet in chapter VII.2.

For $SU(3)$ it suffices to consider only tensors with all their upper indices symmetrized and all their lower indices symmetrized. Thus, the representations of $SU(3)$ are uniquely labeled by two integers (m, n), where m and n denote the number of upper and lower indices. The reason is that the antisymmetric symbols ε^{ijk} and ε_{ijk} carry three indices. We can always trade a pair of lower indices in which the tensor is antisymmetric for one upper index, and similarly for upper indices.

You can see easily that these special properties do not generalize beyond $SU(2)$ and $SU(3)$.

Multiplying representations together

In a course on quantum mechanics you learn how to combine angular momentum. We have already encountered this concept in (5), which when specialized to $SO(3)$, tells us that $3 \otimes 3 = 5 \oplus 1 \oplus 3$, as we noted. This is sometimes described by saying that when we combine two angular momentum $L = 1$ states we obtain $L = 0, 1, 2$. Students are justifiably confused when this procedure is also known as addition of angular momentum.

Given two tensors φ and η of $SU(N)$, with m upper and n lower indices and with m' upper and n' lower indices, respectively, we can consider a tensor T with $(m + m')$ upper and $(n + n')$ lower indices that transforms in the same way as the product $\varphi\eta$. We can then reduce T by the various operations described above. This operation of multiplying two representations together is of course of fundamental importance in physics. In quantum field theory, for example, we multiply fields together to construct the Lagrangian.

As an example, multiply 5^* and 10 in $SU(5)$. To reduce $T_k^{ij} = \varphi_k\eta^{ij}$ we separate out the trace $\varphi_k\eta^{kj}$ (which transforms as a 5) after which there is nothing more we can do. Thus,

$$5^* \otimes 10 = 5 \oplus 45 \qquad (23)$$

As another example, consider $10 \otimes 10$: $\varphi^{ij}\eta^{kl}$. It is easiest to write η^{kl} equivalently as a tensor with three lower indices $\varepsilon_{mnhkl}\eta^{kl}$. The product $10 \otimes 10$ then carries two upper and three lower indices and we will write it as T_{mnh}^{ij}. Taking traces, we separate out T_{mij}^{ij}, which we recognize as 5^*, and the traceless part of T_{mnj}^{ij}, which we recognize as 45^* (see above), thus obtaining:

$$10 \otimes 10 = 5^* \oplus 45^* \oplus 50^* \qquad (24)$$

As exercises you can work out

$$5 \otimes 5 = 10 \oplus 15 \qquad (25)$$

and

$$5 \otimes 5^* = 1 \oplus 24 \qquad (26)$$

You should recognize the 24 as the adjoint.

In physics we are often called upon to multiply a tensor by itself. Statistics then plays a role. For instance, $SU(5)$ grand unification contains a scalar field φ^i transforming as 5. Because of Bose statistics, the product $\varphi^i \varphi^j$ contains only the 15.

Restriction to subgroup

To explain the next group theoretic concept, let me take a physical example. The $SU(3)$ of Gell-Mann and Ne'eman transforms the three quarks u, d, and s into linear combinations of each other. It contains as a subgroup the isospin $SU(2)$ of Heisenberg, which transforms u and d, but leaves s alone. In other words, upon restriction to the subgroup $SU(2)$ the irreducible representation 3 of $SU(3)$ decomposes as

$$3 \to 2 \oplus 1 \tag{27}$$

Consider an irreducible representation with dimension d of some group G. When we restrict our attention to a subgroup H, the set of d objects will in general decompose into n subsets, containing $d_1, d_2, \ldots, d_n$ objects, such that the objects of each subset only transform among themselves under the action of H. This makes obvious sense since there are fewer transformations in H than in G.

The decomposition of the fundamental or defining representation specifies how the subgroup H is embedded in G. Since all representations may be built up as products of the fundamental representation, once we know how the fundamental representation decomposes, we know how all representations decompose. For example, in $SU(3)$

$$3 \otimes 3^* = 8 \oplus 1 \tag{28}$$

while in $SU(2)$

$$(2 \oplus 1) \otimes (2 \oplus 1) = (3 \oplus 1) \oplus 2 \oplus 2 \oplus 1 \tag{29}$$

Comparing (28) and (29) we learn that

$$8 \to 3 \oplus 1 \oplus 2 \oplus 2. \tag{30}$$

Alternatively, we can simply look at the tensors involved. Consider φ^i of $SU(3)$ where the index i takes on the value 1, 2, 3. Let the index μ takes on the value 1, 2. Obviously, $\varphi^i = \{\varphi^\mu, \varphi^3\}$ corresponds to an explicit display of (27). Then $\varphi^i_j = \{\bar{\varphi}^\mu_\nu, \varphi^\mu_3, \varphi^3_\mu, \varphi^3_3\}$, where the bar on $\bar{\varphi}^\mu_\nu$ is to remind us that it is traceless. This corresponds precisely to (30).

Actually, $SU(3)$ also contains the larger subgroup $SU(2) \otimes U(1)$, where the $U(1)$ is generated by the traceless hermitean matrix

$$\begin{pmatrix} -1 & 0 & 0 \\ 0 & -1 & 0 \\ 0 & 0 & 2 \end{pmatrix}$$

We can then write (27) as $3 \to (2, -1) \oplus (1, 2)$, where the notation is almost self-explanatory. Thus, $(2, -1)$ denotes a 2 under $SU(2)$ with "charge" -1 under $U(1)$.

In the text, we will decompose various representations of $SU(5)$ and $SO(10)$. Everything we do there will simply be somewhat more elaborate versions of what we did here.

More on $SO(4)$, $SO(3,1)$, and $SO(2,2)$

In chapter II.3 you learned that acting on the two objects ψ_α with $\alpha = 1, 2$ in the spinor representation $(\frac{1}{2}, 0)$, the generators of rotation and boost are represented by $J_i = \frac{1}{2}\sigma_i$ and $iK_i = \frac{1}{2}\sigma_i$, respectively. I remind you that the equal sign means "represented by." For most purposes (for example, classifying quantum fields) and at the level of rigor of this book, it suffices to think of the Lie algebra generated by commuting J_i and K_i. Occasionally, however, it is useful to contemplate the actual group with group elements $e^{i\vec{\theta}\cdot\vec{J}}$ and $e^{i\vec{\varphi}\cdot\vec{K}}$.

In the spinor representation $(\frac{1}{2}, 0)$ the group elements are represented by $e^{i\vec{\theta}\cdot\frac{\vec{\sigma}}{2}}$ and $e^{\vec{\varphi}\cdot\frac{\vec{\sigma}}{2}}$. While $e^{i\vec{\theta}\cdot\frac{\vec{\sigma}}{2}}$ is special unitary, the 2 by 2 matrix $e^{\vec{\varphi}\cdot\frac{\vec{\sigma}}{2}}$, bereft of the i, is merely special but not unitary. (Incidentally, to verify these and subsequent statements, since you understand rotation thoroughly, you could, without loss of generality, choose

$\vec{\varphi}$ to point along the third axis, in which case $e^{\vec{\varphi} \cdot \frac{\vec{\sigma}}{2}}$ is diagonal with elements $e^{\frac{\varphi}{2}}$ and $e^{-\frac{\varphi}{2}}$. Thus, while the matrix is not unitary, its determinant is manifestly equal to 1.) This set of matrices defines the multiplicative group $SL(2, C)$, consisting of all 2 by 2 complex-valued matrices with unit determinant.

Let us count the number of generators of this group. Two conditions on the determinant (real part $= 1$, imaginary part $= 0$) cut the four complex entries containing eight real numbers down to six numbers, which accounts for the six generators of the Lorentz group $SO(3, 1)$.

To exhibit the map explicitly, we extend the earlier discussion showing that $SU(2)$ covers $SO(3)$. Consider the most general 2 by 2 hermitean matrix

$$X_M = x^0 I - \vec{x} \cdot \vec{\sigma} = \begin{pmatrix} x^0 - x^3 & x^1 - ix^2 \\ x^1 + ix^2 & x^0 + x^3 \end{pmatrix} \tag{31}$$

By explicit computation, $\det X_M = (x^0)^2 - \vec{x}^2$. (To see this instantly, choose $\vec{x}$ to point along the third axis and invoke rotational invariance.) Now consider $X_M' = L^\dagger X_M L$, with L an element of $SL(2, C)$. Manifestly, $\det X_M' = \det X_M$ and thus the transformation preserves $(x^0)^2 - \vec{x}^2$ and hence corresponds to Lorentz transformations. Since L and $-L$ give the same transformation $x \rightarrow x'$, we see that $SL(2, C)$ double covers $SO(3, 1)$. Mathematicians say that $SO(3, 1) = SL(2, C)/Z_2$. If L is also unitary, then $x^{0'} = x^0$ and the transformation is a rotation. The $SU(2)$ subgroup of $SL(2, C)$ double covers the rotation subgroup $SO(3)$ of the Lorentz group $SO(3, 1)$, that is, $SO(3) = SU(2)/Z_2$.

Incidentally, if we introduce an i at a strategic location and define the 2 by 2 matrix $X_E = x^4 I + i\vec{x} \cdot \vec{\sigma}$, regarding $(\vec{x}, x^4)$ as a 4-dimensional vector, we have $\det X_E = (x^4)^2 + \vec{x}^2$, the Euclidean length squared of the 4-vector. (Once again, choose $\vec{x}$ to point along the third axis so that X_E is a diagonal matrix with elements $x^4 \pm ix^3$.) Since $e^{i\theta \frac{\vec{\sigma}}{2}} = \cos \frac{\theta}{2} + i \sin \frac{\theta}{2}(\hat{\theta} \cdot \sigma)$ with $\hat{\theta}$ a unit vector in the θ direction (to see this, once again choose $\vec{\theta}$ to point along the 3rd axis), we see that $X_E/((x^4)^2 + \vec{x}^2)^{\frac{1}{2}}$ is an element of $SU(2)$. (We will come back to this observation in the next section.) Thus, for any two elements U and V of $SU(2)$, the matrix $X_E' = V^\dagger X_E U$ can also be decomposed in the form $X_E' = x'^4 I + i\vec{x}' \cdot \vec{\sigma}$. Evidently, $\det X_E' = \det X_E$. Thus the transformation preserves $(x^4)^2 + \vec{x}^2$ and describes an element of $SO(4)$. This shows explicitly that $SO(4)$ is locally isomorphic to $SU(2) \otimes SU(2)$. If $V = U$, we have a rotation, and if $V^\dagger = U$, the Euclidean analog of a boost.

Note that while the rotation group $SO(3)$ is compact, the Lorentz group $SO(3, 1)$ is not, since the range of the boost parameters $\vec{\varphi}$ is unbounded. In contrast, the group $SO(4)$ is compact and thus can be covered by a compact group, namely, $SU(2) \otimes SU(2)$, but the noncompact group $SO(3, 1)$ cannot be.

At this point, having done $SO(4)$ and $SO(3, 1)$, I might as well (with a wink toward the nuts who complained that this book is not encyclopedic enough) throw in the group $SO(2, 2)$ for use in part N. Let us strip the Pauli matrix σ^2 (kind of a "troublemaker" or at least an odd man out) of his i and define (just for this paragraph)

$$\sigma^2 \equiv \begin{pmatrix} 0 & -1 \\ 1 & 0 \end{pmatrix}$$

Any real 2 by 2 matrix X_H could be decomposed as $X_H = x^4 I + \vec{x} \cdot \vec{\sigma}$. Now $\det X_H = (x^4)^2 + (x^2)^2 - (x^3)^2 - (x^1)^2$, the quadratic form of a spacetime with two time and two space coordinates. The set of all linear transformations (with unit determinant) on (x^1, x^2, x^3, x^4) that preserve this quadratic form defines the group $SO(2, 2)$.

Introduce the multiplicative group $SL(2, R)$ consisting of all 2 by 2 real-valued matrices with unit determinant. For any two elements L_l and L_r of this group, consider the transformation $X_H' = L_l X_H L_r$. Evidently, $\det X_H' = \det X_H$. This shows explicitly that the group $SO(2, 2)$ is locally isomorphic to $SL(2, R) \otimes SL(2, R)$. Although two-timing theories are bound to be trouble, we could use $SO(2, 2)$ formally in computing scattering amplitudes, as we will see in chapter N.3.

Topological quantization of helicity

As promised, let us go back to the observation in the previous section that the matrix $X_E/((x^4)^2 + \vec{x}^2)^{\frac{1}{2}}$ is an element of $SU(2)$. Define $w^A \equiv x^A/((x^4)^2 + \vec{x}^2)^{\frac{1}{2}}$ for $A = 1, 2, 3, 4$. An arbitrary element of $SU(2)$ can be written as $U = w^4 I + i\vec{w} \cdot \vec{\sigma}$, with $\det U = 1 = (w^4)^2 + \vec{w}^2$. The 4-dimensional unit vector $w = (w^4, \vec{w})$ traces out the 3-sphere S^3, the surface of the 4-ball B^4 living in 4-dimensional Euclidean space. Thus the group manifold of $SU(2)$ is S^3.

Next, recall that $SU(2)$ double covers the rotation group $SO(3)$, or in plain talk, two elements U and $-U$ of $SU(2)$ corresponds to the same rotation. Thus the group manifold of $SO(3)$ is S^3/Z_2, that is, the 3-sphere with antipodal points identified.

Consider closed paths in $SO(3)$. Starting at some point P on S^3, wander off a bit and come back to P. The path you traced can evidently be continuously shrunk to a point. But suppose you go off to the other side of the world and arrive at $-P$, the antipodal point of P. You also trace a closed path in $SO(3)$ since P and $-P$ correspond to the same element of $SO(3)$, but this closed path obviously cannot be shrunk to a point. On the other hand, if after arriving at $-P$ you keep going and eventually return to P, then the entire path you traced can be continuously shrunk to a point. Using the language of homotopy groups introduced in chapter V.7, we say that $\Pi_1(SO(3)) = Z_2$: there are two topologically inequivalent classes of paths in the 3-dimensional rotation group.

Now we can go back and tie up a loose end in chapter III.4. Back in school you learned that the nonlinear algebraic structure of the Lie algebra $[J_i, J_j] = i\epsilon_{ijk}J_k$ enforces quantization of angular momentum. But the little group for a massless particle is merely $O(2)$. In the "rich man's approach" to gauge invariance, how do we get the helicity of the photon and the graviton quantized?

The answer is that we invoke topological, rather than algebraic, quantization. A rotation through 4π is represented by $e^{i4\pi h}$ on the helicity h state of the massless particle, but the path traced out by this rotation can be continuously shrunk to a point. Hence, we must have $e^{i4\pi h} = 1$ and $h = 0, \pm\frac{1}{2}, \pm 1, \ldots$.

Appendix C | Feynman Rules

Here we gather the Feynman rules given in various chapters.

Draw all possible diagrams. Label each line with a momentum. If applicable, also label each line with an incoming and an outgoing Lorentz index (for a line describing a vector field), with an incoming and an outgoing internal index (for a line describing a field transforming under an internal symmetry), so on and so forth. Momentum is conserved at each vertex. Momenta associated with internal lines are to be integrated over with the measure $\int [d^4 p/(2\pi)^4]$. A factor of (-1) is to be associated with each closed fermion loop. External lines are to be amputated. For an incoming fermion line write $u(p, s)$ and for an outgoing fermion line $\bar{u}(p', s')$. For an incoming antifermion, write $\bar{v}(p, s)$, and for an outgoing antifermion, $v(p', s')$. If there are symmetry transformations leaving the diagram invariant, then we have to worry about the infamous symmetry factors. Since I don't trust the compilations in various textbooks I work out the symmetry factors from scratch, and that is what I advise you to do.

Scalar field interacting with Dirac field

$$\mathcal{L} = \bar{\psi}(i\gamma^\mu \partial_\mu - m)\psi + \frac{1}{2}[(\partial\varphi)^2 - \mu^2 \varphi^2] - \frac{\lambda}{4!}\varphi^4 + f\varphi\bar{\psi}\psi \tag{1}$$

Scalar propagator:

$$\frac{i}{k^2 - \mu^2 + i\varepsilon} \tag{2}$$

Scalar vertex:

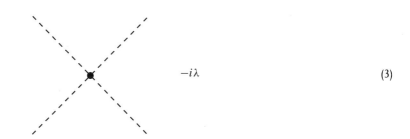

$$-i\lambda \tag{3}$$

Fermion propagator:

$$\frac{i}{\not{p} - m + i\varepsilon} = i\frac{\not{p} + m}{p^2 - m^2 + i\varepsilon} \tag{4}$$

Scalar fermion vertex:

$$if \tag{5}$$

Initial external fermion:

$$u(p, s) \tag{6}$$

Final external fermion:

$$\bar{u}(p, s) \tag{7}$$

Initial external antifermion:

$$\bar{v}(p, s) \tag{8}$$

Final external antifermion:

$$v(p, s) \tag{9}$$

Vector field interacting with Dirac field

$$\mathcal{L} = \bar{\psi}(i\gamma^\mu(\partial_\mu - ieA_\mu) - m)\psi - \tfrac{1}{4}F_{\mu\nu}F^{\mu\nu} - \tfrac{1}{2}\mu^2 A_\mu A^\mu \tag{10}$$

Vector boson propagator:

$$\frac{i}{k^2 - \mu^2}\left(\frac{k_\mu k_\nu}{\mu^2} - g_{\mu\nu}\right) \tag{11}$$

Photon propagator (with ξ an arbitrary gauge parameter):

$$\frac{i}{k^2}\left[(1 - \xi)\frac{k_\mu k_\nu}{k^2} - g_{\mu\nu}\right] \tag{12}$$

Vector boson fermion vertex:

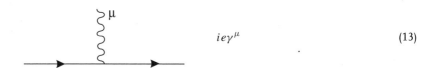

$$ie\gamma^{\mu} \qquad (13)$$

Initial external vector boson:

$$\varepsilon_{\mu}(k) \qquad (14)$$

Final external vector boson:

$$\varepsilon_{\mu}(k)^{*} \qquad (15)$$

Nonabelian gauge theory

Gauge boson propagator:

$$\frac{i}{k^2}\left[(1-\xi)\frac{k_\mu k_\nu}{k^2} - g_{\mu\nu}\right]\delta_{ab} \qquad (16)$$

Ghost propagator:

$$\frac{i}{k^2}\delta_{ab} \qquad (17)$$

Cubic interaction between the gauge bosons:

$$gf^{abc}[g_{\mu\nu}(k_1-k_2)_\lambda + g_{\nu\lambda}(k_2-k_3)_\mu + g_{\lambda\mu}(k_3-k_1)_\nu] \qquad (18)$$

Quartic interaction between the gauge bosons:

$$-ig^2[f^{abe}f^{cde}(g_{\mu\lambda}g_{\nu\rho} - g_{\mu\rho}g_{\nu\lambda}) \\ + f^{ade}f^{cbe}(g_{\mu\lambda}g_{\nu\rho} - g_{\mu\nu}g_{\rho\lambda}) \\ + f^{ace}f^{bde}(g_{\mu\nu}g_{\lambda\rho} - g_{\mu\rho}g_{\nu\lambda})] \qquad (19)$$

Gauge boson coupling to the ghost field:

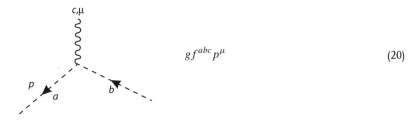

$$g f^{abc} p^\mu \qquad (20)$$

Cross sections and decay rates

Given the Feynman amplitude $\mathcal{M}$ for a process $p_1 + p_2 \to k_1 + k_2 + \cdots + k_n$ the differential cross section is given by

$$d\sigma = \frac{1}{|\vec{v}_1 - \vec{v}_2| \mathcal{E}(p_1)\mathcal{E}(p_2)} \frac{d^3 k_1}{(2\pi)^3 \mathcal{E}(k_1)} \cdots \frac{d^3 k_n}{(2\pi)^3 \mathcal{E}(k_n)} (2\pi)^4 \delta^{(4)}\left(p_1 + p_2 - \sum_{i=1}^n k_i\right) |\mathcal{M}|^2 \qquad (21)$$

Here $\vec{v}_1$ and $\vec{v}_2$ denote the velocities of the incoming particles. The energy factor $\mathcal{E}(p) = 2\sqrt{\vec{p}^2 + m^2}$ for bosons and $\mathcal{E}(p) = \sqrt{\vec{p}^2 + m^2}/m$ for fermions come from the different normalization of the creation and annhilation operators in chapters I.8 and II.2.

For a decay of a particle of mass M the differential decay rate in its rest frame is given by

$$d\Gamma = \frac{1}{2M} \frac{d^3 k_1}{(2\pi)^3 \mathcal{E}(k_1)} \cdots \frac{d^3 k_n}{(2\pi)^3 \mathcal{E}(k_n)} (2\pi)^4 \delta^{(4)}\left(P - \sum_{i=1}^n k_i\right) |\mathcal{M}|^2 \qquad (22)$$

Gamma matrices

Identities for the trace of a product of an even number of gamma matrices:

$$\text{tr}\gamma^{\mu}\gamma^{\nu} = 4\eta^{\mu\nu} \tag{1}$$

$$\text{tr}\gamma^{\mu}\gamma^{\nu}\gamma^{\lambda}\gamma^{\sigma} = 4(\eta^{\mu\nu}\eta^{\lambda\sigma} - \eta^{\mu\lambda}\eta^{\nu\sigma} + \eta^{\mu\sigma}\eta^{\nu\lambda}) \tag{2}$$

We define the totally antisymmetric symbol $\varepsilon^{\mu\nu\lambda\sigma}$ by $\varepsilon^{0123} = +1$ (note $\varepsilon_{0123} = -1$). Then with our definition $\gamma^5 \equiv i\gamma^0\gamma^1\gamma^2\gamma^3$, we have

$$\text{tr}\gamma^5\gamma^{\mu}\gamma^{\nu}\gamma^{\lambda}\gamma^{\sigma} = -4i\varepsilon^{\mu\nu\lambda\sigma} \tag{3}$$

Identities that follow from the basic Clifford identity:

$$\gamma^{\mu}\,\slashed{p}\gamma_{\mu} = -2\,\slashed{p} \tag{4}$$

$$\gamma^{\mu}\,\slashed{p}\,\slashed{q}\gamma_{\mu} = 4p\cdot q \tag{5}$$

$$\gamma^{\mu}\,\slashed{p}\,\slashed{q}\,\slashed{r}\gamma_{\mu} = -2\,\slashed{r}\,\slashed{q}\,\slashed{p} \tag{6}$$

I leave it to you to derive these identities. For example, to obtain (4) keep moving γ^{μ} to the right in the expression $\gamma^{\mu}\,\slashed{p}\gamma_{\mu} = (2p^{\mu} - \slashed{p}\gamma^{\mu})\gamma_{\mu} = 2\,\slashed{p} - 4\,\slashed{p} = -2\,\slashed{p}$.

Evaluating Feynman diagrams

Over the years, a number of tricks and identities have been developed for evaluating the integrals associated with Feynman diagrams.

Let us evaluate

$$I = \int \frac{d^4k}{(2\pi)^4} \frac{1}{(k^2 - m^2 + i\varepsilon)^3} = \int \frac{d^3k}{(2\pi)^3} \int \frac{dk_0}{2\pi} \frac{1}{[k_0^2 - (\vec{k}^2 + m^2) + i\varepsilon]^3}$$

Focus on the k_0 integral. Draw where the poles are in the complex k_0-plane and you will see that the integration contour can be rotated anticlockwise so that [we denote the integrand by $f(k_0)$]

$$\int_{-\infty}^{+\infty} dk_0 f(k_0) = \int_{-i\infty}^{+i\infty} dk_0 f(k_0) = i \int_{-\infty}^{+\infty} dk_4 f(ik_4) \tag{7}$$

where in the last step we define $k_0 = i k_4$ (corresponding to the Wick rotation mentioned in chapters I.2 and V.2.) Thus,

$$I = i(-1)^3 \int \frac{d_E^4 k}{(2\pi)^4} \frac{1}{(k_E^2 + m^2)^3}$$

where $d_E^4 k$ is the integration element in Euclidean 4-dimensional space and $k_E^2 \equiv k_4^2 + \vec{k}^2$ the square of a Euclidean 4-vector. The infinitesimal ε can now be set equal to zero. We can integrate immediately over the three angles since the integrand does not depend on them. You can look up the angular element in Euclidean space in a book, but we will use a neat trick instead.

I will do the more general d-dimensional integral $H = \int d^d k F(k^2)$, where $k^2 = k_1^2 + k_2^2 + \cdots + k_d^2$ and F can be any function as long as the integral converges. (I now drop the subscript E; the context makes clear that we are in Euclidean space.) We can of course set d equal to 4 at the end. The result for arbitrary d will be useful to us in regularizing dimensionally (chapter III.1).

We imagine integrating over the $(d - 1)$ angular variables to obtain $H = C(d) \int_0^\infty dk$ $k^{d-1} F(k^2)$. To determine $C(d)$ we will do the integral $J = \int d^d k e^{-\frac{1}{2} k^2}$ in two different ways. Using (I.2.8) we have $J = (\sqrt{2\pi})^d$. Alternatively,

$$J = C(d) \int_0^\infty dk\, k^{d-1} e^{-\frac{1}{2} k^2} = C(d) 2^{\frac{d}{2}-1} \int_0^\infty dx\, x^{\frac{d}{2}-1} e^{-x} = C(d) 2^{\frac{d}{2}-1} \Gamma(\frac{d}{2})$$

where we changed integration variables and recognized the integral representation of the gamma function $\Gamma(z+1) \equiv \int_0^\infty dx\, x^z e^{-x}$. (Recall that upon integration by parts we obtain $\Gamma(z+1) = z\Gamma(z)$, so that $\Gamma(n) = (n-1)!$ for n an integer.) Therefore $C(d) = 2\pi^{d/2} / \Gamma(d/2)$ and

$$\int d^d k F(k^2) = \frac{2\pi^{d/2}}{\Gamma(d/2)} \int_0^\infty dk\, k^{d-1} F(k^2) \tag{8}$$

Setting $d = 1$ in (8) we determine $\Gamma(\frac{1}{2}) = \pi^{\frac{1}{2}}$, and setting $F(k^2) = \delta(k - 1)$ we see that the area of the $(d - 1)$-dimensional sphere is equal to $C(d)$, thus recovering various results you learned in school about circles and spheres: $C(2) = 2\pi$ and $C(3) = 4\pi$.

The new result you need as a budding field theorist is for $d = 4$:

$$\int d^4 k F(k^2) = \pi^2 \int_0^\infty dk^2 k^2 F(k^2) \tag{9}$$

So finally we have

$$I = \frac{-i}{16\pi^2} \int_0^\infty dk^2 k^2 \frac{1}{(k^2 + m^2)^3} = \frac{-i}{16\pi^2} \frac{1}{2m^2} \tag{10}$$

We have derived the basic formula for doing Feynman integrals:

$$\int \frac{d^4 k}{(2\pi)^4} \frac{1}{(k^2 - m^2 + i\varepsilon)^3} = \frac{-i}{32\pi^2 m^2} \tag{11}$$

(With the telltale $i\varepsilon$ we have evidently moved back to Minkowski space.) As an exercise you can go through the same steps to find

$$\int^\Lambda \frac{d^4 k}{(2\pi)^4} \frac{1}{(k^2 - m^2 + i\varepsilon)^2} = \frac{i}{16\pi^2} \left[\log\left(\frac{\Lambda^2}{m^2}\right) - 1 + \cdots \right] \tag{12}$$

Here a cutoff is needed, which we introduce by setting the upper limit in the integral over k^2 in the analog of (10) to Λ^2. As a check, differentiate (12) with respect to m^2 to recover (11). As another exercise show that

$$\int^\Lambda \frac{d^4 k}{(2\pi)^4} \frac{k^2}{(k^2 - m^2 + i\varepsilon)^2} = \frac{-i}{16\pi^2} \left[\Lambda^2 - 2m^2 \log\left(\frac{\Lambda^2}{m^2}\right) + m^2 + \cdots \right] \tag{13}$$

In (12) and (13) $(\cdots)$ denote terms that vanish for $\Lambda^2 \gg m^2$. In some texts, the (-1) in (12) is dropped by absorbing it into Λ^2. But then we have to be careful to adjust (13) accordingly if it appears in the same calculation.

A useful identity in combining denominators is

$$\frac{1}{x_1 x_2 \dots x_n} = (n-1)! \int_0^1 \int_0^1 \dots \int_0^1 d\alpha_1 d\alpha_2 \dots d\alpha_n$$

$$\delta\left(1 - \sum_j^n \alpha_j\right) \frac{1}{(\alpha_1 x_1 + \alpha_2 x_2 + \dots + \alpha_n x_n)^n} \tag{14}$$

For $n = 2$,

$$\frac{1}{xy} = \int_0^1 d\alpha \frac{1}{[\alpha x + (1-\alpha)y]^2} \tag{15}$$

and for $n = 3$,

$$\frac{1}{xyz} = 2 \int_0^1 \int_0^1 \int_0^1 d\alpha d\beta d\gamma \, \delta(\alpha + \beta + \gamma - 1) \frac{1}{(\alpha x + \beta y + \gamma z)^3} \tag{16}$$

$$= 2 \int\int_{\text{triangle}} d\alpha d\beta \frac{1}{[z + \alpha(x-z) + \beta(y-z)]^3}$$

where the integration region is the triangle in the α-β plane bounded by $0 \le \beta \le 1 - \alpha$ and $0 \le \alpha \le 1$.

Appendix E | Dotted and Undotted Indices and the Majorana Spinor

We develop the dotted and undotted notation introduced in chapter II.3 for further use in discussing supersymmetry in chapter VIII.4 and in part N. In essence, the appearance of undotted and dotted indices can be traced back to the fact that the algebra of the Lorentz group $SO(3, 1)$, with the generators $\vec{J} + i\vec{K}$ and $\vec{J} - i\vec{K}$, breaks up into two pieces, each isomorphic to the algebra of $SU(2)$. The absence or presence of the dot allows us to keep track of which $SU(2)$ we are talking about.

Here I will use extensively results from chapter II.3 and from the exercises (do them!) there without bothering to write them down again here.

In the Weyl basis of chapter II.1

$$\gamma^\mu = \begin{pmatrix} 0 & \sigma^\mu \\ \bar{\sigma}^\mu & 0 \end{pmatrix} \tag{1}$$

where $\sigma^\mu = (I, \vec{\sigma})$ and $\bar{\sigma}^\mu = (I, -\vec{\sigma})$. Knowing that γ^μ acts on

$$\Psi = \begin{pmatrix} \psi_\alpha \\ \bar{\chi}^{\dot{\alpha}} \end{pmatrix}$$

we see that σ^μ and $\bar{\sigma}^\mu$ carry indices as follows:

$$(\sigma^\mu)_{\alpha\dot{\alpha}} \text{ and } (\bar{\sigma}^\mu)^{\dot{\alpha}\alpha} \tag{2}$$

This is consistent with what you know: the Lorentz vector transforms like $(\frac{1}{2}, \frac{1}{2})$ and thus straddles the two $SU(2)$'s. The matrices σ^μ and $\bar{\sigma}^\mu$ mix dotted and undotted indices. We will make good use of this observation later.

Let us check that the Lorentz transformation property of the Dirac spinor Ψ is consistent with what was discussed in chapter II.1. There we learned that $\Psi \to e^{-\frac{i}{4}\omega_{\mu\nu}\Sigma^{\mu\nu}}\Psi$, where $\Sigma^{\mu\nu} \equiv \frac{i}{2}[\gamma^\mu, \gamma^\nu]$. (We want to use the symbol $\sigma^{\mu\nu}$ for some other quantity, hence the change of notation.) Using (1) we obtain

$$\Sigma^{\mu\nu} = 2i \begin{pmatrix} \sigma^{\mu\nu} & 0 \\ 0 & \bar{\sigma}^{\mu\nu} \end{pmatrix}$$

where $\sigma^{\mu\nu} \equiv \frac{1}{4}(\sigma^\mu\bar{\sigma}^\nu - \sigma^\nu\bar{\sigma}^\mu)$ and $\bar{\sigma}^{\mu\nu} \equiv \frac{1}{4}(\bar{\sigma}^\mu\sigma^\nu - \bar{\sigma}^\nu\sigma^\mu)$. From (2) we see that these two matrices carry indices as follows:

$$(\sigma^{\mu\nu})_\alpha{}^\beta \text{ and } (\bar{\sigma}^{\mu\nu})^{\dot{\alpha}}{}_{\dot{\beta}} \tag{3}$$

Again, this reflects the fact that the antisymmetric tensor (such as the electromagnetic field $F_{\mu\nu}$) transforms like $(1, 0) + (0, 1)$.

The matrices $\sigma^{\mu\nu}$ and $\bar{\sigma}^{\mu\nu}$ may seem alien, but recall that they are manufactured out of the familiar Pauli matrices and so they are simply Pauli matrices (what else could they be?) themselves. In particular,

$$\sigma^{0i} = -\bar{\sigma}^{0i} = -\frac{1}{2}\sigma^i \quad \text{and} \quad \sigma^{ij} = \bar{\sigma}^{ij} = -\frac{i}{2}\,\varepsilon^{ijk}\sigma^k$$

Note that these relations are consistent with $(\sigma^{\mu\nu})^\dagger = -(\bar{\sigma}^{\mu\nu})$, which in turn follows from $(\Sigma^{\mu\nu})^\dagger = \gamma^0\Sigma^{\mu\nu}\gamma^0$.

Mother Nature is kind to the students of quantum field theory. The relativistic spinor Ψ breaks up into two 2-component spinors acted on by the Pauli matrices. What you learned in nonrelativistic quantum mechanics continues to be relevant here.

Thus under an infinitesimal Lorentz transformation

$$\psi_\alpha \to \left(I + \tfrac{1}{2}\omega_{\mu\nu}\sigma^{\mu\nu}\right)_\alpha^{\ \beta} \psi_\beta \tag{4}$$

and

$$\bar{\chi}^{\dot\alpha} \to \left(I + \tfrac{1}{2}\omega_{\mu\nu}\bar{\sigma}^{\mu\nu}\right)^{\dot\alpha}_{\ \dot\beta} \bar{\chi}^{\dot\beta} \tag{5}$$

You should check that it all works out according to plan. Everything is consistent with what we learned in chapter II.3, in particular, that boosts act oppositely on $(\frac{1}{2}, 0)$ and $(0, \frac{1}{2})$, but rotations act the same.

Thus far, on the spinor fields ψ_α and $\bar{\chi}^{\dot\alpha}$, the dotted indices always live upstairs and the undotted indices downstairs. What would get them to change floors? Charge conjugation.

Recall from chapter II.1 that the charge conjugated field is defined by $\Psi^c \equiv C\bar{\Psi}^T$ [where T denotes transpose, $\bar{\Psi}$ means $\Psi^\dagger\gamma^0$, and $C^{-1}\gamma^\mu C = -(\gamma^\mu)^T$.] In the Weyl basis, we can choose

$$C = \zeta\gamma^0\gamma^2 = \zeta \begin{pmatrix} -\sigma_2 & 0 \\ 0 & \sigma_2 \end{pmatrix} \tag{6}$$

The condition $(\Psi^c)^c = \Psi$ implies $|\zeta| = 1$. We choose $\zeta = -i$. Explicitly,

$$\Psi^c = \begin{pmatrix} i\sigma_2\bar{\chi}^* \\ -i\sigma_2\psi^* \end{pmatrix}$$

We now introduce some notation, the wisdom of which will soon become clear. Given ψ_α and $\bar{\chi}^{\dot\alpha}$, define

$$\bar{\psi}_{\dot\alpha} \equiv (\psi_\alpha)^* \text{and} \chi^\alpha \equiv (\bar{\chi}^{\dot\alpha})^* \tag{7}$$

Weird, complex conjugation puts on a dot and a bar.

We raise and lower undotted indices as follows: $\psi_\alpha = \varepsilon_{\alpha\beta}\psi^\beta$ and $\psi^\beta = \varepsilon^{\beta\gamma}\psi_\gamma$ which implies that $\varepsilon_{\alpha\beta}\varepsilon^{\beta\gamma} = \delta_\alpha^{\ \gamma}$. Thus, if we choose

$$\varepsilon_{\alpha\beta} = \begin{pmatrix} 0 & 1 \\ -1 & 0 \end{pmatrix} = (i\sigma_2)_{\alpha\beta}$$

then

$$\varepsilon^{\beta\gamma} = \begin{pmatrix} 0 & -1 \\ 1 & 0 \end{pmatrix} = (-i\sigma_2)^{\beta\gamma}$$

We are forced to define $\varepsilon_{12} = +1$ and $\varepsilon^{12} = -1$ to have opposite signs, a fact to keep in mind.

You should realize by now that what we are doing can again be traced back to that peculiar fact about Pauli matrices (appendix B):

$$(i\sigma_2)\sigma_i^*(-i\sigma_2) = -\sigma_i \tag{8}$$

or equivalently

$$\sigma_2\sigma_i^T\sigma_2 = -\sigma_i \tag{9}$$

an identity in one guise or another familiar from quantum mechanics. We have used it again and again, in appendix B and in the text (for example, in connection with Majorana masses and with the Higgs field). From (8) we have $(i\sigma_2)\sigma^{\mu*}(-i\sigma_2) = \bar{\sigma}^\mu$ and hence

$$(i\sigma_2)(\sigma^{\mu\nu})^*(-i\sigma_2) = \bar{\sigma}^{\mu\nu} \tag{10}$$

Analogously, we raise and lower dotted indices as follows: $\bar{\psi}_{\dot\alpha} = \varepsilon_{\dot\alpha\dot\beta}\bar{\psi}^{\dot\beta}$ and $\bar{\psi}^{\dot\beta} = \varepsilon^{\dot\beta\dot\gamma}\bar{\psi}_{\dot\gamma}$. Referring to (7) we see that $\varepsilon_{\dot\alpha\dot\beta}$ is numerically the same as $\varepsilon_{\alpha\beta}$, and $\varepsilon^{\dot\beta\dot\gamma}$ is numerically the same as $\varepsilon^{\beta\gamma}$.

You now see the rationale of these apparently capricious choices: we can now write

$$\Psi^c = \begin{pmatrix} \chi_\alpha \\ \bar{\psi}^{\dot\alpha} \end{pmatrix} \tag{11}$$

Referring to

$$\Psi = \begin{pmatrix} \psi_\alpha \\ \bar{\chi}^{\dot\alpha} \end{pmatrix} \tag{12}$$

we see that the point of the notation is that ψ_α and χ_α transform in the same way and are the same kind of creature (and similarly for $\bar{\chi}^{\dot\alpha}$ and $\bar{\psi}^{\dot\alpha}$.)

We now come to the all-important concept of a Majorana spinor. Ettore Majorana, a brilliant physicist, mysteriously disappeared early in his career. Fermi supposedly described Majorana as "a towering giant without any common sense."[1]

Given a Dirac spinor Ψ, if $\Psi = \Psi^c$, then Ψ is said to be a Majorana spinor.

Comparing (12) and (11), we see that a Majorana spinor has the form

$$\Psi_M = \begin{pmatrix} \psi_\alpha \\ \bar{\psi}^{\dot\alpha} \end{pmatrix} \tag{13}$$

An obvious remark but a handy mnemonic: Given a Weyl spinor ψ_α we can construct a Majorana spinor, and given two Weyl spinors we can construct a Dirac spinor: one Weyl equals one Majorana, and two Weyls equal one Dirac.

Incidentally, another way of seeing that complex conjugation puts on a dot is that (see chapter II.3) conjugation interchanges $\vec{J} + i\vec{K}$ and $\vec{J} - i\vec{K}$.

The point to remember is simply that given a spinor λ_α, then $\lambda_{\dot\alpha}$ transforms like $(\lambda_\alpha)^*$. You should verify this, keeping in mind (10).

The utility of the notation is similar to that of the covariant and contravariant (or upper and lower) indices in special and general relativity. We always contract an upper index with a lower index. Here we have the additional rule that an undotted upper index can only be contracted with an undotted lower index, but never with a dotted lower index, (obviously, since they belong to different algebras.) It is easy to verify these rules. For example, let us show that $\eta^\alpha\psi_\alpha$ is invariant. Using (4) we proceed with laboriously careful pedagogy:

$$\eta^\alpha \to \eta'^\alpha = \varepsilon^{\alpha\beta}\eta'_\beta = \varepsilon^{\alpha\beta}(e^{\frac{1}{2}\omega\sigma})_\beta{}^\gamma\eta_\gamma = \varepsilon^{\alpha\beta}(e^{\frac{1}{2}\omega\sigma})_\beta{}^\gamma\varepsilon_{\gamma\rho}\eta^\rho = (e^{-\frac{1}{2}\omega\sigma^T})^\alpha{}_\rho\eta^\rho \tag{14}$$

where we used once again the identity (9). Then $\eta^\alpha\psi_\alpha \to \eta(e^{-\frac{1}{2}\omega\sigma^T})^T(e^{\frac{1}{2}\omega\sigma})\psi = \eta\psi$, which is indeed an invariant.

In special and general relativity we raise and lower indices with the metric, which is of course symmetric. Here we raise and lower indices with the antisymmetric ε symbol and as a result signs pop up here and there. For example, $\eta^\alpha\psi_\alpha = \varepsilon^{\alpha\beta}\eta_\beta\psi_\alpha = \eta_\beta(-\varepsilon^{\beta\alpha})\psi_\alpha = -\eta_\beta\psi^\beta$. Contrast this with the scalar product of two vectors $v^\mu w_\mu = v_\mu w^\mu$. If we want to suppress indices and write $\eta\psi$, we must decide once and for all what that means. The standard convention is to define

$$\eta\psi \equiv \eta^\alpha\psi_\alpha \tag{15}$$

and not $\eta_\beta\psi^\beta$. This rule is sometimes stated by saying that in contracting undotted indices we always go from the northwest to the southeast, and never from southeast to northwest. As we learned in chapter II.5, spinor fields are to be treated as anticommuting Grassman variables under the path integral, so that $-\eta_\beta\psi^\beta = \psi^\beta\eta_\beta$. We end up with the nice rule $\eta\psi = \psi\eta$.

---

[1] M. Gell-Mann, private communication. Incidentally, the name Ettore corresponds to Hector in English.

Similarly, we define

$$\bar{\chi}\bar{\xi} \equiv \bar{\chi}_{\dot\alpha}\bar{\xi}^{\dot\alpha} = \bar{\xi}\,\bar{\chi} \tag{16}$$

In contracting dotted indices we always go from southwest to northeast. Of course, none of this "Santa Barbara to Cambridge" convention is needed if the indices are displayed explicitly.

Just as in special and general relativity, where the upper and lower indices are very useful in telling us whether expressions we write down make sense, the undotted and dotted upper and lower indices allow us to see immediately that $\eta\psi$ and $\eta\sigma^{\mu\nu}\psi$ make sense, but that $\eta\sigma^{\mu}\psi$ does not. [Look at (2) and (3) and notice the kind of indices that appear.] The notation of course just codifies in a convenient way the group theory fact that $(\frac{1}{2}, 0) \otimes (\frac{1}{2}, 0) = (0, 0) \oplus (1, 0)$, namely, that out of two Weyl spinors we can make a scalar and a tensor but not a vector.

As always, notation should be driven by physics and computational convenience (which is intimately connected to elegance).

To gain familiarity with the dotted and undotted 2-component notation, you should work out some of the identities in the exercises. These identities are useful when working with supersymmetric field theories.

Exercises

E.1   Show that $\eta\sigma^{\mu\nu}\psi = -\psi\sigma^{\mu\nu}\eta$ and $\bar{\chi}\bar{\sigma}^{\mu}\psi = -\psi\sigma^{\mu}\bar{\chi}$.

E.2   Show that $(\theta\varphi)(\bar{\chi}\bar{\xi}) = -\frac{1}{2}(\theta\sigma^{\mu}\bar{\xi})(\bar{\chi}\bar{\sigma}_{\mu}\varphi)$.

E.3   Show that $\theta^{\alpha}\theta_{\beta} = \frac{1}{2}(\theta\theta)\delta^{\alpha}_{\beta}$. [Hint: simply evaluate the two sides for all possible cases.]

Solutions to Selected Exercises

Part I

I.3.1 From the text we have for $x^0 = 0$,

$$
\begin{aligned}
D(x) &= -i \int \frac{d^3k}{(2\pi)^3 2\sqrt{\vec{k}^2 + m^2}} e^{-i\vec{k}\cdot\vec{x}} \\
&= -\frac{i}{2(2\pi)^2} \int_0^\infty \frac{dk\, k^2}{\sqrt{\vec{k}^2 + m^2}} \int_{-1}^{+1} d(\cos\theta) e^{-ikr\cos\theta} \\
&= -\frac{1}{2(2\pi)^2 r} \int_0^\infty \frac{dk\, k}{\sqrt{\vec{k}^2 + m^2}} (e^{ikr} - e^{-ikr}) = -\frac{1}{8\pi^2 r} \int_{-\infty}^\infty \frac{dk\, k}{\sqrt{\vec{k}^2 + m^2}} e^{ikr} \\
&= \frac{i}{8\pi^2 r} \frac{\partial}{\partial r} \int_{-\infty}^\infty \frac{dk}{\sqrt{\vec{k}^2 + m^2}} e^{ikr}
\end{aligned}
$$

The integrand in $I \equiv \int_{-\infty}^\infty (dk/\sqrt{\vec{k}^2 + m^2}) e^{ikr}$ has a cut along the imaginary axis going from im to $i\infty$ (and another cut we don't care about.) So fold the contour around the cut and change variable to $k = i(m + y)$:

$$
\begin{aligned}
I &= 2 \int_0^\infty dy\, e^{-(y+m)r} \frac{1}{\sqrt{(y+m)^2 - m^2}} \\
&= 2 \int_1^\infty du\, e^{-mru} \frac{1}{\sqrt{u^2 - 1}} \\
&= 2 \int_0^\infty dt\, e^{-mr\cosh t}.
\end{aligned}
$$

At this point you can look in a table and find that this is some Bessel function and read off the large r behavior, but it is more stylish to press on and descend steeply: we obtain

$$D(x) = -\frac{im}{4\pi^2 r} \int_0^\infty dt\,(\cosh t)e^{-mr\cosh t}$$

$$= -\frac{im}{4\pi^2 r} \int_0^\infty d(\sinh t)e^{-mr\cosh t}$$

$$= -\frac{im}{4\pi^2 r} \int_0^\infty ds\,e^{-mr\sqrt{s^2+1}} \simeq -\frac{im}{4\pi^2 r}\int_0^\infty ds\,e^{-mr(1+\frac{1}{2}s^2)}$$

$$= \frac{-im^2}{4\pi^2}\left(\frac{\pi}{2(mr)^3}\right)^{\frac{1}{2}}e^{-mr},$$

using the Gaussian integral from the appendix of chapter I.2.

I.3.2 We evaluate

$$D(x) = \int \frac{d^2k}{(2\pi)^2}\frac{e^{ikx}}{k^2 - m^2 + i\varepsilon}$$

by contours as in the text and obtain

$$D(x) = -i\int\frac{dk}{(2\pi)2\omega_k}[e^{-i(\omega_k t - kx)}\theta(x^0) + e^{i(\omega_k t - kx)}\theta(-x^0)]$$

For $x^0 = 0$, we recognize the integral

$$D(x) = -i\int_{-\infty}^{+\infty}\frac{dk}{(2\pi)2\sqrt{\vec{k}^2 + m^2}}e^{-ikx}$$

as a Bessel function from exercise I.3.1:

$$D(x) = \frac{-i}{2\pi}K_0(m|x|) \rightarrow \frac{-i}{2\pi}\sqrt{\frac{\pi}{2m|x|}}e^{-m|x|}$$

with the expected exponential decay for large x.

I.7.2 Expanding and keeping only the desired terms

$$Z(J) \rightarrow C\left\{1 + \frac{1}{2!}\left(-\frac{i}{4!}\lambda\right)^2 \iint d^4w_1 d^4w_2 \left[\frac{\delta}{i\delta J(w_1)}\right]^4 \left[\frac{\delta}{i\delta J(w_2)}\right]^4\right.$$

$$\left.\frac{1}{6!}\left[-\frac{i}{2}\iint d^4x d^4y\, J(x)D(x-y)J(y)\right]^6\right\}$$

Just keep on differentiating.

I.7.4 Write $k_1 = (\sqrt{k^2 + m^2}, 0, 0, k)$ and $k_2 = (\sqrt{k^2 + m^2}, 0, 0, -k)$. Then $E = 2\sqrt{k^2 + m^2} \geq 2m$. Physically, a pair of mesons can be produced when $E \geq 2m$.

I.8.1 Do the k^0 integral on the left-hand side of (I.8.14): $\int dk^0\delta((k^0)^2 - \omega_k^2)\theta(k^0)f(k^0, \vec{k})$, where $\omega_k \equiv +\sqrt{\vec{k} + m^2}$. Using (I.2.12) and picking up the positive root because of the step function, we obtain $\int_0^\infty dk^0(\delta(k^0 - \omega_k)/(2k^0))f(k^0, \vec{k}) = f(\omega_k, \vec{k})/(2\omega_k)$.

To verify the invariance explicitly, boost in the x direction and drop the subscript on ω_k: $k^x \rightarrow \sinh\phi\,\omega + \cosh\phi\,k^x$ and $\omega \rightarrow \cosh\phi\,\omega + \sinh\phi\,k^x$. Then, using $\omega^2 = (k^x)^2 + \cdots$ and hence $\omega d\omega = k^x dk^x$, we have $dk^x \rightarrow (\sinh\phi\,(k^x/\omega) + \cosh\phi)dk^x$. Hence $dk^x/\omega \rightarrow dk^x/\omega$.

I.8.2 Clearly, only the terms $aa^\dagger$ and $a^\dagger a$ in H contributes to $< \vec{k}'|H|\vec{k} >$. Extract these two types of terms in

$$\int d^D x \varphi(x)^2$$

$$= \int d^D x \iint \frac{d^D q}{\sqrt{(2\pi)^D 2\omega_q}} \frac{d^D q'}{\sqrt{(2\pi)^D 2\omega_{q'}}} [a(\vec{q})a^\dagger(\vec{q}')e^{-i(\omega_q t - \vec{q}\cdot\vec{x})} e^{i(\omega_{q'} t - \vec{q}'\cdot\vec{x})} + \text{h.c.}]$$

$$= \int \frac{d^D q}{2\omega_q} [a(\vec{q})a^\dagger(\vec{q}) + a^\dagger(\vec{q})a(\vec{q})]$$

and so H is for our purposes effectively equal to $\int d^D q \frac{\omega_q}{2} [a(\vec{q})a^\dagger(\vec{q}) + a^\dagger(\vec{q}) \, a(\vec{q})]$, which upon using the commutation relation is equal to $\int d^D q \frac{\omega_q}{2} [\delta^{(D)}(\vec{0}) + 2a^\dagger(\vec{q})a(\vec{q})]$. We recognize the first term as the vacuum calculated in the text. Note that the definition of the delta function $(2\pi)^D \delta^{(D)}(\vec{k}) = \int d^D x e^{i\vec{k}\cdot\vec{x}}$ implies $\delta^{(D)}(\vec{0}) = [1/(2\pi)^D] \int d^D x = V/(2\pi)^D$. Thus, subtracting off the vacuum energy, we have H effectively equal to $\int d^D q \omega_q a^\dagger(\vec{q})a(\vec{q})$, which just says that a mode of momentum $\vec{q}$ carries energy ω_q. In particular, using the commutation relation twice we have $< \vec{k}'|H|\vec{k} >= \delta^{(D)}(\vec{k}' - \vec{k})\omega_k$. The energy of a particle of momentum $\vec{k}$ is ω_k relative to the vacuum.

I.8.4 $Q = \int d^D x J_0(x) = \int d^D x (\varphi^\dagger i\partial_0 \varphi - i(\partial_0 \varphi^\dagger)\varphi)$. Focus on the first term:

$$\int d^D x \iint \frac{d^D k'}{\sqrt{(2\pi)^D 2\omega_{k'}}} \frac{d^D k}{\sqrt{(2\pi)^D 2\omega_k}}$$

$$[a^\dagger(\vec{k}')e^{i(\omega_{k'} t - \vec{k}'\cdot\vec{x})} + b(\vec{k}')e^{-i(\omega_{k'} t - \vec{k}'\cdot\vec{x})}]\omega_k [a(\vec{k})e^{-i(\omega_k t - \vec{k}\cdot\vec{x})} - b^\dagger(\vec{k})e^{i(\omega_k t - \vec{k}\cdot\vec{x})}]$$

Note that $i\partial_0$ brings down a factor of ω_k and produces a relative sign between a and $b^\dagger$. As in exercise I.8.2 the integral over x produces a delta function that collapses the two k integrals into one, giving

$$\int d^D k \frac{1}{2}(a^\dagger(\vec{k})a(\vec{k}) - b(\vec{k})b^\dagger(\vec{k}) - a^\dagger(-\vec{k})b^\dagger(\vec{k})e^{2i\omega_k t} + b(-\vec{k})a(\vec{k})e^{-2i\omega_k t})$$

The second term in $-i(\partial_0 \varphi^\dagger)\varphi$ in $J_0(x)$ is just the hermitean conjugate of the first term $\varphi^\dagger i\partial_0 \varphi$. Thus, adding the hermitean conjugate of what we have just obtained, we find

$$Q = \int d^D k [a^\dagger(\vec{k})a(\vec{k}) - b(\vec{k})b^\dagger(\vec{k})]$$

$$= \int d^D k [a^\dagger(\vec{k})a(\vec{k}) - b^\dagger(\vec{k})b(\vec{k})] + \delta^{(D)}(\vec{0}) \int d^D k$$

The infinite additive constant is to be subtracted out much like the vacuum energy. In some texts a normal ordering operation, denoted by a pair of colons, is defined as follows: If you see $: (\cdots) :$ you are instructed to move all the creation operators in the expression $(\cdots)$ to the left of the annihilation operators. In other words, by fiat $: b(\vec{k})b^\dagger(\vec{k}) := b^\dagger(\vec{k})b(\vec{k})$. The current is then defined by $J_\mu(x) \equiv : (\varphi^\dagger i\partial_\mu \varphi - i(\partial_\mu \varphi^\dagger)\varphi) :$. Since the normal ordered current differs from the naively defined current by a c-number the most crucial property of the current, namely current conservation $\partial_\mu J^\mu = 0$, is not affected. This is of course just a formal way of saying that the value of the charge in the vacuum state is to be subtracted. In any case, the result $Q = \int d^D k [a^\dagger(\vec{k})a(\vec{k}) - b^\dagger(\vec{k})b(\vec{k})]$ shows that a and b annihilate positive and negative charges, respectively.

I.10.2 We have (repeated indices summed)

$$R_{aa'}R_{bb'}iD_{a'b'}(x) = \int D\varphi R_{aa'}\varphi_{a'}(x) R_{bb'}\varphi_{b'}(0)e^{iS}$$

But we can change the integration variable from φ to $R\varphi$. Since the action S and the measure $D\varphi$ are both invariant under $SO(N)$ rotations, this is equal to $\int D\varphi \varphi_a(x)\varphi_b(0)e^{iS} = iD_{ab}(x)$. Thus, we obtain $D_{ab} = R_{aa'}R_{bb'}D_{a'b'}$. The properties of the rotation group are such that the only solution of this equation is D_{ab} proportional to δ_{ab}.

I.10.3 The field φ transforms as a symmetric traceless tensor (see appendix B) under $SO(3)$, that is, with all indices displayed, $\varphi_{ab} \to R_{aa'}R_{bb'}\varphi_{a'b'} = R_{aa'}\varphi_{a'b'}R^T_{b'b} = (R\varphi R^T)_{ab}$. As suggested in the hint, writing φ as a 3 by 3 symmetric traceless matrix we have $\varphi \to R\varphi R^T$ and thus the invariants are (up to quartic order in φ) $\text{tr}(\partial_\mu\varphi)^2$, $\text{tr}\,\varphi^2$, $\text{tr}\,\varphi^4$, and $(\text{tr}\,\varphi^2)^2$. Remarkably, you can prove that $\text{tr}\,\varphi^4$ and $(\text{tr}\,\varphi^2)^2$ actually amount to only one invariant by diagonalizing

$$
\varphi = \begin{pmatrix} \alpha & 0 & 0 \\ 0 & \beta & 0 \\ 0 & 0 & -(\alpha+\beta) \end{pmatrix}
$$

You can see by computation $\text{tr}\,\varphi^4$ and $(\text{tr}\,\varphi^2)^2$ are both proportional to $[\alpha^2 + \beta^2 + (\alpha+\beta)^2]^2$. Thus, if we restrict ourselves to quartic terms the Lagrangian $\mathcal{L} = \frac{1}{2}\text{tr}(\partial_\mu\varphi)^2 - \frac{1}{2}m^2\,\text{tr}\,\varphi^2 - \lambda(\text{tr}\,\varphi^2)^2$ actually has an $SO(5)$ symmetry (since φ has 5 components.) This is an example of what is known as "accidental symmetry." Convince yourself that this holds only to quartic order in φ.

I.11.2 Varying $g^{\mu\rho}g_{\rho\lambda} = \delta^\mu_\lambda$ we have $(\delta g^{\mu\rho})g_{\rho\lambda} = -g^{\mu\rho}(\delta g_{\rho\lambda})$, which upon multiplication by $g^{\lambda\nu}$ becomes $\delta g^{\mu\nu} = -g^{\mu\rho}(\delta g_{\rho\lambda})g^{\lambda\nu}$. You may recognize this as just the statement $\delta M^{-1} = -M^{-1}(\delta M)M^{-1}$ for a matrix M. To evaluate δg we use the important identity $\det M = e^{\text{Tr}\log M}$, which you can prove easily by diagonalizing M with a similarity transformation. The left hand side is equal to the product of the eigenvalues of M, while the right hand side is equal to the exponential of the sum of logarithms of the eigenvalues. {You can define the logarithm of a matrix by expanding $\log[I + (M - I)]$ in a power series in $(M - I)$.}

Thus, $\delta \det M = (\det M)\,\text{tr}\,M^{-1}\delta M$ and so $\delta g = gg^{\nu\mu}\delta g_{\mu\nu}$. We are now ready to vary

$$
S = \int d^4x\sqrt{-g}\,\frac{1}{2}(g^{\mu\nu}\partial_\mu\varphi\partial_\nu\varphi - m^2\varphi^2) \equiv \int d^4x\sqrt{-g}\,\mathcal{L}
$$

Plugging in, we have

$$
\delta S = \int d^4x\sqrt{-g}[\frac{1}{2}g^{\nu\mu}\delta g_{\mu\nu}\mathcal{L} - g^{\mu\rho}(\delta g_{\rho\lambda})g^{\lambda\nu}\frac{1}{2}\partial_\mu\varphi\partial_\nu\varphi]
$$

Thus,

$$
T^{\mu\nu} = -\frac{2}{\sqrt{-g}}\frac{\delta S}{\delta g_{\mu\nu}} = g^{\mu\rho}g^{\nu\lambda}\partial_\rho\varphi\partial_\lambda\varphi - g^{\mu\nu}\mathcal{L}
$$

In the flat spacetime limit

$$
T^{00} = (\partial_0\varphi)^2 - \mathcal{L} = \frac{1}{2}((\partial_0\varphi)^2 + (\vec{\nabla}\varphi)^2 + m^2\varphi^2)
$$

precisely the energy density as promised.

I.11.3 Using the expression for $T^{\mu\nu}$ from the preceding exercise, we have

$$
P^i = \int d^3x\,T^{0i} = -\int d^3x\,\partial_0\varphi\partial_i\varphi
$$

and

$$
[P^i, \varphi(x)] = -\int d^3y[\partial_0\varphi(y), \varphi(x)]\partial_i\varphi(y) = i\partial_i\varphi(x)
$$

Thus, combined with the fact that $P^0 = H$, we have $[P^\mu, \varphi(x)] = -i\partial^\mu\varphi(x)$, which just reflects the fact that P^μ and x^ν are conjugate variables.

I.11.4 Evaluating $T_{\mu\nu} = -F_{\mu\lambda}F^\lambda_{\ \nu} - \eta_{\mu\nu}\mathcal{L}$, we have

$$
T_{ij} = -F_{i\lambda}F^\lambda_{\ j} + \frac{1}{2}\delta_{ij}(\vec{E}^2 - \vec{B}^2) = -E_iE_j + F_{ik}F_{jk} + \frac{1}{2}\delta_{ij}(\vec{E}^2 - \vec{B}^2)
$$

Since $F_{ik}F_{jk} = \varepsilon_{ikm}\varepsilon_{jkn}B_m B_n = \delta_{ij}\vec{B}^2 - B_i B_j$, we obtain the announced result. Note that $\delta_{ij}T_{ij} = \frac{1}{2}(\vec{E}^2 + \vec{B}^2) = T_{00}$ and hence $T = 0$.

Part II

II.1.1   Continuing the hint, we have

$$\delta(\bar{\psi}\gamma^\mu\gamma^5\psi) = \bar{\psi}\frac{i}{4}\omega_{\lambda\rho}[\sigma^{\lambda\rho}, \gamma^\mu\gamma^5]\psi = \bar{\psi}\frac{i}{4}\omega_{\lambda\rho}[\sigma^{\lambda\rho}, \gamma^\mu]\gamma^5\psi$$

since γ^5 anticommutes with gamma matrices and hence commutes with the product of two gamma matrices. Inserting $[\sigma^{\lambda\rho}, \gamma^\mu]$ as given in the text we have $\delta(\bar{\psi}\gamma^\mu\gamma^5\psi) = \omega^\mu_\lambda \bar{\psi}\gamma^\lambda\gamma^5\psi$, which is precisely how a vector transforms. Under parity $\bar{\psi}\gamma^\mu\gamma^5\psi \rightarrow \bar{\psi}\gamma^0\gamma^\mu\gamma^5\gamma^0\psi$ which equals $\bar{\psi}\gamma^5\gamma^0\psi = -\bar{\psi}\gamma^0\gamma^5\psi$ for $\mu = 0$, and $\bar{\psi}\gamma^0\gamma^i\gamma^5\gamma^0\psi = \bar{\psi}\gamma^i\gamma^5\psi$ for $\mu = i$. The time component flips sign while the spatial components do not. Thus, the behavior under parity is opposite to that of a normal vector: $\bar{\psi}\gamma^\mu\gamma^5\psi$ is an axial vector. The other cases proceed similarly.

II.1.2   From $\psi_L = \frac{1}{2}(1-\gamma^5)\psi$ and $\psi_R = \frac{1}{2}(1+\gamma^5)\psi$, we find $\bar{\psi}_L = \psi_L^\dagger\gamma^0 = \psi^\dagger\frac{1}{2}(1-\gamma^5)\gamma^0 = \bar{\psi}\frac{1}{2}(1+\gamma^5)$ and $\bar{\psi}_R = \bar{\psi}\frac{1}{2}(1-\gamma^5)$. We then just use the properties of P_L and P_R repeatedly. For example, $\bar{\psi}_L\psi_R = \bar{\psi}\frac{1}{2}(1+\gamma^5)\psi$ and $\bar{\psi}_R\psi_L = \bar{\psi}\frac{1}{2}(1-\gamma^5)\psi$, or equivalently, $\bar{\psi}\psi = \bar{\psi}_L\psi_R + \bar{\psi}_R\psi_L$ and $\bar{\psi}\gamma^5\psi = \bar{\psi}_L\psi_R - \bar{\psi}_R\psi_L$. As another example, $\bar{\psi}_L\gamma^\mu\psi_L = \bar{\psi}\frac{1}{2}(1+\gamma^5)\gamma^\mu\frac{1}{2}(1-\gamma^5)\psi = \bar{\psi}\gamma^\mu\frac{1}{2}(1-\gamma^5)\psi$ and $\bar{\psi}_R\gamma^\mu\psi_R = \bar{\psi}\gamma^\mu\frac{1}{2}(1+\gamma^5)\psi$. Note that various combinations vanish, for example, $\bar{\psi}_L\psi_L = 0$, $\bar{\psi}_L\gamma^\mu\psi_R = 0$, and so on. Complete the exercise.

II.1.3–4 In the appropriate basis the Dirac equation becomes

$$\begin{pmatrix} E-m & p\sigma_3 \\ -p\sigma_3 & -E-m \end{pmatrix}\begin{pmatrix} \phi \\ \chi \end{pmatrix} = 0$$

that is, $(E-m)\phi + p\sigma_3\chi = 0$ and $-p\sigma_3\phi - (E+m)\chi = 0$. The second equation informs us that $\chi = -[p/(E+m)]\sigma_3\phi$. For a slow electron $\chi \simeq -(p/2m)\sigma_3\phi$, so that χ is smaller than ϕ by the factor $p/2m$. The first equation then reduces to $(E - m - p^2/2m)\phi = 0$, which just reminds us of the relation between energy and momentum in the nonrelativistic limit.

II.1.5   In the Weyl basis the Dirac equation for a relativistic electron moving along the 3-axis $E(\gamma^0 - \gamma^3)\psi = 0$ becomes

$$\begin{pmatrix} 0 & I-\sigma^3 \\ I+\sigma^3 & 0 \end{pmatrix}\begin{pmatrix} \psi_L \\ \psi_R \end{pmatrix} = 0$$

Since

$$\sigma^{12} \equiv \frac{i}{2}[\gamma^1, \gamma^2] = -\frac{i}{2}[\sigma^1, \sigma^2] \otimes I = \sigma^3 \otimes I = \begin{pmatrix} \sigma^3 & 0 \\ 0 & \sigma^3 \end{pmatrix}$$

under a rotation around the 3-axis, $\psi_L \rightarrow e^{-(i/4)\omega\sigma^3}\psi_L = e^{+(i/4)\omega}\psi_L$ while $\psi_R \rightarrow e^{-(i/4)\omega\sigma^3}\psi_R = e^{-(i/4)\omega}\psi_R$. Indeed, the left and right handed fields rotate in opposite directions.

II.1.6   In the Weyl basis, the Dirac equation $\gamma \cdot pu = 0$ becomes $\sigma^\mu p_\mu\eta = 0$, and $\bar{\sigma}^\mu p_\mu\chi = 0$, with

$$u = \begin{pmatrix} \chi \\ \eta \end{pmatrix}$$

The solutions are

$$\eta = \begin{pmatrix} p^1 - ip^2 \\ p^0 - p^3 \end{pmatrix} \quad \text{and} \quad \eta = \begin{pmatrix} p^0 + p^3 \\ p^1 + ip^2 \end{pmatrix}$$

for the two possible helicities. The corresponding solutions for χ may be obtained by $\vec{p} \leftrightarrow -\vec{p}$. We have $\bar{u}u = p \cdot p = 0$. For a particle moving in the $+3$ direction, $\eta = 0$ and

$$\chi = 2E \begin{pmatrix} 0 \\ 1 \end{pmatrix}$$

and

$$\eta = 2E \begin{pmatrix} 1 \\ 0 \end{pmatrix}$$

and $\chi = 0$. The Lorentz vector $\bar{u}\gamma^\mu u = (2E)^2(1, 0, 0, 1)$. (What other direction could it point in?) For a particle moving in the -3 direction, η and χ exchange roles. This exercise shows explicitly that for massless particles we can use 2-component spinors. (What happens if parity is broken?)

II.1.8   In either the Dirac or the Weyl basis, $(\psi_c)_c = \gamma^2(\gamma^2\psi^*)^* = \psi$.

II.1.9   It is easiest to work in either the Dirac or the Weyl basis. Let ψ be left handed, that is $(1 + \gamma^5)\psi = 0$. Then $(1 - \gamma^5)\psi_c = (1 - \gamma^5)\gamma^2\psi^* = \gamma^2(1 + \gamma^5)\psi^* = 0$ since γ^5 is real.

II.1.10 $\psi C\psi \rightarrow \psi e^{-\frac{i}{4}\omega_{\lambda\rho}(\sigma^{\lambda\rho})^T} C e^{-\frac{i}{4}\omega_{\mu\nu}\sigma^{\mu\nu}}\psi = \psi C\psi$ since $(\sigma^{\lambda\rho})^T C = -C\sigma^{\lambda\rho}$.

II.1.12   Under parity or reflection in a mirror, $x^1 \rightarrow x^1$ and $x^2 \rightarrow -x^2$. Choose $\gamma^0 = \sigma^3$, $\gamma^0\gamma^1 = \sigma^1$, and $\gamma^0\gamma^2 = \sigma^2$. Multiply the Dirac equation $(i\gamma^\mu\partial_\mu - m)\psi = 0$ by γ^0 and write $[i(\partial_0 + \gamma^0\gamma^i\partial_i) - \gamma^0 m]\psi = 0$. Then multiplying by σ^1 reverses the sign of the ∂_2 term, but also the mass term. I leave it to you to discuss time reversal.

II.2.1   Apply Noether for the transformation

$$\psi \rightarrow e^{i\theta}\psi = (1 + i\theta)\psi$$

Then

$$\frac{\delta\mathcal{L}}{\delta(\partial_\mu\psi)}\delta\psi + \frac{\delta\mathcal{L}}{\delta(\partial_\mu\bar{\psi})}\delta\bar{\psi} = \bar{\psi}i\gamma^\mu(i\theta\psi)$$

Note that formally $\mathcal{L}$ does not depend on $\partial_\mu\bar{\psi}$. Thus, up to overall factors we can choose $J^\mu = \bar{\psi}\gamma^\mu\psi$ with the corresponding charge $Q = \int d^3x \bar{\psi}\gamma^0\psi$ into which we plug (II.2.10)

$$\psi(x) = \int \frac{d^3p}{(2\pi)^{3/2}(E_p/m)^{1/2}} \sum_s [b(p, s)u(p, s)e^{-ipx} + d^\dagger(p, s)v(p, s)e^{ipx}]$$

At this point, the calculation pretty much parallels what you did in exercise I.8.4. The integration $\int d^3x$ over space produces a delta function that sets the momentum variables in ψ and in $\bar{\psi}$ equal to one another. The new feature here is that we encounter objects such as $\bar{u}\gamma^0 u$. Invoking Lorentz invariance and referring to the rest frame form of u and v we have $\bar{u}(p, s)\gamma^\mu u(p, s') = \delta_{ss'}p^\mu/m$, $\bar{u}(p, s)\gamma^\mu v(p, s') = 0$, and so on. We obtain

$$Q = \int \frac{d^3p}{(2\pi)^3(E_p/m)} \sum_s [b^\dagger(p, s)b(p, s) + d(p, s)d^\dagger(p, s)]$$

As in exercise I.8.4 we have to move the creation operator $d^\dagger$ to the left of the annihilation operator d and subtract off an infinite constant. Thus, finally

$$Q = \int \frac{d^3p}{(2\pi)^3(E_p/m)} \sum_s [b^\dagger(p, s)b(p, s) - d^\dagger(p, s)d(p, s)]$$

showing clearly that b annihilates a negative charge and d a positive charge.

To calculate $[Q, \psi(0)] = \int d^3x [\bar{\psi}(x)\gamma^0\psi(x), \psi(0)]$ we use the identity $[AB, C] = A\{B, C\} - \{A, C\}B$ and the canonical anticommutation relation (II.2.4). We find $[Q, \psi(0)] = -\psi(0)$, thus showing that b and $d^\dagger$ must carry the same charge.

II.3.4 The desired equations are $\gamma^\mu\Psi_{\alpha\mu} = 0$ (this takes out 4 components since α takes on 4 values) and $(\not{p} - m)^\beta_\alpha \Psi_{\beta\mu} = 0$ (for each μ this takes out 2 components and so altogether $4 \times 2 = 8$ components.) Thus, $16 - 4 - 8 = 4$ components as desired. Another way of saying this is that $\gamma^\mu\Psi_{\alpha\mu}$ is a Dirac spinor and hence the spin $\frac{1}{2}$ part of the vector-spinor $\Psi_{\alpha\mu}$.

II.6.4 It is good practice to be as symmetrical as one can in calculations. So define $p_3 \equiv -P_1$ and $p_4 \equiv -P_2$ and add the 6 (not 3) combinations appearing in the definitions of s, t, and u, thus obtaining

$$2(s + t + u) = (p_1 + p_2)^2 + (p_3 + p_4)^2 + (p_3 + p_1)^2 + (p_4 + p_2)^2 + (p_4 + p_1)^2 + (p_3 + p_2)^2$$

$$= 3\sum_{i=1}^{4} m_i^2 + 2(p_1 \cdot p_2 + p_3 \cdot p_4 + p_1 \cdot p_3 + p_2 \cdot p_4 + p_1 \cdot p_4 + p_2 \cdot p_3)$$

The second group of terms on the right-hand side collect into $(\sum_{i=1}^{4} p_i)^2 - \sum_{i=1}^{4} m_i^2$. (Obviously, we have for convenience changed notation slightly, setting $m_3 = M_1$ and $m_4 = M_2$.)

II.6.5 Referring to (C.11) we see that in $d\sigma$ the factor

$$\frac{1}{|\vec{v}_1 - \vec{v}_2|\mathcal{E}(p_1)\mathcal{E}(p_2)} \frac{1}{(2\pi)^3\mathcal{E}(k_1)} \cdots \frac{1}{(2\pi)^3\mathcal{E}(k_n)}(2\pi)^4$$

reduces to $\frac{1}{2}(m/E)^4[1/(2\pi)^2]$. Integrating the factor $d^3P_1 d^3P_2 \delta^{(4)}(p_1 + p_2 - P_1 - P_2)$ over $\vec{P}_2$ we knock off 3 of the delta functions, leaving us with $d\Omega dP_1 P_1^2 \delta(2E - E_1)$, and so the integral over P_1 gives $d\Omega \frac{1}{2}E^2$. Finally, the factor containing the "real physics" is $\frac{1}{2}\sum_s \sum_s |\mathcal{M}|^2 = (e^4/4m^4)f(\theta)$. Multiplying the three factors together and dividing by $d\Omega$, we obtain $d\sigma/d\Omega = (\frac{1}{2})^5[e^4/(2\pi)^2](1/E^2)f(\theta)$, as given in the text. Note that m cancels out as expected. We should be able to take the limit $m \to 0$ compared to the energies without the cross section either blowing up or vanishing.

II.6.7 $$\Gamma = \frac{|\mathcal{M}|^2}{2M} \int \frac{d^3k}{(2\pi)^3 2\omega} \frac{d^3k'}{(2\pi)^3 2\omega'}(2\pi)^4 \delta^4(k + k' - q)$$

Knock off the $\vec{k}'$ integral and do the angular part of the $\vec{k}$ integral to obtain

$$\Gamma = \frac{|\mathcal{M}|^2}{8\pi M} \int \frac{dk k^2}{\omega\omega'}\delta(\sqrt{k^2 + m^2} + \sqrt{k'^2 + m^2} - M)$$

Using (I.2.12), we evaluate the integral as $(k^2/\omega\omega')(1/(\frac{k}{\omega} + \frac{k}{\omega'})) = k/M$. Solving $\sqrt{k^2 + m^2} + \sqrt{k'^2 + m^2} = M$ for k we obtain the stated result.

Part III

III.1.2 The amplitude should become nonanalytic when both denominators of the integrand

$$(k^2 - m^2 + i\varepsilon)((K - k)^2 - m^2 + i\varepsilon)$$

vanish, namely when $k^2 = m^2$ and $(K - k)^2 = m^2$. But we found the condition in exercise I.7.4, namely that $K^2 \geq 4m^2$. Referring to (III.1.14)

$$\mathcal{M} = \frac{i\lambda^2}{32\pi^2}\int_0^1 d\alpha \log\left[\frac{\Lambda^2}{\alpha(1 - \alpha)K^2 - m^2 + i\varepsilon}\right]$$

we see that the log has a cut starting at $K^2 = m^2/\alpha(1 - \alpha)$. As α ranges from 0 to 1, the minimum value of $m^2/\alpha(1 - \alpha)$ is attained at $\alpha = \frac{1}{2}$. So indeed, the cut starts at $K^2 = 4m^2$.

III.1.3   Under the indicated change, $\log \Lambda \to \log e^{\varepsilon} \Lambda = \log \Lambda + \varepsilon$, and so $\delta \mathcal{M} = -i\delta\lambda + iC\lambda^2 3(2\varepsilon) + O(\lambda^3)$. Thus, $\delta \mathcal{M} = 0$ implies $\delta\lambda = 6C\lambda^2\varepsilon + O(\lambda^3) = 6C\lambda^2\delta \log \Lambda + O(\lambda^3)$ giving the stated result for $\Lambda(d\lambda/d\Lambda)$.

III.2.1   For $\int d^d x (\partial\varphi)^2$ to be dimensionless, we need $[\varphi] = (d-2)/2$. Thus $[\varphi^n] = n(d-2)/2$ and so in order for $\int d^d x \lambda_n \varphi^n$ to be dimensionless, we must have $[\lambda_n] = n(2-d)/2 + d$.

III.3.3   When we set $m = 0$, the integrand is manifestly a linear combination of γ matrices. The integral cannot produce a term independent of the γ matrices, which is what B is. For electrodynamics, the integral is changed to

$$(ie)^2 i^2 \int \frac{d^4 k}{(2\pi)^4} \frac{1}{k^2} [(1-\xi)\frac{k_\mu k_\nu}{k^2} - g_{\mu\nu}]\gamma^\mu \frac{\not{p} + \not{k} + m}{(p+k)^2 - m^2} \gamma^\nu$$
$$\equiv A(p^2) \not{p} + B(p^2)$$

When $m = 0$, the integrand is a linear combination of the product of three γ matrices, which can only reduce to one γ matrix, not to none. Incidentally, an alternative way of seeing the results stated here is to recall from chapter II.1 that with $m = 0$ the Lagrangian is invariant under the chiral transformation $\psi \to e^{i\theta\gamma^5}\psi$.

III.3.4   This essentially follows from $D = 4 - B_E - \frac{3}{2}F_E$ for $B_E = 0$ and $F_E = 2$. Then $D = 1$ but the linear divergence is reduced to logarithmic divergence by the symmetry argument given in the text.

III.5.2   Basically, you have already done this problem in exercises II.1.3 and II.1.4. You merely have to replace E and $\vec{p}$ by $\partial/\partial t$ and $\vec{\nabla}$ (see also chapter III.6).

III.5.3   In nonrelativistic quantum mechanics, the scattering amplitude is given in the Born approximation by i times the Fourier transform of the potential: $i \int d^3 x e^{i\vec{k}\cdot\vec{x}} U(\vec{x})$. The scattering amplitude owing to the exchange of a scalar meson of mass m is just $i/(k^2 - m^2) \simeq -i/(\vec{k}^2 + m^2)$. Thus, we just repeat the calculation in chapter I.4, obtaining

$$U(\vec{x}) = -\int \frac{d^3 k}{(2\pi)^3} \frac{e^{i\vec{k}\cdot\vec{x}}}{\vec{k}^2 + m^2} = -\frac{1}{4\pi r} e^{-mr}$$

III.6.1 $\bar{u}(p')(\not{p}'\gamma^\mu + \gamma^\mu \not{p})u(p) = 2m\bar{u}(p')\gamma^\mu u(p)$ by the equation of motion, but using $\gamma^\mu\gamma^\nu = \frac{1}{2}\{\gamma^\mu, \gamma^\nu\} + \frac{1}{2}[\gamma^\mu, \gamma^\nu] = \eta^{\mu\nu} - i\sigma^{\mu\nu}$ we can also write $(\not{p}'\gamma^\mu + \gamma^\mu \not{p}) = (p' + p)^\mu + i\sigma^{\mu\nu}(p' - p)_\nu$. We thus obtain the Gordon decomposition.

III.6.2   We compute

$$q_\mu \bar{u}(p')[\gamma^\mu F_1(q^2) + \frac{i\sigma^{\mu\nu}q_\nu}{2m} F_2(q^2)]u(p) = \bar{u}(p') \not{q} u(p) F_1(q^2)$$
$$= \bar{u}(p')(\not{p}' - \not{p})u(p)F_1(q^2)$$
$$= \bar{u}(p')(m - m)u(p)F_1(q^2) = 0$$

where the first and third equality follows from the antisymmetry of σ and the equation of motion, respectively.

III.7.1   Proceeding as in the text but living in d−dimensional spacetime, we obtain $i\Pi_{\mu\nu}(q) = -i \int \frac{d^d l}{(2\pi)^d} \frac{N_{\mu\nu}}{D}$ where $\frac{1}{D} = \int_0^1 d\alpha \frac{1}{\mathcal{D}}$ with $\mathcal{D} = (l^2 + \alpha(1-\alpha)q^2 - m^2 + i\varepsilon)^2$ as before but with $N_{\mu\nu}$ now effectively equal to $-d((1 - \frac{2}{d})g_{\mu\nu}l^2 + \alpha(1-\alpha)(2q_\mu q_\nu - g_{\mu\nu}q^2) - m^2 g_{\mu\nu})$. Rotating to Euclidean space we see that we have to do the integrals (with $c^2 \equiv m^2 - \alpha(1-\alpha)q^2$) $\int \frac{d_E^d l}{(2\pi)^d} \frac{1}{(l^2+c^2)^2}$ and $\int \frac{d_E^d l}{(2\pi)^d} \frac{l^2}{(l^2+c^2)^2} = \int \frac{d_E^d l}{(2\pi)^d} \frac{1}{(l^2+c^2)} - c^2 \int \frac{d_E^d l}{(2\pi)^d} \frac{1}{(l^2+c^2)^2}$. I did the first of these integrals for you in appendix II in chapter III.1. Generalizing slightly we have $\int_0^\infty dl l^{d-1} \frac{1}{(l^2+c^2)^a} = \frac{1}{2}c^{d-2a} \int_0^1 dx(1-x)^{\frac{d}{2}-1}x^{a-1-\frac{d}{2}}$. I will let you carry on from here.

Part IV

IV.1.1 Write $\vec{\varphi} = (\varphi_1, \varphi_2, \ldots, v + \varphi'_N)$. We compute $\frac{1}{2}\mu^2\vec{\varphi}^2 - (\lambda/4)(\vec{\varphi}^2)^2$ up to $O(\varphi^3)$ and find (upon dropping the $'$)

$$\frac{1}{2}\mu^2(v^2 + 2v\varphi_N + \vec{\varphi}^2) - \frac{\lambda}{4}(v^4 + 4v^2\varphi_N^2 + 4v^3\varphi_N + 2v^2\vec{\varphi}^2)$$

The condition of no linear term in φ_N fixes $v^2 = \mu^2/\lambda$ and so the coefficient of $\vec{\varphi}^2$ is equal to $\frac{1}{2}\mu^2 - \lambda/4(2v^2) = 0$. The $(N-1)$ fields $\varphi_1, \varphi_2, \ldots, \varphi_{N-1}$ are massless.

IV.3.1 We have $\int_0^\infty dk \, \log[(k^2+a^2)/k^2] = \pi a$ and so $V_{\text{eff}}(\varphi) = V(\varphi) + \hbar\sqrt{V''(\varphi)}/2 + O(\hbar^2)$. For $\mathcal{L} = \frac{1}{2}(\partial\varphi)^2 - \frac{1}{2}\omega^2\varphi^2$, the quantum oscillator with φ identified as position, we have $V_{\text{eff}}(0) = \frac{1}{2}\hbar\omega$.

IV.3.3 We have

$$m(\varphi) = f\varphi \text{ in } V_F(\varphi) = 2i \int \frac{d^2p}{(2\pi)^2} \log \frac{p^2-m(\varphi)^2}{p^2}$$

which after Wick rotation becomes

$$-2 \int \frac{d^2p_E}{(2\pi)^2} \log \frac{p_E^2+m(\varphi)^2}{p_E^2} = -\frac{1}{2\pi} \int_0^\infty dx \, \log \frac{x+m(\varphi)^2}{x}$$

After cutting off the integral at Λ^2 and adding a counterterm $B\varphi^2$ we obtain

$$V_F = \frac{1}{2\pi}(f\varphi)^2 \log \frac{\varphi^2}{M^2}$$

IV.3.4 $V_{\text{eff}} = i \sum_{n=1}^\infty \int d^4k/(2\pi)^4 (1/2n)[V''(\varphi)/k^2]^n$. For $V''(\varphi) = \frac{1}{2}\lambda\varphi^2$ the corresponding Feynman diagrams consist of a circle with n V's attached to the circumference, where the $2n$ is the infamous symmetry factor that I tried to avoid talking about in chapter I.7.

IV.4.1 With H an arbitrary p-form,

$$ddH = \frac{1}{(p+1)!} \frac{1}{p!} \partial_\lambda\partial_\nu H_{\mu_1\mu_2\ldots\mu_p} dx^\lambda dx^\nu dx^{\mu_1} dx^{\mu_2} \ldots dx^{\mu_p}$$

$$= \frac{1}{2} \frac{1}{(p+1)!} \frac{1}{p!} [\partial_\lambda, \partial_\nu] H_{\mu_1\mu_2\ldots\mu_p} dx^\lambda dx^\nu dx^{\mu_1} dx^{\mu_2} \ldots dx^{\mu_p} = 0$$

IV.5.1 If you have done all the exercises thus far (see exercise I.10.3), you have already made the acquaintance of the $I = 2$ scalar field transforming as $\varphi^{ab} \to R^{ac}R^{bd}\varphi^{cd} = R^{ac}\varphi^{cd}(R^T)^{db} = (R\varphi R^T)^{ab}$, which thus can be written as a traceless 3 by 3 symmetric matrix $\varphi \to R\varphi R^T$. Now you merely have to write out the covariant derivative $D_\mu\varphi$ (see IV.5.20) explicitly. [Hint: The action of the generators on φ is similar to what is shown in (B.20).]

IV.5.2 $dF = d(dA + A^2) = dAA - AdA$ and $[A, F] = AdA - dAA$ and so $dF + [A, F] = 0$. Explicitly with indices, this reads $\varepsilon^{\mu\nu\lambda\sigma}(\partial_\nu F_{\lambda\sigma} + [A_\nu, F_{\lambda\sigma}]) = 0$. In the abelian case, we have, for $\mu = 0$, $\varepsilon^{ijk}\partial_i F_{jk} = \vec{\nabla} \cdot \vec{B} = 0$ (recall chapter IV.4!), and for $\mu = i$, $\varepsilon^{ijk}(-\partial_0 F_{jk} + \partial_j F_{0k} - \partial_j F_{k0}) = -\partial_0 B_i + (\vec{\nabla} \times \vec{E})_i = 0$.

IV.5.4 From the general arguments mentioned in the problem we know that $\text{tr } F^2$ must be the "d of something." Now $d \text{ tr } AdA = \text{tr } dAdA$ and $d \text{ tr } \frac{2}{3}A^3 = \frac{2}{3} \text{tr}(dAA^2 - AdAA + A^2dA) = 2 \text{ tr } dAA^2$ but on the other hand $\text{tr } F^2 = \text{tr}(dA + A^2)(dA + A^2) = \text{tr}(dAdA + 2dAA^2)$ since $\text{tr } A^4 = \text{tr } A^3A = -\text{tr } AA^3 = -\text{tr } A^4 = 0$. In electromagnetism, $\text{tr } A^3 = 0$, and $d \text{ tr } AdA$ when written out in elementary notation is just $\partial_\mu(\varepsilon^{\mu\nu\lambda\sigma}A_\nu\partial_\lambda A_\sigma) = \frac{1}{4}\varepsilon^{\mu\nu\lambda\sigma}F_{\mu\nu}F_{\lambda\sigma}$.

IV.5.6 We simply plug in the general expression in the text and obtain

$$\mathcal{L} = -\frac{1}{4g^2} F^a_{\mu\nu} F^{a\mu\nu} + \bar{q}(i\gamma^\mu D_\mu - m)q$$

with the covariant derivative $D_\mu = \partial_\mu - iA_\mu = \partial_\mu - iA^a_\mu T^a$, where T^a ($a = 1, \dots, 8$) are traceless hermitean 3 by 3 matrices. Explicitly, $(A_\mu q)^\alpha = A^a_\mu (T^a)^\alpha_\beta q^\beta$, with $\alpha, \beta = 1, 2, 3$ (see chapter VII.3).

IV.6.3 Observe

$$A^a_\mu \tau^a = \begin{pmatrix} A^3_\mu & A^{1-i2}_\mu \\ A^{1+i2}_\mu & -A^3_\mu \end{pmatrix}$$

with the obvious notation $A^{1\pm i2}_\mu \equiv A^1_\mu \pm iA^2_\mu$. Let $\langle\varphi\rangle = \begin{pmatrix} 0 \\ v \end{pmatrix}$ so that

$$D_\mu \varphi = \partial_\mu \varphi - i(gA^a_\mu \frac{\tau^a}{2} + g'B_\mu \frac{1}{2})\varphi \rightarrow -\frac{i}{2}v \begin{pmatrix} gA^{1-i2}_\mu \\ -gA^3_\mu + g'B_\mu \end{pmatrix}$$

Thus, $|D_\mu\varphi|^2$ contains $v^2[g^2 A^{1+i2}_\mu A^{1-i2}_\mu + (-gA^3_\mu + g'B_\mu)^2]$. The combinations A^{1+i2}_μ, A^{1-i2}_μ, and $(-gA^3_\mu + g'B_\mu)$ acquire mass while $(g'A^3_\mu + gB_\mu)$ remain massless.

IV.7.4 We have

$$\Delta^{\mu\nu}(k_1, k_2) = i \int \frac{d^4 p}{(2\pi)^4} \frac{N^{\mu\nu}}{D} + \{\mu, k_1 \leftrightarrow \nu, k_2\}$$

where

$$N^{\mu\nu} \equiv \mathrm{tr}\, \gamma^5 (\not{p} - \not{q} + M)\gamma^\nu (\not{p} - \not{k}_1 + M)\gamma^\mu (\not{p} + M)$$

Only the term linear in M in $N^{\mu\nu}$ does not vanish, giving $N^{\mu\nu} = 4iM\varepsilon^{\mu\nu\sigma\tau}k_{1\sigma}k_{2\tau}$. Since we are interested only in terms of $O(k_1 k_2)$ we can set $D \rightarrow (p^2 - M^2)^3$ so that

$$\Delta^{\mu\nu}(k_1, k_2) = -8M\varepsilon^{\mu\nu\sigma\tau}k_{1\sigma}k_{2\tau} \int \frac{d^4 p}{(2\pi)^4} \frac{1}{(p^2 - M^2)^3} = \frac{i}{4\pi^2 M}\varepsilon^{\mu\nu\sigma\tau}k_{1\sigma}k_{2\tau}$$

with a dependence on M as stated in the problem. The effect of the regulator, like some unsavory acquaintance, remains even after we have sent him to infinity.

IV.7.5 We will sketch the solution. The details may be found in the lectures given by S. Adler at the 1970 Brandeis Summer School. The point is to imagine a regularization scheme that preserves the various relevant symmetries, namely Lorentz invariance, vector current conservation, and Bose statistics. As you will see, we don't actually have to specify the regularization. By Lorentz invariance, we have

$$\begin{aligned} \Delta^{\lambda\mu\nu}(k_1, k_2) = {} & \varepsilon^{\lambda\mu\nu\sigma}k_{1\sigma}A_1 + \varepsilon^{\lambda\mu\nu\sigma}k_{2\sigma}A_2 + \varepsilon^{\lambda\mu\sigma\tau}k_{1\sigma}k_{2\tau}k^\nu_1 A_3 \\ & + \varepsilon^{\lambda\mu\sigma\tau}k_{1\sigma}k_{2\tau}k^\nu_2 A_4 + \varepsilon^{\lambda\nu\sigma\tau}k_{1\sigma}k_{2\tau}k^\mu_1 A_5 \\ & + \varepsilon^{\lambda\nu\sigma\tau}k_{1\sigma}k_{2\tau}k^\mu_2 A_6 + \varepsilon^{\mu\nu\sigma\tau}k_{1\sigma}k_{2\tau}k^\lambda_1 A_7 \\ & + \varepsilon^{\mu\nu\sigma\tau}k_{1\sigma}k_{2\tau}k^\lambda_2 A_8 \end{aligned}$$

Since the Feynman integral representing $\Delta^{\lambda\mu\nu}$ is superficially linearly divergent, we see that $A_3, \dots, A_8$ are all convergent since we have to pull out three powers of momentum to extract them. In contrast, A_1 and A_2 are logarithmically divergent. But we can relate them to $A_3, \dots, A_8$ by vector current conservation since $0 = k_{1\mu}\Delta^{\lambda\mu\nu} = \varepsilon^{\lambda\nu\sigma\tau}k_{1\sigma}k_{2\tau}(-A_2 + k_1^2 A_5 + k_1 \cdot k_2 A_6)$ and thus $A_2 = k_1^2 A_5 + k_1 \cdot k_2 A_6$. Similarly for A_1. Rationalizing the Feynman integrand and evaluating the trace in the numerator, we can systematically ignore terms that contribute only to A_1 and A_2. Furthermore, Bose statistics gives us relations such as $A_3(k_1^2, k_2^2, q^2) = -A_6(k_2^2, k_1^2, q^2)$.

Part V

V.1.1 We dropped the term $h^2 \partial_0 \theta$ but kept the term $4g^2 \bar\rho h^2$. This requires $\partial_0 \theta \ll g^2 \bar\rho$, that is, $\omega \ll g^2 \bar\rho$, but since in our solution $\omega \sim g\sqrt{\bar\rho/mk}$ this requires $k \ll g\sqrt{m\bar\rho}$, which is consistent with what we assumed about k. Looking at the terms $-2\sqrt{\bar\rho}h\partial_0\theta - 4g^2\bar\rho h^2$ in $\mathcal{L}$ we see that $h \sim \partial_0\theta/(g^2\sqrt{\bar\rho}) \ll \sqrt{\bar\rho}$, which is also consistent.

V.5.1 With $\gamma^5 = \sigma_3$, $\frac{1}{2}(I \pm \gamma^5)$ clearly projects out the top and bottom component of $\psi = \begin{pmatrix} \psi_L \\ \psi_R \end{pmatrix}$, respectively. Everything is formally the same as in chapter II.1, but we can also work things out explicitly in the specific representation given here. Thus, $\bar\psi\psi = \psi^\dagger \sigma_2 \psi = i(\psi_R^\dagger \psi_L - \psi_L^\dagger \psi_R)$ and $\bar\psi\gamma^5\psi = \psi^\dagger \sigma_2 \sigma_3 \psi = i(\psi_R^\dagger \psi_L + \psi_L^\dagger \psi_R)$. Under the transformation $\psi \to e^{i\theta\gamma^5}\psi$, $\psi_L \to e^{i\theta}\psi_L$ and $\psi_R \to e^{-i\theta}\psi_R$, and the massless Dirac Lagrangian

$$\mathcal{L} = i\psi_R^\dagger (\frac{\partial}{\partial t} + v_F \frac{\partial}{\partial x})\psi_R + i\psi_L^\dagger (\frac{\partial}{\partial t} - v_F \frac{\partial}{\partial x})\psi_L$$

clearly does not change.

V.6.1 This of course just follows from Lorentz invariance. We have

$$\partial_t \varphi(\frac{x - vt}{\sqrt{1 - v^2}}) = \frac{-v}{\sqrt{1 - v^2}}\varphi'(\frac{x - vt}{\sqrt{1 - v^2}})$$

and

$$\partial_x \varphi(\frac{x - vt}{\sqrt{1 - v^2}}) = \frac{1}{\sqrt{1 - v^2}}\varphi'(\frac{x - vt}{\sqrt{1 - v^2}})$$

and thus the equation

$$(\partial_t^2 - \partial_x^2)\varphi(\frac{x - vt}{\sqrt{1 - v^2}}) + V'[\varphi(\frac{x - vt}{\sqrt{1 - v^2}})] = 0$$

becomes

$$\varphi''(\frac{x - vt}{\sqrt{1 - v^2}}) - V'[\varphi(\frac{x - vt}{\sqrt{1 - v^2}})] = 0$$

Note that this does not depend on the form of V. For any relativistic theory, the soliton moves like a relativistic particle (obviously!).

V.6.2 The sine-Gordon theory has an infinite number of vacua occurring at $\varphi = (2n + 1)\pi/\beta$. Thus, there exists a whole spectrum of solitons, such that $\varphi(\pm\infty) = (2n_\pm + 1)\pi/\beta$. The topological current is $J^\mu = (\beta/2\pi)\varepsilon^{\mu\nu}\partial_\nu\varphi$ with the corresponding charge $Q = (n_+ - n_-)$. The $Q = 2$ soliton decays into two $Q = 1$ solitons.

V.7.4 $(i/2\pi)\int_{S^1} g\,dg^\dagger = (i/2\pi)\int_{S^1} e^{iv\theta}de^{-iv\theta} = (i/2\pi)\int_{S^1}(-ivd\theta) = (v/2\pi)\int_0^{2\pi} d\theta = v$, which indeed counts the number of times $e^{iv\theta}$ winds around the circle. What mathematicians call the winding number is indeed just the magnetic flux of the physicist.

V.7.5 Within a region small enough so that we can treat $\varphi^a = v\delta^{a3}$ as constant, using $(D_\mu\varphi)^b = \partial_\mu\varphi^b + e\varepsilon^{bcd}A_\mu^c\varphi^d$ we have $(D_\mu\varphi)^1 = evA_\mu^2$ and $(D_\mu\varphi)^2 = -evA_\mu^1$ and thus

$$\mathcal{F}_{\mu\nu} \equiv \frac{F_{\mu\nu}^a \varphi^a}{|\varphi|} - \frac{(1/e)\varepsilon^{abc}\varphi^a(D_\mu\varphi)^b(D_\nu\varphi)^c}{|\varphi|^3}$$

$$\to F_{\mu\nu}^3 + e(A_\mu^2 A_\nu^1 - A_\nu^2 A_\mu^1) = \partial_\mu A_\nu^3 - \partial_\nu A_\mu^3$$

precisely the electromagnetic field strength since A_μ^3 is the massless component of the Yang-Mills field. Let us compute $B_k = \varepsilon_{ijk}\mathcal{F}_{ij}$ far from the magnetic monopole. To calculate the magnetic charge we are interested only in the term of order $1/r^2$ in $\vec{B}$. Since $D_\mu \varphi \to O(1/r^2)$ by construction we can drop the second term in $\mathcal{F}_{ij}$. Thus, we merely have to compute $F_{ij}^a \equiv \partial_i A_j^a - \partial_j A_i^a + e\varepsilon^{abc} A_i^b A_j^c$. Since F_{ij}^a will eventually be contracted with the unit vector $\varphi^a/|\varphi| = x^a/r$, we can effectively drop some of the terms in F_{ij}^a, thus simplifying the computation. We have

$$\partial_i A_j^a = \partial_i (\frac{1}{e}\varepsilon^{ajl}\frac{x^l}{r^2})" = "\frac{1}{e}\varepsilon^{aji}\frac{1}{r^2}$$

and

$$e\varepsilon^{abc} A_i^b A_j^c = \frac{(1/e)\varepsilon^{abc}\varepsilon^{bim}\varepsilon^{cjn}x^m x^n}{r^4}$$

$$= \frac{(1/e)(\delta^{ci}\delta^{am} - \delta^{cm}\delta^{ai})\varepsilon^{cjn}x^m x^n}{r^4} = \frac{1}{er^4}\varepsilon^{ijn}x^a x^n$$

so that

$$\frac{F_{ij}^a \varphi^a}{|\varphi|} = \frac{F_{ij}^a x^a}{r} = \frac{1}{er^3}(-2+1)\varepsilon^{aij}x^a = -\frac{1}{er^3}\varepsilon^{aij}x^a$$

and hence $B_k = -(1/er^2)\hat{x}^k$. The magnetic charge $g = -4\pi/e$.

Our result appears to differ from Dirac's quantization condition (IV.4.10) by a factor of 2. The resolution of this apparent paradox is instructive. In fact, we can always introduce into this theory a field Ψ (which could be a Bose or a Fermi field) transforming in the $I = \frac{1}{2}$ representation with the corresponding covariant derivative $D_\mu \Psi = \partial_\mu \Psi - ie(\frac{1}{2}\tau^a)A_\mu^a \Psi$. The field Ψ carries electric charge $\frac{1}{2}e$. Thus, the fundamental unit of electric charge is actually $\frac{1}{2}e$, not e, and our result $g = -4\pi/e = -2\pi/(e/2)$ is actually nothing but the Dirac quanization condition. (The sign is trivial: just a question of which one we call the monopole and which the antimonopole.)

V.7.7 Plugging in the Ansatz $\varphi^a = (H(r)/er)(x^a/r)$ and $A_i^b = [1 - K(r)]\varepsilon^{bij}(x^j/er^2)$ [so that $H(r) \xrightarrow[r\to\infty]{} evr$ and $K(r) \xrightarrow[r\to\infty]{} 0$ in accordance with the asymptotic behavior (V.7.5) and (V.7.6)] into $M = \int d^3x\{\frac{1}{4}(\vec{F}_{ij})^2 + \frac{1}{2}(D_i\vec{\varphi})^2 + V(\vec{\varphi})\}$ we get M as a functional of H and K. Minimizing M gives (with $H' = dH/dr$ etc) the equations $r^2 H'' = 2HK^2 + (\lambda/e^2)[H^3 - (ev)^2 r^2 H]$ and $r^2 K'' = K(K^2 - 1) + KH^2$. For help, see M. K. Prasad and C. M. Sommerfeld, *Phys. Rev. Lett.* 35: 760, 1975.

V.7.8 The BPS solution corresponds to setting $\lambda = 0$ in the two equations in exercise V.7.7, rendering the equations soluble, with the solution $H(r) = evr(\coth evr) - 1$ and $K(r) = evr/(\sinh evr)$. Ask yourself why $H(r)$ and $K(r)$ approach their asymptotic values exponentially. What determines the length scale?

V.7.9 For help, see B. Julia and A. Zee, *Phys. Rev.* D11: 2227, 1975.

V.7.11 We derived the lower bound for the mass of the magnetic monopole $4\pi v|g| \sim 4\pi(ev)/e^2 \sim M_W/\alpha$.

V.7.12 Near the identity element $g = e^{i\vec{\theta}\cdot\vec{\sigma}} \simeq 1 + i\vec{\theta}\cdot\vec{\sigma}$ and thus $g dg^\dagger \simeq -id\vec{\theta}\cdot\vec{\sigma}$. In a small neighborhood of the identity element the group manifold is locally Euclidean and so

$$\text{tr}(gdg^\dagger)^3 = i\,\text{tr}(\sigma^i\sigma^j\sigma^k)d\theta^i d\theta^j d\theta^k = -12d\theta^1 d\theta^2 d\theta^3$$

is manifestly proportional to the volume element on S^3. For $g = e^{i(\theta_1\sigma_1 + \theta_2\sigma_2 + m\theta_3\sigma_3)}$, $\text{tr}(gdg^\dagger)^3 = -12md\theta^1 d\theta^2 d\theta^3$.

V.7.13 $\int d^4x(\partial_\mu J_5^\mu) = \int d^3x J_5^0|_{t=+\infty} - \int d^3x J_5^0|_{t=-\infty}$. Recalling that $J_5^0 = \psi_R^\dagger \psi_R - \psi_L^\dagger \psi_L$, we see that the two spatial integrals just count the number of right moving fermion quanta minus the number of left moving fermion quanta at $t = \pm\infty$ respectively. So $\int d^4x(\partial_\mu J_5^\mu)$ is an integer. On the other hand, in the text we proved that $\int \text{tr}\,F^2$ is a topological invariant. In other words, with suitable normalization, evidently $1/(4\pi)^2$, the integral $[1/(4\pi)^2] \int d^4x \varepsilon^{\mu\nu\lambda\sigma}\,\text{tr}\,F_{\mu\nu}F_{\lambda\sigma}$ is an integer. Thus, the coefficient $1/(4\pi)^2$ cannot be shifted even a little bit by quantum fluctuations.

Part VI

VI.4.2 The quartic interaction term $(1/2f^2)(\vec{\pi} \cdot \partial\vec{\pi})^2$ in $\mathcal{L}$ gives the amplitude $i(1/2f^2)i^2\delta^{ab}\delta^{cd}(k_1k_3 + k_1k_4 + k_2k_3 + k_2k_4) + \text{permutations} = (i/2f^2)\delta^{ab}\delta^{cd}(k_1 + k_2)^2 + \text{permutations}$ for the 4-pion interaction vertex (where for convenience we have labeled all the momenta as going outward so that $k_1 + k_2 + k_3 + k_4 = 0$).

VI.4.3 After writing $\sigma = v + \sigma'$, we find, as in chapter IV.1, that $\mathcal{L} = -\frac{1}{2}(2\mu^2)\sigma'^2 - \lambda v\sigma'\vec{\pi}^2 - \frac{1}{4}\lambda(\vec{\pi}^2)^2 + \cdots$, where we have displayed only terms relevant for our purposes. The diagrams contributing to four-pion interaction are of two types, those involving the $\lambda(\vec{\pi}^2)^2$ term and those invoking σ' exchange. The former gives for the amplitude $(-\frac{i}{4}\lambda)2 \cdot 2(\delta^{ab}\delta^{cd} + \delta^{ac}\delta^{bd} + \delta^{ad}\delta^{bc})$ while the latter gives $(-i\lambda v)^2\{2i/[(k_1 + k_2)^2 - m_{\sigma'}^2]\}\delta^{ab}\delta^{cd}$. Thus, expanding to first order in momenta squared we find the coefficient of $\delta^{ab}\delta^{cd}$:

$$-i\lambda - 2i\lambda^2v^2\left[-\frac{1}{m_{\sigma'}^2}\right](1 + \frac{(k_1 + k_2)^2}{m_{\sigma'}^2}) = -i\lambda + \frac{2i\lambda^2v^2}{2\mu^2}\left[1 + \frac{(k_1 + k_2)^2}{2\mu^2}\right] = \frac{i\lambda}{2\mu^2}(k_1 + k_2)^2$$

To compare with exercise VI.4.2 we remember that $f^2 = v^2 = \mu^2/\lambda$ so that the amplitude here is also equal to $(i/2f^2)\delta^{ab}\delta^{cd}(k_1 + k_2)^2 + \text{permutations}$, as we had anticipated in the text.

VI.4.4 We will track down factors of 2 carefully but not factors of i and -1. Let us go back to the chiral transformations $\psi \to [1 + i\vec{\theta} \cdot (\vec{\tau}/2)\gamma^5]\psi$ and $\bar{\psi} \to \bar{\psi}[1 + i\vec{\theta} \cdot (\vec{\tau}/2)\gamma^5]$. Thus, $\delta(\bar{\psi}\psi) = \theta^a\bar{\psi}i\gamma^5\tau^a\psi$ and $\delta(\bar{\psi}i\gamma^5\tau^a\psi) = -\theta^a\bar{\psi}\psi$. Hence, for $\mathcal{L} = \bar{\psi}\{i\gamma\partial + g(\sigma + i\vec{\tau} \cdot \vec{\pi}\gamma_5)\}\psi + \mathcal{L}(\sigma, \vec{\pi})$ to be invariant we must have $\delta\sigma = \theta^a\pi^a$ and $\delta\pi^a = -\theta^a\sigma$. Applying Noether's theorem $J_\mu = (\delta\mathcal{L}/\delta\partial_\mu\varphi)\delta\varphi$, we obtain the current $J_{\mu 5}^a = \bar{\psi}i\gamma_\mu\gamma_5(\tau^a/2)\psi + \pi^a\partial_\mu\sigma - \sigma\partial_\mu\pi^a$ written in the text. Comparing the term $\bar{p}i\gamma_\mu\gamma_5 n$ contained in $J_{\mu 5}^{1+i2} \equiv J_{\mu 5}^1 + iJ_{\mu 5}^2$ with the current $J_{5\mu}$ defined in chapter IV.2, we see that $J_{5\mu} = -iJ_{\mu 5}^{1+i2}$. The normalized state $|\pi^-\rangle = (1/\sqrt{2})(|\pi^1\rangle - i|\pi^2\rangle)$ so that $\langle 0|\pi^{1+i2}|\pi^-\rangle = 2/\sqrt{2}$. The current $J_{\mu 5}^{1+i2}$ contains the term $-v\partial_\mu\pi^{1+i2}$ and thus $f = \sqrt{2}v$. Next, we have to work out the pion-nucleon coupling $g_{\pi NN}$ as defined in chapter IV.2. Here $\mathcal{L}$ contains $g\bar{\psi}i\vec{\tau} \cdot \vec{\pi}\gamma_5\psi$, which contains $\sqrt{2}g\bar{p}i\gamma_5 n\pi^-$ since $\pi^{1-i2} = \sqrt{2}\pi^-$. Thus, $g_{\pi NN} = \sqrt{2}g$. Putting it together, we see that $M = gv$ translates to $2M = fg_{\pi NN}$ in agreement with chapter IV.2.

VI.6.1 See figure VI.6.1. From $\Delta h = (d/\cos\theta) \simeq d(1 + \frac{1}{2}\theta^2)$ and $(\partial h/\partial x) = \tan\theta \simeq \theta$, we have $(\partial h/\partial t) \propto \theta^2 \propto (\partial h/\partial x)^2$, thus giving rise to the term $(\lambda/2)(\vec{\nabla}h)^2$. It all goes back to Mr. Pythagoras.

VI.6.2 We integrate the term $\frac{1}{2}\int d^D\vec{x}\, dt[((\partial/\partial t) - \vec{\nabla}^2)h]^2$ in $S(h)$ by parts to obtain $-\frac{1}{2}\int d^D\vec{x}\, dt[h((\partial/\partial t) + \vec{\nabla}^2)((\partial/\partial t) - \vec{\nabla}^2)h]$. Thus, the propagator is the inverse of the operator $(\partial/\partial t + \vec{\nabla}^2)(\partial/\partial t - \vec{\nabla}^2) = \partial^2/\partial t^2 - (\vec{\nabla}^2)^2$, the Fourier transform of which is $-(\omega^2 + k^4)$.

VI.8.3 With $h'(\vec{x}, t) = h(\vec{x} + g\vec{u}t, t) + \vec{u} \cdot \vec{x} + \frac{g}{2}u^2t$, we have $\partial h'/\partial t = \partial h/\partial t + g\vec{u} \cdot \vec{\nabla}h + (g/2)u^2$ and $\vec{\nabla}h' = \vec{\nabla}h + \vec{u}$. Thus the combination $(\partial h/\partial t) - \frac{g}{2}(\vec{\nabla}h)^2$ is invariant, as is (obviously) $\vec{\nabla}^2h$. In other words, $\tilde{S}(h)$ must be constructed out of these two invariant combinations.

VI.8.5 Look at the action $S(h) = \frac{1}{2}\int d^D\vec{x}\, dt\,[(\partial h/\partial t) - \nabla^2h - g(\nabla h)^2/2]^2$. Comparing $\partial h/\partial t - \nabla^2h$ we see that time has the dimension of length squared: $T \sim L^2$. From the term $\int d^D\vec{x}\, dt\,(\partial h/\partial t)^2$ and the fact that S is dimensionless, we have $[h]^2 \sim T^2/(L^DT) \sim 1/L^{D-2}$ and so h has the dimension of $(1/L^{D-2})^{\frac{1}{2}}$. Comparing $\vec{\nabla}^2h$ with $g(\vec{\nabla}h)^2$ we see that g has the dimension of $1/h$, that is, $L^{(D-2)/2}$.

VI.8.7 We are told that $L(dg/dL) = (2 - D)g/2 + (2D - 3)f_Dg^3 + \cdots$. We are assuming that the terms $(\cdots)$ can be neglected. Thus (in what follows a^2 and b^2 are two generic positive numbers) for $D = 1$, $L(dg/dL) = a^2g - b^2g^3$ and g flows toward the fixed point $g^* = a/b$. (Incidentally, the KPZ equation is soluble for $D = 1$ by methods not explained in this text and both z and χ are known exactly.) For $D = 2$, $L(dg/dL) = b^2g^3$ and g flows toward some unknown strong (presumably) coupling fixed point. For $D = 3$, $L(dg/dL) = -a^2g + b^2g^3$. The fixed point $g^* = a/b$ is unstable. For $g < g^*$, g flows toward the trivial (i.e., free, or Gaussian) fixed point. Since the theory at the fixed point is free we know the critical exponents exactly: $z = 2$ and $\chi = (2 - D)/2$. For $g > g^*$, g flows toward some unknown strong (presumably) coupling fixed point.

Part VII

VII.1.1. Set $n^\mu A'_\mu(x) = 0$ with $A'_\mu = U^\dagger A_\mu U + iU^\dagger \partial_\mu U$, so that $n \cdot \partial U(x) = in \cdot A(x)U(x)$. Define $\lambda(x) = r \cdot x/(r \cdot n)$ for any 4-vector r and write $x = \lambda(x)n + x_\perp$, so that $r \cdot x_\perp = 0$. Then

$$U(x) = \mathcal{P}e^{i \int_0^{\lambda(x)} d\sigma n \cdot A(\sigma n + x_\perp)}$$

(with a path ordering) solves the differential equation, since $n \cdot \partial\lambda = 1$ by construction.

VII.1.2 Using the BHC formula given, we have (the V's are clearly irrelevant)

$$U_{ij}U_{jk} = e^{iaA_\mu}e^{iaA_\nu} = e^{ia(A_\mu + A_\nu) - \frac{1}{2}a^2[A_\mu, A_\nu] + a^3 C + a^4 D + O(a^5)}$$

Similarly,

$$U_{kl}U_{li} = e^{-iaA'_\mu}e^{-iaA'_\nu} = e^{-ia(A'_\mu + A'_\nu) - \frac{1}{2}a^2[A_\mu, A_\nu] + a^3 E + a^4 F + O(a^5)} :$$

the prime reminding us that the A_μ and A_ν in this expression is to be evaluated on the "north" and "west" side of the plaquette in figure VII.1.2, respectively, in contrast to the A_μ and A_ν in $U_{ij}U_{jk}$ which are evaluated on the "south" and "east" side, respectively. Here C, D, E, and F denote various commutators, which we drag along merely to show that they eventually drop out in what interests us. (Note how the different terms are associated with different powers of a, as indicated. Note also that in some places we have dropped the prime on A and absorbed the "error" in doing so into terms of higher order in a.) Thus,

$$U_{kl}U_{li} = e^{-ia(A_\mu + A_\nu) - ia^2(\partial_\nu A_\mu - \partial_\mu A_\nu - \frac{1}{2}i[A_\mu, A_\nu]) + a^3 G + a^4 H + O(a^5)}$$

where G and H denote sums of commutators and terms such as $\partial_\nu \partial_\nu A_\mu$ and $\partial_\nu \partial_\nu \partial_\nu A_\mu$. Applying the BHC formula again to the order indicated we have

$$U_{ij}U_{jk}U_{kl}U_{li} = e^{ia^2(\partial_\mu A_\nu - \partial_\nu A_\mu) - a^2[A_\mu, A_\nu] + O(a^4)} = e^{ia^2 F_{\mu\nu} + a^3 I + a^4 J + O(a^5)}$$

with $F_{\mu\nu} = \partial_\mu A_\nu - \partial_\nu A_\mu + i[A_\mu, A_\nu]$. The same remarks on G and H apply to I and J. The Yang-Mills field strength emerges naturally, as we would anticipate. Since the traces of commutators and of A vanish, when we apply the trace all the junk drops out to $O(a^5)$ and we have

$$S(P) = \operatorname{Re} \operatorname{tr}[1 - \tfrac{1}{2}a^4 F_{\mu\nu}F_{\mu\nu} + O(a^5)]$$

By gauge invariance, the corrections must be of even order in a but for our purposes we don't care about them anyway. Evidently, f and g are related by some uninteresting factors of a.

Part VIII

VIII.1.7 $R^{12} = d\omega^{12} = d(-\cos\theta d\varphi) = \sin\theta d\theta d\varphi = 2R^{12}_{\theta\varphi}d\theta d\varphi \implies R^{12}_{\theta\varphi} = \frac{1}{2}\sin\theta$. Since $e^\theta_1 = 1$, $e^\varphi_2 = 1/\sin\theta$, we obtain $R \equiv R^{ab}_{\mu\nu}e^\mu_a e^\nu_b = 2R^{12}_{\theta\varphi}e^\theta_1 e^\varphi_2 = 1$, independent of θ and φ as expected.

Part N

N.1.1. The effective action for an electrically neutral system is given in the point particle limit by $S = \int d\tau(-m + b_E E_\mu E^\mu + b_B B_\mu B^\mu + \cdots)$, with E_μ and B_μ defined in the text. The interaction terms involve two powers of derivatives, which translate into two powers of ω in the scattering amplitude and hence four powers of ω in the scattering cross section. (Note that a possible term like $\int d\tau F_{\mu\nu}F^{\mu\nu}$ can be absorbed into the two terms already displayed.)

N.3.2. As in (III.3.7) we have $3V_3 + 4V_4 = 2I + n$ where I denotes the number of internal lines. The number of loops (III.3.6) $L = I - (V_3 + V_4 - 1)$ is 0 in a tree diagram. Thus $V_3 = n - 2 - 2V_4 \leq n - 2$.

Further Reading

Books on field theory

This is a list of field theory textbooks that I know about. I do not necessarily recommend them all. In food as in books, each has his or her own taste.

T. Banks, *Modern Quantum Field Theory*, Cambridge University Press, New York, 2008.

J. D. Bjorken and S. D. Drell, *Relativistic Quantum Mechanics*, McGraw-Hill, New York, 1964.

————, *Relativistic Quantum Fields*, McGraw-Hill, New York, 1965.

L. S. Brown, *Quantum Field Theory*, Cambridge University Press, New York, 1992.

S. J. Chang, *Introduction to Quantum Field Theory*, World Scientific, Singapore, 1990.

T. P. Cheng and L. F. Li, *Gauge Theory of Elementary Particle Physics*, Clarendon Press, Oxford, 1984.

F. Dyson and D. Derbes, *Advanced Quantum Mechanics*, World Scientific, Singapore, 2007.

R. P. Feynman, *Quantum Electrodynamics*, W. A. Benjamin, New York, 1962.

K. Huang, *Quantum Field Theory*, John Wiley & Sons, New York, 1998.

C. Itzykson and J-B. Zuber, *Quantum Field Theory*, McGraw-Hill, New York, 1980.

T. D. Lee, *Particle Physics and Introduction to Field Theory*, Taylor & Francis, New York, 1981.

V. P. Nair, *Quantum Field Theory*, Springer, New York, 2005.

M. E. Peskin and D. V. Schroeder, *An Introduction to Quantum Field Theory*, Perseus, Reading MA, 1995.

L. H. Ryder, *Quantum Field Theory*, 2nd Ed., Cambridge University Press, New York, 1996.

M. Stednicki, *Quantum Field Theory*, Cambridge University Press, New York, 2007.

G. Sterman, *An Introduction to Quantum Field Theory*, Cambridge University Press, New York, 1993.

S. Weinberg, *Quantum Theory of Fields*, Vols. 1 & 2, Cambridge University Press, New York, 1996.

X. G. Wen, *Quantum Field Theory of Many-Body Systems*, Oxford University Press, New York, 2007.

and finally, of course,

F. Mandl, *Introduction to Quantum Field Theory*, Interscience, New York, 1959.

Books on various topics mentioned

A. A. Abrikosov, L. Gorkov, and A. Dzyaloshinski, *Methods of Quantum Field Theory in Statistical Physics*, Prentice Hall, Englewood Cliffs, NJ, 1963.

S. L. Adler, "Perturbation Theory Anomalies," in: *Lectures on Elementary Particles and Quantum Field Theory*, 1970, Brandeis University Summer Institute in Theoretical Physics, S. Deser et al, ed., MIT Press, Cambridge, 1970.

P. Anderson, *Basic Notions of Condensed Matter Physics*, Benjamin-Cummings, Menlo Park, CA 1984.

D. Bailin and A. Love, *Supersymmetric Gauge Field Theory and String Theory*, IOP Publishing, Bristol and Philadelphia, 1994.

R. Balian and J. Zinn-Justin, eds., *Methods in Field Theory*, North Holland Publishing, Amsterdam, and World Scientific, Singapore, 1981.

A. L. Barabasi and H. E. Stanley, *Fractal Concepts in Surface Growth*, Cambridge University Press, Cambridge, 1995.

D. Budker, S. J. Freedman, and P. H. Bucksbaum, eds., *Art and Symmetry in Experimental Physics: Festschrift for Eugene D. Commins*, American Institute of Physics, New York, 2001.

J. Cardy, *Scaling and Renormalization in Statistical Physics*, Cambridge University Press, New York, 1996.

S. Coleman, *Aspects of Symmetry*, Cambridge University Press, Cambridge, 1985.

J. Collins, *Renormalization*, Cambridge University Press, Cambridge, 1985.

E. D. Commins, *Weak Interactions*, McGraw-Hill, New York, 1973.

E. D. Commins and P. H. Bucksbaum, *Weak Interactions of Leptons and Quarks*, Cambridge University Press, Cambridge, 2000.

M. Creutz, *Quarks, Gluons and Lattices*, Cambridge University Press, Cambridge, 1983.

P. A. M. Dirac, *The Principles of Quantum Mechanics*, Oxford University Press, Oxford, 1935. (On p. 253 he explained why he wanted the equation of motion for the electron to be first order in time derivative.)

A. Dobado et al., *Effective Lagrangians for the Standard Model*, Springer-Verlag, Berlin, 1997.

O.J.P. Éboli et al., *Particle Physics*, World Scientific, Singapore 1992.

R. P. Feynman, *Statistical Mechanics*, Perseus Publishing, Reading, MA, 1998.

R. P. Feynman and A. R. Hibbs, *Quantum Mechanics and Path Integrals*, McGraw-Hill, New York, 1965.

J. M. Figueroa-O' Farrill, *Electromagnetic Duality for Children*, on the World Wide Web 1998.

V. Fitch et al., eds., *Critical Problems in Physics*, Princeton University Press, Princeton, 1997.

M. Gell-Mann and Y. Ne'eman, *The Eightfold Way*, W. A. Benjamin, New York, 1964.

H. B. Geyer, ed., *Field Theory, Topology and Condensed Matter Physics*, Springer, 1995 (A. Zee, "Quantum Hall Fluids.")

M. L. Goldberger and K. M. Watson, *Collision Theory*, Dover, New York, 2004.

N. Goldenfeld, *Lectures on Phase Transitions and the Renormalization Group*, Addison-Wesley, Reading, MA, 1992.

F. Guerra and N. Robotti, *Ettore Majorana: Aspects of His Scientific and Academic Activity*, Springer, New York, 2008.

C. Itzykson and J-M. Drouffe, *Statistical Field Theory*, Cambridge University Press, Cambridge, 1989.

S. Iyanaga and Y. Kawada, eds., *Encyclopedic Dictionary of Mathematics*, MIT Press, Cambridge, 1980.

L. Kadanoff, *Statistical Physics*, World Scientific, Singapore, 2000.

G. Kane and M. Shifman, eds., *The Supersymmetric World: The Beginning of the Theory*, World Scientific, Singapore, 2000.

J. I. Kapusta, *Finite-Temperature Field Theory*, Cambridge University Press, Cambridge, 1989.

L. D. Landau and E. M. Lifschitz, *Statistical Physics*, Addison-Wesley, Reading, MA, 1974.

S. K. Ma, *Modern Theory of Critical Phenomena*, Benjamin/Cummings, Reading, MA, 1976.

H. J. W. Müller-Kirsten and A. Wiedemann, *Supersymmetry*, World Scientific, Singapore 1987.

T. Muta, *Foundations of Quantum Chromodynamics*, World Scientific, Singapore, 1998.

D. I. Olive and P. C. West, eds., *Duality and Supersymmetric Theories*, Cambridge University Press, Cambridge, 1999.

J. Polchinski, *String Theory*, Cambridge University Press, Cambridge, 1998.

J. J. Sakurai, *Invariance Principles and Elementary Particles*, Princeton University Press, Princeton, 1964.

L. Schulman, *Techniques and Applications of Path Integrals*, John Wiley & Sons, New York, 1981.

R. F. Streater and A. S. Wightman, *PCT, Spin Statistics, and All That*, W. B. Benjamin, New York, 1968.

G. 't Hooft, *Under the Spell of the Gauge Principle*, Word Scientific, Singapore, 1994.

G. 't Hooft et al., eds. *Recent Developments in Gauge Theories*, Plenum, New York, 1980.

D. Voiculescu, ed., *Free Probability Theory*, American Mathematical Society, Providence, R.I., 1997.

S. Weinberg, *Gravitation and Cosmology*, John Wiley & Sons, New York, 1972.

C. N. Yang, *Selected Papers 1945–1980 with Commentary*, W. H. Freeman, San Francisco, 1983.

A. Zee, *Unity of Forces in the Universe*, World Scientific, Singapore, 1982.

J.-B. Zuber, ed., *Mathematical Beauty of Physics*, World Scientific, Singapore, 1997.

Some popular books and books on the history of quantum field theory

M. Bartusiak, *Einstein's Unfinished Symphony*, Joseph Henry Press, Washington, D.C., 2000.

I. Duck and E. C. G. Sudarshan, *Pauli and the Spin-Statistics Theorem*, World Scientific, Singapore 1997.

R. P. Feynman, *QED: The Strange Theory of Light and Matter*, Princeton University Press, Princeton, 2006.

D. Kaiser, *Drawing Theories Apart*, University of Chicago Press, Chicago, 2005.

A. I. Miller, *Early Quantum Electrodynamics*, Cambridge University Press, Cambridge, 1994.

L. O'Raifeartaigh, *The Dawning of Gauge Theory*, Princeton University Press, Princeton, 1997.

S. S. Schweber, *QED and the Men Who Made It: Dyson, Feynman, Schwinger, and Tomonaga*, Princeton University Press, Princeton, 1994.

A. Zee, *Fearful Symmetry*, Princeton University Press, Princeton, 1999.

————, *Einstein's Universe*, Oxford University Press, New York, 2001.

————, *Swallowing Clouds*, University of Washington Press, Seattle, 2002.

Further Reading for Part N

In writing a textbook, the author has the luxury of not preparing a detailed scholarly bibliography (unless he or she chooses to follow the example of S. Weinberg, who is, in my opinion, most admirable in this regard). Even more extravagant is the freedom accorded to authors of popular books who in most cases give their unsuspecting and gullible readers the impression that the physics of an entire era was done by two or three greats, individuals worthy of their own personality cults. Presenting recent developments still in flux, I am faced with the dilemma of whether to give proper credit. In scholarly publications, conscientious referencing is of course ethically mandated, but this is a textbook. Fortunately, in this age of omniscient search engines, the reader could easily compile a bibliography more exhaustive than even a myopic humanist used to be able to muster in half a lifetime. I could do the same, but it is of little help to you for me to merely list the names of those responsible for, say, the new way of computing amplitudes using the spinor helicity formalism.[1] Instead, I can best serve the typical reader by listing a few papers and review articles starting from which you can track down the literature to your scholarly heart's desire. To those who feel that they should be mentioned, I apologize and refer you to Glashow's description of a tapestry in the preface.

W. Goldberger and I. Z. Rothstein, arXiv: hep-th/0409156v2.

Z. Bern, L. J. Dixon, D. C. Dunbar, D. A. Kosower, arXiv: hep-ph/9602280.

N. Arkani-Hamed and J. Kaplan, arXiv: hep-th/0801.2385.

E. Witten, arXiv: hep-th/0312171.

---

[1] F. A. Berends, Z. Bern, L. Chang, P. De Causmaecker, L. J. Dixon, D. C. Dunbar, R. Gastmans, W. Giele, J. F. Gunion, R. Kleiss, D. A. Kosower, Z. Kunszt, M. Mangano, A. G. Morgan, S. J. Parke, W. J. Stirling, T. R. Taylor, W. Troost, T. T. Wu, Z. Xu, D. H. Zhang, and many many others. I know how to copy and paste also! Please forgive me if I inadvertently left you off this list.

Index

Page numbers followed by letters f and n refer to figures and notes, respectively.